# *Introductory Engineering Electromagnetics*

# *Introductory Engineering Electromagnetics*

**BRANKO D. POPOVIĆ**

Virginia Polytechnic Institute and State University
and The University of Belgrade

ADDISON-WESLEY PUBLISHING COMPANY

Reading, Massachusetts

Menlo Park, California · London · Don Mills, Ontario

This book is in the
**ADDISON-WESLEY SERIES IN ELECTRICAL ENGINEERING**

*Consulting Editors:*
DAVID K. CHENG
LEONARD A. GOULD
FRED K. MANASSE

 Printed in the United States of America. Published simultaneously in Canada. Library of Congress Catalog Card No. 70-111954.

*To my wife Olya*

*and*

*to the memory of my sister Sophia*

# PREFACE

This book is intended primarily for undergraduate courses in engineering electromagnetics. With an appropriate choice of topics and problems it can also be used for more advanced levels. In the author's experience, courses totalling 75 to 120 hours can be taught from this text. The mathematical prerequisites are the elements of differential and integral calculus and (for the last part of the book) a basic knowledge of complex numbers. A knowledge of vector calculus is helpful, but not necessary, since a short introduction to vectors is provided. Further prerequisites are basic courses in university physics and in circuit theory, although the latter is useful rather than essential.

The book is divided into five parts: (1) the static electric field, (2) the steady electric current, (3) the static magnetic field, (4) the quasi-stationary electromagnetic field, and (5) Maxwell's equations and their application. A similar structure is to be found in many books on electromagnetic fields, but the characteristic feature of the present text is a careful balance of the material between the five parts. The section on the static electric field constitutes approximately thirty per cent of the book; the steady electric current is represented by ten per cent. The static magnetic field and the quasi-stationary electromagnetic field are the most often neglected stepchildren of authors; in the present text they take up almost thirty-five per cent of the book. The final part, Maxwell's equations and their application, could be almost any length, because doors are opened to many specialized subjects; an introductory text such as this can deal with only a few simple cases. Approximately twenty-five per cent of the text is devoted to this topic.

At an advanced level it is undoubtedly advantageous to introduce the general field concepts and Maxwell's equations as early as possible, to make a unified approach to electromagnetic phenomena. However, this method is not so convenient at the undergraduate level: it is likely that a number of students will be unable to follow the basic ideas of the theory, because of an inadequate background. At this level, Coulomb's law is probably the least risky unit to start teaching, since the field concept is introduced in connection with forces in a simple system of two static electric charges, which is easy to understand and assimilate. A similar approach—to derive the field concepts starting from the force law between two charges in motion—seems to be a consistent and educationally promising method of developing, step by step, the ideas and equations of the general, unique electromagnetic field. However, the static magnetic field is only rarely defined starting from the expression for force between *individual* electric charges, by analogy to Coulomb's law. Electromagnetic induction phenomena seem never to have been attacked in a similar manner.

This book is intended to present a novel, unified approach to electromagnetic fields. Basically, a three-term force between pairs of individual elemental charges is postulated and the field concepts are defined from that, following the known pattern used in the static electric field. The author is aware of the dangers implicit in his attempt, but the advantages are considerable. Electromagnetic induction, a most difficult and obscure phenomenon for students, becomes simple and evident. The concept of the "displacement" current in a vacuum can be given a clear physical interpretation. The uniqueness of the electric and magnetic fields as two parts and as the limiting cases of the only really existing unique electromagnetic field becomes obvious. In a sense, the present simple approach is equivalent to the modern advanced way of starting with Coulomb's law and the special theory of relativity.

The theory of electromagnetic fields is probably best studied by first understanding its basic ideas, and then fixing and deepening the knowledge through solving a large number of problems. The fundamental ideas of the theory are relatively simple. It might be said, therefore, that the problems themselves constitute a course in electromagnetism, and the present text is remarkable for having a very large number of problems. By choosing appropriate problems, the teacher can teach courses of different lengths and levels from it. The problems are divided into three groups according to function: test, general, and numerical problems.

The *test problems* usually require more understanding than mathematics. Some of them are even of the "yes-or-no" type. Their task is to develop in students a self-confident approach to the particular material studied. They are also intended to compel the student to reread and reinvestigate concepts explained in the main text.

The student's understanding is fixed and promoted further through the second group, the *general problems.* Some of them, those explaining the pattern of a solving procedure and the more difficult ones, are solved in detail, in an attempt to show the student how the problems *should* be approached. The others are not solved. However, perhaps the most important phase in acquiring confidence is the repeated satisfaction of finding the correct answer. For this reason answers to practically all the general problems are provided, but usually not in the analytical form. Rather, an answer corresponding to a particular set of numerical values (always as practical as possible) is often given. This serves a double purpose: it makes possible a check of the result found, and it gives an idea of the order of magnitude of the quantities considered.

The criterion of choice for the general problems was that the most important problems are those which analyze some practical case (usually in a simplified form). Unfortunately, it was not always possible to adhere to this principle. For example, in the section on the static electric field it is very simple to master the major part of the necessary mathematical technique, and so a certain number of problems in that part of the book are aimed at promoting the understanding of some particular mathematical methods. In other parts, problems of such a nature are rare.

The idea of the order of magnitude is developed to a certain extent through the general problems, but for practical applications a "feeling" for the order of magnitude of physical quantities deserves more attention. A third group of *numerical problems* is therefore added to each chapter. In working them out the student is expected to remember the values of some of the important constants and the order of magnitude of the values of physical quantities considered.

This book was written during the author's stay as a visiting professor at the Virginia Polytechnic Institute (now Virginia Polytechnic Institute and State University), on leave from the University of Belgrade. Without the initial encouragement of Professor W. A. Blackwell, Head of the Department of Electrical Engineering of VPI, and the understanding and help of the Assistant Head, Dr. H. H. Hull, this book would probably never have been written. Several colleagues from VPI helped the author with critical comments and useful suggestions. In this respect the author would like to mention particularly Dr. R. H. Bond. Some of the author's students at VPI corrected the English of the first draft of the book. The author is greatly indebted to all of them.

The author would also like very much to express his gratitude to his professor and friend, Dr. J. V. Surutka of the University of Belgrade. It was the author's good fortune to have had the opportunity to learn from him and to discuss with him the logical foundations of electromagnetic theory.

*Belgrade, Yugoslavia*
*August 1970*

B. D. P.

# CONTENTS

CHAPTER 0

# SHORT SURVEY OF VECTORS

## 0.1 INTRODUCTION

The theory of electromagnetic fields uses as tools certain mathematical concepts. Some of them—such as elements of differential and integral calculus—are indispensable for defining many basic concepts of the theory. Others—such as vectors, complex numbers, and various mathematical transformations—are not indispensable in the above sense, but familiarity with these concepts makes possible a much simpler formal development of the subject. This simplified treatment cannot reveal any new detail of the theory. However, it enables one to concentrate more on the theory's important aspects, rather than to dwell on lengthy mathematical procedures. Concepts of vectors and vector functions play a central role in this formally simplified approach. In fact, the development of electromagnetic theory is today regarded as inseparable from these concepts.

This text uses vectors as a mathematical tool from the very beginning. Very often students are not sufficiently familiar with vectors when they begin their course in electromagnetic fields. For this reason basic ideas of vectors and vector functions are explained in the following few pages. More complicated concepts are explained in the text, as they become necessary.

## 0.2 SCALARS AND VECTORS. SCALAR AND VECTOR FIELDS

Several types of mathematical concepts are utilized to describe various physical quantities encountered in practice. Two of these concepts—scalars and vectors—are used in basic electromagnetic theory.

Many physical quantities are completely specified by a single number, describing the magnitude of the quantity in terms of a convenient unit of the same quantity. Such physical quantities are called *scalar quantities* or *scalars*. For example, temperature at a point of a room is a scalar quantity, because it is completely specified by one number only. Mass density, size of a class, caloric value of a slice of bread, and many other quantities are classified as scalars for the same reason.

There is, however, a wide class of physical quantities that cannot be described by one number only. These can be completely specified only if both magnitude *and* direction of the quantity are given. Such quantities are known as *vector quantities*, or simply *vectors*. For example, the magnitude of the velocity of a tornado is far from being sufficient information about the tornado; knowledge of its direction is extremely useful. Thus the velocity of a tornado, and in fact the velocity of any object, is a vector quantity. Further examples of vector quantities are force on an object, displacement of an object, and acceleration.

Many branches of physics—electromagnetic field theory among them—deal with phenomena which are distributed continuously throughout a certain region of space. In order to describe such a phenomenon mathematically, we must describe it at *every point* of that region. In that case we speak about the *field* of a quantity in the region. The word field thus simply implies that a physical quantity is continuously distributed throughout the region. If the quantity considered is a scalar quantity, we say that the field is a *scalar field.* If the quantity is a vector, we say that the field is a *vector field.* The temperature associated with every point of a room is an example of a scalar field. The mass density of heterogeneous media, the atmospheric pressure on the earth's surface, and the humidity field in a region of the atmosphere are further examples of scalar fields. The velocity field of a fluid or gas—that is, the wholeness of velocities of all particles in a certain region—is a vector field. Further examples of vector fields are the gravitational field of the earth and all fields associated with forces acting in a region of space.

A scalar field can be represented mathematically by a single function of coordinates with respect to some convenient coordinate system. For example, the three axes of a rectangular coordinate system may be imagined along three edges of a corner of a room. The temperature at all points inside the room may be described by a single function of these coordinates, giving the correct value of temperatures at all points.

Vector fields are more difficult to describe mathematically. It will be shown that in the general case a vector field is defined uniquely if three certain scalar fields are known with respect to a convenient coordinate system. Before that, however, it is necessary to define algebraic operations with vector quantities and to describe a few types of coordinate systems that are most often used in electromagnetic theory.

## 0.3 ALGEBRAIC OPERATIONS WITH VECTORS

Displacement of a point for a certain distance along a straight line is perhaps the most obvious physical vector quantity. Although, strictly speaking, all algebraic operations with vectors have to be *defined*, some of them follow in a sense logically when we consider displacement as a model of a vector.

In this text vectors will be designated by boldface type, for example **A**, and their magnitudes by italics, for example $A$. Great confusion may arise if vectors are not designated properly. *It is strongly recommended that the utmost care be exercised from the very beginning to designate all vector quantities in such a way as to distinguish them clearly from scalar quantities.* Usual signs for vector quantities in handwritten material are a small arrow or bar above symbols representing vector quantities, for example $\vec{a}$ or $\bar{a}$.

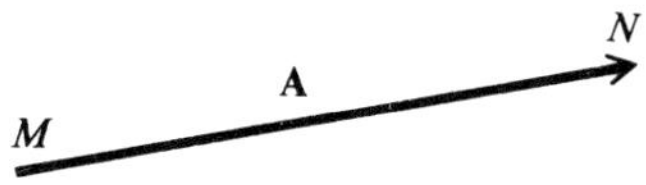


**Fig. 0.1** Graphical representation of a vector.

When we consider displacement along a straight line as a model of a vector quantity, we can first represent vectors graphically by a straight line with an arrowhead (Fig. 0.1). The arrowhead indicates the particular direction of the displacement. The length of the line is proportional to the magnitude of the displacement (or, in the general case, to the magnitude of the vector quantity considered). The starting point $M$ is called the *origin* of the vector, and the endpoint $N$ is called its *head*.

Suppose that a point is first moved along the line $MN$, represented by vector $\mathbf{A}$, and then along $NP$, represented by vector $\mathbf{B}$ (Fig. 0.2). The real displacement was along the broken line $MNP$. But the final position of the point is the same as if the point moved directly from $M$ to $P$ along line $MP$, indicated as vector $\mathbf{C}$. Vector $\mathbf{C}$ thus defined is known as *the sum of vectors* $\mathbf{A}$ and $\mathbf{B}$,*

$$\mathbf{C} = \mathbf{A} + \mathbf{B}. \tag{0.3.1}$$

It should be noted that Eq. (0.3.1) has no meaning unless one has in mind Fig. 0.2. If vectors $\mathbf{A}$ and $\mathbf{B}$ are imagined to have a common origin, point $M$, they constitute two sides of a parallelogram. From Fig. 0.2 it is evident that vector $\mathbf{C} = \mathbf{A} + \mathbf{B}$ may be obtained graphically also as the diagonal $MP$ of the parallelogram $MNPQ$.

From Fig. 0.2 it is also evident that the order in adding two vectors is irrelevant; that is,

$$\mathbf{A} + \mathbf{B} = \mathbf{B} + \mathbf{A}. \tag{0.3.2}$$

Hence vector addition obeys the commutative law.

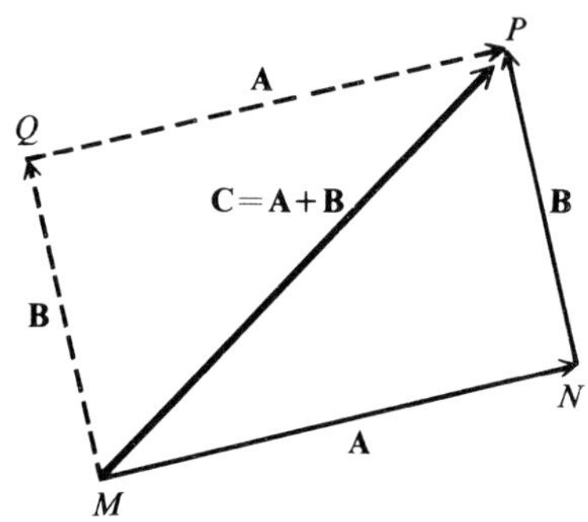


**Fig. 0.2** Vector addition and the commutative law for vector addition.

* Equations in this survey of vectors will be designated by a zero, plus the number of the section.

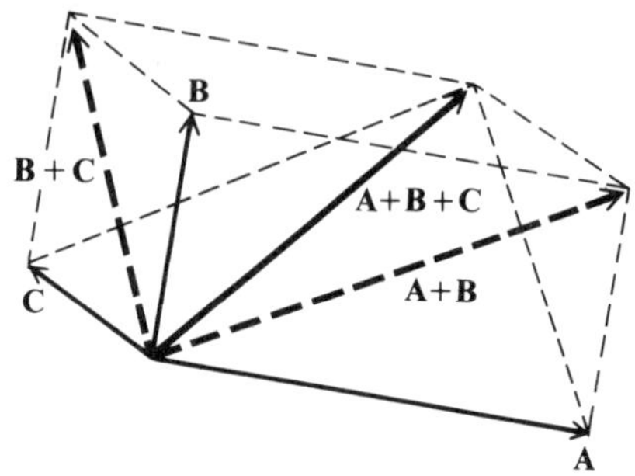


**Fig. 0.3** The associative law for vector addition.

Considering now three arbitrary vectors **A**, **B**, and **C** (Fig. 0.3), it can easily be inferred that

$$(\mathbf{A} + \mathbf{B}) + \mathbf{C} = \mathbf{A} + (\mathbf{B} + \mathbf{C}). \tag{0.3.3}$$

Thus vector addition also satisfies the associative law of addition. That is, the order of partial summing of several vectors does not change the final result.

A vector having the same magnitude as a vector **B**, but having the opposite direction, is denoted conventionally by $-\mathbf{B}$. With this convention, the rule for subtracting the vector **B** from a vector **A** follows directly:

$$\mathbf{A} - \mathbf{B} = \mathbf{A} + (-\mathbf{B}). \tag{0.3.4}$$

This is illustrated in Fig. 0.4.

The product of a scalar quantity $M$ and a vector **A** is obviously again a vector. It has the same direction as **A**, but is $M$ times larger in magnitude. The product of two vector quantities cannot be defined, however, in a manner analogous to the definition of the product of two scalars or the product of a scalar and a vector.

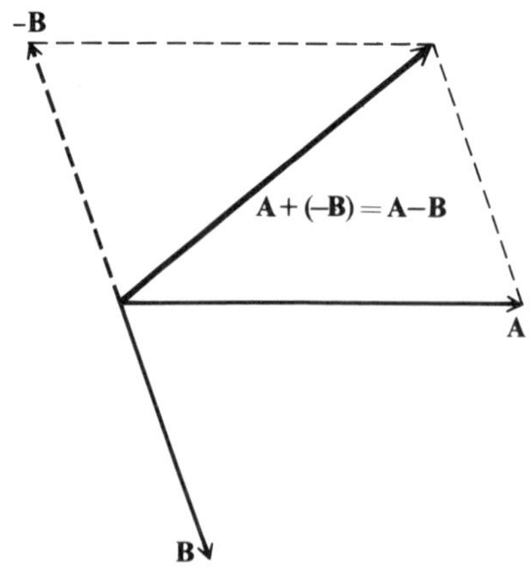


**Fig. 0.4** Subtraction of two vectors.

Indeed, what does it mean to "multiply" two vector quantities, that is, two quantities having both magnitude and direction? Thus a "product" of two vectors does not follow from the definition of a vector quantity; it has to be *defined*. This could be done in many ways. In practical applications it is useful to define two different "products" of two vectors. These are the so-called *scalar product*, or *dot product* of two vectors and the *vector product*, or *cross product* of two vectors.

**0.3.1 Scalar (dot) product of two vectors.** Consider a body sliding over a rough horizontal surface (Fig. 0.5). Suppose that the motion of the body is caused by a constant force **F**, acting in the direction shown. If the body is moved from position 1 to position 2, for a (vector) length **l**, as indicated in Fig. 0.5, the force **F** performs a work

$$W = F_l l = (F \cos \alpha) l = Fl \cos \alpha \tag{0.3.5}$$

according to the definition of the work done by a force. The product on the right side of Eq. (0.3.5) involves magnitudes $F$ and $l$ of *vectors* **F** and **l** and the cosine of the angle between them. But it is *not* a vector quantity, since work is a scalar quantity. This type of product is encountered very often, and the following convenient shorthand notation is very useful. Instead of writing $Fl \cos \alpha$, we conventionally write $\mathbf{F} \cdot \mathbf{l}$. Thus instead of Eq. (0.3.5) we simply write

$$W = \mathbf{F} \cdot \mathbf{l}, \tag{0.3.6}$$

with the understanding that $\mathbf{F} \cdot \mathbf{l}$ stands for $Fl \cos \alpha$. The "product" of two vectors of this type is called the *scalar* or *dot* (because of the dot sign) *product*.

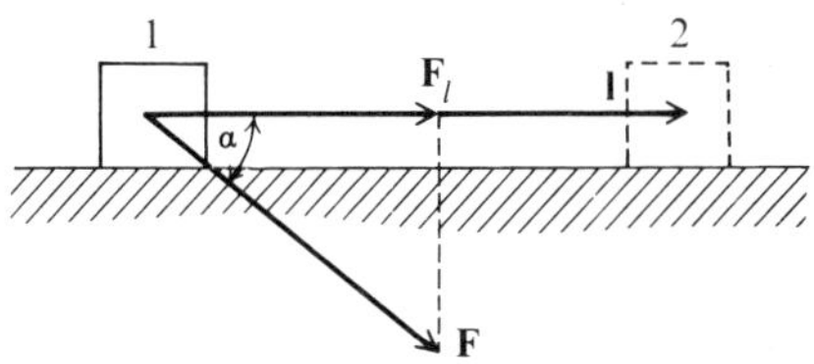


**Fig. 0.5** An illustration of work.

Generally speaking, given two vectors **A** and **B** and the angle $\alpha$ between them, the dot product $\mathbf{A} \cdot \mathbf{B}$ of the two vectors denotes, in fact, the product $AB \cos \alpha$:

$$\boxed{\mathbf{A} \cdot \mathbf{B} = AB \cos \alpha} \,. \tag{0.3.7}$$

In particular, for $\mathbf{B} = \mathbf{A}$ and denoting $\mathbf{A} \cdot \mathbf{A}$ by $\mathbf{A}^2$, we can determine the magnitude $A$ of a vector **A** from the dot product $\mathbf{A} \cdot \mathbf{A}$:

$$\boxed{\mathbf{A} \cdot \mathbf{A} = \mathbf{A}^2 = A^2} \,. \tag{0.3.7a}$$

From the definition of the dot product it follows that

$$\boxed{\mathbf{A} \cdot \mathbf{B} = \mathbf{B} \cdot \mathbf{A}} \,. \tag{0.3.8}$$

Hence the dot product is commutative. It may be shown that the so-called distributive law applies also to scalar multiplication, i.e., that

$$\mathbf{A} \cdot (\mathbf{B} + \mathbf{C}) = \mathbf{A} \cdot \mathbf{B} + \mathbf{A} \cdot \mathbf{C}. \tag{0.3.9}$$

It is important to note that vectors are not defined with respect to any particular coordinate system. Consequently, definition (0.3.7) of the dot product is quite general. Expressions for the dot product in the three coordinate systems used most often will be given later.

**0.3.2 Vector (cross) product of two vectors.** Consider a lever which can rotate about one end, denoted in Fig. 0.6 by $O$. Suppose that a force $\mathbf{F}$ acts at the other end, denoted by $P$, at a distance $\mathbf{r}$ from the axis of rotation. The force $\mathbf{F}$ can be represented as a vector sum of two vectors, vector $\mathbf{F}_r$ parallel to $\mathbf{r}$, and vector $\mathbf{F}_n$ normal to $\mathbf{r}$. $\mathbf{F}_r$ and $\mathbf{F}_n$ are called the *components* of vector $\mathbf{F}$. Component $\mathbf{F}_r$ cannot cause any rotation of the lever. Only component $\mathbf{F}_n$ is responsible for its rotation. The strength with which the lever is urged to rotate is obviously proportional to the magnitudes $F_n$ and $r$ of both $\mathbf{F}_n$ and $\mathbf{r}$. Their product is known as the moment of force $\mathbf{F}$ with respect to the axis of rotation. Denoting the magnitude of the moment by $M$ and the angle between vectors $\mathbf{F}$ and $\mathbf{r}$ by $\alpha$, we have

$$M = F_n r = (F \sin \alpha)\, r = Fr \sin \alpha. \tag{0.3.10}$$

It is very convenient to consider the moment $M$ of a force as a vector quantity. The magnitude of $M$ is given by Eq. (0.3.10). *By definition*, the vector $\mathbf{M}$ is perpendicular to both vectors $\mathbf{F}$ and $\mathbf{r}$. Its direction is *defined* as the direction of advance of a right-hand screw at $O$, if rotated together with the lever. In Fig. 0.6, vector $\mathbf{M}$ is thus directed toward the reader, which is indicated by the sign $\odot$ (the "tip" of the arrow).

By convention, all the above definitions concerning the magnitude and sign of the vector $\mathbf{M}$ are compressed in the following way of designating this type of "product" of two vectors:

$$\mathbf{M} = \mathbf{r} \times \mathbf{F}. \tag{0.3.11}$$

Thus Eq. (0.3.11) implies *by definition* the following. The magnitude of vector $\mathbf{M}$ is equal to $Fr \sin \alpha$. Vector $\mathbf{M}$ is perpendicular to both vectors $\mathbf{r}$ and $\mathbf{F}$. Its direction is obtained by the right-hand-screw rule if the *first* vector (in our case $\mathbf{r}$) is

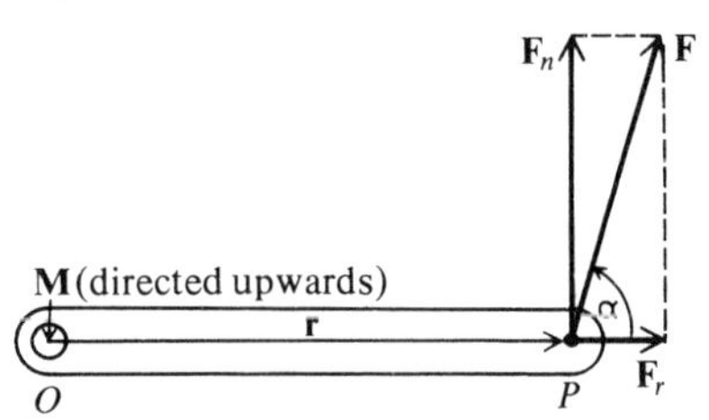


**Fig. 0.6** An illustration of the moment of force **F**.

rotated to coincide with the *second* vector (in our case **F**) *in the shortest way*. Only then would a right-hand screw placed at $O$ advance, if rotated with **r**, in the direction indicated (toward the reader). The "product" of two vectors of the type represented by Eq. (0.3.11) is termed the *vector* or *cross* (because of the cross sign) *product*.

Quite generally, given two vectors **A** and **B**, their cross product **C**,

$$\mathbf{C} = \mathbf{A} \times \mathbf{B}, \tag{0.3.12}$$

by definition is again a vector, perpendicular to both **A** and **B**. Its direction is obtained by the right-hand-screw rule rotating the *first* vector (**A**) to coincide with the *second* vector (**B**) in the shortest way. The magnitude of the cross product is equal to $AB \sin \alpha$. When **A** and **B** represent displacements, $AB \sin \alpha$ is equal to the area of the parallelogram having **A** and **B** as two sides (Fig. 0.7). The definition of the cross product is rather complicated. The student has to be familiar with it before using it as a shorthand notation in developing electromagnetic theory. It is highly recommended that Fig. 0.7 be kept in mind when writing the cross product of two vectors.

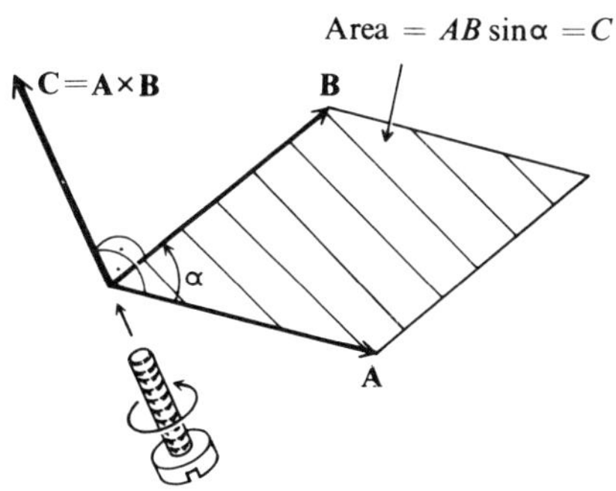


**Fig. 0.7** Right-hand-screw rule for determining direction of the cross product of two vectors.

It follows directly from the definition of the cross product that

$$\boxed{\mathbf{B} \times \mathbf{A} = -\mathbf{A} \times \mathbf{B}}\,. \tag{0.3.13}$$

So the cross product is *not* commutative; that is, $\mathbf{B} \times \mathbf{A} \neq \mathbf{A} \times \mathbf{B}$. It may be shown (see Problem 0.G1) that the distributive law applies to vector multiplication; that is,

$$\mathbf{A} \times (\mathbf{B} + \mathbf{C}) = \mathbf{A} \times \mathbf{B} + \mathbf{A} \times \mathbf{C}. \tag{0.3.14}$$

Just as in the case of the dot product, the cross product is not defined with respect to any particular coordinate system. The expression for the cross product in any coordinate system can be derived starting from the above general definition.

**0.3.3 Products of three vectors.** Recall that we have defined only the product of a scalar and a vector, the scalar product of two vectors, and the vector product of two vectors. It is then easy to conclude that only the following three types of products of three vectors are possible: $\mathbf{A}\,(\mathbf{B}\cdot\mathbf{C})$, $\mathbf{A}\cdot(\mathbf{B}\times\mathbf{C})$ and $\mathbf{A}\times(\mathbf{B}\times\mathbf{C})$. It is understood that the product in the parentheses is to be computed first. The first triple product is obtained by simply multiplying a vector $\mathbf{A}$ with the scalar $\mathbf{B}\cdot\mathbf{C}$. The second represents a dot product of a vector $\mathbf{A}$ with the vector $\mathbf{B}\times\mathbf{C}$, and is known as the *triple scalar product*. The third is a cross product of a vector $\mathbf{A}$ and the vector $\mathbf{B}\times\mathbf{C}$. It is referred to as the *triple vector product*. The last two have many applications in electromagnetic theory, and will be considered now in some detail.

The triple scalar product has the very important property that

$$\boxed{\mathbf{A}\cdot(\mathbf{B}\times\mathbf{C}) = \mathbf{C}\cdot(\mathbf{A}\times\mathbf{B}) = \mathbf{B}\cdot(\mathbf{C}\times\mathbf{A})}\,. \tag{0.3.15}$$

which can be proved as follows. According to the definition of the cross product (Fig. 0.7), the magnitude of the cross product $\mathbf{B}\times\mathbf{C}$ equals the area of the parallelogram formed by vectors $\mathbf{B}$ and $\mathbf{C}$. Hence $\mathbf{A}\cdot(\mathbf{B}\times\mathbf{C})$ equals the volume of the parallelepiped formed by vectors $\mathbf{A}$, $\mathbf{B}$, and $\mathbf{C}$, because $A\cos\beta$ is the height of the parallelepiped shown in Fig. 0.8. It can be demonstrated easily that the volume of the same parallelepiped is obtained by the other two triple scalar products in Eq. (0.3.15).

The triple vector product $\mathbf{A}\times(\mathbf{B}\times\mathbf{C})$ represents a vector normal to both vectors $\mathbf{A}$ and $(\mathbf{B}\times\mathbf{C})$. Since vector $(\mathbf{B}\times\mathbf{C})$ is normal to the plane of vectors $\mathbf{B}$ and $\mathbf{C}$, it follows that vector $\mathbf{A}\times(\mathbf{B}\times\mathbf{C})$ lies in the plane of vectors $\mathbf{B}$ and $\mathbf{C}$. Therefore

$$\mathbf{A}\times(\mathbf{B}\times\mathbf{C}) = a\mathbf{B} + b\mathbf{C}, \tag{0.3.16}$$

where $a$ and $b$ are scalars to be expressed in terms of vectors $\mathbf{A}$, $\mathbf{B}$, and $\mathbf{C}$. It may be proved that $a = \mathbf{A}\cdot\mathbf{C}$ and $b = -\mathbf{A}\cdot\mathbf{B}$ (see Problem 0.G1), so that

$$\boxed{\mathbf{A}\times(\mathbf{B}\times\mathbf{C}) = (\mathbf{A}\cdot\mathbf{C})\,\mathbf{B} - (\mathbf{A}\cdot\mathbf{B})\,\mathbf{C}}\,. \tag{0.3.17}$$

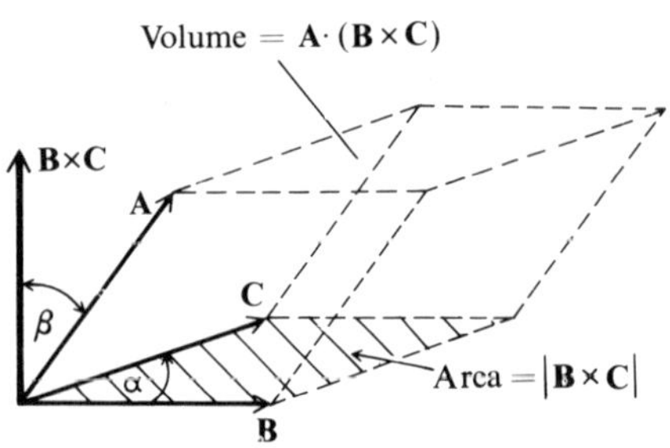


**Fig. 0.8** Geometrical interpretation of the triple scalar product.

This is also an important vector relation, used extensively in the theory of electromagnetic fields.

## 0.4 REPRESENTATION OF VECTORS WITH RESPECT TO COORDINATE SYSTEMS

**0.4.1 Unit vectors.** In vector operations a vector which is of unit magnitude and which has no physical dimensions (i.e., a vector which is a pure number) plays a very important role. Such a vector is referred to as the *unit vector*. According to its definition, the unit vector determines a *direction* in space. Unit vectors are usually denoted by special symbols, like **a**, **u**, **i**, etc. In this text the symbol **u** (with appropriate subscript if necessary) is adopted for unit vectors.

Given an arbitrary vector **A** (which may be any physical vector quantity), we can easily determine the unit vector $\mathbf{u}_A$ in the direction of **A**. From the above definition of the unit vector,

$$\boxed{\mathbf{u}_A = \mathbf{A}/A} \qquad \text{(unit vector in direction of vector } \mathbf{A}). \tag{0.4.1}$$

Hence any vector can be represented as a product of the unit vector in its direction with the magnitude of the vector:

$$\mathbf{A} = A\,\mathbf{u}_A. \tag{0.4.2}$$

**0.4.2 Rectangular, cylindrical, and spherical coordinate systems.** For locating a point in space it is indispensable to have a certain reference system. Any reference system is satisfactory if it enables *unique* description of the position of a point. For example, given the latitude and longitude of any point on the surface of the earth, the position of the point is completely specified. Or if we say that a nail is driven into the north wall of our room, two yards from the floor and two yards from the east wall, the position of the nail is again completely specified. From elemental geometry we know that one datum is sufficient for specifying the position of a point on a line. For specifying the position of a point on a surface (for example the earth's surface) two data are necessary. In the general case of a point in space three data are required. Such data, describing uniquely the position of a point with respect to a certain reference system, are called the *coordinates* of the point. The reference system is referred to as the *coordinate system*.

Many coordinate systems could be devised, but some of them are simpler to use than the others. For this reason only a relatively small number of coordinate systems is used in practice (approximately a dozen). In this text only three are necessary, the so-called rectangular, cylindrical, and spherical coordinate systems.

The *rectangular* or *Cartesian* coordinate system is perhaps the one best known. It consists of three mutually perpendicular straight lines, referred to as the *axes* of the system, intersecting at a single point $O$, called the *origin* of the system. Axes are denoted by symbols $x$, $y$, and $z$. To make possible the unique description of a point, the origin is assumed to divide the axes into positive and negative parts.

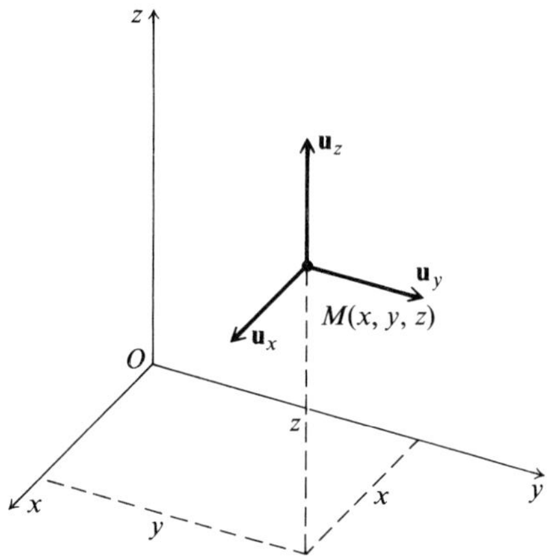


**Fig. 0.9** Rectangular (Cartesian) coordinate system and the corresponding unit vectors.

Axes are usually provided with an arrowhead indicating their positive direction (Fig. 0.9). Conventionally, axes $x$, $y$, and $z$ are always assumed (unless stated otherwise) to be in such a mutual relationship that if the $x$-axis is rotated in the shortest way to coincide with the $y$-axis, the direction of the $z$-axis is obtained by the right-hand-screw rule. Planes defined by any two coordinate axes and planes parallel to these are referred to as *coordinate surfaces.*

Any point in a rectangular coordinate system is in fact defined by the intersection of three coordinate (plane) surfaces. Distances $x$, $y$, and $z$ of these coordinate surfaces from "basic" surfaces $x = 0$, $y = 0$, and $z = 0$, respectively, are used as the coordinates of a point (Fig. 0.9).

Directions in which an arbitrary point $M$ would move if only one coordinate at a time is increased are designated by three unit vectors, $\mathbf{u}_x$, $\mathbf{u}_y$, and $\mathbf{u}_z$. These obviously are parallel to the respective coordinate axes (Fig. 0.9).

In the case of problems characterized by axial symmetry, it is much simpler to use a set of coordinate surfaces different from the planes used in the rectangular system. These coordinate surfaces are: coaxial circular cylinders (the axis of which for simplicity is always chosen to coincide with the axis of symmetry), half-planes emerging from that axis, and planes perpendicular to the axis. Radii of coordinate cylinders are usually denoted by $r$. They represent the first coordinate in this system, known as the *cylindrical coordinate system.*

Although the cylindrical coordinate system is a self-contained and independent system, it is customary to consider it in connection with a Cartesian coordinate system having the $z$-axis coinciding with the axis of coordinate cylinders, as shown in Fig. 0.10. We then take as the second coordinate the angle $\phi$ between the half-planes and the $x$-axis (i.e., the plane $y = 0$). The third coordinate is the distance $z$ of the point from the plane $z = 0$ (Fig. 0.10). Thus according to the definition $0 \leq r < \infty, 0 \leq \phi < 2\pi$, and $-\infty < z < \infty$.

To each point in the cylindrical coordinate system there are attached three

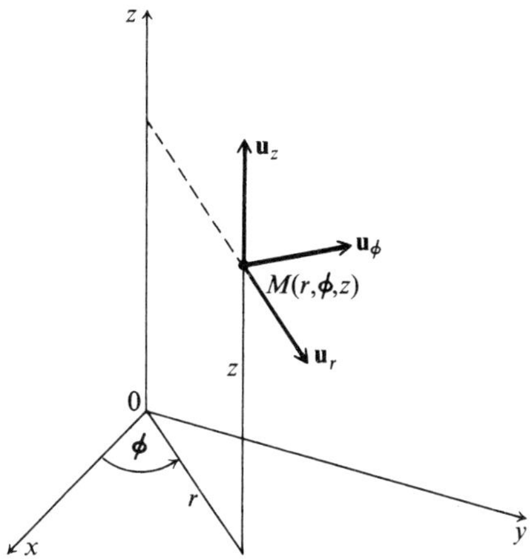


**Fig. 0.10** Cylindrical coordinate system and the corresponding unit vectors.

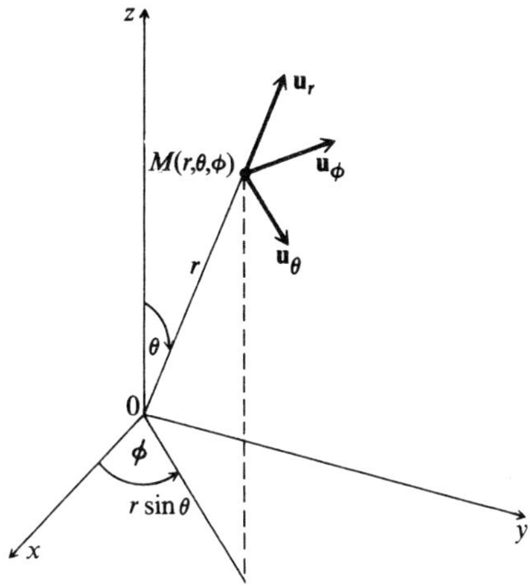


**Fig. 0.11** Spherical coordinate system and the corresponding unit vectors.

unit vectors, $\mathbf{u}_r$, $\mathbf{u}_\phi$ and $\mathbf{u}_z$. They indicate directions of motion of the point when only one of the coordinates $r$, $\phi$, or $z$, respectively, is given an infinitesimal positive increment. We see from Fig. 0.10 that these unit vectors are mutually orthogonal. If sequence $\mathbf{u}_r$, $\mathbf{u}_\phi$, $\mathbf{u}_z$ is adopted, the direction of $\mathbf{u}_z$ is obtained from $\mathbf{u}_r$ and $\mathbf{u}_\phi$ according to the right-hand-screw rule. We shall always adhere to this convention.

The *spherical coordinate system* is very convenient whenever we are dealing with problems having a certain degree of spherical symmetry. Coordinate surfaces in the spherical system, the intersection of which locates a point, are concentric spheres, cones with apexes at the center of the spheres, and half-planes emerging from the axis of the cones. The three coordinates of a point are the radius $r$ of a sphere, the half-angle $\theta$ of the cones, and angle $\phi$ between a half-plane and the $x$-axis. Hence, according to the definitions, we have that $0 \le r < \infty$, $0 \le \theta \le \pi$, $0 \le \phi < 2\pi$.

With each point in the spherical coordinate coordinate system are associated three mutually perpendicular unit vectors $\mathbf{u}_r$, $\mathbf{u}_\theta$, and $\mathbf{u}_\phi$ (Fig. 0.11). They indicate directions in which the point would move if only one of the coordinates—$r$, $\theta$, or $\phi$, respectively—is given an infinitesimal increment. We shall adhere to the sequence $r$, $\theta$, $\phi$ of the three coordinates. Then we determine the direction of $\mathbf{u}_\phi$ from $\mathbf{u}_r$ and $\mathbf{u}_\theta$ according to the right-hand-screw rule.

**0.4.3 Representation of a vector function with respect to a coordinate system.** A vector quantity is completely specified with respect to any coordinate system if we know the origin of the vector and its normal projections on directions of the three unit vectors at that point. These projections of a vector are known as its *vector components*. The magnitudes of the vector components are called *scalar components* of a vector, or simply its components. The scalar components will be designated in this book by the symbol used for the magnitude of the vector, with corresponding subscript. For example, components of a vector **A** are designated by $A_x$, $A_y$, $A_z$ in the Cartesian coordinate system, by $A_r$, $A_\phi$, $A_z$ in the cylindrical coordinate system, and by $A_r$, $A_\theta$, $A_\phi$ in the spherical coordinate system. This is illustrated in Figs. 0.12a–12c. Given the vector components of a vector, the vector itself can be determined as their vector sum. A vector component is written as a product of the scalar component and the corresponding unit vector. In the three coordinate systems considered, the total vector **A** may be represented as

$$\begin{aligned}
\mathbf{A} &= A_x\,\mathbf{u}_x + A_y\,\mathbf{u}_y + A_z\,\mathbf{u}_z \qquad &&\text{(Cartesian system),}\\
\mathbf{A} &= A_r\,\mathbf{u}_r + A_\phi\,\mathbf{u}_\phi + A_z\,\mathbf{u}_z \qquad &&\text{(cylindrical system),}\\
\mathbf{A} &= A_r\,\mathbf{u}_r + A_\theta\,\mathbf{u}_\theta + A_\phi\,\mathbf{u}_\phi \qquad &&\text{(spherical system).}
\end{aligned}$$

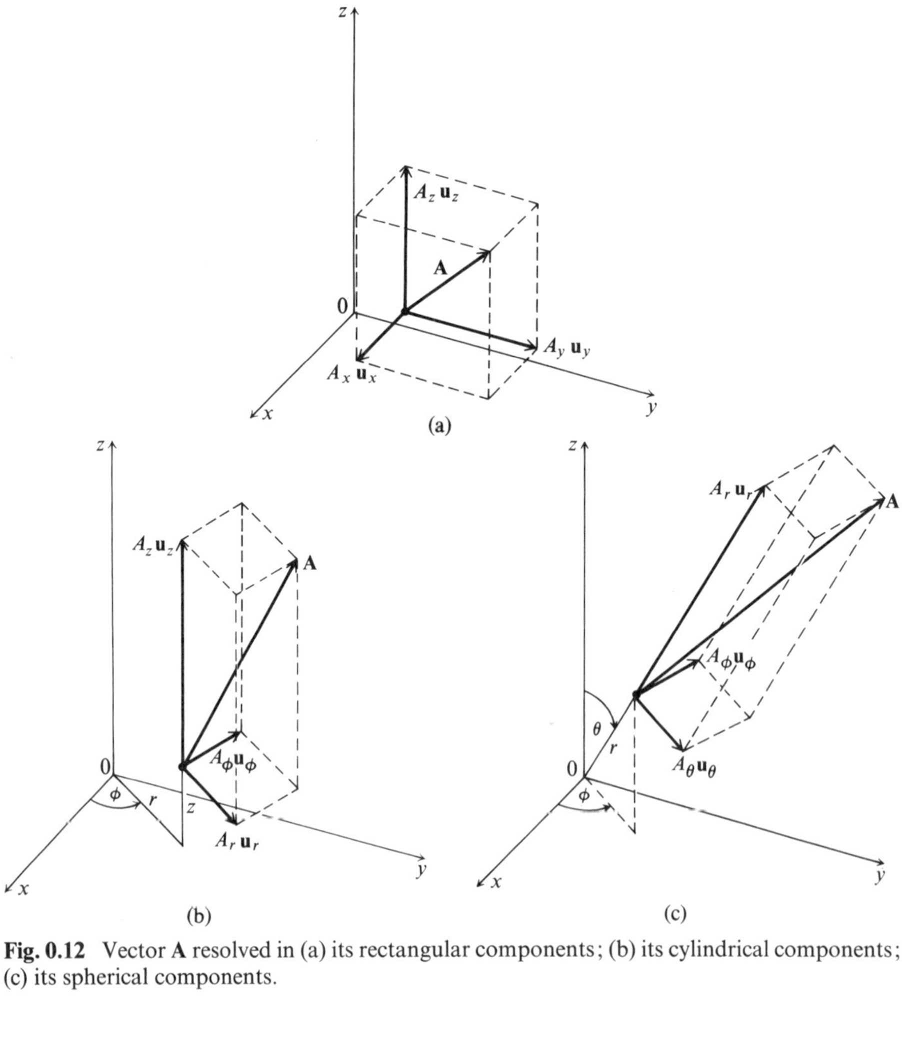


**Fig. 0.12** Vector **A** resolved in (a) its rectangular components; (b) its cylindrical components; (c) its spherical components.

Suppose now that vector **A** is a function of coordinates, i.e., that it represents a vector field. From the above, this vector function is uniquely defined if three independent scalar functions, representing the magnitudes of its components, are given. For example, if a vector function is given with respect to the spherical coordinate system, we have that

$$\mathbf{A}(r, \theta, \phi) = A_r(r, \theta, \phi)\,\mathbf{u}_r + A_\theta(r, \theta, \phi)\,\mathbf{u}_\theta + A_\phi(r, \theta, \phi)\,\mathbf{u}_\phi.$$

$A_r$, $A_\theta$, and $A_\phi$ generally may be completely different functions of coordinates $r$, $\theta$, and $\phi$.

**0.4.4 Algebraic operations with vectors given with respect to a coordinate system.** If a vector is given by its vector components with respect to a coordinate system, it is represented in fact as a sum of three vectors. With this in mind, we can easily perform all algebraic operations with vectors.

Consider two vector functions $\mathbf{A}\,(v_1, v_2, v_3)$ and $\mathbf{B}\,(v_1, v_2, v_3)$. Variables $(v_1, v_2, v_3)$ stand for either $(x, y, z)$, $(r, \phi, z)$, or $(r, \theta, \phi)$. Then (omitting, for simplicity, the parentheses denoting dependence on variables $v_1$, $v_2$, and $v_3$)

$$\mathbf{A} = A_1\,\mathbf{u}_1 + A_2\,\mathbf{u}_2 + A_3\,\mathbf{u}_3, \tag{0.4.3}$$

$$\mathbf{B} = B_1\,\mathbf{u}_1 + B_2\,\mathbf{u}_2 + B_3\,\mathbf{u}_3, \tag{0.4.4}$$

where $\mathbf{u}_1$, $\mathbf{u}_2$, and $\mathbf{u}_3$ stand for any set of three unit vectors.

A very important note on unit vectors is necessary at this point. *Only in the Cartesian coordinate system are unit vectors directed in the same directions at all points.* In the other two coordinate systems considered, and in fact in all other coordinate systems, *unit vectors generally have different directions at different points.* For example, if only $\phi$ in Fig. 0.11 is increased, point $M$ would move along a circle centered on the $z$-axis. All three unit vectors at this new point would have directions that were different with respect to the old ones. Hence if vectors **A** and **B** have origins at different points, $\mathbf{u}_1$, $\mathbf{u}_2$, and $\mathbf{u}_3$ in Eq. (0.4.3) are generally different from $\mathbf{u}_1$, $\mathbf{u}_2$, and $\mathbf{u}_3$ in (0.4.4). Thus only if the two vector functions **A** and **B** refer to the same point, or are given with respect to a Cartesian system, are the corresponding unit vectors in Eqs. (0.4.3) and (0.4.4) identical.

Suppose that vectors **A** and **B** have origins at the same point. In that case all algebraic operations have essentially the same expressions with respect to all three coordinate systems. Thus we first have that

$$\mathbf{A} \pm \mathbf{B} = (A_1 \pm B_1)\,\mathbf{u}_1 + (A_2 \pm B_2)\,\mathbf{u}_2 + (A_3 \pm B_3)\,\mathbf{u}_3. \tag{0.4.5}$$

The three unit vectors are mutually orthogonal. Therefore

$$\mathbf{u}_1 \cdot \mathbf{u}_2 = \mathbf{u}_1 \cdot \mathbf{u}_3 = \mathbf{u}_2 \cdot \mathbf{u}_3 = 0. \tag{0.4.6}$$

Recalling that the magnitude of a unit vector is by definition unity, according to Eq. (0.3.7a) we have

$$\mathbf{u}_1 \cdot \mathbf{u}_1 = \mathbf{u}_2 \cdot \mathbf{u}_2 = \mathbf{u}_3 \cdot \mathbf{u}_3 = 1. \tag{0.4.7}$$

Thus the dot product of the two vectors with origins at the same point is

$$\begin{aligned}\mathbf{A}\cdot\mathbf{B} &= (A_1\,\mathbf{u}_1 + A_2\,\mathbf{u}_2 + A_3\,\mathbf{u}_3)\cdot(B_1\,\mathbf{u}_1 + B_2\,\mathbf{u}_2 + B_3\,\mathbf{u}_3)\\ &= A_1\,B_1 + A_2\,B_2 + A_3\,B_3.\end{aligned} \tag{0.4.8}$$

In Eq. (0.4.8), $A_1$, $A_2$, $A_3$ and $B_1$, $B_2$, $B_3$ stand for the rectangular, cylindrical or spherical components of the two vectors.

In the particular case when $\mathbf{B} = \mathbf{A}$, Eq. (0.4.8) becomes

$$\mathbf{A}\cdot\mathbf{A} = \mathbf{A}^2 = A^2 = A_1^2 + A_2^2 + A_3^2.$$

This shows that the magnitude $A$ of the vector $\mathbf{A}$ can be calculated from its components $A_1$, $A_2$, and $A_3$ with respect to any coordinate system with mutually orthogonal unit vectors as follows:

$$\boxed{A = (A_1^2 + A_2^2 + A_3^2)^{1/2}}\,. \tag{0.4.9}$$

Because the three unit vectors are mutually orthogonal, we have

$$\mathbf{u}_1 \times \mathbf{u}_2 = \mathbf{u}_3, \qquad \mathbf{u}_2 \times \mathbf{u}_3 = \mathbf{u}_1, \qquad \mathbf{u}_3 \times \mathbf{u}_1 = \mathbf{u}_2. \tag{0.4.10}$$

Also, the cross product of any unit vector with itself is zero:

$$\mathbf{u}_1 \times \mathbf{u}_1 = \mathbf{u}_2 \times \mathbf{u}_2 = \mathbf{u}_3 \times \mathbf{u}_3 = 0. \tag{0.4.11}$$

According to Eqs. (0.4.10) and (0.4.11), and taking into account (0.3.13), we thus have that the cross product of two vectors is given by

$$\begin{aligned}\mathbf{A}\times\mathbf{B} &= (A_1\,\mathbf{u}_1 + A_2\,\mathbf{u}_2 + A_3\,\mathbf{u}_3)\times(B_1\,\mathbf{u}_1 + B_2\,\mathbf{u}_2 + B_3\,\mathbf{u}_3)\\ &= \mathbf{u}_1(A_2B_3 - A_3B_2) + \mathbf{u}_2(A_3B_1 - A_1B_3) + \mathbf{u}_3(A_1B_2 - A_2B_1).\end{aligned} \tag{0.4.12}$$

Note the cyclical law 1–2–3, 2–3–1, and 3–1–2 characteristic for positive terms. All the general remarks made in connection with Eq. (0.4.8) apply to Eq. (0.4.12) as well. From Eq. (0.4.12) we see that the cross product of vectors $\mathbf{A}$ and $\mathbf{B}$ having origins at the same point can be written in the form of a determinant:

$$\mathbf{A}\times\mathbf{B} = \begin{vmatrix} \mathbf{u}_1 & \mathbf{u}_2 & \mathbf{u}_3 \\ A_1 & A_2 & A_3 \\ B_1 & B_2 & B_3 \end{vmatrix}\,. \tag{0.4.13}$$

This is perhaps more easy to remember than Eq. (0.4.12).

The triple scalar product of three vectors, say $\mathbf{C}\cdot(\mathbf{A}\times\mathbf{B})$, is readily obtained from Eqs. (0.4.8) and (0.4.12):

$$\begin{aligned}\mathbf{C}\cdot(\mathbf{A}\times\mathbf{B}) &= C_1(A_2B_3 - A_3B_2) + C_2(A_3B_1 - A_1B_3)\\ &\quad + C_3(A_1B_2 - A_2B_1).\end{aligned} \tag{0.4.14}$$

It can also be represented as a determinant, by simply replacing $\mathbf{u}_1$, $\mathbf{u}_2$, and $\mathbf{u}_3$ in the determinant of Eq. (0.4.13) by $C_1$, $C_2$, and $C_3$, respectively.

The triple vector product $\mathbf{C}\times(\mathbf{A}\times\mathbf{B})$ of three vectors $\mathbf{C}$, $\mathbf{A}$, and $\mathbf{B}$ can also be obtained by using Eq. (0.4.12). We first determine the components of vector $\mathbf{A} \times \mathbf{B}$, and then the cross product of vector $\mathbf{C}$ with vector $\mathbf{A} \times \mathbf{B}$ thus found.

**0.4.5 Concluding remarks.** Several times in this survey it was stressed that vector functions are not really connected with any particular coordinate system. However, mathematical operations with vectors are sometimes much more *convenient* to perform with respect to a coordinate system. Since vectors themselves are independent of the choice of the system, one can use the most suitable system for a problem at hand without impairing the generality of the result.

In particular, some vector identities can be proved only with difficulty if no reference to a coordinate system is made, while it is easy to prove them with respect to some convenient system. Most often the rectangular coordinate system is the one used for that purpose. It is very important to understand that, although such a proof is derived with respect to a particular coordinate system, there is no lack of generality as a result. Problem 0.G1 gives two examples of the application of this method.

## PROBLEMS

### Test problems

**0.T1.** Classify the following quantities as either scalar or vector quantities: mass, time, weight, course of a plane, position of a point with respect to another point, acceleration of a body, power of an engine.

**0.T2.** Several vectors, each starting from the end of the preceding one, form a closed polygon. What is the sum of these vectors?

**0.T3.** Discuss whether the definition of subtraction of vectors follows from the definition of vector addition, or whether an additional definition is really indispensable.

**0.T4.** Sketch the dependence of the dot product of two vectors having unit magnitudes on the angle $\alpha$ between them.

**0.T5.** In the diagram of the preceding problem, sketch the dependence of the magnitude of the cross product of the two vectors on the angle $\alpha$ between them.

**0.T6.** Discuss identity (0.3.13).

**0.T7.** State which of the following products of three or more vectors have a meaning: $\mathbf{A} \times (\mathbf{B} \cdot \mathbf{C})$, $(\mathbf{A} \times \mathbf{B}) \cdot \mathbf{C}$, $(\mathbf{A} \times \mathbf{B}) \times (\mathbf{C} \cdot \mathbf{D})$, $(\mathbf{A} \cdot \mathbf{B}) \cdot (\mathbf{C} \times \mathbf{D})$, $[(\mathbf{A} \cdot \mathbf{B}) \cdot \mathbf{C}] \times \mathbf{D}$, $[(\mathbf{A} \times \mathbf{B}) \cdot \mathbf{C}] \times \mathbf{D}$, $[\mathbf{A} \times (\mathbf{B} \times \mathbf{C})] \cdot \mathbf{D}$.

**0.T8.** What is the necessary and sufficient condition that three vectors, $\mathbf{A}$, $\mathbf{B}$, and $\mathbf{C}$, be in the same plane?

**0.T9.** Prove that $(\mathbf{A} \times \mathbf{B}) \cdot \mathbf{A}$ and $(\mathbf{A} \times \mathbf{B}) \cdot \mathbf{B}$ are zero.

**0.T10.** If $\mathbf{A} \cdot \mathbf{C} = \mathbf{B} \cdot \mathbf{C}$, does it mean that $\mathbf{A} = \mathbf{B}$?

**0.T11.** Which of the following sets of coordinates define a point: (1) $x = 2$, $y = -4$, $z = 0$; (2) $r = -4$, $\phi = 0°$, $z = -1$; (3) $r = 3$, $\theta = -90°$, $\phi = 0°$?

**0.T12.** Expand the determinant in Eq. (0.4.13) to prove that it is identical with the right-hand side of Eq. (0.4.12).

## General problems

**0.G1.** Prove Eqs. (0.3.14) and (0.3.17) by expanding left and right sides in the rectangular coordinate system.

**0.G2.** Show that if the origin of a vector is at a point $M(x, y, z)$ and the head is at a point $M'(x', y', z')$, then the unit vector along the vector is

$$\mathbf{u} = \frac{(x' - x)\,\mathbf{u}_x + (y' - y)\,\mathbf{u}_y + (z' - z)\,\mathbf{u}_z}{[(x' - x)^2 + (y' - y)^2 + (z' - z)^2]^{1/2}}\,.$$

**0.G3.** Show that the following relations between unit vectors of rectangular, cylindrical, and spherical coordinate systems are valid:

$$\text{cylindrical}\;\left\{\begin{aligned} \mathbf{u}_r &= \mathbf{u}_x \cos\phi + \mathbf{u}_y \sin\phi \\ \mathbf{u}_\phi &= -\mathbf{u}_x \sin\phi + \mathbf{u}_y \cos\phi \\ \mathbf{u}_z &= \mathbf{u}_z \end{aligned}\right\}\;\text{rectangular}$$

$$\text{spherical}\;\left\{\begin{aligned} \mathbf{u}_r &= \mathbf{u}_x \sin\theta\cos\phi + \mathbf{u}_y \sin\theta\sin\phi + \mathbf{u}_z\cos\theta \\ \mathbf{u}_\theta &= \mathbf{u}_x \cos\theta\cos\phi + \mathbf{u}_y \cos\theta\sin\phi - \mathbf{u}_z\sin\theta \\ \mathbf{u}_\phi &= -\mathbf{u}_x \sin\phi + \mathbf{u}_y\cos\phi \end{aligned}\right\}\;\text{rectangular}$$

**0.G4.** Show that if the coordinates of a point with respect to the cylindrical system are $r$, $\phi$, and $z$, and with respect to the spherical coordinate system they are $r$, $\theta$, and $\phi$, then the coordinates of the same point with respect to the rectangular coordinate system are $x = r\cos\phi$, $y = r\sin\phi$, $z = z$, and $x = r\sin\theta\cos\phi$, $y = r\sin\theta\sin\phi$, $z = r\cos\theta$, respectively.

**0.G5.** At a point having coordinates $(x, y, z)$, $(r, \phi, z)$, and $(r, \theta, \phi)$ with respect to the rectangular, cylindrical, and spherical coordinate systems, the rectangular components $A_x$, $A_y$, and $A_z$ of a vector **A** are known. Prove that the cylindrical components of the vector **A** are

$$\begin{aligned} A_r &= A_x\cos\phi + A_y\sin\phi, \\ A_\phi &= -A_x\sin\phi + A_y\cos\phi, \\ A_z &= A_z, \end{aligned}$$

and that the spherical components are

$$\begin{aligned} A_r &= A_x\sin\theta\cos\phi + A_y\sin\theta\sin\phi + A_z\cos\theta, \\ A_\theta &= A_x\cos\theta\cos\phi + A_y\cos\theta\sin\phi - A_z\sin\theta, \\ A_\phi &= -A_x\sin\phi + A_y\cos\phi. \end{aligned}$$

[*Hint:* Use the results of Problem 0.G3 and recall that one obtains the component of a vector in a certain direction as the dot product of the vector and the unit vector of that direction.]

**0.G6.** Given that $\mathbf{A} = A_x\,\mathbf{u}_x + A_y\,\mathbf{u}_y + A_z\,\mathbf{u}_z$ and $\mathbf{B} = B_x\,\mathbf{u}_x + B_y\,\mathbf{u}_y + B_z\,\mathbf{u}_z$, show that the smaller angle between the two vectors is given by

$$\alpha = \cos^{-1}\left\{\frac{A_x B_x + A_y B_y + A_z B_z}{[(A_x^2 + A_y^2 + A_z^2)(B_x^2 + B_y^2 + B_z^2)]^{1/2}}\right\}.$$

[*Hint:* Use Eq. (0.3.7).]

**0.G7.** The *direction cosines* of a vector are cosines of angles between the vector and the positive directions of the $x$-, $y$-, and $z$-axes. They are usually designed by $l$, $m$, and $n$, respectively. Show that $l^2 + m^2 + n^2 = 1$.

**0.G8.** Show that if $l_A$, $m_A$, and $n_A$ are the direction cosines of a vector **A**, and $l_B$, $m_B$, and $n_B$ those of a vector **B**, the cosine of the smaller angle between the two vectors is $l_A l_B + m_A m_B + n_A n_B$.

## Numerical problems

**0.N1.** Given that $\mathbf{A} = 3\mathbf{u}_x - 5\mathbf{u}_y + 8\mathbf{u}_z$, $\mathbf{B} = 4\mathbf{u}_x + \mathbf{u}_y - 3\mathbf{u}_z$, and $\mathbf{C} = -2\mathbf{u}_x + 3\mathbf{u}_y - 4\mathbf{u}_z$, determine: $\mathbf{A} + \mathbf{B} - \mathbf{C}$, $\mathbf{A}\cdot\mathbf{B}$, $\mathbf{A}\cdot\mathbf{C}$, $\mathbf{B} \times \mathbf{C}$, $\mathbf{A} \times \mathbf{C}$, $\mathbf{A}\cdot(\mathbf{B} \times \mathbf{C})$, $(\mathbf{A}\cdot\mathbf{B})\,\mathbf{C}$, $(\mathbf{A} \times \mathbf{B})\cdot\mathbf{C}$ and $(\mathbf{A} \times \mathbf{B}) \times \mathbf{C}$. Find the smaller angle between **A** and **B**, and **A** and **C**, and magnitudes of the three vectors.

**0.N2.** Given that $\mathbf{A} = 2\mathbf{u}_r - \mathbf{u}_\phi + \mathbf{u}_z$, with origin at the point of the cylindrical coordinate system $(3, 0°, 0)$, and $\mathbf{B} = 2\mathbf{u}_r - \mathbf{u}_\phi + \mathbf{u}_z$, with origin at the point $(5, \pi/2, 5)$, determine: $\mathbf{A} + \mathbf{B}$, $\mathbf{A} - \mathbf{B}$, $\mathbf{A}\cdot\mathbf{B}$, $\mathbf{A} \times \mathbf{B}$, the smaller angle between the two vectors, and their magnitudes.

**0.N3.** Given that $\mathbf{A} = 3\mathbf{u}_r + 2\mathbf{u}_\theta - \mathbf{u}_\phi$ has origin at the point $(1, \pi/2, 0°)$ in the spherical system, and $\mathbf{B} = -2\mathbf{u}_r - 4\mathbf{u}_\theta + 2\mathbf{u}_\phi$ has origin at the point $(3, \pi/2, \pi)$, determine: $\mathbf{A} + \mathbf{B}$, $\mathbf{A} - \mathbf{B}$, $\mathbf{A}\cdot\mathbf{B}$, $\mathbf{A} \times \mathbf{B}$, the smaller angle between the two vectors, and their magnitudes.

**0.N4.** Show that vectors $\mathbf{A} = \mathbf{u}_x + 4\mathbf{u}_y + 3\mathbf{u}_z$ and $\mathbf{B} = 2\mathbf{u}_x + \mathbf{u}_y - 2\mathbf{u}_z$ are perpendicular. Determine the ratio of their magnitudes.

**Suggested reading.** The following list progresses from simpler to more advanced texts. For a full bibliographical description, see the Bibliography.

1. HOLT, C. A., *Introduction to Electromagnetic Fields and Waves*, Chapter 2.
2. BOHN, E. V., *Introduction to Electromagnetic Fields and Waves*, Chapter 1.
3. PLONSEY, R., and R. E. COLLIN, *Principles and Applications of Electromagnetic Fields*, Chapter 1.

# PART I

# THE STATIC ELECTRIC FIELD

CHAPTER 1

# THE STATIC ELECTRIC FIELD IN THE ABSENCE OF DIELECTRICS

## 1.1 INTRODUCTION

The phenomena that we shall devote most of our time to in this text are the interactions between large aggregates of elemental electric charges. It is well known that elemental electric charges are the building-blocks of atoms of all substances. Interactions between charged particles inside individual atoms are the subject of theories of microscopic properties of matter, and thus we shall not consider them here. Rather we shall concentrate on interactions between large ensembles of atoms, interactions that are due to mutual effects between elemental charges inside these ensembles.

Since the number of charged particles inside *any* volume to be considered is supposed to be very large, the concept of infinitesimally small volume in the mathematical sense is not allowed. Instead we impose a restriction on the size of the "infinitesimally small" volume. It has to be very small, but large enough to enclose a vast number of elemental charges. Such an "infinitesimally small" volume is known as a *physically small* volume. The physically small volume will be implied whenever we refer to "infinitesimally" or "very" small volume.

We can give no unique order of magnitude of the physically small volume, since such a magnitude depends on the type of structure considered. However, it seems that a volume of $10^{-6}$ cubic centimeters can be regarded as an upper limit, with a vast amount of reserve in most cases. To put it in numbers, since there are, speaking very roughly, between $10^{22}$ and $10^{24}$ atoms per cubic centimeter of a rigid body, $10^{-6}$ cubic centimeters contain some $10^{16}$ to $10^{18}$ atoms. Although the exact number of atoms varies greatly for different substances, one can appreciate the fact that the above upper limit of the physically small volume is a safe one.

From the microscopic, atomic point of view, physical quantities inside any physically small volume can vary tremendously from point to point. For example, mass density is zero at all points outside elementary particles of matter, but enormous inside them. When we are considering interactions between large aggregates of electric charges, these variations on the atomic, microscopic scale are unimportant. It is enough simply to know the average values of the quantities considered inside a physically small volume. So we consider all physical quantities to be "smoothed out." Such an approach is referred to as the *macroscopic* approach, and the averaged physical quantities are termed *macroscopic* quantities.

This text deals with *macroscopic* phenomena related to interactions between elemental electric charges. Nevertheless, to enable the reader to better understand the mechanism of certain interactions, we shall use a model of the atomic structure of matter. Since we shall use this model only for analyzing macroscopic relations, we shall have no need for a model corresponding to the present sophisticated knowledge of the atomic structure of matter. It is sufficient for our purposes to envision a crude model of the atom, considered as having a central, heavy, positively charged nucleus and a number of light, negatively charged electrons orbiting around it.

Some substances have certain elemental electric charges practically free to move inside them. Such substances are known as *conductors*. Still considering the above-mentioned crude model of the atom, we may say that one type of conductor, known as a metallic conductor, has atoms in which the electrons of the outer atomic shells (the valence or *conduction* electrons) are attached very loosely to the rest of the atom. Conductors also exist in the form of electrolytic solutions and gases, in which there are movable charged particles, called *ions*, which can carry either positive or negative charges.

The other type of substances, known as *insulators* or *dielectrics*, in normal circumstances have practically no freely movable charges in their structures. Substances having only a relatively small number of free charges in their structure are known as *semiconductors*.

Atoms are usually electrically neutral. That is, the total positive charge of the nucleus is equal in magnitude to the total negative charge of its orbiting electrons. Bodies consisting of such neutral atoms are said to be (macroscopically) uncharged. If there is on a body a surplus of either positive or negative elemental charges, it is referred to as a *charged body*.

Macroscopic interactions between aggregates of electric charges which are in an arbitrary state of motion are very complicated. Therefore it is useful to start with the simplest case of these interactions. So in the first part of the book we shall consider macroscopic phenomena relating to distributions of electric charges which are static with respect to the observer. In a macroscopic theory the concept of static charges implies that the total charge inside any physically small volume is constant in time.

The interactions between static macroscopic charges are simplest when the charges are located on conducting bodies in the absence of dielectrics. This chapter will therefore deal with this case. And although this situation is the simplest, it is of a considerable practical importance. It will enable us to define and understand many important concepts necessary for analyzing more general types of interactions between charges. It is particularly important as the first door into the science of electromagnetic fields.

## 1.2 COULOMB'S LAW

Charged bodies of dimensions small in comparison with the distance between them are referred to as *point* charges. It was found experimentally by Coulomb in 1785 that the force between two point charges is directly proportional to the product of their charges and inversely proportional to the square of their distance. Denote the charges on the two bodies by $Q_1$ and $Q_2$ and the distance between them by $r$. The magnitude of this force, known as the *Coulomb* or the *electric force*, can then be calculated from the formula

$$F = K\frac{Q_1Q_2}{r^2}. \tag{1.2.1}$$

Here $K$ is a constant depending on the medium in which the charges are immersed and on the system of units used. In this chapter we shall consider only charges in vacuum. In the MKSA system of units, which is used in this book,* the constant $K$ for vacuum (and approximately for air) is written in the form $K = 1/(4\pi\varepsilon_0)$.† The new constant, $\varepsilon_0$, is usually called the *permittivity of free space*.

Recall that by convention we associate the minus sign with the electron (negative) type of charge, and the plus sign with the other (positive) type of charge. It has been found experimentally that two charges repel each other if they are of the same sign and attract each other if they are of opposite signs. Using vector notation, we can also express this fact in mathematical form. With reference to Fig. 1.1, the force $\mathbf{F}_{12}$ by which charge $Q_1$ acts on charge $Q_2$ is

$$\boxed{\mathbf{F}_{12} = \frac{1}{4\pi\varepsilon_0}\frac{Q_1Q_2}{r^2}\mathbf{u}_{12}} \quad \text{(Coulomb's law).} \tag{1.2.2}$$

Here $\mathbf{u}_{12}$ is the unit vector directed from charge $Q_1$ to charge $Q_2$. Force $\mathbf{F}_{21}$ on charge $Q_1$ is obtained by replacing $\mathbf{u}_{12}$ in Eq. (1.2.2) by $-\mathbf{u}_{12}$, or $\mathbf{u}_{21}$.

In the MKSA system of units the unit of charge is the *coulomb* (abbreviated C). It is approximately equal to $6.242 \times 10^{18}$ electron charges, taken with the positive sign. The permittivity of free space is found to have a value

$$\boxed{\varepsilon_0 = 8.854 \times 10^{-12} \simeq \frac{1}{36\pi \times 10^9}\frac{\text{coulomb}^2}{\text{newton}\cdot\text{m}^2}}. \tag{1.2.3}$$

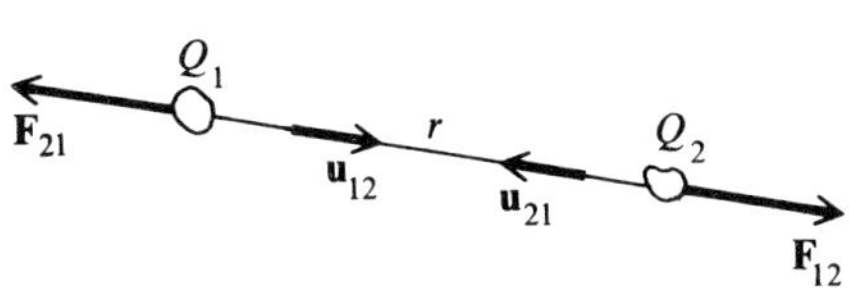


**Fig. 1.1** Two "point" charges and reference directions of vectors entering the vector form of Coulomb's law.

---

* A short account of the MKSA system is given in Appendix 1.

† We write $4\pi$ for convenience. When it is introduced in $K$ it drops out of many important formulas to be derived later.

(The *newton* is the unit of force in the MKSA system.) For reasons that will become apparent in Section 1.14, the equivalent unit *farad per meter* (F/m) of $\varepsilon_0$ is frequently used instead of coulomb$^2$/(newton $\cdot$ m$^2$).

It has also been shown experimentally that if there are more than two "point" charges, the total force on any one of them can be obtained as a vector sum of partial Coulomb forces (1.2.2). This is known as the *principle of superposition.*

Coulomb's law is a macroscopic experimental law. Hence two points should be appreciated: First, the domain of its validity, strictly speaking, is limited to macroscopic structures. It has been proved, however, that Coulomb's law is also valid for phenomena which are on an atomic scale, if distances between elemental charges are larger than approximately $10^{-12}$ cm. Second, as with any physical law, Coulomb's law has been verified only within limits of experimental error. It has been shown by an indirect method that if we suppose the force to be proportional to $r^{-n}$, then $n = 2 \pm 10^{-9}$. Therefore we can use Coulomb's law (1.2.2) with complete confidence as a starting point for a macroscopic theory of phenomena relating to systems of static electric charges.

*Problems referring to Section 1.2:* 1.T1 to 1.T4, 1.G1, 1.G2, 1.N1 to 1.N3.

## 1.3 CONCEPT OF THE STATIC ELECTRIC FIELD

Suppose that an observer wishes to determine experimentally whether there are stationary charged bodies around him. For that purpose he has to be equipped with a small charged body, which we shall call simply a *test charge*, and he has to consider electric forces on it. Suppose that he finds that at every point in his laboratory the test charge experiences a certain force, but that there are no bodies inside the laboratory which are charged. The observed forces are certainly due to some unknown charges outside the laboratory. His experimental evidence does not show him where they are. So he can only conclude that his laboratory is in the "field of interest" of some electric charges. In order to be more concise in the future, he introduces a new name for that experience, and says that in his laboratory there exists an *electric field.* Thus it is said that an electric field exists in a region of space whenever a test charge introduced in the region is acted on by an electric force.

To characterize the electric field at each point quantitatively, without reference to a specific test charge, we can divide the force acting on a small test charge by the magnitude of the charge. The ratio is always independent of the test charge, which is a direct consequence of Coulomb's law (1.2.2) and of the principle of superposition. Denoting the test charge by $\Delta Q$, we have

$$\frac{\text{Force on } \Delta Q \text{ at a point}}{\Delta Q} = \text{vector quantity independent of } \Delta Q.$$

This ratio is called the *electric field strength* or the *electric field intensity*, and is denoted by **E**:

$$\mathbf{E} = \frac{\mathbf{F}_{\text{on } \Delta Q}}{\Delta Q} \qquad \text{(definition of } \mathbf{E}\text{).} \qquad (1.3.1)$$

From Eq. (1.3.1), we see that the direction of **E** is the direction of the force on a *positive* test charge at the point considered. In the general case the electric field intensity **E** can be considered mathematically as a (vector) function of coordinates with respect to some convenient coordinate system.

According to Eq. (1.3.1), the unit of the electric field intensity in the MKSA system is the newton per coulomb. For reasons that will become apparent in Section 1.6, the equivalent unit *volt per meter* (V/m) is usually used instead.

Let us determine the electric field intensity at a point of the electric field surrounding a point charge $Q$. Changing the $Q_2$ in Eq. (1.2.2) to $\Delta Q$ and the $Q_1$ to $Q$, we find that the electric field intensity at a distance $r$ from the charge $Q$ is

$$\mathbf{E} = \frac{\mathbf{F}_{\text{on } \Delta Q}}{\Delta Q} = \frac{1}{4\pi\varepsilon_0}\frac{Q}{r^2}\mathbf{u} \qquad (\mathbf{E} \text{ of a point charge}). \qquad (1.3.2)$$

According to Eq. (1.2.2), the unit vector **u** is directed *away* from the charge $Q$. The electric field intensity of a point charge is a vector directed along imaginary straight lines emerging from the charge. If the charge is positive **E** is directed away from it, and if the charge is negative **E** is directed toward it (Fig. 1.2).

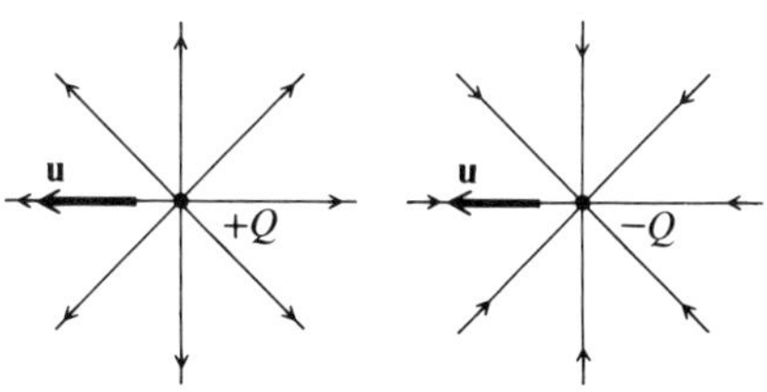


**Fig. 1.2** Graphical representation of the electric field of positive and negative point charges.

Using the superposition principle and Eq. (1.3.2), we can determine the electric field intensity due to $n$ point charges $Q_1, Q_2, \ldots, Q_n$. Consider a point $A$ at distances $r_1, r_2, \ldots, r_n$ from these charges. Denoting by $\mathbf{u}_{1A}, \mathbf{u}_{2A}, \ldots, \mathbf{u}_{nA}$ the unit vectors directed *from* charges $Q_1, Q_2, \ldots, Q_n$ *toward* the point $A$, we can readily find the electric field intensity to be

$$\mathbf{E}_A = \sum_{k=1}^{n} \frac{1}{4\pi\varepsilon_0}\frac{Q_k}{r_k^2}\mathbf{u}_{kA}. \qquad (1.3.3)$$

*Problem referring to Section 1.3:* 1.N4.

## 1.4 ELECTRIC FIELD OF CONTINUOUS DISTRIBUTION OF CHARGES

The real sources of all electric fields are always discrete elemental electric charges. However, macroscopic theory usually deals with a vast number of elemental charges distributed in a finite region of space. Since only the macroscopic properties of these charges are important, the concept of the *volume charge density* $\rho$ is defined as follows:

$$\rho = \frac{\text{sum of all charges inside a physically small volume } \Delta v}{\Delta v}. \tag{1.4.1}$$

The volume charge density is of course a function of the point of observation. If $\rho$ is known in advance, or determined in some appropriate manner, the charge inside any small volume $\Delta v$ can be determined as

$$\Delta Q = \rho \, \Delta v. \tag{1.4.2}$$

Consider the whole region occupied by charges to be divided into such small volumes. According to Eq. (1.3.3), it is possible to obtain the field intensity at any point as the sum of field intensities due to charges inside all volumes $\Delta v$:

$$\mathbf{E} = \sum_{v} \frac{1}{4\pi\varepsilon_0} \frac{\rho \, \Delta v}{r^2} \mathbf{u}. \tag{1.4.3}$$

The summation sign refers to the volume $v$ occupied by the charges. The unit vector $\mathbf{u}$ is directed from the volume element $\Delta v$ toward the point at which the electric field intensity $\mathbf{E}$ is being determined. For $\Delta v$ small enough, Eq. (1.4.3) becomes the volume integral over the region $v$ (Fig. 1.3):

$$\mathbf{E} = \frac{1}{4\pi\varepsilon_0} \int_v \frac{\rho \, dv}{r^2} \mathbf{u}. \tag{1.4.4}$$

Hence, given $\rho$ as a function of coordinates, we can always (at least numerically) determine the electric field intensity $\mathbf{E}$. It should be noted that in most cases the distribution of charges is *not* known. So Eqs. (1.4.3) and (1.4.4) have a rather limited practical usefulness.

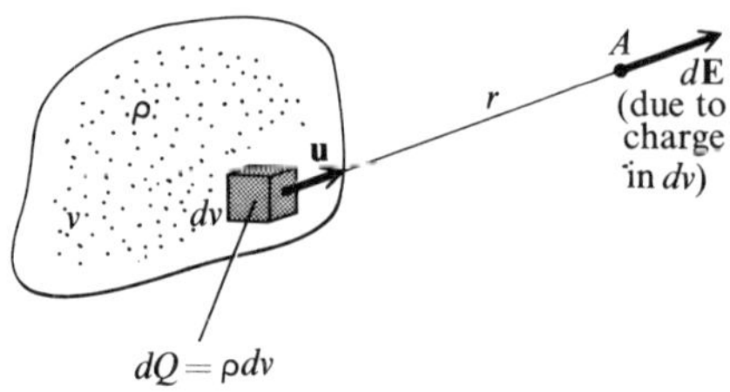


**Fig. 1.3** A volume distribution of charges.

Elemental charges are always distributed within a certain volume. In some important cases, for example in the case of charged conducting bodies, elemental charges are distributed in a very thin layer on the surface of the body (Section 1.9). It is then convenient to consider charges to be spread over the very surface, and to introduce the concept of the *surface charge density* $\rho_s$, which is defined as the total charge on a surface element $\Delta S$ divided by the area $\Delta S$, with $\Delta S$ approaching zero:

$$\rho_s = \lim_{\Delta S \to 0} \frac{\Delta Q_{\text{on } \Delta S}}{\Delta S}. \tag{1.4.5}$$

Just as in the case of Eq. (1.4.4), we can determine the electric field intensity resulting from such a distribution of charges as a surface integral over all charged surfaces of the system (Fig. 1.4):

$$\mathbf{E} = \frac{1}{4\pi\varepsilon_0} \int_S \frac{\rho_s dS}{r^2} \mathbf{u}. \tag{1.4.6}$$

Suppose finally that charges are distributed along very thin wires (of radii small in comparison with distances from the wire to the points of observation). It is convenient to introduce the concept of the *linear charge density* $\rho_l$. The linear charge density is defined as the ratio of charge $\Delta Q$ on an element $\Delta l$ of the wire to the length $\Delta l$ of the element, for $\Delta l$ tending to zero:

$$\rho_l = \lim_{\Delta l \to 0} \frac{\Delta Q_{\text{on } \Delta l}}{\Delta l}. \tag{1.4.7}$$

The electric field intensity resulting from such a distribution of charges can be calculated according to the formula (Fig. 1.5),

$$\mathbf{E} = \frac{1}{4\pi\varepsilon_0} \int_C \frac{\rho_l \, dl}{r^2} \mathbf{u}. \tag{1.4.8}$$

As an example of the application of formula (1.4.6), let us determine the electric field intensity on the axis of a circular disk of radius $a$, uniformly charged over its surface with a surface charge density $\rho_s$. (Such a distribution is possible only if

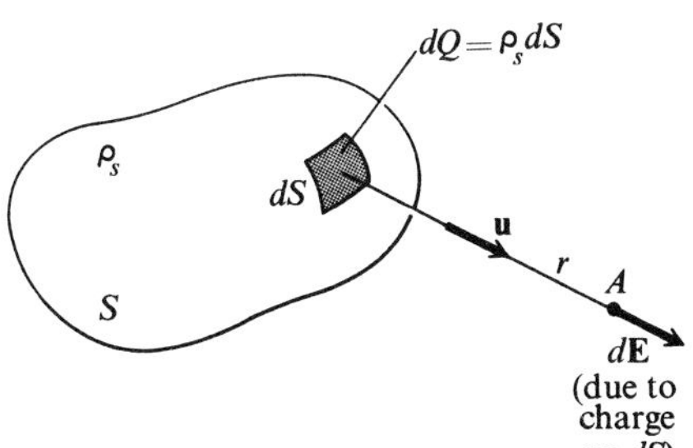


**Fig. 1.4** Charges distributed over a surface.

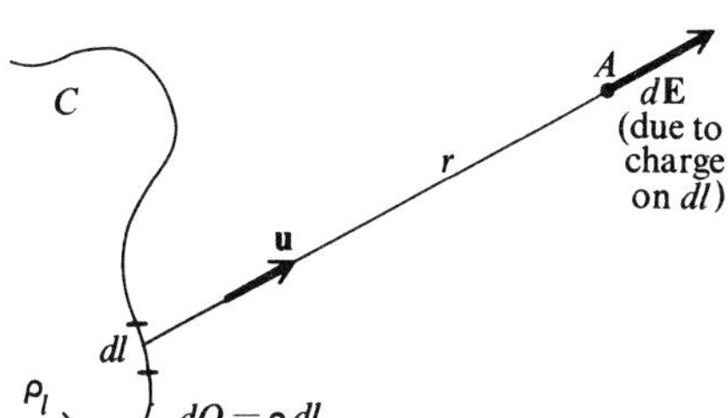


**Fig. 1.5** A line distribution of charges.

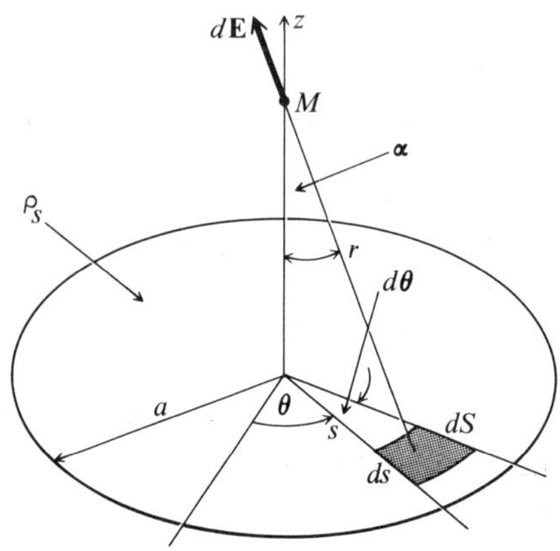


**Fig. 1.6** A thin circular disk charged uniformly over its surface.

the disk is made of an insulating material; see Problem 1.G33.) The geometry is represented in Fig. 1.6.

The magnitude of the electric field intensity vector at a point $M$ on the disk axis due to the charge $\rho_s\, dS$ on the surface element $dS = s\, d\theta\, ds$ is

$$dE = \frac{\rho_s\, dS}{4\pi\varepsilon_0 r^2} = \frac{\rho_s\, s\, d\theta\, ds}{4\pi\varepsilon_0(z^2 + s^2)}.$$

The direction of the vector $d\mathbf{E}$ is shown in Fig. 1.6. By symmetry, on the disk axis (the $z$-axis), there exists only the $z$-component of the electric field intensity. So only the $z$-component of $d\mathbf{E}$ is required:

$$dE_z = dE \cos\alpha = \frac{\rho_s\, z\, s\, d\theta\, ds}{4\pi\varepsilon_0(z^2 + s^2)^{3/2}},$$

because $\cos\alpha = z/r = z/(z^2 + s^2)^{1/2}$. Hence the total electric field intensity at $M$ is

$$E_z = \frac{\rho_s z}{4\pi\varepsilon_0}\int_0^{2\pi} d\theta \int_0^a \frac{s\, ds}{(z^2 + s^2)^{3/2}} = \frac{\rho_s}{2\varepsilon_0}\left[1 - \frac{z}{(z^2 + a^2)^{1/2}}\right]. \quad (1.4.9)$$

*Problems referring to Section 1.4:* 1.G3 to 1.G7, 1.N5.

## 1.5 LINES OF FORCE

As an aid to visualizing the structure of the electric field associated with a particular distribution of charges, it is helpful to draw a set of lines having the property that the electric field intensity vector $\mathbf{E}$ is tangential to these lines at every point. We add to these lines small arrow signs, indicating the direction of $\mathbf{E}$ along them. From such a picture alone we can determine the direction in which the electric force would act on a positive test charge at any point of the field. These lines are

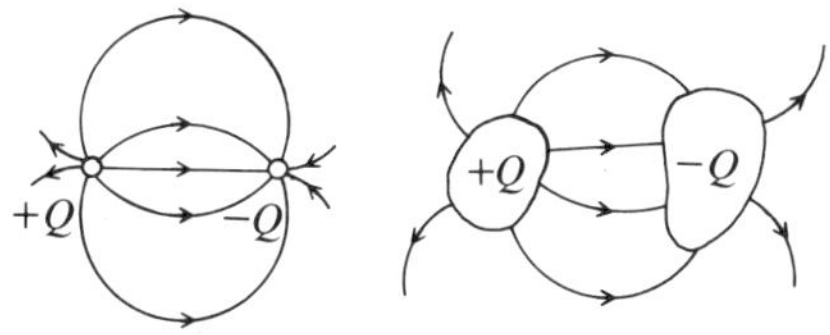


**Fig. 1.7** Examples of representing the electric field by lines of force.

called the *lines of force*, or the *lines of the electric field intensity*. Examples of the lines of force of the electric field of positive and negative point charges are shown in Fig. 1.2. It follows directly from the definition of the lines of force that these lines emerge from positive charges and end on negative charges. Further examples of the electric field represented by lines of force are shown in Fig. 1.7.

## 1.6 ELECTROSTATIC POTENTIAL

**1.6.1 Introduction. Line integral of the electric field intensity vector.** To describe quantitatively the force properties of a static electric field, we introduced the concept of the electric field intensity vector. In this section it will be shown that the electric field intensity vector created by static charges satisfies a certain mathematical relation. When we use this relation, a new description of the static electric field is possible. To understand this, we must make a short mathematical digression.

Consider two points, $A$ and $B$, in an arbitrary *static* electric field, that is, in a field produced by static charges (Fig. 1.8). Suppose that a test charge $\Delta Q$ is initially at $A$. Let the *electric* forces move the test charge $\Delta Q$ from $A$ to $B$ along an arbitrary path $AaB$. Note that the electric field intensity vector generally *varies* from one point to another. So the work done by the electric forces in moving $\Delta Q$ along a small element $\Delta l$ of the path is *approximately* equal to the product of $\Delta l$ and the projection of the force $\Delta Q\mathbf{E}$, acting on the charge at a point on the element $\Delta l$, in the direction of $\Delta l$. That is, designating the work done by the electric forces by $\Delta W$, we have $\Delta W \simeq \Delta Q\, E\, \Delta l \cos\alpha$ (Fig. 1.8). If we consider the element $\Delta l$ of the path as a *vector* pointed in the direction of the motion of the test charge, the elemental work done by the electric forces can be represented simply as

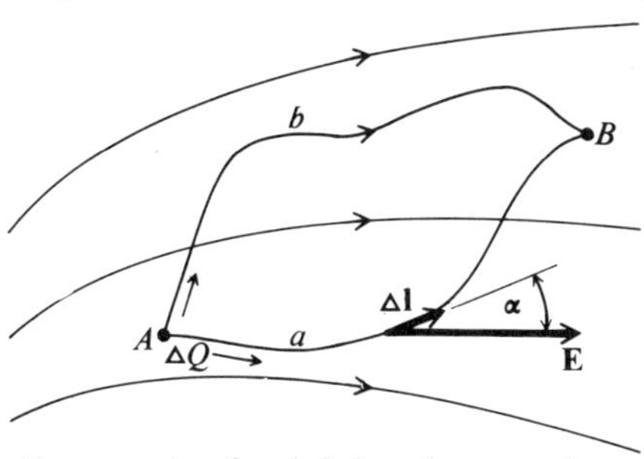


**Fig. 1.8** Construction for defining electrostatic potential.

$\Delta W \simeq \Delta Q\mathbf{E}\cdot\Delta\mathbf{l}$. The total work $W$ done by the electric forces in moving $\Delta Q$ from $A$ to $B$ is approximately equal to the sum of all these elemental works:

$$W \simeq \sum_A^B \Delta W = \sum_A^B \Delta Q\mathbf{E}\cdot\Delta\mathbf{l}. \tag{1.6.1}$$

To obtain this work more precisely, it is necessary to decrease $\Delta\mathbf{l}$, that is, to increase the number of terms in the sum on the right. For, if $\Delta\mathbf{l}$ is small enough, the electric field intensity $\mathbf{E}$ will be practically constant along it. Let $\Delta\mathbf{l}$ be infinitesimally small. Then the right-hand side of Eq. (1.6.1) represents an infinite sum of infinitely small terms. It is then referred to as the *line integral* of force $\Delta Q\mathbf{E}$, and is written as

$$W = \int_A^B \Delta Q\mathbf{E}\cdot d\mathbf{l}. \tag{1.6.2}$$

Thus the line integral is defined as the limit of the sum in Eq. (1.6.1) for elements $\Delta\mathbf{l}$ of the path between $A$ and $B$ assumed sufficiently small. The integral in Eq. (1.6.2) is a scalar integral. Theoretically, it can be determined, provided that $\mathbf{E}$ and the path between $A$ and $B$ are known functions of coordinates. Determining this integral exactly is often a very intricate matter. But to determine a line integral approximately, we can always start with its summation form (1.6.1).

The line integral of the electric field intensity vector of static electric charges has a simple and important property. It is a function of the starting and end points, but not of the particular path between them. To prove this, suppose that the work done by the electric forces in moving $\Delta Q$ from $A$ to $B$ is different if $\Delta Q$ is moved along $AaB$ from what it is if $\Delta Q$ is moved along $AbB$ (Fig. 1.8). Let us say that

$$\left\{\int_A^B \Delta Q\mathbf{E}\cdot d\mathbf{l}\right\}_{\text{along } a} > \left\{\int_A^B \Delta Q\mathbf{E}\cdot d\mathbf{l}\right\}_{\text{along } b}.$$

Now we can obtain the work done by the electric forces in moving $\Delta Q$ from $B$ to $A$ simply by changing the sign in Eq. (1.6.2), because when we go from $B$ to $A$ the element $d\mathbf{l}$ is in the direction opposite to that assumed earlier. Provided that the above inequality is valid, if we let $\Delta Q$ move along $AaB$ and then back along $BbA$, at $A$ it will have energy *different* from zero. We can take this surplus energy from the charge in some appropriate manner and let it again go around $AaBbA$, to provide us with more energy. This is obviously inconsistent with the law of conservation of energy. For when $\Delta Q$ is back at $A$, the whole system is in exactly the same state as before. The charges producing the field are static and unchanging, and $\Delta Q$ is back at the starting point. Hence the work done by the electric forces must be the same along both paths. Since $AaB$ and $AbB$ are two arbitrary paths, we have quite generally that in the *static* electric field

$$\int_A^B \Delta Q\mathbf{E}\cdot d\mathbf{l} = \text{constant independent of the path between } A \text{ and } B. \tag{1.6.3}$$

A direct corollary of this conclusion is that the work done by the electric forces in moving a test charge $\Delta Q$ around a closed contour $C$ is zero; that is,

$$\Delta Q \oint_C \mathbf{E} \cdot d\mathbf{l} = 0. \tag{1.6.4}$$

An extremely important relation follows, which must be satisfied by the electric field intensity vector of any real *static* electric field:

$$\boxed{\oint_C \mathbf{E} \cdot d\mathbf{l} = 0} \quad \text{(consequence of law of conservation of energy).} \tag{1.6.5}$$

**1.6.2 Definition of the electrostatic potential.** If Eq. (1.6.3) is divided by $\Delta Q$, it becomes independent of the magnitude of the test charge:

$$\int_A^B \mathbf{E} \cdot d\mathbf{l} = \text{const.} \tag{1.6.6}$$

This shows that the work done by the electric forces per unit test charge is a function of the terminal points of the path only, and not of the path itself. But it also represents an integral relation which the electric field intensity vector function $\mathbf{E}$ created by *static* electric charges must satisfy in all cases. We shall see in Section 11.2 that this is not true for time-varying fields.

The property (1.6.6) of electrostatic fields can be used to describe them from the standpoint of energy. For that purpose let us choose *any* point $R$, in or out of the electrostatic field. We shall refer to $R$ as the *reference point*. Let us determine the line integral of the electric field intensity $\mathbf{E}$ *from* each point of the field *to* the reference point $R$. These integrals, since they are independent of the path chosen from the field points to the reference point $R$, have unique values. They represent the work that would be done by electric forces in moving a unit test charge from field points to the point $R$. In this manner we could, figuratively speaking, attach to every point of the field a small card with information as to how much work would be done by the electric forces in moving the unit test charge from that point to the reference point $R$. This situation is sketched in Fig. 1.9, in which the work per unit charge from one point to $R$ is specified as 2.35 J/C (J stands for *joule*, the

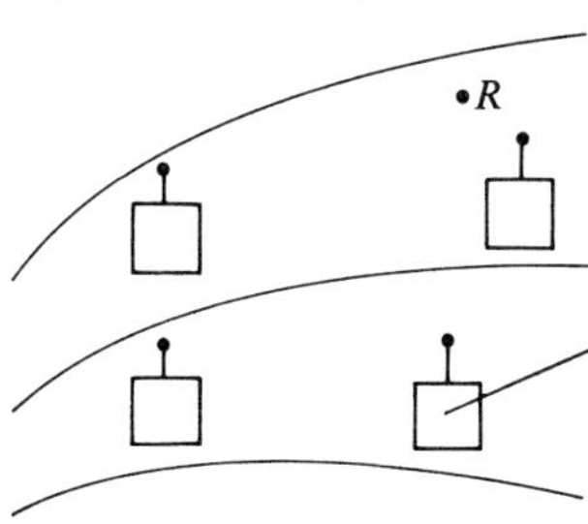


**Fig. 1.9** Description of the static electric field in terms of electrostatic potential.

unit of work and energy in the MKSA system). So we can describe the electric field in terms of a scalar rather than a vector function, the electric field intensity **E**.

Mathematically speaking, vector functions are more complicated than scalar functions, and it is precisely this that makes it more advantageous to describe the electric field in terms of energy than in terms of force. Instead of dealing with the vector field of the electric field intensity, we deal with a certain scalar field. We shall demonstrate later that the force properties of a static electric field are easily derivable from its energy properties. So this simpler, scalar description of the static electric field provides, directly or indirectly, complete data about the field.

The line integral

$$\boxed{V_A = \int_A^R \mathbf{E}\cdot d\mathbf{l}} \qquad \text{(definition of electrostatic potential)} \qquad (1.6.7)$$

is referred to as the *electrostatic potential at the field point A with respect to the reference point R*. The electrostatic potential at some other point, say $B$, with respect to the same reference point is

$$V_B = \int_B^R \mathbf{E}\cdot d\mathbf{l}.$$

Using Eq. (1.6.5), we can easily prove that the difference of potential at two points, $A$ and $B$, is

$$\boxed{V_A - V_B = \int_A^R \mathbf{E}\cdot d\mathbf{l} - \int_B^R \mathbf{E}\cdot d\mathbf{l} = \int_A^B \mathbf{E}\cdot d\mathbf{l}}\ . \qquad (1.6.8)$$

Note that the reference point $R$ does not appear on the right-hand side of Eq. (1.6.8). Thus the difference of potential at any two points is independent of the position of the reference point.

If Eq. (1.6.7) is evaluated for a general field point $A(x, y, z)$,* the scalar function

$$V(x, y, z) = \int_{A(x,y,z)}^R \mathbf{E}\cdot d\mathbf{l} \qquad (1.6.9)$$

is called the *electrostatic potential function*, or often simply the *potential*. It should be noted that the electrostatic potential is, by definition, *always* given with respect to a certain reference point.

Let us determine the potential of the electric field due to a single point charge. This serves a double purpose: It illustrates the method of evaluating the potential if the electric field intensity is a known function, and also provides a basis for evaluating the electrostatic potential of an arbitrary distribution of charges.

* Coordinates $(x, y, z)$ indicate only *a* set of coordinates, not necessarily a rectangular set. We shall use this convention frequently.

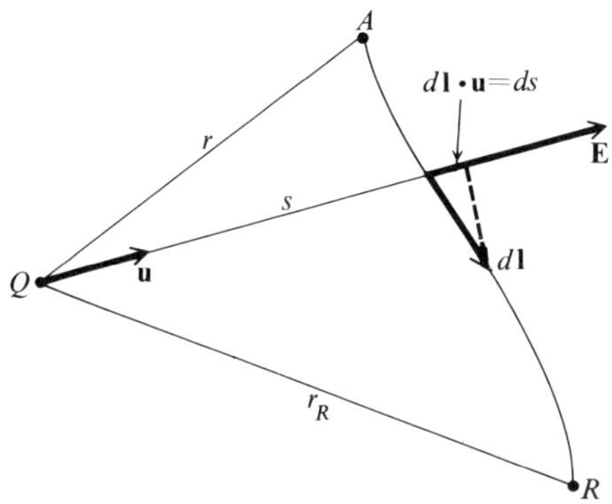


**Fig. 1.10** Construction for determining potential of the field created by a point charge.

Suppose that the reference point $R$ is at a distance $r_R$ from a point charge $Q$. Let us evaluate the potential of a point $A$ at a distance $r$ from the charge (Fig. 1.10). Using Eqs. (1.6.7) and (1.3.2), we have

$$V_A = \int_A^R \mathbf{E}\cdot d\mathbf{l} = \frac{Q}{4\pi\varepsilon_0}\int_r^{r_R}\frac{\mathbf{u}\cdot d\mathbf{l}}{s^2} = \frac{Q}{4\pi\varepsilon_0}\int_r^{r_R}\frac{ds}{s^2} = \frac{Q}{4\pi\varepsilon_0}\left(\frac{1}{r}-\frac{1}{r_R}\right). \qquad (1.6.10)$$

The location of the reference point is completely arbitrary, so we can conventionally take it at infinity. Assuming $r_R$ in (1.6.10) to be infinite, we obtain the electrostatic potential function of a point charge $Q$ with respect to the reference point at infinity (dropping the subscript $A$ as unecessary):

$$\boxed{V(r) = \frac{Q}{4\pi\varepsilon_0 r}} \qquad \text{(single point charge, reference point at infinity).} \qquad (1.6.11)$$

It is obvious that we can adopt the convention of the reference point being at infinity whenever charges are located inside a finite volume.

Using the same notation as in Figs. 1.3 to 1.5, we can determine the electrostatic potential due to an arbitrary distribution of charges. Suppose that the potential is created by a set of point charges, and by volume, surface, and line charge distributions. Then

$$V = \sum_{k=1}^{n}\frac{Q_k}{4\pi\varepsilon_0 r_k} + \int_v \frac{\rho\, dv}{4\pi\varepsilon_0 r} + \int_S \frac{\rho_s\, dS}{4\pi\varepsilon_0 r} + \int_C \frac{\rho_l\, dl}{4\pi\varepsilon_0 r} \qquad (1.6.12)$$

(reference point at infinity). Equation (1.6.12) is of course valid only if there are no charges at infinity.

According to Eq. (1.6.9), the unit of the potential $V$ is equal to the unit of the electric field intensity times one meter. Since the potential is a quantity used extensively in practice, it is given a separate name, the *volt* (abbreviated V). Therefore the unit of the electric field intensity is the volt per meter (V/m), as mentioned in Section 1.3.

To illustrate the application of Eq. (1.6.12), let us determine the electrostatic potential at point $M$ of Fig. 1.6. The only term existing in this case is the surface integral. Since $dS = s\,d\theta\,ds$,

$$V(z) = \int_0^{2\pi}\int_0^a \frac{\rho_s\,s\,d\theta\,ds}{4\pi\varepsilon_0 r}$$

$$= \frac{\rho_s}{4\pi\varepsilon_0}\int_0^{2\pi} d\theta \int_0^a \frac{s\,ds}{\sqrt{z^2+s^2}} = \frac{\rho_s}{2\varepsilon_0}\left(\sqrt{z^2+a^2} - z\right). \quad (1.6.13)$$

In the following section it will be shown that the electric field intensity at $M$ can be obtained from $V(z)$ by a simple differentiation.

*Problems referring to Section 1.6:* 1.T5 to 1.T8, 1.G8 to 1.G10.

## 1.7 EVALUATION OF ELECTRIC FIELD INTENSITY FROM ELECTROSTATIC POTENTIAL. GRADIENT OF A SCALAR FUNCTION

Equation (1.6.8) defines the differences of potential at any pair of points in a field. In particular, this equation can be used to evaluate the electric field intensity from the electrostatic potential, if the latter is known. It is obvious from Eq. (1.6.9) that, since the potential is determined as an *integral* of the electric field intensity, the electric field intensity must be a function of *derivatives* of the potential. Recall that in the general case the potential is a scalar function of *three* variables (three coordinates of a field point).

Let us write Eq. (1.6.8) for the case of two adjacent points $A$ and $B$, $B$ being determined with respect to $A$ by a differential vector length $d\mathbf{l}$ (Fig. 1.11). The potential difference is hence infinitesimally small, say $dV$, and instead of the integral (sum), we have only one term. Assuming that the potential *rises* in the direction of $d\mathbf{l}$, Eq. (1.6.8) yields in this case

$$V_A - V_B = V - (V + dV) = -dV = \mathbf{E}\cdot d\mathbf{l} = E\cos(\mathbf{E}, d\mathbf{l})\,dl = E_l\,dl. \quad (1.7.1)$$

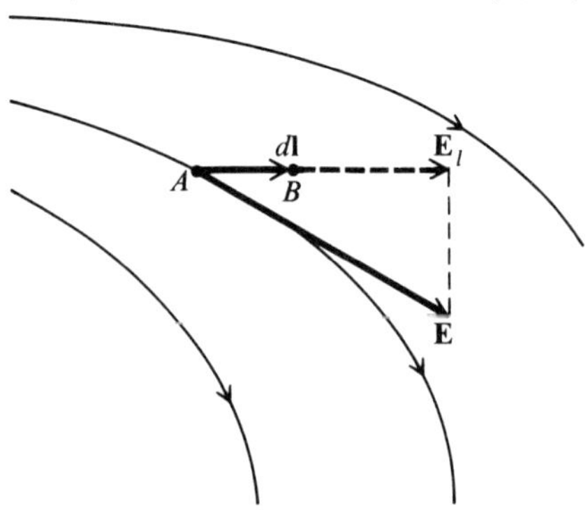


**Fig. 1.11** Difference of potentials at points $A$ and $B$ is $\mathbf{E}\cdot d\mathbf{l} = E_l\,dl$.

Here $E_l$ denotes the component of $\mathbf{E}$ in the direction of $d\mathbf{l}$. So

$$E_l = -\frac{dV}{dl}. \tag{1.7.2}$$

Now $E_l$ is the component of the *vector* $\mathbf{E}$ in the direction defined by the element $d\mathbf{l}$. Therefore $dV/dl$ can also be regarded as a component of a certain vector function. This vector function is referred to as the *gradient* of $V$, and is designated as grad $V$. So the component $\text{grad}_l\, V$ of vector grad $V$ in the direction defined by $d\mathbf{l}$ is

$$\boxed{\text{grad}_l\, V = \frac{dV}{dl}} \quad \text{(definition of component of gradient in direction } dl\text{).} \tag{1.7.3}$$

Using this definition, we can determine components of the gradient of a scalar function $V$ in the directions of three unit vectors of any coordinate system. For example, in a rectangular system,

$$\text{grad}_x\, V = \frac{\partial V}{\partial x}, \qquad \text{grad}_y\, V = \frac{\partial V}{\partial y}, \qquad \text{grad}_z\, V = \frac{\partial V}{\partial z},$$

from which the total vector grad $V$ is

$$\text{grad}\, V = \frac{\partial V}{\partial x}\mathbf{u}_x + \frac{\partial V}{\partial y}\mathbf{u}_y + \frac{\partial V}{\partial z}\mathbf{u}_z \quad \text{(rectangular coordinate system).} \tag{1.7.4}$$

Partial rather than ordinary derivative signs are used because $V$ is a function of all three coordinates.

In a rectangular coordinate system a shorthand notation for gradient is often used. First the so-called *del* operator $\nabla$ is defined as

$$\nabla = \frac{\partial}{\partial x}\mathbf{u}_x + \frac{\partial}{\partial y}\mathbf{u}_y + \frac{\partial}{\partial z}\mathbf{u}_z. \tag{1.7.5}$$

Next a convention is adopted that "multiplying" a scalar function by the del operator yields the right-hand side of Eq. (1.7.4):

$$\nabla V = \left(\frac{\partial}{\partial x}\mathbf{u}_x + \frac{\partial}{\partial y}\mathbf{u}_y + \frac{\partial}{\partial z}\mathbf{u}_z\right) V = \frac{\partial V}{\partial x}\mathbf{u}_x + \frac{\partial V}{\partial y}\mathbf{u}_y + \frac{\partial V}{\partial z}\mathbf{u}_z. \tag{1.7.6}$$

It should be noted that the del operator (1.7.5) is defined with respect to the Cartesian coordinate system. But it is often used as a short notation for gradient in general.

With the above conventions, we can represent the electric field intensity vector in terms of the potential function $V$ as

$$\boxed{\mathbf{E} = -\text{grad}\, V}\,. \tag{1.7.7}$$

This relation is often written in the symbolic form $\mathbf{E} = -\nabla V$. Thus, if potential $V$ is known as a function of coordinates, the electric field intensity vector can be obtained as a simple sum of certain derivatives of $V$, multiplied by appropriate unit vectors.

**1.7.1 Equipotential surfaces.** Surfaces whose points are at the same potential are called *equipotential surfaces*. Consider two arbitrary adjacent points of an equipotential surface. The potential difference (1.7.1) in that case is zero. This means that at every point of an equipotential surface the electric field intensity is normal to the surface. In other words, *lines of force are always normal to equipotential surfaces*. According to Eq. (1.7.7), grad $V$ is therefore directed in the direction of the largest increase of the potential at a point.

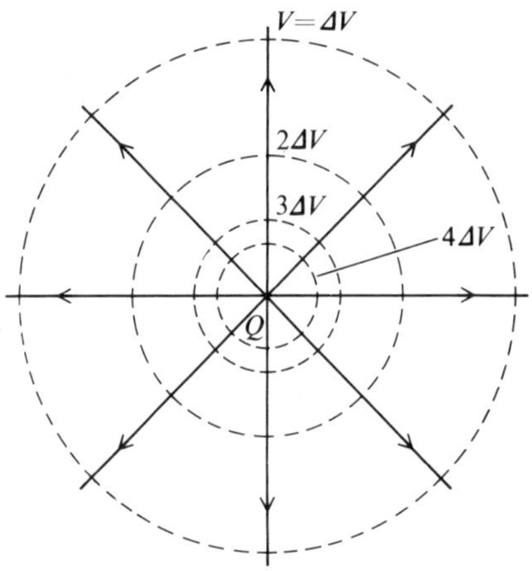


**Fig. 1.12** Equipotential lines of a point charge. Equipotential surfaces are spheres centered at the charge.

Just as lines of force are useful for visualizing the electric field, so are equipotential surfaces. We adopt the convention of drawing equipotential surfaces such that the difference of potential between any two adjacent surfaces is the same. It is evident from Eq. (1.7.2) that the electric field intensity is larger where equipotential surfaces are closer, and vice versa. Vector $\mathbf{E}$ is normal to equipotential surfaces, and is directed toward the adjacent surface of *smaller* potential.

For example, according to Eq. (1.6.11), the equipotential surfaces of the electric field surrounding an isolated point charge are spheres. When we adopt a potential difference $\Delta V$ between adjacent equipotential surfaces, it is not difficult to determine the successive values $r_n$ $(n = 1, 2, \ldots)$ of their radii. The zero equipotential surface corresponds to $r_0 = \infty$. According to Eq. (1.6.11), the equipotential surface having potential $n\,\Delta V$ has a radius $r_n = Q/4\pi\varepsilon_0 n\Delta V$. Shown in Fig. 1.12 are equipotential lines of a point charge depicted according to that rule. These lines represent intersections of equipotential spheres with a plane containing the point charge.

*Problems referring to Section 1.7:* 1.T9, 1.G11 to 1.G13.

## 1.8 GAUSS' LAW

**1.8.1 Flux of a vector function. Concept of the solid angle.** In Section 1.6 it was shown that the electric field intensity vector of any real static electric field must satisfy the integral relation (1.6.5). It will now be shown that the electric field intensity **E** must satisfy another integral relation, the so-called *Gauss' law*.

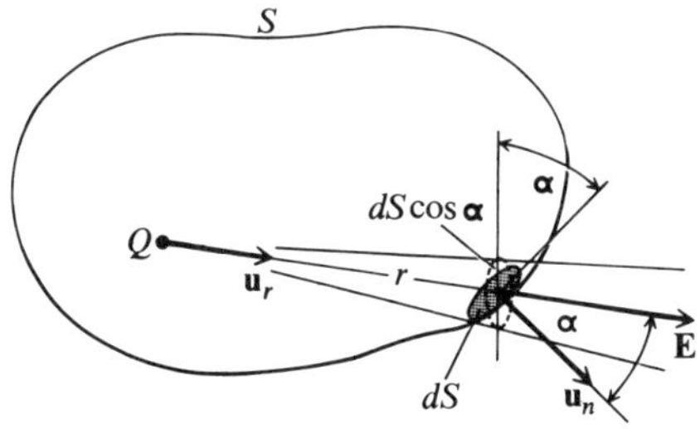


**Fig. 1.13** A point charge $Q$ enclosed by a surface $S$.

Consider an arbitrary *closed* surface $S$, enclosing a single point charge $Q$ (Fig. 1.13). The unit vector $\mathbf{u}_n$ normal to any closed surface is *defined* to be directed *from* the surface *outward*. This is a very important convention. We shall adhere to it without exception. With this convention, we can associate a unique unit vector with each element $dS$ of the surface $S$. Therefore we can also regard element $dS$ of the surface $S$ as a *vector* quantity $d\mathbf{S} = dS\,\mathbf{u}_n$. Thus by definition $d\mathbf{S}$ is a vector of magnitude $dS$ and *perpendicular* to $dS$.

Let us evaluate the (infinite) sum of all infinitesimal scalar products $\mathbf{E}\cdot d\mathbf{S}$ over the surface $S$, that is, the integral

$$\Psi_E = \lim_{\Delta S\to 0}\sum_S \mathbf{E}\cdot\Delta\mathbf{S} = \oint_S \mathbf{E}\cdot d\mathbf{S}. \tag{1.8.1}$$

An integral of this type is referred to as a *surface integral*. $\Psi_E$ as defined by Eq. (1.8.1) is called the *flux* of the electric field intensity vector through surface $S$. The same term is also used for the integral of this type if the surface is not closed. (The term *flux*, meaning *rate of flow*, originated in fluid mechanics. In the case of the electrostatic field, no physical meaning analogous to flow of some fluid can be associated with the flux of the electric field intensity.) Note that, according to definition (1.8.1), the flux of a vector function through a surface can be determined theoretically if the surface and the vector function are known as functions of coordinates. If the integration involved is impossible to perform analytically, we can always find the approximate value of the integral as the sum in Eq. (1.8.1), with $\Delta S$ sufficiently small.

From Eq. (1.3.2) and Fig. 1.13, we have

$$\mathbf{E}\cdot d\mathbf{S} = \frac{Q}{4\pi\varepsilon_0}\frac{\mathbf{u}_r\cdot d\mathbf{S}}{r^2} = \frac{Q}{4\pi\varepsilon_0}\frac{dS\cos\alpha}{r^2}. \tag{1.8.2}$$

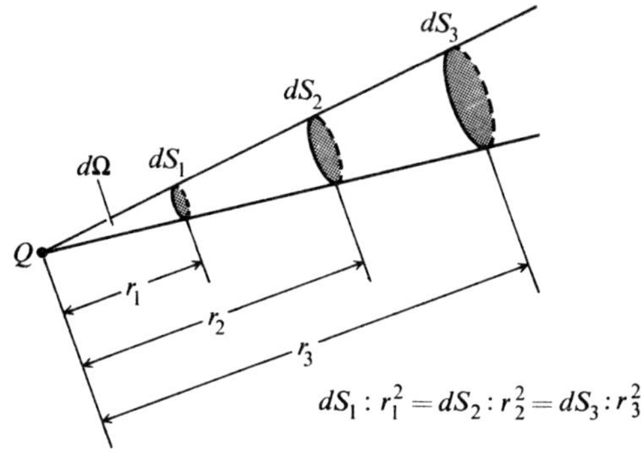


**Fig. 1.14** The ratio $dS_k/r_k^2$ is a constant, known as the elemental solid angle $d\Omega$ subtended at $Q$ by surfaces $dS_1$, $dS_2$, and $dS_3$.

Now $dS \cos \alpha$ represents the elemental surface obtained when $dS$ is projected on the plane normal to radius $r$. Since $dS$ is very small, $dS \cos \alpha$ is equal to the elemental surface cut out from an imaginary sphere of radius $r$ by rays emerging from charge $Q$ and touching the contour line of the elemental surface $dS$ (Fig. 1.13).

We now need the concept of the *solid angle*. It is defined as follows. Suppose that several cones are cut out of the same conical surface by spheres of different radii centered at the apex of the cone (Fig. 1.14). The ratio of their base surfaces to the square of the corresponding radius is a constant. This constant is a measure of the cone "opening." It is referred to as the *solid angle* $d\Omega$ subtended by the surface $dS$ at the point where charge $Q$ is situated. If the solid angle is not very small, as above, its value is defined to be numerically equal to the surface cut out by a cone from the sphere of unit radius centered at the apex of the cone. The unit of the solid angle is the *steradian*. For example, the solid angle subtended by an infinite plane at a point on (or off) its surface is numerically equal to the surface of half the unit sphere, or to $2\pi$ steradians. Similarly, the "full" solid angle is equal to $4\pi$ steradians. Note that a solid angle is formed by a cone of *any* shape, not necessarily by one that has a circular cross-section.

**1.8.2 Proof of Gauss' law.** Using the concept of the solid angle, we can write Eq. (1.8.2) in the form

$$\mathbf{E} \cdot d\mathbf{S} = \frac{Q}{4\pi\varepsilon_0} \, d\Omega. \tag{1.8.3}$$

Let us integrate—that is, sum—$\mathbf{E} \cdot d\mathbf{S}$ given by Eq. (1.8.3) over the *closed* surface $S$. Since $Q$ is *inside* the surface, the sum of all $d\Omega$ is equal to the full solid angle $4\pi$. So

$$\oint_S \mathbf{E} \cdot d\mathbf{S} = \frac{Q}{4\pi\varepsilon_0} \, 4\pi = \frac{Q}{\varepsilon_0}. \tag{1.8.4}$$

Suppose that inside $S$ there exists an arbitrary charge distribution of total charge $Q_{\text{total}}$ (the sum of all point charges, and all volume, surface, and line charges). From Eq. (1.8.4) and the principle of superposition, it follows that

$$\oint_S \mathbf{E} \cdot d\mathbf{S} = \frac{Q_{\text{total enclosed by } S}}{\varepsilon_0} \quad \text{(Gauss' law).} \tag{1.8.5}$$

This integral relation is satisfied by the electric field intensity vector created by any distribution of static charges situated in vacuum. It is referred to as *Gauss' law.*

It is easy to show that Eq. (1.8.5) remains valid even if the surface $S$ is such that the elemental cone cuts it more than once, and even if there are charges *outside* the surface. Refer to Fig. 1.15, and note that $\alpha_2$ and $\alpha_1'$ are obtuse angles. Consider charge $Q$ first. According to Eqs. (1.8.2) and (1.8.3), the contribution to the total flux through $dS_2$ is equal to that through or $dS_1$ or through $dS_3$, but of opposite sign. In the case of charge $Q'$, the flux through $dS_1'$ is equal to that through $dS_2'$, but of opposite sign. Thus, if a charge is enclosed by the surface, the flux of $\mathbf{E}$ created by it through the surface is always given by Eq. (1.8.5). The flux of the electric field intensity due to charges outside the surface is zero.

Gauss' law, Eq. (1.8.5), cannot be used for some new description of the electrostatic field, as Eq. (1.6.5) can. But it is extremely useful in solving some important electrostatic problems in a simple manner. Two distinct types of problems can be solved. First, suppose that a distribution of charges has such a high degree of symmetry that the direction of the electric field intensity vector is known in advance at every point. Then Gauss' law (1.8.5) is an equation sufficient to determine the one unknown: the magnitude of $\mathbf{E}$. Second, by the help of Gauss' law, one can easily prove some very general properties of the static electric field. In this section we shall solve only a few illustrative, but important, problems of the first type.

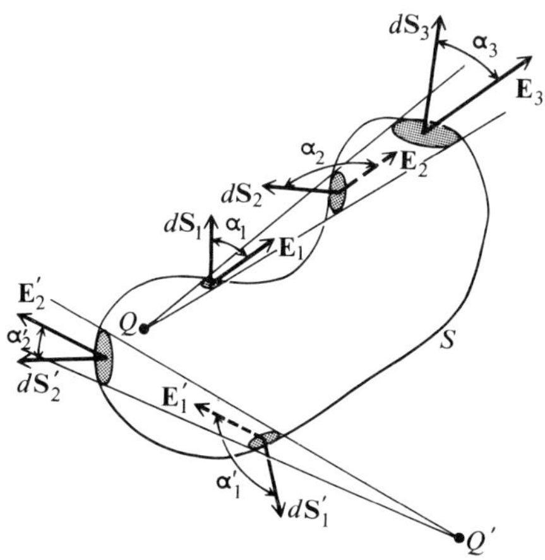


**Fig. 1.15** Construction for proving Gauss' law.

**1.8.3 Electric field and potential of uniform distribution of charges over a spherical surface.** Consider a sphere of radius $a$ uniformly charged over its surface by a total charge $Q$. Owing to symmetry the following statements concerning the electric field intensity vector can be made: (1) The lines of force are radial, and (2) the magnitude of the electric field intensity vector is a function of the distance $r$ only of a field point from the center of the sphere.

Let us apply Gauss' law to an imaginary spherical surface $S$ of radius $r > a$ (Fig. 1.16). According to Eq. (1.8.5) and the abovementioned properties of vector $\mathbf{E}$, we have

$$\oint_S \mathbf{E} \cdot d\mathbf{S} = \oint_S E(r)\, dS = E(r) \oint_S dS = E(r) 4\pi r^2 = \frac{Q}{\varepsilon_0},$$

from which

$$E(r) = \frac{Q}{4\pi\varepsilon_0 r^2} \qquad \text{for} \qquad r > a. \tag{1.8.6}$$

Therefore, outside a uniformly charged sphere, the electric field intensity is the same as if the total charge $Q$ were located at the center of the sphere.

Now let us apply Gauss' law to an imaginary spherical surface $S'$ of radius $r < a$ (Fig. 1.16). By exactly the same procedure as before, we obtain

$$E(r) = 0 \qquad \text{for} \qquad r < a, \tag{1.8.7}$$

because the charge enclosed by surface $S'$ is zero.

Since the electric field intensity vector outside the sphere is the same as that of a point charge at the center of the sphere, the potential in both cases is the same

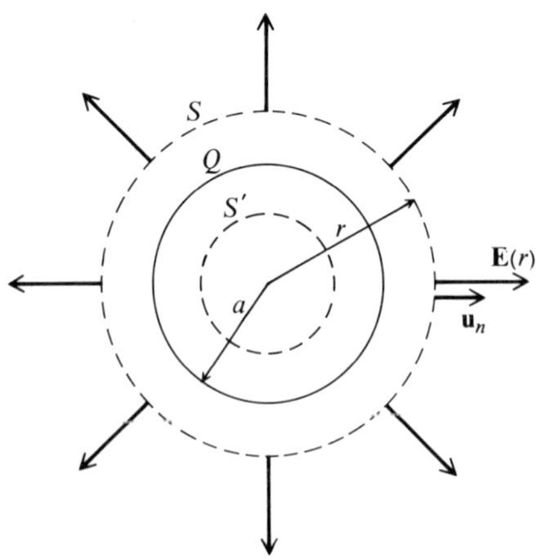


Fig. 1.16 A uniformly charged sphere.

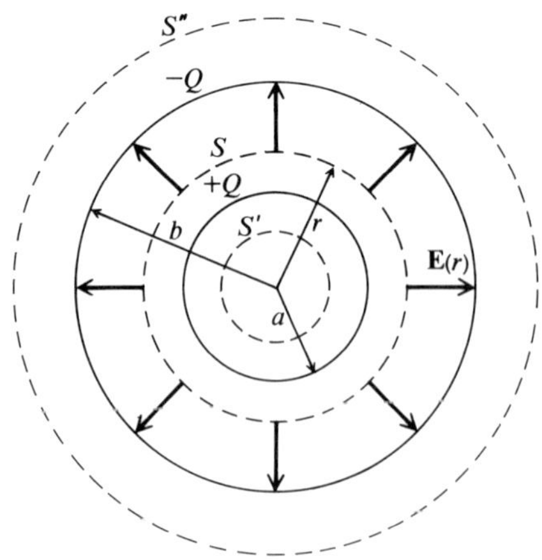


Fig. 1.17 Two oppositely charged concentric spheres.

as well. The potential outside the sphere is therefore given by Eq. (1.6.11). In particular, the potential at a point on the surface of the sphere is

$$V(a) = \frac{Q}{4\pi\varepsilon_0 a}. \tag{1.8.8}$$

Since inside the sphere $\mathbf{E} = 0$, all points inside the sphere are at the potential equal to that on its surface.

Let us consider two concentric uniformly charged spheres having the same total charge $Q$, but of opposite signs (Fig. 1.17). We apply Gauss' law successively to imaginary spherical surfaces $S$, $S'$, and $S''$, and thus we obtain that between the spheres

$$E(r) = \frac{Q}{4\pi\varepsilon_0 r^2} \qquad \text{for} \qquad a < r < b. \tag{1.8.9}$$

Inside the inner sphere and outside the outer sphere the electric field intensity is zero,

$$E(r) = 0 \qquad \text{for} \qquad r < a \qquad \text{and} \qquad r > b,$$

since $S'$ and $S''$ both enclose a zero *total* charge.

**1.8.4 Electric field and potential of uniform distribution of charges over an infinite circular cylinder.** Imagine an infinite circular cylinder of radius $a$, uniformly charged over its surface by charges of density $\rho_s$. By symmetry the lines of force are radial, and the electric field intensity is a function only of the distance $r$ of the field point from the cylinder axis (Fig. 1.18).

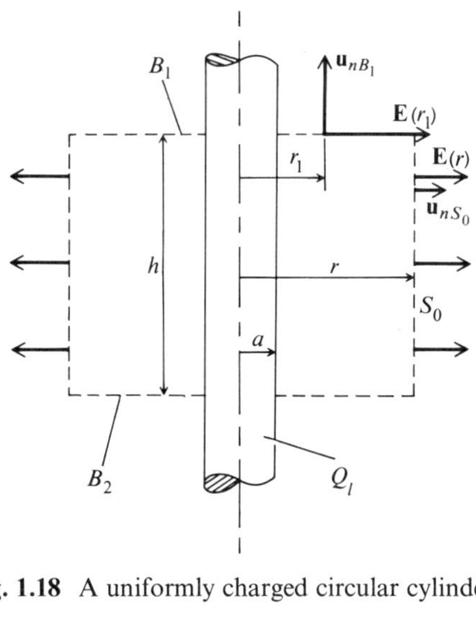


**Fig. 1.18** A uniformly charged circular cylinder.

Consider the imaginary cylinder shown of height $h$ and radius $r$. On the bases $B_1$ and $B_2$ the electric field intensity $\mathbf{E}(r)$ and the surface element $\mathbf{u}_{nB}\,dS$ are normal to one another, so that their scalar product is zero. On the curved surface, $\mathbf{E}(r)$ and the vector surface element $\mathbf{u}_{nS_0}dS$ are parallel. Using Gauss' law, we thus obtain

$$\oint_S \mathbf{E}\cdot d\mathbf{S} = \int_{S_0} E(r)\,dS = E(r)\int_{S_0} dS = E(r)2\pi rh = \frac{\rho_s 2\pi ah}{\varepsilon_0},$$

so that

$$E(r) = \frac{2\pi a\rho_s}{2\pi\varepsilon_0 r} = \frac{Q_l}{2\pi\varepsilon_0 r} \quad \text{for} \quad r > a. \tag{1.8.10}$$

Here $Q_l$ represents the charge per unit length of the cylinder. By the same argument as in the previous example, we can conclude that

$$E(r) = 0 \quad \text{for} \quad r < a. \tag{1.8.11}$$

The potential at a point $M$ which is at a distance $r$ ($r > a$) from the cylinder axis, with respect to a reference point at a distance $r_R$ from the axis, is

$$V(r) = \int_M^R E(r)\mathbf{u}_r\cdot d\mathbf{l} = \int_r^{r_R} E(r)\,dr = \frac{Q_l}{2\pi\varepsilon_0}\int_r^{r_R}\frac{dr}{r} = \frac{Q_l}{2\pi\varepsilon_0}\ln\frac{r_R}{r}. \tag{1.8.12}$$

The same expression is valid for a line charge, since $V(r)$ is not a function of the radius $a$ of the cylinder. In particular, at the surface of the cylinder,

$$V(a) = \frac{Q_l}{2\pi\varepsilon_0}\ln\frac{r_R}{a}. \tag{1.8.13}$$

It should be noted that the distance $r_R$ of the reference point from the axis of the cylinder cannot be taken to be infinite, because there are charges at infinity.

Suppose that there are two coaxial cylinders of radii $a$ and $b$, charged uniformly with charges $Q_l$ and $-Q_l$ per unit length. We can readily show that

$$E(r) = \frac{Q_l}{2\pi\varepsilon_0 r} \quad \text{for} \quad a < r < b, \tag{1.8.14}$$

and

$$E(r) = 0 \quad \text{for} \quad r < a \quad \text{and} \quad r > b.$$

**1.8.5 Electric field of charges uniformly distributed over an infinite plane surface.** Consider an infinite plane surface uniformly charged with a surface charge density $\rho_s$. Owing to symmetry, the lines of force are normal to the surface. The electric field intensity is a function of the distance $x$ from the surface only (Fig. 1.19), and $E(x) = E(-x)$.

Let us apply Gauss' law to the cylinder the cross-section of which is indicated in dashed lines in Fig. 1.19. Note that the flux of $\mathbf{E}$ is zero through the curved

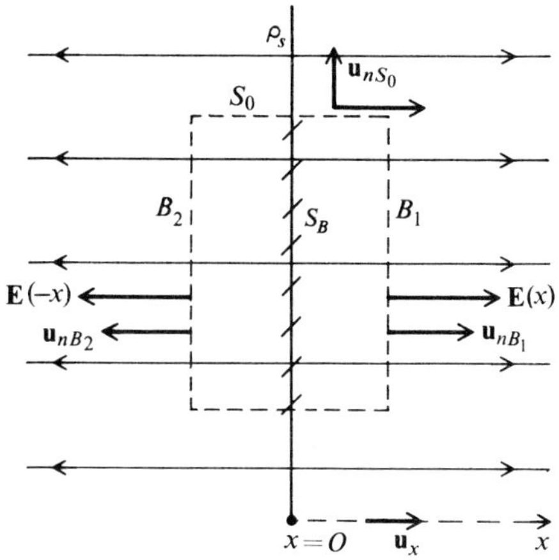


**Fig. 1.19** An uniformly charged plane.

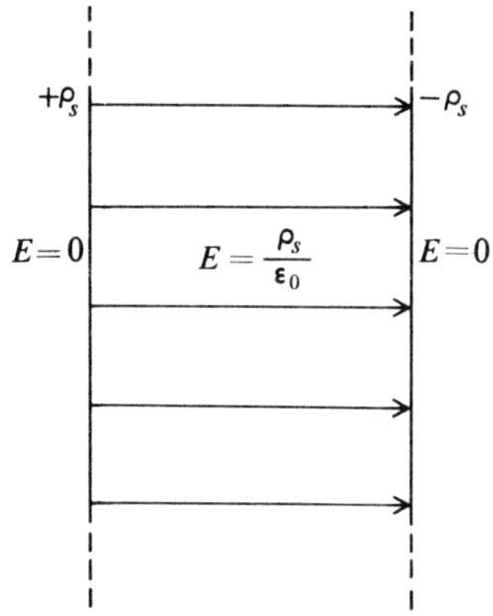


**Fig. 1.20** Two oppositely charged parallel planes.

surface $S_0$, and that the cylinder encloses the charge $\rho_s S_B$, located on the cross-hatched portion of the plane shown in Fig. 1.19. Therefore

$$\oint_S \mathbf{E}\cdot d\mathbf{S} = \int_{B_1} \mathbf{E}\cdot d\mathbf{S} + \int_{B_2} \mathbf{E}\cdot d\mathbf{S} = 2E(x)S_B = \frac{\rho_s S_B}{\varepsilon_0},$$

from which we obtain

$$E(x) = \frac{\rho_s}{2\varepsilon_0}. \tag{1.8.15}$$

The electric field intensity has the same magnitude at every point, however far from the plane. This is obviously the result of the assumption that the charged plane is infinite.

If we have two parallel, infinite planes charged uniformly by surface charge densities $\rho_s$ and $-\rho_s$ (Fig. 1.20), the electric field intensity between the planes is given by

$$E = \frac{\rho_s}{\varepsilon_0}, \tag{1.8.16}$$

and is zero outside.

A word of explanation is here necessary to explain why we consider an "infinite" charged cylinder and an "infinite" charged plane. Obviously, such cases are never encountered in practice. However, they are excellent *approximations* to some real systems. For example, thin long charged wires of circular cross section can be approximated with a high degree of accuracy by the idealized infinitely long cylinder considered. The electric field between the plates of a real parallel-plate capacitor is practically the same as that between the two "infinite" charged planes everywhere except near edges. Note that in both cases the exact solution is extremely difficult to find, if indeed it can be found at all. This illus-

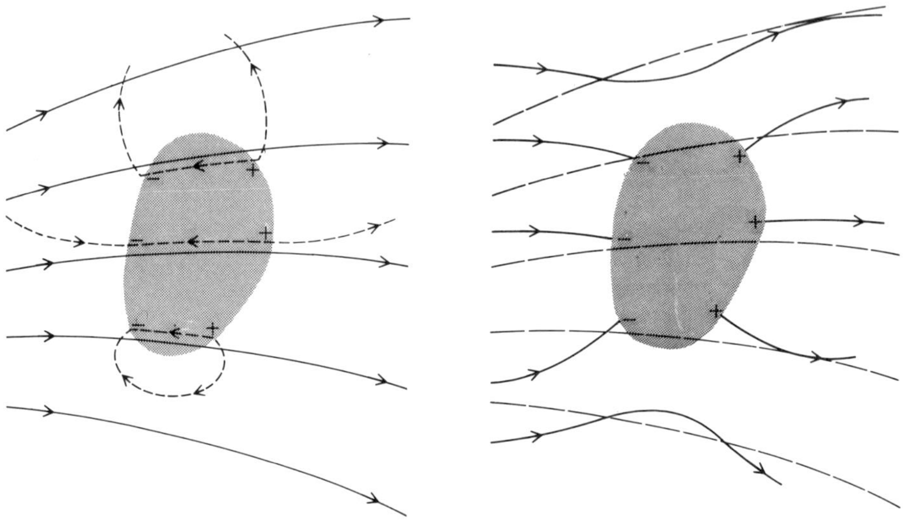

—— Lines of primary field
---- Lines of secondary field

(a)

---- Lines of primary field
—— Lines of resultant field

(b)

**Fig. 1.21** (a) A conducting body introduced into the electric field. (b) Resultant field corresponding to (a).

trates clearly the significance of analyzing idealized or simple cases. They can be always used as approximations to some more complicated structures that have practical importancc.

*Problems referring to Section 1.8:* 1.T10, 1.T11, 1.G14 to 1.G22, 1.N6, 1.N7.

## 1.9 BEHAVIOR OF CONDUCTORS IN THE STATIC ELECTRIC FIELD

As mentioned in Section 1.1, conductors are defined as substances having elemental charges free to move in response to an applied field. In ordinary circumstances these charges cannot leave the surface of a conductor. Recalling this, we can deduce how a piece of a conducting material behaves when it is introduced into an electrostatic field.

Consider an uncharged piece of a conductor. Suppose that in some appropriate manner we switched on the electric field. We can imagine that, at the very first instant, thc piece of the conductor is "permeated" with this electric field. As a consequence, there is an electric force that exists on every elemental charge inside the conductor. Free charges inside it start an organized drift. Positive charges move in the direction of the field, negative ones in the opposite direction. We can readily conclude from Fig. 1.21a that the secondary electric field of these "induced" charges is, in essence, opposite to the primary electric field which caused

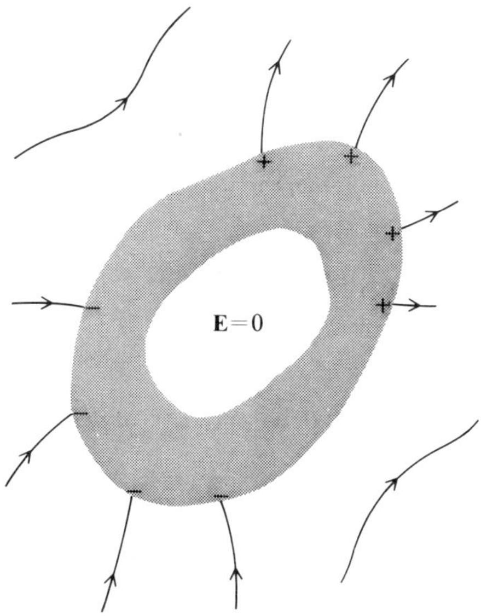


**Fig. 1.22** A shell of conducting material is a perfect protection of the volume enclosed from the external static electric fields.

the charges to move. Ultimately, when electrostatic equilibrium is attained, this secondary field must cancel the primary field at each point inside the conducting body. Otherwise the free charges would keep on moving. Hence *a static electric field cannot exist inside a conductor*. Sketched in Fig. 1.21b is the resultant field corresponding to the primary and secondary field shown in Fig. 1.21a.

We can now prove that the charge density of induced charges is zero inside the conducting body, which means that all induced charges are located in a very thin layer on the surface. Let us apply Gauss' law to *any* closed surface $S$ completely within the conducting body. At every point of the surface the electric field intensity $\mathbf{E}$ is zero, and consequently the flux of $\mathbf{E}$ through it is identically zero. Since the surface $S$ is completely arbitrary, it follows from Eq. (1.8.5) that there can be no net charge inside a conducting body.

An important example of the application of this fact is the so-called *electrostatic shielding*. Figure 1.22 shows a hollow uncharged conductor in an external electrostatic field. Since the electric field vanishes in the conducting wall, however thin the wall may be, no charges can be induced on the inner surface. Therefore the electric field intensity in the cavity is zero. So, in the case of the static electric field, a very thin conducting shell suffices for *complete* electrostatic shielding. We shall see that this is not the case with electric fields that vary in time.

The *resultant* electric field intensity on the surface of a conductor (the vector sum of the external electric field intensity and that due to the induced charges) is normal to the surface. This can be shown as follows. Consider the narrow rectangular contour 12341 shown in Fig. 1.23. Note that an electric field exists only

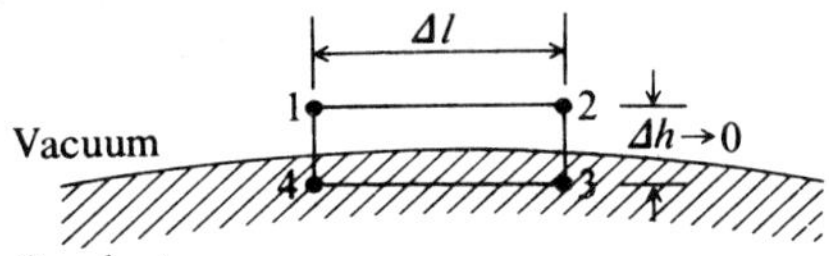


**Fig. 1.23** Contour for proving that the tangential component of **E** at the surface of a conductor is zero.

in vacuum. Applying Eq. (1.6.5) to the contour, we have

$$\oint_{12341} \mathbf{E} \cdot d\mathbf{l} = E_t \Delta l = 0,$$

where $E_t$ is the possible tangential component of **E** at the conductor surface. Therefore

$$\boxed{E_t = 0} \quad \text{(boundary condition for } E_t \text{ at the surface of a conductor).} \qquad (1.9.1)$$

Thus the electric field intensity vector of a static electric field is always normal to the surface of a conductor. Hence it follows that the surface of any conductor in the static electric field is an equipotential surface. As there is no field inside a conducting body, all points of a conductor are at the same potential, equal to that at its surface.

The same conclusions are also valid if a conducting body is charged. The charge is again distributed in a thin layer on the surface, because no electric field, and consequently no charges, can exist inside a conductor.

Let us now consider the case of a small charged body which is introduced into the cavity of an uncharged conducting body (Fig. 1.24). The primary field due to charge $Q$ on the small body will induce on the *inner* (cavity) surface of the hollow conductor charges of the opposite sign. These charges are so distributed that their

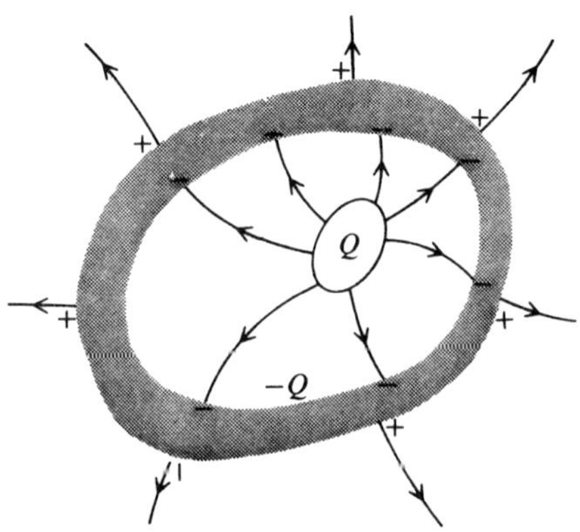


**Fig. 1.24** A charged body inside a conducting shell.

electric field cancels the electric field of charge $Q$ in the conducting wall. To obtain the total induced charge, apply Gauss' law to a surface $S$ (not shown) situated completely inside the conductor and enclosing the cavity. We obtain

$$\oint_S \mathbf{E} \cdot d\mathbf{S} = 0 = \frac{Q + Q_{\text{induced}}}{\varepsilon_0},$$

so that $Q_{\text{induced}} = -Q$.

However, the hollow conductor was supposed to be uncharged. The charge $-Q_{\text{induced}} = +Q$ remained uncompensated, since $Q_{\text{induced}}$ is distributed on the inner surface. How is this uncompensated charge distributed? We have proved that there can be no charges inside a conductor. So the uncompensated charge $+Q$ is spread over the outer surface of the conductor. Note that the field of the charge $Q$ and of the induced charge $-Q$ on the inner surface is limited to the cavity. The charge $+Q$ on the outer surface is therefore distributed in a manner which is *independent* of the position of the charged body inside the cavity.

Let us determine the relation between the normal (only existing) component $E_n$ of the electric field intensity on the conductor surface and the surface charge density $\rho_s$ at the same point. For that purpose, apply Gauss' law to the thin elemental cylinder $S$ shown in Fig. 1.25. Note that one base of the cylinder is *inside* the conductor. By Gauss' law, we have

$$\oint_S \mathbf{E} \cdot d\mathbf{S} = \Delta S\, E_n = \frac{\rho_s \Delta S}{\varepsilon_0},$$

from which

$$\boxed{E_n = \frac{\rho_s}{\varepsilon_0}} \quad \text{(boundary condition for } E_n \text{ on conductor surface).} \quad (1.9.2)$$

This is referred to as the *boundary condition* for the normal component of vector $\mathbf{E}$ at the conductor surface. Recalling that $E_n = -\partial V/\partial n$, we can also write

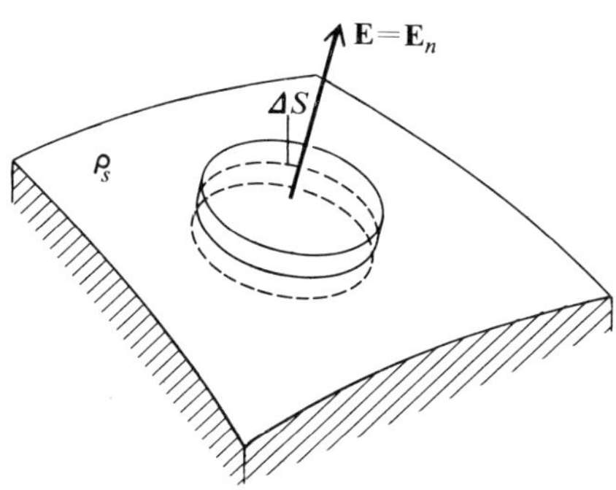


**Fig. 1.25** A Gaussian surface for proving that $E_n = \rho_s/\varepsilon_o$.

Eq. (1.9.2) in the form

$$-\frac{\partial V}{\partial n} = \frac{\rho_s}{\varepsilon_0}. \tag{1.9.3}$$

*Problems referring to Section 1.9:* 1.T12 to 1.T20, 1.G23 to 1.G33.

## 1.10 TUBES OF FLUX OF THE ELECTRIC FIELD INTENSITY

When we use an appropriate convention in drawing the lines of force, these lines can indicate not only the direction of the electric field vector at a point, but also its magnitude.

Consider two charged conducting bodies, and imagine a tubular surface $S_T$ formed by a bundle of lines of force (Fig. 1.26). There is no "leakage" of flux through the tubular surface $S_T$, since $\mathbf{E}$ is tangential to it at every point. The flux of the electric field intensity is consequently the same through each cross section of the tube. For this reason such tubes are called the *tubes of constant flux of the electric field intensity vector*.

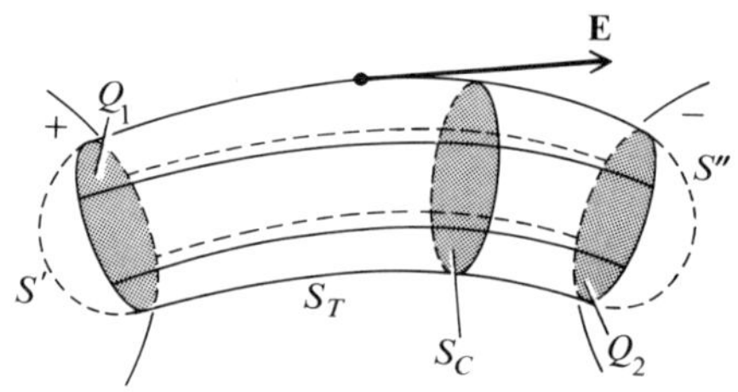


**Fig. 1.26** A tube of flux of the electric field intensity vector **E**.

It is easy to prove two important properties of the tubes of the electric field intensity. First, let us apply Gauss' law to the closed surface composed of the tubular surface $S_T$ and of the two end surfaces, $S'$ and $S''$, lying *entirely* inside the conductors. We find that charges $Q_1$ and $Q_2$ on which the tube of the electric flux terminates are of equal magnitude, but have opposite signs:

$$\oint_{S'+S_T+S''} \mathbf{E}\cdot d\mathbf{S} = 0 = \frac{Q_1 + Q_2}{\varepsilon_0}, \qquad \text{from which} \qquad Q_2 = -Q_1. \tag{1.10.1}$$

The flux of $\mathbf{E}$ through the closed surface is zero, because on $S'$ and $S''$ $\mathbf{E}$ vanishes ($\mathbf{E}$ inside a conductor is zero), and $\mathbf{E}$ is tangential to $S_T$ at every point.

Second, let us apply Gauss' law to the closed surface formed by $S'$, a part of $S_T$ and $S_c$ (Fig. 1.26). We obtain that the flux of the electric field intensity through an arbitrary cross section of the tube of flux considered is equal to $Q_1/\varepsilon_0$:

$$\oint_{S'+S_T+S_c} \mathbf{E}\cdot d\mathbf{S} = \int_{S_c} \mathbf{E}\cdot d\mathbf{S} = \frac{Q_1}{\varepsilon_0}. \tag{1.10.2}$$

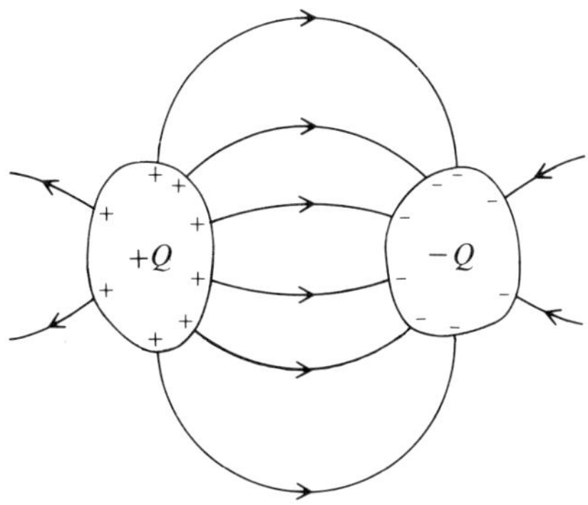


**Fig. 1.27** An example of representing tubes of flux of vector **E** by lines of force.

The entire volume of a region in which the electric field exists can be divided into tubes of *equal* flux. Each tube can then be represented by a single *line* of force (say the line of force along its axis). Since the tubes are supposed to be of *equal* flux, the charges on which they end are also equal. We can represent these charges by a single plus or minus sign. If these conventions are adopted, the magnitude of the electric field intensity is proportional to the density of the lines of force at a point, and the charge density is proportional to the density of the plus and minus signs. We shall always adhere to this method of representing the electric field and the density of charges. Figure 1.27 shows an example of representing lines of force and distribution of charges using these conventions.

*Problems referring to Section 1.10:* 1.T21, 1.T22.

## 1.11 DIFFERENTIAL FORM OF GAUSS' LAW. DIVERGENCE OF A VECTOR FUNCTION

Gauss' law, Eq. (1.8.5), relates certain *integrals* of the electric field intensity and the electric charge density. It is therefore referred to as the *integral* form of Gauss' law. We have seen that Gauss' law in the integral form is a very useful tool for analyzing the electrostatic field. But even more useful is its *differential* form, obtained as follows.

Consider a volume distribution of charges of density $\rho$. We can then write Gauss' law (1.8.5) in the form

$$\oint_S \mathbf{E} \cdot d\mathbf{S} = \frac{1}{\varepsilon_0} \int_v \rho \, dv. \tag{1.11.1}$$

Now, for $v$ small enough, say equal to $\Delta v$, the right-hand side of Eq. (1.11.1) is simply $\rho \, \Delta v/\varepsilon_0$:

$$\oint_{\Delta S} \mathbf{E} \cdot d\mathbf{S} = \frac{\rho}{\varepsilon_0} \Delta v \qquad (\Delta v \text{ very small}). \tag{1.11.2}$$

If $\Delta v$ tends to zero, so does the flux of **E** through the surface $\Delta S$ enclosing $\Delta v$.

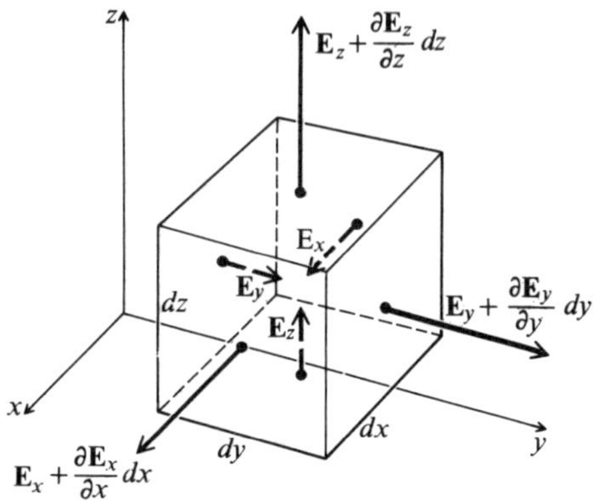


**Fig. 1.28** Volume element for determining divergence of **E** in a rectangular coordinate system.

But the limit of the ratio of the left-hand side and $\Delta v$ generally does not vanish:

$$\lim_{\Delta v \to 0} \frac{\oint_{\Delta S} \mathbf{E} \cdot d\mathbf{S}}{\Delta v} = \frac{\rho}{\varepsilon_0}. \tag{1.11.3}$$

Essentially, this is Gauss' law applied to the immediate vicinity of a field point. It is therefore sometimes called the *point form* of Gauss' law. We shall show that Eq. (1.11.3) represents a partial differential equation of the first order. It is therefore more frequently referred to as the *differential form* of Gauss' law.

Expressions of the form of the left-hand side of Eq. (1.11.3) are encountered in all branches of physics that deal with vector fields. The whole expression is conventionally shortened into div **E** (read as "the divergence of **E**"):

$$\boxed{\operatorname{div} \mathbf{E} = \lim_{\Delta v \to 0} \frac{\oint_{\Delta S} \mathbf{E} \cdot d\mathbf{S}}{\Delta v}} \quad \text{(definition of divergence).} \tag{1.11.4}$$

Consequently, we can write Eq. (1.11.3) as:

$$\boxed{\operatorname{div} \mathbf{E} = \frac{\rho}{\varepsilon_0}} \quad \text{(differential form of Gauss' law).} \tag{1.11.5}$$

We can determine divergence of a vector function with respect to any coordinate system from the definition (1.11.4). As an example, let us determine div **E** with respect to the Cartesian coordinate system. The volume element $dv$ in that case is (see Fig. 1.28)

$$dv = dx\, dy\, dz. \tag{1.11.6}$$

We obtain the flux of **E** through the surface enclosing the elemental volume $dv$ as follows. Recall that the positive normal to a closed surface is always directed away from that surface. Referring to Fig. 1.28, we have

$$\oint_{dS} \mathbf{E}\cdot d\mathbf{S} = -E_x\,dy\,dz + \left(E_x + \frac{\partial E_x}{\partial x}\,dx\right)dy\,dz - E_y\,dx\,dz$$

$$+\left(E_y + \frac{\partial E_y}{\partial y}\,dy\right)dx\,dz - E_z\,dx\,dy + (E_z + \frac{\partial E_z}{\partial z}\,dz)\,dx\,dy$$

$$= \left(\frac{\partial E_x}{\partial x} + \frac{\partial E_y}{\partial y} + \frac{\partial E_z}{\partial z}\right)dx\,dy\,dz.$$

So according to Eq. (1.11.4), with respect to the rectangular coordinate system,

$$\operatorname{div}\mathbf{E} = \frac{\partial E_x}{\partial x} + \frac{\partial E_y}{\partial y} + \frac{\partial E_z}{\partial z}. \tag{1.11.7}$$

Equation (1.11.5) thus indeed represents a partial differential equation of the first order.

Since in the rectangular coordinate system

$$\mathbf{A}\cdot\mathbf{B} = A_x B_x + A_y B_y + A_z B_z, \tag{1.11.8}$$

using the del operator, defined by Eq. (1.7.6), we can write Eq. (1.11.7) in a more compact form:

$$\operatorname{div}\mathbf{E} = \nabla\cdot\mathbf{E}. \tag{1.11.9}$$

Representation of div **E** as the dot "product" of the del operator and vector **E** is true only in the Cartesian coordinate system. Very often $\nabla\cdot\mathbf{E}$ is also used as a short notation for div **E**. We then write the differential form (1.11.5) of Gauss' law as $\nabla\cdot\mathbf{E} = \rho/\varepsilon_0$.

Let us substitute $\rho/\varepsilon_0$ given by Eq. (1.11.5) into Eq. (1.11.1). We obtain the relation between the flux of a vector function **E** through a closed surface $S$ and the integral of its divergence over the volume $v$ enclosed by $S$:

$$\boxed{\oint_S \mathbf{E}\cdot d\mathbf{S} = \int_v \operatorname{div}\mathbf{E}\,dv} \qquad \text{(divergence theorem).} \tag{1.11.10}$$

This relation is known as Gauss' or *divergence theorem*. It is valid for any vector function that is continuous and differentiable inside $S$ and on $S$. It is an important theorem of vector analysis, and we shall refer to it frequently.

Observe that the partial differential equation (1.11.5) represents a single scalar equation. Therefore it is not in general sufficient to enable one to determine *three* scalar components of the electric field intensity vector **E**. Only if vector **E**, by symmetry, has only *one* component, depending on only *one* coordinate, we can use

Eq. (1.11.5) to determine that one unknown. In that case it is usually easier to use its integral-form counterpart, Eq. (1.11.1).

*Problems referring to Section 1.11:* 1.T23, 1.G34, 1.G35.

## 1.12 POISSON'S AND LAPLACE'S EQUATIONS

Combining the differential form (1.11.5) of Gauss' law with Eq. (1.7.7), we obtain a partial differential equation of the second order, with the electrostatic potential $V$ as the unknown:

$$\operatorname{div} \mathbf{E} = \operatorname{div}(-\operatorname{grad} V) = \frac{\rho}{\varepsilon_0}.$$

This equation is known as *Poisson's equation*:

$$\boxed{\operatorname{div} \operatorname{grad} V = -\frac{\rho}{\varepsilon_0}} \qquad \text{(Poisson's equation).} \tag{1.12.1}$$

In the particular case in which all charges are distributed on surfaces of conducting bodies, the volume charge density $\rho$ is zero at all points. Then Eq. (1.12.1) becomes

$$\boxed{\operatorname{div} \operatorname{grad} V = 0} \qquad \text{(Laplace's equation).} \tag{1.12.2}$$

This equation is called *Laplace's equation.*

From Eqs. (1.7.4) and (1.11.7), it follows that with respect to the Cartesian coordinate system Eqs. (1.12.1) and (1.12.2) have the forms

$$\frac{\partial^2 V}{\partial x^2} + \frac{\partial^2 V}{\partial y^2} + \frac{\partial^2 V}{\partial z^2} = -\frac{\rho}{\varepsilon_0} \tag{1.12.3}$$

and

$$\frac{\partial^2 V}{\partial x^2} + \frac{\partial^2 V}{\partial y^2} + \frac{\partial^2 V}{\partial z^2} = 0. \tag{1.12.4}$$

In the Cartesian coordinate system one can make use of the del operator and write formally

$$\operatorname{div} \operatorname{grad} V = \nabla \cdot (\nabla V) = (\nabla \cdot \nabla)\, V = \nabla^2 V.$$

The symbol $\nabla^2$ is called the *Laplacian operator*. With the Laplacian operator,

$$\nabla^2 = \nabla \cdot \nabla = \frac{\partial^2}{\partial x^2} + \frac{\partial^2}{\partial y^2} + \frac{\partial^2}{\partial z^2}, \tag{1.12.5}$$

Poisson's and Laplace's equations (1.12.3) and (1.12.4) can be written in the short forms $\nabla^2 V = -\rho/\varepsilon_0$ and $\nabla^2 V = 0$, respectively. These equations are strictly valid only with respect to the Cartesian coordinate system. However, they are often used in other systems, implying that the symbol $\nabla^2$ really stands for the symbol "div grad."

Poisson's equation (1.12.1) is the most general equation which must be satisfied by the electrostatic potential function $V$. In practice one often encounters cases in which charges are located only on surfaces of conductors. The electrostatic potential function in the space between conductors is then determined by Laplace's equation (1.12.2). Our attention in the following section will be restricted to the properties of Laplace's equation and its solutions.

*Problems referring to Section 1.12:* 1.G36 to 1.G41.

## 1.13 PROPERTIES OF LAPLACE'S EQUATION AND ITS SOLUTIONS

**1.13.1 Introduction. Uniqueness theorem.** The electrostatic potential function $V$ in vacuum between charged conducting bodies must satisfy Laplace's equation (1.12.2). Therefore this equation alone suffices for determining $V$. Boundary surfaces between conductors and vacuum are equipotential. It is necessary to determine the particular solution of Laplace's equation having constant, usually given values on the conductor surfaces. These conditions are known as *boundary conditions*. Most often it is not possible to find the required solution starting directly from Laplace's equation.

Suppose that we found in some indirect manner a certain function, satisfying both Laplace's equation *and* the prescribed boundary conditions. The process leading to such a solution might be very illogical. For example, we could simply *try* some of the known functions to see whether they satisfied these requirements. Obviously, we could not be sure in advance that this would be the *only* solution to the problem. Such a trial-and-error method, if used, would require as a working hypothesis that *there is only one solution of Laplace's equation satisfying a given set of boundary conditions*. This statement *is* true, and is known as the *theorem of uniqueness of solutions of Laplace's equation*. The importance of the uniqueness theorem cannot be sufficiently stressed. If a solution is found in *any* manner, according to this theorem it is the only possible one. The uniqueness theorem is the basis of most methods of solving static electric field problems, and therefore its meaning must be understood and remembered. The proof of the theorem, which follows, does not add anything to the clarity of its meaning. Therefore it may be omitted if desired.

**1.13.2 Proof of the uniqueness theorem.*** Consider an arbitrary system of charged conducting bodies. Assume that we know either the potentials of all bodies, or the normal derivatives of the potential at all conductor surfaces. Suppose that, contrary to the uniqueness theorem, two different functions, $V_1$ and $V_2$, satisfying

* This section may be omitted without loss of continuity.

Laplace's equation,

$$\nabla^2 V_1 = 0, \qquad \nabla^2 V_2 = 0, \tag{1.13.1}$$

and the prescribed boundary conditions are possible. Then the difference potential $V_d = V_1 - V_2$ as well as the normal derivative $\partial V_d/\partial n$ of the difference potential must be zero on all conductor surfaces. In addition, according to Eq. (1.13.1),

$$\nabla^2 (V_1 - V_2) = \nabla^2 V_d = 0. \tag{1.13.2}$$

Let us apply the divergence theorem (1.11.10) to a particular vector function $V_d \nabla V_d$. We have

$$\oint_S V_d \nabla V_d \cdot d\mathbf{S} = \int_v \nabla \cdot (V_d \nabla V_d)\, dv = \int_v [(\nabla V_d)^2 + V_d \nabla^2 V_d]\, dv$$
$$= \int_v (\nabla V_d)^2\, dv \tag{1.13.3}$$

[according to Eq. (1.13.2), $\nabla^2 V_d = 0$]. We must evaluate the surface integral for all surfaces limiting the volume $v$ inside which the field exists. These include the surface $S_0$ enclosing the whole system, and surfaces $S_1, S_2, \ldots, S_n$ of the conducting bodies limiting the volume $v$ *inside* (Fig. 1.29).

Imagine $S_0$ as a very large sphere. The potential $V$ on it is proportional to $1/r$, and $\nabla V$ to $1/r^2$, because charge in the domain is seen from far points as a point charge. This is true if the total charge of the system is different from zero. If it is zero, both the potential and its derivative decrease with $r$ still faster. The area of $S_0$, however, increases only as $r^2$. So the integral over $S_0$ tends to zero as $r$ increases.

The surface integrals over $S_1, S_2, \ldots, S_n$ are also zero, for both $V_d$ and

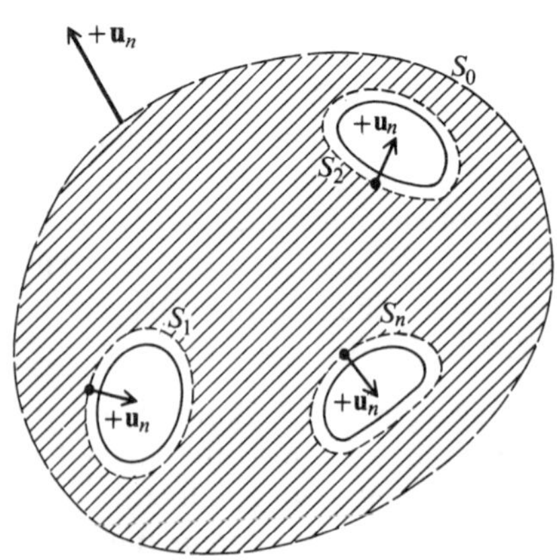


**Fig. 1.29** Construction for proving the uniqueness theorem.

$\nabla V_d = \partial V_d/\partial n$ on them are zero by hypothesis. Hence Eq. (1.13.3) becomes

$$\int_v (\nabla V_d)^2 \, dv = 0. \tag{1.13.4}$$

Since the integrand is positive (the square of a function), this equation can be satisfied only if $\nabla V_d = 0$. Therefore

$$V_d = V_1 - V_2 = \text{constant}. \tag{1.13.5}$$

Now, since on the conductor surfaces $V_1 = V_2$, the constant in Eq. (1.13.5) is zero. Thus the electrostatic potential function that satisfies Laplace's equation and *either* the condition that it is a given constant on conductor surfaces *or* that its normal derivative on the surfaces has prescribed value, is unique. It is not difficult to prove that the uniqueness theorem is in fact valid under a less severe restriction on $\partial V/\partial n$. It is sufficient that the total charge on conductors is specified, i.e., that

$$\oint_{S_k} (\partial V/\partial n) \, dS = -Q_k/\varepsilon_0, \qquad k = 1, 2, \ldots, n,$$

are known.

**1.13.3 Superposition principle for solutions of Laplace's equation.** The superposition principle for solutions of Laplace's equation is another important property of this equation. Essentially, this principle is a consequence of the linear character of Laplace's equation, and can be proved as follows.

Let $V = a_1 V_1 + a_2 V_2$, where $a_1$ and $a_2$ are arbitrary constants. Then

$$\text{grad } V = \text{grad } (a_1 V_1 + a_2 V_2) = a_1 \text{ grad } V_1 + a_2 \text{ grad } V_2.$$

It follows that if $V = a_1 V_1 + a_2 V_2$, then

$$\begin{aligned} \text{div grad } V &= \text{div } (a_1 \text{ grad } V_1 + a_2 \text{ grad } V_2) \\ &= a_1 \text{ div grad } V_1 + a_2 \text{ div grad } V_2. \end{aligned} \tag{1.13.6}$$

Suppose that both $V_1$ and $V_2$ are solutions of Laplace's equation (1.12.2). From Eq. (1.13.6), the sum $V = a_1 V_1 + a_2 V_2$ is also a solution of the same equation. Generally speaking, if $V_1, V_2, \ldots, V_n$ are solutions of Laplace's equation, their sum with *arbitrary* constant coefficients $a_1, a_2, \ldots, a_n$,

$$V = a_1 V_1 + a_2 V_2 + \cdots + a_n V_n, \tag{1.13.7}$$

is also a solution of Laplace's equation. This statement is known as the *superposition principle for solutions of Laplace's equation*.

Combined with the uniqueness theorem, the superposition principle represents a very powerful tool for solving electrostatic problems. We can form a series of solutions of Laplace's equation of the type (1.13.7), and then try to determine

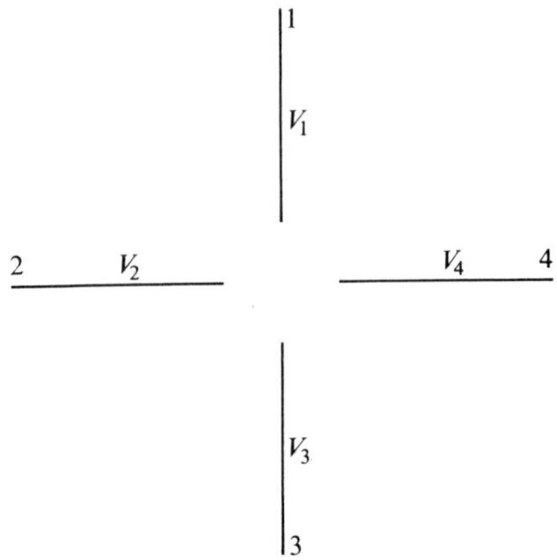


**Fig. 1.30** A two-dimensional electrostatic system.

the coefficients $a_k$ so that the boundary conditions for a particular problem are satisfied. Now any function of coordinates can be represented as a series (finite, infinite, or perhaps in the form of an integral, i.e., infinite sum of infinitesimally small functional terms) of convenient functions with appropriate coefficients. Using a series of functions satisfying Laplace's equation, we can theoretically find the solution to any electrostatic problem. If the solution thus obtained satisfies the prescribed boundary conditions, according to the uniqueness theorem it is the only possible solution.

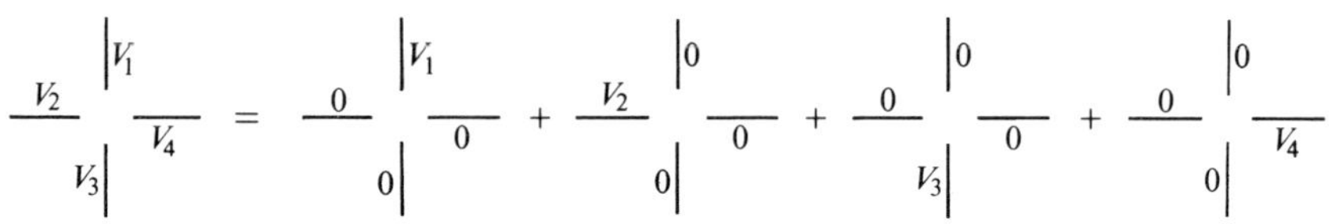

**Fig. 1.31** A schematic explanation of how the superposition principle is used for solving the system of Fig. 1.30.

We shall make use of this method extensively in Section 4.3. To illustrate how useful it can be, let us consider a system of four thin identical metal strips, as shown in Fig. 1.30. Suppose that we wish to determine the potential at every point when these strips are at potentials $V_1$, $V_2$, $V_3$ and $V_4$. It is sufficient to determine the potential function only for the case $V_1 = V_1$, $V_2 = V_3 = V_4 = 0$. We obtain the same type of function if we imagine the other three to be successively at their respective potentials, assuming the potential of the rest to be zero. We can readily find the resultant potential function as the sum of these four partial functions (Fig. 1.31).

*Problems referring to Section 1.13:* 1.T24, 1.T25, 1.G42.

## 1.14 RELATION BETWEEN ELECTROSTATIC POTENTIAL AND CHARGES ON CONDUCTING BODIES

**1.14.1 The case of an isolated conductor.** Consider an isolated charged conducting body. Distribution of charges over the surface of the body is such that the surface is equipotential and there is no electric field inside. Suppose that we increase the surface charge density at each point of the surface $n$ times. Again the conductor surface is equipotential, only the potential is different. According to the uniqueness theorem, this conversely implies that the $n$-times-larger charge on the body *must* be distributed in the same manner as before. So *the relative distribution of charges on an isolated conducting body is independent of the total charge on the body.* From Eq. (1.6.12) it follows that the potential of an isolated conducting body is directly proportional to the total charge on the body. Or the total charge on an isolated conductor is directly proportional to the conductor potential. Let us denote the total charge on the conducting body by $Q$ and its corresponding potential by $V$. Then

$$\boxed{Q = CV} \qquad \text{(definition of capacitance, isolated body)}, \qquad (1.14.1)$$

where $C$ is a constant depending on the shape of the body, but not on $Q$ or $V$. The constant $C$ obviously characterizes the "capacity" of an isolated body with respect to the charge accumulated on it per unit of potential. It is referred to as the *capacitance* of that body.

In addition to defining the capacitance of an isolated body, Eq. (1.14.1) shows a useful pattern for determining the capacitance. We assume a certain charge $Q$ on the body, determine the corresponding electric field intensity around the body, and then determine its potential as the line integral of **E** from the surface of the body to the reference point (infinity). We then obtain the capacitance of the body as $C = Q/V$.

As a simple example, let us calculate the capacitance of an isolated conducting sphere of radius $a$. The potential of the sphere corresponding to a charge $Q$ on it is determined in Section 1.8.3, Eq. (1.8.8). The capacitance of the sphere is

$$C = \frac{Q}{V(a)} = 4\pi\varepsilon_0 a. \qquad (1.14.2)$$

According to Eq. (1.14.1), the unit of capacitance in the MKSA system is the coulomb per volt (C/V). Since capacitance is a very important and often-used quantity, this unit is given a special name, the *farad* (abbreviated F). From Eq. (1.14.2) one can see that $\varepsilon_0$ can be given in units of farad per meter (F/m), as mentioned in Section 1.2.

The capacitance of one farad is much larger than most capacitances encountered in practice. Therefore submultiples of the farad, the *microfarad* (abbreviated $\mu$F, equal to $10^{-6}$ F) and the *micromicrofarad* or *picofarad* (abbreviated pF, equal to $10^{-12}$ F) are frequently used instead.

**1.14.2 The case of two bodies with equal charges of opposite sign (the capacitor).** A system encountered very often in practice consists of two conducting bodies with equal charges of opposite sign. Charges on the two bodies of equal magnitude but of opposite sign are obtained if the bodies are connected to a device called a *source of electromotive force*, which "pumps" charges of one sign from one body to another. So if a charge $Q$ is taken away from one body and transferred to the other body, there is a charge $-Q$ left on the first body (Fig. 1.32). Owing to a very frequent appearance in practice, such a system has a particular name, a *capacitor*. The two bodies of a capacitor are usually referred to as its *electrodes*.

A very simple relation exists between the charge $Q$ of a capacitor and the potential difference between its electrodes. It is usually written in the form

$$\boxed{Q = C(V_1 - V_2)} \qquad \text{(definition of capacitance).} \tag{1.14.3}$$

The factor $C$ is a constant which depends on the geometry of the capacitor, but not on $Q$ or $(V_1 - V_2)$. It is referred to as the *capacitance* of the capacitor.

To prove that a relation of the form (1.14.3) must be valid for a capacitor, let us consider two isolated bodies carrying charges $Q$ and $-Q$. By the same arguments as used for an isolated body, we can easily conclude that relative distribution of charges is independent of the particular charge $Q$. Consequently, the electric field intensity vector created by charges $Q$ and $-Q$ on the capacitor electrodes is at all points directly proportional to $Q$. The potential difference between the electrodes, that is, the line integral of $\mathbf{E}$ from one electrode to the other, is therefore also proportional to $Q$. So a relation of the form (1.14.3) must hold.

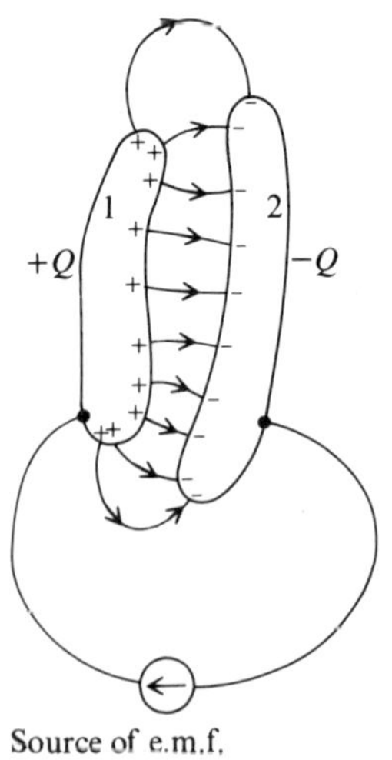


**Fig. 1.32** A capacitor: 1 and 2 are capacitor electrodes.

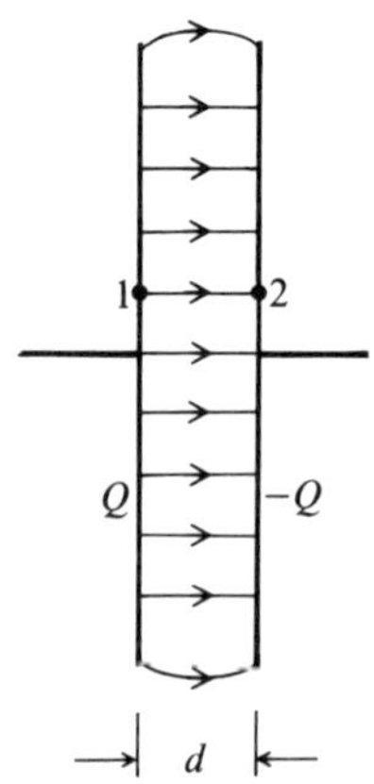


**Fig. 1.33** A parallel-plate capacitor.

Equation (1.14.3) represents the defining equation of the capacitance $C$ of a capacitor. It can also be used for evaluating $C$ according to the following simple pattern:

1) Assume charges $Q$ and $-Q$ on the capacitor electrodes.
2) Determine the electric field intensity due to these charges.
3) Determine the potential difference $(V_1 - V_2)$ between the electrodes as the line integral of $\mathbf{E}$ between them.
4) Determine $C$ as $C = Q/(V_1 - V_2)$.

This simple procedure is of course not the only one possible, and cannot always be used. It is applicable, however, to all the simple cases that we shall consider of determining the capacitance.

As an example, let us evaluate the capacitance of the so-called *parallel-plate capacitor*. It consists of two conducting parallel plates placed close together (Fig. 1.33). It is assumed that the distance $d$ between the plates is much smaller than their linear dimensions. The electric field intensity $\mathbf{E}$ between the plates is then approximately the same as in the case when they are infinite (Section 1.8.5) everywhere except near the edges. Assuming the area of each plate to be $S$, the electric field intensity vector due to charges $Q$ and $-Q$ on the plates is perpendicular to the plates, and of magnitude

$$E = \frac{\rho_s}{\varepsilon_0} = \frac{Q}{S\varepsilon_0}.$$

Hence the potential difference between the plates is

$$V_1 - V_2 = \int_1^2 \mathbf{E} \cdot d\mathbf{l} = Ed = \frac{Q}{S\varepsilon_0} d.$$

According to Eq. (1.14.3), the capacitance of the parallel-plate capacitor is

$$C = \varepsilon_0 \frac{S}{d} \qquad \text{(parallel-plate capacitor).} \tag{1.14.4}$$

**1.14.3 Parallel and series connections of capacitors.** Capacitors are, in practice, frequently connected in different ways. The most important connections of a group of capacitors are the *parallel* and the *series connections*.

The usual symbol for capacitors is two thick parallel lines. Figure 1.34 shows a group of capacitors connected in parallel, and Fig. 1.35 a group of capacitors connected in series. Both parallel and series connections of capacitors are two-terminal networks, each having a certain equivalent capacitance. The equivalent capacitances can be determined as follows.

Referring to the parallel connection shown in Fig. 1.34, the potential difference between the electrodes of all the capacitors is the same, equal to that between terminals $A$ and $B$. The total charge accumulated in all capacitors corresponding to a potential difference $V_A - V_B$ between $A$ and $B$ is

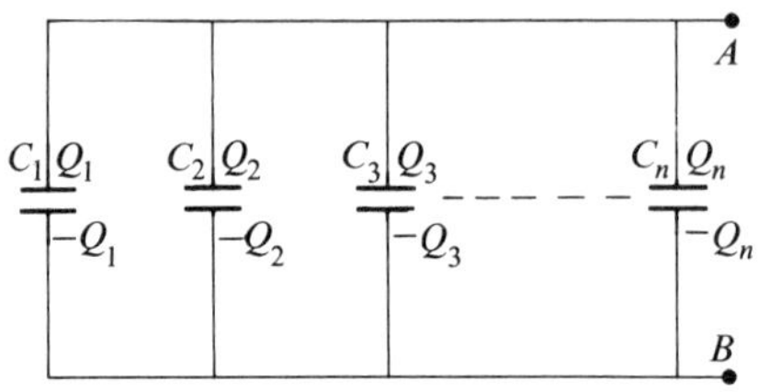


**Fig. 1.34** Parallel connection of capacitors.

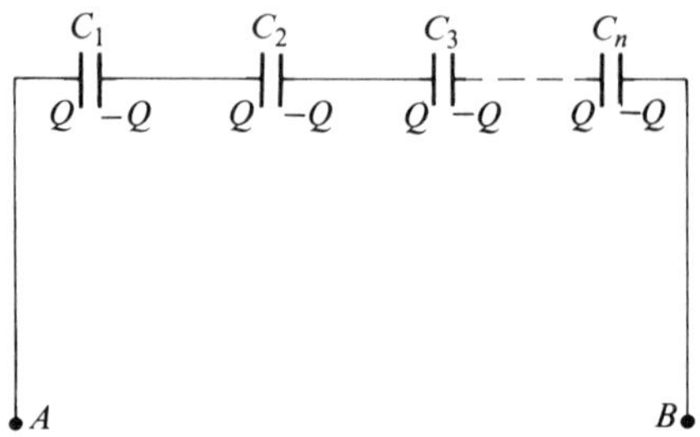


**Fig. 1.35** Series connection of capacitors.

$$
\begin{aligned}
Q_{\text{total}} = Q_1 + Q_2 + \cdots + Q_n &= C_1(V_A - V_B) + C_2(V_A - V_B) \\
&\quad + \cdots + C_n(V_A - V_B) \\
&= (C_1 + C_2 + \cdots + C_n)(V_A - V_B).
\end{aligned}
$$

So the equivalent capacitance of the parallel connection of capacitors is

$$
C_{\text{eq}} = \frac{Q_{\text{total}}}{V_A - V_B} = C_1 + C_2 + \cdots + C_n. \tag{1.14.5}
$$

In the case of the series connection shown in Fig. 1.35, the potential differences between the electrodes of the capacitors are generally different. But the charge on all capacitors is the same. This is obvious from the following reasoning: If capacitors are not connected to a source, all electrodes are uncharged. So the total charge on the—for example—right electrode of $C_1$ and the left electrode of $C_2$ is zero. Now suppose that when a source is connected, a charge $-Q$ is induced on the right electrode of $C_1$ by the charge $+Q$ on its left electrode. A charge $+Q$ must be then induced on the left electrode of $C_2$. The same argument applies to all other plates.

Denoting this common charge by $Q$, we obtain the potential difference between points $A$ and $B$ as the sum of potential differences across all capacitors,

$$
V_A - V_B = \frac{Q}{C_1} + \frac{Q}{C_2} + \cdots + \frac{Q}{C_n}.
$$

The equivalent capacitance of the series connection is hence defined by the expression

$$
\frac{1}{C_{\text{eq}}} = \frac{V_A - V_B}{Q} = \frac{1}{C_1} + \frac{1}{C_2} + \cdots + \frac{1}{C_n}. \tag{1.14.6}
$$

*Problems referring to Section 1.14:* 1.G43 to 1.G52, 1.N8 to 1.N14.

## PROBLEMS

### Test problems

**1.T1.** Discuss the statement that Eq. (1.2.2) shows indeed not only the magnitude but also the correct direction of force $\mathbf{F}_{12}$.

**1.T2.** Would the vector form (1.2.2) of Coulomb's law be possible if plus and minus signs were not associated with the two types of charges? For example, suppose that they were denoted by subscripts $A$ and $B$ instead of by plus and minus signs. (This problem is intended to show how important proper conventions are for simplifying the mathematical description of physical phenomena.)

**1.T3.** Is it possible to *derive* the principle of superposition of Coulomb's forces, starting from Coulomb's law?

**1.T4.** Prove that there can be no net electric force on an isolated charged body due to its charge only.

***1.T5.** Does it follow from Section 1.6.1 that Coulomb's inverse-square law can be derived from the principle of conservation of energy?

**1.T6.** Prove Eqs. (1.6.5) and (1.6.8).

**1.T7.** If instead of the reference point $R_1$ some other reference point $R_2$ is adopted, the potential $V$ at *all* points would increase by precisely the value of the line integral of $\mathbf{E}$ from $R_1$ to $R_2$. Prove this statement.

**1.T8.** Given that $C$ is an arbitrary closed contour, determine $\oint_C dl$ and $\oint_C d\mathbf{l}$.

**1.T9.** Using Eq. (1.7.2), deduce Eq. (1.4.9) from Eq. (1.6.13).

**1.T10.** Visualize the solid angle subtended by several objects in your room at one of your eyes, and estimate their magnitude in steradians.

***1.T11.** In Section 1.8 it is shown that Gauss' law is a consequence of Coulomb's inverse-square law. Is the converse also true, if it is supposed that $\mathbf{E}$ is proportional to $Q\,r^{-n}\,\mathbf{u}$, and $n$ is a constant to be determined?

**1.T12.** A small charged body is placed near a large uncharged *conducting* body. Is there a force on the small body? If so, what is its direction?

***1.T13.** A small charge $+Q_1$ is situated in the electric field of a charge $+Q_2$ distributed over a *conducting* body. Is the force on $Q_1$ repulsive at all points of the field?

**1.T14.** Two point charges are placed near a conducting body. Why cannot Coulomb's law be applied in this case to determine the *total* forces on the point charges?

**1.T15.** Does Coulomb's law give the correct result in the preceding problem if it is intended only for determining the component of the total force due to the other point charge?

***1.T16.** What is the net force on an uncharged conducting body in a uniform external electric field? [*Hint:* Use the result of Problem 1.T4 and the principle of superposition.]

**1.T17.** A point charge $Q$ is situated inside an uncharged conducting spherical shell. The outer radius of the shell is $a$, and its thickness is $d$. Charge $Q$ is at a distance $b$ from the center of the

---

* An asterisk designates problems that are more difficult or complicated than average.

sphere. Determine the electric field intensity at points outside the sphere, and sketch the lines of force inside the sphere.

**1.T18.** A point charge $Q$ is situated at the center of a hollow conducting sphere. Determine the force on the point charge when the system is introduced into a uniform electric field of intensity $\mathbf{E}_0$.

**1.T19.** Using Eq. (1.8.5), prove that there can be no charges inside a conductor.

**1.T20.** An electric charge $Q$ is moved slowly around in a room having conducting walls. Is it possible to detect this motion outside the room by means of some electric field meter?

**1.T21.** Sketch the lines of force and distribution of charge in a transversal cross section of two parallel cylindrical *conductors* of radii $a$, whose axes are $3a$ apart, given that the charge per unit length on the cylinders is of equal magnitude but of opposite sign.

**1.T22.** A long straight charged conductor of radius $a$ is placed parallel to an infinite conducting plane, its axis being at a height $3a/2$ from the plane. Sketch the lines of force and distribution of charges on both the conductor and the plane, in a cross-section of the system normal to the conductor.

**1.T23.** Given that $S$ is an arbitrary closed surface, determine $\oint_S dS$ and $\oint_S d\mathbf{S}$. [*Hint:* make use of Eq. (1.11.10).]

**1.T24.** Prove that if $V_1$ and $V_2$ are solutions of Laplace's equation, their product is not generally a solution to that equation.

**1.T25.** Suppose that the potential is determined for a sphere uniformly charged throughout its volume. Propose a method, based on the principle of superposition, for determining potential of uniform volume distribution of charges distributed inside a spherical *shell* of inner and outer radii $a$ and $b$, respectively.

## General problems

**1.G1.** Two small balls of masses $m$ have a charge $Q$ each, and are suspended at a common point by a thin light conducting filament of length $l$. Assuming the charges to be located approximately at the centers of the balls, find the angle $\alpha$ between the filaments. Suppose that $\alpha$ is small. (Such a system can be used as a primitive device for measuring charge and potential, and is called an *electroscope*.)

*Answer.** For $Q = 10^{-11}$ C, $l = 0.1$ m and $m = 10^{-5}$ kg, $\alpha = 0.705°$.

**1.G2.** Two conducting balls of equal radii are charged with charges $Q_1$ and $Q_2$. The distance $d$ between the centers of the balls is much larger than their radii. The balls are brought into contact, then placed back in their original positions. Determine the force between the two balls in the two cases.

*Answer.* For $Q_1 = 2 \times 10^{-10}$ C, $Q_2 = 8 \times 10^{-10}$ C, and $d = 10$ cm, the force before the two balls are brought into contact is $F_1 = 1.44 \times 10^{-7}$ N (newtons), and after they have been brought into contact it is $F_2 = 2.25 \times 10^{-7}$ N.

---

* Answers to problems are often *not* given in the form of a complete analytical expression. Rather the value of this expression for some special case is given, so that the student can check his general result, substituting in it the particular quantities given in the answer.

***1.G3.** The disk shown in Fig. 1.6 is charged with a surface charge density $\rho_s$, which is a function of the distance $s$ from the center of the disk of the form $\rho_s = As$, where $A$ is a constant. Determine the electric field intensity at the point $M$ on the disk axis shown in the figure.

*Answer.* For $a = z = 0.1$ m and $A = 10^{-7}$ C/m$^3$, $E = 98.8$ V/m.

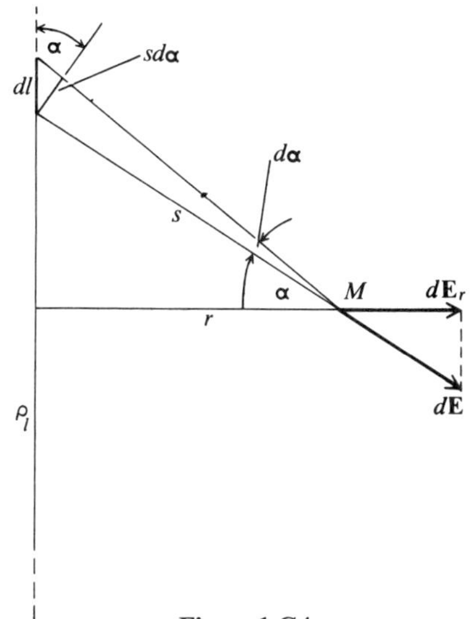


**Figure 1.G4**

**1.G4.** A very long thin straight conductor is charged with a charge $\rho_l$ per unit length. Determine the electric field intensity in the vicinity of the conductor.

*Solution.* The charge on an element $dl$ of the conductor creates at a point $M$ the electric field intensity (Fig. 1.G4) of magnitude.

$$dE = \frac{\rho_l\, dl}{4\pi\varepsilon_0 s^2}\,.$$

Owing to symmetry, the resultant electric field intensity is radial. Let us therefore determine the radial component $dE_r$ of $d\mathbf{E}$:

$$dE_r = dE\cos\alpha = \frac{\rho_l\, dl\cos\alpha}{4\pi\varepsilon_0 s^2}\,.$$

Since $s = r/\cos\alpha$ and $dl = s\,d\alpha/\cos\alpha$,

$$dE_r = \frac{\rho_l\cos\alpha\, d\alpha}{4\pi\varepsilon_0 r}\,,$$

so that

$$E_r = \frac{\rho_l}{4\pi\varepsilon_0 r}\int_{-\pi/2}^{\pi/2}\cos\alpha\, d\alpha = \frac{\rho_l}{2\pi\varepsilon_0 r}\,.$$

**1.G5.** Determine the electric field intensity due to a uniformly charged circular loop of radius $a$ and total charge $Q$, at points on the loop axis normal to its plane.

*Answer.* For $Q = 10^{-11}$ C, $a = 5$ cm, and at a distance 5 cm from the center of the loop, $E = 12.7$ V/m.

**1.G6.** A hemispherical shell of radius $a$ is charged uniformly over its surface by a total charge $Q$. Determine the electric field intensity at the center of the sphere, one-half of which is the shell.

*Answer.* For $Q = 10^{-10}$ C and $a = 10$ cm, $E = 45$ V/m.

**1.G7.** Determine the electric field intensity of an infinite, plane uniformly charged with surface charge density $\rho_s$, starting from Eq. (1.4.6).

*Answer.* $E = \rho_s/2\varepsilon_0$, directed away from the plane.

**1.G8.** Prove that an electrostatic field with parallel lines of force, but with **E** not constant in planes perpendicular to the lines, cannot exist.

***1.G9.** The source of an electrostatic field is a volume charge distribution of finite charge density $\rho$, distributed in a finite region of space. Prove that the electrostatic potential has a finite value at all points.

*Proof.* From expression (1.6.12), the electrostatic potential due to a volume distribution of charge is

$$V = \frac{1}{4\pi\varepsilon_0}\int_v \frac{\rho\, dv}{r}, \tag{a}$$

where $r$ is the distance of the volume element $dv$ from the point at which the potential is being determined. The potential given by (a) is obviously finite at all points at which $\rho = 0$. At a point at which $\rho \neq 0$, the ratio $\rho/r$ tends to an infinitely large value when $r$ becomes smaller and smaller. It has to be proved that the electrostatic potential function $V$ at these points is nevertheless finite.

Consider a point $M$ and suppose that the charge density at $M$ is $\rho$. Imagine a small sphere $S$ of radius $a$ centered at $M$. The potential at $M$ can be considered as the sum of potentials created at $M$ by charges outside the small sphere $S$, which is finite (as explained above), and by charges inside $S$. Hence it is necessary to prove that the potential at $M$ due to charges inside $S$ is finite.

If the sphere $S$ is sufficiently small, the charge density in the sphere can be considered to be equal to the charge density $\rho$ at $M$. If this is not the case, it will be supposed that $\rho$ denotes the largest value of charge density inside $S$.

Let us divide the sphere $S$ into elemental concentric spherical shells of radii $r$ $(0 \leq r \leq a)$ and thicknesses $dr$. Each of these shells creates at its interior a constant potential, equal to the potential at its surface, given by Eq. (1.8.8):

$$dV = \frac{\rho\, 4\pi r^2\, dr}{4\pi\varepsilon_0 r} = \frac{\rho\, r\, dr}{\varepsilon_0}.$$

The potential at the center of the sphere due to all charges inside $S$ is equal to the sum (that is, integral) of potentials created by all elemental spherical shells at their centers:

$$V = \int_0^a \frac{\rho\, r\, dr}{\varepsilon_0} = \frac{\rho a^2}{2\varepsilon_0}. \tag{b}$$

It was assumed that the charge density $\rho$ was finite at all points. The potential given by (b) is therefore finite even if we consider $\rho$ to represent the maximum value of charge density inside $S$. The electrostatic potential at all points of an electric field due to a finite distribution of charges of a finite density is a finite quantity.

***1.G10.** Prove that the electrostatic potential has a finite value at all points on a finite surface $S$ charged with a finite charge density $\rho_s$.

**1.G11.** Determine the expression for the gradient of a scalar function $V$ with respect to the cylindrical coordinate system.

*Solution.* The component of the gradient of a scalar function $V$ in a certain direction is given by Eq. (1.7.3). The elemental lengths in the directions of the unit vectors $\mathbf{u}_r$, $\mathbf{u}_\phi$, and $\mathbf{u}_z$ of the cylindrical system are $dr$, $r\,d\phi$, and $dz$, respectively (see Section 0.4.2 for details on the cylindrical coordinate system). Hence the components of the gradient are

$$\text{grad}_r\, V = \frac{\partial V}{\partial r}, \qquad \text{grad}_\phi\, V = \frac{1}{r}\frac{\partial V}{\partial \phi}, \qquad \text{grad}_z\, V = \frac{\partial V}{\partial z},$$

from which

$$\text{grad}\, V = \frac{\partial V}{\partial r}\mathbf{u}_r + \frac{1}{r}\frac{\partial V}{\partial \phi}\mathbf{u}_\phi + \frac{\partial V}{\partial z}\mathbf{u}_z.$$

**1.G12.** Determine the gradient of a scalar function $V$ with respect to the spherical coordinate system. (For details on the spherical coordinate system, see Section 0.4.2.)

*Answer.*
$$\text{grad}\, V = \frac{\partial V}{\partial r}\mathbf{u}_r + \frac{1}{r}\frac{\partial V}{\partial \theta}\mathbf{u}_\theta + \frac{1}{r\sin\theta}\frac{\partial V}{\partial \phi}\mathbf{u}_\phi.$$

**1.G13.** Determine the electrostatic potential, and hence the electric field intensity, at points on the axis normal to the plane of the charged circular loop of Problem 1.G5.

**1.G14.** Using Gauss' law, determine the electric field intensity inside and outside a sphere of radius $a$ uniformly charged with charge density $\rho$.

*Solution.* Owing to the symmetrical distribution of charges, the lines of the electric field intensity are radial, as shown in Fig. 1.G14. The magnitude of the electric field intensity is a

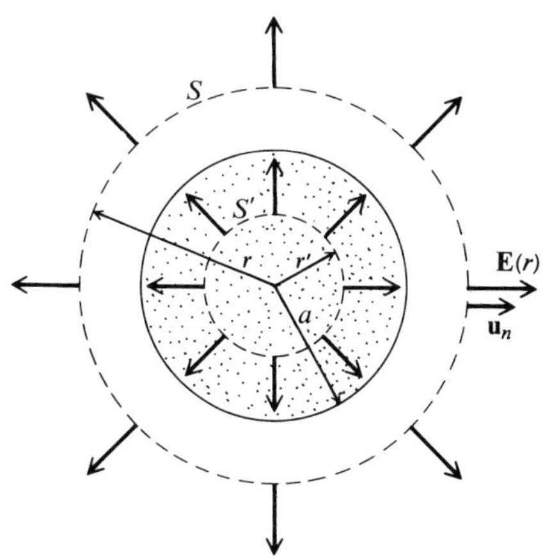


**Figure 1.G14**

function of the distance $r$ only of a point from the center of the sphere. When we apply Gauss' law to an imaginary spherical surface $S$ of radius $r$ larger than $a$, we obtain

$$\oint_S \mathbf{E}(r)\cdot d\mathbf{S} = \oint_S E(r)\, dS = E(r)\oint_S dS = E(r)\, 4\pi r^2 = \frac{1}{\varepsilon_0}\frac{4}{3}a^3\pi\rho = \frac{Q}{\varepsilon_0}.$$

Here $Q$ denotes the total charge inside the sphere. Hence

$$E(r) = \frac{Q}{4\pi\varepsilon_0 r^2} \qquad \text{for} \qquad r \geq a. \tag{a}$$

Now we apply Gauss' law to an imaginary sphere $S'$ of radius $r'$ smaller than $a$, to obtain

$$\oint_{S'} \mathbf{E}(r')\cdot d\mathbf{S}' = E(r')\, 4\pi r'^2 = \frac{1}{\varepsilon_0}\int_{v'} \rho\, dv' = \frac{1}{\varepsilon_0}\rho\int_{v'} dv' = \frac{4r'^3\pi}{3\varepsilon_0}\rho,$$

from which we get

$$E(r') = \frac{\rho}{3\varepsilon_0}r' \qquad \text{for} \qquad r' \leq a. \tag{b}$$

The sketch of the dependence of $E$ on $r$ is left to the reader.

**1.G15.** A uniform charge distribution of charge density $\rho$ is limited by two parallel planes at a distance $d$ from each other. Using Gauss' law, determine the electric field intensity inside and outside the charge layer.

*Answer.* When we take the origin at the middle of the layer and direct the $x$-axis normal to the end planes, outside the layer $\mathbf{E}$ is constant, and inside the layer $\mathbf{E}$ is a linear function of $x$.

**1.G16.** Repeat the preceding problem, given that $\rho$ is a function of $x$, of the form $\rho = A(x^2 - d^2/4)$, where $A$ is a constant. The $x$-axis is perpendicular to the end planes, and the origin is at the middle of the layer.

*Answer.* Outside the charged layer, $E_x$ is constant. Inside the layer, for $x = \pm d/4$, $E_x(\pm d/4) = \mp 11\, A\, d^3/192\,\varepsilon_0$. At all points $E_y = E_z = 0$.

**1.G17.** Inside an infinite right circular cylindrical surface of radius $a$ are distributed charges of volume charge density $\rho$. Using Gauss' law, determine the electric field intensity inside and outside the surface as a function of the distance $r$ from the axis of the cylinder.

*Answer.* $E_r(a/2) = E_r(2a) = \rho a/4\varepsilon_0$.

**1.G18.** A volume distribution of charges is bounded by a spherical surface of radius $a$. The volume charge density inside the sphere is given by $\rho(r) = \rho_0(a - r)/a$, where $r$ is the distance of a point from the center of the sphere. Using Gauss' law, determine the electric field intensity inside and outside the sphere.

*Answer.* $E_r(a/2) = 5\rho_0 a/48\varepsilon_0$; $E_r(2a) = \rho_0 a/48\varepsilon_0$.

**1.G19.** Under certain circumstances the electric field intensity in the atmosphere has a value $E_1$ at the surface of the earth, and a value $E_2$ at a height $h$ above the surface. In both cases the direction of the electric field intensity vector is toward the earth. Determine the average value of the volume charge density in the atmosphere below height $h$.

*Answer.* For $E_1 = 300$ V/m, $E_2 = 20$ V/m and $h = 1400$ m, $\rho_{\text{ave}} = +1.77 \times 10^{-12}$ C/m$^3$.

**1.G20.** Prove that the electrostatic potential cannot have a maximum or a minimum value, except at points occupied by positive and negative charges, respectively.

*Proof.* Suppose that at a point $M$ the potential has a maximum value with respect to neighboring points. Then, according to Eq. (1.7.2), the electric field intensity at all points on the surface of a small sphere $S$ enclosing the point $M$ must be directed outward. Consequently the flux of the electric field intensity through $S$ must be positive. According to Gauss' law, however, this means that $S$ encloses a certain *positive* charge, however small $S$ is. This can be true only if, at $M$, there is a positive charge of nonvanishing density. Similar argument leads to the conclusion that the potential at a field point can have a minimum value only if, at that point, there exists a nonvanishing negative charge density.

**1.G21.** Prove *Earnshaw's theorem:* A stationary system of charges cannot be in a stable equilibrium under their own influence.

*Proof.* Suppose that a small positive charge in a system of stationary charges is in stable equilibrium at a field point $M$. The electric field intensity in the vicinity of point $M$ due to all *other* charges must then be directed toward $M$. This requires a minimum of the electrostatic potential due to all charges which are *not* at $M$. In the preceding problem this was proved to be impossible.

**1.G22.** Prove that if charge density inside a spherical surface $S$ is zero at every point, then the average potential on the sphere $S$ is equal to the potential at its center, whatever its radius.

*Proof.* The average potential $V_{\text{ave}}$ on the sphere $S$ of radius $r$ is

$$V_{\text{ave}} = \frac{\oint_S V\,dS}{S} = \frac{\oint_S V\,dS}{4\pi r^2}.$$

In order to prove that $V_{\text{ave}}$ is independent of $r$, we must show that $\partial V_{\text{ave}}/\partial_r = 0$. It is hard to prove this by differentiating the right-hand side of the above equation, because not only the denominator but also $V$ and $S$ are functions of $r$. However, if we introduce the (constant) radius $r$ under the integral sign, we obtain

$$V_{\text{ave}} = \frac{1}{4\pi}\oint_S V\frac{dS}{r^2} = \frac{1}{4\pi}\oint_S V\,d\Omega.$$

Here $d\Omega$ is the elemental solid angle subtended by $dS$ at the center of the sphere. The only function of $r$ on the right-hand side of the last equation is $V$. Hence

$$\frac{\partial V_{\text{ave}}}{\partial r} = \frac{1}{4\pi}\oint_S \frac{\partial V}{\partial r}\,d\Omega.$$

If we write $dS/r^2$ instead of $d\Omega$, and if we note that $\partial V/\partial r = -E_r$, we obtain

$$\frac{\partial V_{\text{ave}}}{\partial r} = -\frac{1}{4\pi r^2}\oint_S E_r\,dS = -\frac{1}{4\pi r^2}\oint_S \mathbf{E}\cdot d\mathbf{S}.$$

We assume that the sphere $S$ encloses no charges. By Gauss' law the last integral therefore vanishes, which means that $V_{\text{ave}}$ is a constant, independent of $r$. To determine this constant, imagine the sphere to be very small. Then $V_{\text{ave}}$ is equal to the potential at the center of the sphere. The proof is thereby completed.

This conclusion provides a simple basis for approximate numerical evaluation of the electrostatic potential function between conductors at given potentials (see Section 5.3).

**1.G23.** A soap bubble of radius $R$ and wall thickness $a$ is at a potential $V$ with respect to the reference point at infinity. Determine the potential of a spherical drop obtained when the bubble explodes. Consider the bubble and the drop as conductors.

*Answer.* For $R = 2$ cm. $a = 10^{-4}$ cm, and $V = 1000$ V, the potential of the drop is $V_{\text{drop}} \simeq 18{,}800$ V.

**1.G24.** A certain volume of a liquid conductor is sprayed into $N$ equal spherical drops. Then, by some appropriate method, each drop is given a potential $V$ with respect to the reference point at infinity. Finally, all these small drops are combined into a large spherical drop. Determine the potential of the large drop.

*Answer.* For $N = 10^6$ and $V = 10$ V, the potential of the large drop is $V_1 = 100{,}000$ V.

**1.G25.** Two small conducting spheres of radii $a$ and $b$ are connected by a very thin, flexible conductor of length $d$. The total charge of the system is $Q$. Assuming that $d$ is much larger than both $a$ and $b$, determine the force $F$ on the wire. Charges may be considered to be located on the two spheres only, and to be distributed practically uniformly over their surfaces.

*Answer.* For $a = 1$ cm, $b = 2$ cm, $d = 1$ m, and $Q = 10^{-10}$ C, the force $F = 2 \times 10^{-11}$ N.

**1.G26.** Two small conducting balls of radii $a$ and $b$ are charged with charges $Q_a$ and $Q_b$, and are placed at a distance $d$ ($d \gg a, b$) apart. Suppose that the balls are connected with a thin conducting wire. What will be the direction of flow of positive charges through the wire? Discuss the question for various values of $Q_a$, $Q_b$, $a$, and $b$.

**1.G27.** Prove that if a small hole is drilled through a thin charged conducting shell, the electric field intensity at the center of the hole is equal to half the electric field intensity at that point before the hole was drilled. [*Hint:* Consider the electric field intensity just above and just below the conductor surface as a sum of the electric field intensities due to the surface charge in the immediate vicinity of the point considered, and due to all other charges of the system.]

**1.G28.** Inside a spherical conducting shell of radius $b$ is a conducting sphere of radius $a$ ($a < b$), charged with a charge $Q_a$. What is the potential of the shell: (1) if it is uncharged? (2) if it is charged with a charge $Q_b$? Does the potential depend on the position of the sphere inside the shell? Will it change if we move the sphere into contact with the inner surface of the shell?

**1.G29.** Suppose that the shell in the preceding problem is connected by a thin conducting wire to the reference point of the potential. Determine: (1) its charge, (2) the electrostatic potential function outside the shell.

**1.G30.** A conducting sphere of radius $a$ carries a charge $Q_1$. Concentric with the sphere there is a spherical shell of inner radius $b$ ($b > a$) and outer radius $c$, carrying a charge $Q_2$. Determine the electric field intensity and the electrostatic potential at every point of the system. Sketch the dependence of $E$ and $V$ on the distance $r$ from the common center.

***1.G31.** An isolated conducting ellipsoid of semi-axes $a$, $b$, and $c$ carries a charge $Q$. Determine the distribution of charges on the surface of the ellipsoid.

*Solution.* Since the ellipsoid is conducting, the total charge $Q$ is distributed over its surface. The charge distribution must be such that inside the ellipsoid the electric field intensity is zero at all points.

Consider an arbitrary point $M$ inside the ellipsoid, and divide the ellipsoid volume into pairs of elemental cones with apexes at that point. The electric field intensity at $M$ will be zero if, at $M$, charges on elemental surfaces cut out by these cone pairs create field intensities of equal

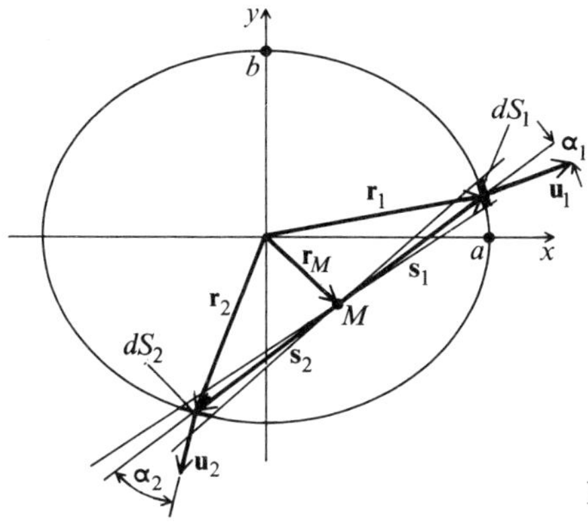


**Figure 1.G31**

magnitude but opposite direction. Indicated in Fig. 1.G31 is one pair of such cones. Denote the elemental surfaces by $dS_1$ and $dS_2$, the surface charge densities at $dS_1$ and $dS_2$ by $\rho_{s1}$ and $\rho_{s2}$, and distances of $dS_1$ and $dS_2$ from $M$ by $s_1$ and $s_2$. According to the above reasoning, it must be true that

$$\frac{\rho_{s1}\, dS_1}{4\pi\varepsilon_0 s_1^2} = \frac{\rho_{s2}\, dS_2}{4\pi\varepsilon_0 s_2^2}.$$

Since the elemental solid angle of elemental cones is

$$d\Omega = \frac{dS_1 \cos\alpha_1}{s_1^2} = \frac{dS_2 \cos\alpha_2}{s_2^2},$$

the above condition becomes

$$\frac{\rho_{s1}}{\cos\alpha_1} = \frac{\rho_{s2}}{\cos\alpha_2}. \tag{a}$$

Here $\alpha_1$ and $\alpha_2$ denote the angles between vectors $\mathbf{s}_1$ and $\mathbf{s}_2$ and the unit vectors $\mathbf{u}_1$ and $\mathbf{u}_2$ normal to the elemental surfaces $dS_1$ and $dS_2$.

With respect to the coordinate system shown in Fig. 1.G31, the equation of the ellipsoid surface is

$$f(x, y, z) = \frac{x^2}{a^2} + \frac{y^2}{b^2} + \frac{z^2}{c^2} - 1 = 0. \tag{b}$$

The unit vector $\mathbf{u}$ normal to the surface of the ellipsoid at a point having coordinates $x$, $y$ and $z$ is given by

$$\mathbf{u} = \left(\frac{\operatorname{grad} f}{|\operatorname{grad} f|}\right)_{f=0} = \frac{2(\mathbf{u}_x x/a^2 + \mathbf{u}_y y/b^2 + \mathbf{u}_z z/c^2)}{2(x^2/a^4 + y^2/b^4 + z^2/c^4)^{1/2}} = (\mathbf{u}_x x/a^2 + \mathbf{u}_y y/b^2 + \mathbf{u}_z z/c^2)\, h, \tag{c}$$

where

$$h = (x^2/a^4 + y^2/b^4 + z^2/c^4)^{-1/2}. \tag{d}$$

We shall now show that $h$ represents the normal distance from the origin to the plane tangential to the surface of the ellipsoid at the point considered.

The equation of the plane tangential to the ellipsoid surface at a point $M'(x', y', z')$ on the surface has the form

$$(\mathbf{r} - \mathbf{r}')\cdot\mathbf{u}' = 0. \tag{e}$$

Here $\mathbf{r}'$ is the position vector of $M'$, $\mathbf{u}'$ the unit vector normal to the surface at $M'$, and $\mathbf{r}$ the position vector of a point on the tangential plane. We can readily understand that this equation represents the equation of the tangential plane at $M'$. Note only that $(\mathbf{r} - \mathbf{r}')$ is in that plane, and that $\mathbf{u}'$ is normal to the plane. So the above equation must be satisfied.

The normal distance from the tangential plane to the origin is equal to $\mathbf{r}\cdot\mathbf{u}'$, or according to Eq. (e), to $\mathbf{r}'\cdot\mathbf{u}'$. Since $M'$ is on the surface of the ellipsoid, from Eqs. (c) and (b) we obtain

$$\begin{aligned}\mathbf{r}'\cdot\mathbf{u}' &= (\mathbf{u}_x x' + \mathbf{u}_y y' + \mathbf{u}_z z')\cdot(\mathbf{u}_x x'/a^2 + \mathbf{u}_y y'/b^2 + \mathbf{u}_z z'/c^2)\,h' \\ &= (x'^2/a^2 + y'^2/b^2 + z'^2/c^2)\,h' = h',\end{aligned}$$

which had to be proved.

Let us now determine $\cos\alpha_1$ and $\cos\alpha_2$ as they enter into Eq. (a). First, the unit vector of the direction from $dS_1$ to $dS_2$ is

$$\mathbf{u}_s = \frac{\mathbf{s}_1 - \mathbf{s}_2}{s_1 + s_2} = \frac{\mathbf{r}_1 - \mathbf{r}_M - (\mathbf{r}_2 - \mathbf{r}_M)}{s_1 + s_2} = \frac{1}{s_1 + s_2}\left[\mathbf{u}_x(x_1 - x_2) + \mathbf{u}_y(y_1 - y_2) + \mathbf{u}_z(z_1 - z_2)\right].$$

So according to Eqs. (b) and (c),

$$\cos\alpha_1 = \mathbf{u}_1\cdot\mathbf{u}_s = \frac{h_1}{s_1 + s_2}(1 - x_1x_2/a^2 - y_1y_2/b^2 - z_1z_2/c^2),$$

$$\cos\alpha_2 = -\mathbf{u}_2\cdot\mathbf{u}_s = \frac{h_2}{s_1 + s_2}(1 - x_2x_1/a^2 - y_2y_1/b^2 - z_2z_1/c^2).$$

The condition (a) of the electrostatic equilibrium of charges on the surface of the conducting ellipsoid becomes

$$\frac{\rho_{s1}}{h_1} = \frac{\rho_{s1}}{h_2} = \text{constant} = A.$$

The point $M$ and the pair of elemental cones were chosen arbitrarily. Hence this relation must be valid for any point on the surface of the ellipsoid. Accordingly, the surface charge density on a charged isolated ellipsoid is proportional to the distance $h$ from the plane tangential to the ellipsoid surface at the point considered, to the center of the ellipsoid,

$$\rho_s = A\,h. \tag{f}$$

The constant $A$ is a function of the dimensions of the ellipsoid and the total charge on it. We can determine it as follows.

The total charge on the ellipsoid is

$$Q = \oint_S \rho_s\,dS = A\oint_S h\,dS,$$

where $S$ denotes the area of the ellipsoid. Imagine the whole ellipsoid to be divided into elemental cones, with apexes at the center of the ellipsoid and with bases $dS$. Each of these cones has a volume $dv = h\,dS/3$, since $h$ represents the normal distance from the cone apex to the plane of its base, that is, the height of the elemental cone. Hence $h\,dS = 3\,dv$, so that the surface integral in the last equation represents the threefold volume of the ellipsoid:

$$Q = A\,3\int_v dv = 3Av.$$

Since the volume of an ellipsoid $v = 4\pi abc/3$, the constant $A$ is

$$A = \frac{Q}{3v} = \frac{Q}{4\pi abc}.$$

According to Eqs. (f) and (d), the surface charge density distribution function on the surface of the ellipsoid is finally given by

$$\rho_s = \frac{Q}{4\pi abc\sqrt{x^2/a^4 + y^2/b^4 + z^2/c^4}}. \tag{g}$$

**1.G32.** Starting with the result of the preceding problem, determine the charge distribution on a very thin straight conducting wire of circular cross section, of radius $R$ and length $L(L \gg R)$, carrying a charge $Q$.

*Solution.* Let $a = b$ in Eq. (g) of the preceding problem, and suppose that $a$ tends to zero. We obtain the distribution of charges on the surface of a very thin straight conductor of length $2c$, carrying a charge $Q$, as follows:

$$\rho_s = \lim_{a\to 0} \frac{Q}{4\pi a^2 c\sqrt{(x^2 + y^2)/a^4 + z^2/c^4}} = \frac{Q}{4\pi c\sqrt{x^2 + y^2}}.$$

According to the assumption, $L \gg R$. So distribution of charge is practically the same in the case of the conducting thin wire, if we only change $2c$ to $L$ and $\sqrt{x^2 + y^2}$ to $R$:

$$\rho_s = \frac{Q}{2\pi LR}.$$

The charge per unit length of the wire

$$\rho_l = 2\pi R\rho_s = \frac{Q}{L}.$$

In the case of a very thin conducting wire, the charge is distributed along its length almost uniformly. Only near its ends is the surface charge density very large. We see this from Eq. (f) of the preceding problem, recalling that the distance $h$ from the tangential plane at points along the thin ellipsoid is practically equal to $R$, and at its ends $h = L/2$. Note that the area of surface carrying such a large charge density is very small. The effect of the ends on the distribution of charges can therefore usually be neglected.

Based on this conclusion, we can develop an approximate method of determining the capacitance of thin wires (see Problem 1.G48).

**1.G33.** Starting with the result (g) of Problem 1.G31, determine the charge distribution on the surface of a very thin flat circular conducting disk of radius $a$, carrying a charge $Q$.

*Answer.* $\rho_s = Q/(4\pi a\sqrt{a^2 - r^2})$, where $r$ is the distance from the center of the disk. Note that the distribution of charges in this case is far from uniform.

**1.G34.** Determine the divergence of a vector function $\mathbf{A}$ with respect to a cylindrical coordinate system.

*Solution.* The divergence of a vector function is defined by Eq. (1.11.4). Figure 1.G34 shows the volume element in a cylindrical coordinate system. Referring to explanations given in

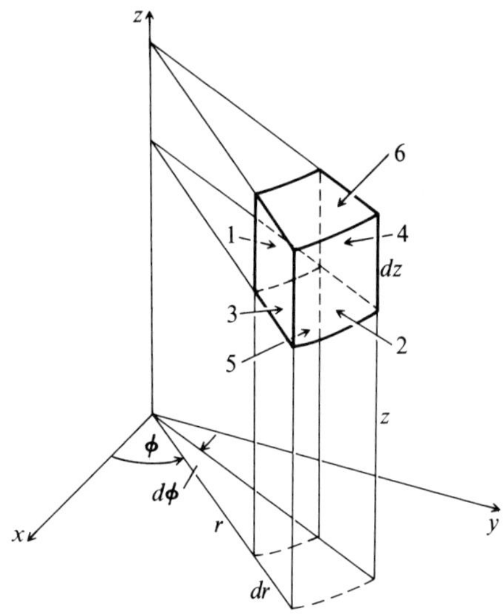


**Figure 1.G34**

Section 0.4.2, this volume element is enclosed by cylinders $r$ = constant and $r + dr$ = constant, by half-planes $\phi$ = constant and $\phi + d\phi$ = constant, and by planes $z$ = constant and $z + dz$ = constant. Neglecting infinitesimal quantities of order higher than the first, these six coordinate surfaces define an infinitesimal rectangular parallelepiped of sides $dr$, $r\,d\phi$, and $dz$. We obtain the flux of vector **A** through the pair of elemental surfaces 1 and 2 (see Fig. 1.G34) of the elemental parallelepiped as follows. Note that the flux of $A_r$ through surface 1 is inward flux (that is, negative), and that the flux through surface 2 is outward flux (that is, positive). Therefore

$$(\mathbf{A}\cdot d\mathbf{S})_{1+2} = -A_r\, r\, d\phi\, dz + \left[A_r\, r\, d\phi\, dz + \frac{\partial}{\partial r}(A_r\, r\, d\phi\, dz)\, dr\right] = -\frac{\partial}{\partial r}(rA_r)\, dr\, d\phi\, dz.$$

We obtain the flux through pairs of elemental surfaces (3 + 4) and (5 + 6) in a similar manner. The result is

$$(\mathbf{A}\cdot d\mathbf{S})_{3+4} = \frac{\partial A_\phi}{\partial \phi}\, dr\, d\phi\, dz, \qquad (\mathbf{A}\cdot d\mathbf{S})_{5+6} = \frac{\partial A_z}{\partial z}\, r\, dr\, d\phi\, dz.$$

The total flux through the elemental parallelepiped is the sum of the three partial fluxes determined above:

$$(\mathbf{A}\cdot d\mathbf{S})_{1+2+\cdots+6} = \left[\frac{1}{r}\frac{\partial}{\partial r}(r\, A_r) + \frac{1}{r}\frac{\partial A_\phi}{\partial \phi} + \frac{\partial A_z}{\partial z}\right] r\, dr\, d\phi\, dz.$$

Since the volume enclosed by the elemental parallelepiped is $dv = r\, dr\, d\phi\, dz$,

$$\operatorname{div}\mathbf{A} = \lim_{\Delta v\to 0}\frac{\oint_S \mathbf{A}\cdot d\mathbf{S}}{\Delta v} = \frac{(\mathbf{A}\cdot d\mathbf{S})_{1+2+\cdots+6}}{r\, dr\, d\phi\, dz} = \frac{1}{r}\frac{\partial}{\partial r}(r\, A_r) + \frac{1}{r}\frac{\partial A_\phi}{\partial \phi} + \frac{\partial A_z}{\partial z}.$$

**1.G35.** Determine the divergence of a vector function **A** with respect to the spherical coordinate system.

*Answer.*

$$\operatorname{div} \mathbf{A} = \frac{1}{r^2}\frac{\partial}{\partial r}(r^2 A_r) + \frac{1}{r \sin\theta}\frac{\partial}{\partial \theta}(\sin\theta\, A_\theta) + \frac{1}{r\sin\theta}\frac{\partial A_\phi}{\partial \phi}.$$

**1.G36.** Determine the Laplacian (divergence of the gradient) of a scalar function $V$ with respect to the cylindrical coordinate system.

*Answer.* According to the results of Problems 1.G11 and 1.G34,

$$\operatorname{div} \operatorname{grad} V = \frac{1}{r}\frac{\partial}{\partial r}\left(r\frac{\partial V}{\partial r}\right) + \frac{1}{r^2}\frac{\partial^2 V}{\partial \phi^2} + \frac{\partial^2 V}{\partial z^2}.$$

**1.G37.** Prove that the Laplacian of a scalar function $V$ with respect to the spherical coordinate system is

$$\operatorname{div} \operatorname{grad} V = \frac{1}{r^2}\frac{\partial}{\partial r}\left(r^2\frac{\partial V}{\partial r}\right) + \frac{1}{r^2 \sin\theta}\frac{\partial}{\partial \theta}\left(\sin\theta\frac{\partial V}{\partial \theta}\right) + \frac{1}{r^2\sin^2\theta}\frac{\partial^2 V}{\partial \phi^2}.$$

**1.G38.** Two large parallel plates at a distance $d$ from each other are at potentials zero and $V_0$. Starting with Laplace's equation, determine the electrostatic potential function and therefrom the electric field intensity between the plates.

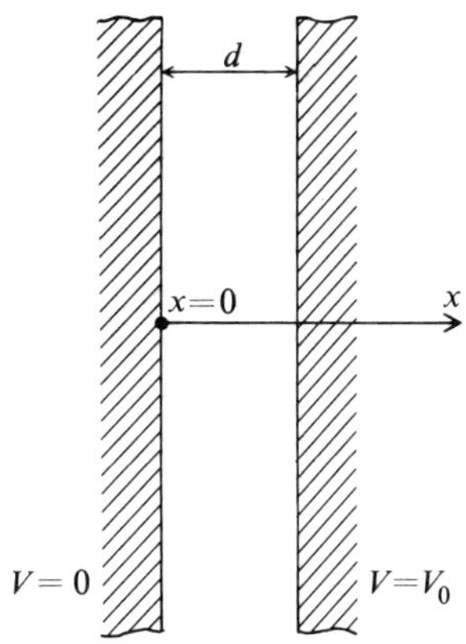


**Figure 1.G38**

*Solution.* We shall determine the electrostatic potential function between the plates at points far from their edges. By reference to Fig. 1.G38, Laplace's equation (1.12.4) reduces in this case to $d^2V/dx^2 = 0$, from which we obtain

$$V(x) = C_1 x + C_2. \tag{a}$$

Here $C_1$ and $C_2$ are constants to be determined so that the prescribed boundary conditions are satisfied. Since $V = 0$ for $x = 0$, and $V = V_0$ for $x = d$, from (a) it follows that $C_2 = 0$ and $C_1 = V_0/d$. Hence

$$V(x) = \frac{V_0}{d}x.$$

The electric field intensity

$$\mathbf{E} = -\operatorname{grad} V = -\mathbf{u}_x \frac{dV}{dx} = -\mathbf{u}_x \frac{V_0}{d}.$$

(Is the negative sign correct or not?)

This example represents one of a few cases in which the electrostatic potential can be determined by direct integration of Laplace's equation.

**1.G39.** The inner conductor of a coaxial cable is at potential $V_0$, and the outer at potential zero. The radii of the conductors are $a$ and $b$ ($a < b$). Starting with Laplace's equation, determine the electrostatic potential function, and hence the electric field intensity vector, between the conductors.

**1.G40.** Repeat the preceding problem for the case of two concentric conducting spherical shells of radii $a$ and $b$ (the spherical capacitor).

**1.G41.** The electrostatic potential function between two parallel very large plates is given by

$$V(x) = A\,x^{4/3} + B\,x + C \qquad (A, B, C \text{ are constants}),$$

where $x$ is the distance of a point between the plates from one of the plates. Determine the volume density of the charges creating such a potential function. [*Hint:* Make use of Poisson's equation.]

**1.G42.** Starting with Laplace's equation (1.12.4), prove that the electrostatic potential function cannot have a relative maximum or minimum in a charge-free region.

**1.G43.** Determine the capacitance of a coaxial cable, with conductors of radii $a$ and $b$, per unit length of the cable.

*Solution.* Follow the pattern given in Section 1.14.2 of determining the capacitance, and suppose that the inner conductor carries a charge $Q_l$ and the outer conductor a charge $-Q_l$ per unit length. The electric field intensity due to these charges was determined in Section 1.8.2, Eq. (1.8.14). The potential difference between the cable conductors corresponding to charges $Q_l$ and $-Q_l$ is hence

$$V_a - V_b = \int_a^b \frac{Q_l}{2\pi\varepsilon_0 r}\,dr = \frac{Q_l}{2\pi\varepsilon_0}\ln\frac{b}{a},$$

so that the capacitance of the cable per unit length is

$$C_l = \frac{Q_l}{V_a - V_b} = \frac{2\pi\varepsilon_0}{\ln(b/a)}.$$

**1.G44.** Determine the capacitance of a spherical capacitor whose electrode radii are $a$ and $b$.

*Answer.* For $a = 5$ cm and $b - 10$ cm, $C = 11.1$ pF.

**1.G45.** The outer conductor of a coaxial cable has a fixed radius $b$. Determine radius $a$ of the inner conductor so that, for a given potential difference between the conductors, the electric field intensity at its surface is minimum. Determine the capacitance per unit length of such a cable.

*Solution.* According to Eq. (1.8.14), if conductors carry charges $Q_l$ and $-Q_l$ per unit length, the electric field intensity at the surface of the inner conductor is

$$E(a) = \frac{Q_1}{2\pi\varepsilon_0 a}.$$

Charge $Q_l$ can be expressed by the potential difference $(V_a - V_b)$ and the cable capacitance per unit length (Problem 1.G43),

$$Q_l = C_l (V_a - V_b) = \frac{2\pi\varepsilon_0(V_a - V_b)}{\ln(b/a)}.$$

Combining the two equations we obtain

$$E(a) = \frac{V_a - V_b}{a \ln(b/a)}.$$

Radius $a$ resulting in the minimum value of $E(a)$ is obtained from the equation

$$\frac{\partial E(a)}{\partial a} = (V_a - V_b)\left(-\frac{1}{a^2 \ln^2(b/a)}\right)\left[\ln\frac{b}{a} + a\frac{1}{b/a}\left(-\frac{b}{a^2}\right)\right] = 0,$$

from which

$$\ln\frac{b}{a} = 1 \qquad \text{or} \qquad \frac{b}{a} = e = 2.7182\ldots \tag{a}$$

Capacitance per unit length of such a coaxial cable (with dielectric vacuum or air) is $C_l = 2\pi\varepsilon_0 = 556$ pF/m.

**1.G46.** For given $b$, determine $a$ in Problem 1.G44 so that the electric field intensity on the surface of the inner electrode for a given potential difference between the electrodes is minimum. Determine the capacitance of such a capacitor.

*Answer.* For $b = 10$ cm, the capacitance $C = 11.1$ pF.

***1.G47.** Starting with the result of Problem 1.G33, determine the capacitance of a very thin conducting flat circular disk of radius $a$.

*Solution.* In Problem 1.G33, we determined the distribution of charges over the surface of a very thin flat circular conducting disk carrying a total charge $Q$. In order to determine its capacitance, we must determine its potential due to the charge $Q$. Since the disk is conducting, the potential is the same at all points of its surface. So it suffices to find the potential at any one of these points. For that purpose let us determine the potential of a point $M$ on the axis normal

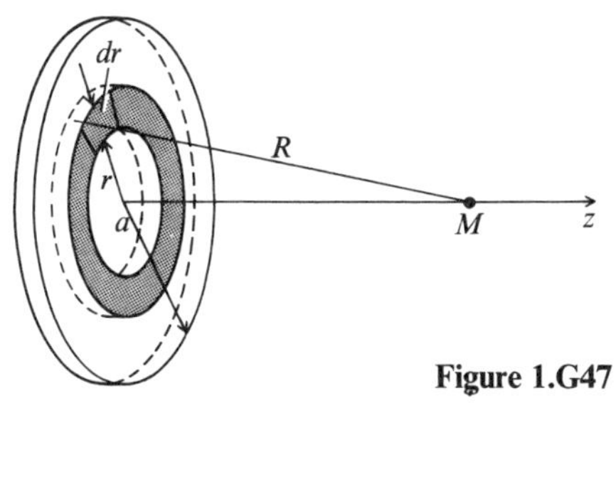


**Figure 1.G47**

to the disk (Fig. 1.G47). We obtain the potential of the disk if we allow the point $M$ to approach the surface of the disk.

The potential at $M$ is

$$V = \frac{1}{4\pi\varepsilon_0} \int_S \frac{\rho_s \, dS}{R},$$

where $R$ represents the distance from the surface element $dS$ to point $M$. Let $dS = 2(2\pi r \, dr)$ (we take into account immediately that the charge is distributed over *both* faces of the disk). According to the result of Problem 1.G33,

$$V = \frac{Q}{4\pi\varepsilon_0 a} \int_0^a \frac{r \, dr}{\sqrt{(a^2 - r^2)(z^2 + r^2)}}, \tag{a}$$

When we introduce a new variable, $v = z^2 + r^2$, $dv = 2\,r\,dr$, the above integral becomes

$$\frac{1}{2} \int_{z^2}^{z^2+a^2} \frac{dv}{\sqrt{v(z^2 + a^2 - v)}} = \tfrac{1}{2} \arcsin \frac{v - (z^2 + a^2)/2}{(z^2 + a^2)/2} \Bigg|_{v=z^2}^{v=z^2+a^2}$$

$$= \frac{1}{2}\left(\frac{\pi}{2} - \arcsin \frac{z^2 - a^2}{z^2 + a^2}\right).$$

Now let $z$ tend to zero. The value of the integral in (a) is then $\pi/2$, so that (a) becomes

$$V = \frac{Q}{4\pi\varepsilon_0 a} \frac{\pi}{2} = \frac{Q}{8\varepsilon_0 a}.$$

Hence the capacitance of the disk is

$$C = \frac{Q}{V} = 8\varepsilon_0 a.$$

***1.G48.** Starting with the result of Problem 1.G32 (i.e., that the charge along a thin straight wire is distributed very nearly uniformly), determine the approximate expression for the capacitance of a thin rod of radius $R$ and length $L$ ($L \gg R$).

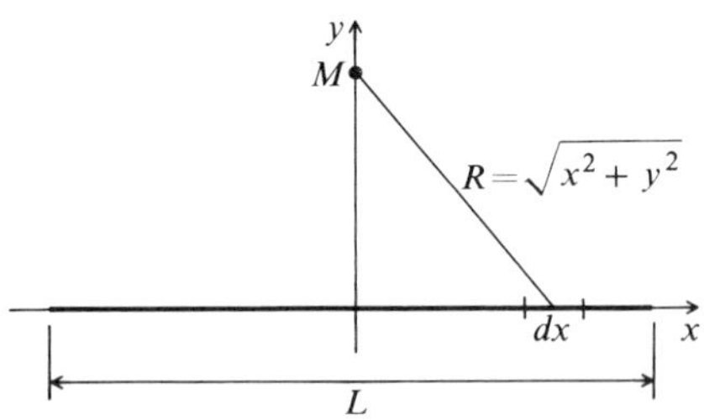


**Figure 1.G48**

*Solution.* To determine the capacitance of the rod, we must determine its potential when it carries a charge $Q$. Assume approximately that the charge $Q$ is distributed along the axis of the rod. The potential at a point $M$ due to this line charge (see Fig. 1.G48) is

$$V = \frac{1}{4\pi\varepsilon_0} \int_{-L/2}^{L/2} \frac{Q}{L} \frac{dx}{\sqrt{x^2 + y^2}} = \frac{Q}{4\pi\varepsilon_0 L} \ln \frac{L/2 + \sqrt{(L/2)^2 + y^2}}{-L/2 + \sqrt{(L/2)^2 + y^2}}.$$

For $y = R$, we obtain

$$V_{\text{rod}} = \frac{Q}{4\pi\varepsilon_0 L} \ln \frac{L/2 + \sqrt{(L/2)^2 + R^2}}{-L/2 + \sqrt{(L/2)^2 + R^2}}.$$

Since $L \gg R$, the numerator in the above equation is approximately equal to $L$. By expanding the square root in the denominator and retaining only the first two terms, we obtain

$$-L/2 + \sqrt{(L/2)^2 + R^2} = (L/2)\left[-1 + \sqrt{1 + (2R/L)^2}\right] \simeq (L/2)\tfrac{1}{2}(2R/L)^2 = R^2/L.$$

Consequently, for a very thin rod,

$$V_{\text{rod}} = \frac{Q}{4\pi\varepsilon_0 L} \ln \frac{L^2}{R^2} = \frac{Q}{2\pi\varepsilon_0 L} \ln \frac{L}{R},$$

from which we obtain the capacitance of the rod:

$$C = \frac{Q}{V_{\text{rod}}} = \frac{2\pi\varepsilon_0 L}{\ln(L/R)}. \tag{a}$$

This formula can be used to compute the capacitance of long, thin isolated wires.

***1.G49.** Show that equipotential surfaces due to a line charge of length $2c$ are confocal rotational ellipsoids with foci at the ends of the line charge. From this, deduce the capacitance of an isolated prolate ellipsoid of semiaxes $a$ and $b$ with focal distance $c = \sqrt{a^2 - b^2}$. By an appropriate approximation obtain the capacitance of a thin rod of radius $R$ and length $L$, given by (a) of the preceding problem.

**1.G50.** Determine the capacitance per unit length between two parallel wires of radii $R$ and distance between axes of the wires $D$, given that $D \gg R$ (the *two-wire line*).

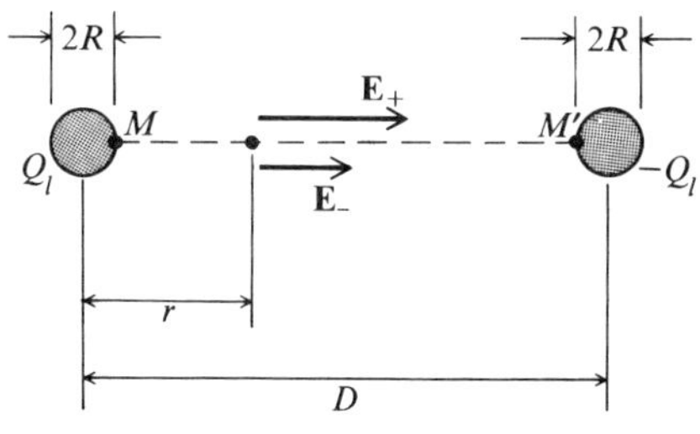


**Figure 1.G50**

*Solution.* In accord with the pattern explained in Section 1.14.2, suppose that the wires carry charges $Q_l$ and $-Q_l$ per unit length. If $D \gg R$, distribution of charges over the conductor surfaces is practically uniform. Hence the electric field intensity due to charges on the two wires is given by Eq. (1.8.10). To determine the potential difference between the two wires carrying charges $Q_l$ and $-Q_l$ per unit length, it is sufficient to determine the potential difference between any two points on the two wires. The simplest case is obtained if we consider points $M$ and $M'$ (Fig. 1.G50) and determine the line integral of the total electric field intensity along the straight line joining them. The electric field intensities $\mathbf{E}_+$ and $\mathbf{E}_-$ are in the direction indicated. Their magnitudes are

$$E_+ = \frac{Q_l}{2\pi\varepsilon_0 r} \quad \text{and} \quad E_- = \frac{Q_l}{2\pi\varepsilon_0(D - r)}.$$

Hence

$$V_M - V_{M'} = \int_R^{D-R} (E_+ + E_-)\,dr = \frac{Q_l}{2\pi\varepsilon_0}\int_R^{D-R}\left(\frac{1}{r} + \frac{1}{D-r}\right)dr$$

$$= \frac{Q_l}{2\pi\varepsilon_0}\left[\ln\frac{D-R}{R} - \ln\frac{D-(D-R)}{D-R}\right] = \frac{Q_l}{\pi\varepsilon_0}\ln\frac{D-R}{R}.$$

Since $D \gg R$, the value of the capacitance per unit length of the two-wire line is

$$C_l = \frac{Q_l}{V_M - V_{M'}} = \frac{\pi\varepsilon_0}{\ln(D/R)}. \quad \text{(a)}$$

**1.G51.** A thick uncharged metal plate is introduced between the electrodes of a parallel-plate capacitor parallel to the capacitor electrodes. The distance between the electrodes is $d$, and the thickness of the plate is $a$ ($a < d$). Is it possible to define the capacitance of the parallel-plate capacitor in this case? If so, determine the capacitance, if area of the plates is $S$.

***1.G52.** Suppose that by some appropriate method the electrostatic potential function $V(x, y, z)$ has been found in a system of two conductors, one within the *closed* cavity of the other, having prescribed values of the potentials $V_1$ and $V_2$ on the surfaces of the two conductors. Is it possible to determine the capacitance between the two conductors from $V(x, y, z)$? If so, determine the general expression for the capacitance.

## Numerical problems

**1.N1.** Determine the electric force between an electron and a nucleus carrying equal charges (of opposite signs) $Q = e = 1.602 \times 10^{-19}$ C. Assume that the distance between them is $d = 0.5 \times 10^{-10}$ m. Compare this force with the gravitational force of attraction $F = \gamma m_e m_n/d^2$, where $\gamma = 6.67 \times 10^{-11}$ N m$^2$/kg$^2$, the electron mass $m_e = 9.11 \times 10^{-31}$ kg, and the mass $m_n$ of the nucleus is 1837 times larger.

**1.N2.** If it were possible to charge two small bodies at a distance of 1 m apart with charges of 1 C each, what would be the force between them?

**1.N3.** Two small bodies at a distance of 1 cm carry a charge of $Q = 10^{-12}$ C each. How many electron charges do we have to add to each body in order to change the force acting between them by 0.0001%?

**1.N4.** Determine the electric field intensity at a point at a distance $r = 1$ cm from a "point" charge $Q = 10^{-11}$ C.

**1.N5.** What is the number of electron charges in one cubic millimeter if the volume density of charges at a certain point is $\rho = 10^{-6}$ C/m$^3$?

**1.N6.** In normal circumstances the air acts as a very good insulator up to electric field intensities of approximately $E_{\max} = 3 \times 10^6$ V/m. For larger values of $E$, it acts like a conductor. Determine the maximum possible charge on a spherical conductor of radius $a = 10$ cm situated in air. Determine the potential of the sphere in that case.

**1N.7.** The radius of the earth is approximately $6.37 \times 10^6$ m. Considering the earth as a large conducting sphere surrounded by air, what is the maximum charge that can exist on it? What is the potential of the earth in that case?

**1.N8.** Determine the capacitance of the earth. (See the preceding problem.)

**1.N9.** Determine the capacitance of a parallel-plate capacitor with plates of surfaces $S = 0.01\ \text{m}^2$ and distance between the plates $d = 0.1$ mm. What is the maximum potential difference between the capacitor electrodes if the dielectric is air?

**1.N10.** Determine the capacitance per unit length of a two-wire line in air whose conductors have radii of $R = 5$ mm; the distance between the axes of the wires is $D = 0.5$ m. (See Problem 1.G50.)

**1.N11.** Determine the capacitance of a thin disk of radius $a = 5$ cm. (See Problem 1.G47.)

**1.N12.** Determine the capacitance of a straight piece of wire of radius $a = 1$ mm and length $L = 10$ m. (See Problem 1.G48.)

**1.N13.** Determine the capacitance per unit length of a coaxial cable with conductors of radii $a = 0.5$ cm and $b = 1.5$ cm. (See Problem 1.G43.)

**1.N14.** The potential difference between plates of the parallel-plate capacitor in Problem 1.N9 is 100 V. The distance between the plates is then changed from 0.1 mm to 1 mm, the source being disconnected. What is the new potential difference? Explain the result.

**Suggested reading.** The list given here progresses from simpler to more advanced texts. For a full bibliographical description, see the Bibliography.

1. KRAUS, J. D., *Electromagnetics*, Chapter 1.
2. SCOTT, W. T., *The Physics of Electricity and Magnetism*, Chapter 1.
3. PANOFSKY, W. K. H., and M. PHILLIPS, *Classical Electricity and Magnetism*, Chapter 1.

CHAPTER 2

# THE STATIC ELECTRIC FIELD IN THE PRESENCE OF DIELECTRICS

## 2.1 THE BEHAVIOR OF DIELECTRICS IN A STATIC ELECTRIC FIELD

As mentioned in Section 1.1, insulators or *dielectrics* are substances that in ordinary circumstances do not possess free electric charges. This does not mean, however, that they cannot modify an electric field into which they are introduced.

In order to understand the mechanism by which a dielectric body influences an electric field into which it is introduced, consider a single atom of a dielectric. Suppose that it is electrically neutral, and that it does not create any appreciable electric field at far points. Of course, a short-range electric field of atomic dimensions necessarily exists. When observed from *distant* points, the atom may be considered to consist of two concentric, uniformly charged spheres, carrying equal charges of opposite signs (Fig. 2.1a). Such a system produces a zero electric field at all points outside the outer shell (see Section 1.8.1), and therefore also at distant points.*

Let us now introduce this atom into an external electric field. The positive and negative charged shells will be acted on by forces of equal magnitude, but opposite direction. Inside the atom there are certain elastic forces which tend to keep it in its neutral form; that is, to keep the two imaginary charged spheres concentric. So only a certain displacement of the two spheres in opposite directions will occur, depending on the intensity of the external field (Fig. 2.1b). Since the centers of the two spheres are not at the same point, the electric field at distant points is not zero any more. It is now equal to the field produced by two point charges at a distance $d$ from one another (Fig. 2.1b). Note that the distance $d$ between the centers of the two spheres is much smaller than the distance between the point of observation and the atom. It is not possible, of course, to represent this fact in a reasonable scale in the figure.

Such a system of two closely spaced "point" charges of equal magnitude but of opposite sign is called an *electric dipole*. The secondary electric field due to a neutral atom in an external electric field is the same as that created by a dipole. Consider a piece of a dielectric substance introduced into an external electric field.

* We shall see later that there are dielectric substances of a different kind, for which a different model has to be used. Although the mechanism is somewhat different, the final effect on the macroscopic field is the same.

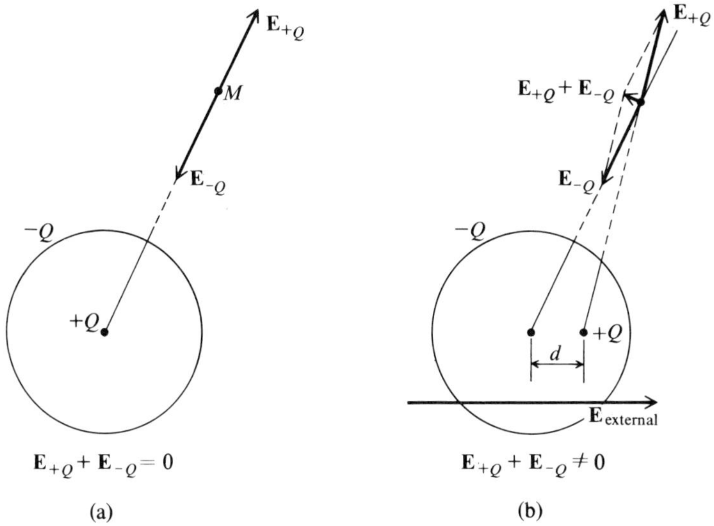


**Fig. 2.1** (a) A model of a neutral atom. (b) The same atom in an external electric field.

The secondary electric field due to the "deformed" atoms inside the dielectric is identical to the electric field of an aggregate of corresponding electric dipoles situated in a vacuum. In that case we say that the substance is *polarized.* The process of creating electric dipoles out of electrically neutral atoms and molecules is referred to as *polarization.*

One can imagine polarization of a wide class of dielectric substances according to the simplified model explained above. But there is also another class of dielectric substances, the molecules of which are dipoles even if there is no applied electric field. Such "permanent" dipole molecules are called *polar molecules.* (Molecules that are not dipoles in ordinary circumstances are known as *nonpolar molecules.*) In the absence of an external electric field, these dipoles are oriented in space statistically, so that again no macroscopic electric field is observed under normal circumstances. When the external field is applied, the force on the positive dipole charge is in the direction of the field, and that on the negative dipole charge in the opposite direction. This results in a torque that tends to align all dipoles. The thermal motion of molecules prevents complete alignment. Nevertheless a piece of a polarized substance consisting of polar molecules creates a secondary electric field which is again the same as the electric field of a large aggregate of electric dipoles.

Consequently, all dielectric substances introduced into an electric field can be considered to be equivalent to a vast number of tiny electric dipoles. The secondary macroscopic electric field due to a polarized substance can be regarded as the electric field of these dipoles *situated in vacuum.* An analysis of the behavior of dielectric substances in an electrostatic field reduces therefore to the analysis of a system of electric dipoles considered in vacuum. Methods of analysis of systems of electric charges in vacuum were explained in the preceding chapter. We have only two additional tasks: First, we have to determine the potential and the electric field due to a single electric dipole and those due to a large aggregate of elemental electric dipoles. Second, we must determine *experimentally* the connection between the state of polarization of different substances and the applied electric field.

*Problems referring to Section 2.1:* 2.T1 to 2.T7.

## 2.2 THE POTENTIAL AND ELECTRIC FIELD OF AN ELECTRIC DIPOLE AND OF AGGREGATE OF DIPOLES

Let us first determine the potential due to a dipole consisting of charges $Q$ and $-Q$ at a distance $d$ apart, at a point which is at a distance $r$ ($r \gg d$) from the center of the dipole (Fig. 2.2). Since there are only two "point" charges $Q$ and $-Q$, potential $V$ at point $M$ is given by

$$V = \frac{Q}{4\pi\varepsilon_0 r_+} + \frac{-Q}{4\pi\varepsilon_0 r_-} = \frac{Q}{4\pi\varepsilon_0}\left(\frac{1}{r_+} - \frac{1}{r_-}\right) = \frac{Q}{4\pi\varepsilon_0}\,\frac{r_- - r_+}{r_- r_+} \simeq \frac{Q\,d\cos\theta}{4\pi\varepsilon_0 r^2}. \quad (2.2.1)$$

Let us direct the vector $\mathbf{d}$ from the negative to the positive dipole charge, as shown in Fig. 2.2. The product

$$\boxed{\mathbf{p} = Q\,\mathbf{d}} \qquad \text{(definition of dipole moment)} \qquad (2.2.2)$$

represents a convenient vector quantity for describing the dipole, since it shows both its direction and its charge. In fact, $\mathbf{p}$ defines only their product, but from Eq. (2.2.1) precisely this product determines the potential, and hence also the electric field created by the dipole. The vector $\mathbf{p}$ is called the *dipole moment*. We

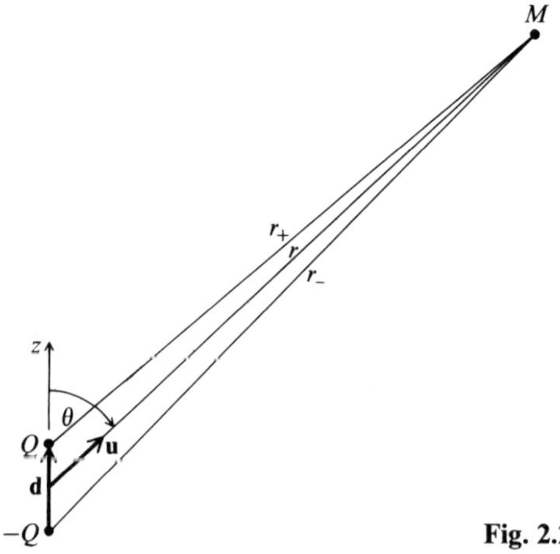


**Fig. 2.2** An electric dipole.

next stipulate that the unit vector **u** be directed from the dipole toward the point $M$ (Fig. 2.2). With these conventions, we can write the expression for the electrostatic potential (2.2.1) of a dipole in the form

$$V = \frac{\mathbf{p}\cdot\mathbf{u}}{4\pi\varepsilon_0 r^2}. \tag{2.2.3}$$

We can obtain the components of the electric field intensity from Eq. (2.2.1). Let the sketched coordinate system be spherical. We obtain the only existing components $E_r$ and $E_\theta$ of the electric field intensity vector from the result of Problem 1.G12:

$$E_r = -\mathrm{grad}_r\, V = -\frac{\partial V}{\partial r} = \frac{p\cos\theta}{2\pi\varepsilon_0 r^3}, \tag{2.2.4}$$

$$E_\theta = -\mathrm{grad}_\theta\, V = -\frac{1}{r}\frac{\partial V}{\partial \theta} = \frac{p\sin\theta}{4\pi\varepsilon_0 r^3}. \tag{2.2.5}$$

Now consider an aggregate of elemental electric dipoles representing a polarized substance. Since we are dealing with a macroscopic theory, we are not interested in the position and moment of each dipole. For macroscopic purposes, it is enough to know their average density at a point, where the term *density* is understood in the following sense: We consider a physically small volume $dv$ of the polarized substance, and sum *vectorially* the dipole moments of all dipoles inside $dv$. The *density of the electric dipole moment* is then given by

$$\boxed{\mathbf{P} = \frac{\sum_{dv} \mathbf{p}}{dv}} \qquad \text{(definition of polarization vector).} \tag{2.2.6}$$

The density (2.2.6) of the dipole moment at a point, which we denoted by **P**, is also referred to as the *polarization vector*.

From the definition (2.2.6) of the polarization vector **P**, the electric dipole moment $d\mathbf{p}$ of an elemental volume $dv$ enclosing a point at which the polarization vector is **P** is given by

$$d\mathbf{p} = \mathbf{P}\, dv. \tag{2.2.7}$$

Consequently, according to Eq. (2.2.3), the electrostatic potential due to the polarization of the element considered is

$$dV = \frac{\mathbf{P}\cdot\mathbf{u}\, dv}{4\pi\varepsilon_0 r^2}.$$

Hence the potential due to all elemental dipoles inside a volume $v$ is

$$V_p = \frac{1}{4\pi\varepsilon_0}\int_v \frac{\mathbf{P}\cdot\mathbf{u}}{r^2}\, dv. \tag{2.2.8}$$

The subscript $p$ in $V_p$ denotes that the potential is due to the polarized substance.

Therefore, *if* the polarization vector is found in some appropriate manner, the potential, and hence the electric field intensity created by the polarized substance, could be—at least theoretically— readily determined. Polarization **P** at a point is obviously some function of the *total* electric field intensity at that point, due to *all* charges. That is, due to the primary sources of the electric field *and* to the electric dipoles to which the influence of the polarized matter can be reduced. Thus we are confronted with a strange situation. The total electric field is a function of the polarization vector at each point of a polarized substance. Conversely, we can determine the polarization vector at a point if, in addition to the intrinsic properties of the particular dielectric considered, we know the total electric field intensity at that point. However, equilibrium must be attained. Our task is to determine this equilibrium, which we can do only if we determine, in some appropriate manner, the dependence of the polarization **P** on the total electric field intensity.

Experimental evidence shows that the polarization vector **P** for the majority of substances, with both polar and nonpolar molecules, under normal circumstances is proportional to the total electric field intensity:

$$\mathbf{P} = \chi_e \varepsilon_0 \, \mathbf{E} \qquad \text{(definition of electric susceptibility).} \qquad (2.2.9)$$

The constant $\chi_e$ is referred to as the *electric susceptibility*, and is a characteristic of every dielectric. For convenience, the constant of proportionality is written in the form $\chi_e \varepsilon_0$. It is obvious that the mechanism of polarization in the case of substances with nonpolar molecules is of internal atomic nature, and consequently cannot depend on temperature. In substances with polar molecules the thermal vibration or motion of polarized molecules prevents their alignment under the influence of the electric field, so that the electric susceptibility decreases with temperature. In the case of polar molecules, nonpolar polarization is also present. For substances with polar molecules, the electric susceptibility is given as the sum of a constant term and a term inversely proportional to the absolute temperature.

In some crystalline structures, such as quartz, strong internal forces interfere with the forces due to the external electric field. As a result, **P** is not in the direction of **E** unless **E** is in some preferred directions. Such dielectrics are called *anisotropic* dielectrics. However, we shall deal only with *isotropic* dielectrics, that is, with those which are polarizable in the same manner in all directions.

*Problems referring to Section 2.2:* 2.T8, 2.G1.

## 2.3 CHARGE DISTRIBUTION EQUIVALENT TO A POLARIZED DIELECTRIC

In the preceding section we showed that from the standpoint of the macroscopic theory a polarized substance can be regarded as an aggregate of elemental electric dipoles situated in vacuum. We shall now show that the electric field of these

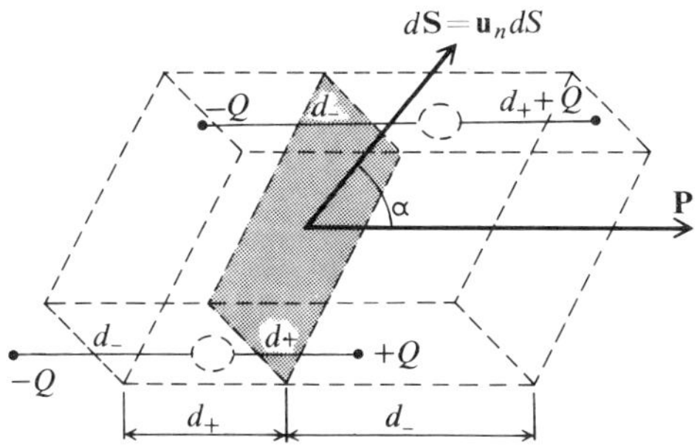


**Fig. 2.3** In the process of polarization positive charges are moved for a distance $d_+$ and negative ones for a distance $d_-$.

elemental dipoles can also be ascribed to a certain equivalent volume and surface distribution of charges, situated in vacuum. This equivalent charge distribution is derivable from the polarization vector **P**.

Consider an imaginary surface element $d\mathbf{S}$ situated in a polarized dielectric. Let the polarization vector in the neighborhood of $d\mathbf{S}$ be **P**, and the angle between $d\mathbf{S}$ and **P** be $\alpha$ (Fig. 2.3). Denote by $d$ the distance between charges $+Q$ and $-Q$ of the dipoles representing the polarized dielectric. In the process of polarization the surface $d\mathbf{S}$ is crossed by a certain number of positive charges in one direction, and by a certain number of negative charges in the opposite direction. Assume that the distance which the positive charges have moved from the equilibrium point is $d_+$, and that the distance which the negative charges have moved (in the opposite direction) is $d_-$. Then $d_+ + d_- = d$. It is obvious that surface element $d\mathbf{S}$ was crossed in its *positive* direction by displaced *positive* charges of all molecules which were to the left of $dS$ up to the distance $d_+$, and that it was crossed in its *negative* direction by displaced *negative* charges of all molecules which were to the right of $dS$ up to the distance $d_-$. Figure 2.3 shows only one representative dipole of both kinds, the dashed circle representing the position of the molecule prior to polarization. Denote by $N$ the number of molecules per unit volume, and note that the crossing of the surface in the negative direction by negative charges is equivalent to the crossing of the surface in the positive direction by positive charges. The *total* charge that crossed the surface $d\mathbf{S}$ in its positive direction in the process of polarization is

$$dQ = N\,Q\,d_+\,dS\cos\alpha - N\,(-Q)\,d_-\,dS\cos\alpha = N\,Q\,d\,dS\cos\alpha = N\,Q\,\mathbf{d}\cdot d\mathbf{S}.$$

However, $NQ\mathbf{d} = \mathbf{P}$, by definition (2.2.6) and (2.2.2) of the polarization vector **P**, so that

$$dQ = \mathbf{P}\cdot d\mathbf{S} \tag{2.3.1}$$

If we now consider a very small imaginary *closed* surface $\Delta S$, the total *positive* charge that *left* the volume enclosed by the surface $\Delta S$ in the process of polarization, according to Eq. (2.3.1), is

$$\Delta Q = \oint_{\Delta S} \mathbf{P} \cdot d\mathbf{S}. \tag{2.3.2}$$

Consequently, the charge that remained inside $\Delta S$ is equal to $-\Delta Q$. The density $\rho_p$ of that charge is

$$\rho_p = -\frac{\Delta Q}{\Delta v} = -\frac{\oint_{\Delta S} \mathbf{P} \cdot d\mathbf{S}}{\Delta v}. \tag{2.3.3}$$

Now $\Delta S$ is assumed to be very small. According to the definition (1.11.4) of the divergence of a vector function, the density $\rho_p$ of charges accumulated at a point inside a dielectric in the process of polarization is

$$\boxed{\rho_p = -\operatorname{div} \mathbf{P}} \qquad \text{(volume density of bound charges).} \tag{2.3.4}$$

The density $\rho_p$ is *not* the density of free electric charges, which could be taken away, but rather the mean density of charges firmly bound to their respective atoms. These charges result from the fact that in the process of polarization the balance of positive and negative charges in a physically small volume might be disturbed. Due to possible different polarization from one point to another, more charges might leave the elemental volume than enter it. Charges thus created in a polarized dielectric are called *polarization charges*. They are also often referred to as *bound charges*, to stress that they are not freely movable.

Consider now the boundary surface of a dielectric and a vacuum. Assume that the polarization vector $\mathbf{P}$ of the polarized dielectric near the surface makes an angle $\alpha$ with the unit vector $\mathbf{u}_n$ normal to the surface (Fig. 2.4). From the figure

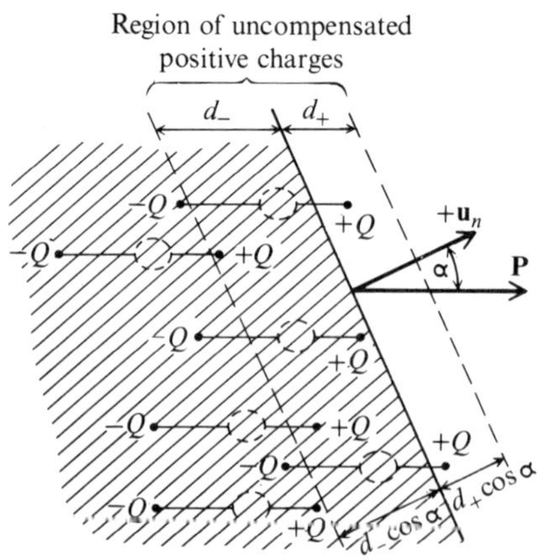


**Fig. 2.4** Boundary surface between a dielectric and a vacuum.

we see that the positive ends of dipoles "protrude" from the surface up to a distance $d_+ \cos\alpha$ (measured along $\mathbf{u}_n$). Also, inside the dielectric up to a distance $d_- \cos\alpha$ (measured along $\mathbf{u}_n$), there exist only positive ends of dipoles. Further to the left, positive and negative ends of dipoles are found in equal number, and their macroscopic field does not exist.

So uncompensated bound charges exist only in the thin layer indicated in Fig. 2.4. Given that the number of dipoles per unit volume is $N$, charge accumulated in this layer per unit area is

$$\rho_{sp} = N(d_+ + d_-)\,Q\cos\alpha = N\,d\,Q\cos\alpha = P\cos\alpha = \mathbf{P}\cdot\mathbf{u}_n. \qquad (2.3.5)$$

Note that the thickness of this layer is of the order of the radius of the dielectric molecules. That is, it is extremely thin, and charges can be regarded to be distributed over the very surface of the dielectric. By analogy with $\rho_p$ given by Eq. (2.3.4), this surface charge density is referred to as the *surface density of polarization* or of *bound charges*.

For the case in which $dS$ is on the boundary surface of two different dielectrics (Fig. 2.5), the charge distributed on the boundary surface has a density

$$\rho_{sp} = \mathbf{P}_1\cdot\mathbf{u}_{n1} + \mathbf{P}_2\cdot\mathbf{u}_{n2}, \qquad (2.3.6)$$

where $\mathbf{P}_1$ and $\mathbf{P}_2$ are polarization vectors in dielectrics 1 and 2, respectively, and $\mathbf{u}_{n1}$ and $\mathbf{u}_{n2}$ are unit vectors normal to the boundary surface and directed *from* dielectrics 1 and 2, respectively. Since $\mathbf{u}_{n2} = -\mathbf{u}_{n1}$, we may write Eq. (2.3.6) in the

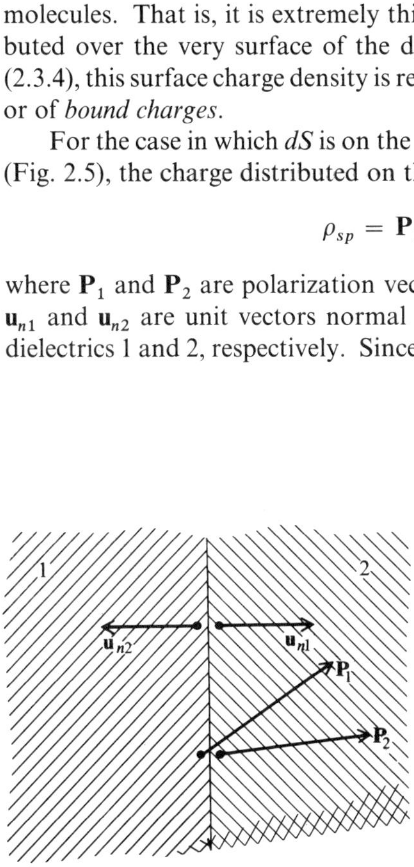


**Fig. 2.5** Boundary surface between two dielectrics. Note that positive normals are directed away from the respective dielectric.

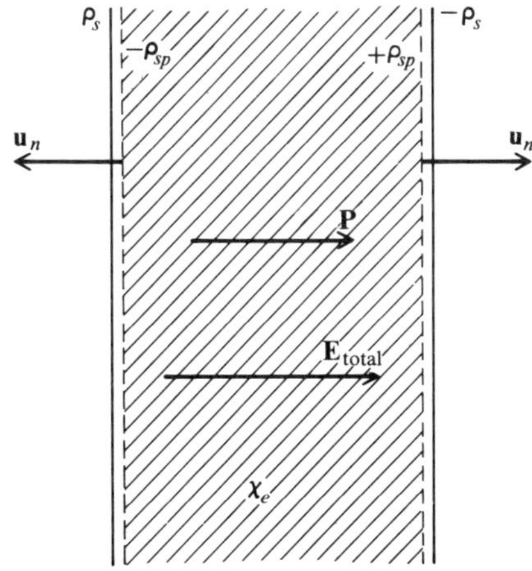


**Fig. 2.6** The total electric field in a parallel-plate capacitor is due to free charges on the capacitor electrodes and to polarization charges on the two surfaces of the dielectric.

form

$$\boxed{\rho_{sp} = (\mathbf{P}_1 - \mathbf{P}_2)\cdot\mathbf{u}_{n1}} \qquad \text{(surface density of bound charges).} \tag{2.3.7}$$

Hence a polarized piece of a dielectric is equivalent to charges of a volume density given by Eq. (2.3.4) and to surface charges of density given by Eq. (2.3.7).

As an example, let us consider the case of a parallel-plate capacitor filled with some dielectric of electric susceptibility $\chi_e$ (Fig. 2.6). Neither the total electric field intensity between the plates nor the polarization vector is known. But it is evident that the total electric field intensity **E**, and hence also the polarization vector **P**, are normal to the plates, and are of the same magnitude throughout the dielectric. Since **P** is a constant vector, its divergence (which implies differentiation!) is zero. So, according to Eq. (2.3.4), the volume density of polarization charges is zero. According to Eq. (2.3.5), on the two end surfaces of the dielectric there exist surface distributions of polarization charges, of surface density

$$\rho_{sp} = |\mathbf{P}\cdot\mathbf{u}_n| = P. \tag{2.3.8}$$

The positive polarization charge is on the end surface of the dielectric adjacent to the negative plate, and vice versa (Fig. 2.6). The *total* electric field intensity is due to charge layers of both free and polarization charges. So, according to Eq. (1.8.16),

$$E_{\text{total}} = \frac{\rho_s - \rho_{sp}}{\varepsilon_0}. \tag{2.3.9}$$

The total electric field intensity is *smaller* when the dielectric is present than when it is not.

Now, according to Eq. (2.3.8), $\rho_{sp} = P$. From the other side, according to Eq. (2.2.9), $P = \chi_e\varepsilon_0 E_{\text{total}}$. Hence, from Eq. (2.3.9), we obtain that in this case

$$\varepsilon_0 E_{\text{total}} = \rho_s - \chi_e\varepsilon_0 E_{\text{total}}$$

or

$$E_{\text{total}} = \frac{\rho_s}{(1 + \chi_e)\varepsilon_0} < \frac{\rho_s}{\varepsilon_0} = E_0. \tag{2.3.10}$$

Here $E_0$ designates the electric field intensity between the plates in the absence of a dielectric. The potential difference between the plates is $E_{\text{total}}d$. So the capacitance per unit surface of the capacitor plates is

$$C_s = \frac{\rho_s}{E_{\text{total}}d} = \frac{(1 + \chi_e)\varepsilon_0}{d} > \frac{\varepsilon_0}{d} = C_{s0}, \tag{2.3.11}$$

where $C_{s0}$ denotes the capacitance per unit surface in the absence of a dielectric.

Historically, the relations (2.3.10) and (2.3.11) were among the first quantitative relations describing the macroscopic influence of a dielectric substance. Particularly, Eq. (2.3.11) represented a basis for experimental determination of the electric susceptibility $\chi_e$. Indeed, the whole phenomenological analysis of electrostatic fields in the presence of dielectric substances can be started from these relations. However, insight into the mechanism of the influence of a dielectric substance on the electrostatic field in which it is introduced is lost, and many concepts necessarily remain unclear. Perhaps one of the most striking examples of this fact is connected with the definition and interpretation of the so-called *electric displacement vector*. This vector will be defined and interpreted in the next section, according to the adopted microscopic approach to macroscopic electric fields in the presence of dielectric substances.

*Problems referring to Section 2.3:* 2.T9 to 2.T16, 2.G2, 2.G3.

## 2.4 GENERALIZED FORM OF GAUSS' LAW. THE ELECTRIC DISPLACEMENT VECTOR

We can obtain the electric field intensity created by both free and polarization charges according to the same formulas, derived for the electrostatic field of charges situated in vacuum. Consequently, *all* deductions discussed in the first chapter remain valid in this case as well, if we replace $\rho$ by $(\rho + \rho_p)$ and $\rho_s$ by $(\rho_s + \rho_{sp})$. In particular, Gauss' law in the integral form (1.11.1) now becomes

$$\oint_S \mathbf{E} \cdot d\mathbf{S} = \frac{1}{\varepsilon_0} \int_v (\rho + \rho_p)\, dv,$$

or, since $\rho_p = -\operatorname{div} \mathbf{P}$,

$$\oint_S \varepsilon_0 \mathbf{E} \cdot d\mathbf{S} + \int_v \operatorname{div} \mathbf{P}\, dv = \int_v \rho\, dv. \tag{2.4.1}$$

When we transform the volume integral on the left-hand side into a surface integral, using the divergence theorem (1.11.10), this becomes

$$\oint_S (\varepsilon_0 \mathbf{E} + \mathbf{P}) \cdot d\mathbf{S} = \int_v \rho\, dv. \tag{2.4.2}$$

Equation (2.4.2) represents the generalized Gauss' law in integral form. We obtain the generalized Gauss' law in differential form by first transforming the surface integral in Eq. (2.4.2) into a volume integral,

$$\int_v \operatorname{div} (\varepsilon_0 \mathbf{E} + \mathbf{P})\, dv = \int_v \rho\, dv. \tag{2.4.3}$$

Since $v$ is arbitrary, it follows that

$$\operatorname{div} (\varepsilon_0 \mathbf{E} + \mathbf{P}) = \rho. \tag{2.4.4}$$

The sum $(\varepsilon_0 \mathbf{E} + \mathbf{P})$ has the remarkable property that its flux through any closed surface is equal to the total *free* charge enclosed by the surface. This means that its sources are only free electric charges. Consequently, this sum appears to be a very convenient function for analyzing electrostatic fields in the presence of dielectric substances. (We have mentioned that the determination of polarization charges, for a given distribution of free charges, is a very difficult task.) The sum $(\varepsilon_0\mathbf{E} + \mathbf{P})$ is therefore given a particular name, the *electric displacement vector*, and denoted by $\mathbf{D}$:

$$\boxed{\mathbf{D} = \varepsilon_0\mathbf{E} + \mathbf{P}} \qquad \text{(definition of electric displacement vector; valid generally).} \tag{2.4.5}$$

Equation (2.4.5) represents the most general—and always valid—connection between vectors **D**, **E**, and **P**. Using Eq. (2.4.5), we can write the generalized integral and differential forms of Gauss' law (2.4.2) and (2.4.4) in the form

$$\boxed{\oint_S \mathbf{D}\cdot d\mathbf{S} = \int_v \rho\, dv} \qquad \text{(integral form of generalized Gauss' law)} \tag{2.4.6}$$

and

$$\boxed{\operatorname{div}\mathbf{D} = \rho} \qquad \text{(differential form of generalized Gauss' law).} \tag{2.4.7}$$

Although it is perhaps early to refer to nonstatic fields at this point, it should be stressed that Eqs. (2.4.7) and (2.4.4), together with their integral equivalents (2.4.6) and (2.4.2), remain valid for electric fields varying with time in an arbitrary manner. Indeed, Eq. (2.4.7) represents one of the four differential equations, the so-called Maxwell's equations, which all electromagnetic fields satisfy. However, we have a long way ahead to cover in order to derive and appreciate the importance of these all-mighty equations.

A special—but the most common—case in practice is the case that exists when **P** is proportional to **E**, their relation being given by Eq. (2.2.9). In that case

$$\mathbf{D} = \varepsilon_0\mathbf{E} + \mathbf{P} = \varepsilon_0(1 + \chi_e)\mathbf{E}. \tag{2.4.8}$$

The sum $(1 + \chi_e)$ is referred to as the *dielectric constant* or *relative permittivity*, and is designated by $\varepsilon_r$,

$$\varepsilon_r = 1 + \chi_e. \tag{2.4.9}$$

Next we refer to the product $\varepsilon_0\varepsilon_r$ as the *permittivity* of the dielectric considered, and designate it by $\varepsilon$,

$$\varepsilon = \varepsilon_0(1 + \chi_e) = \varepsilon_0\varepsilon_r \qquad \text{(definition of permittivity).} \tag{2.4.10}$$

Hence we may write Eq. (2.4.8) in the form

$$\mathbf{D} = \varepsilon_r\varepsilon_0\mathbf{E} = \varepsilon\mathbf{E} \qquad \text{(linear dielectrics only).} \tag{2.4.11}$$

Values of the relative permittivity $\varepsilon_r$ of some dielectrics are given in Table 2.1.

**Table 2.1.** Relative permittivity $\varepsilon_r$ of some dielectrics (representative values)

| | | | | | |
|---|---|---|---|---|---|
| Vacuum | 1.00000 | Distilled water at 0°C | 88.0 | Mica | 6.0 |
| Air (at 1 atm) | 1.00059 | Distilled water at 20°C | 80.0 | Rubber | 2–3.5 |
| Transformer oil | 2.24 | Glass | 3.8–6.8 | Nylon | 3.5 |

One can see that the permittivity of air is practically equal to that of vacuum. Distilled water is a typical example of dielectric with polar molecules; its relative permittivity is high, and changes substantially with temperature.

Finally, according to Eq. (2.4.9), the relation (2.2.9) between the polarization vector and the electric field intensity in the case of a linear dielectric can be written in the form

$$\mathbf{P} = (\varepsilon_r - 1)\varepsilon_0\mathbf{E} = (\varepsilon - \varepsilon_0)\mathbf{E} \qquad \text{(linear dielectrics only).} \qquad (2.4.12)$$

**2.4.1 Interpretation of the displacement vector D.*** According to definition (2.4.5), the electric displacement vector **D** represents a sum of two quantities which are completely different physically, although expressible by the same unit. We have shown that the polarization vector **P** represented the density of electric dipole moments at a point, created by establishing the electric field inside a dielectric. The polarization vector may be also interpreted as being a quantity of charges that have crossed the surface normal to the direction of polarization, per unit area (Section 2.3). Hence we measure it in coulombs per square meter ($C/m^2$). The product $\varepsilon_0\mathbf{E}$ is measured by the same unit, $C/m^2$, but its physical meaning is completely different. While **P** represents *local* density of dipole moments, $\varepsilon_0\mathbf{E}$ represents the electric field intensity (multiplied by a physical constant $\varepsilon_0$) at a point due to *all* electric charges of the system (free and polarization charges). Hence if we wish to stand on firm ground we have to consider the electric displacement vector **D** as a hybrid physical quantity. It consists of two physical quantities of completely different physical meaning, and is introduced only as a convenient quantity for analyzing electric fields in the presence of dielectric substances. If we interpret **D**, the electric displacement vector, more freely, it can be given a very interesting meaning. Let us stress again that at this point we are leaving the firm, unique standpoint, and are interpreting the displacement vector in a way that cannot be justified experimentally at the present level of our knowledge of electric fields.

Vector **P** represents, as mentioned, the quantity of real electric charges which crossed the surface normal to the direction of **P**, per unit surface. It can be also *imagined* that the first term $\varepsilon_0\mathbf{E}$ in the definition (2.4.5) of the displacement vector **D** represents likewise a quantity of some unknown type of electric charges, which crossed the surface normal to vector **E**, per unit area. These, imaginary submicroscopic electric charges must be present not only within dielectrics, but also in

* This section may be omitted without loss of continuity.

vacuum (free space), since the term $\varepsilon_0 \mathbf{E}$ does not depend on the state of polarization at a certain point. Essentially, this was the meaning attributed to the term $\varepsilon_0 \mathbf{E}$ by the founder of the field theory of the electric and magnetic phenomena, a great physicist of the second half of the nineteenth century, James Clerk Maxwell.

Now, if $\mathbf{P}$ varies with time, there will be a certain motion (of microscopic dimensions) of *real* electric charges inside atoms of a dielectric, i.e., a kind of electric current. Maxwell imagined that if $\mathbf{E}$ varies with time in vacuum, a motion of these submicroscopic charges takes place, and that the effects due to their motion must be the same as the effects of the motion of charges in a dielectric. Strangely enough, this hypothesis led Maxwell to one of the most brilliant achievements of theoretical physics of all time: the prediction of the existence of electromagnetic waves. Maxwell predicted these waves some twenty years before they were detected experimentally.

From the above short and rather free description of Maxwell's ideas, one may understand why Maxwell named $\mathbf{D}$ the electric displacement vector, and the term $\varepsilon_0 \mathbf{E}$ the *electric displacement in vacuum*. However, it will be shown in Chapter 13 that Maxwell's hypothesis is in fact unnecessary for predicting the existence of electromagnetic waves. So at the present time, let us regard the terms *electric displacement in vacuum* and *displacement current* simply as names of certain physical quantities, not as names describing the physical meaning of these quantities.

*Problems referring to Section 2.4:* 2.G4 to 2.G14, 2.N1 to 2.N5.

## 2.5 THE ELECTRIC FIELD INSIDE A HOMOGENEOUS DIELECTRIC

If the permittivity at all points of a dielectric is the same, we say that the dielectric is *homogeneous*. In that case $\mathbf{D} = \varepsilon \mathbf{E}$, where $\varepsilon$ is independent of coordinates. Therefore Eq. (2.4.7) becomes

$$\operatorname{div} \mathbf{D} = \operatorname{div} \varepsilon \mathbf{E} = \varepsilon \operatorname{div} \mathbf{E} = \rho,$$

from which we obtain

$$\operatorname{div} \mathbf{E} = \frac{\rho}{\varepsilon}. \tag{2.5.1}$$

So the volume density of polarization charges, according to Eqs. (2.3.4) and (2.4.12), is given by

$$\rho_p = -\operatorname{div} \mathbf{P} = -\operatorname{div} [(\varepsilon - \varepsilon_0)\mathbf{E}] = -(\varepsilon - \varepsilon_0) \operatorname{div} \mathbf{E} = -\rho \frac{\varepsilon - \varepsilon_0}{\varepsilon}. \tag{2.5.2}$$

The total charge density at a point—i.e., the sum of the densities of free and polarization charges—is

$$\rho_{\text{total}} = \rho + \rho_p = \rho - \rho \frac{\varepsilon - \varepsilon_0}{\varepsilon} = \frac{\varepsilon_0}{\varepsilon} \rho = \frac{\rho}{\varepsilon_r}. \tag{2.5.3}$$

Hence at a point inside a homogeneous dielectric of relative permittivity $\varepsilon_r$, where the density of the free charges is $\rho$, the total charge density is smaller than $\rho$, owing to the induced polarization charges of opposite sign. Since this total charge is to be considered in vacuum, instead of Eq. (2.5.1) we may write

$$\operatorname{div} \mathbf{E} = \frac{\rho_{\text{total}}}{\varepsilon_0} = \frac{\rho}{\varepsilon_r \varepsilon_0}. \tag{2.5.4}$$

This equation is identical to Eq. (2.5.1), and in fact can be directly deduced from it, but in a purely formal manner.

Therefore the electric field intensity due to a volume distribution of charges of density $\rho$ in a homogeneous dielectric can be considered as the field intensity due to a volume distribution of charges of equivalent density $\rho/\varepsilon_r$ *situated in vacuum*. In Chapter 1 we formally solved the problem of determining the electrostatic field in such a case. For example, the electric field intensity due to a very small spherical charged cloud, given by Eq. (1.3.2), in this case becomes

$$\mathbf{E} = \frac{1}{4\pi\varepsilon_0} \frac{Q/\varepsilon_r}{r^2} \mathbf{u},$$

or

$$\mathbf{E} = \frac{1}{4\pi\varepsilon_r\varepsilon_0} \frac{Q}{r^2} \mathbf{u} = \frac{1}{4\pi\varepsilon} \frac{Q}{r^2} \mathbf{u}. \tag{2.5.5}$$

Since the whole analysis of the electrostatic field in vacuum was based on Eq. (1.3.2), it follows that all formulas derived there are applicable in this case if we simply change $\varepsilon_0$ to $\varepsilon$. We recommend, however, that the reader bear in mind the fact that changing $\varepsilon_0$ to $\varepsilon$ implicitly means taking into account the density of polarization charges, and then considering the whole system of charges to be situated in vacuum.

If the charges are distributed on surfaces of conducting bodies embedded in a homogeneous dielectric, we can readily show that the total influence of the dielectric on the electric field of these charges can be reduced to layers of polarization charges at the surfaces of charged bodies. These charges have signs opposite to those of the charges on the bodies. The whole system of charges is then to be considered in vacuum. To show this, suppose that the surface charge density of free charges at a point of a charged conductor is $\rho_s$. If we formally change $\varepsilon_0$ in Eq. (1.9.2) to $\varepsilon$, we obtain

$$E = E_n = \frac{\rho_s}{\varepsilon}. \tag{2.5.6}$$

The surface charge density of the polarization charges at the boundary surface dielectric/conductor is given by Eq. (2.3.5), in which the positive normal is directed away from the dielectric. Hence

$$\rho_{sp} = +\mathbf{P}\cdot\mathbf{u}_{n\,\text{diel}} = -\mathbf{P}\cdot\mathbf{u}_{n\,\text{cond}} = -P_n = -(\varepsilon - \varepsilon_0)E_n = -\rho_s\frac{\varepsilon - \varepsilon_0}{\varepsilon}. \quad (2.5.7)$$

The total surface charge density is

$$\rho_{s\,\text{total}} = \rho_s + \rho_{sp} = \rho_s - \rho_s\frac{\varepsilon - \varepsilon_0}{\varepsilon} = \frac{\varepsilon_0}{\varepsilon}\rho_s = \frac{\rho_s}{\varepsilon_r}. \quad (2.5.8)$$

Therefore, in the case of free charges spread over the surface of a conducting body embedded in a homogeneous dielectric, the surface charge density of the free charges is $\varepsilon_r$ times larger than the total charge on the conductor surface (free and polarization charge). Again we can formally obtain the influence of the polarized dielectric, according to Eq. (2.5.8), by simply changing the permittivity of vacuum $\varepsilon_0$ to the permittivity $\varepsilon$ of the dielectric considered, in all formulas valid for the field of charged conducting bodies in vacuum.

The most immediate consequence of this fact is that the force between two small charged bodies situated in vacuum is $\varepsilon_r$ times larger than that which would be obtained in a dielectric of relative permittivity $\varepsilon_r$. This fact can be seen from Coulomb's law (1.2.2). Historically, this was the first observed influence of dielectric substances on the static electric field.

Another important consequence is the following. Since the free charge is larger than the total charge on the surface of a conductor, for a given free charge the electric field intensity at all points—and hence also the potential difference between two electrodes of a capacitor—is $\varepsilon_r$ times larger for the dielectric vacuum than for the dielectric of permittivity $\varepsilon = \varepsilon_r\varepsilon_0$. Hence the capacitance of any capacitor is $\varepsilon_r$ times larger for a dielectric of relative permittivity $\varepsilon_r$ than for dielectric vacuum (air):

$$C = \frac{Q_{\text{free}}}{(V_1 - V_2)_{\text{diel}}} = \frac{Q_{\text{free}}}{(V_1 - V_2)_{\text{air}}/\varepsilon_r} = \varepsilon_r\frac{Q_{\text{free}}}{(V_1 - V_2)_{\text{air}}} = \varepsilon_r C_0, \quad (2.5.9)$$

where $C_0$ denotes the capacitance of the capacitor with dielectric vacuum (air). That was shown to be true in the special case of a parallel-plate capacitor [Fig. 2.6, Eq. (2.3.11)].

## 2.6 BOUNDARY CONDITIONS

Consider the interface between two different dielectric media. At adjacent points on two sides of the boundary surface, vectors **E** and **D** generally have components both tangential and normal to the surface. A simple relation exists between the corresponding components. These relations are known as *boundary conditions.*

Let us derive the boundary condition for tangential components of the electric field intensity: Consider two infinitely close points $A_1$ and $A_2$ on the two sides of the interface between two dielectrics. Let us apply Eq. (1.6.5) to a small rectangular contour $A_1B_1B_2A_2A_1$ shown in Fig. 2.7. When we neglect contributions to the integral along $B_1B_2$ and $A_2A_1$, since these are supposed to be vanishingly small, it

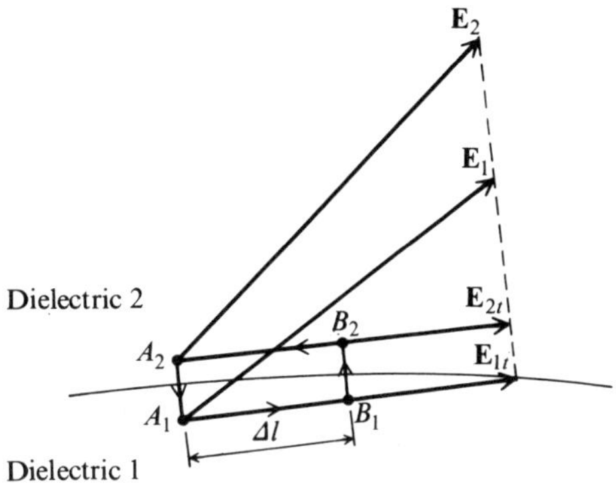


**Fig. 2.7** Contour for proving that $E_{1t} = E_{2t}$.

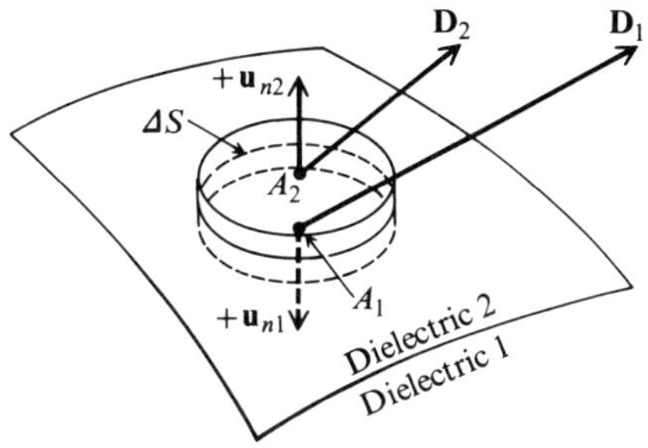


**Fig. 2.8** Gaussian surface for proving that $D_{1n} = D_{2n}$.

follows that

$$\oint_{A_1B_1B_2A_2A_1} \mathbf{E}\cdot d\mathbf{l} = E_{1t}\Delta l - E_{2t}\Delta l = 0,$$

from which

$$\boxed{E_{1t} = E_{2t}} \quad \text{(boundary condition for tangential components of } \mathbf{E}\text{).} \quad (2.6.1)$$

When the two dielectrics are linear, $\mathbf{E}_1 = \mathbf{D}_1/\varepsilon_1$ and $\mathbf{E}_2 = \mathbf{D}_2/\varepsilon_2$, $\varepsilon_1$ and $\varepsilon_2$ being permittivities of the two dielectrics. From Eq. (2.6.1), we find that the boundary condition for tangential components of vector **D** is

$$\frac{D_{1t}}{\varepsilon_1} = \frac{D_{2t}}{\varepsilon_2} \quad \text{(linear dielectrics).} \quad (2.6.2)$$

Suppose now that there are no free charges on the boundary surface. Let us apply the generalized integral form (2.4.6) of Gauss' law to a very thin cylinder (a "pillbox") with plane base surfaces $\Delta S$ parallel to the boundary surface and the curved surface normal to it (Fig. 2.8). Neglecting the flux of **D** through the vanishingly small curved surface, we have

$$\oint_S \mathbf{D}\cdot d\mathbf{S} = (-D_{1n} + D_{2n})\Delta S = 0,$$

from which

$$\boxed{D_{1n} = D_{2n}} \quad \text{(boundary condition for normal components of } \mathbf{D}\text{, no free charges on the interface).} \quad (2.6.3)$$

When the two dielectrics are linear, with permittivities $\varepsilon_1$ and $\varepsilon_2$, according to Eq. (2.6.3) the boundary condition for normal components of **E** is

$$\varepsilon_1 E_{1n} = \varepsilon_2 E_{2n} \qquad \text{(linear dielectrics).} \tag{2.6.4}$$

For the case in which there is a surface distribution of charges on the interface of density $\rho_s$, the "pillbox" encloses a charge $\rho_s \Delta S$, so that we have

$$\boxed{D_{2n} - D_{1n} = \rho_s} \qquad \text{(free charges on the interface).} \tag{2.6.5}$$

Finally, let us apply the generalized integral form of Gauss' law to a very thin cylinder one of whose plane base surfaces $\Delta S$ is in the conductor and the other of which is in the dielectric (Fig. 1.25, p. 47). We obtain that, at the point of the conductor surface at which the free surface charge density is $\rho_s$, the normal component, $D_n$, of $\mathbf{D}$ is given by

$$\boxed{D_n = \rho_s} \qquad \text{(boundary surface of conductor and dielectric).} \tag{2.6.6}$$

Of course, $D_t = 0$ in this case. Since $D_n = \varepsilon E_n$, we also have

$$E_n = \frac{\rho_s}{\varepsilon}. \tag{2.6.7}$$

If the two dielectrics are linear, a simple relation exists between the angles $\alpha_1$ and $\alpha_2$ which the electric field intensities $\mathbf{E}_1$ and $\mathbf{E}_2$ (or the electric displacement vectors $\mathbf{D}_1$ and $\mathbf{D}_2$) make with the normal to the boundary surface (Fig. 2.9). This is known as the *law of refraction* of lines of vectors $\mathbf{E}$ and $\mathbf{D}$. According to Fig. 2.9 and Eqs. (2.6.1) through (2.6.4), we have

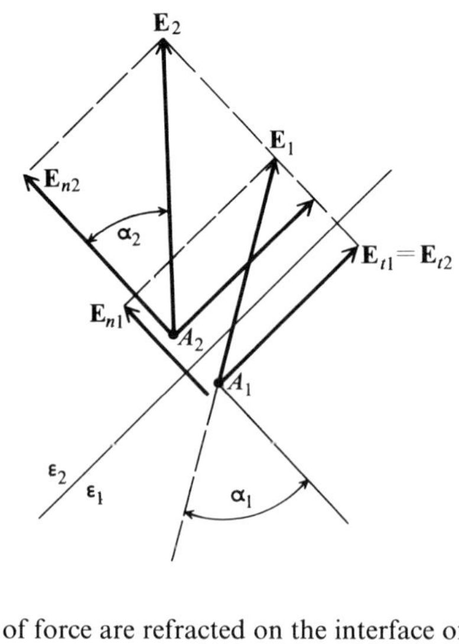


**Fig. 2.9** Lines of force are refracted on the interface of two dielectrics.

$$\frac{\tan \alpha_1}{\tan \alpha_2} = \frac{E_{t1}/E_{n1}}{E_{t2}/E_{n2}} = \frac{E_{n2}}{E_{n1}} = \frac{D_{n2}/\varepsilon_2}{D_{n1}/\varepsilon_1} = \frac{\varepsilon_1}{\varepsilon_2}. \tag{2.6.8}$$

Finally we obtain the following boundary conditions for the electrostatic potentials $V_1$ and $V_2$ at points $A_1$ and $A_2$:

$$\varepsilon_1 \frac{\partial V_1}{\partial n} = \varepsilon_2 \frac{\partial V_2}{\partial n} \tag{2.6.9}$$

and

$$V_1 = V_2. \tag{2.6.10}$$

The first equation follows directly from Eq. (2.6.4). The second results from the fact that if points $A_1$ and $A_2$ are quite close, as assumed, the line integral of the electric field intensity between them is vanishingly small, **E** being finite at all points.

*Problems referring to Section 2.6:* 2.T17, 2.T18, 2.G15 to 2.G18.

## 2.7 LINES OF THE ELECTRIC DISPLACEMENT VECTOR

As in the case of lines of the electric field intensity, lines of the electric displacement vector are defined as lines to which vector **D** is tangential at every point. Next we can define tubes of equal flux of vector **D** in the same manner as we defined the tubes of equal flux of the electric field intensity (Section 1.10). Using the same

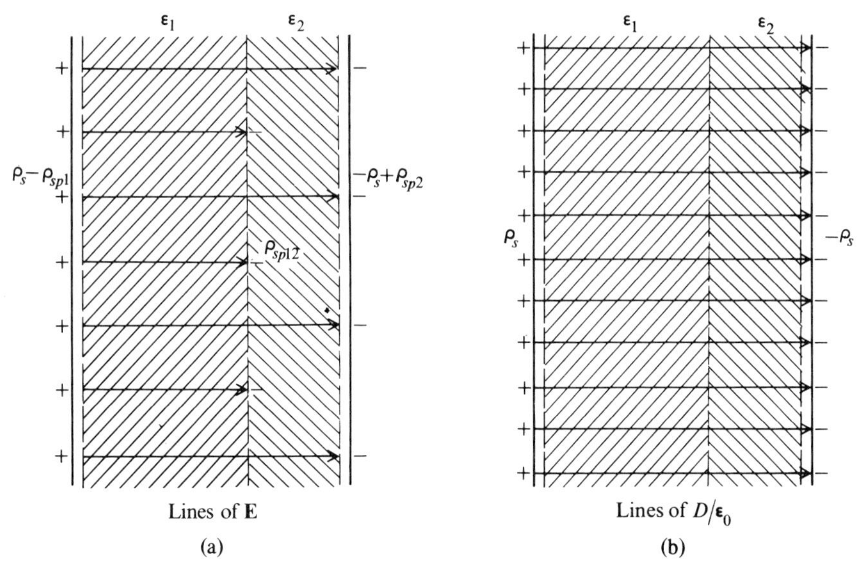


**Fig. 2.10** (a) If a dielectric is not homogeneous, the lines of vector **E** are discontinuous. (b) Lines of vector **D** are continuous even if the dielectric is inhomogeneous.

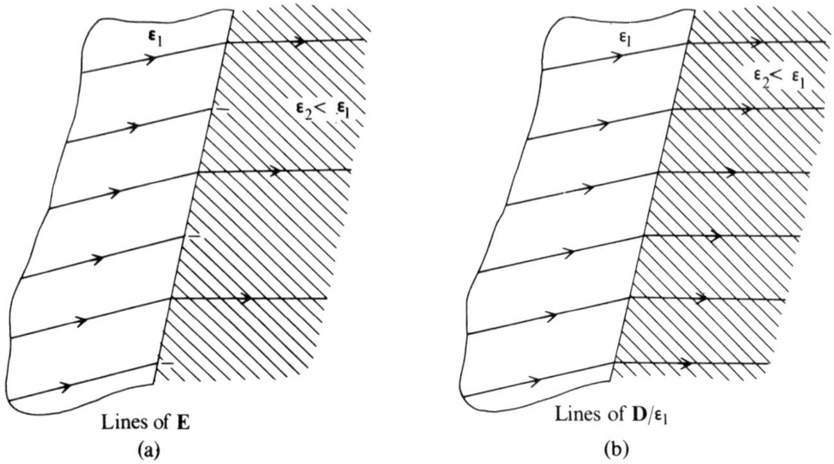


**Fig. 2.11** (a) A more general case of that illustrated in Fig. 2.10a. (b) A more general case of that illustrated in Fig. 2.10b.

arguments and starting from the generalized integral form of Gauss' law, we can prove that these tubes of flux end on electric charges of equal magnitudes but of opposite signs, whatever the dielectric between the charged bodies. This is not the case with tubes of flux of the electric field intensity, since sources of the electric field intensity are not only free charges, but also polarization charges. This fact is best illustrated in parts (a) and (b) of Fig. 2.10, which show the lines of the electric field intensity and of the electric displacement vector (each representing a tube of flux of the respcctive vector) in a parallel-plate capacitor with two dielectric slabs of permittivities $\varepsilon_1$ and $\varepsilon_2$. The lines of **D** are continuous, beginning at positive free charges and ending on negative free charges. The lines of **E** are discontinuous, since on the boundary surface of the two dielectrics, as well as on the end surfaces of the dielectrics, there appear layers of polarization charges, which are also sources of the electric field intensity.

A more general case, illustrating the same fact, is shown in Figs. 2.11a and 2.11b. From these two examples, it is obvious that lines of the electric displacement vector **D** are more convenient for representing the electric field in the presence of dielectrics than lines of the electric field intensity vector **E**.

*Problems referring to Section 2.7:* 2.T19 to 2.T21.

## 2.8 SOME PHYSICAL PROPERTIES OF DIELECTRICS

The main macroscopic characteristic of a dielectric in electromagnetic theory is its permittivity. But real dielectric substances have other physical properties also, which in some instances are of considerable practical importance. A brief account of some of these properties is presented below.

**2.8.1 Dielectric strength.** One of the most important characteristics of a dielectric is the lowest value of the electric field intensity for which a dielectric loses its insulating properties and becomes a conductor. That particular value of the electric field intensity for a given dielectric is referred to as its *dielectric strength.*

To every dielectric there corresponds a certain critical value of the electric field intensity. For smaller field intensities than the critical one, the dielectric behaves as an insulator, and for larger field intensities it behaves as a conductor. To understand this, recall that the force between electrons and nuclei in an insulator is such that under *ordinary* circumstances elecrons and nuclei are tightly bound together. However, if we introduce a dielectric into an electrostatic field of sufficient intensity, the electric force acting on some of the electrons may be large enough to tear these electrons away from the rest of the atom. Free electric charges appear in the structure of the dielectric, and large electric currents may result. This is known as the *dielectric breakdown.*

It may also happen that the electric field intensity is large enough only in certain regions of a dielectric. Then only sparking or steady discharging currents (known as *corona*) occur in those regions; this sparking is not necessarily followed by a general dielectric breakdown. Representative values of dielectric strength of some common dielectrics are given in Table 2.2.

**Table 2.2.** Dielectric strength of some common dielectrics in V/m (representative values)

| | | | |
|---|---|---|---|
| Air | $3 \times 10^6$ | Porcelain | $20 \times 10^6$ |
| Glass | $10 \times 10^6$ | Rubber | $40 \times 10^6$ |
| Transformer oil | $15 \times 10^6$ | Mica | $50 \times 10^6$ |

In almost all instances sparking and dielectric breakdown are highly undesirable. There are, however, a number of practical applications in which discharging of electricity by means of sparking or corona is promoted rather than suppressed. For example, an airplane may be charged, perhaps due to friction (see Section 2.8.4), during its flight. This charging may cause a spark and may even lead to a fire when the airplane lands. A very simple device may prevent excessive accumulation of charge. If some sharp edges are provided on the plane, the electric field intensity in the vicinity of such edges will be much higher than the dielectric strength of air, for even a moderate charge on the plane. The charge will leak through these conducting regions back into the air. Lightning rods work on the same principle: High electric field intensities near the sharp end of a lightning rod stimulate the discharge of lightning through the rod (which is always grounded) to the ground, thereby preventing the discharge of the lightning through nearby objects.

**2.8.2 Dielectric relaxation.** In the case of some substances which have polar molecules, the alignment of dipoles under the influence of an external electric field, and also the return to their initial statistical distribution after the field is switched off, require a certain period of time. This time is referred to as the *relaxation time*. The relaxation time varies greatly for different substances, but it may be of a duration of hours or days. Several important practical consequences result from this fact.

First, complete charging and discharging of capacitors with such a dielectric take some time. For example, if such a capacitor is discharged for only a short period of time, it may happen that considerable potential difference—enough to be dangerous—appears across its terminals hours and even days after this short, incomplete discharge (see Problem 2.T24).

Also, in the case of varying electric fields, the polarization lags with respect to the electric field intensity. In addition to the undesirable losses resulting from "friction" during change of polarization, the relationship between **P** and **E** is not linear any more, but is of the form shown in Fig. 9.20 (p. 320), if we replace $B$ by $P$ and $H$ by $E$. This phenomenon is referred to as the *electric hysteresis*.

**2.8.3 Permanently polarized dielectrics (electrets).** When some types of waxes and some other substances are either exposed to very high electric fields, or are exposed in a molten state to an electric field and then allowed to solidify in the field, permanent polarization of these substances occurs, which may last for a long period of time. Such permanently polarized pieces of dielectrics are called *electrets*. The electric field of an electret is very similar to the magnetic field of a permanent magnet. The concept of an electret is very useful for understanding the relation between vectors **D**, **P**, and **E** (see Problems 2.T22 through 2.T24, 2.G3 and 2.G19).

**2.8.4 Triboelectricity.** Historically, the very first experiment with electric phenomena was performed in ancient times by rubbing a piece of amber with some type of cloth. It turned out that both the piece of amber and the cloth attracted light pieces of various substances. We know today that this attraction is due to electric charges appearing on the surfaces of the amber and cloth, and that this phenomenon, the so-called *triboelectricity* (which in Greek means the electricity created by rubbing), is not limited to amber and cloth. Also, rubbing promotes triboelectricity but is not necessary for producing it.

The following simple model suffices to explain, roughly, the mechanism of triboelectricity. Let us consider two different insulators. The forces keeping the outer electrons from escaping from a molecule are of different magnitudes for the two substances. Hence, if they are brought into contact, at the surface of the contact the outer electrons in the molecules may be acted on by an additional force due to nearby molecules of a different type. This force might possibly tear these outer electrons away from the original molecule. Consequently, the two pieces of insulator, after they are separated, will be charged. Rubbing promotes this effect only because it enables a larger number of molecules of the two substances to come into close contact.

*Problems referring to Section 2.8:* 2.T22 to 2.T24, 2.G19 to 2.G23, 2.N6.

## PROBLEMS

### Test problems

**2.T1.** An atom acquires a dipole moment proportional to the electric field intensity **E** of the external field, $\mathbf{p} = \alpha\mathbf{E}$ ($\alpha$ is often referred to as the *polarizability*). Determine the total force acting on the atom when it is introduced in a *uniform* electric field of intensity **E**.

**2.T2.** Repeat the preceding problem for a case in which the atom is introduced into the field of a point charge $Q_1$. Determine only the direction of the force, and not its magnitude.

**2.T3.** A small body—either dielectric or conducting—is introduced into a nonuniform electric field; in which direction (qualitatively) does the force act on the body?

**2.T4.** Two point charges are placed near a piece of dielectric. Explain why Coulomb's law cannot be used to determine the *total* force on the two charges.

**2.T5.** A small charge is placed near a dielectric body. Will there be a force acting between the charge and the body? Explain.

**2.T6.** A closed surface $S$ situated in vacuum encloses a total charge $Q$ *and* a piece of a dielectric substance. Using a sound physical argument, prove that in this case also flux of the electric field intensity vector **E** through $S$ is $Q/\varepsilon_0$.

**2.T7.** Arbitrary pieces of dielectrics and conductors carrying a total charge $Q$ are introduced through an opening in a hollow, uncharged conducting body. The opening is then closed. Using a physical argument and Gauss' law for a vacuum, prove that the charge appearing at the outer surface of the body is exactly equal to $Q$.

**2.T8.** At a point of a polarized dielectric, the number of dipoles per unit volume is $N$, and their moment is **p**. What is the polarization vector at that point?

**2.T9.** Explain in your own words why the capacitance of a capacitor filled with a dielectric is larger than the capacitance of the same capacitor without the dielectric.

**2.T10.** A positive point charge is placed in air near the interface of air and a liquid dielectric. Will the surface of the dielectric be deformed? If you think that it will be deformed, then will it raise or sink? What if the charge is negative?

***2.T11.** Is there a mechanical pressure acting on a boundary surface between two different dielectrics situated in an electrostatic field? Explain.

**2.T12.** Prove that the total polarization charge on any piece of a dielectric substance is zero.

**2.T13.** Inside an isolated uncharged hollow spherical conductor situated in vacuum is placed a point charge $Q$ at a distance $d$ from the center of the conductor. The cavity is, in addition, filled with an inhomogeneous dielectric. Determine the electric field intensity outside the conductor.

***2.T14.** Prove that a charged conductor situated in an inhomogeneous *linear* dielectric has a potential proportional to its charge. [*Hint:* consider the polarized dielectric as an aggregate of dipoles situated in vacuum.]

***2.T15.** Prove that a relation of the type (1.14.3) is valid for a system of two charged conductors situated in an inhomogeneous *linear* dielectric.

***2.T16.** Discuss Problems 2.T14 and 2.T15 for a case in which the dielectric is *not* linear.

**2.T17.** Prove that, in a needle-like cavity which is parallel to the electric field intensity vector inside a homogeneous dielectric, the electric field intensity is the same as in the dielectric.

**2.T18.** Prove that, in a disk-like cavity with faces normal to the electric field intensity $\mathbf{E}_{\text{diel}}$ in a homogeneous dielectric, $\mathbf{E}_{\text{cavity}} = \varepsilon \mathbf{E}_{\text{diel}}/\varepsilon_0$.

**2.T19.** Would the electric field in the capacitor of Fig. 2.10a be changed if a thin conducting uncharged foil were introduced between dielectrics 1 and 2? Explain.

**2.T20.** A dielectric slab of permittivity $\varepsilon = 2\varepsilon_0$ is situated in vacuum in an external uniform electric field. Field lines are perpendicular to the faces of the slab. Sketch the lines of the resultant vectors **E** and **D**.

**2.T21.** Repeat the preceding problem, given that lines of the external field make an angle of 45° with the normal to the slab.

**2.T22.** A disk-shaped electret is polarized perpendicular to the two plane faces. Sketch the lines of vectors **E** and **D** created by the polarization of the electret. [*Hint:* Sources of **E** are free *and* polarization charges; sources of **D** are free charges only; outside the electret $\mathbf{D} = \varepsilon_0\mathbf{E}$, inside the electret $\mathbf{D} = \varepsilon_0\mathbf{E} + \mathbf{P}$.]

**2.T23.** What happens if the two faces of the electret of the preceding problem are covered with thin conducting foils? If the foils are then short-circuited by a wire, will there be an electric current of short duration through the wire?

**2.T24.** Using the logic of the preceding problem, explain the following: A capacitor with a dielectric having a long relaxation time was charged and then short-circuited (that is, the electrodes were connected with a conducting wire) for a period of time shorter than the relaxation time. A potential difference will be created later between the electrodes of the capacitor.

## General problems

**2.G1.** Determine the magnitude of the force acting on the atom of Problem 2.T2.

*Solution.* Suppose that the atom is placed at a distance $r$ from the point charge $Q_1$. In that case it is situated in an external field

$$E_1 = \frac{Q_1}{4\pi\varepsilon_0 r^2}, \tag{a}$$

so that its dipole moment is

$$p = \alpha E_1 = \frac{\alpha Q_1}{4\pi\varepsilon_0 r^2}. \tag{b}$$

Since $p = Qd$ (we assume that $Q$ is known), from (b) it follows that

$$d = \frac{p}{Q} = \frac{\alpha Q_1}{4\pi\varepsilon_0 r^2 Q}. \tag{c}$$

Suppose that $Q_1$ is positive. Then charge $-Q$ will be (say) at a distance $(r - d/2)$, and charge $+Q$ at a distance $(r + d/2)$ from the point charge $Q_1$. Hence the total *attractive* force on the polarized atom is, according to (a) through (c),

$$F = F_{-Q} - F_{+Q} = \frac{Q_1 Q}{4\pi\varepsilon_0(r - d/2)^2} - \frac{Q_1 Q}{4\pi\varepsilon_0(r + d/2)^2}. \tag{d}$$

Now $d \ll r$. Hence (d) can be written in the form

$$F \approx \frac{Q_1 Q}{4\pi\varepsilon_0 r^2}\left(1 + 2\frac{d}{2r}\right) - \frac{Q_1 Q}{4\pi\varepsilon_0 r^2}\left(1 - 2\frac{d}{2r}\right) = \frac{Q_1 Q d}{2\pi\varepsilon_0 r^3}. \tag{e}$$

Substituting value (c) of $d$ into (e), we find that the (attractive) force on the dipole is finally

$$F = \frac{\alpha Q_1^2}{8\pi^2\varepsilon_0^2 r^5}. \tag{f}$$

**2.G2.** A homogeneous dielectric sphere is uniformly polarized throughout its volume with density of dipole moment **P**. Determine the distribution of volume and surface polarization charges.

**2.G3.** A thin circular dielectric disk, of radius $a$ and thickness $d$, is permanently polarized with a dipole moment per unit volume **P**, parallel to the axis of the disk which is normal to its plane faces. Determine the electrostatic potential at a point on the axis of the disk.

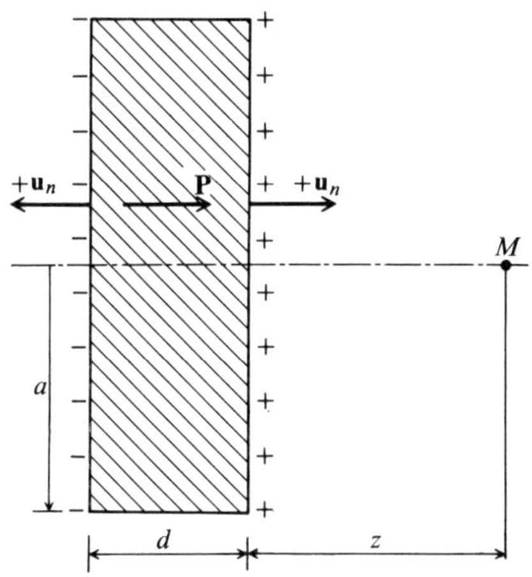


**Figure 2.G3**

*Solution.* According to Eq. (2.3.7), the field of the array of dipoles in the permanently polarized disk can be reduced to the field due to surface distribution of charges considered to be situated in vacuum, of surface density (Fig. 2.G3)

$$\rho_{sp} = \mathbf{P}\cdot\mathbf{u}_n = \pm P.$$

We can calculate the potential at $M$ created by the two circular layers of polarization charges by using Eq. (1.6.13), only changing $\rho_s$ to $\rho_{sp}$ and taking into account charges on both faces of the disk:

$$\begin{aligned} V_M &= \frac{\rho_{sp}}{2\varepsilon_0}(\sqrt{z^2 + a^2} - z) - \frac{\rho_{sp}}{2\varepsilon_0}[\sqrt{(z + d)^2 + a^2} - (z + d)] \\ &= \frac{\rho_{sp}}{2\varepsilon_0}[d + \sqrt{z^2 + a^2} - \sqrt{(z + d)^2 + a^2}]. \end{aligned} \tag{a}$$

When $d \ll z$, then

$$\sqrt{(z+d)^2+a^2} = \sqrt{z^2+a^2+2zd} = \sqrt{z^2+a^2}\sqrt{1+\frac{2zd}{z^2+a^2}}$$
$$\approx \sqrt{z^2+a^2}\left(1+\frac{zd}{z^2+a^2}\right),$$

so that, in that case, (a) becomes

$$V_M = \frac{\rho_{sp}d}{2\varepsilon_0}\left(1-\frac{z}{\sqrt{z^2+a^2}}\right).$$

**2.G4.** The dielectric of a parallel-plate capacitor with plates of surface $S$ consists of two slabs of permittivities $\varepsilon_1$ and $\varepsilon_2$ and thicknesses $d_1$ and $d_2$ (Fig. 2.G4a). Determine: (1) the capacitance of such a capacitor, (2) the ratio of the electric field intensity in slab 1 and the field that would exist if the capacitor were filled with dielectric 1 only, for the same potential difference between the plates, and (3) the polarization charges and their densities at all points, given that the potential difference across the capacitor is $V$.

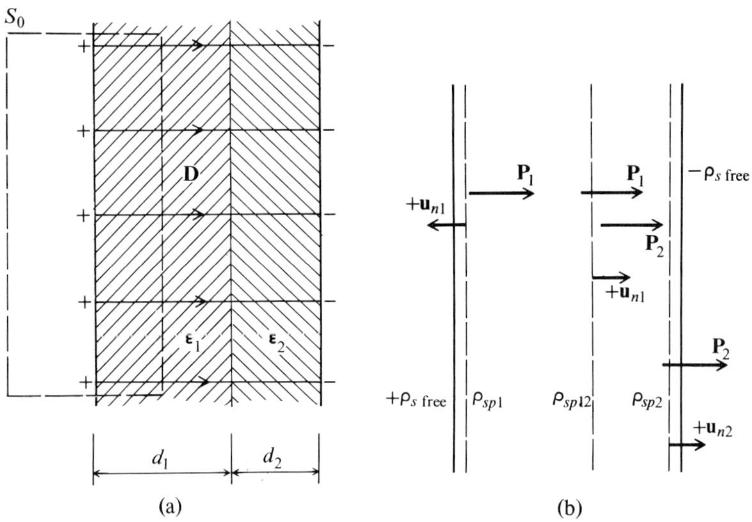


**Figure 2.G4**

*Solution.* (1) We shall determine the capacitance of the capacitor according to the pattern explained in connection with Eq. (1.14.3). In accord with that procedure, suppose that the capacitor plates carry charges $Q$ and $-Q$. Since free charges exist only on the capacitor plates, lines of the electric displacement vector are continuous through the capacitor and are normal to the plates (Fig. 2.G4a). Applying the generalized integral form (2.4.6) of Gauss' law [since dielectrics are present, the form (1.8.5) of Gauss' law is *not* applicable in this case] to the surface $S_0$ shown in Fig. 2.G4a, we obtain directly that $DS = Q$, whence $D = Q/S$. Now in slab 1, $D = \varepsilon_1 E_1$, and in slab 2, $D = \varepsilon_2 E_2$. The electric field intensities in the two slabs are

$$E_1 = \frac{D}{\varepsilon_1} = \frac{Q}{\varepsilon_1 S}, \qquad E_2 = \frac{D}{\varepsilon_2} = \frac{Q}{\varepsilon_2 S}. \tag{a}$$

Hence the potential difference between the plates corresponding to the charge $Q$ and $-Q$ on them is

$$V = E_1\, d_1 + E_2\, d_2 = Q\left(\frac{d_1}{\varepsilon_1 S} + \frac{d_2}{\varepsilon_2 S}\right),$$

whence the capacitance of the capacitor is

$$C = \frac{Q}{V} = \frac{1}{\dfrac{d_1}{\varepsilon_1 S} + \dfrac{d_2}{\varepsilon_2 S}}. \tag{b}$$

From (b), we obtain

$$\frac{1}{C} = \frac{d_1}{\varepsilon_1 S} + \frac{d_2}{\varepsilon_2 S} = \frac{1}{C_1} + \frac{1}{C_2},$$

where $C_1 = \varepsilon_1 S/d_1$ and $C_2 = \varepsilon_2 S/d_2$. The capacitance of the capacitor is thus equal to the equivalent capacitance of two capacitors formed by slabs 1 and 2 imagined to be connected in series.

(2) According to (a) and (b), the electric field intensity in slab 1 corresponding to the potential difference $V$ between the electrode is

$$E_1 = \frac{Q}{\varepsilon_1 S} = \frac{CV}{\varepsilon_1 S} = \frac{V}{d_1 + \dfrac{\varepsilon_1}{\varepsilon_2} d_2}. \tag{c}$$

For the case in which the whole capacitor is filled with any homogeneous dielectric, the electric field intensity $E$ corresponding to the potential difference $V$ is

$$E = \frac{V}{d_1 + d_2}. \tag{d}$$

Dividing (c) by (d), we obtain

$$\frac{E_1}{E} = \frac{d_1 + d_2}{d_1 + \dfrac{\varepsilon_1}{\varepsilon_2} d_2}. \tag{e}$$

From (e) it follows that, if $\varepsilon_2 > \varepsilon_1$, the electric field intensity in slab 1 is larger than if the dielectric is homogeneous, of say permittivity $\varepsilon_1$. Consequently, if a capacitor is used for high voltages, such that the electric field intensity is of the order of dielectric strength of the dielectrics, one should avoid a multilayer dielectric with dielectrics of very different permittivities (compare Problem 2.N3).

(3) Since the polarization of the slab is uniform, the volume density of the polarization charges is zero at all points. Polarization charges exist at the surfaces of contact of slab 1 and the left plate, of the two slabs, and of slab 2 and the right plate (Fig. 2.G4b). Their respective surface densities according to Eqs. (2.3.7), (a), and (b) are

$$\rho_{sp1} = -P_1 = -(\varepsilon_1 - \varepsilon_0)E_1 = -\varepsilon_2 \frac{(\varepsilon_1 - \varepsilon_0)V}{d_1\varepsilon_2 + d_2\varepsilon_1},$$

$$\rho_{sp12} = P_1 - P_2 = (\varepsilon_1 - \varepsilon_0)E_1 - (\varepsilon_2 - \varepsilon_0)E_2 = \varepsilon_0 \frac{(\varepsilon_1 - \varepsilon_2)V}{d_1\varepsilon_2 + d_2\varepsilon_1},$$

$$\rho_{sp2} = +P_2 = +(\varepsilon_2 - \varepsilon_0)E_2 = \varepsilon_1 \frac{(\varepsilon_2 - \varepsilon_0)V}{d_1\varepsilon_2 + d_2\varepsilon_1}.$$

It is left to the reader to prove that the resultant electric field intensities $E_1$ and $E_2$ in the two slabs are due to free charges on the two plates and to the above calculated surface polarization charges, and that the total polarization charge is zero (as it must always be).

**2.G5.** Repeat the preceding problem for the case of a coaxial cable with a two-layer dielectric. The radius of the inner conductor is $a$, the radius of the outer conductor is $b$, and the permittivity is $\varepsilon_1$ for $a < r < c$, and $\varepsilon_2$ for $c < r < b$, $r$ being the distance from the cable axis.

*Answer.* For $\varepsilon_1 = \varepsilon_2$, or for $c = a$, or for $c = b$, the result must be the same as that for a cable with a homogeneous dielectric.

**2.G6.** Repeat Problem 2.G4 for the case of a spherical capacitor with two layers of dielectrics. The radius of the inner electrode is $a$, the radius of the outer electrode is $b$, and the permittivity is $\varepsilon_1$ for $a < r < c$, and $\varepsilon_2$ for $c < r < b$, $r$ denoting the distance from the center of the capacitor.

*Answer.* For $\varepsilon_1 = \varepsilon_2$, or for $c = a$, or for $c = b$, the result must be the same as that for a spherical capacitor with a homogeneous dielectric.

**2.G7.** A point charge $Q$ is located at the center of a dielectric sphere of radius $a$ and permittivity $\varepsilon$. Outside the sphere there is a vacuum. Determine the electric field intensity inside and outside the sphere, and the potential of its surface.

*Solution.* When we apply the generalized integral form (2.4.6) of Gauss' law, we find that the electric displacement vector at all points is given by

$$D = \frac{Q}{4\pi r^2},$$

where $r$ denotes the distance of a point considered from the center of the sphere. Hence the electric field intensity inside the sphere is

$$E = \frac{D}{\varepsilon} = \frac{Q}{4\pi\varepsilon r^2},$$

and outside the sphere it is

$$E_0 = \frac{D}{\varepsilon_0} = \frac{Q}{4\pi\varepsilon_0 r^2}.$$

The potential of the sphere surface is the same as the potential at a distance $a$ from a point charge $Q$ situated in vacuum:

$$V = \int_a^\infty E_0\, dr = \frac{Q}{4\pi\varepsilon_0 a}.$$

This is obvious, since the total polarization charge in the dielectric sphere is zero, and is distributed symmetrically with respect to its center, where the point charge $Q$ is located.

**2.G8.** Inside a dielectric sphere of radius $a$ and permittivity $\varepsilon$ are distributed free charges of volume density $\rho = kr$, where $k$ is a constant and $r$ the distance of a point inside the sphere from its center. Outside the sphere there is a vacuum with no charges. Determine the potential at the center of the sphere (with respect to the reference point at infinity) and also the volume and surface density of the polarization charges.

*Solution.* In order to determine the potential at the center of the sphere, one must determine the electric field intensity at all points. Apply the generalized integral form (2.4.6) of Gauss' law to an imaginary spherical surface $S$ of radius $r < a$. Since $\mathbf{D}$ is radial (by symmetry), we have

$$\oint_S \mathbf{D} \cdot d\mathbf{S} = 4\pi r^2 D(r) = \int_v \rho \, dv = \int_0^r (kr')(4\pi r'^2 \, dr') = k4\pi \frac{r^4}{4},$$

from which we obtain

$$D(r) = k\frac{r^2}{4} \qquad \text{for} \qquad r < a.$$

In a similar manner, we find that

$$D(r) = \frac{ka^4}{4r^2} \qquad \text{for} \qquad r > a.$$

Hence the electric field intensity is

$$E(r) = \frac{D(r)}{\varepsilon} = \begin{cases} kr^2/4\varepsilon & \text{for} \quad r < a \\ ka^4/4\varepsilon_0 r^2 & \text{for} \quad r > a. \end{cases} \tag{a}$$

The potential of the center of the sphere is

$$V = \int_0^\infty E(r)\, dr = \int_0^a E(r)\, dr + \int_a^\infty E(r)\, dr = \frac{k}{4\varepsilon}\int_0^a r^2\, dr + \frac{ka^4}{4\varepsilon_0}\int_a^\infty \frac{dr}{r^2}$$

$$= \frac{ka^3}{12\varepsilon} + \frac{ka^3}{4\varepsilon_0} = \frac{ka^3}{12\varepsilon\, \varepsilon_0}(\varepsilon_0 + 3\varepsilon).$$

According to Eqs. (2.3.4), (2.4.12), and the answer to Problem 1.G35, the volume density of the polarization charges inside the sphere is

$$\rho_p = -\operatorname{div} \mathbf{P} = -\operatorname{div}[(\varepsilon - \varepsilon_0)E(r)\mathbf{u}_r] = -\frac{1}{r^2}\frac{d}{dr}[r^2(\varepsilon - \varepsilon_0)E(r)]$$

$$= -\frac{\varepsilon - \varepsilon_0}{r^2} k \frac{r^3}{\varepsilon} = -\frac{\varepsilon - \varepsilon_0}{\varepsilon} kr. \tag{b}$$

The surface density of polarization charges on the surface of the sphere, according to Eq. (2.3.5), is

$$\rho_{sp} = P(a) = (\varepsilon - \varepsilon_0)ka^2/4\varepsilon. \tag{c}$$

It is left to the reader to prove that the sum of all polarization charges determined above is zero, as well as that the electric field intensity given by (a) may be considered to be due to *all* charges: free charges, of density $\rho = kr$, as well as volume and surface distribution of polarization charges, given by (b) and (c).

**2.G9.** The dielectric in a parallel-plate capacitor is not homogeneous. The permittivity varies linearly from the value $\varepsilon_1$ at the surface of one plate to the value $\varepsilon_2$ at the surface of the other plate. The distance between the plates is $d$, and their area is $S$. Determine the capacitance of this capacitor and the density of the polarization charges, given that the plates are charged with charges $Q$ and $-Q$. Ignore the fringing effect.

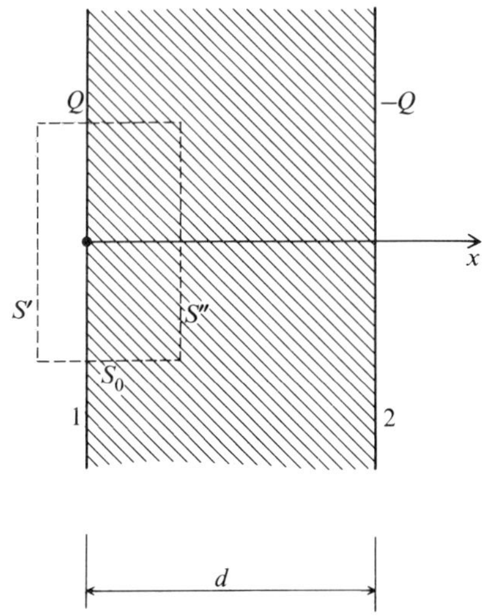


**Figure 2.G9**

*Solution.* By symmetry, the electric field intensity vector is normal to the plates at all points. Let us apply the generalized integral form (2.4.6) of Gauss' law to a cylindrical surface with bases $S'$ and $S''$ parallel to the plates, one inside and the other outside the capacitor, and with the curved surface normal to the plates (Fig. 2.G9). We easily find that $D = Q/S$ at all points. The electric field intensity is given by

$$E(x) = \frac{D}{\varepsilon(x)} = \frac{Q}{S\left(\varepsilon_1 + \dfrac{\varepsilon_2 - \varepsilon_1}{d}x\right)},$$

according to the assumed dependence of permittivity on coordinate $x$. Hence the potential difference between the plates when they are charged with charges $Q$ and $-Q$ is

$$V_1 - V_2 = \int_0^d E(x)\,dx = \frac{Q\,d}{S(\varepsilon_2 - \varepsilon_1)} \ln\frac{\varepsilon_2}{\varepsilon_1},$$

from which we obtain the capacitance of the capacitor:

$$C = \frac{Q}{V_1 - V_2} = \frac{S(\varepsilon_2 - \varepsilon_1)}{d \ln \dfrac{\varepsilon_2}{\varepsilon_1}}.$$

The surface density of polarization charges at the surface of plate 1 is

$$\rho_{sp1} = -P(0) = -[\varepsilon(0) - \varepsilon_0]E(0) = -(\varepsilon_1 - \varepsilon_0)\frac{Q}{S\varepsilon_1},$$

and at the surface of plate 2

$$\rho_{sp2} = +P(d) = (\varepsilon_2 - \varepsilon_0)\frac{Q}{S\varepsilon_2}.$$

The volume density of polarization charges is

$$\rho_p(x) = -\operatorname{div} \mathbf{P} = -\frac{d}{dx}\{[\varepsilon(x) - \varepsilon_0]E(x)\} = -\frac{Q\varepsilon_0(\varepsilon_2 - \varepsilon_1)}{S\,d\left(\varepsilon_1 + \dfrac{\varepsilon_2 - \varepsilon_1}{d}x\right)^2}.$$

It is left to the reader to prove that the total polarization charge is zero.

**2.G10.** Repeat the preceding problem for the case of a coaxial cable with conductors of radii $a$ and $b$ ($b > a$), with permittivity given by $\varepsilon(r) = \varepsilon b/r$, $r$ being the distance of a point inside the cable from the axis of the cable.

*Answer.* For $\varepsilon = \varepsilon_0$ and $b/a = 3$, the capacitance per unit length of the cable is $C_l = 83.5$ pF/m.

**2.G11.** A spherical capacitor with electrodes of radii $a$ and $b$ ($b > a$) is filled with an inhomogeneous dielectric whose permittivity ($\varepsilon(r)$ is a function of the distance $r$ from the center of the capacitor only. Given that the permittivity on the inner electrode is $\varepsilon$, determine $\varepsilon(r)$ in such a manner that the electric field intensity is the same at all points. Next determine the capacitance of such a capacitor, and the density of polarization charges at all points.

*Answer.* For $\varepsilon = 4\varepsilon_0$, $a = 2$ cm, and $b = 4$ cm, the capacitance of the capacitor is $C = 8.89$ pF.

**2.G12.** In Problems 2.G9 through 2.G11, it is shown that the volume distribution of polarization charges also exist in the case of a linear but *inhomogeneous* dielectric with *no* volume density of free charges. Prove that this is true in a general case, and determine the volume density of polarization charges.

*Solution.* According to Eqs. (2.4.7) and (2.4.11), for the case in which permittivity is a function of coordinates we obtain

$$\operatorname{div} \mathbf{D} = \operatorname{div} \varepsilon\mathbf{E} = \varepsilon \operatorname{div} \mathbf{E} + \mathbf{E}\cdot\operatorname{grad} \varepsilon = \rho. \tag{a}$$

From Eqs. (2.3.4) and (2.4.5), it follows that

$$\rho_p = -\operatorname{div} \mathbf{P} = -\operatorname{div} (\mathbf{D} - \varepsilon_0\mathbf{E}) = -\operatorname{div} \mathbf{D} + \varepsilon_0 \operatorname{div} \mathbf{E}. \tag{b}$$

Now if $\rho = 0$, from (a) it follows that

$$\operatorname{div} \mathbf{E} = -\mathbf{E}\cdot\frac{\operatorname{grad} \varepsilon}{\varepsilon}. \tag{c}$$

Since $\rho = 0$, by hypothesis, $\operatorname{div} \mathbf{D} = \rho = 0$ as well, so that from (b) and (c) we finally have

$$\rho_p = -\mathbf{E}\cdot\frac{\varepsilon_0}{\varepsilon}\operatorname{grad} \varepsilon. \tag{d}$$

It is left as an exercise for the reader to evaluate the volume density of polarization charges in Problems 2.G9 through 2.G11, starting from (d) of this problem.

***2.G13.** The permittivity of an infinite dielectric medium is given as the following function of the distance $r$ from the center of symmetry: $\varepsilon(r) = \varepsilon_0(1 + a/r)$. A small conducting sphere of radius $R$, carrying a charge $Q$, is centered at $r = 0$. Determine the electrostatic potential function $V(r)$ and the volume density of polarization charges at all points.

*Solution.* Since the system is symmetrical with respect to the point $r = 0$, using the generalized integral form (2.4.6) of Gauss' law we readily find the electric displacement vector (which is radial) to be

$$D(r) = \frac{Q}{4\pi r^2} \qquad \text{for} \qquad r > R,$$

from which we obtain

$$E(r) = \frac{Q}{4\pi\varepsilon(r)r^2} = \frac{Q}{4\pi\varepsilon_0 r(a + r)} \qquad \text{for} \qquad r > R. \tag{a}$$

According to (a), the potential at a distance $r$ from the center of symmetry, with respect to a reference point at a distance $r_R$, is

$$V(r) = \int_r^{r_R} E(r)\, dr = \int_r^{r_R} \frac{Q}{4\pi\varepsilon_0 r(a + r)}\, dr.$$

Now $1/[r(a + r)] = 1/ar - 1/[a(a + r)]$, so that the potential becomes

$$V(r) = \frac{Q}{4\pi\varepsilon_0 a}\left(\ln\frac{r_R}{r} - \ln\frac{a + r_R}{a + r}\right) = \frac{Q}{4\pi\varepsilon_0 a}\ln\frac{r_R(a + r)}{r(a + r_R)}.$$

If the reference point is chosen at infinity, $r_R = \infty$, we obtain

$$V(r) = \frac{Q}{4\pi\varepsilon_0 a}\ln\frac{a + r}{r}.$$

According to (a), according to (d) of the preceding problem, and according to the expression for gradient with respect to the spherical coordinate system (Problem 1.G12), the volume density of the polarization charges is found to be

$$\rho_p(r) = -E(r)\frac{\varepsilon_0}{\varepsilon(r)}\frac{d\varepsilon(r)}{dr} = \frac{aQ}{4\pi r^2(a + r)^2}.$$

The total polarization charge distributed throughout the dielectric is

$$Q_p = \int_R^{\infty} \rho_p(r) 4\pi r^2\, dr = \frac{aQ}{a + R}. \tag{b}$$

On the surface of the charged sphere the surface density of polarization charges according to Eqs. (2.3.5) and (a) is

$$\rho_{sp} = \mathbf{P}(R)\cdot\mathbf{u}_{n\,\text{diel}} = -P(R) = -[\varepsilon(R) - \varepsilon_0]E(R) = -\frac{aQ}{4\pi R^2(a + R)},$$

so that the total polarization charge on the boundary surface of the sphere and the dielectric is

$$Q_p(R) = 4\pi R^2 \rho_{sp} = -\frac{aQ}{a + R}. \tag{c}$$

Comparing (b) and (c), we see that the total polarization charge is zero, as it should be.

**2.G14.** Shown in Fig. 2.G14 is what is known as *capacitor bushing*, which is used to insulate a high-potential conductor $A$ at its passage through the grounded wall $W$. The cross-hatched surfaces represent very thin dielectric sheets of permittivity $\varepsilon$, and the thicker lines represent conducting foils placed between these sheets.

Referring to Fig. 2.G14, prove that the electric field intensity throughout the bushing is approximately the same, provided that $a_1 d_1 = a_2 d_2 = \cdots = a_5 d_5$.

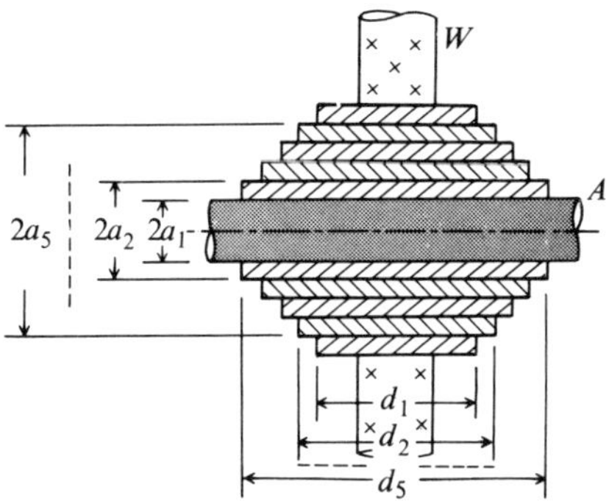


Figure 2.G14

**2.G15.** The dielectric in a parallel-plate capacitor consists of two slabs, of permittivities $\varepsilon_1$ and $\varepsilon_2$, as shown in Fig. 2.G15a. The capacitor is connected to a potential difference $V$. Determine: (1) the electric field intensities and the electric displacement vectors in both slabs; (2) the surface charge density of free charges on the electrodes; (3) the surface charge density of polarization charges on the surfaces of the slabs; (4) the capacitance of the capacitor. Neglect the fringing effect.

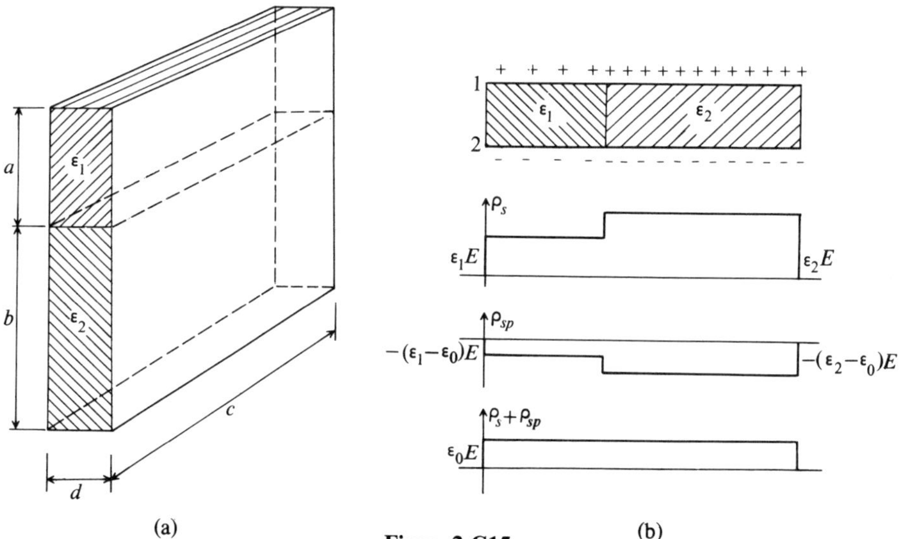


Figure 2.G15

*Solution.* The electric field intensity vector **E** is obviously normal to the plates at all points. Since

$$\int_1^2 \mathbf{E} \cdot d\mathbf{l} = E\,d = V,$$

the electric field intensity is the same in the two slabs,

$$E_1 = E_2 = E = \frac{V}{d}. \tag{a}$$

The boundary condition (2.6.1) for tangential components of the electric field intensity at the boundary surface of the dielectrics is seen to be satisfied.

Since $\mathbf{D}_1 = \varepsilon_1 \mathbf{E}_1$ and $\mathbf{D}_2 = \varepsilon_2 \mathbf{E}_2$, from (a) it follows that

$$D_1 = \varepsilon_1 \frac{V}{d}, \qquad D_2 = \varepsilon_2 \frac{V}{d}. \tag{b}$$

According to Eq. (2.6.6), the surface charge density of free charges on the plates is

$$\rho_{s1} = D_1 = \varepsilon_1 \frac{V}{d} \tag{c}$$

on the parts of the electrodes in contact with dielectric 1, and

$$\rho_{s2} = D_2 = \varepsilon_2 \frac{V}{d} \tag{d}$$

on the parts of the electrodes in contact with dielectric 2.

The surface charge density of polarization charges on the surfaces of slab 1 is

$$\rho_{sp1} = P_1 = (\varepsilon_1 - \varepsilon_0)E = (\varepsilon_1 - \varepsilon_0)\frac{V}{d},$$

and on the surfaces of slab 2,

$$\rho_{sp2} = P_2 = (\varepsilon_2 - \varepsilon_0)\frac{V}{d}.$$

The polarization charges have, of course, signs that are opposite to those of the free charges on the adjacent electrode. It is seen that

$$\rho_{s1} - \rho_{sp1} = \rho_{s2} - \rho_{sp2} = \varepsilon_0 \frac{V}{d},$$

so that the *total* surface charge density (the sum of the free and polarization charges) is constant over the surfaces of the electrodes.

Figure 2.G15b shows the distribution of free and polarization charges, assuming that $\varepsilon_1 < \varepsilon_2$.

The capacitance of the capacitor is

$$C = \frac{Q}{V} = \frac{\rho_{s1}ac + \rho_{s2}bc}{V} = \frac{(\varepsilon_1 ac + \varepsilon_2 bc)E}{V} = \varepsilon_1 \frac{ac}{d} + \varepsilon_2 \frac{bc}{d},$$

since $E = V/d$. The capacitance of the capacitor is the same as if the two capacitors, formed by the two slabs, are independent and connected in parallel.

**2.G16.** Repeat the preceding problem for a spherical capacitor whose inner electrode is placed on a cone of permittivity $\varepsilon$, as shown in Fig. 2.G16.

*Answer.* For $\varepsilon = \varepsilon_0$, or for $\alpha = 2\pi$, or for $\alpha = 0$, the result must be equal to that for the case of a spherical capacitor with a homogeneous dielectric.

**2.G17.** Repeat problem 2.G15 for a coaxial cable whose inner conductor is placed on a conical slab. Figure 2.G16 applies in this case also, if it is considered as the cross section of the cable.

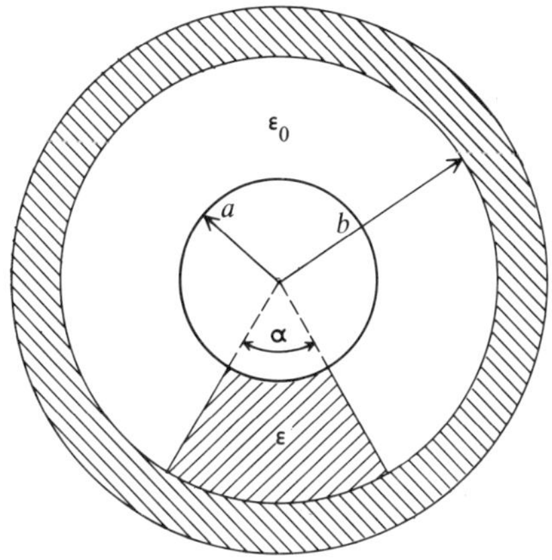


**Figure 2.G16**

**2.G18.** For the coaxial cable of Problem 2.G5, determine the radius $c$ of the cylindrical boundary surface of the two dielectric layers so that the maximal electric field intensity in the two layers is the same.

*Answer.* For $a = 2$ cm, $\varepsilon_1 = 4\varepsilon_0$, and $\varepsilon_2 = 2\varepsilon_0$, $c = 4$ cm.

***2.G19.** A homogeneous dielectric sphere of radius $a$ is polarized uniformly throughout its volume, the polarization vector at all points being **P** (a spherical electret). Determine the electric field intensity and the electric displacement vector both outside and inside the sphere. Sketch the lines of vectors **E** and **D**. Outside the electret there is a vacuum.

*Solution.* The electric field of the polarized sphere is due to the polarization charges spread over the sphere surface only. According to Eq. (2.3.4), if **P** is a constant vector (as assumed), the volume density of polarization charges is zero. The surface charge density of polarization charges on the sphere surface, according to Fig. 2.G19a and Eq. (2.3.5), is

$$\rho_{sp} = \mathbf{P} \cdot \mathbf{u}_n = P \cos \theta. \tag{a}$$

Starting from the determined distribution (a) of the polarization charges, one could determine the potential and the electric field intensity using general formulas (1.6.12) and (1.4.6). However, one can obtain the result more quickly by making use of the following reasoning.

Imagine two spherical distributions of charges of uniform volume densities $\rho$ and $-\rho$. Suppose that, if the centers of the two distributions coincide, each of the elemental positive charges constituting the positive sphere coincides exactly with one of the elemental negative charges constituting the negative sphere. Hence, in this case, the total (even microscopic) electric field would be zero.

Imagine now that the centers of the two spheres are separated by a very small distance $d$. Then the pairs of formerly coinciding elemental charges would likewise be separated by the distance $d$, resulting thereby in elemental electric dipoles. So the two slightly eccentric charged spheres may be looked on as a cloud of identical, evenly spaced dipoles, that is, as a spherical, uniformly polarized electret. Conversely, the electric potential and the field inside and outside the spherical electret is the same as in the case of the two uniformly charged, slightly eccentric spheres. In order to profit from this conclusion, one must establish the connection between the polarization vector **P** of the electret, the total charge $\pm Q$ of the two spheres, and the distance by which one must imagine the centers of the spheres to be separated.

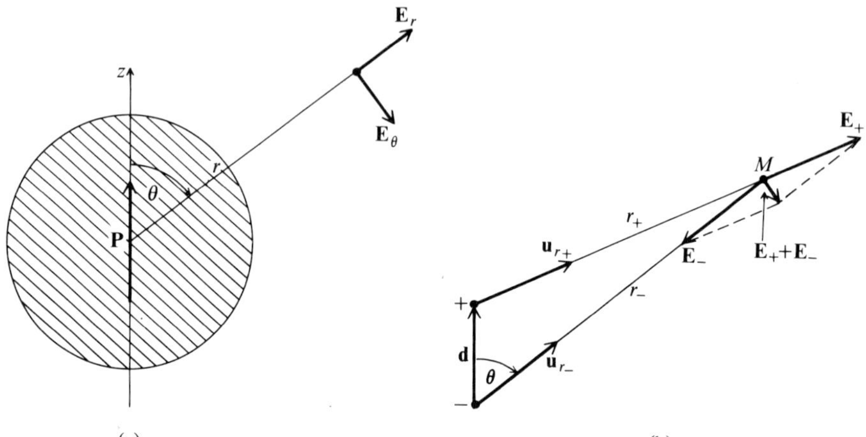


**Figure 2.G19**

Suppose that the number of charges in each of the two spheres is $N$, that their individual charge is $\pm q$, and that positive charges are displaced with respect to negative charges by a small distance $\mathbf{d}$. Since dipole moments $q\mathbf{d}$ of the elemental dipoles thus obtained are the same, the density of the dipole moment inside the sphere, i.e., the polarization of the two displaced spheres, will be equal to the polarization of the electret if

$$\frac{Nq\,\mathbf{d}}{\frac{4}{3}a^3\pi} = \frac{Q\,\mathbf{d}}{\frac{4}{3}a^3\pi} = \mathbf{P}. \tag{b}$$

Now outside the two spheres the electric field intensity is the same as that of a dipole of moment $Q\mathbf{d}$, since the field of each sphere is the same as if its total charge is concentrated at its center [Eq. (a) of Problem 1.G14]. Therefore, according to (b) and according to expressions (2.2.4) and (2.2.5) for the spherical components of the electric field intensity of a dipole, we have that outside the spherical electret (see Fig. 2.G19a)

$$E_r = \frac{Q\,d\cos\theta}{2\pi\varepsilon_0 r^3} = \frac{2a^3P\cos\theta}{3\varepsilon_0 r^3}, \tag{c}$$

$$E_\theta = \frac{Q\,d\sin\theta}{4\pi\varepsilon_0 r^3} = \frac{a^3P\sin\theta}{3\varepsilon_0 r^3}. \tag{d}$$

If the point is inside the electret, we make use of formula (b) of Problem 1.G14. According to that formula and Fig. 2.G19b, the electric field intensity at a point inside the two oppositely charged, slightly displaced spheres is

$$\mathbf{E}_{\text{in}} = \mathbf{E}_+ + \mathbf{E}_- = \frac{\rho}{3\varepsilon_0}r_+\mathbf{u}_{r+} - \frac{\rho}{3\varepsilon_0}r_-\mathbf{u}_{r-} - \frac{\rho}{3\varepsilon_0}(r_+\mathbf{u}_{r+} - r_-\mathbf{u}_{r-}) = -\frac{\rho}{3\varepsilon_0}\mathbf{d}. \tag{e}$$

Replacing $\rho$ in (e) by the ratio of the total charge $Q$ inside the spheres and the volume of the spheres $4a^3\pi/3$, and taking into account (*b*), we finally obtain

$$\mathbf{E}_{\text{in}} = -\frac{1}{3\varepsilon_0}\frac{Q}{\frac{4}{3}a^3\pi}\mathbf{d} = -\frac{\mathbf{P}}{3\varepsilon_0}. \tag{f}$$

Hence the electric field intensity inside the uniformly polarized spherical electret is *uniform*, in the direction opposite to **P**, and is independent of the radius of the sphere.

The electric displacement vector outside the electret is

$$\mathbf{D}_{\text{out}} = \varepsilon_0 \mathbf{E}_{\text{out}} \qquad \text{for} \qquad r > a,$$

and inside the electret

$$\mathbf{D}_{\text{in}} = \varepsilon_0 \mathbf{E}_{\text{in}} + \mathbf{P} = \tfrac{2}{3}\mathbf{P}. \tag{g}$$

The sketch of the lines of vectors **E** and **D** is left as an exercise for the reader.

***2.G20.** Consider a point $M$ inside a uniformly polarized dielectric of relative permittivity $\varepsilon_r$. The electric field intensity inside the dielectric is $\mathbf{E}_0$. Determine the electric field intensity at $M$ due to the polarization of a small sphere centered at $M$, and that due to all other charges of the system (the so-called *effective* polarizing electric field intensity). [*Hint:* Make use of formula (f) of the preceding problem, and also the principle of superposition.]

*Answer.* For $\varepsilon_r = 4$, the ratio of the effective polarizing electric field intensity $\mathbf{E}_{\text{eff}}$ and the electric field intensity $\mathbf{E}_0$ actually existing at $M$ is $\mathbf{E}_{\text{eff}}/\mathbf{E}_0 = 2$.

**2.G21.** The two-wire line of Problem 1.G50 is situated in air. Determine the approximate maximum value of the potential difference between the conductors, subject to the condition that there is no sparking (corona) at the surface of the conductors.

*Solution.* Ignoring the electric field intensity due to the charge on the other conductor, the electric field intensity at the surface of both conductors, for charge $Q_l$ per unit length, is

$$E_{\text{max}} = \frac{Q_l}{2\pi\varepsilon_0 R}.$$

Taking into account (a) of Problem 1.G50, we obtain

$$E_{\text{max}} = \frac{V}{2R \ln (D/R)}.$$

Since this electric field must be less than the dielectric strength $E_{\text{diel str}}$ of air, it follows that the required maximum value of the applied potential difference between the line conductors is

$$V_{\text{max}} = 2RE_{\text{diel str}} \ln \frac{D}{R}.$$

**2.G22.** The potential difference, in air, between wires of a two-wire line of given distance $D$ between conductors is $V$. Given that $E_{\text{diel str}}$ is the dielectric strength of air, determine the approximate minimum value of the wires' radius, subject to the condition that no corona surround the wires.

*Answer.* For $D = 2.5$ m, $V = 110$ kV (1 kV = 1000 V), $E_{\text{diel str}} = 3 \times 10^6$ V/m, the radius $R \simeq 2.8$ mm.

***2.G23.** The radii of conductors of a coaxial cable with air dielectric are $a$ and $b$, where $a < b$. Determine the maximum value of the potential difference between the conductors for which a complete breakdown of the air dielectric still does not occur. The dielectric strength of air is $E_{\text{diel str}}$.

*Solution.* According to the solution to Problem 1.G45, the maximal electric field intensity for a given potential difference $V$ between cable conductors is at the surface of the inner conductor,

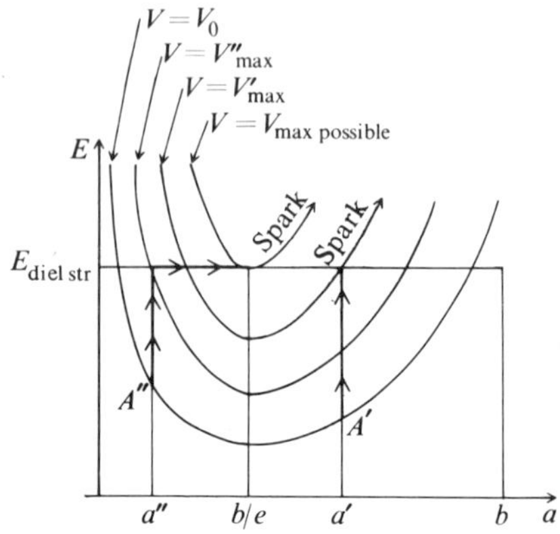


**Figure 2.G23**

and is given by

$$E_{max} = E(a) = \frac{V}{a \ln (b/a)}.$$

It was also shown there that for given $b$ the minimum field intensity $E(a)$ is obtained if $a = b/e$ ($e = 2.7182\ldots$, the natural base of logarithms).

We can represent $E(a)$ graphically as a function of $a$, for a given set of values of $V$, as in Fig. 2.G23; this figure shows some representative curves necessary for the analysis.

We must distinguish between two cases: (1) when $a < b/e$, and (2) when $a > b/e$. Let us consider the second case first.

If $a = a' > b/e$, increasing the potential difference $V$ from $V = V_0$ to the value

$$V'_{max} = a' E_{diel\ str} \ln \frac{b}{a'} \tag{a}$$

causes the point $A'$ to move from curve $V = V_0$ upward to the particular point at which $E(a') = E_{diel\ str}$ (curve $V = V'_{max}$). If we increase the potential difference still more, the air around the inner conductor will become ionized, which corresponds to increasing the inner diameter of the cable. But if $a'$ is thus made larger, we see from Fig. 2.G23 that the electric field intensity is also increased. Hence an avalanche process—the breakdown of the dielectric—will occur if $V$ is even slightly larger than $V'_{max}$, as given by (a)

If $a = a'' < b/e$, increasing the potential difference $V$ from $V = V_0$ to the value

$$V''_{max} = a'' E_{diel\ str} \ln \frac{b}{a''} \tag{b}$$

causes the point $A''$ to move from curve $V = V_0$ upward to the particular point at which $E(a'') = E_{diel\ str}$ (curve $V = V''_{max}$). In this case also an increase of $V$ above value (b) results in an ionized layer around the inner conductor. However, this does not result in a breakdown, since according to Fig. 2.G23 an increase in $a''$ in this case results in a *decrease* in $E(a'')$, so that the process will eventually stop. If we increase the potential difference $V$ still more, the point $A''$ moves to the right along the straight line $E(a) = E_{diel\ str}$, to the point at which the total radius of the inner conductor ($a''$ *plus* the thickness of the ionized layer) is equal to $b/e$. This is the maximum possible potential difference, $V_{\text{max possible}}$, to which the cable can be connected,

$$V_{\text{max possible}} = \frac{b}{e} E_{\text{diel str}}.$$

A further increase of the potential difference will initiate the breakdown of the air throughout the cable, as in the first case.

## Numerical problems

**2.N1.** A parallel-plate capacitor, with dielectric transformer oil and distance between the plates $d = 1$ cm, is connected to a source of potential difference $V = 10{,}000$ V. Determine the density of surface free and polarization charges in the capacitor, and its capacitance per unit area of the electrodes.

**2.N2.** Calculate the capacitance of the capacitor described in Problem 2.G4, given that $\varepsilon_1 = 2\varepsilon_0$, $\varepsilon_2 = 3\varepsilon_0$, $d_1 = 4$ mm, $d_2 = 2$ mm and $S = 0.05$ m$^2$.

**2.N3.** The dielectric in a parallel-plate capacitor is mica. Owing to an imperfect contact between the mica and the plates, there is a tiny air gap between the mica and both plates. Given that the thickness of the mica is $d_1 = 1$ mm, and the total thickness of both air gaps is $d_2 = 0.1$ mm, determine the electric field intensity in the air gaps in terms of the electric field intensity that would exist in the mica without gaps. Assume that the latter electric field intensity is designed to be $10 \times 10^6$ V/m. What is the electric field intensity in the air gaps? Deduce that in the case of such high values of the electric field intensity, a liquid dielectric may be strongly recommended. (See Problem 2.G4.)

**2.N4.** Determine the free and polarization surface charge densities for the capacitor of Problem 2.N2, given that the capacitor is connected to a potential difference of 100 V.

**2.N5.** The surface charge density of free charges on the plates of the capacitor of Problem 2.G9 is $Q/S = 10^{-8}$ C/m$^2$, $\varepsilon_1 = 2\varepsilon_0$, $\varepsilon_2 = 4\varepsilon_0$, and $d = 2$ mm. Determine the surface and volume density of the polarization charges, and the total polarization charge distributed throughout the dielectric (excluding surface polarization charges).

**2.N6.** Determine the approximate maximum value of the potential difference to which a two-wire line situated in air can be connected. The distance between the wires is $D = 3$ m, and their radii are $R = 5$ mm. What is the corresponding charge per unit length of the wires? (See Problem 2.G21.)

**Suggested reading.** The list given here progresses from simpler to more advanced texts. For a full bibliographical description, see the Bibliography.

1. REITZ, J. R., and F. J. MILFORD, *Foundations of Electromagnetic Theory*, Chapter 4.
2. SCOTT, W. T., *The Physics of Electricity and Magnetism*, Chapter 3.
3. SLATER, J. C., and N. H. FRANK, *Electromagnetism*, Chapter IV.

CHAPTER 3

# ENERGY AND FORCES IN THE STATIC ELECTRIC FIELD

## 3.1 POTENTIAL ENERGY OF A GIVEN DISTRIBUTION OF CHARGES

Let us consider a system of charges distributed in a known manner in a finite region $v$ of space. Suppose that their density is $\rho$ (a function of coordinates) and that the electrostatic potential with respect to the reference point at infinity is $V$ (again a function of coordinates). If charges distributed on the surfaces of conductors are also present, we can consider them as volume distribution of charges of very large densities, as in reality they are.

Generally speaking, a certain elemental force acts on every elemental volume carrying either free or polarization charges. In the case of free charges, a certain amount of work has evidently to be done in placing the charges in their present positions. The total work done *by external forces*, i.e. the total work done *against* the forces of the electric field, in establishing the system of charges, can be calculated as follows.

Suppose a small part $d\rho\, dv$ of the total charge $\rho\, dv$ inside an elemental volume $dv$ is allowed to recede to the reference point under the influence of the electric forces created by the other charges of the system. According to the definition of the electrostatic potential, the work of the electric forces is given by

$$dW_{\text{el.f.}} = V\, d\rho\, dv.$$

If we now wish to move the charge $d\rho\, dv$ back to its original position, some work has to be done by external, nonelectric forces *against* the electric forces. This work will be exactly equal to the work previously done by the electric forces:

$$dW_{\text{ext.f.}} = V\, d\rho\, dv. \tag{3.1.1}$$

Equation (3.1.1) is the basis for evaluating the total work done against electric forces during the establishment of the system. If we denote by $\rho'$ and $V'$ the charge density and potential at a given point of the system at any instant during its creation by bringing charges from infinity, we shall find that according to Eq. (3.1.1) the total work done against the electric forces in assembling the system of charges is

$$W_{\text{ext.f.}} = \int_v \int_o^\rho V'\, d\rho'\, dv, \tag{3.1.2}$$

where $\rho$ is the final value of the charge density at a point inside $dv$.

The total work (Eq. 3.1.2) done by the external forces against the electric forces cannot be a function of the particular process by which the final state is reached. This is proved as follows. Suppose that this statement were not correct. We could then choose two different processes using correspondingly different amounts of work to attain the *same* final state of the system. We would assemble the system by the process requiring less energy and then let it disintegrate by the other process, which requires more energy to assemble and thus provides more energy in disintegrating. By such a procedure we would be able to obtain any amount of energy we desired, but this is impossible. Thus, by the law of conservation of energy, only the final state of the system is important and the process of assembling the system has no influence on the total work done against the electric forces.

For this reason, in calculating the work done by the external forces, we can choose any convenient process to assemble the system. We shall take the simplest one, in which the charge density $\rho'$ at all points corresponding to any of the temporary states is proportional to the final density:

$$\rho' = k\rho, \qquad 0 \le k \le 1. \tag{3.1.3}$$

In this case the potential $V'$ at all points corresponding to any of the temporary states of the system is also proportional to the final potential $V$, with the same (variable) constant of proportionality:

$$V' = kV, \qquad 0 \le k \le 1. \tag{3.1.4}$$

Combining Eqs. (3.1.2) to (3.1.4), we find that the total work done by the external forces against the electric forces in establishing any system of static charges is given by

$$W_{\text{ext.f.}} = \int_v \int_{k=0}^{1} kV\rho \, dv \, dk = \int_v V\rho \, dv \int_{k=0}^{1} k \, dk = \frac{1}{2} \int_v V\rho \, dv. \tag{3.1.5}$$

The work given in Eq. (3.1.5), being done by the external forces against the electric forces only, is exactly equal to the energy located in the system. It can be used potentially at a later instant by allowing the system to disintegrate. Hence, the potential energy of a system of charges, which we shall simply refer to as the *electrostatic energy* of the system and designate by $W_e$, is

$$\boxed{W_e = \frac{1}{2} \int_v V\rho \, dv} \quad \text{(potential energy of a static system of charges).} \tag{3.1.6}$$

If the charges of the system are located in thin layers on surfaces $S_1, S_2, \ldots, S_n$ of conductors $1, 2, \ldots, n$, Eq. (3.1.6) becomes

$$W_e = \frac{1}{2} \int_{S_1+S_2+\cdots} V\rho_s \, dS = \frac{1}{2} \sum_{j=1}^{n} V_j \int_{S_j} \rho_s \, dS = \frac{1}{2} \sum_{j=1}^{n} V_j Q_j, \tag{3.1.7}$$

since the surface of each conductor is equipotential.

A simple, but important, special case of Eq. (3.1.7) is the electrostatic energy located in a charged capacitor, in which case $j = 2$ and $Q_2 = -Q_1 = -Q$, so that

$$W_e = \tfrac{1}{2}(V_1 Q - V_2 Q) = \tfrac{1}{2}(V_1 - V_2)\,Q \qquad \text{(capacitor).} \qquad (3.1.8)$$

Using the relation given in Eq. (1.14.3) we can write Eq. (3.1.8) in the following equivalent forms

$$W_e = \tfrac{1}{2}(V_1 - V_2)\,Q = \tfrac{1}{2}C(V_1 - V_2)^2 = Q^2/2C \qquad \text{(capacitor).} \quad (3.1.9)$$

It should be noted that the expressions derived above for the electrostatic energy of the system represent only the potential, or free, energy of the system, i.e. that part of the total energy which can be obtained if the system disintegrates. The energy required to create the elemental charges $d\rho\, dv$ is *not* included in these expressions. We are assuming that these charges were created at infinity, before the system was assembled.

*Problems referring to Section 3.1:* 3.T1 to 3.T6, 3.G1 to 3.G10, 3.N1 to 3.N2.

## 3.2 DISTRIBUTION OF ENERGY IN THE STATIC ELECTRIC FIELD

In the preceding section, the energy of a system of charges was considered as the *potential* energy, i.e. energy resulting from their relative positions. However, it is also possible to consider this energy as distributed throughout the field with a certain energy *density*.

To understand the basis of such an approach, let us consider first a mechanical system consisting of a rigid body pressed into a rubber slab. If released, the body would jump out from the rubber slab and possibly perform some mechanical work. Where was the energy located in this case?

According to an approach analogous to that used in the preceding section, the energy can be considered to be due to the special *position* of the body (i.e. because it was pressed into the rubber). An equivalent approach is to consider the energy as located throughout the stressed rubber and to be connected with its elastic properties.

In the case of static electric fields it is possible to imagine that a certain amount of energy is associated with the "elastic" properties of polarized pieces of dielectric substances. But no obvious physical argument can explain how energy is stored *in a vacuum*, i.e. in those parts of space which are not occupied by dielectric bodies (for example, the case of two charged conductors situated in a vacuum). Nevertheless, we shall now prove that, if energy is distributed throughout the field, by analogy with the above mechanical case, it is necessary to consider also parts of fields in vacuum as seats of the electrostatic energy.

We transform the expression (3.1.6) for the electrostatic energy of a system of charges as follows. First, according to Eq. (2.4.7), $\rho = \operatorname{div} \mathbf{D}$. Next, formally applying the differential del operator ($\nabla$), we have

$$\nabla \cdot (V\mathbf{D}) = V\nabla \cdot \mathbf{D} + \mathbf{D} \cdot \nabla V,$$

from which, by rearranging terms and writing the corresponding symbols instead of $\nabla$, we have

$$V\rho = V \operatorname{div} \mathbf{D} = \operatorname{div} (V\mathbf{D}) - \mathbf{D} \cdot \operatorname{grad} V = \operatorname{div} (V\mathbf{D}) + \mathbf{D} \cdot \mathbf{E}. \tag{3.2.1}$$

By introducing (3.2.1) into (3.1.6) we obtain

$$W_e = \frac{1}{2} \int_v \operatorname{div} (V\mathbf{D})\, dv + \frac{1}{2} \int_v \mathbf{D} \cdot \mathbf{E}\, dv. \tag{3.2.2}$$

Volume $v$ in Eqs. (3.2.2) and (3.1.6) must enclose all charges, but is otherwise arbitrary, since integrating Eq. (3.1.6) in the region with no charges ($\rho = 0$) does not affect the final result at all. Therefore, we can assume $v$ to be in the form of a large sphere centered somewhere in the system. The first integral in Eq. (3.2.2) can be transformed into a surface integral [using the divergence theorem (Eq. 1.11.10)] over the surface $S$ of the sphere. Now, over large distances $V$ decreases at least as $1/r$ and $D$ as $1/r^2$. Since $S$ increases as $r^2$, the first integral for $r$ increasing without limit is zero, and Eq. (3.2.2) becomes

$$\boxed{W_e = \int_v \tfrac{1}{2}\mathbf{E} \cdot \mathbf{D}\, dv} \quad \text{(energy of the static electric field expressed as an integral over the whole field),} \tag{3.2.3}$$

where $v$ denotes the *whole* space occupied by the field. It is obvious from Eq. (3.2.3) that the expression

$$\boxed{\frac{dW_e}{dv} = \tfrac{1}{2}\mathbf{E} \cdot \mathbf{D}} \quad \text{(density of energy in static electric field)} \tag{3.2.4}$$

can be interpreted as density of energy located in the field. If thus interpreted, Eq. (3.2.4) shows that energy is located at all points where the field exists, and hence also in a vacuum.

It is not possible to prove this hypothesis experimentally, as can be done in the analogous mechanical case. We can cut out a piece of stressed rubber and prove that a certain amount of energy is obtained if the piece is allowed to assume its unstressed shape. But we cannot "cut out" a piece of vacuum around a point in the static electric field.

In Chapter 15 we shall prove that in the case of *time-varying* fields energy *is* located at all points of the field, even in a vacuum. Hence, although the hypothesis that energy is distributed in the static electric field according to Eq. (3.2.4) cannot be proved experimentally, it can be regarded at least as the limiting case of a hypothesis which has been proved experimentally (for time-varying fields).

A simple illustration of the general conclusion that Eqs. (3.2.3) and (3.1.6) must always yield the same result is a parallel-plate capacitor with plates of surface $S$, distance between plates $d$, and a dielectric vacuum (air). According to Eqs. (3.1.9) and (1.14.4), assuming that the potential difference between the plates is $(V_1 - V_2)$,

$$W_e = \tfrac{1}{2}C(V_1 - V_2)^2 = \tfrac{1}{2}\varepsilon_0 \frac{S}{d}(V_1 - V_2)^2 = \tfrac{1}{2}\varepsilon_0 Sd\left(\frac{V_1 - V_2}{d}\right)^2 = \tfrac{1}{2}EDv,$$

where $v$ denotes the volume, $Sd$, of the part of space where the field exists.

*Problems referring to Section 3.2:* 3.G1 to 3.G6, 3.N3.

## 3.3 FORCES ON CHARGED CONDUCTORS IN VACUUM

It is much easier to determine the forces acting on charged conductors in vacuum than to determine forces in the presence of dielectrics. In the first case, only the forces on free charges distributed over conductor surfaces have to be taken into account. In the second case, additional forces appear, which act on polarization charges and on polarized substances throughout the dielectric. Forces on charged conducting bodies situated in a vacuum will therefore be considered first.

**3.3.1 Pressure on surfaces of conductors.** The electric field intensity $\mathbf{E}$ at the surface of a conductor and the surface charge density of free charges at the same point are connected by the relation (1.9.2):

$$E = E_n = \rho_s/\varepsilon_0. \tag{3.3.1}$$

Starting from this, it is possible to determine easily the pressure on the conductor surface.

Consider part of the surface of a charged conductor situated in a vacuum. Inside the conductor the electric field intensity is zero and outside the conductor it is perpendicular to the surface, as shown in Fig. 3.1. Now, the total electric field intensity both inside and outside the conductor is the sum of the electric field intensities created by the surface charges in the immediate neighborhood of the point being considered and all the other charges in the system. The first component, which we shall designate by $\mathbf{E}_{\rho_s}$, is perpendicular to the surface, but directed

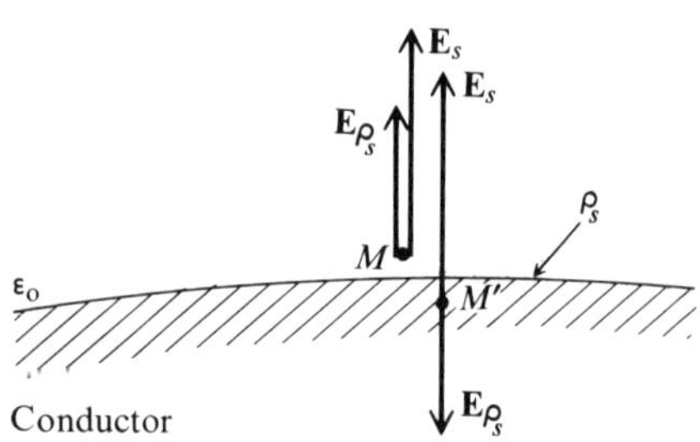


**Fig. 3.1** Components of the field intensity vector near the surface of a charged conductor.

*away* from it on both sides (for positive $\rho_s$). This is because, for two points $M$ and $M'$ close to the surface, the surface charge is equivalent to a plane charged sheet. The second component, which we shall denote by $\mathbf{E}_s$, is obviously of the same magnitude and direction at points $M$ and $M'$. Hence, according to Fig. 3.1 we find that

$$\text{at } M'\text{:} \quad -\mathbf{E}_{\rho_s} + \mathbf{E}_s = 0, \qquad \text{and at } M\text{:} \quad \mathbf{E}_{\rho_s} + \mathbf{E}_s = \mathbf{E}, \tag{3.3.2}$$

from which we get

$$\mathbf{E}_s = \mathbf{E}_{\rho_s} = \mathbf{E}/2. \tag{3.3.3}$$

Consequently, the pressure on the conductor surface for a dielectric vacuum according to Eqs. (3.3.1) and (3.3.3) is

$$p = dF/dS = \rho_s E_s = \rho_s(\rho_s/2\varepsilon_0) = \rho_s^2/2\varepsilon_0. \tag{3.3.4}$$

Since $\rho_s = D = \varepsilon_0 E$ on the surface of a conductor, Eq. (3.3.4) may be written in the form

$$p = \tfrac{1}{2}\varepsilon_0 E^2 = \tfrac{1}{2}\mathbf{E}\cdot\mathbf{D} = \frac{1}{2}\frac{D^2}{\varepsilon_0}. \tag{3.3.5}$$

Therefore, the pressure on the surface of a conductor is always directed toward a vacuum; it is proportional to $\rho_s^2$ (and thus does not depend on the sign of $\rho_s$) and is equal to the expression which in Section 3.2 we named the density of energy located in the field. It will be shown (Section 3.5) that this fact is not accidental, but follows from the law of conservation of energy and the assumption that energy is located in the field.

**3.3.2 Forces on charged conductors in a vacuum.** The total forces on charged conductors situated in a vacuum can be determined either by integrating Eq. (3.3.5) for pressure over the surface $S_0$ of a conductor,

$$\mathbf{F} = \oint_{S_0} p\, d\mathbf{S} = \frac{1}{2}\oint_{S_0} (\mathbf{E}\cdot\mathbf{D})\, d\mathbf{S}, \tag{3.3.6}$$

or by applying Coulomb's law. In both cases, however, it is essential that the distribution of charges over the conducting bodies be known in advance. This is a serious limitation of the method, since the exact distribution of charges is often not known.

In a few cases, these forces can be determined easily. For example, for two parallel thin conductors at a distance $d$, carrying charges $Q_l$ and $-Q_l$ per unit length, the electric field intensity due to charge on one conductor at points where the other is situated is given by Eq. (1.8.10). So, by direct application of Coulomb's law, we obtain the (attractive) force per unit length of the conductors:

$$F_l = E_s Q_l = Q_l^2/2\pi\varepsilon_0 d. \tag{3.3.7}$$

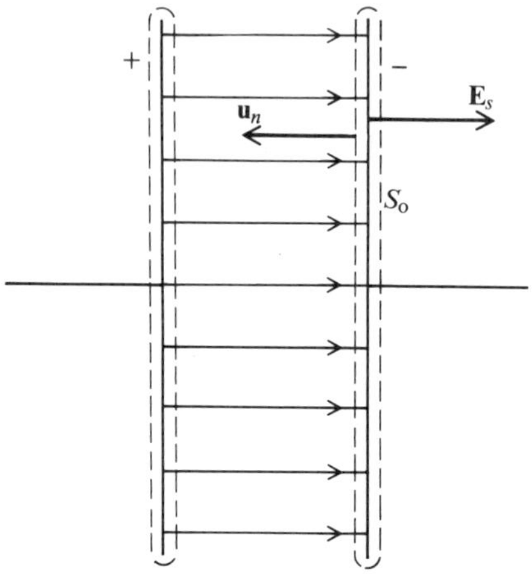


**Fig. 3.2** Construction for determining the total force on the electrode of a parallel-plate capacitor.

It should be noted that Eq. (3.3.6) cannot be applied in this case, since the exact distribution of charges over the conductor surfaces has not been determined.

As a second example, let us determine the force acting on one plate of a parallel-plate capacitor with plates of surface $S$ and a distance $d$ between them. Let the charges on the plates be $Q$ and $-Q$. The force in this case can be determined by either method. Since the pressure on the *inner* surface of each plate, according to Eq. (1.8.16) and ignoring the fringing effect, is

$$p = \tfrac{1}{2}\varepsilon_0 E^2 = \frac{1}{2}\frac{\rho_s^2}{\varepsilon_0} = \frac{1}{2}\frac{Q^2}{S^2\varepsilon_0},$$

and on the outer surface is $p = 0$ (theoretically, the field-free region), Eq. (3.3.6) yields

$$\mathbf{F} = \frac{1}{2}\frac{Q^2}{S^2\varepsilon_0}\mathbf{S} = \frac{1}{2}\frac{Q^2}{S^2\varepsilon_0}S\,\mathbf{u}_n = \frac{1}{2}\frac{Q^2}{\varepsilon_0 S}\mathbf{u}_n. \tag{3.3.8}$$

The vector $\mathbf{u}_n$ represents the positive (outward) unit vector normal to the inner surfaces of the plates (Fig. 3.2). Hence the force is attractive, as expected.

Equation (3.3.8) can also be easily obtained if we bear in mind that each plate is situated in the electric field of the charges on the other. The intensity $E_s$ of that field according to Eq. (1.8.15) is

$$E_s = \frac{1}{2}\frac{\rho_s}{\varepsilon_0} = \frac{1}{2}\frac{Q}{\varepsilon_0 S}$$

(see Fig. 3.2). Since the total charge on each plate is $Q$, Eq. (3.3.8) follows immediately.

*Problems referring to Section 3.3:* 3.T7 to 3.T8, 3.G17, 3.N4 to 3.N5.

## 3.4 QUALITATIVE DISCUSSION OF ELECTROSTATIC FORCES IN DIELECTRICS

**3.4.1 Introduction.** When, in addition to charged conducting bodies, dielectric substances are also present in the field, determining the forces acting in the system becomes extremely involved. On the one hand, it is necessary first to determine the exact distribution of all charges, and even the polarization vector and the field at all points, because a force exists also on a small polarized volume of a dielectric with no net polarization charge, if it is situated in a nonuniform electric field. On the other hand, real dielectrics are compressible. Hence permittivity, and consequently also the polarization charges, are a function of the stresses in the dielectric. This phenomenon is called *electrostriction*. We are again in a vicious circle: volume forces in a dielectric are functions of the polarization charges, and the polarization charges are functions (to a certain extent) of stresses in the dielectric, i.e. of the volume forces.

Although the above physical picture is a necessary basis for determining the exact distribution of forces in an electrostatic system, it is fortunately not necessary for determining the total forces acting on the separate bodies of the system. In Section 3.5 it will be shown that total forces and torques can be easily calculated (at least theoretically), by the *method of virtual displacements*. In practice we deal, almost without exception, with these total forces and torques, and knowledge of the exact distribution of forces is usually unimportant. In addition, the electrostatic forces encountered in practice are very small, and their application is mainly restricted to rather delicate electrical measurements.

For these reasons, the distribution of forces throughout a dielectric will be considered only qualitatively, to obtain some idea where such forces really act, and only the total forces, pressures, and torques will be calculated in Sections 3.5 and 3.6, by the method of virtual displacements.

**3.4.2 Volume forces in a homogeneous polarized uncharged dielectric.** It was proved in Section 2.5 that, in a homogeneous polarized dielectric with no free

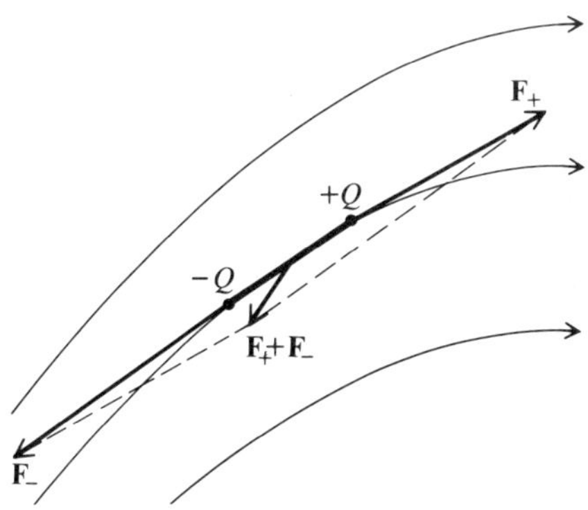


**Fig. 3.3** An electric dipole in an inhomogeneous electric field.

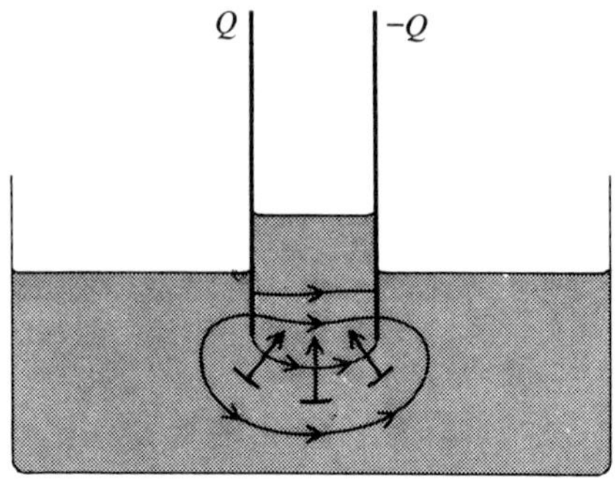


**Fig. 3.4** Forces acting on dipoles in the fringe-field region push the liquid dielectric upward.

volume charges, there can be no polarization charges. One may be tempted to conclude that, consequently, there can be no volume forces inside such a dielectric. This reasoning will be shown to be incorrect.

Let us consider a dipole of a homogeneous dielectric situated in a nonuniform electric field. In the case of isotropic dielectrics, which are the most common type, the dipole moment is in the direction of the field, as shown in Fig. 3.3. Although the situation is very exaggerated, it is obvious that a resultant force is acting on the dipole, directed generally toward points of higher field. This force can even be at right angles to the electric field intensity at that point.

For example, if the plates of a parallel-plate capacitor are partly immersed in a homogeneous fluid dielectric, a force exists on each dipole in the fringe-field region, tending to push it into the region of the higher field. Consequently, the pressure in the dielectric between the plates is higher, resulting usually in a small, but measurable, rise of the dielectric surface (Fig. 3.4). The expression for calculating this pressure will be derived in Section 3.6.2.

*Problems referring to Section 3.4:* 3.T9 to 3.T11.

## 3.5 GENERAL METHOD OF DETERMINING FORCES IN THE STATIC ELECTRIC FIELD

As indicated in the two preceding sections, the determination of forces in an electrostatic field as the integral of volume and surface forces is an extremely involved task, since it is very difficult to determine the exact distribution of these forces. However, if only *total* forces and torques on bodies in the static electric field are required, a general method for their determination is available. This method is based on the *principle of virtual displacements*, which in turn is based on the law of conservation of energy. The idea underlying the method is as follows.

In an electrostatic field, all charged bodies and dielectrics are by definition static. This implies that the electric forces acting on the bodies are balanced by exactly equal mechanical forces—otherwise the bodies would not be stationary. Now, imagine that electric forces moved a charged or a dielectric body for an infinitesimal distance $\delta x$. If the component of the total electric force **F** on the body in the direction $\delta x$ is $F_x$, this imaginary (virtual) displacement must be accompanied by an elemental (virtual) amount of work $\delta W$ done by the force $F_x$, such that $\delta W = F_x \, \delta x$. Hence, if we can determine $\delta W$ as a function of $x$, we can also determine $F_x$:

$$F_x = \delta W / \delta x. \tag{3.5.1}$$

Determination of the virtual work $\delta W$ performed during the virtual displacement can be reduced to determining the accompanying change in the energy stored in the field in two special cases: (1) when the *charges* on all bodies during the virtual displacement stay constant, and (2) when the *potentials* of all conducting bodies during the virtual displacement stay constant.

Suppose first that the charges on all conducting bodies are kept constant; i.e. the conductors are not connected to any sources capable of changing their charge when moved. Consequently, no energy is supplied to the system, and the virtual work done by the electric forces according to the law of conservation of energy must be accompanied by an equal amount of change of energy of the system, $\delta W_e$. We are assuming that no loss of energy accompanies the virtual displacement. Therefore we may write

$$\delta W = -\delta W_e. \tag{3.5.2}$$

The minus sign is necessary: since we know that the energy of the system was *larger* before the displacement, the increment of energy of the system must be negative. Combining Eqs. (3.5.1) and (3.5.2), we find that the component $F_x$ (in a direction $x$) of the electric force **F** on any body of the system (Fig. 3.5) may be calculated as

$$\boxed{F_x = -\frac{\delta W_e}{\delta x}} \quad \text{(charges kept constant).} \tag{3.5.3}$$

If torque on a body due to the electric forces is to be evaluated rather than the total force, the same procedure applies. The virtual work $\delta W$ done by the component $T_z$ of the total torque **T** in rotating a body for a virtual angle $\delta\phi$ around, say, the $z$-axis of rotation (Fig. 3.6) is

$$\delta W = T_z \, \delta\phi.$$

Combining this with Eq. (3.5.2) we obtain

$$T_z = -\frac{\delta W_e}{\delta\phi} \quad \text{(charges kept constant).} \tag{3.5.4}$$

Suppose now that the potentials of all the conducting bodies of the system are kept constant during the virtual displacement; i.e., all the conducting bodies are connected to sources which supply charges necessary to keep their potentials

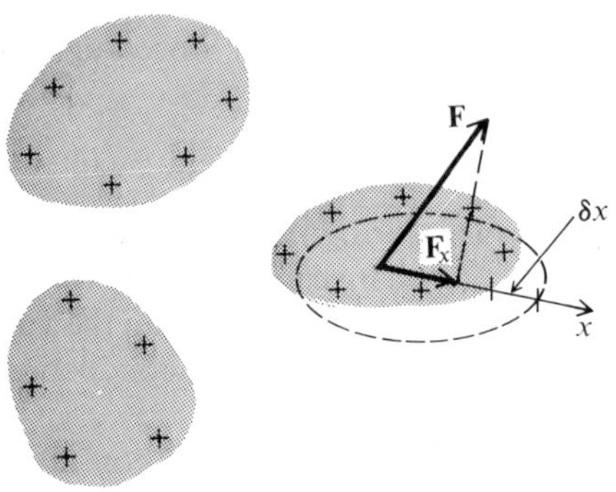


Fig. 3.5 Imaginary displacement of a body under the action of the electrostatic force **F**.

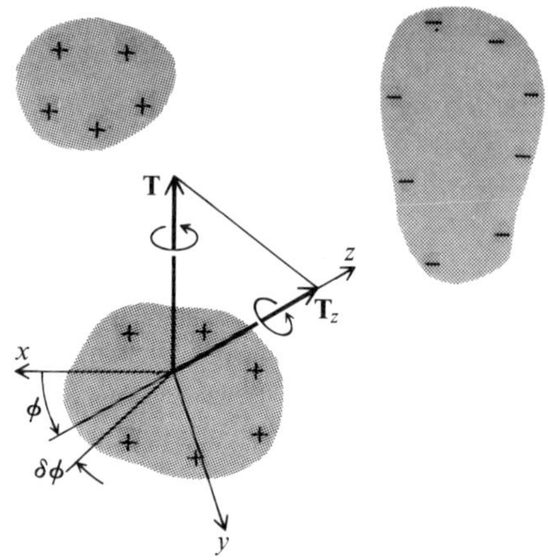


Fig. 3.6 Imaginary rotation of a body for angle $\delta\phi$.

constant, in spite of the somewhat different geometry of the system after the displacement. Consequently, the sources must in this case do a certain amount of work. If the potentials of conducting bodies are designated by $V_1, V_2, \ldots, V_n$ and the respective supply of charges during the virtual displacement by $\delta Q_1, \delta Q_2, \ldots, \delta Q_n$, according to the definition of the potential the *total* work $\delta W_s$ done by all sources is

$$\delta W_s = \sum_{j=1}^{n} V_j \, \delta Q_j. \tag{3.5.5}$$

Instead of Eq. (3.5.2), the law of conservation of energy (again assuming no losses) now reads

$$\delta W_s = \delta W + \delta W_e. \tag{3.5.6}$$

According to Eq. (3.1.7), the increase in energy of the system if the potentials of all conducting bodies are kept constant and their charges changed for $\delta Q_j$, $j = 1, 2, \ldots, n$, is

$$\delta W_e = \frac{1}{2} \sum_{j=1}^{n} V_j \, \delta Q_j. \tag{3.5.7}$$

From Eqs. (3.5.5) and (3.5.7) it is evident that $\delta W_s = 2\delta W_e$, so that Eq. (3.5.6) becomes $\delta W = \delta W_e$. This means that *exactly* one-half of the energy supplied by the generators is used for performing the mechanical work during the displacement. The energy of the system must increase by the same amount. From this and Eq. (3.5.1) we therefore deduce that

$$\boxed{F_x = +\frac{\delta W_e}{\delta x}} \quad \text{(potentials of conducting bodies kept constant).} \tag{3.5.8}$$

The analogous expression to Eq. (3.5.4) for the torque $T_z$ acting on a body is

$$T_z = +\frac{\delta W_e}{\delta \phi} \qquad \text{(potentials of conducting bodies kept constant).} \quad (3.5.9)$$

As an example of the application of Eqs. (3.5.3) and (3.5.8) let us determine the force acting on a plate of a charged parallel-plate capacitor with dielectric vacuum. (This was determined in Section 3.3.2 by a different method.)

According to Eqs. (3.1.9) and (1.14.4), the energy located in the capacitor for constant $Q$ can be expressed in the form

$$W_e = \frac{1}{2}\frac{Q^2}{C} = \frac{1}{2}\frac{Q^2}{\varepsilon_0 S/d} = \frac{1}{2}\frac{Q^2}{\varepsilon_0 S}d,$$

so that, using Eq. (3.5.3), we obtain

$$F_d = -\frac{\delta W_e}{\delta d} = -\frac{1}{2}\frac{Q^2}{\varepsilon_0 S}.$$

The minus sign indicates that the force $F_d$ tends to *decrease* the distance $d$ between the plates (which obviously is correct).

If the potential difference $(V_1 - V_2)$ between the plates is assumed to be constant, from Eqs. (3.1.9) and (1.14.4) we have

$$W_e = \tfrac{1}{2}C(V_1 - V_2)^2 = \tfrac{1}{2}\varepsilon_0\frac{S}{d}(V_1 - V_2)^2,$$

so that, using Eq. (3.5.8), we find that

$$F_d = +\frac{\delta W_e}{\delta d} = -\tfrac{1}{2}\varepsilon_0\frac{S}{d^2}(V_1 - V_2)^2 = -\tfrac{1}{2}\varepsilon_0 SE^2.$$

This is the same as before, since $\varepsilon_0 E^2 = D^2/\varepsilon_0 = Q^2/\varepsilon_0 S^2$.

*Problems referring to Section 3.5:* 3.G11 to 3.G16, 3.N8 to 3.N10.

## 3.6 DETERMINATION OF PRESSURE ON BOUNDARY SURFACES

**3.6.1 Pressure on surfaces of charged conductors.** Consider a charged conductor situated in a homogeneous dielectric of permittivity $\varepsilon$. Imagine that the electric forces push a small surface $\Delta S$ of the conductor (Fig. 3.7) for a small length $\delta x$ in a direction $x$, and assume that the charge on the conductor is kept constant. The accompanying decrease of energy stored in the field is exactly equal to the energy stored in the elemental volume $\delta v = \Delta S\, \delta x \cos\alpha$ before it was "annexed" by the conductor, or

$$\delta W_e = -\tfrac{1}{2}\varepsilon E^2\, \Delta S\, \delta x \cos\alpha.$$

Hence, according to Eq. (3.5.3), the force acting on the surface element $\Delta S$ in the

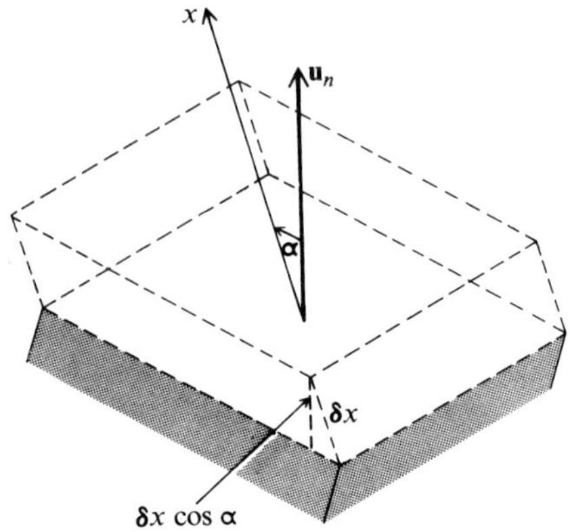


**Fig. 3.7** A virtual displacement of a conductor surface.

direction of the $x$-axis is

$$\Delta F = -\frac{\delta W_e}{\delta x} = \tfrac{1}{2}\varepsilon E^2 \, \Delta S \cos \alpha. \tag{3.6.1}$$

According to Eq. (3.6.1), the component of force tangential to the surface ($\alpha = 90°$) is zero, so that on the surface of a conductor there is only a normal pressure:

$$p = \frac{\Delta F}{\Delta S} = \tfrac{1}{2}\varepsilon E^2 = \tfrac{1}{2}\mathbf{E} \cdot \mathbf{D}. \tag{3.6.2}$$

This expression was obtained in Section 3.3.1 for the special case of a dielectric vacuum. From the process of derivation it is evident that Eq. (3.6.2) represents in fact the total pressure acting on *both* the surface of the conductor and the surface of the dielectric with which it is in contact.

**3.6.2 Pressure on boundary surfaces of two dielectrics.** There is no tangential force acting on the boundary surface of two dielectrics. A virtual displacement of the boundary surface parallel to the surface is not accompanied by a change in the electrostatic energy, since the system is exactly the same as before. Consequently, according to either (3.5.3) or (3.5.8), the tangential component of the force is zero.

The normal component of force at the boundary surface depends upon the direction of the electric field intensity vector in both dielectrics. Since the force corresponding to any direction of vector **E** may be obtained if the force corresponding to tangential and normal components of **E** is found, these two cases will be considered in turn.

Consider first the case when the lines of force are parallel to the boundary surface (Fig. 3.8). If we suppose that under the influence of the pressure the boundary surface moved for a distance $\delta x$, as shown in Fig. 3.8, it is simplest to assume that the potential difference between the two conductors shown is unchanged. Then

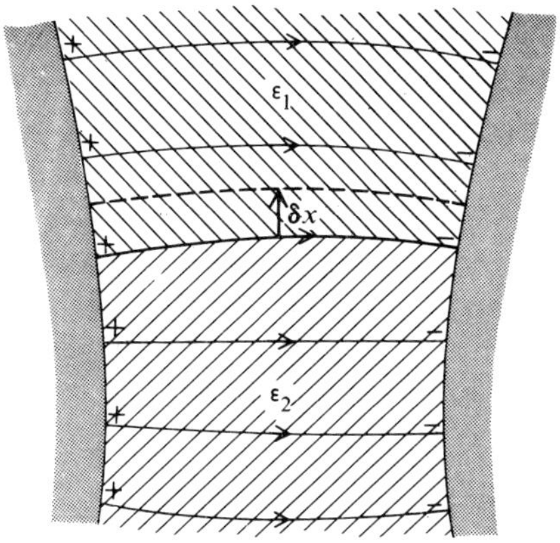


**Fig. 3.8** An interface between two dielectrics parallel to the lines of force.

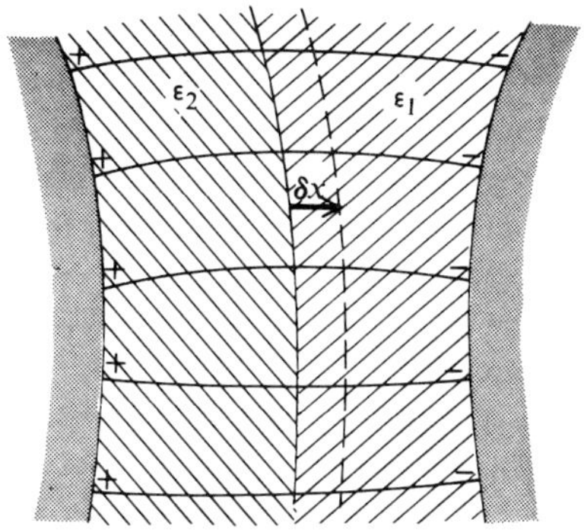


**Fig. 3.9** An interface between two dielectrics perpendicular to the lines of force.

the electric field intensity in both dielectrics will not be changed after the virtual displacement. [According to the boundary condition (2.6.1), the tangential components of **E** must be equal. If the potential difference remained the same, **E** would also remain the same on both sides of the boundary surface.] Taking a small portion $\Delta S$ of the boundary surface and noting that $E_1 = E_2 = E_t$, we find that the change in energy located in the field is

$$\delta W_e = (\tfrac{1}{2}\varepsilon_2 E_2^2 - \tfrac{1}{2}\varepsilon_1 E_1^2)\,\Delta S\,\delta x = \tfrac{1}{2}(\varepsilon_2 - \varepsilon_1)\,E_t^2\,\Delta S\,\delta x.$$

According to Eq. (3.5.8), the force per unit area, assumed to act from the dielectric of permittivity $\varepsilon_2$ toward the dielectric of permittivity $\varepsilon_1$, is

$$p_{E_t} = \frac{1}{\Delta S}\frac{\delta W_e}{\delta x} = \tfrac{1}{2}(\varepsilon_2 - \varepsilon_1)\,E_t^2 \qquad \text{(directed from 2 to 1).} \qquad (3.6.3)$$

The pressure tends to push the boundary surface toward the dielectric having the smaller permittivity.

Suppose now that the field vectors are normal to the boundary surface (Fig. 3.9). Imagine that, under the influence of the pressure on the boundary surface, the dielectric of permittivity $\varepsilon_2$ "annexed" a small portion of the dielectric of permittivity $\varepsilon_1$. According to the boundary condition (2.6.3), $D_1 = D_2 = D_n$, so that it is simplest to assume the free charges on the two conductors remained the same, in which case there will be no change of **D**. Therefore, for determining the pressure in this case it is simplest to apply Eq. (3.5.3).

The virtual change in the energy located in the field resulting from the virtual displacement $\delta x$ of a small part $\Delta S$ of the boundary surface in the direction indicated is

$$\delta W_e = \left(\frac{1}{2}\frac{D_2^2}{\varepsilon_2} - \frac{1}{2}\frac{D_1^2}{\varepsilon_1}\right)\Delta S\,\delta x = \frac{1}{2}\left(\frac{1}{\varepsilon_2} - \frac{1}{\varepsilon_1}\right)D_n^2\,\Delta S\,\delta x.$$

According to Eq. (3.5.3) the pressure on the surface, assumed to be in the direction indicated, is

$$p_{D_n} = -\frac{1}{\Delta S}\frac{\delta W_e}{\delta x} = \frac{1}{2}\left(\frac{1}{\varepsilon_1} - \frac{1}{\varepsilon_2}\right) D_n^2 \qquad \text{(directed from 2 to 1).} \qquad (3.6.4)$$

In this case also, the pressure acts in the direction of the dielectric of smaller permittivity.

Noting that $\frac{1}{2}\varepsilon E^2 = \frac{1}{2}\varepsilon E_t^2 + \frac{1}{2}D_n^2/\varepsilon$, from Eqs. (3.6.3) and (3.6.4) it may be deduced that the pressure on the boundary surface in the general case is

$$\boxed{p = \tfrac{1}{2}(\varepsilon_2 - \varepsilon_1)(E_t^2 + D_n^2/\varepsilon_1\varepsilon_2)} \qquad \text{(directed from 2 to 1).} \qquad (3.6.5)$$

As an example, consider the pressure on the surface of the liquid dielectric in Fig. 3.4. If $\varepsilon$ denotes the permittivity of the dielectric, the upward pressure on the surface of the liquid dielectric between the plates will be

$$p = \tfrac{1}{2}(\varepsilon - \varepsilon_0)\, E_t^2 = \tfrac{1}{2}(\varepsilon - \varepsilon_0)\left(\frac{V_1 - V_2}{d}\right)^2,$$

where $(V_1 - V_2)$ denotes the potential difference between the plates. It is obvious, however, that the origin of this pressure is *not* on the boundary surface, since polarization charges are not present nor is the field nonuniform. This clearly shows that the derived expressions for pressure enable us to evaluate the pressure acting on the boundary surface without necessarily locating correctly the forces to which this pressure is equivalent.

*Problems referring to Section 3.6:* 3.T12, 3.G17 to 3.G21, 3.N11.

## PROBLEMS

### Test problems

**3.T1.** Discuss whether a system of charged bodies can have zero total electrostatic energy.

***3.T2.** Can the electrostatic energy of a system of charges be negative?

**3.T3.** Derive Eq. (3.1.8) for energy stored in a charged capacitor with capacitance $C$, assuming that it is charged from zero to $Q$ and $-Q$ by transferring the charge from one electrode to the other in amounts $dQ$.

**3.T4.** Capacitors of capacitance $C_1, C_2, \ldots, C_n$ are connected (1) in parallel and (2) in series. Determine the energy of the system in both cases, given that the system is connected to a source of potential difference $V$.

**3.T5.** A parallel-plate capacitor, with dielectric air and plate area $S$, is charged with a fixed charge $Q$. If the plates are moved a distance $(d_0 + d)$ apart, the original distance being $d_0$, what is the change in the electrostatic energy stored in the capacitor? Explain the result.

**3.T6.** A capacitor with dielectric air and capacitance $C_0$ is charged with a charge $Q$. The space between the electrodes is then filled with a liquid dielectric of permittivity $\varepsilon$. Determine the change in the electrostatic energy stored in the capacitor. Explain the result.

**3.T7.** Is it possible to have a system of three point charges which are in equilibrium under the influence of their own mutual forces? Is the equilibrium stable or unstable?

**3.T8.** Small pieces of polarized dielectrics and uncharged conductors, e.g. particles of dust and smoke, may be approximated by dipoles oriented in the direction of the electric field intensity. If the electric field is not uniform, is there a force on the dipoles? Could this phenomenon be used to filter dust and smoke particles from the air?

**3.T9.** Explain where the seat of the electrostatic forces is in Figs. 3.G13, 3.G14, 3.G18, and 3.G19.

***3.T10.** A system of charged conductors is situated in a homogeneous dielectric. The density of polarization charges on conductor surfaces is proportional to the density of free charges at the same points. What is the electric field inside conducting bodies and on conductor surfaces due to *polarization* charges, if conductor surfaces are imagined to be *inside* the surface layer of polarization charges?

***3.T11.** From the conclusion of the preceding problem deduce that the total force between conducting bodies proper is not influenced by the presence of a *homogeneous* dielectric.

**3.T12.** Prove that the *total* pressure on the boundary surface of a conductor and a dielectric is really the sum of pressures acting on the surface of the conductor and the surface of the dielectric, assuming that a very thin air layer exists between them.

## General problems

**3.G1.** An isolated conducting sphere of radius $a$ is charged with a charge $Q$ and is situated in a homogeneous dielectric of permittivity $\varepsilon$. Prove that the same result, $W_e = Q^2/8\pi\varepsilon a$, for the electrostatic energy stored in the system is obtained by using either Eq. (3.1.6) or Eq. (3.2.3).

**3.G2.** A dielectric sphere of radius $a$ and permittivity $\varepsilon$ is situated in a vacuum and is charged throughout its volume by a volume density of free charges $\rho = \rho_0\, a/r$, $r$ being the distance from the center of the sphere. Determine the electrostatic energy of the system.

*Solution.* According to Eq. (3.1.6), to determine the energy stored in the system it is necessary to know, in addition to the charge density $\rho$ at all points, the potential at these points. The potential function $V$ can be determined as follows.

Since the electric field is radial (because of symmetry), using the generalized integral form of Gauss' law we obtain

$$E_{\text{inside}}(r) = \rho_0 a/2\varepsilon \quad \text{for} \quad r < a \quad \text{and} \quad E_{\text{outside}}(r) = \rho_0 a^3/2\varepsilon_0 r^2 \quad \text{for} \quad r > a.$$

The potential at a point at a distance $r(r < a)$ from the center of the sphere is

$$V(r) = \int_r^a E_{\text{inside}}\, dr + \int_a^\infty E_{\text{outside}}\, dr = \frac{\rho_0 a}{2\varepsilon}(a - r) + \frac{\rho_0 a^2}{2\varepsilon_0},$$

from which we obtain the energy of the system:

$$W_e = \frac{1}{2}\int_0^a \frac{\rho_0 a}{r}\left[\frac{\rho_0 a}{2\varepsilon}(a - r) + \frac{\rho_0 a^2}{2\varepsilon_0}\right] 4\pi r^2\, dr = \frac{\rho_0^2 a^5 \pi}{6\varepsilon} + \frac{\rho_0^2 a^5 \pi}{2\varepsilon_0}.$$

It is left to the reader to check that the same result is obtained according to Eq. (3.2.3).

**3.G3.** Repeat the preceding problem, given that free charges inside the dielectric sphere are of constant density $\rho_0$.

*Answer.* For $\varepsilon = \varepsilon_0$, $W_e = 12\rho_0^2\pi a^5/45\varepsilon_0$.

**3.G4.** Inside a hollow metal sphere, of inner radius $b$ and outer radius $c$, is a metal sphere of radius $a$. The centers of the two spheres coincide, and the dielectric is air. If the inner sphere carries a charge $Q_1$ and the outer sphere a charge $Q_2$, what is the energy stored in the system?

*Answer.* For $Q_1 = -Q_2$, energy $W_e = Q_1^2(b - a)/8\pi\varepsilon_0 ab$.

**3.G5.** A spherical capacitor with dielectric air, of electrode radii $a$ and $b$, is connected to a source of potential difference $V$. Determine the electrostatic energy of the system.

**3.G6.** Repeat the preceding problem for the case of a coaxial cable of length $l$, with dielectric of permittivity $\varepsilon$ and conductors of radii $a$ and $b$.

**3.G7.** A variable capacitor is made of a set of parallel metal plates between which another set of plates can be inserted. The two sets of plates represent the two electrodes of the capacitor and the capacitance can be varied by pushing the movable plates between the stationary plates to a greater or lesser degree. The minimal capacitance of the system is $C_{\text{min}}$ and the maximal capacitance $C_{\text{max}}$, where $C_{\text{max}} = kC_{\text{min}}$ and $k$ is a constant.

With the movable plates in the position corresponding to the maximal capacitance, the capacitor is charged with a charge $Q$ and the source is disconnected. The movable plates are then moved to the position of minimal capacitance. Determine the energy of the system in each case and explain the result.

*Solution.* Since, during the change of capacitance, the charge $Q$ is supposed to remain constant, it is most natural to use the last expression in Eq. (3.1.9) to determine the energy. Hence,

$$W_{C_{\text{max}}} = \frac{1}{2}\frac{Q^2}{C_{\text{max}}}; \qquad W_{C_{\text{min}}} = \frac{1}{2}\frac{Q^2}{C_{\text{min}}},$$

from which we obtain

$$W_{C_{\text{min}}}/W_{C_{\text{max}}} = C_{\text{max}}/C_{\text{min}} = k > 1.$$

During the motion of movable plates the source was disconnected. How then is an increase of energy in the system possible? The answer is simple. During the motion of the plates a certain amount of work had to be done *against* the forces of the field, thus increasing the energy of the field. Physically, it may also be stated that in the latter case the charges on the plates are pressed closer together than in the initial state, for which obviously a certain amount of work must be performed.

**3.G8.** A coaxial cable of length $L$ and conductors of radii $a$ and $b$, if first filled with a liquid dielectric of permittivity $\varepsilon$, and then for a short time connected to a source of potential difference $V$. After the source is disconnected the liquid dielectric is drained from the cable. Determine: (1) the potential difference $V_1$ between the cable conductors after the dielectric is drained from the cable; (2) the energies $W_{e1}$ and $W_{e2}$ stored in the system before and after the dielectric is drained. Ignore the fringing effect.

*Answer.* $V_1 = \varepsilon V/\varepsilon_0$; $W_{e2} = \varepsilon W_{e1}/\varepsilon_0$.

***3.G9.** Prove *Thomson's theorem:* the distribution of static charges on conductors is such that the energy of the system of charged conductors is minimal.

*Proof.* There is no tangential electric force on charges spread over surfaces of conductors. Consequently, no energy is required in the first approximation to move any small patch of charges an infinitesimal distance along the conductor surface. If this is true for *any* small patch of charges on *any* conductor of the system, it follows that charges are distributed over surfaces of conductors in such a manner that the energy of the system is stationary; i.e. the first variation of energy corresponding to a slightly disturbed charge distribution is zero.

However, if we imagine that the patch of charges is moved farther in any direction, the repulsive forces of the adjacent charges of the same sign would require some work to be done against them, the energy located in the field thereby increasing. Consequently, energy corresponding to the real distribution of charges over surfaces of charged bodies is minimal.

***3.G10.** Prove that, if an uncharged conductor, or a conductor at potential zero, is introduced in an electrostatic field of charges distributed on conducting bodies, the energy of the system decreases.

*Proof.* Consider a system of $n$ conductors carrying constant charges $Q_1, Q_2, \ldots, Q_n$. Denote by $V_1, V_2, \ldots, V_n$ the potentials of conductors corresponding to the initial state of the system. The energy of the system prior to introducing the uncharged body, which we shall denote by subscript $(n + 1)$, according to (3.1.7) is

$$W_e = \frac{1}{2}\sum_{j=1}^{n} V_j Q_j.$$

Now, when the $(n + 1)$th body is introduced, the energy of the system is

$$W'_e = \frac{1}{2}\sum_{j=1}^{n} V'_j Q_j + \tfrac{1}{2}V_{n+1}Q_{n+1}.$$

(Since the charges are constant, the potential of all bodies must have changed.) If *either* $V_{n+1}$ or $Q_{n+1}$ are zero, the last term vanishes, so that no energy is supplied to the system by introducing the $(n + 1)$th body and the energy of the system is either less than, or equal to, the initial energy, that is $W'_e \leq W_e$. However, during the introduction of the $(n + 1)$th body from infinity the forces on the induced charge were always in the direction of the increasing field, that is, toward the system of charged conductors. Hence the forces of the field performed a certain amount of work. Since the energy stored in the system was the only source of energy available, it must have decreased during the process.

**3.G11.** The potential difference between the conductors of a long, straight, two-wire line, of radius $R$ and distance between axes $D$ ($D \gg R$), is $V$. Determine the intensity and direction of the force acting on the conductors per unit length.

*Solution.* We shall determine the force acting on the two conductors by both direct application of Coulomb's law and the principle of virtual displacements.

The solution using the first method is found in Section 3.3.2, Eq. (3.3.7). Changing $d$ in that equation to $D$ and expressing $Q_l$ by the potential difference between conductors $V$ and the capacitance per unit length $C_l$ of the two-wire line [Eq. (a) of Problem 1.G50], we obtain the expression for the (attractive) force per unit length between conductors of the line:

$$F_l = \frac{\pi\varepsilon_0 V^2}{2D(\ln D/R)^2}. \tag{a}$$

To calculate the force by Eq. (3.5.8), it is necessary to determine the energy of the system per unit length of the line, when the potential difference $V$ between the conductors is known (and is supposed to stay constant during virtual displacement). From Eqs. (3.1.9) and (a) of Problem 1.G50 the energy per unit length of the line is

$$W_{el} = \tfrac{1}{2}C_l V^2 = \frac{\pi\varepsilon_0 V^2}{2 \ln D/R}. \tag{b}$$

According to (3.5.8) the force per unit length acting on the conductors is

$$F_l = +\frac{\delta W_{el}}{\delta D} = -\frac{\pi\varepsilon_0 V^2}{2(\ln D/R)^2}\frac{1}{D/R}\frac{1}{R} = -\frac{\pi\varepsilon_0 V^2}{2D(\ln D/R)^2},$$

which is the same as (a) above. The minus sign indicates that the force tends to decrease the distance $D$ between the two conductors; i.e. that it is attractive.

It is left as an exercise for the reader to prove that the same result is obtained by using Eq. (3.5.3). (In this case the other equivalent expression for energy should be used.)

**3.G12.** The inner conductor of a coaxial cable can slide through the cylindrical cavity formed by a solid dielectric of the cable. The permittivity of the dielectric is $\varepsilon$, the radius of the inner (movable) conductor is $a$, and the radius of the outer conductor is $b$ (Fig. 3.G12). Determine the magnitude and the direction of the force acting on the inner conductor, assuming that the potential difference between the cable conductors is $V$. Ignore the fringing effect.

*Solution.* The force on the inner conductor may be calculated by either Eq. (3.5.3) or Eq. (3.5.8). Energy stored in the part of the cable of length $l$ is

$$W_e = \tfrac{1}{2}C_l l V^2 = \frac{1}{2}\frac{2\pi\varepsilon l}{\ln b/a}V^2.$$

Applying Eq. (3.5.8) to this expression for energy we find

$$F = \frac{\delta W_e}{\delta l} = \frac{\pi\varepsilon V^2}{\ln b/a}.$$

We leave it as an exercise for the reader to show that the same result is obtained using Eq. (3.5.3).

**3.G13.** The end of a coaxial cable is closed by a dielectric piston of permittivity $\varepsilon$, as shown in Fig. 3.G13. The radii of the cable conductors are $a$ and $b$ and the dielectric in the other part of the cable is air. What is the magnitude and direction of the axial force acting on the dielectric piston, if the potential difference between the conductors is $V$?

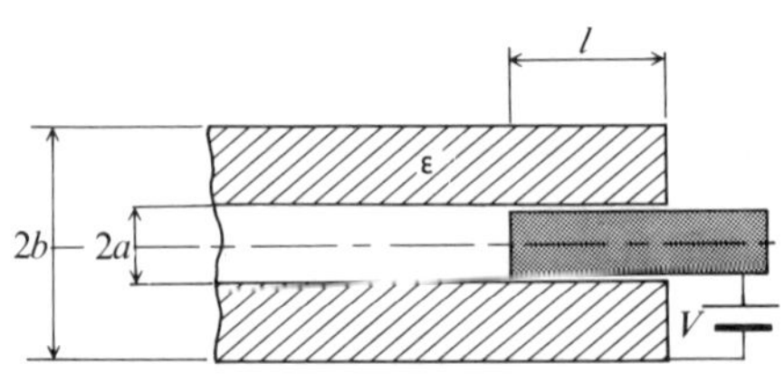


Figure 3.G12

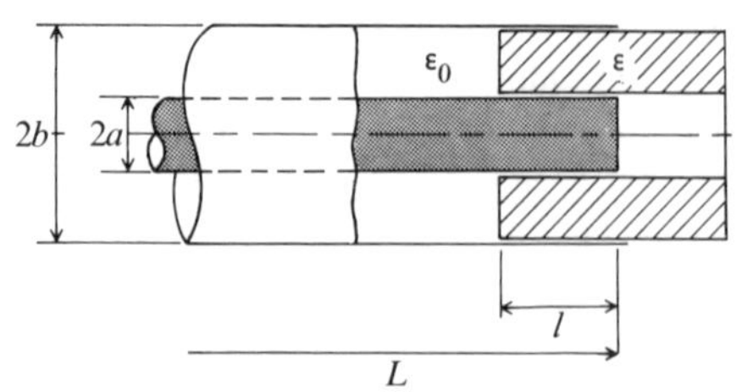


Figure 3.G13

*Solution*. Suppose that the total length of the cable is $L$. Then the energy stored in the cable is

$$W_e = \tfrac{1}{2}CV^2 = \frac{1}{2}\,\frac{2\pi\varepsilon_0(L - l)}{\ln b/a}\,V^2 + \frac{1}{2}\,\frac{2\pi\varepsilon l}{\ln b/a}\,V^2.$$

Using Eq. (3.5.8) we obtain

$$F = \frac{\delta W_e}{\delta l} = \frac{\pi V^2(\varepsilon - \varepsilon_0)}{\ln b/a}.$$

**3.G14.** One end of an open coaxial cable is immersed vertically in a liquid dielectric of unknown permittivity, as shown in Fig. 3.G14. The cable is connected to a source of potential difference $V$. The electrostatic forces draw the liquid dielectric into the cable for a height $h$ above the level of the dielectric outside the cable. The radius of the inner conductor is $a$, that of the outer conductor $b$, and the mass density of the liquid dielectric is $\rho_m$. Determine the relative permittivity of the liquid dielectric.

*Answer*. For $V = 1000$ V, $h = 3.29$ cm, $a = 0.5$ mm, $b = 1.5$ mm, and $\rho_m = 1$ gr/cm$^3$, the relative permittivity $\varepsilon_r = 81$.

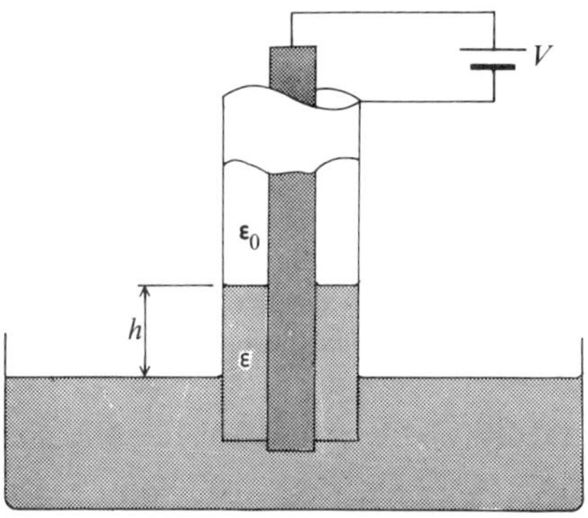


**Figure 3.G14**

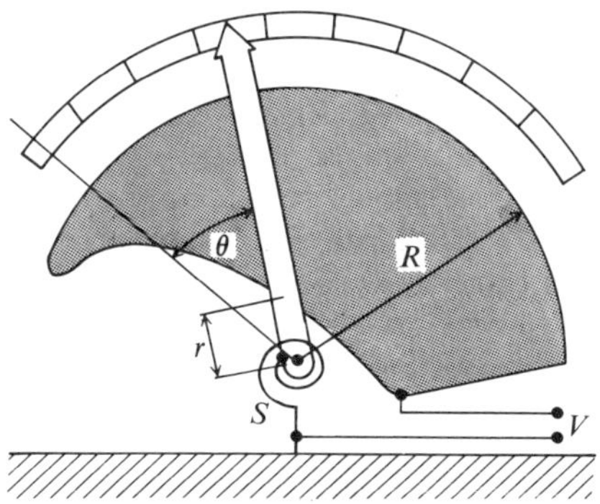


**Figure 3.G15**

**3.G15.** The needle electrometer (apparatus for measuring potential difference) consists of a thin metal pointer above a stationary metal plate (Fig. 3.G15). The outer edge of the plate is a circle of radius $R$, centered at the axis of rotation of the pointer. The inner edge of the plate is determined by an arbitrary function $r = r(\theta)$. The capacitance between the plate and the pointer is proportional to the length of the pointer above the plate, $C = C_l(R - r)$, where $C_l$ is a constant.

Determine the torque of electrostatic forces on the pointer, given that the electrometer is connected to a potential difference $V$. If this torque is counterbalanced by a spiral spring $S$ of torsion constant $k$, show that it is possible to determine $r = r(\theta)$ in such a manner that the deflection of the pointer is proportional to the potential difference $V$.

*Solution*. According to Eq. (3.5.9) the torque acting on the pointer is

$$T = +\frac{\delta W_e}{\delta\theta} = \tfrac{1}{2}V^2\frac{dC}{d\theta} = -\tfrac{1}{2}V^2 C_l\frac{dr}{d\theta},$$

since $C = C_l(R - r)$. If this torque is balanced by the spiral spring, the following equation holds

$$k\theta = -\tfrac{1}{2}C_l V^2 \frac{dr}{d\theta},$$

from which

$$dr = -\frac{2k}{C_l V^2}\theta\, d\theta. \tag{a}$$

If the deflection angle $\theta$ is proportional to the applied potential difference $V$ we can put $V = k'\theta$. In that case (a) becomes

$$dr = -\frac{2k}{C_l k'^2}\frac{d\theta}{\theta},$$

whence

$$r(\theta) = \frac{2k}{C_l k'^2}\ln\frac{\theta_0}{\theta}, \tag{b}$$

$\theta_0$ representing an integration constant. From (b) we can see that it is possible to make a needle electrometer with a deflection angle proportional to the applied potential difference, since it is possible to construct the inner edge of the plate according to (b), if $\theta_0$ is chosen in an appropriate manner.

**3.G16.** Figure 3.G16 shows the so-called quadrant electrometer. When the potential difference $V$ is zero the rotatable plates $A$ are situated in quadrants out of the staionary plates. The torque of the electrostatic forces is counterbalanced by a spiral spring $S$ with a torsion constant $k'$. The capacitance between the rotatable and the stationary plates is proportional to the deflection angle $\alpha$, $C = k''\alpha$. Determine $\alpha$ as a function of the applied potential difference $V$. What is the maximal potential difference that can be measured by the electrometer? (Ignore possible friction.)

*Answer.* For $k' = 0.0318$ Nm/radian, $k'' = 500$ pF/radian and $V = 10{,}000$ V, the deflection angle $\alpha = \pi/4$ radians.

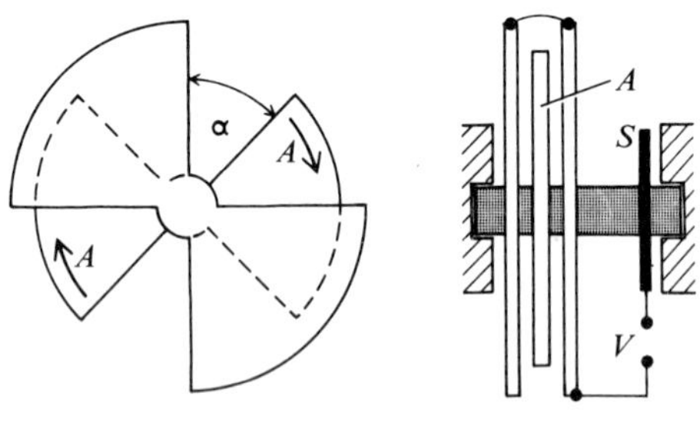


Figure 3.G16

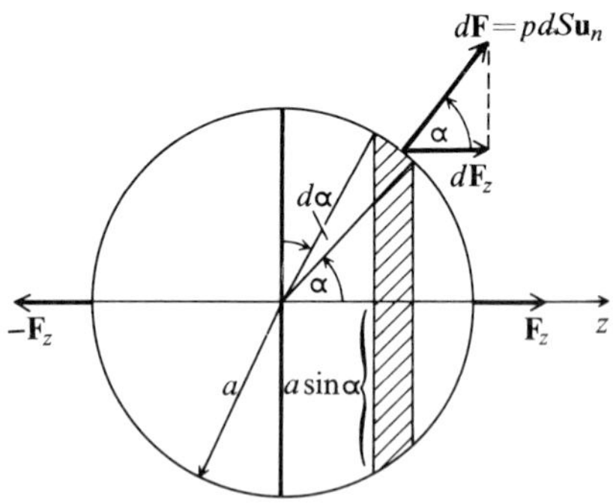


Figure 3.G17

**3.G17.** A conducting sphere of radius $a$ is cut into two halves, which are pressed together by a spring inside the sphere. The sphere is situated in air and is charged with a charge $Q$. Determine the force on the spring due to the charge on the sphere.

*Solution.* The pressure acting on the surface of the sphere is given by Eqs. (3.6.2) or (3.3.5).

Since the electric field intensity at the surface of the sphere is $E = Q/4\pi\varepsilon_0 a^2$, Eq. (3.6.2) or (3.3.5) yields

$$p = \tfrac{1}{2}\varepsilon_0 E^2 = \frac{Q^2}{32\pi^2\varepsilon_0 a^4}. \tag{a}$$

By reference to Fig. 3.G17 we see that the force on one-half of the sphere is

$$F_z = \int_0^{\pi/2} p\, 2\pi a \sin\alpha\, a\, d\alpha \cos\alpha = pa^2\pi.$$

or, according to (a),

$$F_z = Q^2/32\pi\varepsilon_0 a^2.$$

**3.G18.** The ends of the plates of a parallel-plate capacitor are immersed vertically in a liquid dielectric of permittivity $\varepsilon$ and mass density $\rho_m$. The distance between the electrodes is $d$ and the dielectric above the liquid is air (Fig. 3.G18). Determine $h$, the rise in the level of the liquid dielectric between the plates when these are connected to a source of potential difference $V$. Ignore the fringing effect and other side effects.

*Solution.* The electric field intensity is the same in air and in the dielectric, $E = V/d$. The pressure acting on the surface of the dielectric between the plates according to Eq. (3.6.3) is

$$p = \tfrac{1}{2}(\varepsilon - \varepsilon_0)\, E^2 = \tfrac{1}{2}(\varepsilon - \varepsilon_0)\,\frac{V^2}{d^2},$$

and is directed upward. The total electric force acting on the dielectric surface is

$$F = pda = \tfrac{1}{2}(\varepsilon - \varepsilon_0)\,\frac{aV^2}{d}, \tag{a}$$

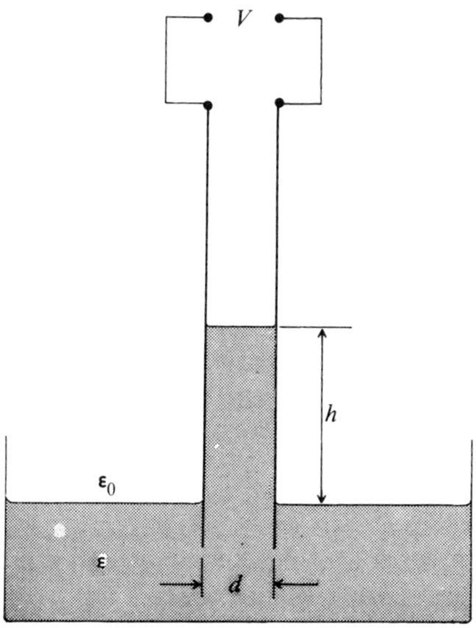


Figure 3.G18

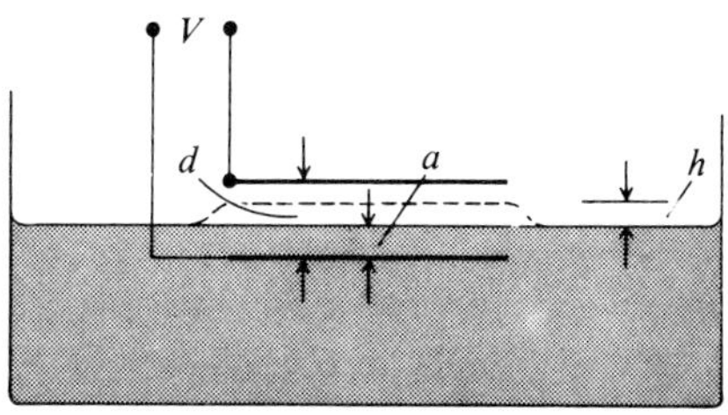


Figure 3.G19

$a$ denoting the width of the plates. In the stationary state this force is counterbalanced by the weight $F_g$ of the dielectric above the outer level,

$$F_g = \rho_m adhg, \tag{b}$$

$g$ representing the strength of the earth gravitational field ($g = 9.81$ m/s$^2$). Equating (a) and (b) we obtain

$$h = \frac{(\varepsilon - \varepsilon_0)V^2}{2d^2\rho_m g}. \tag{c}$$

**3.G19.** One of the electrodes of a parallel-plate capacitor is immersed in a liquid dielectric of unknown permittivity and mass density $\rho_m$ and the other electrode is in the air above the surface of the dielectric (Fig. 3.G19). The distance between the plates is $d$ and the depth of the dielectric above the immersed plate is $a$. When the capacitor is connected to a source of potential difference $V$, the level of the liquid between the plates is for $h$ higher than the level outside the capacitor. Determine the permittivity of the liquid dielectric.

*Solution.* When the electrodes are connected to a source of potential difference $V$, the surface of the liquid dielectric between the plates is acted on by an upward pressure. The intensity of this pressure, according to Eq. (3.6.4), is

$$p = \frac{1}{2}\left(\frac{1}{\varepsilon_0} - \frac{1}{\varepsilon}\right) D^2 = \frac{1}{2}\,\frac{\varepsilon - \varepsilon_0}{\varepsilon_0\varepsilon}\, D^2 = \tfrac{1}{2}(\varepsilon - \varepsilon_0)\frac{\varepsilon_0}{\varepsilon}\, E^2, \tag{a}$$

where $E$ is the electric field intensity in the air. The level of the dielectric will be pushed upward, until the total force on the surface of the dielectric is counterbalanced by the weight of the liquid above the surrounding level. If $S$ is the surface of the plates this condition yields

$$pS = \rho_m hSg. \tag{b}$$

Substituting (a) into (b), we obtain

$$\tfrac{1}{2}(\varepsilon - \varepsilon_0)\frac{\varepsilon_0}{\varepsilon}\, E^2 = \rho_m hg. \tag{c}$$

Since $V = E(d - a - h) + E'(a + h)$, $E'$ denoting the electric field intensity in the liquid dielectric, and $\varepsilon_0 E = \varepsilon E'$, the electric field intensity $E$ in the air is

$$E = \frac{\varepsilon V}{\varepsilon(d - a - h) + \varepsilon_0(a + h)}.$$

Equation (c) thus becomes

$$\tfrac{1}{2}(\varepsilon - \varepsilon_0)\frac{\varepsilon_0\varepsilon V^2}{[\varepsilon(d - a - h) + \varepsilon_0(a + h)]^2} = \rho_m hg.$$

Solving this equation for $\varepsilon$ we find

$$\varepsilon = \varepsilon_0\frac{(2rst + 1) + \sqrt{(2rst + 1)^2 + 4rt^2(1 - rs^2)}}{2(1 - rs^2)},$$

where $r = 2\rho_m gh/\varepsilon_0 V^2$, $s = d - a - h$, and $t = a + h$.

**3.G20.** Solve Problems 3.G13 and 3.G14 starting from the expression for pressure acting on the surface of the dielectric.

**3.G21.** One branch of a U-shaped dielectric tube filled with a liquid dielectric of unknown permittivity is situated between the plates of a parallel-plate capacitor. The potential difference between the capacitor plates is $V$ and the distance is $d$. The cross-section of the U-tube is a very thin rectangle, with the larger side parallel to the electric field intensity vector in the charged capacitor. The dielectric in the tube above the liquid dielectric is air, and the density of mass of the liquid dielectric is $\rho_m$. Assuming that $h$ is the measured difference between the levels of the liquid dielectric in the two branches of the U-tube, determine the permittivity of the dielectric.

*Answer.* For $V = 10{,}000$ V, $d = 1$ cm, $\rho_m = 0.9$ gr/cm$^3$ and $h = 1$ mm, the relative permittivity $\varepsilon_r \simeq 3$.

## Numerical problems

**3.N1.** An isolated conducting sphere of radius $a = 10$ cm is situated in air and is charged with a charge $Q = 10^{-10}$ C. Determine the energy used for charging the sphere.

**3.N2.** A capacitor of capacitance $C = 10{,}000$ pF is connected to a source of potential difference $V = 1000$ V. Determine the energy stored in the capacitor.

**3.N3.** As we noted in Section 2.8.1, the highest possible field intensity in air is approximately $E_{\max} = 3 \times 10^6$ V/m. Determine the corresponding energy density located in the field. Compare this with the density of chemical energy of a cumbustible gas and air mixed, with an energy density of 0.5 J/cm$^3$.

**3.N4.** Calculate the pressure on the conducting sphere of Problem 3.N1.

**3.N5.** Calculate the force per unit length of the conductors of a two-wire line situated in air, if the conductor radii are $R = 1$ mm, the distance between them $D = 1$ m, and the potential difference between the conductors $V = 10{,}000$ V.

**3.N6.** If an isolated conducting sphere of radius $a = 5$ cm is situated in air, what is the total force on one-half of the sphere corresponding to the maximal possible charge on it?

**3.N7.** A parallel-plate capacitor with dielectric air has plates of area $S = 100$ cm$^2$ and the distance between them $d = 1$ mm. Determine the maximum possible force on the plates of the capacitor due to the charge on them.

**3.N8.** Calculate the force on the inner cable conductor of Problem 3.G12, for $V = 1000$ V, $\varepsilon = 3\varepsilon_0$, and $b/a = e = 2.718\ldots$.

**3.N9.** Calculate the force on the dielectric piston of Problem 3.G13 for the same values as in Problem 3.N8.

**3.N10.** If, in Problem 3.G15, $C_l = 1$ pF/cm, $V = 1000$ V and $dr/d\theta = -5$ cm/radian, what is the torque $T$ on the pointer?

**3.N11.** Calculate the pressure on the surface of the dielectric and the rise of the level of the dielectric in Problem 3.G18, given that $\varepsilon = 3\varepsilon_0$, $V = 1000$ V, $d = 1$ mm, and $\rho_m = 0.9$ gr/cm$^3$ (i.e., 900 kg/m$^3$).

**Suggested reading.** The list given here progresses from simpler to more advanced texts. For full bibliographical description, see the Bibliography.

1. CARTER, G. W., *The Electromagnetic Field in Its Engineering Aspects*, Chapter 4.
2. CHESTON, W. B., *Elementary Theory of Electric and Magnetic Fields*, Chapter 4.
3. ABRAHAM, M., and R. BECKER, *The Classical Theory of Electricity and Magnetism*, Chapter V.

CHAPTER 4

# ANALYTICAL METHODS OF SOLVING ELECTROSTATIC PROBLEMS

## 4.1 INTRODUCTION

To solve an electrostatic problem usually implies determining the electrostatic potential $V$ at all points, since the electric field intensity and all other quantities of interest may then be determined easily.

There are several typical types of problem concerned with known initial quantities. The first type is that in which the distribution of free and possible polarization charges at all points is known in advance. The procedure for finding the potential in these cases is almost straightforward, but unfortunately few practical problems are of this type. Problems solved in Chapters 1 and 2 using Gauss' law fall into this category.

Often we encounter a different type of problem: the distribution of charge is known in advance at some points and on some conducting bodies, while the other conducting bodies are known either to carry a certain total charge or to be at a given potential. Problems of this type may be attacked in some important cases by what is called the method of images, which is explained in the following section.

Finally, the type of problem most often encountered in practice is that in which the potentials of all conducting bodies are known. In this case the solution reduces to determining the particular solution of Laplace's equation which satisfies the prescribed boundary conditions, i.e. which has the prescribed (constant) values on the surfaces of conducting bodies. Except in some very simple cases (three of them are discussed in Problems 1.G38 to 1.G40) solution of these, known as *boundary value problems*, is usually rather intricate mathematically. One of a few existing analytical methods for their solution is presented in Section 4.3.

## 4.2 THE METHOD OF ELECTRIC IMAGES

**4.2.1 Introduction.** Speaking quite generally, we can say that the method of electric images implies determining an *imaginary* distribution of charges *inside* a conducting body of an electrostatic system, which would produce *outside* the body exactly the same field as that produced by the actual free charges over its surface. For example, consider a point charge $Q$ placed near a conducting body, as shown in Fig. 4.1. Suppose that in some manner we concluded that the electric field at all points at the surface of the conducting body would be the same as before, that is, perpendicular to the surface, if we removed the body and introduced a certain

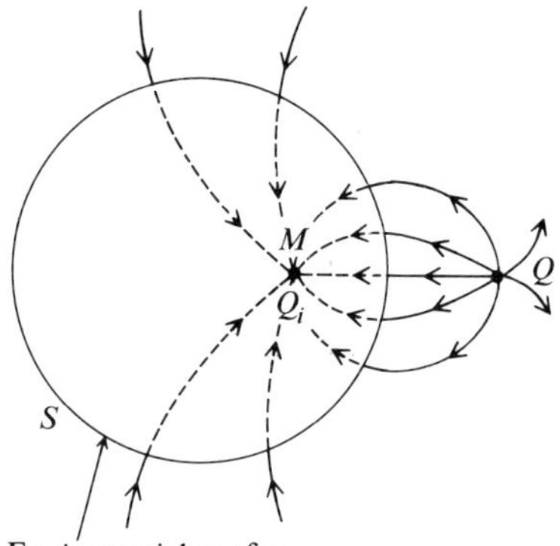


**Fig. 4.1** An illustration of the method of images.

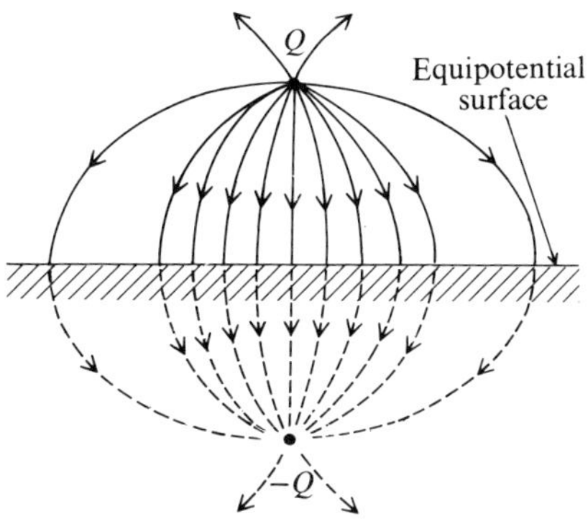


**Fig. 4.2** Image charge of the point charge $Q$ near a plane conducting surface.

charge $Q_i$ at a point $M$ inside the volume occupied earlier by the body. By virtue of the uniqueness theorem, the electric field outside the imaginary surface $S$ (which earlier was the surface of the conducting body) must be exactly the same as before. For boundary conditions on $S$ are the same in both cases (the electric field intensity is perpendicular to $S$ at all points, i.e. $S$ is equipotential), and the only existing charge outside is $Q$. The very complicated problem of finding the exact distribution of induced charges on the surface of the conductor, and of the electric field due to these charges, is hence reduced to a much simpler problem of determining the field due to a single point charge $Q_i$. Charge $Q_i$, together with its position at $M$, is referred to as the *electric image* of charge $Q$ in the surface $S$ of the conductor.

If the influence of the induced charges cannot be reduced to the field of a single charge, but rather to a certain number of charges, or even to an equivalent volume charge distribution inside $S$, these charges are again referred to as the image charges. But when the image charges are not in the form of some simple charge distribution, the method of images loses its main advantage—its simplicity. For this reason it can be applied in practice only if the conductor or dielectric boundaries are of a simple geometrical form, such as plane, cylindrical, or spherical. These three cases will be considered in turn.

**4.2.2 Images in the case of plane conducting surface.** Let us first consider a single point charge $Q$ situated above a conducting plane (Fig. 4.2). The image charge of the charge $Q$ in the conducting plane is a point charge $-Q$ placed with respect to the plane symmetrically to $Q$. For, if the plane is removed and $-Q$ introduced instead (Fig. 4.2), the field *above* the plane is the same as before (the field in reality does not exist below the plane). Consequently, the field of all the charges induced on the conducting plane may be reduced to the field of a single charge $-Q$, symmetrical to $Q$ with respect to the conducting plane. The same reasoning applies to any given distribution of charges above a conducting plane, as illustrated in Fig. 4.3.

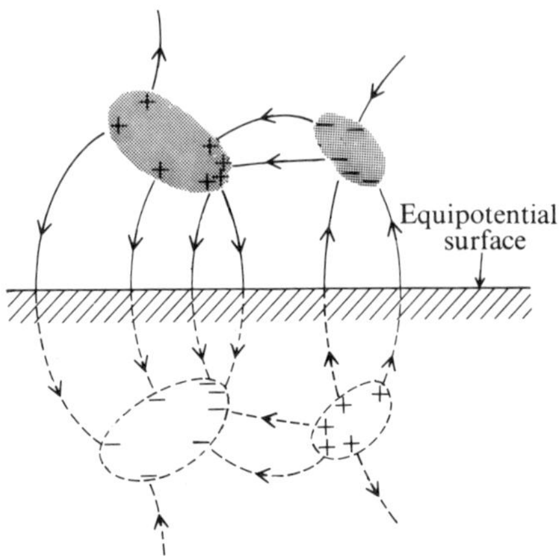


**Fig. 4.3** Image charges of a system of charges near a plane conducting surface.

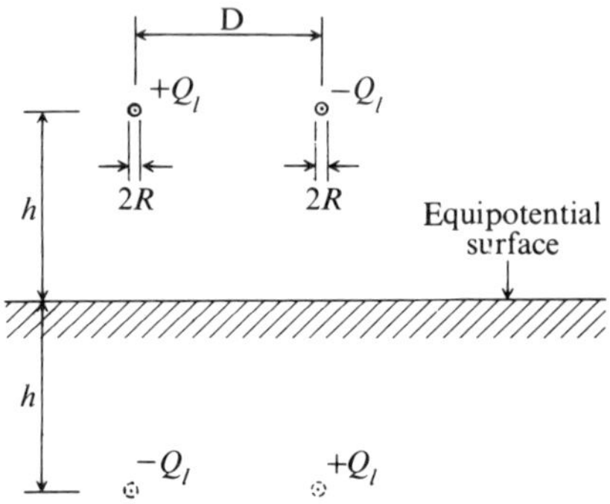


**Fig. 4.4** A two-wire line running parallel to the conducting surface of the earth and its image.

One of the most important practical applications of the method of images in the case of a plane conducting surface is that of charged bodies above the surface of the earth, regarded as being approximately plane. For example, let us consider a two-wire line running parallel to the surface of the earth (Fig. 4.4), and suppose that both $h$ and $D$ are much larger than $R$ (which is usually the case). We wish to determine the capacitance of the line, per unit length, *in the presence of the earth's surface.* Following the pattern for determining the capacitance described in connection with Eq. (1.14.3), we suppose first that charges $+Q_l$ and $-Q_l$ exist per unit length of the two conductors. But these charges induce certain charges on the earth's surface, which have some influence on the potential difference between the two conductors. The influence of these induced charges (of unknown distribution) can be reduced to that of images of the two charged conductors in the surface of the earth, as indicated in Fig. 4.4. The required capacitance of the line per unit length can now be easily determined (see Problem 4.G8). The method obviously is also applicable if the conductors are at different heights (see Problem 4.G10).

Let us now consider a point charge $Q$ placed near two conducting half-planes, perpendicular to one another (Fig. 4.5). If the conducting half-planes are removed and the three image charges shown introduced instead, the surface where the conducting half-planes were remains equipotential. So the field in that part of space where it really exists (in the quadrant with charge $Q$) is the same as before.

**4.2.3 Images in a conducting cylinder.** Consider a line charge $\rho_l$ running parallel to a conducting cylinder of radius $R$ and carrying a charge $Q_l = -\rho_l$ per unit length. The cross-section of the system is shown in Fig. 4.6. If the electric field of charges induced on the surface of the cylinder can be reduced to the field of a certain image line charge $\rho_l'$, by symmetry the image line charge $\rho_l'$ must be somewhere in the plane of the line charge $\rho_l$ and the axis of the cylinder, as indicated in

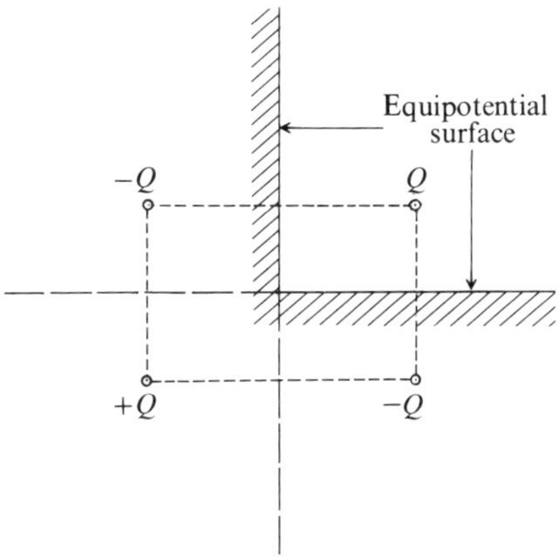


**Fig. 4.5** Image charges of a point charge in a 90° conducting corner.

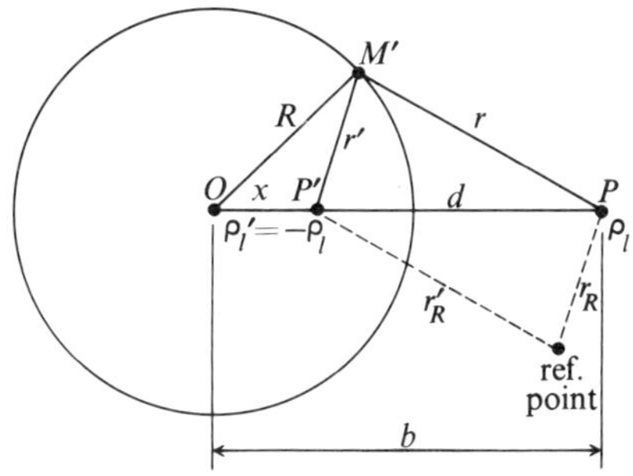


**Fig. 4.6** The image of the line charge $\rho_l$ running parallel to a conducting circular cylinder is the line charge $-\rho_l$.

Fig. 4.6. The magnitude of the charge $\rho_l'$ and its position $x$ with respect to the cylinder axis are unknown. If possible, we must determine these so that the surface of the cylinder is equipotential.

Now, according to Gauss' law, the total charge enclosed by a cylinder coinciding with the conductor surface equals the flux of the electric field intensity through the cylinder, multiplied by $\varepsilon_0$. If the field is assumed to be identical in the real case and that, when the cylinder is removed and the image line charge introduced instead, the flux of the electric field intensity is also the same, it follows that $\rho_l'$ must be equal to $Q_l$, that is to $-\rho_l$, as indicated in Fig. 4.6.

According to Eq. (1.8.12), the potential at a point $M'$ at a distance $r$ from a line charge $\rho_l$, with respect to a reference point at a distance $r_R$ from the line charge, is

$$V = \int_r^{r_R} \frac{\rho_l}{2\pi\varepsilon_0 r}\,dr = \frac{\rho_l}{2\pi\varepsilon_0} \ln \frac{r_R}{r}. \tag{4.2.1}$$

The distance $r_R$ cannot be taken to be infinite, since in this case there *are* charges at infinity. If we denote by $r_R'$ and $r'$ the corresponding distances from the line charge $\rho_l' = -\rho_l$, then the potential at the same point due to $\rho_l'$ is

$$V' = \frac{\rho_l'}{2\pi\varepsilon_0} \ln \frac{r_R'}{r'} = -\frac{\rho_l}{2\pi\varepsilon_0} \ln \frac{r_R'}{r'}.$$

So the total potential at $M'$ is

$$V_{\text{total}} = V + V' = \frac{\rho_l}{2\pi\varepsilon_0}\left(\ln \frac{r_R}{r} - \ln \frac{r_R'}{r'}\right) = \frac{\rho_l}{2\pi\varepsilon_0} \ln \left(\frac{r_R}{r_R'}\frac{r'}{r}\right). \tag{4.2.2}$$

From (4.2.2) it is clear that the equation of *any* equipotential surface due to the two equal line charges of opposite signs is $r'/r$ = constant. Hence, if the distance $x$ of the image line charge $\rho_l'$ from the axis of the cylinder can be determined so that $r'/r$ = constant for all points on the surface of the cylinder, the image line charge is completely determined.

From Fig. 4.6, if triangles $OM'P$ and $OP'M'$ are similar (if $x$ is chosen appropriately, this can always be the case for a fixed point $M'$) then

$$r'/r = R/b = x/R.$$

[$b$ stands for $(x + d)$, that is, for the distance from the line charge $\rho_l$ to the cylinder axis.] Thus, equation $r'/r$ = constant can be satisfied for any point on the cylinder surface if

$$x = R^2/b. \tag{4.2.3}$$

The image line charge $\rho'_l$ of the line charge $\rho_l$ is therefore completely determined.

Now, by symmetry, the cylindrical surface of radius $R$, centered at a distance $x$ to the *right* of the line charge $\rho_l$ is also equipotential (Fig. 4.7). Imagine that this cylindrical surface is made conductive, say by covering it with a very thin conducting foil. The boundary condition requiring that the surface of the conducting foil is equipotential is satisfied, because the cylindrical surface was equipotential before the foil was introduced. The only difference will be that inside and outside the cylindrical foil charges will be induced, of total magnitude per unit length $-\rho_l$ and $\rho_l$ respectively. If we now imagine that the charges inside the cylinder are allowed to cancel out, nothing will be changed outside the cylinder. We are therefore left with the important practical case of two parallel circular cylinders carrying equal charges of opposite signs (a two-wire line with wire radius comparable to the distance between their axis). The electric field outside two such conductors is therefore the same as the field of the two line charges situated at $P$ and $P'$.

This fact enables us to make an easy determination of the capacitance per unit length of the two cylinders. To determine the potential difference between the cylinders corresponding to charges $Q_l$ and $-Q_l$ per unit length, it suffices to determine the potential difference between any two points on their surfaces, since these

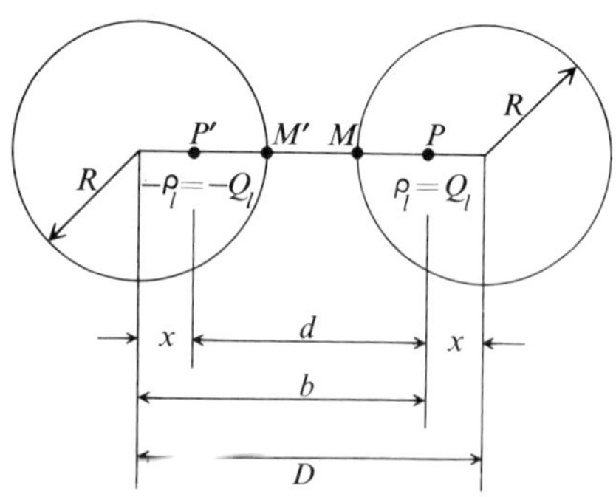


**Fig. 4.7** Electric field of two oppositely charged conducting cylinders can be reduced to the field of two line charges inside them.

surfaces are equipotential. Choosing for simplicity points $M$ and $M'$, from Eq. (4.2.2) and the notation shown in Figs. 4.7 and 4.6, it follows that

$$V_M - V_{M'} = \frac{Q_l}{2\pi\varepsilon_0} \ln \frac{r_R(b - R)}{r'_R(R - x)} - \frac{Q_l}{2\pi\varepsilon_0} \ln \frac{r_R(R - x)}{r'_R(b - R)} = \frac{Q_l}{\pi\varepsilon_0} \ln \frac{b - R}{R - x},$$

from which the capacitance per unit length is

$$C_l = \frac{Q_l}{V_M - V_{M'}} = \frac{\pi\varepsilon_0}{\ln\left[(b - R)/(R - x)\right]}. \tag{4.2.4}$$

From Fig. 4.7 and Eq. (4.2.3), $b = D - x = D - R^2/b$. Solving this for $b$ we find

$$b = D/2 + \sqrt{(D/2)^2 - R^2}. \tag{4.2.5}$$

Substituting this value of $b$ into Eq. (4.2.4), after some rearrangements we obtain the capacitance $C_l$ per unit length between the two conductors, expressed in terms of the data usually given, the radius $R$ of the conductors and the distance $D$ between their axes:

$$C_l = \frac{\pi\varepsilon_0}{\ln\left[(D/2R) + \sqrt{(D/2R)^2 - 1}\right]}. \tag{4.2.6}$$

We can readily see that, if $D \gg R$, Eq. (4.2.6) becomes Eq. (a) of Problem 1.G50.

**4.2.4 Images in a conducting sphere.** Consider a point charge $Q$ at a distance $b$ from the center of a conducting sphere of radius $R$ and at potential zero (Fig. 4.8). Some charges are obviously induced on the outer surface of the sphere. The system is symmetrical with respect to the line $OQ$. If the electric field of these induced charges can be reduced to a single point charge $Q'$ inside the sphere, the sphere being removed, $Q'$ must be on this line (Fig. 4.8). Both $Q'$ and its distance $x$ from the center of the sphere are as yet unknown.

If the real sphere is removed and the influence of the charges induced on its surface replaced by the image $Q'$ of the charge $Q$, the (now geometrical) surface of the

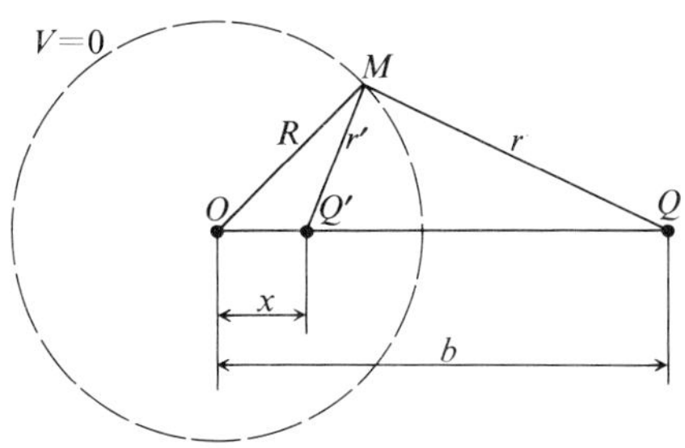


**Fig. 4.8** Image of a point charge in a conducting sphere at zero potential.

sphere must remain at potential zero. Otherwise the boundary conditions are not the same in the two cases. It follows that

$$\frac{Q}{4\pi\varepsilon_0 r} + \frac{Q'}{4\pi\varepsilon_0 r'} = 0, \qquad \text{whence} \qquad Q' = -Q\frac{r'}{r}. \tag{4.2.7}$$

If this equation is to be valid for any $r'/r$, this ratio must be a constant. This is true if triangles $OMQ$ and $OQ'M$ are similar, since then

$$r'/r = x/R = R/b. \tag{4.2.8}$$

From Eqs. (4.2.7) and (4.2.8) we can see that both the magnitude and position of the image charge $Q'$ can be determined:

$$Q' = -QR/b, \tag{4.2.9}$$

$$x = R^2/b. \tag{4.2.10}$$

Suppose now that the sphere is at some given potential $V$ rather than at potential zero. In that case, the images of $Q$ in the sphere are its image $Q'$, defined by Eqs. (4.2.9) and (4.2.10), and a point charge $Q'' = 4\pi\varepsilon_0 RV$ at the center of the sphere. The image $Q'$ together with $Q$ makes the sphere at potential zero, and the image $Q''$ adjusts its potential to the desired value.

Suppose that, instead of its potential, the total charge $Q_s$ on the sphere is given. The field outside the sphere should be the same when the charge on its surface is replaced by a few point charges inside it. According to Gauss' law, the total charge of all images must be exactly equal to $Q_s$. The potential produced by them and by the external charge $Q$ must be constant over the surface of the sphere. This can be attained if the charge $Q'$ (given by Eq. 4.2.9) is placed at the distance $x$, Eq. (4.2.10), from the center of the sphere, and a charge $Q'' - Q_s - Q'$ at the center of the sphere.

These results can be used for solving several important practical problems: a conducting (uncharged or charged) sphere in a uniform external electric field (see Problem 4.G17), the approximate determination of capacitance and the electric field of two metal spheres charged with equal charges of opposite signs (the "sparker"; see Problem 4.G15), etc.

*Problems referring to Section 4.2:* 4.T1 to 4.T8, 4.G1 to 4.G17, and 4.N1 to 4.N8.

## 4.3 DIRECT SOLVING OF LAPLACE'S EQUATION

**4.3.1 Introduction.** Suppose that the potentials of all conductors of a system are given, and that there are no free charges in the dielectric surrounding the conductors. To determine the potential at all points it is necessary to find a function which satisfies Laplace's equation and has the prescribed constant values on the surfaces of the conductors.

Three of the simplest cases of this type are solved in Problems 1.G38 to 1.G40. They are all characterized by the potential being a function of one coordinate only of an appropriately chosen coordinate system. The general form of Laplace's equation, a partial differential equation, reduces to an ordinary differential equation, but, unfortunately, such cases are very few in number. In the majority of problems it is necessary to find a solution to the general form of Laplace's equation, subject to some prescribed boundary conditions.

From the mathematical point of view this is a very difficult class of problem. Certain problems can be solved by using a special technique, known as the method of separation of variables. According to this method, the potential function is assumed to be in the form of a product of functions of *one* of (in the general case) three coordinates. It can be shown that, in eleven coordinate systems, this yields three ordinary differential equations instead of Laplace's partial differential equation. Solutions to most of these equations are known. We have to combine these solutions so as to satisfy the boundary conditions of the problem at hand. If we succeed, by virtue of the uniqueness theorem, the solution obtained is the right one.

The separation of variables in Laplace's equation may be performed easily in rectangular, cylindrical, and spherical coordinate systems. In the other eight coordinate systems, the mathematics involved is more difficult, but the ideas underlying the method are the same. We shall illustrate the method in connection with the three aforementioned coordinate systems only.

**4.3.2 Separation of variables in a rectangular coordinate system.** According to Eq. (1.12.4), Laplace's equation with respect to a rectangular coordinate system has the form

$$\frac{\partial^2 V}{\partial x^2} + \frac{\partial^2 V}{\partial y^2} + \frac{\partial^2 V}{\partial z^2} = 0. \tag{4.3.1}$$

Following the procedure of the method of separation of variables outlined above, let us assume that $V(x, y, z)$ can be represented in the form

$$V(x, y, z) = X(x)\, Y(y)\, Z(z), \tag{4.3.2}$$

where $X(x)$ is a function of $x$ only, $Y(y)$ of $y$ only, and $Z(z)$ of $z$ only. Substituting this form of the potential function $V(x, y, z)$ into Eq. (4.3.1), we have

$$Y(y)\, Z(z) \frac{d^2 X(x)}{dx^2} + X(x)\, Z(z) \frac{d^2 Y(y)}{dy^2} + X(x)\, Y(y) \frac{d^2 Z(z)}{dz^2} = 0.$$

If we divide this equation by $X(x)\, Y(y)\, Z(z)$, it becomes

$$\frac{1}{X(x)} \frac{d^2 X(x)}{dx^2} + \frac{1}{Y(y)} \frac{d^2 Y(y)}{dy^2} + \frac{1}{Z(z)} \frac{d^2 Z(z)}{dz^2} = 0. \tag{4.3.3}$$

Ordinary derivatives are used instead of partial derivatives, because $X(x)$, $Y(y)$, and $Z(z)$ are functions of $x$, $y$, and $z$ only.

Each of the three terms on the left of Eq. (4.3.3) is a function of only one coordinate. Since Eq. (4.3.3) must be satisfied for *all* values of $x$, $y$, and $z$, all three must be equal to a constant. To understand this, let us suppose that Eq. (4.3.3) is satisfied for a particular set of values of $x$, $y$, and $z$. If we now keep $x$ and $y$ constant and change $z$, the third term in Eq. (4.3.3) will generally change. The equation would be satisfied for *all* $x$, $y$, and $z$ only if the third term is *not* changed when $z$ is changed, that is, unless it is a constant. The same argument applies to the first two terms.

Hence, Eq. (4.3.3) represents in fact three ordinary differential equations:

$$\frac{1}{X(x)}\frac{d^2X(x)}{dx^2} = k_x^2, \tag{4.3.4}$$

$$\frac{1}{Y(y)}\frac{d^2Y(y)}{dy^2} = k_y^2, \tag{4.3.5}$$

$$\frac{1}{Z(z)}\frac{d^2Z(z)}{dz^2} = k_z^2, \tag{4.3.6}$$

with constants $k_x^2$, $k_y^2$, and $k_z^2$ satisfying the condition

$$k_x^2 + k_y^2 + k_z^2 = 0. \tag{4.3.7}$$

From this we can see that at least one of the three constants must be negative.

The solutions of Eqs. (4.3.4) to (4.3.6) are

$$X(x) = A_x e^{k_x x} + B_x e^{-k_x x}, \quad \text{or} = A_x \sin k_x x + B_x \cos k_x x \quad \text{if} \quad k_x^2 < 0, \tag{4.3.8}$$

$$Y(y) = A_y e^{k_y y} + B_y e^{-k_y y}, \quad \text{or} = A_y \sin k_y y + B_y \cos k_y y \quad \text{if} \quad k_y^2 < 0, \tag{4.3.9}$$

$$Z(z) = A_z e^{k_z z} + B_z e^{-k_z z}, \quad \text{or} = A_z \sin k_z z + B_z \cos k_z z \quad \text{if} \quad k_z^2 < 0. \tag{4.3.10}$$

This can be proved by substitution. $A_x, B_x, \ldots, B_z$ represent arbitrary constants, which have to be determined, together with constants $k_x$, $k_y$, and $k_z$ so that the boundary conditions of the problem considered are satisfied.

Usually one function of the above type cannot alone be made to satisfy all prescribed boundary conditions. But the sum of any number of these functions is again a solution of Esq. (4.3.4) to (4.3.6), i.e. to Laplace's equation. Therefore we try to find the desired solution in the form of a sum (finite or infinite) of solutions (4.3.8) through (4.3.10), with appropriately determined constants. We shall illustrate this process by an example.

Consider a two-dimensional problem, such as shown in Fig. 4.9. The two parallel plates at potential zero extend to the right to infinity.

For $y = 0$ and $y = a$ the potential is zero. Since $e^{\pm k_y y}$ cannot be zero, we must take the solution of Eq. (4.3.9) of $Y(y)$ in the form of trigonometric functions. Next, since in our case $k_z = 0$, it follows from Eq. (4.3.7) that

$$k_x^2 = -k_y^2 = k^2 \qquad (k\text{—a real number}). \tag{4.3.11}$$

We can now write solution (4.3.2) in the form

$$V(x, y) = (A_x e^{kx} + B_x e^{-kx})(A_y \sin ky + B_y \cos ky). \tag{4.3.12}$$

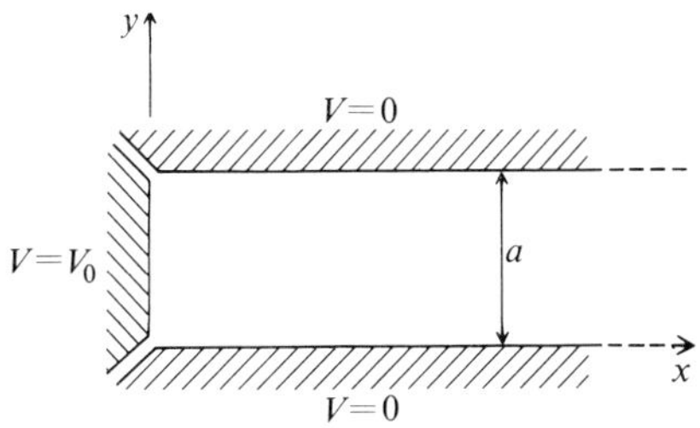


**Fig. 4.9** A two-dimensional potential problem.

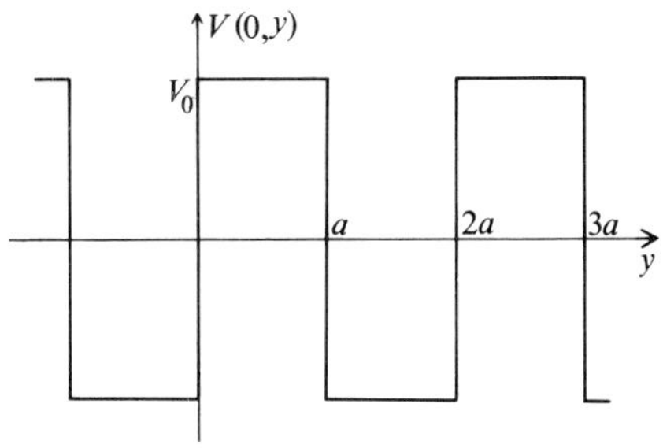


**Fig. 4.10** $V(0,y)$ of the system of Fig. 4.9 is represented by the section of the periodic function shown in the region from 0 to $a$.

For $x \to \infty$, $V$ obviously tends to zero. This can be only if $A_x = 0$. From the other side, since for $y = 0$ we have $V(x, 0) = 0$, it follows that $B_y = 0$. The condition that $V(x, a) = 0$ is then satisfied if $ka = n\pi$, $n = 1, 2, \ldots$, or $k = n\pi/a$, $n = 1, 2, \ldots$. The solution, Eq. (4.3.12), subject to the boundary conditions at infinity and at $y = 0$ and $y = a$, therefore becomes

$$V_n(x, y) = C_n e^{-(n\pi/a)x} \sin (n\pi/a)y, \qquad n = 1, 2, \ldots \tag{4.3.13}$$

(Constant $A_xA_y$ is shortened into a new constant $C_n$.) Consequently, an infinite number of functions $V_n$, $n$ being any positive integer, satisfies the boundary conditions for $y = 0$, $y = a$, and $x = \infty$. But it is obvious that none of them satisfies the last boundary condition, that $V = V_0$ for $x = 0$, and for any $y$ from zero to $a$.

Since all functions of Eq. (4.3.13) are solutions to Laplace's equation, their sum is also a solution. Let us try to determine constants $C_n$, $n = 1, 2, \ldots$, so that

$$V(x, y) = \sum_{n=1}^{\infty} V_n(x, y) = \sum_{n=1}^{\infty} C_n e^{-(n\pi/a)x} \sin (n\pi/a)y$$

$$= \begin{cases} V_0 \text{ for } x = 0 \text{ and } 0 < y < a \\ 0 \text{ for } x = 0, \text{ and } y = 0 \text{ and } y = a \end{cases} \tag{4.3.14}$$

i.e. that

$$V(0, y) = \sum_{n=1}^{\infty} C_n \sin (n\pi/a)y = \begin{cases} V_0 \text{ for } 0 < y < a \\ 0 \text{ for } y = 0 \text{ and } y = a \end{cases} \tag{4.3.15}$$

This question may be solved using basic concepts of trigonometric or Fourier series. What we need in this case is the Fourier series for the function shown in Fig. 4.10. For $0 < y < a$ it represents precisely the function $V(0, y)$ specified by Eq. (4.3.15). To obtain coefficients $C_n$, let us multiply Eq. (4.3.15) by

$\sin(m\pi/a)y\,dy$, where $m$ is a positive integer, and then integrate the expression thus obtained from 0 to $2a$:

$$\sum_{n=1}^{\infty} C_n \int_0^{2a} \sin\frac{n\pi}{a}y \sin\frac{m\pi}{a}y\,dy = \int_0^{2a} V(0, y) \sin\frac{m\pi}{a}y\,dy. \tag{4.3.16}$$

Now,

$$\int_0^{2a} \sin\frac{n\pi}{a}y \sin\frac{m\pi}{a}y\,dy = \frac{1}{2}\int_0^{2a}\left[\cos\frac{(n-m)\pi}{a}y - \cos\frac{(n+m)\pi}{a}y\right]dy$$
$$= \begin{cases} 0 & \text{for} \quad m \neq n \\ a & \text{for} \quad m = n \end{cases} \tag{4.3.17}$$

so that

$$C_n = \frac{1}{a}\int_0^{2a} V(0, y) \sin\frac{n\pi}{a}y\,dy. \tag{4.3.18}$$

In the case of $V(0, y)$ represented in Fig. 4.10, Eq. (4.3.18) yields

$$C_n = \frac{V_0}{a}\int_0^{a} \sin\frac{n\pi}{a}y\,dy + \frac{-V_0}{a}\int_a^{2a} \sin\frac{n\pi}{a}y\,dy = \frac{1 + \cos 2n\pi - 2\cos n\pi}{n\pi}V_0$$

or

$$C_{2n+1} = \frac{4V_0}{(2n+1)\pi}, \qquad C_{2n} = 0, \qquad n = 0, 1, 2, \ldots \tag{4.3.19}$$

The coefficients $C_n$ in Eq. (4.3.15) are thus determined so that Eq. (4.3.15) is satisfied. The final potential function (4.3.14) is given by

$$V(x, y) = \frac{4V_0}{\pi}\sum_{n=0}^{\infty}\frac{\sin\frac{(2n+1)\pi}{a}y}{2n+1}e^{-\frac{(2n+1)\pi}{a}x} \tag{4.3.20}$$

**4.3.3 Separation of variables in a cylindrical coordinate system.*** According to the result of Problem 1.G36, Laplace's equation in a cylindrical coordinate system has the form

$$\frac{1}{r}\frac{\partial}{\partial r}\left(r\frac{\partial V}{\partial r}\right) + \frac{1}{r^2}\frac{\partial^2 V}{\partial\phi^2} + \frac{\partial^2 V}{\partial z^2} = 0. \tag{4.3.21}$$

Following the procedure outlined for separating the variables, let us assume that $V(r, \phi, z)$ can be represented in the form

$$V(r, \phi, z) = R(r)\,\Phi(\phi)\,Z(z). \tag{4.3.22}$$

* This section may be omitted without loss of continuity.

Substituting Eq. (4.3.22) into Eq. (4.3.21) and dividing the equation thus obtained by $R(r)\,\Phi(\phi)\,Z(z)$, we have

$$\frac{1}{rR(r)}\frac{d}{dr}\left[r\frac{dR(r)}{dr}\right]+\frac{1}{r^2\Phi(\phi)}\frac{d^2\Phi(\phi)}{d\phi^2}+\frac{1}{Z(z)}\frac{d^2Z(z)}{dz^2}=0. \tag{4.3.23}$$

By the use of the same argument as we used in connection with Eq. (4.3.3), this equation can be satisfied for *any z* only if the last term on the left side is a constant, or

$$\frac{1}{Z(z)}\frac{d^2Z(z)}{dz^2}=k_z^2. \tag{4.3.24}$$

With Eq. (4.3.24), Eq. (4.3.23) becomes

$$r^2\left\{\frac{1}{rR(r)}\frac{d}{dr}\left[r\frac{dR(r)}{dr}\right]+k_z^2\right\}+\frac{1}{\Phi(\phi)}\frac{d^2\Phi(\phi)}{d\phi^2}=0. \tag{4.3.25}$$

Following the same line of reasoning, we find that the second term in this equation must be a constant,

$$\frac{1}{\Phi(\phi)}\frac{d^2\Phi(\phi)}{d\phi^2}=-k_\phi^2. \tag{4.3.26}$$

The minus sign is necessary because $\Phi(\phi)$ can only be a periodic function of coordinate $\phi$, that is, $\Phi(\phi+2n\pi)=\Phi(\phi)$. Equation (4.3.25) thus becomes

$$\frac{1}{rR(r)}\frac{d}{dr}\left[r\frac{dR(r)}{dr}\right]-\frac{k_\phi^2}{r^2}+k_z^2=0. \tag{4.3.27}$$

The solutions of Eqs. (4.3.24) and (4.3.26) are of the form

$$Z(z)=A_ze^{k_zz}+B_ze^{-k_zz} \tag{4.3.28}$$

$$\Phi(\phi)=A_\phi\sin k_\phi\phi+B_\phi\cos k_\phi\phi. \tag{4.3.29}$$

After we have performed the indicated differentiation and made some simple rearrangements, Eq. (4.3.27) becomes

$$\frac{d^2R(r)}{dr^2}+\frac{1}{r}\frac{dR(r)}{dr}+\left(k_z^2-\frac{k_\phi^2}{r^2}\right)R(r)=0. \tag{4.3.30}$$

This is known as *Bessel's differential equation*. Its solutions, which cannot be expressed in the form of a finite series of elementary functions, are called *Bessel functions*. A detailed account of the Bessel equation and of Bessel functions may be found elsewhere.* Here it is sufficient to mention that the values of Bessel functions are tabulated in extensive tables, like the values of sine and cosine functions.

---

* See, for example, W. R. Smythe, *Static and Dynamic Electricity*, McGraw-Hill, 1968 (third edition).

Thus no difficulties whatsoever appear when we have to solve numerically a problem involving Bessel functions.

A particular, but common, case in practice is when the potential is not a function of $z$. In that case we have to put $k_z = 0$. The Bessel equation (4.3.30) then reduces to

$$r^2 \frac{d^2R(r)}{dr^2} + r\frac{dR(r)}{dr} - k_\phi^2 R(r) = 0. \tag{4.3.31}$$

By a simple check we can show that the solution of this equation is

$$R(r) = A_r r^{k_\phi} + B_r r^{-k_\phi} \qquad (k_z = 0). \tag{4.3.32}$$

In fact, from Eq. (4.3.24) we can see that $Z(z)$ is not necessarily independent of $z$ if $k_z = 0$, but rather that

$$Z(z) = A_z z + B_z \qquad (k_z = 0). \tag{4.3.33}$$

If the potential is independent of $z$, we have to put $A_z = 0$.

We will illustrate the application of the method of separation of variables in the case of the cylindrical coordinate system by a case where the potential is not a function of $z$. Consider a thin, infinite, conducting, circular cylinder of radius $a$, split into two halves, as shown in Fig. 4.11, and suppose that the upper half of the cylinder is at a potential $V_0$, and the lower at potential $-V_0$. Because of our choice of the coordinate system, the potential is a function of coordinates $r$ and $\phi$ only. In addition, it is an odd function of $\phi$, that is, $V(r, -\phi) = -V(r, \phi)$. According to Eqs. (4.3.22), (4.3.32), and (4.3.29), the solution of Laplace's equation in this case is

$$V_k(r, \phi) = (A_r r^{k_\phi} + B_r r^{-k_\phi})\, A_\phi \sin k_\phi \phi = (C_k r^k + C'_k r^{-k}) \sin k\phi. \tag{4.3.34}$$

($C_k$ and $C'_k$ are new constants, and $k_\phi$ is replaced by $k$ for simplicity.) It is not possible to choose the constants $C_k$ and $C'_k$ in Eq. (4.3.34) so that the boundary con-

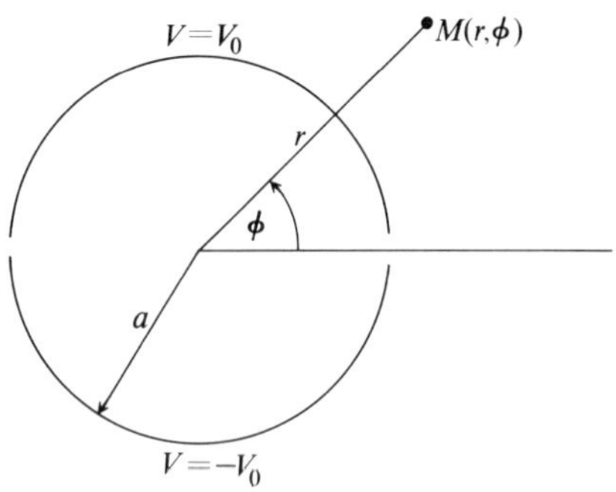


**Fig. 4.11** A two-dimensional potential problem. By symmetry the potential function is an odd function in $\phi$, that is, $V(r, -\phi) = -V(r, \phi)$.

ditions are satisfied at $r = a$ for all $\phi$. But it is again possible to form a series solution

$$V(r, \phi) = \sum_{k=1}^{\infty} (C_k r^k + C'_k r^{-k}) \sin k\phi, \tag{4.3.35}$$

and to try to determine constants $C_k$ and $C'_k$, $k = 1, 2, \ldots$, so that the boundary conditions are satisfied.

Let us first consider points inside the cylinder. Since the potential must be finite at all points for $r < a$, it follows that inside the cylinder we have to put $C'_k = 0$. The remaining coefficients $C_k$ are to be determined so that

$$[V(r, \phi)]_{r \leq a} = \sum_{k=1}^{\infty} C_k r^k \sin k\phi = \begin{cases} +V_0 \text{ for } 0 < \phi < \pi, r = a \\ -V_0 \text{ for } \pi < \phi < 2\pi, r = a \\ 0 \text{ for } \phi = 0 \text{ and } \phi = \pi, r = a \end{cases} \tag{4.3.36}$$

This function is represented in Fig. 4.10 if we change $y$ to $\phi$ and $a$ to $\pi$. Consequently, $C_n$ given by Eq. (4.3.19) represents $C_k a^k$ in our case, so that the final solution for points inside the cylinder is

$$[V(r, \phi)]_{r \leq a} = \frac{4V_0}{\pi} \sum_{k=0}^{\infty} \frac{r^{2k+1} \sin (2k + 1)\phi}{(2k + 1)a^{2k+1}}. \tag{4.3.37}$$

If we now consider the potential outside the cylinder, we have to put $C_k = 0$, since the potential must be zero for $r \to \infty$. We try to determine constants $C'_k$ so that

$$[V(r, \phi)]_{r \geq a} = \sum_{k=1}^{\infty} C'_k r^{-k} \sin k\phi = \begin{cases} +V_0 \text{ for } 0 < \phi < \pi, r = a \\ -V_0 \text{ for } \pi < \phi < 2\pi, r = a \\ 0 \text{ for } \phi = 0 \text{ and } \phi = \pi, r = a \end{cases} \tag{4.3.38}$$

The series is of the same form as before. The values of $C'_k$ are hence obtained by equating $C'_k a^{-k}$ to $C_n$ given by Eq. (4.3.19). Thus

$$[V(r, \phi)]_{r \geq a} = \frac{4V_0}{\pi} \sum_{k=0}^{\infty} \frac{a^{2k+1} \sin (2k + 1)\phi}{(2k + 1)r^{2k+1}}. \tag{4.3.39}$$

**4.3.4 Separation of variables in a spherical coordinate system.*** According to the result of Problem 1.G37, Laplace's equation in the spherical coordinate system has the form

$$\frac{\partial}{\partial r}\left(r^2 \frac{\partial V}{\partial r}\right) + \frac{1}{\sin \theta} \frac{\partial}{\partial \theta}\left(\sin \theta \frac{\partial V}{\partial \theta}\right) + \frac{1}{\sin^2\theta} \frac{\partial^2 V}{\partial \phi^2} = 0. \tag{4.3.40}$$

* This section may be omitted without loss of continuity.

Assuming that

$$V(r, \theta, \phi) = R(r)\,\Theta(\theta)\,\Phi(\phi), \tag{4.3.41}$$

Eq. (4.3.40) becomes

$$\frac{1}{R(r)}\frac{d}{dr}\left[r^2\frac{dR(r)}{dr}\right] + \frac{1}{\sin\theta\,\Theta(\theta)}\frac{d}{d\theta}\left[\sin\theta\frac{d\Theta(\theta)}{d\theta}\right] + \frac{1}{\sin^2\theta\,\Phi(\phi)}\frac{d^2\Phi(\phi)}{d\phi^2} = 0. \tag{4.3.42}$$

If Eq. (4.3.42) is to be satisfied for all $r$, the first term on the left must be a constant,

$$\frac{1}{R(r)}\frac{d}{dr}\left[r^2\frac{dR(r)}{dr}\right] = k_r. \tag{4.3.43}$$

Hence Eq. (4.3.42) becomes, after multiplying by $\sin^2\theta$,

$$\left\{k_r\sin^2\theta + \frac{\sin\theta}{\Theta(\theta)}\frac{d}{d\theta}\left[\sin\theta\frac{d\Theta(\theta)}{d\theta}\right]\right\} + \frac{1}{\Phi(\phi)}\frac{d^2\Phi(\phi)}{d\phi^2} = 0. \tag{4.3.44}$$

The term within the braces is a function of $\theta$, and the second term of $\phi$ only. If the equation is to be satisfied for all $\theta$ and $\phi$, both must be equal to a constant. Since with respect to coordinate $\phi$ the potential must be periodic, that is, $V(r, \theta, \phi) = V(r, \theta, \phi + 2n\pi)$, $n = \pm 1, \pm 2, \ldots$, we put

$$\frac{1}{\Phi(\phi)}\frac{d^2\Phi(\phi)}{d\phi^2} = -k_\phi^2. \tag{4.3.45}$$

The term within the braces in Eq. (4.3.44) must consequently be equal to $k_\phi^2$. After making simple rearrangements, we obtain

$$\sin\theta\frac{d}{d\theta}\left[\sin\theta\frac{d\Theta(\theta)}{d\theta}\right] + (k_r\sin^2\theta - k_\phi^2)\,\Theta(\theta) = 0. \tag{4.3.46}$$

We can easily check that, if we write the arbitrary constant $k_r$ in the form

$$k_r = n(n + 1), \tag{4.3.47}$$

where $n$ is a new constant, the solution of Eq. (4.3.43) is

$$R(r) = A_r r^n + B_r r^{-(n+1)}. \tag{4.3.48}$$

If, for the sake of simplicity, we change $k_\phi$ to $m$, the solution of Eq. (4.3.45) becomes

$$\Phi(\phi) = A_\phi\sin m\phi + B_\phi\cos m\phi \qquad (m \equiv k_\phi). \tag{4.3.49}$$

If the potential function is to be single-valued with respect to coordinate $\phi$, the constant $m$ must be an integer:

$$k_\phi = m = 0, \pm 1, \pm 2, \ldots \tag{4.3.50}$$

Substituting Eqs. (4.3.47) and (4.3.50) into Eq. (4.3.46) we obtain

$$\sin\theta \frac{d}{d\theta}\left[\sin\theta \frac{d\Theta(\theta)}{d\theta}\right] + [n(n+1)\sin^2\theta - m^2]\Theta(\theta) = 0. \tag{4.3.51}$$

This equation is called the *Legendre differential equation*, and its solutions are referred to as Legendre functions. A detailed account of the Legendre equation and of Legendre functions may be found elsewhere;* we shall not introduce these special functions. We shall illustrate the solution of Laplace's equation in the spherical coordinate system by a couple of examples not involving Legendre functions.

Let us determine the potential function in the case of a conducting uncharged sphere of radius $a$ situated in an external uniform electric field of intensity $E_0$ (Fig. 4.12). Assuming that $V = 0$ in plane $\theta = \pi/2$, the potential of the external field is

$$V_0(r, \theta) = \int_M^R \mathbf{E}_0 \cdot d\mathbf{l} = -E_0 r \cos\theta. \tag{4.3.52}$$

The total potential is obviously not a function of the coordinate $\phi$. Hence we have to put $m = 0$. According to Eq. (4.3.48) the solution of Eq. (4.3.41) thus becomes

$$V_n(r, \theta) = [A_n r^n + B_n r^{-(n+1)}]\,\Theta_n(\theta). \tag{4.3.53}$$

$\Theta_n(\theta)$ is a solution of the Legendre equation (4.3.51) for $m = 0$:

$$\frac{d}{d\theta}\left[\sin\theta \frac{d\Theta(\theta)}{d\theta}\right] + n(n+1)\sin\theta\,\Theta(\theta) = 0. \tag{4.3.54}$$

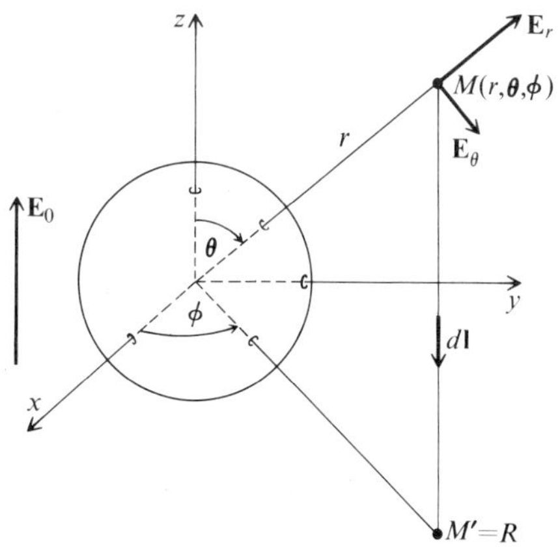


**Fig. 4.12** A conducting uncharged sphere in a uniform external field.

---

* For example, see footnote on p. 153.

At a large distance from the sphere, that is for large $r$, the potential must approach that given by Eq. (4.3.52). It hence follows that $A_1 = -E_0$ and $A_n = 0$ for $n \neq 1$, where it is assumed that $n$ is positive. It is easy to check that the solution of Eq. (4.3.54) for $n = 1$ is $\cos\theta$, $\Theta_1(\theta) = \cos\theta$. So the potential $V(r, \theta)$ can be represented in the form

$$V(r, \theta) = \left(-E_0 r + \frac{B_1}{r^2}\right)\cos\theta + \sum_{n \neq 1} B_n r^{-(n+1)}\Theta_n(\theta). \tag{4.3.55}$$

Now, on the surface of the sphere $V(a, \theta) = 0$, since it was assumed to be zero at its equator. So Eq. (4.3.55) yields

$$V(a, \theta) = 0 = \left(-E_0 a + \frac{B_1}{a^2}\right)\cos\theta + \sum_{n \neq 1} B_n a^{-(n+1)}\Theta_n(\theta). \tag{4.3.56}$$

This equation is satisfied if $B_1 = E_0 a^3$ and $B_n = 0$ for $n \neq 1$. According to the uniqueness theorem, this is the only possible solution. The total potential in this case is thus

$$V(r, \theta) = -E_0(r - a^3/r^2)\cos\theta. \tag{4.3.57}$$

The corresponding components of the electric field intensity are

$$E_r(r, \theta) = -\frac{\partial V}{\partial r} = E_0\left(1 + 2\frac{a^3}{r^3}\right)\cos\theta, \tag{4.3.58}$$

$$E_\theta(r, \theta) = -\frac{1}{r}\frac{\partial V}{\partial \theta} = -E_0\left(1 - \frac{a^3}{r^3}\right)\sin\theta. \tag{4.3.59}$$

At the surface of the sphere

$$E_r(a, \theta) = 3E_0\cos\theta, \qquad E_\theta(a, \theta) = 0. \tag{4.3.60}$$

In Fig. 4.13, the lines of the electric field intensity and the equipotential lines of a conducting sphere in a uniform external electric field are shown.

Another problem which can be solved by the same simple technique is that of a dielectric sphere in a uniform external electric field. Let the radius of the sphere be $a$, its permittivity $\varepsilon_2$, and the permittivity of the surrounding medium $\varepsilon_1$, and suppose that the intensity of the external field is $E_0$ (Fig. 4.14). The general solution for the potential in this case is also given by Eq. (4.3.53). Since at large distances the potential must approach that given by Eq. (4.3.52), solution $V_1(r, \theta)$ for points *outside* the sphere is in the form Eq. (4.3.55). Inside the sphere the potential must be zero for $r = 0$. So, if we suppose that $n > 0$, all $B_n$ inside the sphere must be zero. A possible solution for potential $V_2(r, \theta)$ for points inside the sphere is thus

$$V_2(r, \theta) = \sum A_n r^n \Theta_n(\theta), \qquad n > 0. \tag{4.3.61}$$

On the surface of the sphere the potentials given in Eqs. (4.3.61) and (4.3.55) must be the same, since the electric field intensity is finite everywhere. The equation

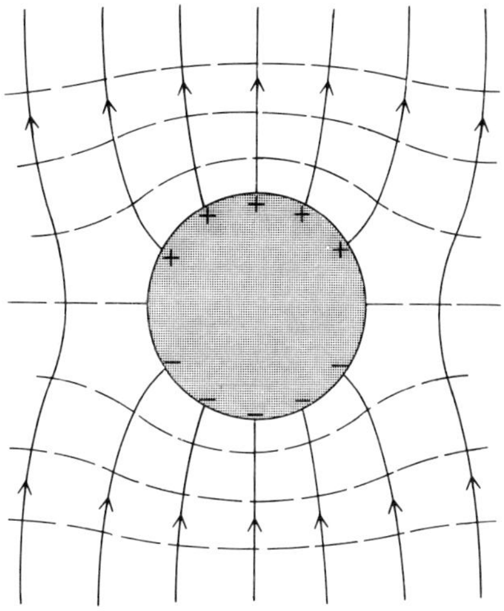

**Fig. 4.13** Lines of force (solid) and equipotential lines (dashed) in the vicinity of an uncharged conducting sphere in a uniform field.

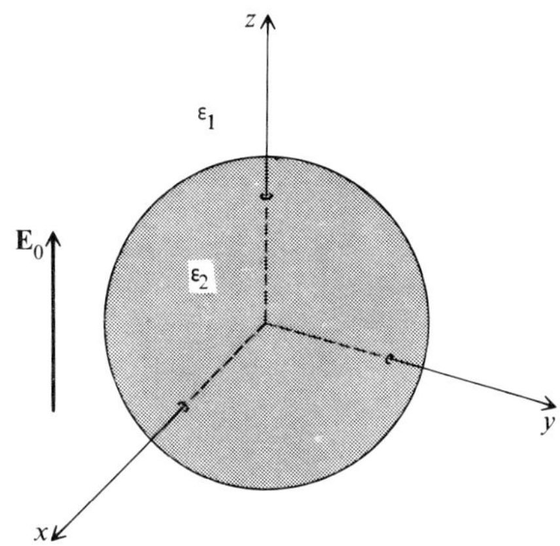


**Fig. 4.14** A dielectric sphere in a uniform external field.

$$\left(-E_0 a + \frac{B_1}{a^2}\right)\cos\theta + \sum_{n \neq 1} B_n a^{-(n+1)}\Theta_n(\theta) = \sum A_n a^n \Theta_n(\theta) \qquad (4.3.62)$$

corresponds to the condition $V_1(a, \theta) = V_2(a, \theta)$. Since this must hold for any $\theta$, and since $\Theta_1(\theta) = \cos\theta$, we have

$$-E_0 a + \frac{B_1}{a^2} = A_1 a, \qquad \text{and} \qquad B_n a^{-(n+1)} = A_n a^n \qquad \text{for} \qquad n \neq 1. \qquad (4.3.63)$$

In addition, the normal components of the electric displacement vector at the surface of the sphere must be the same in both dielectrics:

$$\left[-\varepsilon_1 \frac{\partial V_1}{\partial r}\right]_{r=a} = \left[-\varepsilon_2 \frac{\partial V_2}{\partial r}\right]_{r=a}. \qquad (4.3.64)$$

This yields

$$\varepsilon_1\left[\left(E_0 + \frac{2B_1}{a^3}\right)\cos\theta + \sum_{n>1}(n+1)B_n a^{-(n+2)}\Theta_n(\theta)\right] = -\varepsilon_2 \sum_{n>1} n A_n a^{n-1}\Theta_n(\theta),$$

from which

$$\varepsilon_1\left(E_0 + \frac{2B_1}{a^3}\right) = -\varepsilon_2 A_1, \qquad \text{and}$$

$$\varepsilon_1(n+1)B_n a^{-(n+2)} = -\varepsilon_2 n A_n a^{n-1} \qquad \text{for} \qquad n \neq 1. \qquad (4.3.65)$$

From Eqs. (4.3.63) and (4.3.65), the constants $A_1$ and $B_1$ are

$$B_1 = \frac{\varepsilon_2 - \varepsilon_1}{\varepsilon_2 + 2\varepsilon_1} a^3 E_0, \tag{4.3.66}$$

$$A_1 = \left(\frac{\varepsilon_2 - \varepsilon_1}{\varepsilon_2 + 2\varepsilon_1} - 1\right) E_0 = -\frac{3\varepsilon_1}{\varepsilon_2 + 2\varepsilon_1} E_0, \tag{4.3.67}$$

and $A_n = B_n = 0$ for $n \neq 1$. Equations (4.3.61) and (4.3.55) for potential thus finally become

$$V_2(r, \theta) = -\frac{3\varepsilon_1}{\varepsilon_2 + 2\varepsilon_1} E_0 r \cos\theta, \qquad r \geq a \tag{4.3.68}$$

$$V_1(r, \theta) = -\left(r - \frac{\varepsilon_2 - \varepsilon_1}{\varepsilon_2 + 2\varepsilon_1} \frac{a^3}{r^2}\right) E_0 \cos\theta, \qquad r \geq a. \tag{4.3.69}$$

According to Eq. (4.3.68) the electric field inside the sphere is uniform, in the direction of $\mathbf{E}_0$, and of magnitude

$$E_2 = \frac{3\varepsilon_1}{\varepsilon_2 + 2\varepsilon_1} E_0, \qquad r < a. \tag{4.3.70}$$

Outside the sphere the components of the electric field intensity are

$$E_{1r} = -\frac{\partial V_1}{\partial r} = \left(1 + 2\frac{\varepsilon_2 - \varepsilon_1}{\varepsilon_2 + 2\varepsilon_1} \frac{a^3}{r^3}\right) E_0 \cos\theta, \qquad r > a, \tag{4.3.71}$$

$$E_{1\theta} = -\frac{1}{r}\frac{\partial V_1}{\partial \theta} = -\left(1 - \frac{\varepsilon_2 - \varepsilon_1}{\varepsilon_2 + 2\varepsilon_1} \frac{a^3}{r^3}\right) E_0 \sin\theta, \qquad r > a. \tag{4.3.72}$$

Comparing Eqs. (4.3.71) and (4.3.72) with Eqs. (2.2.4) and (2.2.5) we can see that the field intensity in dielectric 1 is the sum of $E_0$ and the field intensity created by a dipole of moment

$$p = 4\pi\varepsilon_1 \frac{\varepsilon_2 - \varepsilon_1}{\varepsilon_2 + 2\varepsilon_1} a^3 E_0 \tag{4.3.73}$$

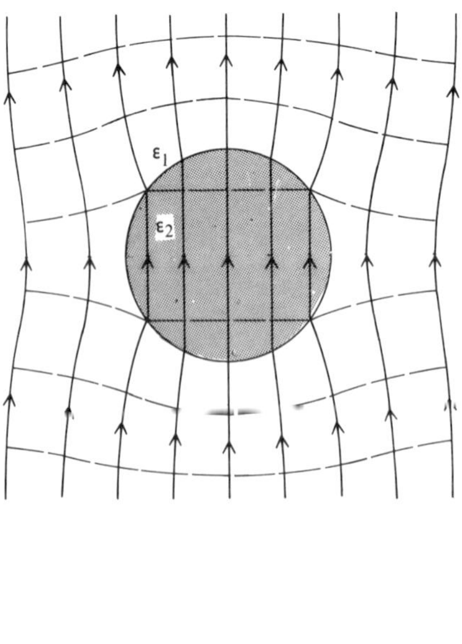


**Fig. 4.15** Lines of vector **D** (solid) and equipotential lines (dashed) in the vicinity of a dielectric sphere in a uniform field.

situated at the center of the sphere. We have to imagine the whole space to be filled with a dielectric of permittivity $\varepsilon_1$.

In Fig. 4.15 the lines of the electric displacement vector and the equipotential lines in this case are shown.

*Problems referring to Section 4.3:* 4.T9 to 4.T12, 4.G18 to 4.G20, 4.N9.

## PROBLEMS

### Test problems

**4.T1.** A point charge $Q$ is placed at a distance $d$ from a large earthed conducting plane. Determine the force on charge $Q$.

**4.T2.** Determine the image charges in the case analogous to that shown in Fig. 4.5, if the angle of the conducting corner is 60° instead of 90°.

***4.T3.** Prove that a finite number of images in the case of a point (or line) charge in a conducting corner is obtained only if the angle of the corner is $\pi/n$, $n$ being any integer. Sketch the image charges for $n = 1, 2, 3$, and 4.

***4.T4.** Discuss the possibility of using the method of images in the case of a conducting corner of angle 120°.

**4.T5.** A point charge $Q$ is placed between two parallel conducting planes. Sketch the sequence of image charges in both planes.

**4.T6.** A point charge $Q$ is placed at a distance $b$ from the center of a grounded conducting sphere of radius $R$ ($R < b$). Determine the force by which the charge $Q$ acts on the induced charges on the surface of the sphere. Sketch the lines of force.

***4.T7.** Repeat the preceding problem if the sphere is (1) uncharged, (2) carrying a charge $Q_1$, and (3) at a given potential $V$. Sketch the lines of the electric field intensity vector in the three cases.

***4.T8.** Describe in words how you would solve the problem of two spheres at equal and opposite potentials, by using an infinite number of images in both spheres.

**4.T9.** Suppose that the electrostatic system shown in Fig. 4.9 is also closed from the right, with a conducting strip at potential $V_0$ (or $-V_0$) at $x = b$. Discuss the values of the constants $k_x^2$ and $k_y^2$ in this case and the types of functions $X(x)$ and $Y(y)$ convenient to form a series solution satisfying the boundary conditions.

***4.T10.** Write the solution of Laplace's equation in the cylindrical coordinate system corresponding to the general form of Eq. (4.3.22) of separated variables, for the particular case when the problem has an axial symmetry and the potential is not a function of the coordinate $z$. Start from the general differential equations (4.3.24), (4.3.26), and (4.3.30).

**4.T11.** Determine the general solution of Eq. (4.3.40) for problems having spherical symmetry.

**4.T12.** Deduce from Eqs. (4.3.70) to (4.3.72) the largest value of the electric field intensity inside and outside a small air bubble in a dielectric of relative permittivity $\varepsilon_r$, at a point where the external electric field intensity is $E_0$. Using the result obtained, discuss the influence of accidental small air bubbles in a liquid dielectric on the quality of the dielectric.

## General problems

**4.G1.** A point charge $Q$ resides at a distance $h$ from a grounded conducting plane. Determine the distribution of induced charges on the plane, and the total induced charge.

*Solution.* As explained in Section 4.2.2, the electric field due to the induced charges on the plane may be reduced to the field of the image charge $-Q$, as shown in Fig. 4.G1a. The total electric field intensity at any point above the plane is equal to the vector sum of intensities due to the two charges. Thus at a point $M$ on the plane vector $\mathbf{E}$ is normal to the plane, and of intensity

$$E(x, 0) = E_z(x, 0) = 2\frac{Q}{4\pi\varepsilon_0 r^2}\cos\alpha = \frac{Q}{2\pi\varepsilon_0 r^2}\frac{h}{r} = \frac{Qh}{2\pi\varepsilon_0 r^3}.$$

The density of charges induced on the surface is hence

$$\rho_s = -\varepsilon_0 E_z(x, 0) = -\frac{Qh}{2\pi r^3} = -\frac{Qh}{2\pi(x^2 + h^2)^{3/2}}.$$

Fig. 4.G1b shows the density of the induced charges as a function of the distance $x$ from the point on the plane below the charge $Q$.

Distribution of the induced charges being symmetrical with respect to the $z$ axis, the total induced charge is

$$Q_{\text{induced}} = \int_0^\infty \rho_s 2\pi x\, dx = -Qh\int_0^\infty \frac{x\, dx}{(x^2 + h^2)^{3/2}} = -Q.$$

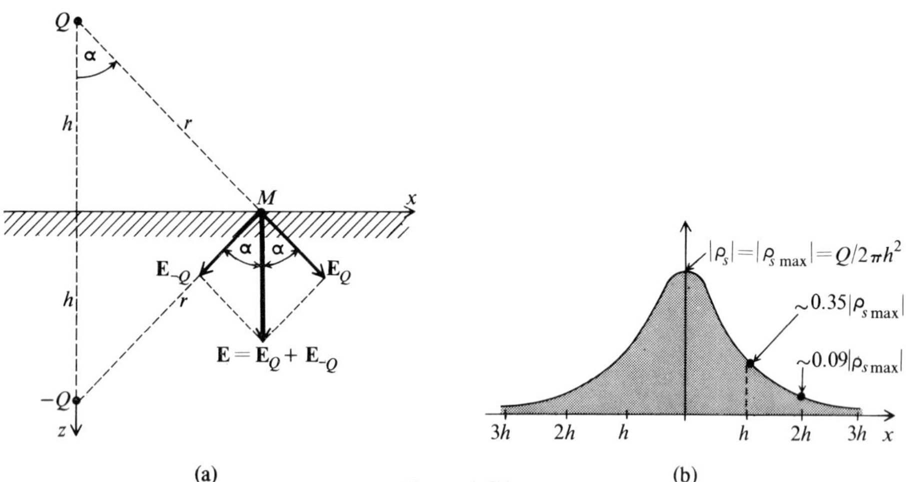


Figure 4.G1

**4.G2.** A small uncharged ball is suspended by an insulating spring at a height $h$ above a grounded conducting plane. The ball is then charged with an unknown charge $Q$. This results in attracting the ball towards the plane for a distance $d$. If a force $f$ is necessary to produce the extension of the spring for unit length, determine the charge $Q$. The medium is air.

*Answer.* For $h = 5$ cm, $d = 1$ cm, and $f = 10^{-6}$ N/m, $Q = 8.43 \times 10^{-11}$ C.

**4.G3.** Determine the force $F_l$ per unit length on a thin conductor charged with a charge $Q_l$ per unit length, running at a height $h$ parallel to the plane surface of the earth.

*Answer.* For $Q_l = 10^{-10}$ C/m and $h = 10$ m, $F_l = 9 \times 10^{-12}$ N/m.

**4.G4.** Determine the distribution of induced charges on the earth's surface from the preceding problem, as well as the total charge induced, per unit length in the direction of the conductor.

***4.G5.** Determine the force on the point charge $Q$ shown in Fig. 4.G5.

*Answer.* The force is directed towards the corner. For $Q = 10^{-12}$ C and $d = 10$ cm its intensity is $13.4 \times 10^{-13}$ N.

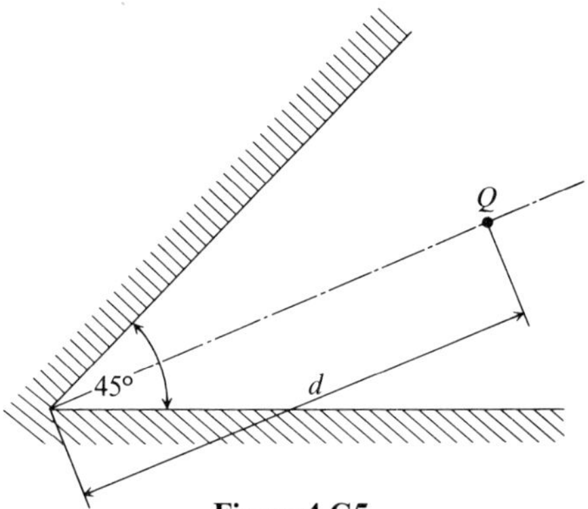


**Figure 4.G5**

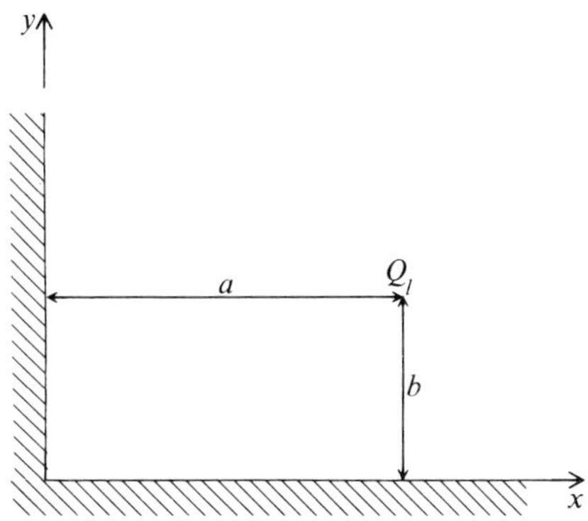


**Figure 4.G6**

**4.G6.** A thin conductor carrying a charge $Q_l$ per unit length runs parallel to a grounded conducting corner, as shown in Fig. 4.G6. Determine the $x$ and $y$ components $F_{xl}$ and $F_{yl}$ of the force acting on the conductor, per unit length.

*Answer.* For $a = b$, $F_{xl} = F_{yl} = -Q_l^2/8\pi\varepsilon_0 a$.

**4.G7.** A grounded conductor of radius $R$ runs parallel to the plane surface of the earth, at a height $h$ ($h \gg R$). An external uniform electric field (due, for example, to a large charged cloud) of intensity $E_0$ exists around the conductor. Vector $\mathbf{E}_0$ is directed towards the earth. Determine the charge per unit length induced on the conductor, and the total electric field intensity at points below the conductor and at the conductor surface.

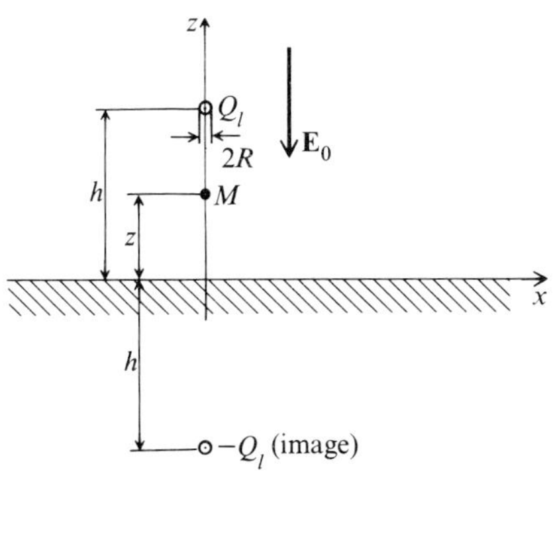


**Figure 4.G7**

*Solution.* Since the conductor is grounded, it will become charged under the influence of the field $E_0$. Suppose that its charge (which is obviously negative, but as yet unknown) per unit length is $Q_l$. Then the total electric field intensity above the earth may be regarded as the sum of $E_0$ and the field due to the charge $Q_l$ and its image $-Q_l$.

Referring to Fig. 4.G7, we see that, at a point $M$ below the conductor, both $\mathbf{E}_{\text{conductor}}$ and $\mathbf{E}_{\text{image}}$ are directed towards the earth. The total electric field intensity $E_t$ at $M$ is hence

$$E_{tM}(z) = E_0 + E_{\text{conductor}} + E_{\text{image}} = E_0 + \frac{Q_l}{2\pi\varepsilon_0(h - z)} + \frac{Q_l}{2\pi\varepsilon_0(h + z)}. \tag{a}$$

The unknown charge $Q_l$ can be determined by using the condition that the conductor is grounded, that is, at potential zero:

$$V_{\text{conductor}} = \int_{h-R}^{0} E_t\,(-dz) = \int_0^{h-R} E_t\,dz = \int_0^{h-R} E_0\,dz + \frac{Q_l}{2\pi\varepsilon_0}\int_0^{h-R}\frac{dz}{h-z}$$

$$+ \frac{Q_l}{2\pi\varepsilon_0}\int_0^{h-R}\frac{dz}{h+z} = E_0(h - R) + \frac{Q_l}{2\pi\varepsilon_0}\left[\ln\frac{h}{h-(h-R)} + \ln\frac{h+(h-R)}{h}\right]$$

$$\simeq E_0 h + \frac{Q_l}{2\pi\varepsilon_0}\ln\frac{2h}{R} = 0,$$

whence

$$Q_l = -\frac{2\pi\varepsilon_0 E_0 h}{\ln 2h/R}. \tag{b}$$

Substituting (b) into (a) we obtain

$$E_{tM}(z) = E_0\left[1 - \frac{2h^2}{(h^2 - z^2)\ln 2h/R}\right]. \tag{c}$$

From this we see that the field intensity below the grounded conductor is smaller than $E_0$. Such conductors are suspended at the tops of towers carrying power transmittion lines, for partially shielding the lines from the atmospheric electrostatic field.

The electric field intensity at the surface of the conductor is largest at the point $z = h + R$:

$$E_t(h + R) = E_0\left[1 - \frac{2h^2}{(h^2 - 2hR - R^2 - h^2)\ln 2h/R}\right] \simeq E_0\left(1 + \frac{h}{R\ln 2h/R}\right). \tag{d}$$

Since $h \gg R$, it is seen that the electric field intensity at the surface of the conductor is much larger than $E_0$. Consequently, a corona will appear at the conductor surface even for very small values of $E_0$. A conducting path through the air is thus initiated for a possible lightning discharge through the grounded conductor. The conductor hence acts primarily as a lightning arrester.

**4.G8.** A two-wire line of conductor radius $R$ and distance between them $D$ runs at a height $h\,(h \gg R)$, parallel to the plane conducting surface of the earth (Fig. 4.4). Determine the capacitance of the line in the presence of the earth.

*Solution.* The potential of both wires, and hence also the potential difference between them for given charges $Q_l$ and $-Q_l$ per unit length of the conductors, is different from those in the case of an isolated two-wire line. An additional potential is created by the induced charges on the surface of the earth. As explained in Section 4.2.2, the influence of the induced charges can be reduced to image charges, indicated in Fig. 4.4.

Let us take the earth's surface as the reference surface of the potential. According to Eq. (4.2.1) for the potential of a line charge, the potential of the positive conductor is then

$$V_+ \simeq \frac{Q_l}{2\pi\varepsilon_0}\ln\frac{h}{R} - \frac{Q_l}{2\pi\varepsilon_0}\ln\frac{h}{D} - \frac{Q_l}{2\pi\varepsilon_0}\ln\frac{h}{2h} + \frac{Q_l}{2\pi\varepsilon_0}\ln\frac{h}{\sqrt{D^2+(2h)^2}}$$

$$= \frac{Q_l}{2\pi\varepsilon_0}\ln\frac{D}{R\sqrt{1+(D/2h)^2}}. \tag{a}$$

The potential of the negative conductor similarly is

$$V_- \simeq -\frac{Q_l}{2\pi\varepsilon_0}\ln\frac{D}{R\sqrt{1+(D/2h)^2}}.$$

The capacitance per unit length of the two-wire line above the surface of the earth therefore is

$$C_l = \frac{Q_l}{V_+ - V_-} = \frac{\pi\varepsilon_0}{\ln\{D/[R\sqrt{1+(D/2h)^2}]\}} \tag{b}$$

[Compare this result with (a) of Problem 1.G50.]

**4.G9.** Derive the expression for capacitance per unit length between the earth and a conductor of radius $R$ running at a height $h$ ($h \gg R$) parallel to the plane conducting surface of the earth.

**4.G10.** Repeat Problem 4.G8 in the case of a two-wire line of conductors of different radii $R_1$ and $R_2$ and at different heights $h_1$ and $h_2$ from the ground. The distance $D$ between the wires, $h_1$, and $h_2$ are all much larger than $R_1$ and $R_2$.

*Answer*. For $h_1 = h_2$ and $R_1 = R_2$ the result must become (b) of Problem 4.G8.

**4.G11.** One of the conductors of the two-wire line shown in Fig. 4.4 is earthed. Determine the capacitance of the other conductor with respect to the first conductor and the earth. [*Hint:* for a given charge per unit length of the unearthed conductor the induced charge per unit length on the earthed conductor is determined from the condition that its potential (due to real *and* image charges) is zero.]

*Answer*. For $D = 2$ m, $h = 10$ m, and $R = 5$ mm the required capacitance per unit length is $C_l \simeq 7.25$ pF/m.

**4.G12.** Assume that the two wires shown in Fig. 4.4 are at the same potential, i.e. that they are connected together by a conductor. Determine the capacitance per unit length between the two wires and the earth. [Hint: suppose that a charge $+Q_l/2$ exists on the two wires. Proceed further following the procedure outlined in Problem 4.G8.]

***4.G13.** Determine an approximate expression for capacitance between a thin metal rod of length $L$ and radius $R$ ($L \gg R$) residing at a height $h$ ($h \gg R$) parallel to the surface of the earth. Assume that a charge $Q$ is distributed along the rod very nearly uniformly.

*Solution*. The rod is assumed to be very thin. Hence, if the rod is charged, the potential at all points outside the rod will be approximately the same as if the charge is distributed uniformly along its axis. The potential of the rod with charge $Q$ is created by both charge $Q$ and the image of the rod. Since the rod is conducting, the potential is the same at all points on its surface. It

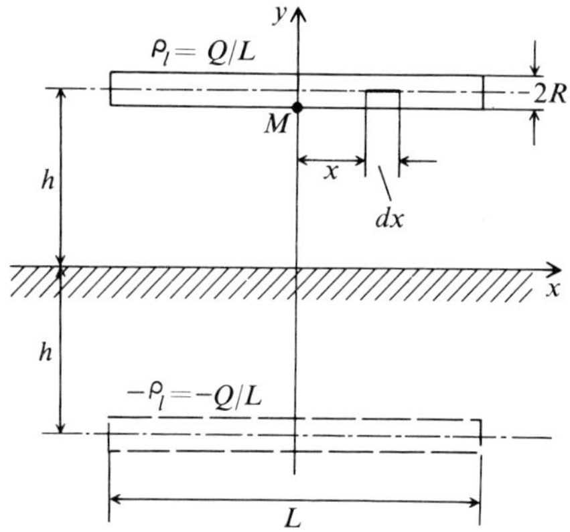


**Figure 4.G13**

is simplest to determine it at the point $M$ at the surface (Fig. 4.G13). Assuming line charges $\rho_l = Q/L$ and $-\rho_l$ along the axis of the rod and of its image, we have

$$V \simeq \frac{1}{4\pi\varepsilon_0}\int_{-L/2}^{L/2}\frac{\rho_l\,dx}{\sqrt{x^2 + R^2}} - \frac{1}{4\pi\varepsilon_0}\int_{-L/2}^{L/2}\frac{\rho_l\,dx}{\sqrt{x^2 + (2h)^2}}. \tag{a}$$

Both integrals in (a) are of the same type. They can be reduced to an ordinary type of integral by a simple change of variable. For example, let

$$u = x + \sqrt{x^2 + R^2}, \qquad du = dx + x\,dx/\sqrt{x^2 + R^2} = u\,dx/\sqrt{x^2 + R^2} \tag{b}$$

The first integral in (a) then becomes

$$\int_{-L/2}^{L/2}\frac{dx}{\sqrt{x^2 + R^2}} = \int_{u_1}^{u_2}\frac{du}{u} = \ln\frac{u_2}{u_1}, \tag{c}$$

where, according to (b),

$$u_1 = -L/2 + \sqrt{(L/2)^2 + R^2}, \qquad u_2 = L/2 + \sqrt{(L/2)^2 + R^2}. \tag{d}$$

The value of the second integral in (a) is obtained by simply changing $R^2$ to $(2h)^2$.

In the case of the first integral $L \gg R$, so that approximately

$$u_1 \simeq -\frac{L}{2} + \frac{L}{2}\left(1 + \frac{1}{2}\frac{4R^2}{L^2}\right) = \frac{R^2}{L^2}, \qquad u_2 \simeq 2L/2 = L.$$

The approximate value of the potential of the rod is therefore

$$V \simeq \frac{\rho_l}{2\pi\varepsilon_0}\ln\frac{L^2}{R^2} - \frac{\rho_l}{2\pi\varepsilon_0}\ln\frac{L/2 + \sqrt{(L/2)^2 + (2h)^2}}{-L/2 + \sqrt{(L/2)^2 + (2h)^2}}.$$

The capacitance of the rod in the presence of the earth

$$C = \frac{Q}{V} = \frac{\rho_l L}{V} = \frac{2\pi\varepsilon_0 L}{\ln\dfrac{L^2(\sqrt{L^2 + 16h^2} - L)}{R^2(\sqrt{L^2 + 16h^2} + L)}}.$$

**4.G14.** A thick conducting cylinder of radius $R$ runs parallel to the surface of the earth. The axis of the cylinder is at a height $h$ above the earth. Determine the capacitance per unit length between the cylinder and the earth.

*Answer.* Twice as large as that given by Eq. (4.2.6) if $D$ is replaced by $2h$.

***4.G15.** Determine approximately the capacitance between two spheres of radius $R$ and distance between their centers $b$ ($b > 2R$). Determine the maximum value of the electric field at the surfaces of the spheres, if the potential difference between them is $2V$.

*Solution.* Following the general pattern of the method of images, let us determine the system of charges inside *imaginary* spheres of radius $R$ shown in Fig. 4.G15 which makes the two spheres equipotential, say at potentials $V$ and $-V$.

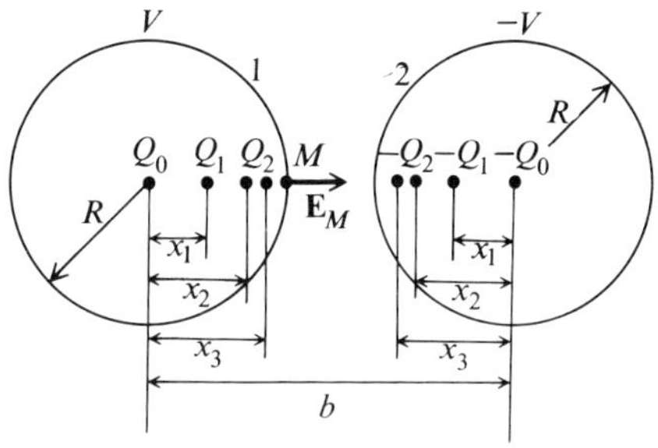


**Figure 4.G16**

First, imagine a charge $Q_0 = 4\pi\varepsilon_0 RV$ at the center of the left sphere (denoted by 1), and charge $-Q_0$ at the center of the right sphere (denoted by 2). The potential of both spheres due *only* to the charge inside them is then just as required. But the other charge makes them non-equipotential. However, according to Section 4.2.4, when the image charges $Q_1$ and $-Q_1$, of magnitudes given by Eq. (4.2.9),

$$Q_1 = -(-Q_0 \frac{R}{b}) = Q_0 \frac{R}{b} = 4\pi\varepsilon_0 \, R \, V \frac{R}{b},$$

are placed in both spheres at a distance $x_1 = R^2/b$ from their centers, the situation is improved. The charge $Q_0$ inside the sphere 1, the charge $-Q_0$ outside it and the charge $Q_1$ inside it make its surface equipotential, at precisely potential $V$. The same is valid for sphere 2. But charges $-Q_1$ and $+Q_1$ outside the two spheres disturb their equipotential character. To compensate for these charges, we add further images $+Q_2$ inside sphere 1, and $-Q_2$ inside sphere 2, of magnitudes

$$Q_2 = -(-Q_1 \frac{R}{b - x_1}) = Q_1 \frac{R}{b - x_1} = 4\pi\varepsilon_0 \, R \, V \frac{R}{b} \frac{R}{b - x_1},$$

and at a distance $x_2 = R^2/(b - x_1)$ from their centers. Obviously, this process has to be continued indefinitely. Charges $Q_n$ are to be placed inside sphere 1, and charges $-Q_n$ inside sphere 2, of magnitudes

$$Q_n = Q_{n-1} \frac{R}{b - x_{n-1}} = 4\pi\varepsilon_0 \, R \, V \prod_{k=1}^{n} \frac{R}{b - x_{k-1}} \tag{a}$$

(the symbol $\prod$ implies product), at distances

$$x_n = \frac{R^2}{b - x_{n-1}}, \qquad n = 1, 2, \ldots, \qquad x_0 = 0, \tag{b}$$

from the centers of the spheres.

According to Gauss' law the total charge on real metal spheres is equal to the sum of all charges $Q_n$:

$$Q_{\rm I} = \sum_{n=0}^{\infty} Q_n; \qquad Q_{\rm II} = -\sum_{n=0}^{\infty} Q_n. \tag{c}$$

The capacitance between the two spheres according to (a) and (c) is

$$C = \frac{Q_{\rm I}}{2V} = 2\pi\varepsilon_0 R \sum_{n=0}^{\infty} \left\{ \prod_{k=1}^{n} \frac{R}{b - x_{k-1}} \right\}, \qquad x_0 = 0. \tag{d}$$

The product function within the braces of (d) is smaller than $[R/(b - R)]^n$, $x_n$ being always smaller than $R$, according to (b). Since $b > 2R$, the series in (d) is a rapidly converging series, so that a few terms only are usually sufficient for practical purposes.

Finally, the electric field intensity at the point $M$ of sphere 1 is equal to the sum of intensities at $M$ due to all image point charges, inside both spheres. According to (a),

$$E_M = \frac{1}{4\pi\varepsilon_0} \sum_{n=0}^{\infty} \frac{Q_n}{(R - x_n)^2} + \frac{1}{4\pi\varepsilon_0} \sum_{n=0}^{\infty} \frac{Q_n}{(b - R - x_n)^2}$$

$$= R\,V \sum_{n=0}^{\infty} \left\{ \prod_{k=1}^{n} \frac{R}{b - x_{k-1}} \right\} \left\{ \frac{1}{(R - x_n)^2} + \frac{1}{(b - R - x_n)^2} \right\}. \tag{e}$$

$V$ represents *half* the potential difference between the spheres.

The same formula is valid for all points along the line $MM'$, if only $R$ in the parentheses $(R - x_n)^2$ and $(b - R - x_n)^2$ is changed to the distance $r$ of the point considered between $M$ and $M'$ from the center of sphere 1.

***4.G16.** Repeat the preceding problem for two spheres of different radii.

***4.G17.** Analyze the potential and the electric field of an *uncharged* sphere of radius $a$ in a uniform external electric field of intensity $E_0$, starting from the case of a point charge $Q$ and an uncharged sphere. [*Hint:* consider the case when $Q$ recedes to infinity, but simultaneously increases in magnitude, so that the field due to it around the sphere is $E_0$.]

*Answer.* Given by Eqs. (4.3.57) to (4.3.59).

***4.G18.** Determine the potential function in the two-dimensional system shown in Fig. 4.G18.

*Solution.* The potential $V$ being zero for $y = 0$ and $y = a$, the solution of $Y(y)$ in Eq. (4.3.9) must be of sinusoidal form. The constant $k_y^2$ being negative, the constant $k_x^2$ must be positive, $k_x^2 = -k_y^2 = k^2$. The solution of Eq. (4.3.8) is in exponential form. Next, since the potential is a symmetric function with respect to $y$ axis, the constants $A_x$ and $B_x$ in Eq. (4.3.8) must be equal. The solution is thus

$$V(x, y) = A_x(e^{kx} + e^{-kx})(A_y \sin ky + B_y \cos ky). \tag{a}$$

Now, for $y = 0$ the potential $V(x, 0) = 0$, whence $B_y = 0$. Also, for $y = a$ the potential $V(x, a) = 0$, whence $ka = n\pi$, or $k = n\pi/a$, $n = 1, 2, \ldots$. Taking into account that $(e^{kx} + e^{-kx}) = 2 \cosh kx$, solution (a) becomes

$$V_n(x, y) = A_n \cosh \frac{n\pi}{a} x \sin \frac{n\pi}{a} y, \qquad n = 1, 2, \ldots. \tag{b}$$

Obviously, the boundary conditions that $V(-b, y) = V(b, y) = V_0$ cannot be satisfied by one term of the form (b) only. It is therefore necessary to determine the constants $A_n$ so that

$$V(\pm b, y) = \sum_{n=1}^{\infty} A_n \cosh \frac{n\pi}{a} b \sin \frac{n\pi}{a} y = \begin{cases} V_0 \text{ for } 0 < y < a \\ 0 \quad \text{for } y = 0 \text{ and } y = a \end{cases}$$

This problem is the same as that of determining the constants $C_n$ in Eq. (4.3.15). The final result for $V(x, y)$ is

$$V(x, y) = \frac{4V_0}{\pi} \sum_{n=0}^{\infty} \frac{\cosh \dfrac{(2n + 1)\pi}{a} x \sin \dfrac{(2n + 1)\pi}{a} y}{(2n + 1) \cosh \dfrac{(2n + 1)\pi}{a} b}.$$

**4.G19.** A very long uncharged cylindrical conductor with a circular cross-section of radius $R$ is perpendicular to the lines of a uniform external electric field of intensity $E_0$. Determine the total potential and electric field intensity, assuming that the reference plane of the potential is the plane $\phi = \pm\pi/2$ (Fig. 4.G19).

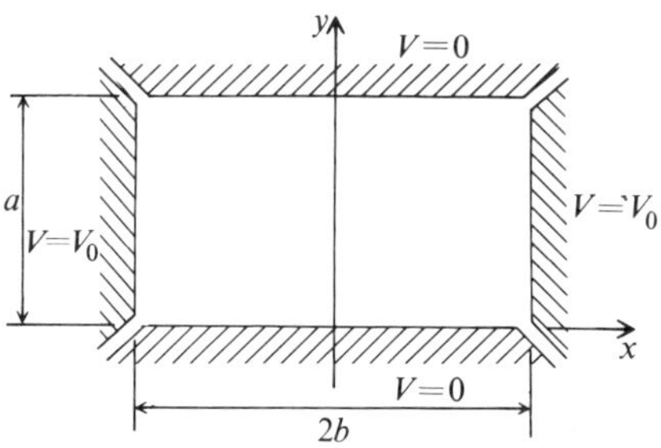


**Figure 4.G18**

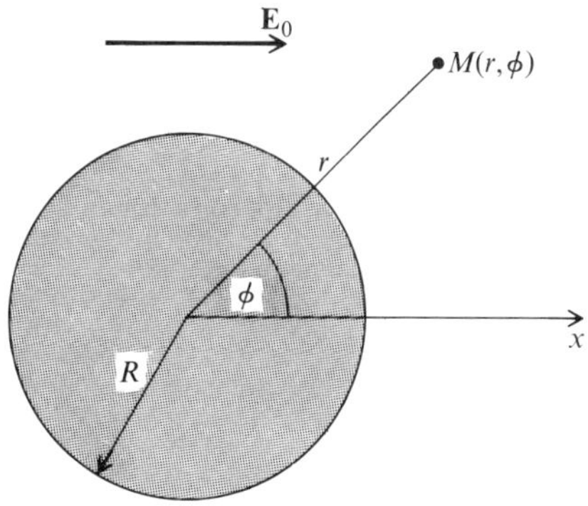


**Figure 4.G19**

*Solution.* The potential in this case is independent of $z$, so that according to Eqs. (4.3.22), (4.3.29), and (4.3.32) the potential function is of the form

$$V(r, \phi) = (A_\phi \sin k_\phi \phi + B_\phi \cos k_\phi \phi)\,(A_r r^{k_\phi} + B_r r^{-k_\phi}). \tag{a}$$

Now, the potential is firstly an even function of $\phi$, so that we have to put $A_\phi = 0$. Secondly, the potential is a periodic function of $\phi$, in the sense that $V(r, \phi + 2n\pi) = V(r, \phi)$, $n$ being any integer, so that $k_\phi$ can only be an integer. Thus (a) becomes

$$V_n(r, \phi) = (A_n r^n + B_n r^{-n}) \cos n\phi, \qquad n = 1, 2, \ldots. \tag{b}$$

$A_n$ and $B_n$ are new constants. But for $r$ large, $V_n(r, \phi)$ must become the potential $V_0$ of the external uniform field,

$$V_0 = -E_0 r \cos \phi, \tag{c}$$

so that $A_1 = -E_0$ and $A_n = 0$ for $n \neq 1$. Consequently the expression (b) for the potential becomes

$$V_n(r, \phi) = -E_0 r \cos \phi + B_n r^{-n} \cos n\phi, \qquad n = 1, 2, \ldots. \tag{d}$$

The solution could in general be of the form $\Sigma V_n(r, \phi)$. But in this case the required boundary condition that $V(R, \phi) = 0$ is satisfied with a single term. Indeed, the equation

$$V_n(R, \phi) = -E_0 R \cos \phi + B_n R^{-n} \cos n\phi = 0$$

can be satisfied for all $\phi$ only if $n = 1$ and $B_n = B_1 = E_0 R^2$. The final expression for the potential is thus

$$V(r, \phi) = -E_0(r - R^2/r) \cos \phi. \tag{e}$$

Determination of the components $E_r$ and $E_\phi$ of the electric field intensity and the sketch of the lines of force and equipotential lines are left to the reader.

***4.G20.** A very long dielectric cylinder of radius $R$ and permittivity $\varepsilon$ resides in a vacuum normal to the lines of a uniform external electric field of intensity $E_0$. Determine the total potential and the components of the electric field intensity inside and outside the dielectric cylinder. Referring to Fig. 4.G19, assume that the potential is zero for $\phi = \pm\pi/2$.

*Answer.* For $\varepsilon \to \infty$ the expression for the potential must become that given by (e) of the preceding problem. Explain the physical basis of such a check.

## Numerical problems

**4.N1.** A conducting charged sphere of radius $a = 1$ cm resides in air at a height $h = 10$ cm above a conducting plane surface. Determine approximately the largest possible charge on the sphere and the corresponding magnitude of the force on it.

**4.N2.** Calculate the electric field intensity at the surface of the grounded conductor described in Problem 4.G7 in terms of $E_0$, if $h = 20$.m and $R = 5$ mm. What will happen if $E_0 = 10^4$ V/m?

**4.N3.** Calculate the capacitance per unit length of the two-wire line of Problem 4.G8 for $R = 3$ mm, $D = 50$ cm and the following heights of the line above ground: 25 cm, 1 m and 10 m.

**4.N4.** Determine the capacitance of a thin straight wire of length $L = 10$ m and radius $R = 3$ mm, residing in air at a height $h = 10$ m and $h = 1$ m parallel to the plane conducting surface of the earth (see Problem 4.G13).

**4.N5.** Calculate the capacitance per unit length of a two-wire line with conductors of radius $R = 1$ cm, if the distance $D$ between their axes is 5 cm, 10 cm and 1 m, using in all three cases the exact and the approximate formula (valid for $D \gg R$).

**4.N6.** A conducting sphere of radius $R = 10$ cm carries a charge $Q_1 = +10^{-10}$ C. Determine the total force on a point charge $Q_0 = 10^{-11}$ C when at distances $b = 12$ cm, 15 cm, and 30 cm from the center of the sphere.

**4.N7.** Calculate the capacitance between two spheres of radius $R = 5$ cm with centers at a distance $D = 15$ cm, using the first 2, 3, and 4 image charges on both spheres. (See Problem 4.G15.)

**4.N8.** Calculate the maximum electric field intensity at the surfaces of the two spheres of the preceding problem, if the potential difference between them is $2V = 10{,}000$ V, using also the first 2, 3, and 4 image charges in both spheres.

**4.N9.** Sketch the potential function represented by the series in Eq. (4.3.20) at $x = 0$, using the first three terms of the series.

**Suggested reading.** The list given here progresses from simpler to more advanced texts. For a full bibliographical description, see the Bibliography.

1. PLOSNEY, R. and R. E. COLLIN, *Principles and Applications of Electromagnetic Fields*, Chapter 4.
2. WEBER, E., *Electromagnetic Fields, Theory and Applications*, Chapters 7 and 8.
3. RAMO, S., J. R. WHINNERY and T. VAN DUZER, *Fields and Waves in Communication Electronics*, Chapter 3.
4. SMYTHE, W. R., *Static and Dynamic Electricity*, Chapters IV and V.

CHAPTER 5

# APPROXIMATE METHODS OF SOLVING ELECTROSTATIC PROBLEMS

## 5.1 INTRODUCTION

In the preceding chapters, several methods for the analytical solution of certain classes of electrostatic problems were explained. The solutions were exact or almost exact, but the geometry of the conducting bodies considered was without exception very simple. They were either conducting planes, or corners, or spheres, etc.

However, in practice electrostatic systems are often of such a complicated geometry that an analytical solution is not feasible. For a rough estimate it is possible in many cases to approximate a real system by some geometrically simpler system, which can be solved analytically. For a more precise analysis, however, it is necessary to develop some approximate methods for the solution. In what follows two such approximate methods are described in some detail, and the logic of a few others is explained briefly.

## 5.2 THE GRAPHICAL METHOD OF SOLVING ELECTROSTATIC PROBLEMS

By using the graphical method we can attack two types of electrostatic problems, two-dimensional problems of arbitrary geometry and problems having axial symmetry. The methods vary somewhat in the two cases, and each will be considered separately.

**5.2.1 The graphical method of solving two-dimensional problems (the method of curvilinear squares).** As mentioned in Section 1.7.1, the equipotential surfaces can be conveniently drawn so that the potential difference between any two adjacent equipotential surfaces is the same. Similarly, in Sections 1.10 and 2.7 we adopted the convention of drawing the lines of the electric field intensity and electric displacement vectors so that each line represented a tube of constant flux of the two vectors, the same for the whole drawing. (It should be recalled that, in the case of isotropic dielectrics, the lines of both vectors **E** and **D** are orthogonal to the equipotential surfaces.) In a two-dimensional case, these conventions require that the same potential difference exists between the equipotential lines in the transverse

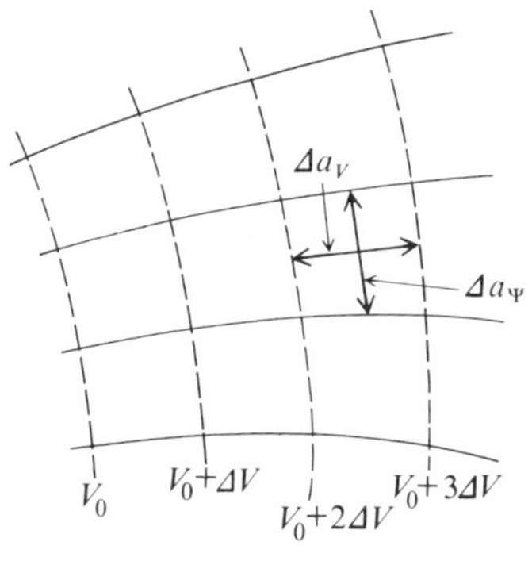


**Fig. 5.1** A two-dimensional electrostatic field.

cross-section of the system, and that the lines of vectors **E** or **D** represent tubes of equal flux of these vectors, per unit length along the system.

Let us suppose that the potential difference $\Delta V$ between two adjacent equipotential surfaces is very small, and that the flux $\Delta\Psi_D$ of vector **D** represented by each line of vector **D** is chosen to be also very small. Let us denote by $\Delta a_V$ the geometrical distance between two adjacent equipotential surfaces, and by $\Delta a_\Psi$ the geometrical distance between two adjacent lines of the electric displacement vector **D** at the same point. Both $\Delta a_V$ and $\Delta a_\Psi$ are different at different points of the field (Fig. 5.1). We can write the following simple approximate relations. First, the approximate magnitude of the electric field intensity (not taking into account its direction, which is known if the plot of equipotential lines is given) is

$$E \simeq \Delta V/\Delta a_V. \tag{5.2.1}$$

If we assume that the permittivity at the point considered is $\varepsilon$, the magnitude of the electric displacement vector at the same point is then approximately

$$D = \varepsilon E \simeq \varepsilon \Delta V/\Delta a_V. \tag{5.2.2}$$

The flux $\Psi_D$ of the vector **D**, per unit length along the two-dimensional system, represented by each line shown in Fig. 5.1, is therefore approximately

$$\Delta\Psi_{lD} \simeq D\,\Delta a_\Psi \simeq \varepsilon \frac{\Delta V}{\Delta a_V} \Delta a_\Psi. \tag{5.2.3}$$

From Eq. (5.2.3) it follows that the ratio $\Delta a_\Psi$ and $\Delta a_V$ must be very nearly constant throughout the field,

$$\frac{\Delta a_\Psi}{\Delta a_V} \simeq \frac{\Delta\Psi_{lD}}{\varepsilon\Delta V}, \tag{5.2.4}$$

provided that $\Delta V$ and $\Delta\Psi_{lD}$ are sufficiently small. Now, the lines of **D** and the equipotential surfaces are normal to each other. According to Eq. (5.2.4) the small curvilinear rectangles of the map formed by the two families of curves must be very nearly similar, the ratio of their sides being approximately constant. For given $\Delta\Psi_{ID}$, $\Delta V$, and $\varepsilon$ this ratio is determined by Eq. (5.2.4). We can make the sketch of the field when this ratio is assumed.

No other restrictions on $\Delta\Psi_{ID}$ and $\Delta V$ have to be imposed except that they are small enough. But it is extremely convenient in practice to adopt $\Delta\Psi_{ID}$ and $\Delta V$ so that the ratio $\Delta a_{\Psi}/\Delta a_V$ is unity. In that case, the curvilinear rectangles become simply curvilinear squares. The method of plotting the two families of curves is based on a visual estimate. It is obviously much more convenient to draw curvilinear squares than similar curvilinear rectangles. If we adopt this, our final sketch of the equipotential lines and of the lines of the electric displacement vector must be such, that the geometrical figures obtained are approximately curvilinear squares (except possibly at some very special points of the field). For this reason we often refer to the method as the *method of curvilinear squares.*

The logical procedure for determining the equipotential lines and the lines of the electric displacement vector using this method is the following:

1. Sketch roughly equipotential lines, bearing in mind that the potential difference between them should be equal. Make use of any existing symmetry as much as possible. Begin to draw the lines where the field is of the simplest form (say, in the region of uniform field).
2. Try to draw a set of curves normal to the sketched equipotential lines and the conducting boundaries so that the geometrical figures obtained are as similar as possible to curvilinear squares.
3. Repeat the procedure several times to obtain better curvilinear squares. For that purpose, either erase the obviously erroneous parts of the sketch or use another sheet of paper. Transparent paper is very convenient for this. Where necessary, make a denser net of lines.

Almost no other general hints can be given. In fact, only a certain amount of experience can help in making the sketches faster and more precise. As an example, Fig. 5.2 is a sketch of equipotential lines and of the lines of the electric displacement vector between two parallel strip conductors charged with equal charges of opposite signs.

Once we have obtained a sufficiently accurate field plot, we can calculate any desired quantity from it. We know the potential function, and from this we can approximately determine the electric field intensity and the density of the charges on the surfaces of the conductors. Hence also we can calculate the capacitance per unit length between two cylindrical conductors of some irregular cross-sections.

For example, the electric field intensity at any point of a two-dimensional system is given by Eq. (5.2.1). Suppose that the number of lines of the electric displacement vector emerging from the contour of a conductor is $N_{\Psi}$. According

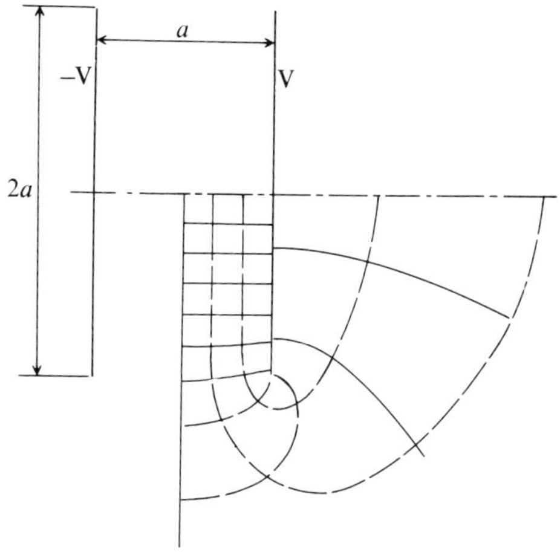


Fig. 5.2 Plot of curvilinear squares for a two-dimensional system.

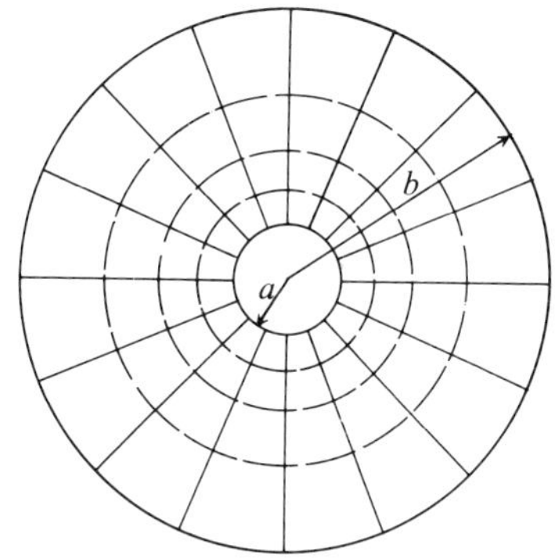


Fig. 5.3 Plot of curvilinear squares for a cross-section of a coaxial cable.

to the generalized form of Gauss' law and Eq. (5.2.3), the charge $Q_l$ on that conductor per unit length is then

$$Q_l = N_\Psi \Delta\Psi_{lD} = N_\Psi \, \varepsilon \frac{\Delta a_\Psi}{\Delta a_V} \Delta V. \tag{5.2.5}$$

If the total potential difference $V$ between two isolated cylindrical conductors is divided in the plot into $N_V$ equal parts, $\Delta V = V/N_V$. Let us assume that the method of plotting the field was the method of curvilinear squares. From Eq. (5.2.5) and the definition of the capacitance, it follows that the capacitance between two isolated cylindrical conductors per unit length is

$$C_l = \frac{Q_l}{V} = \varepsilon \frac{N_\Psi}{N_V} \frac{\Delta a_\Psi}{\Delta a_V} = \varepsilon \frac{N_\Psi}{N_V}. \tag{5.2.6}$$

If $\Delta a_\Psi / \Delta a_V \neq 1$, the first expression rather than the second is valid.

For example, in Fig. 5.2 we chose the value $N_V = 6$. The field plot resulted in $N_\Psi = 21$. If the dielectric between the strips is air, the capacitance per unit length between them is approximately

$$C_l = 8.85 \times 10^{-12} \frac{21}{6} = 31 \text{ pF/m}.$$

This differs considerably from the capacitance calculated without taking into account the fringing effect:

$$C_{l\,\text{apr}} = \varepsilon_0 \frac{2a}{a} = 2\varepsilon_0 = 17.71 \text{ pF/m}.$$

What is the degree of precision of this approximate method? The error is obviously smaller for smaller curvilinear squares. But, even with a moderate size of squares and with almost no experience, the error need be no more than ten to fifteen per cent. To illustrate this, let us take the case where the capacitance per unit length of the coaxial cable shown in Fig. 5.3 is calculated by using Eq. (5.2.6) and the exact formula (see Problem 1.G43). Since $N_\Psi = 16$ and $N_V = 4$, according to Eq. (5.2.6), we have

$$C_l \simeq \varepsilon \frac{16}{4} = 4\varepsilon.$$

According to the exact formula, since $b/a \simeq 5.1$, we have

$$C_l = \frac{2\pi\varepsilon}{\ln b/a} \simeq \frac{2\pi\varepsilon}{\ln 5.1} \simeq 3.85\varepsilon.$$

The difference between the approximate and the exact value is less than five per cent.

### 5.2.2 The graphical method of solving axially symmetrical problems.*

Axially symmetrical fields can also be solved approximately by the graphical method. The plot of the field is made in a plane containing the axis of symmetry. We make use again of Eqs. (5.2.1) and (5.2.2). Each line of the electric displacement vector now represents the flux through a circular strip, with a radius equal to the distance $r$ of the point considered from the axis of symmetry (Fig. 5.4). Thus, instead of Eq. (5.2.3), we have

$$\Delta\Psi_D \simeq D\,\Delta a_\Psi\, 2\pi r = \varepsilon\,\Delta a_\Psi\, 2\pi r \frac{\Delta V}{\Delta a_V}, \tag{5.2.7}$$

from which

$$\frac{\Delta a_\Psi}{\Delta a_V} \simeq \frac{1}{r}\frac{\Delta\Psi_D}{2\pi\varepsilon\Delta V}. \tag{5.2.8}$$

Observe that $\Delta\Psi_D$ and $\Delta V$ are constants for a given plot. The ratio $\Delta a_\Psi/\Delta a_V$ in a correct plot must therefore vary with the distance $r$ of the curvilinear rectangle considered from the axis of symmetry. Assuming that

$$\Delta\Psi_D/2\pi\varepsilon\Delta V = a_0, \tag{5.2.9}$$

where $a_0$ is a certain constant length, (5.2.8) becomes

$$\frac{\Delta a_\Psi}{\Delta a_V} \simeq \frac{a_0}{r} \quad \text{or} \quad \Delta a_\Psi \simeq \frac{a_0}{r}\Delta a_V. \tag{5.2.10}$$

For $r = a_0$, the curvilinear rectangles should be approximately curvilinear squares.

* This section may be omitted without loss of continuity.

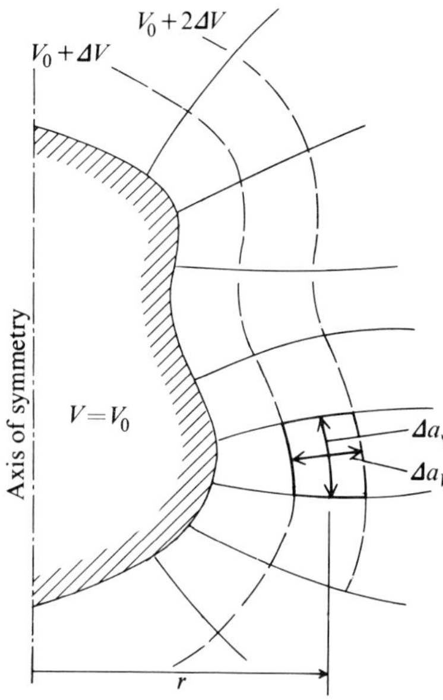


**Fig. 5.4** An axially symmetrical electrostatic field.

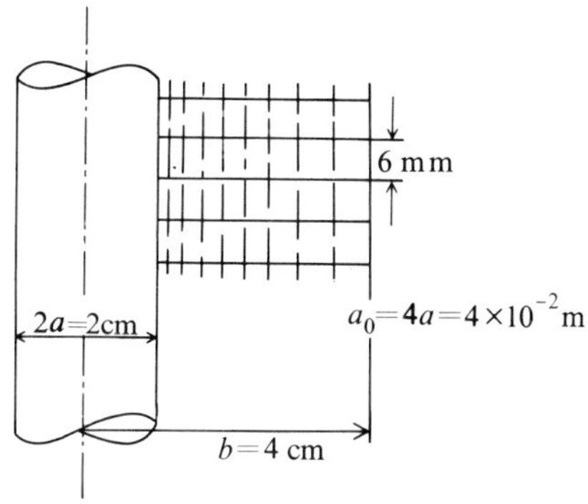


**Fig. 5.5** A coaxial cable with $b/a = 4$ considered as an axially symmetrical system.

The plot of an axially symmetrical electrostatic field is hence more involved than the plot in the case of two-dimensional fields. If such a plot is determined, we can obtain the same informations as before from the plot. For example, the total charge on an axially symmetrical conductor, in analogy with Eq. (5.2.5) and according to Eq. (5.2.7), is

$$Q = N_\Psi \Delta\Psi_D = N_\Psi\, \varepsilon\, 2\pi r \frac{\Delta a_\Psi}{\Delta a_V}\, \Delta V.$$

If we have two bodies charged with equal charges of opposite signs, their capacitance is given by

$$C = \frac{Q}{V} = \frac{N_\Psi \varepsilon 2\pi r(\Delta a_\Psi/\Delta a_V)\Delta V}{N_V \Delta V} = \varepsilon \frac{N_\Psi}{N_V} 2\pi r \frac{\Delta a_\Psi}{\Delta a_V}.$$

According to Eq. (5.2.10), this becomes

$$C = 2\pi\varepsilon a_0 \frac{N_\Psi}{N_V}. \tag{5.2.11}$$

The constant $a_0$ represents an adopted distance (in meters!) from the axis of symmetry at which curvilinear rectangles are approximately curvilinear squares.

Consider a simple example. Figure 5.5 shows the plot of the field inside a part of a coaxial cable, assuming that $a_0 = b$, i.e. that the rectangles are roughly squares at $r = b$. Note that $\Delta a_\Psi/\Delta a_V \simeq 2$ at $r = a_0/2 = b/2$, and that $\Delta a_\Psi/\Delta a_V \simeq 4$ at $r = a_0/4 = a$.

The approximate value of the capacitance per unit length of such a cable is obtained according to Eq. (5.2.11). We assumed that $a_0 = 4$ cm and $\Delta a_\Psi \simeq 6$ mm. Since the plot resulted in $N_V = 9$,

$$C_l \simeq 2\pi\varepsilon a_0 \frac{1/(\Delta a_\Psi)}{N_V} = 2\pi\varepsilon\, 4 \times 10^{-2} \frac{167}{9} \simeq 2\pi\varepsilon \times 0.745.$$

According to the exact formula, for $b/a = 4$,

$$C_l = 2\pi\varepsilon/\ln(b/a) = 2\pi\varepsilon/\ln 4 \simeq 2\pi\varepsilon \times 0.741.$$

The difference between the two results is very small.

**5.2.3 The graphical method in the case of inhomogeneous dielectric.*** The procedure explained above for plotting both two-dimensional and axially symmetrical fields is valid only if the permittivity is constant. The procedure can also be used if several dielectrics are present. In that case, however, the method requires different ratios of $\Delta a_\Psi/\Delta a_V$ for different dielectrics. We can see this from Eqs. (5.2.4) and (5.2.8). In addition, the lines of vector **D** must refract at the boundary surfaces of two different dielectrics according to the law given by Eq. (2.6.8). Thus, although the method is applicable in these cases as well, it becomes much more intricate, and we shall not spend more time on it here.†

*Problems referring to Section 5.2:* 5.T1 to 5.T4, 5.G1 to 5.G8.

## 5.3 THE NUMERICAL SOLUTION OF LAPLACE'S EQUATION

Numerical solutions of many types of differential equations are today possible as a result of the extremely high storage possibilities of modern digital computers. Laplace's equation can be solved numerically in many cases, and this method of solving it is becoming of ever-increasing importance in engineering practice. The logical basis for such a solution is described below.

The numerical solutions of Laplace's equation are used in practice mainly for solving two-dimensional problems, but the procedure is also possible in the case of three-dimensional fields.

To solve Laplace's equation numerically, we represent it in the approximate form of finite differences instead of true differentials. Consider a point $M(x, y)$ in a two-dimensional field, and suppose that the potential at $M$ is $V(x, y)$. The potential at the neighbouring points $M_1$, $M_2$, $M_3$, and $M_4$ shown in Fig. 5.6 is obtained approximately if $V(x, y)$ is expanded in a Taylor series and only the

* This section may be omitted without loss of continuity.

† For a more detailed description of graphical field plotting and for further references see E. Weber, *Electromagnetic Fields*, Vol. 1—"Mapping of fields," John Wiley & Sons, 1950, pp. 197–206.

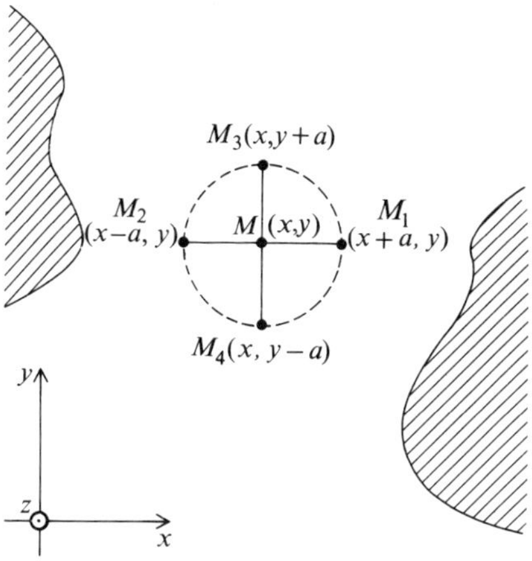


**Fig. 5.6** Construction for determining the finite difference approximation to Laplace's equation.

three first terms are retained:

$$V(x + a, y) = V(x, y) + a\frac{\partial V}{\partial x} + \frac{a^2}{2}\frac{\partial^2 V}{\partial x^2} + \text{terms of higher order.}$$

In the same way

$$V(x - a, y) = V(x, y) - a\frac{\partial V}{\partial x} + \frac{a^2}{2}\frac{\partial^2 V}{\partial x^2} + \text{terms of higher order.}$$

If we add these two equations, the first derivative $\partial V/\partial x$ cancels out. The second partial derivative $\partial^2 V/\partial x^2$ can thus be expressed approximately by the value of $V$ at $M$, and at the neighbouring points $M_1$ and $M_2$:

$$\frac{\partial^2 V}{\partial x^2} \simeq \frac{V(x + a, y) + V(x - a, y) - 2V(x, y)}{a^2}. \tag{5.3.1}$$

Similarly, the approximate value of the second partial derivative $\partial^2 V/\partial y^2$ of $V$ is

$$\frac{\partial^2 V}{\partial y^2} \simeq \frac{V(x, y + a) + V(x, y - a) - 2V(x, y)}{a^2}. \tag{5.3.2}$$

According to Eqs. (5.3.1) and (5.3.2) the approximate, finite difference expression of the Laplacian of the two-dimensional scalar potential $V(x, y)$ is

$$\nabla^2 V = \frac{\partial^2 V}{\partial x^2} + \frac{\partial^2 V}{\partial y^2} \simeq \frac{1}{a^2}[V(x + a, y) + V(x - a, y) + V(x, y + a) + V(x, y - a) - 4V(x, y)].$$

Hence the approximate, finite difference form of Laplace's equation $\nabla^2 V = 0$ is

$$V(x, y) \simeq \tfrac{1}{4}[V(x + a, y) + V(x - a, y) + V(x, y + a) + V(x, y - a)]. \quad (5.3.3)$$

This equation can also be obtained from the result of Problem 1.G22. According to that result, the potential at the center of any sphere enclosing no charges is exactly equal to the average value of the potential over the sphere surface. In the two-dimensional case considered, $V(x, y)$ is hence approximately equal to the average value of the potential at six symmetric points at the surface of the sphere of radius $a$, or

$$V(x, y, z) \simeq \tfrac{1}{6}\,[V(x + a, y, z) + V(x - a, y, z) + V(x, y + a, z) + V(x, y - a, z) + V(x, y, z + a) + V(x, y, z - a)].$$

Since the field is two-dimensional, the potential is not a function of $z$, so that Eq. (5.3.3) follows directly.

To demonstrate how Eq. (5.3.3) is used for the approximate numerical solution of two-dimensional electrostatic problems, let us consider the system shown in Fig. 5.7. We can write the equation (5.3.3) for every point of the square net shown, that is, for points 1, 2, 3, . . . , 171. We thus obtain 171 linear equations with 171 values $V_1, V_2, V_3, \ldots, V_{171}$ of the potential at these points as unknowns:

$$\begin{aligned} V_1 &= \tfrac{1}{4}\,(V_2 + V_{\mathrm{I}} + V_{\mathrm{II}} + V_{20}), \\ V_2 &= \tfrac{1}{4}\,(V_3 + V_1 + V_{\mathrm{II}} + V_{21}), \\ &\cdots\cdots\cdots\cdots\cdots \\ V_{171} &= \tfrac{1}{4}\,(V_{\mathrm{III}} + V_{170} + V_{152} + V_{\mathrm{IV}}). \end{aligned} \quad (5.3.4)$$

($V_{\mathrm{I}}$, $V_{\mathrm{II}}$, $V_{\mathrm{III}}$, and $V_{\mathrm{IV}}$ are given potentials of the four walls.)

Laplace's partial differential equation is reduced to a system of linear equations, but with 171 unknowns! It is obvious that for determining the 171 values of the potential at the 171 points shown we cannot easily solve this system of equations using any standard procedure, say the method of determinants. Digital

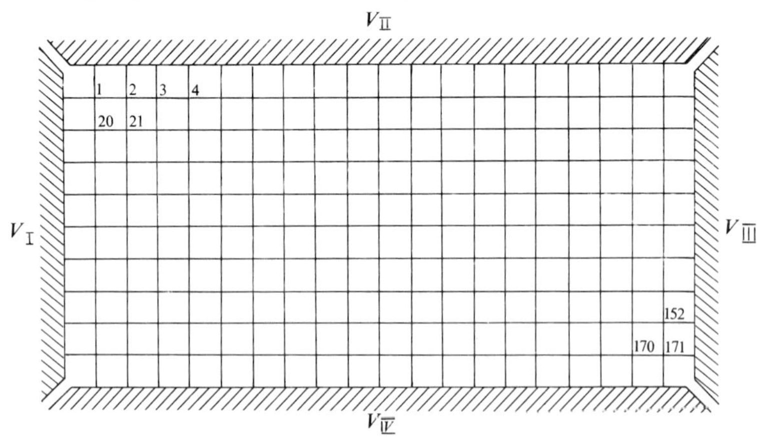


**Fig. 5.7** A two-dimensional potential problem with the square net for determining the potential function using numerical method.

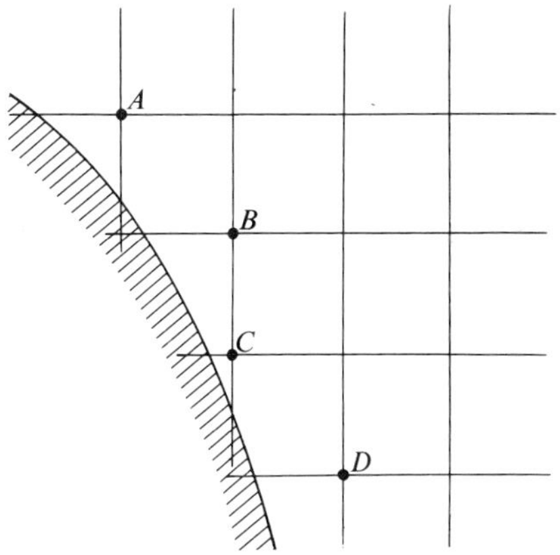


**Fig. 5.8** A case when neighbouring points in a net are not at equal distances.

computers can solve systems of linear equations with a large number of unknowns, but their capacity is nevertheless restricted. Consequently, some simpler approximate method must be used for solving such systems of linear equations with a large number of unknowns, as in Eq. (5.3.4). The following iterative method may be applied to that end.

Suppose that by some auxiliary means (say, by a rough plot of the field, or even by an educated guess) we assign to all 171 nodes of the net in Fig. 5.7 some approximate values $V_1^{(0)}, V_2^{(0)}, \ldots, V_{171}^{(0)}$ of the potential. Since they are only approximate values, none of the equations (5.3.4) will in general be satisfied. However, we can calculate a second set of approximate values $V_1^{(1)}, V_2^{(1)}, \ldots, V_{171}^{(1)}$, using the first set of approximate values and Eq. (5.3.4):

$$V_1^{(1)} = \tfrac{1}{4}[V_2^{(0)} + V_{\mathrm{I}} + V_{\mathrm{II}} + V_{20}^{(0)}], \text{ etc.}$$

The new values of the potential thus obtained will in general be better approximations to the true values [i.e. to the solutions of Eqs. (5.3.4], because the exact values of the potential of the four walls are taken into account. We next introduce these improved values on the right of Eqs. (5.3.4), and obtain the second improved set $V_1^{(2)}, V_2^{(2)}, \ldots, V_{171}^{(2)}$ of the potential at the 171 points. After several such iteration cycles no considerable changes of values of the potential are presumably to be obtained by repeated cycles. The task of evaluating the approximate values of the potential at the 171 points may then be regarded as completed.

The logic of the iterative method is very simple. Unfortunately, many iteration cycles are usually necessary to obtain reasonably accurate results. To make the process converge more rapidly, several refinements of the method have been devised. These include, for example, the use of a finer net only where the changes of potential are more pronounced, the possible extrapolation of values of the potential from a sequence of values obtained by several iteration cycles, etc. In addition, the boundary surfaces may not coincide with the nodes of the net, as in the case shown in Fig. 5.8. At points $A$, $B$, $C$, and $D$ in that figure we must write an equation analogous to Eq. (5.4.3), but valid in the case when one or more of the adjacent points of the net are at unequal distances from these points (see Problem 5.G10).

For further details on the numerical solution of Laplace's equation the reader is referred to more specialized literature.*

*Problems referring to Section 5.3:* 5.T7, 5.G9 to 5.G13.

## 5.4 ANALOG METHODS FOR THE APPROXIMATE SOLUTION OF ELECTROSTATIC PROBLEMS

**5.4.1 Introduction.** The electrostatic potential function in charge-free regions is a solution of Laplace's partial differential equation, subject to a particular set of boundary conditions. In Section 1.13.2, it was shown that only one solution simultaneously satisfying Laplace's equation and the prescribed set of boundary conditions is possible. Obviously, from the mathematical point of view it is irrelevant if the function to be determined is the electrostatic potential or any other function of coordinates satisfying Laplace's equation, subject to the same set of boundary conditions. For example, Laplace's equation is also satisfied by the electric potential in the case of current flow in a homogeneous conductor. The two-dimensional Laplace's equation is also approximately satisfied by the displacement of a thin elastic membrane deformed from its plane by supports and weights, provided that the displacements are small. The three-dimensional Laplace's equation is exactly satisfied by the velocity potential existing at the points of an incompressible fluid flow, provided that certain conditions are met. The streamlines in this case correspond to the lines of the electric displacement vector. The uniqueness property of solutions of Laplace's equation insures that for an analogous set of boundary conditions the solutions in all these cases are essentially the same. This conclusion provides the basis of analog methods for solving electrostatic problems. The logic of some of them is described below.

**5.4.2 The current-flow analogy.** It was mentioned that the electric potential in the case of current flow in a homogeneous conductor is also a function satisfying Laplace's equation. (This will be demonstrated in the next chapter.) Consequently, by determining experimentally the distribution of potential in the analogous current-flow case, we can find an approximate solution to some electrostatic problems. For a complete understanding of this method knowledge of the electric current fields is indispensable. We have therefore postponed a more detailed explanation of the method until Section 6.7.2. Here we need only mention that in practice only two-dimensional and axially symmetrical problems can be analyzed by this method.

---

* For example, G. E. Forsythe and W. R. Wasow, *Finite Difference Methods for Partial Differential Equations*, John Wiley & Sons, 1960.

**5.4.3 The elastic membrane analogy.** It may be shown that the exact partial differential equation governing the vertical displacement $z$ of a horizontal elastic membrane is

$$\frac{\partial^2 z}{\partial x^2} + \frac{\partial^2 z}{\partial y^2} + \frac{\partial^2 z}{\partial x^2}\left(\frac{\partial z}{\partial y}\right)^2 + \frac{\partial^2 z}{\partial y^2}\left(\frac{\partial z}{\partial x}\right)^2 - \frac{\partial^2 z}{\partial x \partial y}\frac{\partial z}{\partial x}\frac{\partial z}{\partial y} = 0.$$

We can see that, if the displacements are small, more precisely if the inclination of the deformed membrane with respect to the horizontal plane is small everywhere, the first two terms only may be retained. Displacement $z$ in such cases approximately satisfies Laplace's equation in two dimensions. With the above restriction, we can obtain the physical model of any two-dimensional electrostatic field, displacement $z$ corresponding to the potential $V$. For example, Fig. 5.9 shows the cross-section of a simple model of the distribution of the potential in the case of two conductors at potentials $V = +V_0$ and $V = -V_0$ inside a circular tube at zero potential. For further details about this method the reader is referred to other texts.*

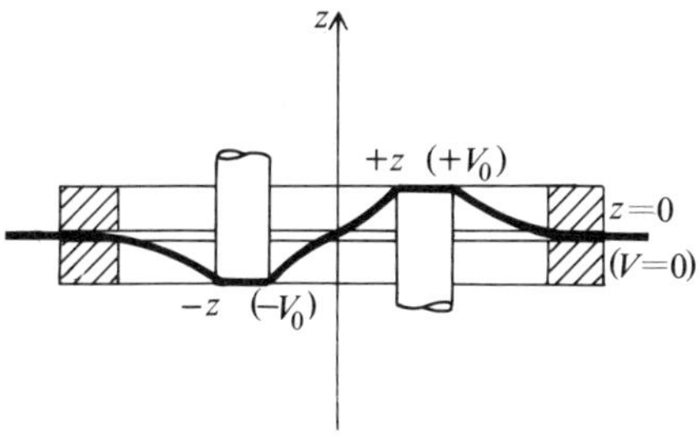


**Fig. 5.9** Elastic membrane analogy of a shielded two-wire line.

**5.4.4 The fluid-flow analogy.** In an incompressible homogeneous fluid the velocity vector $\mathbf{v}$ of the fluid particles must satisfy the equation

$$\operatorname{div} \mathbf{v} = 0. \tag{5.4.1}$$

If this were not so, the density of the fluid would become progressively larger at points where (5.4.1) is not satisfied. This is impossible if the fluid is supposed to be incompressible. If, in addition, the flow is such that along any closed line

$$\oint_C \mathbf{v} \cdot d\mathbf{l} = 0, \tag{5.4.2}$$

the velocity vector $\mathbf{v}$ precisely satisfies the same equations as the electric field intensity vector $\mathbf{E}$ in a charge-free region. Thus, the streamlines correspond to the lines of vector $\mathbf{E}$ in the analogous electrostatic case.

---

* See, for example, P. Moon and D. E. Spencer, *Field Theory for Engineers*, Van Nostrand, 1961, pp. 26–28, or E. Weber, *op. cit.*, pp. 193–194.

Equation (5.4.1) is satisfied for all fluids, since their compressibility is generally very small. To satisfy the other necessary equation (5.4.2), a special technique is required. The velocity vector of a fluid flow between two parallel, closely spaced planes very nearly satisfies Eq. (5.4.2). We can obtain very clear pictures of the streamlines of the velocity field, corresponding to the lines of force in the analogous electrostatic case, by using this method with colored fluids.*

*Problems referring to Section 5.4:* 5.T8 to 5.T10.

## PROBLEMS

### Test problems

**5.T1.** Discuss the method of curvilinear squares if applied for plotting the whole field in the case of a two-dimensional electrostatic system with an inhomogeneous dielectric. Is it possible to use it in this case?

**5.T2.** Can the method of curvilinear squares be used for plotting the field of two identical spheres charged with equal charges of opposite signs?

***5.T3.** Explain why the graphical method cannot be used to obtain the field plot in some cross-section of an arbitrary three-dimensional field.

**5.T4.** In the case of the two-dimensional system shown in Fig. 5.2, it is not possible to draw all equipotential lines and all lines of vector **D** in full length, since the drawing is of limited dimensions. Does that fact interfere with the calculation of capacitance according to Eq. (5.2.6)?

**5.T5.** Is it possible to use a system of finite difference equations in the case of a three-dimensional system? If you think that it is possible, explain why such a method is not very often used in practice in those cases.

***5.T6.** Is it possible to adapt the numerical method described in Section 5.3 for solving axially symmetrical problems? Discuss the question quite generally, bearing in mind Laplace's equation in cylindrical coordinates.

**5.T7.** Suppose that the system of difference equations in Eq. (5.3.4) is solved exactly, i.e. that values $V_1, V_2, \ldots, V_{171}$ are found which satisfy the system exactly. Are these values of the potential the *exact* values of the potential at these points?

**5.T8.** Is it necessary that the ends of the two rods in Fig. 5.9, representing the electrodes, be flat as shown, or may they also be rounded?

**5.T9.** What would be the elastic membrane model for the electrostatic system of two isolated parallel bus bars at potentials $V$ and $-V$? What if the two bus bars are inside a tube of arbitrary cross-section and at zero potential?

**5.T10.** Where would you inject and where take a fluid from between two closely spaced parallel plates to make the fluid flow between the plates analogous to the lines of force of a two-wire lines with thin conductors?

---

* See, for example, W. H. Hayt, *Engineering Electromagnetics*, 2nd edition, McGraw-Hill, 1967, p. 166. For a more detailed description of the method and for further references see E. Weber, *op. cit.*, p. 194, and P. Moon and D. E. Spencer, *op. cit.*, pp. 21 – 25.

## General problems

**5.G1.** The distance $d$ between two parallel conducting strips of width $a$ is $d = a/10$. Construct a plot of curvilinear squares and determine the capacitance per unit length of this strip-line. Since $d \ll a$, the result should almost be the same as that according to the approximate analytical formula in which thc fringing effect is ignored.

**5.G2.** Repeat the preceding problem if $d = a$.

**5.G3.** The distance between the axes of two parallel conductors of radius $R$ is $3R$. Construct a plot of curvilinear squares for some convenient value of $R$ (say $R = 2$ cm), and determine the capacitance per unit length of this line. Check the result thus found by calculating the capacitance using the exact formula in Eq. (4.2.6).

**5.G4.** Two parallel bus bars of square cross-section with sides $a$ are at a distance $a$ apart. Construct a plot of curvilinear squares for some convenient value of $a$ (say, $a = 3$ cm). Compute the capacitance per unit length of such a line. It should not be too different from the capacitance per unit length between two conductors of circular cross-section of, say, radius $a/2$, and with distance between their axes $2a$.

**5.G5.** Construct a plot of curvilinear squares and compute the approximate value of the capacitance per unit length of two parallel strips shown in Fig. 5.G5.

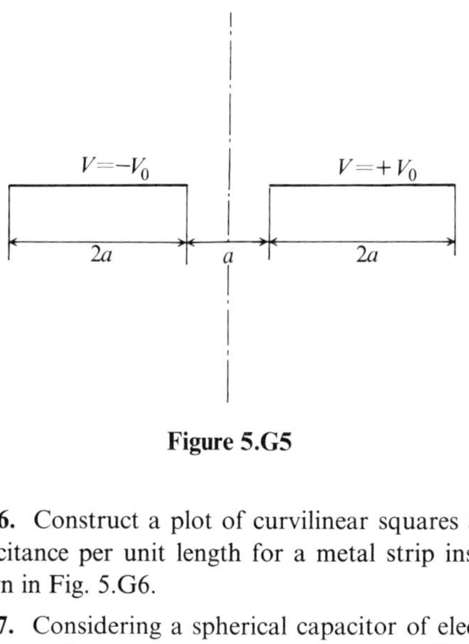


Figure 5.G5

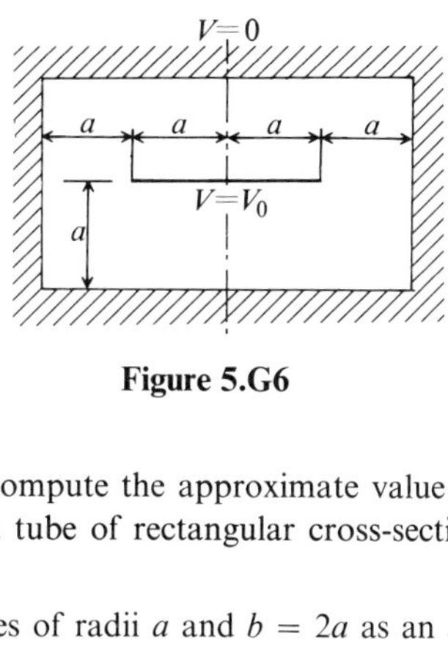


Figure 5.G6

**5.G6.** Construct a plot of curvilinear squares and compute the approximate value of the capacitance per unit length for a metal strip inside a tube of rectangular cross-section, as shown in Fig. 5.G6.

**5.G7.** Considering a spherical capacitor of electrodes of radii $a$ and $b = 2a$ as an axially symmetrical system, construct the plot of curvilinear rectangles, assuming that in Eqs. (5.2.10) and (5.2.11) $a_0 = a$. This means that, at the distance $a$ from the axis of symmetry, $\Delta a_\Psi \simeq \Delta a_V$, at the distance $a/2$, $\Delta a_\Psi \simeq 2\Delta a_V$, and at the distance $b = 2a$, $\Delta a_\Psi \simeq \Delta a_V/2$. These facts are of considerable help in constructing the plot in axially symmetrical cases. Determine the approximate value of the capacitance from the plot, and compare with the exact value of the capacitance.

***5.G8.** A coaxial cable of conductors of radii $a$ and $b = 3a$ has a two-layer dielectric, of permittivities $\varepsilon_1$ (the inner layer) and $\varepsilon_2 = \varepsilon_1/2$ (the outer layer). The thickness of both dielectric layers is $a$. Construct a plot of curvilinear rectangles, assuming that they are square-shaped in dielectric of permittivity $\varepsilon_1$ [then in the other dielectric, according to Eq. (5.2.4) and since $\varepsilon_2 = \varepsilon_1/2$, $\Delta a_\Psi/\Delta a_V \simeq 2$]. Find the approximate value of capacitance per unit length, substituting either $\varepsilon_1$ or $\varepsilon_2$ into Eq. (5.2.6) instead of $\varepsilon$. Note that if $\varepsilon_2$ is introduced, $\Delta a_\Psi/\Delta a_V = 2$, not 1, so that the same result is obtained. Compare the capacitance thus obtained with the exact value of the capacitance.

**5.G9.** Suppose we determined the approximate values of the potential at the nodes of a very coarse square net. Explain how it is possible to obtain from them the approximate values of the potential at the nodes of a net with squares of half the size. This procedure is very convenient for determining more precisely the potential in domains where the potential changes very much from node to node.

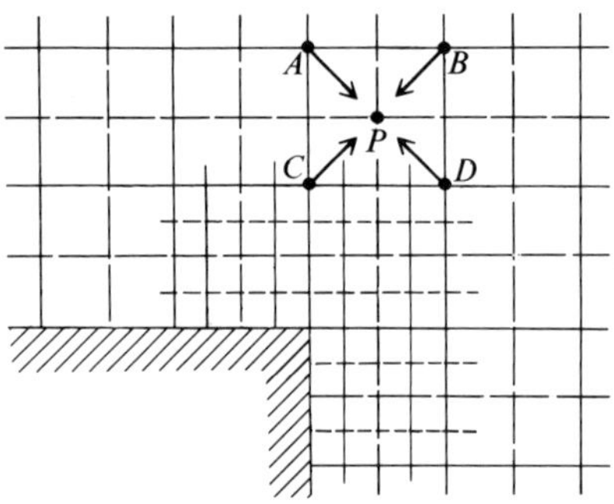


**Figure 5.G9**

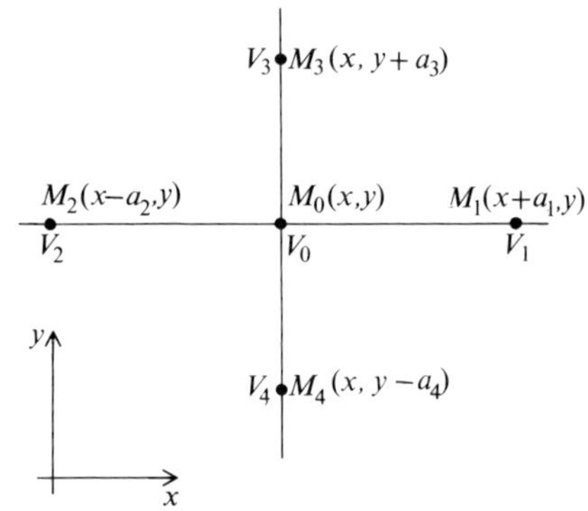


**Figure 5.G10**

*Solution.* Consider the coarse net shown in Fig. 5.G9 in solid lines. Suppose that we determined the values of the potential at the nodes of the net using the iterative or any other process. Obviously, in the vicinity of the conducting corner shown the potential changes very much from node to node. To determine the potential in this region more precisely, we construct around the corner a net twice as fine, and in the immediate surroundings of the corner a net four times as fine, as shown in Fig. 5.G9. We first determine the approximate values of the potential at the nodes of the second net, using the known values of the potential and an expression analogous to Eq. (5.3.3). For example, at the point $P$

$$V_P \simeq \tfrac{1}{4}(V_A + V_B + V_C + V_D).$$

The values of the potential thus obtained must be improved by using the iterative process before we repeat the same procedure for the still finer net shown.

**5.G10.** Determine the expression analogous to Eq. (5.3.3) if the distances of points where the potential is given from the point where it is being determined are unequal, rather than equal (see Fig. 5.G10).

*Solution.* By analogy with equations from which the approximate expressions for the second derivatives in Eqs. (5.3.1) and (5.3.2) were derived for $a_1 = a_2 = a_3 = a_4$, we can approximately write

$$V_1 = V(x + a_1, y) \simeq V_0 + a_1 \frac{\partial V}{\partial x} + \frac{a_1^2}{2} \frac{\partial^2 V}{\partial x^2}, \tag{a}$$

$$V_2 = V(x - a_2, y) \simeq V_0 - a_2 \frac{\partial V}{\partial x} + \frac{a_2^2}{2} \frac{\partial^2 V}{\partial x^2}. \qquad \text{(b)}$$

Multiplying (a) by $a_2$, (b) by $a_1$, and adding the results, we have

$$a_2 V_1 + a_1 V_2 \simeq V_0(a_1 + a_2) + \frac{\partial^2 V}{\partial x^2} \frac{a_1 a_2(a_1 + a_2)}{2},$$

from which

$$\frac{\partial^2 V}{\partial x^2} \simeq \frac{2(a_2 V_1 + a_1 V_2)}{a_1 a_2(a_1 + a_2)} - \frac{2V_0}{a_1 a_2}. \qquad \text{(c)}$$

We similarly find that

$$\frac{\partial^2 V}{\partial y^2} \simeq \frac{2(a_4 V_3 + a_3 V_4)}{a_3 a_4(a_3 + a_4)} - \frac{2V_0}{a_3 a_4}. \qquad \text{(d)}$$

Adding (c) and (d), and noting that the sum of the left-hand sides is approximately zero if $V$ satisfies Laplace's equation, we finally have

$$V_0 \simeq \frac{a_3 a_4}{(a_1 + a_2)(a_1 a_2 + a_3 a_4)}(a_2 V_1 + a_1 V_2) + \frac{a_1 a_2}{(a_3 + a_4)(a_1 a_2 + a_3 a_4)}(a_4 V_3 + a_3 V_4). \qquad \text{(e)}$$

For $a_1 = a_2 = a_3 = a_4$, (e) reduces to Eq. (5.3.3).

**5.G11.** Determine the equipotential lines in the two-dimensional system shown in Fig. 5.G11a. Use the numerical method.

*Solution.* Because of symmetry it is sufficient to determine the potential in the crosshatched part of Fig. 5.G11a. In Fig. 5.G11b, this region is shown on a larger scale, with a net of rather large squares. To obtain some rough estimate, we determine first the approximate values of the potential at the nodes of this coarse net.

At all nodes of the net are shown the adopted initial values of the potential. To explain how the process is performed for points on the two axes of symmetry, the symmetric points

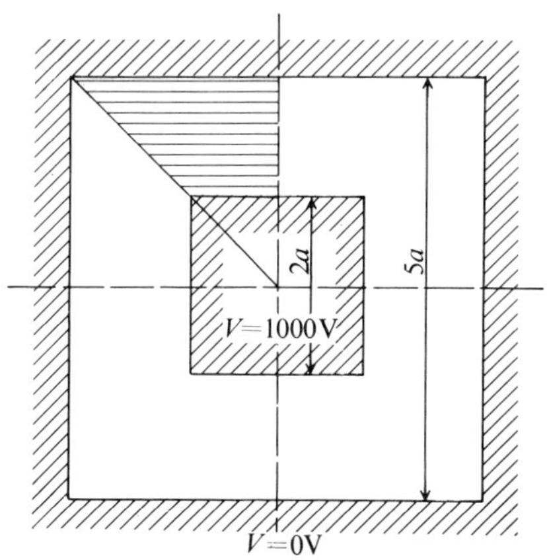


**Figure 5.G11a**

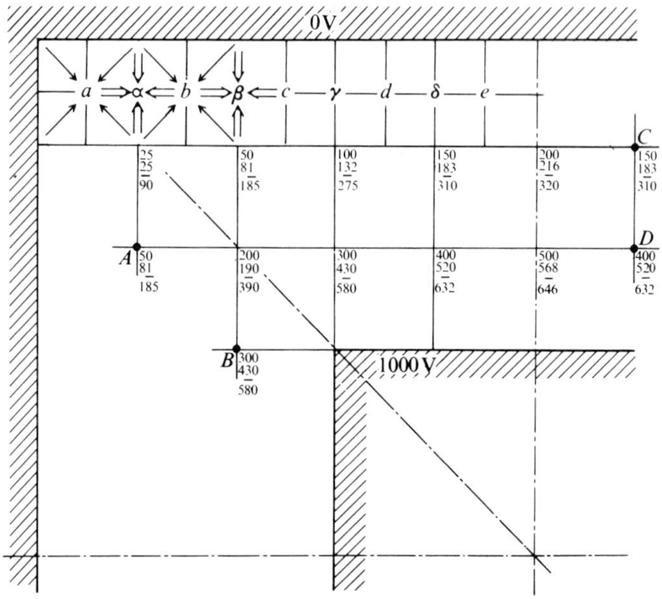


Figure 5.G11b

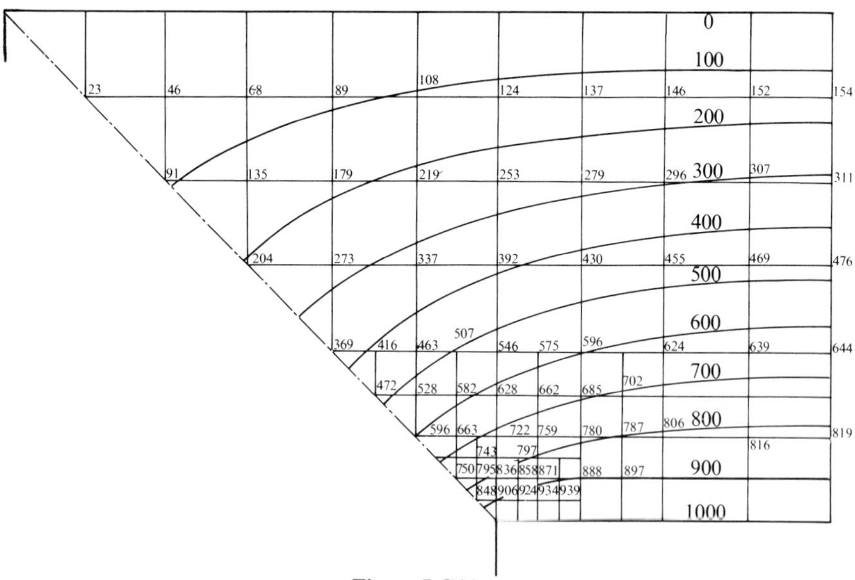


Figure 5.G11c

$A$, $B$, $C$, and $D$ are shown with the corresponding initial values of the potential. Whenever, during the process, these points are to be taken into account, we assume that their potential is equal to the potential of the symmetric point.

Below the initial values of the potential the first improved values are shown. If the calculations are performed using a small desk calculator, it is advisable either to erase or to cross out the old values, to eliminate possible confusion. The last values shown are obtained after seven iteration cycles. We then adopt these values as approximate values, and a net twice as fine is constructed. The approximate, initial values at the nodes of the finer net must first be determined at the points $a, b, c, \ldots$, shown in Fig. 5.G11b. As an example, single arrows indicate which points are to be taken into account for determining the potential at points $a$ and $b$. When the potential is thus obtained at all points $a, b, c, \ldots$, the potential at points $\alpha, \beta, \gamma, \ldots$ can be found. As an example, double arrows indicate which points should be taken into account for determining the approximate values of the potential at points $\alpha$ and $\beta$.

The values of the potential thus found are taken as initial values, and the iteration cycles are repeated several times to obtain the final values indicated in Fig. 5.G11c. For greater accuracy the net is subdivided into a finer net twice more, as shown in the figure. At the end the equipotential lines, with potential difference $\Delta V = 100$ V between the adjacent lines are sketched.

**5.G12.** Assuming that in Fig. 5.G6 $V = 1000$ V, determine the potential at the nodes of a net with the squares of its sides $a/4$. Then, where necessary, use a net which is twice as fine.

**5.G13.** Starting with a net having squares of sides $a/8$, determine the approximate values of the potential for the two-dimensional system shown in Fig. 5.G13. Then, where necessary, use a net which is twice as fine.

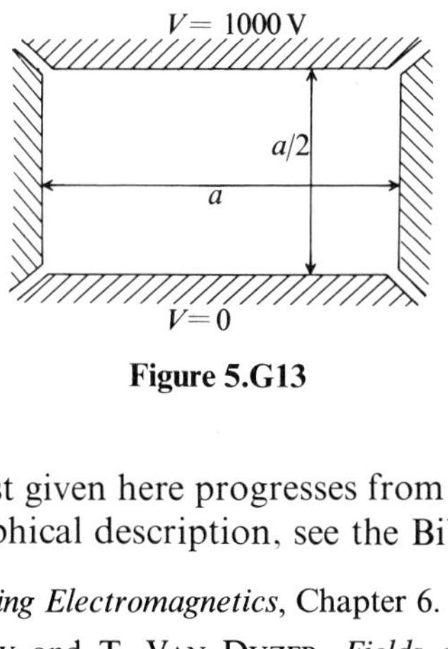


**Figure 5.G13**

**Suggested reading.** The list given here progresses from simpler to more advanced texts. For a full bibliographical description, see the Bibliography.


1. HAYT, W. H., Jr., *Engineering Electromagnetics*, Chapter 6.
2. RAMO, S., J. R. WHINNERY and T. VAN DUZER, *Fields and Waves in Communication Electronics*, pp. 159–172.

# PART II

# THE STEADY ELECTRIC CURRENT

CHAPTER 6

# STEADY ELECTRIC CURRENT IN SOLID AND LIQUID CONDUCTORS

## 6.1 INTRODUCTION

In the preceding chapters we have considered the static electric field, that is, the field due to static electric charges. In this, and in the next chapters, we shall deal with cases where the continuous distributions of charges are macroscopically in a steady state of motion. Such a steady state is obtained when large aggregates of elemental electric charges are moving so, that the macroscopic picture of the charge motion is constant in time.

Since the charges are not static any more, a revision of the fundamental law of the theory of the field of static charges (i.e., Coulomb's law of force) is necessary. We have to investigate experimentally whether the force between pairs of moving charges can still be described by Coulomb's law alone, or whether, perhaps, some additional terms will have to be introduced into it. In Chapters 8 and 11 we shall see that, whenever electric charges are in motion with respect to an observer, additional terms in the force law *are* necessary. Fortunately, in the case of the continuous distribution of charges in a *steady* state of motion, the forces of this new type on *individual* elemental electric charges are usually negligible.

The steady motion of charges is most often canalized by suitable conductors. We noticed in Section 1.1 that by the term "conductor" we meant any substance having throughout its volume a relatively large number of charges free to move inside it. These charges cannot leave the conductor in normal circumstances.

The most important class of conductors is that of metallic ones. According to the crude, but for us adequate, model described in Section 1.1, metallic conductors are characterized by a very loose connection between the valence, or conduction electrons, and the rest of an atom. In normal circumstances, these electrons move at random from one atom to another inside the metal. This random motion produces no average drift of electrons. However, in response to even a very weak electric field an organized motion of electrons, superimposed on their random motion, will result. Such an organized motion of elementary charged particles is referred to as the *electric current*. The electrons will be accelerated by the electric field, but sooner or later they will collide with an atom, the consequences of which will be twofold. First, the average velocity of the electrons in this directed motion will be limited, in fact very small. Second, the electrons will transfer to the atoms a part of the acquired kinetic energy. This will be manifested in vibration of the atoms which is known to represent thermal energy. The electric current in any conductor is thus necessarily accompanied by the generation of heat. This is known as *Joule's heat*.

The mechanism of conduction in liquid solutions is somewhat different. A certain percentage of the molecules is known to be dissociated into two oppositely charged parts, called *ions*. If an electric field is established in such a liquid, the positive ions will move in the direction of the field, and the negative ions in the opposite direction. The electric current in liquid conductors is thus made up of two streams of charged particles moving in opposite directions, and generally with different average velocities. Generation of heat will also occur, as with metallic conductors. In addition, the electric current in liquid solutions is always accompanied by chemical changes in the liquid, a process we call *electrolysis*.

The number of free electric charges per unit volume of both metallic and liquid conductors is extremely large, for we must remember that there are some $10^{22}$ to $10^{24}$ atoms per cubic centimeter of a rigid body. In $10^{-6}$ cubic centimeters of a metallic conductor there is therefore a huge number, approximately $10^{16}$ to $10^{18}$ electrons. The electric current in metals and liquids can thus be visualized *macroscopically* as a flow of some "electric fluid". Historically, the fluid model was the first model of the electric current.

## 6.2 CURRENT DENSITY AND CURRENT INTENSITY

In the preceding section a qualitative description of the electric current was given. Quantitatively, the electric current, that is, the flow of a continuous distribution of electric charges, is represented by two physical quantities. The *current density vector* characterizes the flow of charges *at a point*; the *current intensity* describes the rate of flow of electric charges *through a macroscopic surface*.

Suppose that at a point inside a conductor there are $N$ free charges per unit volume, each carrying a charge $Q$, moving (at the point considered) with an average velocity $\mathbf{v}$. The *current density* $\mathbf{J}$ at that point is defined as:

$$\boxed{\mathbf{J} = NQ\mathbf{v}} \qquad \text{(definition of current density).} \tag{6.2.1}$$

We can see that according to this definition the current density vector of negative charges $-Q$ with average velocities $-\mathbf{v}$ is the same as the current density vector of positive charges $+Q$ with average velocities $+\mathbf{v}$:

$$\mathbf{J}_{-Q} = N(-Q)(-\mathbf{v}) = NQ\mathbf{v} = \mathbf{J}_{+Q}. \tag{6.2.2}$$

The definition in Eq. (6.2.1) of the current density vector thus implies that a stream of negative charges in one direction is equivalent to the stream of these charges taken with the positive sign and moving in the opposite direction.

In the general case, the electric current may be made up of a continuous distribution of different charges, $Q_1, Q_2, \ldots, Q_n$, of different numbers $N_1, N_2, \ldots, N_n$, per unit volume, and with different velocities $\mathbf{v}_1, \mathbf{v}_2, \ldots, \mathbf{v}_n$, at the point con-

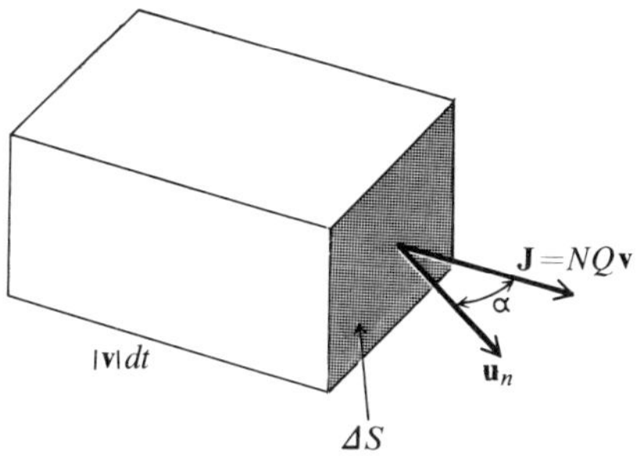


**Fig. 6.1** A small planar surface in a current field.

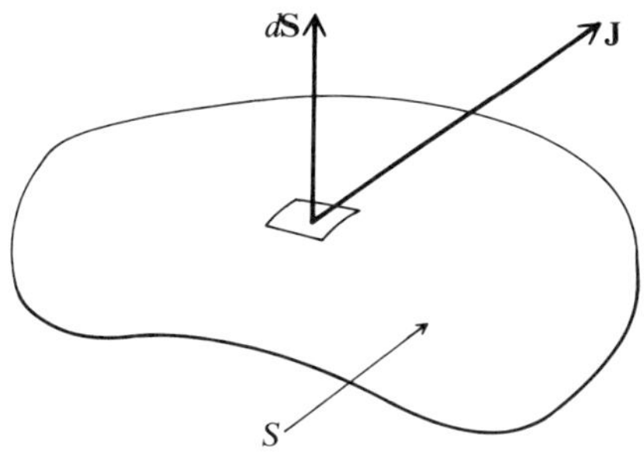


**Fig. 6.2** Illustration for defining current intensity through a macroscopic surface.

sidered. The current density is then defined as the vector sum of the partial current densities:

$$\mathbf{J}_{\text{total}} = N_1Q_1\mathbf{v}_1 + N_2Q_2\mathbf{v}_2 + \cdots + N_nQ_n\mathbf{v}_n. \tag{6.2.3}$$

Let us consider a small planar surface $\Delta S$ in a stream of charged particles constituting the electric current, and suppose that the current density vector at that point is $\mathbf{J} = NQ\mathbf{v}$ (Fig. 6.1). During a small time interval $\Delta t$, the surface $\Delta S$ will be traversed by the charge $\Delta Q$ occupying the volume $\Delta S(v\Delta t)\cos\alpha$ to the left of $\Delta S$, as shown in Fig. 6.1. The charge $\Delta Q$ that crosses $\Delta S$ in the time interval $\Delta t$ is thus

$$\Delta Q = NQv\Delta S\Delta t\cos\alpha. \tag{6.2.4}$$

The *electric current intensity* through the surface, or, simply, the current intensity, is defined as the charge that crosses the surface per unit time. The adopted symbol for current intensity is $I$ in the case of steady currents, and $i(t)$, or simply $i$, in the case of time-varying currents. According to this definition the current intensity $\Delta I$ through $\Delta S$ in Fig. 6.1 is

$$\Delta I = \Delta Q/\Delta t = NQv\Delta S\cos\alpha. \tag{6.2.5}$$

Consider the small surface $\Delta S$ as a vector quantity, $\Delta\mathbf{S} = \mathbf{u}_n\Delta S$. Then Eq. (6.2.5) can be represented in the vector form:

$$\Delta I = NQ\mathbf{v}\cdot\Delta\mathbf{S} = \mathbf{J}\cdot\Delta\mathbf{S}. \tag{6.2.6}$$

From Eq. (6.2.6) the total current intensity $I$ through an arbitrary surface $S$, that is, the charge which crosses the surface per unit time, is equal to the flux of the current density vector $\mathbf{J}$ through the surface (Fig. 6.2):

$$\boxed{I = \frac{dQ}{dt} = \int_S \mathbf{J}\cdot d\mathbf{S}} \qquad \text{(definition of current intensity).} \tag{6.2.7}$$

According to this definition, the current intensity through a surface may be either positive or negative. Its sign depends upon the choice of the positive normal

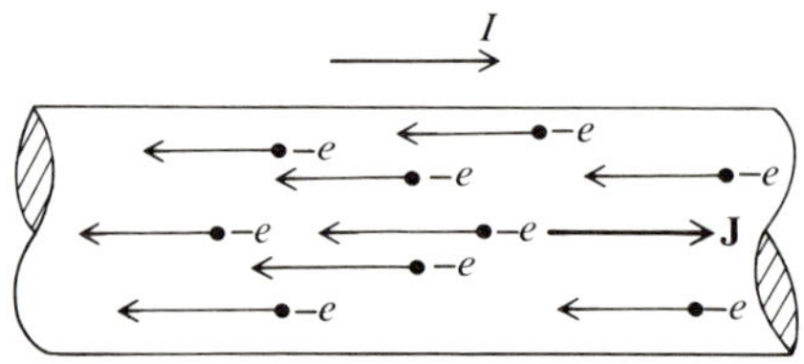


**Fig. 6.3** In metallic conductors current density and current intensity are in the opposite direction to the flow of electrons.

to the surface. By the definition, Eq. (6.2.7), the current intensity is a *scalar* quantity. It is customary, however, particularly in wire conductors, to refer to the direction of current, which is assumed to be in the direction of the current density vector. According to this convention, the direction of the current in a metal wire is opposite to the direction of the flow of the free electrons which constitute the current (Fig. 6.3).

The unit of the current intensity in the MKSA system is *coulomb per second*. Since this is a very important quantity, a separate name, the *ampere* (abbreviation A) is used instead of C–s. From Eq. (6.2.7) it follows that the unit for the current density is *ampere per square meter* ($A/m^2$).

Very often in practice, particularly in electronics, submultiples of the ampere are used, the *milliampere* (mA, equal to $10^{-3}$ A) and the *microampere* ($\mu$A, equal to $10^{-6}$ A). Similarly, since current density of 1 $A/m^2$ is a very small density for most practical applications, larger units for the current density are very often used instead. These are $A/mm^2$ (equal to $10^6$ $A/m^2$) and $A/cm^2$ (equal to $10^4$ $A/m^2$).

The lines of the current density vector **J**, or *current flow lines*, are defined as geometrical lines having the property that vector **J** is tangential to such a line at all points. Any tubular surface formed by a bunch of current flow lines has the obvious property that the flux of vector **J**, that is, the electric current intensity, is the same through any cross-section of the tube. For this reason such a tube is referred to as a tube of current. Conventionally it is always assumed that each line of current flow in a drawing represents a tube of current, and that the whole current field is divided into tubes of *equal* current.

*Problems referring to Section 6.2:* 6.N1 to 6.N2.

### 6.3 THE CONTINUITY EQUATION

There is no experimental evidence that electric charges can be either created or destroyed in macroscopic quantities. This fact is known as the law of conservation of electric charges. The continuity equation is the mathematical form of this law, and is one of the basic equations of electromagnetic theory.

To derive the continuity equation, let us consider a closed surface $S$, enclosing at an instant of time a total charge $Q$ (Fig. 6.4). $Q$ is part of a cloud of charges of density $\rho$ (a function of position and time), distributed in space and moving in

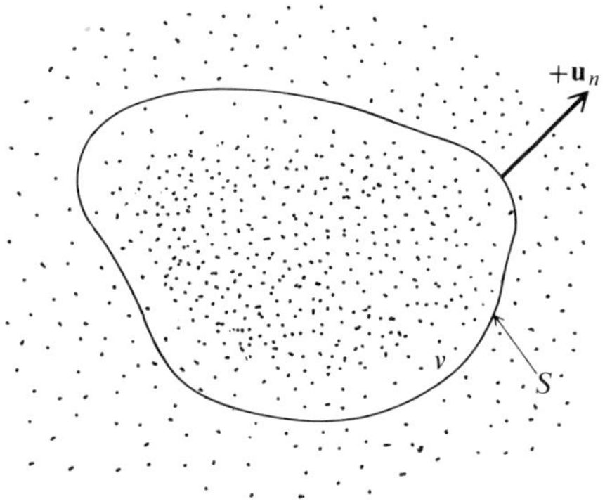


**Fig. 6.4** An imaginary surface enclosing a volume $v$ out of a cloud of charges.

an arbitrary manner. The total current $i_s(t)$ through a fixed surface $S$, with reference to the outward positive normal to $S$, is by definition

$$i_s(t) = \oint_S \mathbf{J} \cdot d\mathbf{S}, \qquad (6.3.1)$$

where $\mathbf{J}$ is the current density vector at $d\mathbf{S}$. Now, the current $i_s(t)$ represents by definition the *positive* charge that crosses $S$ per unit time in the *positive* (outward) direction. But, according to the law of conservation of charges, the outward flow of charges must be accompanied by exactly the same *decrease* of the charge enclosed by $S$. Thus $i_s(t)$ can also be written in the form

$$i_s(t) = \frac{dQ_{\text{crossing } S}}{dt} = -\frac{dQ_{\text{inside } S}}{dt} = -\frac{d}{dt}\int_v \rho \, dv. \qquad (6.3.2)$$

Combining Eqs. (6.3.1) and (6.3.2) we have

$$\boxed{\oint_S \mathbf{J} \cdot d\mathbf{S} = -\frac{d}{dt}\int_v \rho \, dv} \quad \text{(integral form of the continuity equation).} \qquad (6.3.3)$$

This equation is known as the *continuity equation* in integral form. To obtain its differential form we can modify Eq. (6.3.3) as follows. First, by using the divergence theorem (1.11.10), we can transform the surface integral of $\mathbf{J}$ into a volume integral of div $\mathbf{J}$ over the volume $v$ enclosed by $S$. The integral on the right-hand side refers to the volume $v$ which is assumed to be fixed. The time derivative therefore refers only to $\rho$. Since $\rho$ is a function of space coordinates as well as of time, it is customary to write the partial derivative sign $\partial/\partial t$ instead of the ordinary sign $d/dt$. In the light of these remarks Eq. (6.3.3) becomes

$$\int_v \text{div } \mathbf{J} \, dv = -\int_v \frac{\partial \rho}{\partial t} \, dv, \qquad \text{or} \qquad \int_v \left(\text{div } \mathbf{J} + \frac{\partial \rho}{\partial t}\right) dv = 0. \qquad (6.3.4)$$

This equation must hold for an arbitrary volume $v$. This is possible only if the function under the integral sign vanishes. Hence we obtain

$$\boxed{\operatorname{div} \mathbf{J} + \frac{\partial \rho}{\partial t} = 0} \qquad \text{(differential form of the continuity equation).} \qquad (6.3.5)$$

If the electric current is stationary, the picture of current flow, and hence also the density of charges at all points, does not vary in time. In that case $\partial\rho/\partial t = 0$, so that the integral and differential forms of the continuity equation become

$$\boxed{\oint_S \mathbf{J} \cdot d\mathbf{S} = 0} \qquad \text{(steady current),} \qquad (6.3.6)$$

$$\boxed{\operatorname{div} \mathbf{J} = 0} \qquad \text{(steady current).} \qquad (6.3.7)$$

Equations (6.3.6) and (6.3.7) may both be regarded as mathematical expressions of the fact that current flow-lines close on to themselves in the case of steady currents. If we apply Eq. (6.3.6) to a complete tubular surface of *any* tube of current, the current through the surface of the tube is zero, so that the tube closes on to itself. At the same time, this means that the current density vector $\mathbf{J}$ has no sources. Electric charges cannot, of course, be continuously created at any point of a steady current flow!

*Problems referring to Section 6.3:* 6.T1 to 6.T3.

## 6.4 ON THE MECHANISM OF CURRENT FLOW IN SOLID AND LIQUID CONDUCTORS. SOURCES (GENERATORS) OF ELECTROMOTIVE FORCE

**6.4.1 On the mechanism of current flow.** Let us consider two conductors charged with equal charges of opposite signs and embedded in a solid or liquid *dielectric* (Fig. 6.5). In normal circumstances, no charges will leave the conductor surface. Suppose, however, that a small charge $+\Delta Q$ left the surface of the positive conductor and entered the dielectric. Under the influence of the electric field due to charges $Q$ and $-Q$ on the two bodies, the charge $\Delta Q$ will be accelerated in the

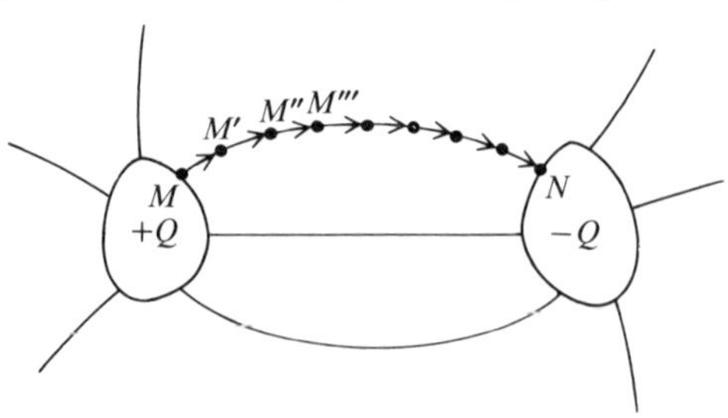


**Fig. 6.5** Two electrodes embedded in a dielectric.

direction of the electric field intensity at $M$. But after a very short path, it will collide, say at a point $M'$, with an atom of the dielectric, and stop. It will next be accelerated in the direction of the electric field intensity at $M'$, but, after a very short path, it will collide with some other atom of the dielectric, and stop again. The process continues until it reaches a point $N$ on the negatively charged body. There it neutralizes with a small fraction $-\Delta Q$ of the total charge $-Q$ on the body. It is obvious that *macroscopically the path of* $\Delta Q$ *will coincide with the line of force from* $M$ *to* $N$, as shown in Fig. 6.5.

Suppose now that, instead of the dielectric, the two charged bodies are embedded in some poor conductor. Free charges available throughout the poor conductor will follow exactly the same type of paths as before. The chains of moving free charges will macroscopically follow the lines of the electric field intensity. The two charged bodies will be, however, invaded by charges with the opposite signs. Thus, sooner or later, the total charge on the two bodies will be neutralized, and the electric field in the poor conductor will vanish. The propelling force on the free charges in the poor conductor disappears, and the current flow is stopped.

To obtain a steady electric current in the poor conductor, it is obviously necessary to keep constant the charges $+Q$ and $-Q$ on the bodies, that is, it is necessary to take away constantly the charge that reaches one body through the poor conductor and to transfer it to the other body. Equilibrium would thus be attained.

**6.4.2 Generators of electromotive force.** To take away, for example, *positive* charges from the *negative* body and to transfer them to the *positive* body, it is necessary for some work to be done, *against* the forces of the electric field of charges $Q$ and $-Q$. To obtain a steady electric current a certain source of energy is hence essential. Such sources of energy must act on the electric charges by some *nonelectric forces*, that is, forces not due to stationary electric charges. These sources are referred to as the *sources or generators of electromotive force*.

The nature of nonelectric forces on electric charges inside generators may be very diverse. Here it is sufficient for us to explain qualitatively the two types of these forces most often encountered.

Historically, the first kind of source of electromotive force capable of producing steady currents of a sufficient intensity and for a significant period of time was the chemical cell. The operation of the chemical cell may be very roughly explained as follows. Consider a metal body immersed in a conducting solution of some chemical compound. The metal body is referred to as the *electrode*, and the solution as the *electrolyte*. In the thin layer of contact between the electrode and the electrolyte the electric charges are acted on by certain forces. These forces have different magnitudes and even different directions for different pairs of electrodes and electrolytes. Thus, if two electrodes of different, suitable materials are immersed in the same electrolyte, these forces could be made to act one from the electrode towards the electrolyte, and the other from the electrolyte towards the other electrode. If the electrodes are connected by a metallic conductor, these two thin layers of contact will act as pumps, pushing the electric charges in the same direction.

The other type of force on the electric charges appears throughout a body moving in a magnetic field, a phenomenon which will be explained in the next part of the book. We will merely note here that these forces act on electric charges at *every* point of a body moving in a magnetic field.

Thus, in general, free electric charges at any point may be acted on simultaneously by electric and by nonelectric forces. The former are due to the electric field of distributed electric charges; the latter are *not* produced by these charges. The nonelectric forces are usually referred to as the *external* or *impressed* forces. If we denote the impressed force on a charge $Q$ at a point by $\mathbf{F}_i$, and the electric field intensity at that point by $\mathbf{E}$, the total force on the charges passing the point considered is then

$$\mathbf{F}_{\text{total}} = Q\mathbf{E} + \mathbf{F}_i. \tag{6.4.1}$$

Now, $\mathbf{F}_i$ can always be represented as a product of $Q$ and a certain vector $\mathbf{E}_i$. $\mathbf{E}_i$ thus defined is *not* any real electric field intensity, since it is not due to stationary electric charges. By definition, only it gives the correct force on the charges if multiplied by their charge. This equivalent electric field intensity is usually referred to as the *external* or *impressed* electric field intensity. In the light of these remarks Eq. (6.4.1) may be written in the form

$$\mathbf{F}_{\text{total}} = Q(\mathbf{E} + \mathbf{E}_i). \tag{6.4.2}$$

The impressed electric field intensity $\mathbf{E}_i$ exists only inside the sources, while the real electric field intensity $\mathbf{E}$, due to the distributed charges, exists both inside and outside the sources.

**6.4.3 The electromotive force of a generator.** All generators used in practice have two common characteristics. Firstly, they are provided with two terminals (usually metallic), known as the electrodes; secondly, there is no current inside the generator if there is no current through the terminal electrodes, as otherwise the generator would transform all its energy into heat inside itself.

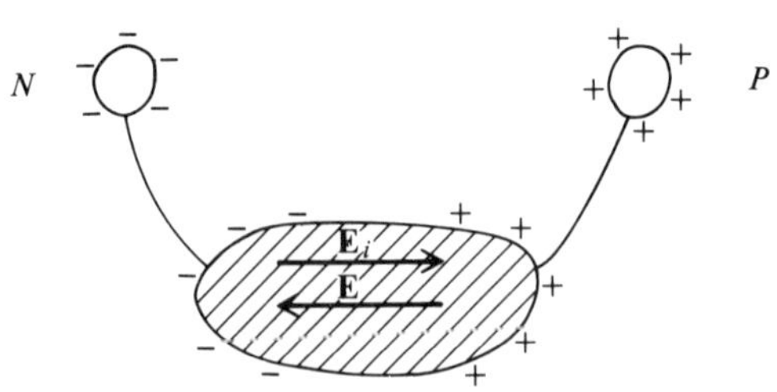


**Fig. 6.6** An open-circuited source of electromotive force.

Such a generator is represented diagrammatically in Fig. 6.6. Let us suppose that the generator is "open-circuited," as shown in the figure. In that case, the electrodes will be charged, as indicated, and the distribution of the charges is such that the electric field intensity $\mathbf{E}$ due to these charges is *exactly* equal in magnitude and opposite to $\mathbf{E}_i$ at all points inside the source. The net force on the charges inside the source must be zero, since it is assumed that no current exists inside an open-circuited generator.

The *electromotive force* $\mathscr{E}$ of a generator is defined as the line integral of the impressed field $\mathbf{E}_i$ from the negative to the positive electrode:

$$\boxed{\mathscr{E} = \int_N^P \mathbf{E}_i \cdot d\mathbf{l}} \qquad \text{(definition of the electromotive force).} \tag{6.4.3}$$

Now, since there are no currents inside the generator, $\mathbf{E}_i = -\mathbf{E}$ at all points, so that

$$\mathscr{E} = \int_N^P \mathbf{E}_i \cdot d\mathbf{l} = -\int_N^P \mathbf{E} \cdot d\mathbf{l} \qquad \text{(through source).} \tag{6.4.4}$$

The electric field intensity $\mathbf{E}$ is due to the stationary distribution of the charges. This field is hence the same as that of static charges. So we have

$$\oint \mathbf{E} \cdot d\mathbf{l} = \underbrace{\int_N^P \mathbf{E} \cdot d\mathbf{l}}_{\text{through source}} + \underbrace{\int_P^N \mathbf{E} \cdot d\mathbf{l}}_{\text{through dielectric}} = 0, \tag{6.4.5}$$

and Eq. (6.4.4) becomes

$$\boxed{\mathscr{E} = \underbrace{\int_P^N \mathbf{E} \cdot d\mathbf{l}}_{\text{through dielectric}} = V_P - V_N} \qquad \begin{array}{l}\text{(alternative definition of}\\ \text{electromotive force;}\\ \text{open-circuited generator).}\end{array} \tag{6.4.6}$$

*The electromotive force of a generator is equal to the potential difference between its open-circuited electrodes.* This fact enables us to make a simple measurement of the electromotive force of any generator.

From Eq. (6.7.6) we can see that the electromotive force has the same unit as the potential difference, that is, the *volt* (V).

**6.4.4 Conclusion.** If we imagine that the electrodes of a generator are connected by a conductor, the following mechanism of the current flow follows from the above explanations. The generator pushes the free charges inside it and creates a certain distribution of quasi-static charges. These charges, though moving, have a constant density at all points, and they produce an electric field at all points of the system. Outside the generator the charges are propelled by forces of this electric field only. Inside the generator the charges are acted on by both the

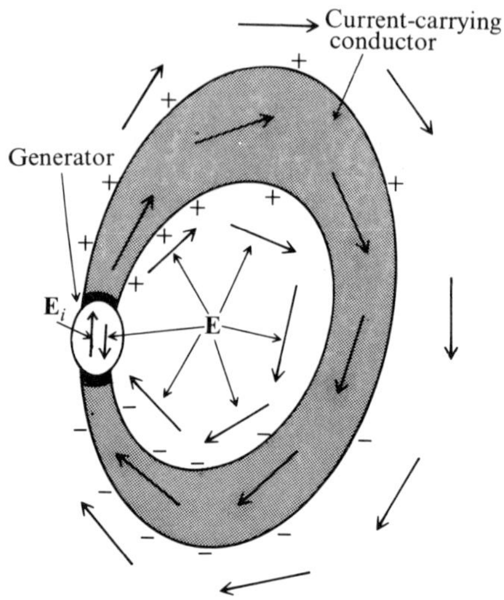


**Fig. 6.7** An illustration of the mechanism of current flow.

electric field of distributed charges and the impressed field, i.e. the external, non-electric forces (Fig. 6.7).

Energy is being converted into heat at *all* points on the current flow, although it is supplied by the generator only. The transfer of energy to different parts of the system is effected by the electric field of the distributed electric charges.

*Problems referring to Section 6.4:* 6.T4, 6.T5.

## 6.5 THE POINT FORM OF OHM'S LAW. CONDUCTIVITY AND RESISTIVITY

The current flow lines of a steady current flow macroscopically coincide with the lines of the *total* force on the moving charges. This is true for all liquid and practically all solid conductors. In these cases the current density $\mathbf{J}$ at all points is in the direction of the *total* force $\mathbf{F}_{\text{total}}$ on charges. According to Eq. (6.4.2), we may quite generally write

$$\boxed{\mathbf{J} = \sigma(\mathbf{E} + \mathbf{E}_i)} \qquad \text{(the point form of Ohm's law; impressed field present).} \tag{6.5.1}$$

The coefficient $\sigma$ in this equation is known as the *conductivity* of the conductor considered. We would expect it to be a function of the properties of the conductor, the temperature, and the current density at a point. It was found experimentally, however, that for solid and liquid conductors the conductivity is not a function of current density, except when the latter is very large. Equation (6.5.1) is called *Ohm's law in point form*.

In particular, for points outside generators Ohm's law (6.5.1) reduces to

$$\boxed{\mathbf{J} = \sigma \mathbf{E}} \quad \text{(point form of Ohm's law; impressed field absent).} \tag{6.5.2}$$

Sometimes it is more convenient to write Eq. (6.5.2) in the form

$$\mathbf{E} = \mathbf{J}/\sigma = \rho \mathbf{J}, \qquad \rho = 1/\sigma, \tag{6.5.3}$$

where $\rho$ (not to be confused with the volume density of charges) is called the *resistivity* of the conductor considered.

Both conductivity $\sigma$ and resistivity $\rho$ are very important characteristics of conductors. According to Eq. (6.5.2) the unit of conductivity is $(A/m^2)/(V/m) = A/Vm$. For reasons which will become apparent in Section 6.7, *siemens per meter* (abbreviation S/m), and very often *mho per meter* (abbreviation usually Mho/m), are used instead. As the unit of resistivity we use the *ohm meter* (abbreviation $\Omega$m). In Table 6.1, representative values for conductivity and resistivity of some common metallic and nonmetallic conductors, and of some common insulators are given.

Resistivity and conductivity are functions of temperature. For a very wide range of temperature $t$ the resistivity may be represented in the form of the series

$$\rho(t) = \rho_0[1 + \alpha(t - t_0) + \beta(t - t_0)^2 + \cdots]. \tag{6.5.4}$$

$\rho_0$ is the known resistivity at a temperature $t_0$, and $\alpha$, $\beta$, . . . are constants to be determined experimentally. If the range of temperature considered is small, Eq. (6.5.4) can be approximated to

$$\rho(t) = \rho_0[1 + \alpha(t - t_0)]. \tag{6.5.5}$$

Coefficient $\alpha$ is referred to as the temperature coefficient. It is generally a function of the reference temperature $t_0$. Values of $\alpha$ are also given in Table 6.1.

**Table 6.1** Resistivity and conductivity of some common materials.

| | $\rho(\Omega m)$, at 20°C | $\sigma$(S/m) | $\alpha(°C^{-1})$, at 20°C |
|---|---|---|---|
| *Conductors* | | | |
| Silver | $0.163 \times 10^{-7}$ | $6.15 \times 10^7$ | $3.7 \times 10^{-3}$ |
| Copper | $0.174 \times 10^{-7}$ | $5.75 \times 10^7$ | $3.9 \times 10^{-3}$ |
| Aluminum | $0.283 \times 10^{-7}$ | $3.54 \times 10^7$ | $3.6 \times 10^{-3}$ |
| Iron | $1.00 \times 10^{-7}$ | $1.00 \times 10^7$ | $4.5 \times 10^{-3}$ |
| Lead | $2.10 \times 10^{-7}$ | $0.48 \times 10^7$ | $4.5 \times 10^{-3}$ |
| Sea water | approx. 0.25 | approx. 4 | |
| Earth | $10^2 - 10^4$ | $10^{-4} - 10^{-2}$ | |
| *Insulators* | | | |
| Fused quartz | $10^{17}$ | $10^{-17}$ | |
| Mica | $10^{15}$ | $10^{-15}$ | |
| Glass | $10^{12}$ | $10^{-12}$ | |
| Transformer oil | $10^{11}$ | $10^{-11}$ | |
| Distilled water | $10^4$ | $10^{-4}$ | |

Even the best insulators have a certain nonzero conductivity. For good insulators, however, conductivity is extremely small, and can usually be ignored.

*Problem referring to Section 6.5:* 6.N3.

### 6.6 JOULE'S LAW

Let us consider a conductor having $N$ free charges $Q$ per unit volume. Let us suppose that if the electric field intensity inside the conductor is $\mathbf{E}$, the average velocity of the free charges is $\mathbf{v}$. The electric force on each charge is $Q\mathbf{E}$. The forces of the electric field in a time interval $dt$ perform work on each charge equal to $Q\mathbf{E}\cdot(\mathbf{v}\,dt)$. The work done by the electric forces in the time interval $dt$ in moving all charges inside a small volume $dv$ is hence

$$dW_e = Q\mathbf{E}\cdot(\mathbf{v}\,dt)N\,dv = NQ\mathbf{v}\cdot\mathbf{E}\,dv\,dt.$$

According to Eq. (6.2.1), this expression can be written in the form

$$dW_e = \mathbf{J}\cdot\mathbf{E}\,dv\,dt. \tag{6.6.1}$$

As was explained in Section 6.1, the motion of free charges inside a conductor under the influence of the electric field is macroscopically uniform rather than accelerated. This is because the charges collide with the atoms of the conductor, and in these collisions transmit to the atoms part of the energy they acquired by acceleration during successive collisions. This part of the energy is transformed into heat, and cannot be returned to the electric field. To maintain the electric current inside a conductor it is necessary constantly to transmit to the charges certain energy, which will be ultimately transformed into heat. Equation (6.6.1) represents precisely the work done in maintaining the current inside a volume $dv$ in a time interval $dt$, that is, the energy transformed in $dv$ into heat during time interval $dt$. The power, or the rate at which the energy of the electric field is transformed into heat, per unit volume, is

$$\boxed{\frac{dP}{dv} = \frac{dW_e/dt}{dv} = \mathbf{J}\cdot\mathbf{E}} \qquad \text{(the point form of Joule's law).} \tag{6.6.2}$$

This expression is known as the *point form of Joule's law*.

Power absorbed from the electric field by a volume $v$ of a current-carrying conductor is

$$\boxed{P = \int_v \mathbf{J}\cdot\mathbf{E}\,dv} \qquad \text{(the integral form of Joule's law).} \tag{6.6.3}$$

This is the *integral form of Joule's law*.

Consider now the volume $v_g$ of a generator. The free charges inside this volume are acted on by some external forces, which, as explained, can be represented by an equivalent impressed electric field intensity $\mathbf{E}_i$. If we change $\mathbf{E}$ to $\mathbf{E}_i$

in Eq. (6.6.2), the power of the external forces per unit volume is obtained:

$$dP_g/dv = \mathbf{J} \cdot \mathbf{E}_i. \tag{6.6.4}$$

The power of the external forces inside the volume $v_g$ is therefore

$$P_g = \int_v \mathbf{J} \cdot \mathbf{E}_i \, dv. \tag{6.6.5}$$

A certain part of power $P_g$ is transformed directly into heat in the generating region. The other, usually much larger, part of $P_g$ is used for propelling the electric charges inside the generator against the electric field of charges accumulated throughout the system. This keeps the energy of the electric field of accumulated charges constant, in spite of conversion of that energy into heat. The ratio of the two parts of the power $P_g$ depends upon the type of generator and the external current-carrying system. A more detailed discussion of this problem is presented in Section 6.8.

*Problems referring to Section 6.6:* 6.T6, 6.N4.

## 6.7 THE GENERAL PROPERTIES OF STEADY CURRENT FIELDS

Suppose that generators of electromotive force are not present in that part of the current field which we are considering. If an impressed electric field is present, $\mathbf{E}$ in Ohm's law has simply to be changed to $(\mathbf{E} + \mathbf{E}_i)$.

The electric field intensity $\mathbf{E}$ of the stationary charges is identical with the electric field intensity of the *static* charges distributed in the same way. Thus *all* relations applying to an electric field of static charges are applicable in this case also. In particular, the line integral of $\mathbf{E}$ around any closed contour is zero,

$$\oint_C \mathbf{E} \cdot d\mathbf{l} = 0, \qquad \text{from which} \qquad \mathbf{E} = -\operatorname{grad} V. \tag{6.7.1}$$

The potential $V$ of the stationary charges in a steady current field is calculated from their distribution in the same way as the electrostatic potential from the distribution of the static charges.

The concept of the permittivity of a conductor is the same as in the case of dielectrics. It is a measure of the polarizability of the atoms of the conductor. In *good* conductors, the permittivity has almost no influence on the current flow, because the density of the polarization charges, proportional to the very small electric field intensity inside a good conductor, is negligible. Hence, the error in assuming that the permittivity of good conductors is $\varepsilon_0$ is also negligible.

If a conductor is poor, the density of the polarization charges may be comparable to the density of the stationary charges. To analyze the current flow in poor conductors we must know the exact value of their permittivity.

In addition to Eq. (6.7.1), the electric field of steady currents also generally satisfies the following equations:

$$\operatorname{div} \mathbf{E} = (\rho + \rho_p)/\varepsilon_0, \tag{6.7.2}$$

$$\rho_p = -\text{div}\,\mathbf{P}, \qquad \mathbf{P} = (\varepsilon - \varepsilon_0)\mathbf{E} \text{ (linear media)}, \tag{6.7.3}$$

$$\mathbf{D} = \varepsilon_0\mathbf{E} + \mathbf{P}, \qquad \text{div}\,\mathbf{D} = \rho. \tag{6.7.4}$$

Finally, we have to add to this system the continuity equation and Ohm's law:

$$\text{div}\,\mathbf{J} = 0, \qquad \mathbf{J} = \sigma\mathbf{E}. \tag{6.7.5}$$

According to Eqs. (6.7.4) and (6.7.5), the density of free charges is

$$\rho = \text{div}\,\varepsilon\mathbf{E} = \text{div}\,\varepsilon(\mathbf{J}/\sigma) = \varepsilon/\sigma\,\text{div}\,\mathbf{J} + \mathbf{J}\cdot\text{grad}\,\varepsilon/\sigma = \mathbf{J}\cdot\text{grad}\,\varepsilon/\sigma. \tag{6.7.6}$$

In an inhomogeneous conductor a distribution of free charges generally exists at all points inside it.

If conductivity $\sigma$ and permittivity $\varepsilon$ are constant (homogeneous conductor), we have that

$$\boxed{\rho = 0} \qquad \text{(at all points of a homogeneous conductor).} \tag{6.7.7}$$

This is a very important conclusion. *Accumulations of electric charges creating the electric field which maintains a steady current in homogeneous conductors cannot be inside the conductors.* Charges can reside only on the boundary surfaces of two different conductors, or of a conductor and an insulator.

**6.7.1 Boundary conditions.** Consider a boundary surface between two conductors of different conductivities $\sigma_1$ and $\sigma_2$ (Fig. 6.8). Applying the continuity equation in integral form (6.3.6) to the surface of the thin disk shown yields

$$J_{1n} = J_{2n}, \qquad \text{or} \qquad \sigma_1 E_{1n} = \sigma_2 E_{2n}. \tag{6.7.8a}$$

The normal components of the electric current density vector are equal on both sides of a boundary surface.

Applying the integral equation in Eq. (6.7.1) to the narrow rectangular contour shown in Fig. 6.8, we obtain

$$E_{1t} = E_{2t}, \qquad \text{or} \qquad J_{1t}/\sigma_1 = J_{2t}/\sigma_2. \tag{6.7.8b}$$

Thus, the tangential components of the electric field intensity are also equal on both sides of a boundary surface.

From Fig. 6.8, Eq. (6.7.8a), and Eq. (6.7.8b) the law of the refraction of current flow lines on the boundary surface follows:

$$\frac{\tan\alpha_1}{\tan\alpha_2} = \frac{J_{1t}/J_{1n}}{J_{2t}/J_{2n}} = \frac{\sigma_1}{\sigma_2}. \tag{6.7.9}$$

The surface density of the free charges on the boundary surface is given by Eq. (2.6.5):

$$\rho_s = D_{2n} - D_{1n} = \varepsilon_2 E_{2n} - \varepsilon_1 E_{1n}. \tag{6.7.10}$$

In the absence of the impressed electric field $\mathbf{E} = \mathbf{J}/\sigma$, and this can be written in

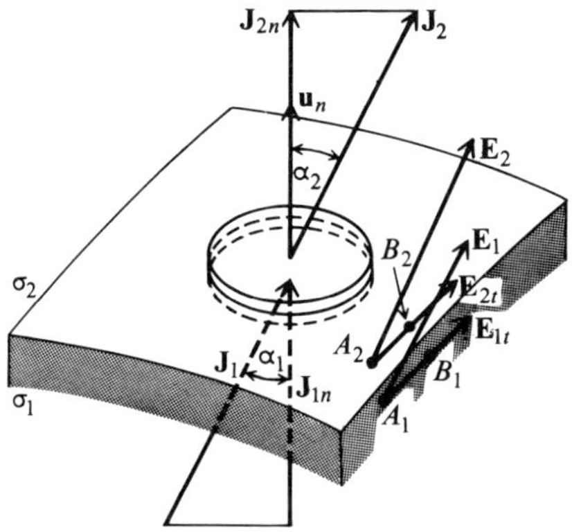


**Fig. 6.8** Gaussian surface and elemental rectangular contour for proving that $J_{1n} = J_{2n}$ and $E_{1t} = E_{2t}$.

the form

$$\rho_s = \frac{\varepsilon_2}{\sigma_2} J_{2n} - \frac{\varepsilon_1}{\sigma_1} J_{1n} = \left(\frac{\varepsilon_2}{\sigma_2} - \frac{\varepsilon_1}{\sigma_1}\right) J_n. \tag{6.7.11}$$

Suppose that medium 1 is a very good conductor, and medium 2 a poor conductor, i.e. that $\sigma_1 \gg \sigma_2$. The boundary conditions expressed by Eqs. (6.7.8a) and (6.7.8b) remain valid in this case, but Eqs. (6.7.9) and (6.7.11) may both be written in an approximate form

$$\left.\begin{aligned} &\tan \alpha_2 \simeq 0, \qquad \text{or} \qquad \alpha_2 \simeq 0 \\ &\rho_s \simeq \frac{\varepsilon_2}{\sigma_2} J_{2n} = \varepsilon_2 E_{2n} = D_{2n} \end{aligned}\right\} \quad \sigma_1 \gg \sigma_2. \tag{6.7.12}$$

The boundary conditions on the surfaces of a good and a poor conductor are approximately the same as on the surface of a conductor in the static electric field. This fact is used as a basis for the experimental analysis of electrostatic fields, which will be explained in the next section. Another consequence is the possibility of defining what we call the resistance and the conductance between two electrodes embedded in a poor conductor. This is discussed in Section 6.7.3.

**6.7.2 The electrolytic trough.** According to Eqs. (6.7.1) and (6.7.2), the potential function $V$ of the stationary charges in a steady current field inside a homogeneous conductor must satisfy Poisson's equation

$$\text{div grad } V = -\rho/\varepsilon. \tag{6.7.13}$$

In regions where the charge density $\rho = 0$, this becomes Laplace's equation,

$$\text{div grad } V = 0. \tag{6.7.14}$$

Let us now consider two electrodes of a very good conductor embedded in a poor conductor. In this case, the boundary conditions on the surfaces of the electrodes are the same as in the electrostatic case, and the potential in both cases satisfies Laplace's equation (6.7.14). According to the uniqueness theorem, the potential functions are hence also the same. Thus, if the distribution of the potential is determined experimentally for a steady current case, the potential is also determined for the corresponding electrostatic case. Conversely, to determine the distribution of the potential of a system of mutually insulated conductors at given potentials (the electrostatic case), we can immerse the whole system into a trough with some very poor liquid conductor, and trace experimentally the equipotential surfaces or lines. The trough containing the liquid conductor (the electrolyte) is referred to as the *electrolytic trough*. This method of analyzing the electrostatic field is known as the *method of current-flow analogy*. Although theoretically the application is general, the method is easily applicable only to two-dimensional and axially symmetrical problems. For details of the electrolytic trough technique the reader is referred to other sources.*

**6.7.3 Conductance and resistance between two electrodes embedded in a poor conductor.** Consider two conducting bodies embedded in a poor homogeneous conductor (or a poor insulator) of conductivity $\sigma$ and permittivity $\varepsilon$, and suppose that they are connected by very thin wires to a generator of electromotive force (Fig. 6.9). A current flowing from the right-hand to the left-hand conductor through the poor insulator results. The propelling force on the charges in the poor insulator is the electric field of the charges accumulated on the surfaces of the electrodes. The electrodes are practically equipotential, as explained in Section 6.7.1. Let us denote the potential difference between the electrodes by $(V_1 - V_2)$, which is proportional to the charge $Q$ on the electrodes, as in the electrostatic case. From this we may define the capacitance between the two conductors, although the insulator is not perfect:

$$C = Q/(V_1 - V_2). \tag{6.7.15}$$

However, through such a capacitor a steady current flows. Let us denote the intensity of this current by $I$. Then, by using Eq. (6.7.12), we may express $I$ in

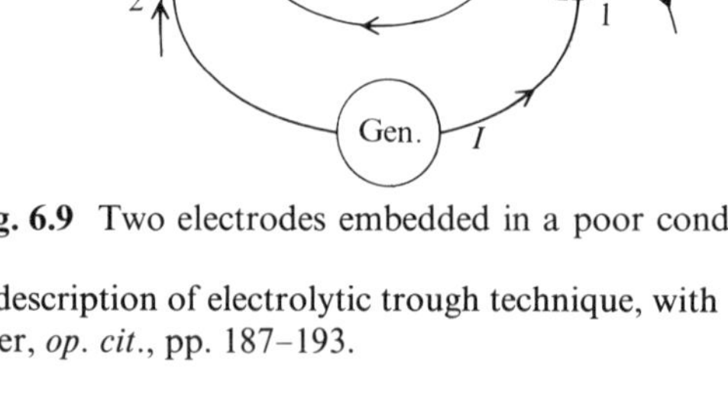


**Fig. 6.9** Two electrodes embedded in a poor conductor.

* A short, but good description of electrolytic trough technique, with further references, may be found in E. Weber, *op. cit.*, pp. 187–193.

terms of $\varepsilon$, $\sigma$, and $Q$ as follows:

$$I = \oint_S \mathbf{J} \cdot d\mathbf{S} = \oint_S J_n \, dS = \frac{\sigma}{\varepsilon} \oint_S \rho_s \, dS = \frac{\sigma}{\varepsilon} Q. \tag{6.7.16}$$

The charge $Q$ is proportional to the potential difference $(V_1 - V_2)$ between the electrodes. The current intensity $I$ is hence also proportional to $(V_1 - V_2)$:

$$\boxed{I = G(V_1 - V_2)} \qquad \text{(definition of conductance).} \tag{6.7.17}$$

The coefficient $G$ is a function of electrode geometry and of the conductivity of the medium only, and is referred to as the *conductance* between the two electrodes. A very simple relation exists between the conductance and the capacitance of a capacitor with a homogeneous imperfect dielectric. From Eqs. (6.7.15) to (6.7.17),

$$\boxed{G = \frac{\sigma}{\varepsilon} C} \qquad \text{(capacitor with an imperfect } \textit{homogeneous} \text{ dielectric).} \tag{6.7.18}$$

The reciprocal value of the conductance is called the *resistance*. It is designated by $R$:

$$\boxed{R = (V_1 - V_2)/I} \qquad \text{(definition of resistance).} \tag{6.7.19}$$

The concepts of resistance and conductance are very much used in practice. According to Eq. (6.7.19), the unit of resistance is volt per ampere (V/A); this unit is given the name *ohm* (abbreviation Ω). The unit of conductance is named *siemens* (abbreviation S) instead of ampere per volt (A/V), but, since it is the reciprocal of the ohm, the name *mho* (abbreviated in different ways) is very often used. (The word "mho" is obtained by spelling "ohm" backwards.

In some simple, but important, cases conductance and resistance may be determined by using the following simple procedure; a parallel procedure is used for computing the capacitance.

1) Imagine a current of intensity $I$ flowing between the electrodes.
2) Determine the current density vector $\mathbf{J}$ at all points (this is possible only in some simple cases, as mentioned).
3) Determine the electric field intensity, $\mathbf{E} = \mathbf{J}/\sigma$.
4) Determine the potential difference between the electrodes as the line integral of $\mathbf{E}$.
5) Determine either $G$ or $R$ according to Eq. (6.7.17) or Eq. (6.7.19).

As an example, let us determine the conductance per unit length between the conductors of a coaxial cable. Suppose that the conductor radii are $a$ and $b$, and that the dielectric is imperfect, of conductivity $\sigma$. Let a current of intensity $I_l$ flow between the conductors per unit length of the cable. The current flow lines are radial by symmetry. The magnitude of the current density vector is a function of the distance $r$ from the cable axis only (Fig. 6.10). According to Eq. (6.2.7), we

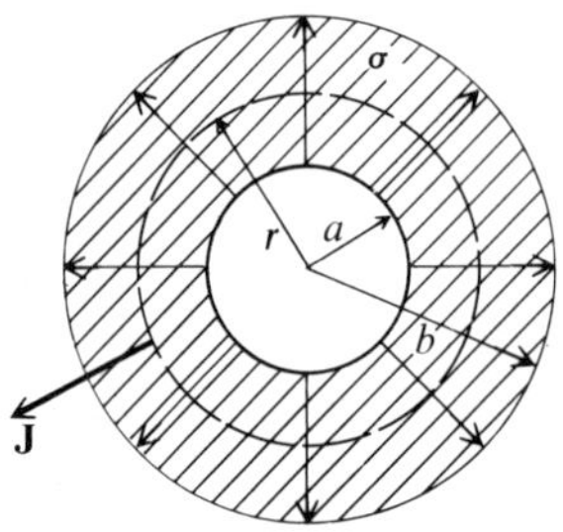


**Fig. 6.10** If the dielectric in a coaxial cable is not perfect, current flows from one conductor to the other through the dielectric.

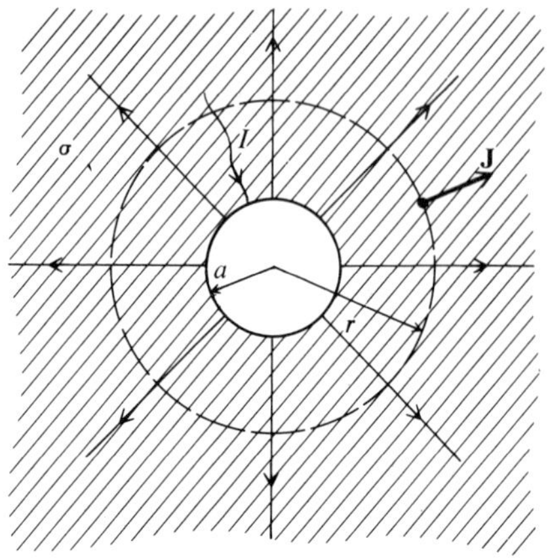


**Fig. 6.11** A spherical electrode embedded in a poor conductor.

hence have

$$I_l = \int_S \mathbf{J} \cdot d\mathbf{S}_l = J(r)2\pi r, \qquad \text{so that} \qquad J(r) = I_l/2\pi r.$$

The electric field intensity is also radial, of magnitude

$$E(r) = J(r)/\sigma = I_l/2\pi r\sigma.$$

The potential difference between the conductors

$$V_1 - V_2 = \int_a^b E(r)\,dr = \frac{I_l}{2\pi\sigma} \ln\frac{b}{a}.$$

The conductance per unit length of the cable is therefore

$$G_l = \frac{I_l}{V_1 - V_2} = \frac{2\pi\sigma}{\ln b/a}.$$

The same result may be obtained directly by combining the result of Problem 1.G43 with Eq. (6.7.18).

Now let us consider a cable of length $L$. The current between the two conductors through the imperfect dielectric is $I_l L$, so that the conductance of the length $L$ of the cable is

$$G = LG_l = \frac{2\pi\sigma L}{\ln b/a}. \tag{6.7.20}$$

As another example, let us determine the resistance of a spherical electrode of radius $a$ embedded in a homogeneous conducting medium of conductivity $\sigma$ (a grounding sphere). Let us suppose that current $I$ flows through the electrode. The current flow lines are radial, and the magnitude of the current density vector is a function of the distance $r$ from the center of the electrode only (Fig. 6.11):

$$J = I/4\pi r^2.$$

The potential of the electrode with respect to the reference point at infinity is

$$V_a = \int_a^\infty E(r)\,dr = \int_a^\infty \frac{J(r)}{\sigma}\,dr = \frac{I}{4\pi\sigma}\int_a^\infty \frac{dr}{r^2} = \frac{I}{4\pi\sigma a}.$$

The resistance of the grounding sphere hence is

$$R = V_a/I = 1/4\pi\sigma a. \tag{6.7.21}$$

**6.7.4 The method of images.** Consider an electrode of an arbitrary shape, buried in a poorly conducting medium (for example, earth) near the plane boundary surface of the poor conductor and insulator. Suppose that a current of intensity $I$ is flowing away from the electrode, and is supplied by a distant generator through a thin, insulated wire (Fig. 6.12). The current flow lines are, as shown, tangential to the boundary surface.

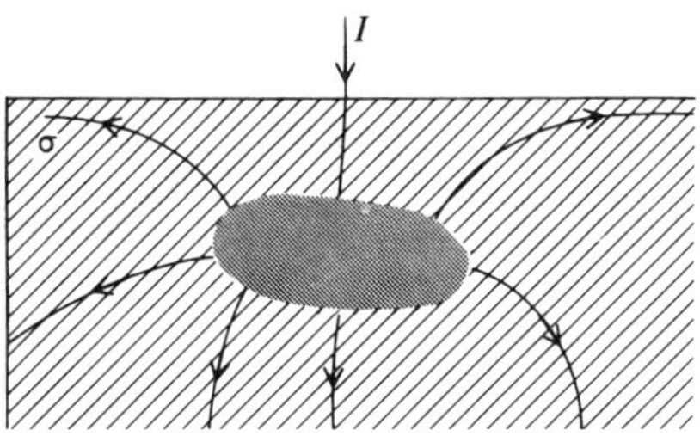


**Fig. 6.12** An electrode buried under the surface of the earth.

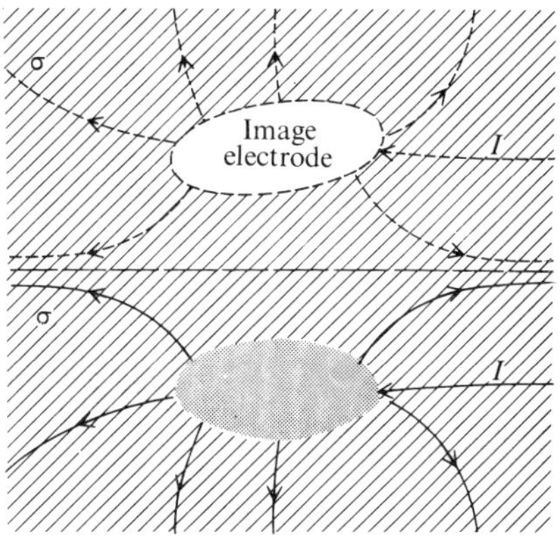


**Fig. 6.13** Two symmetrical electrodes embedded in a homogeneous poor conductor.

Imagine that the whole space is filled with the poor conductor. If an "image" electrode is placed in the upper half-space, as shown in Fig. 6.13, the current flow lines will be tangential to the plane of symmetry (the former boundary surface). From the uniqueness theorem, we know that the current distribution in both halves is exactly the same as in the real case. Thus, the influence of the plane boundary surface of a conducting medium on the distribution of current may be replaced by an image electrode. The current through the image electrode is of the same intensity *and direction* as that through the real electrode. We have to imagine the whole space filled with the same poor conductor. This method is extremely useful in analyzing current flow from electrodes buried under the surface of the earth (see Problems 6.G6 to 6.G17).

*Problems referring to Section 6.7:* 6.T7 to 6.T9, 6.G1 to 6.G17, 6.N5 to 6.N9.

## 6.8 ON TWO-TERMINAL GENERATORS AND RESISTORS

A generator of electromotive force is, in general, a *region* inside a conductor where external, non-electric forces act on free electric charges. Generators encountered in practice always have two equipotential terminals (electrodes). In practical applications the mechanism of propelling the electric charges inside such a generator is irrelevant. The only facts of interest are its electromotive force and its internal resistance (to be explained).

Also, in many practical applications, the current flow is confined to relatively small pieces of a resistive substance. Only rarely is current flow considered in extended conductors, such as currents in the earth, sea water, etc. These pieces of resistive substances are invariably ended by two approximately equipotential surfaces. For this reason they are referred to as the *two-terminal resistors*.

Two-terminal resistors are of many types. Their common characteristic are two equipotential end-surfaces (Fig. 6.14). If we suppose that the resistive material of a resistor is linear (although it may be inhomogeneous), then, to $n$ times larger current through the resistor there corresponds $n$ times larger current density at all points. The electric field intensity, and hence also the potential difference between the equipotential end-surfaces (electrodes) 1 and 2, will also be $n$ times larger. Thus, for any two-terminal linear resistor, the potential difference between the electrodes is proportional to the current intensity through it:

$$\boxed{V_1 - V_2 = RI} \quad \text{(definition of the resistance of a two-terminal linear resistor).} \tag{6.8.1}$$

The coefficient $R$ is referred to as the *resistance* of the resistor. The unit of resistance is the ohm ($\Omega$). The reciprocal value of the resistance is called the conductance, and is designated by $G$.

The power expended in a resistor can be calculated by using Eq. (6.6.2), and by integrating over tubes of equal current intensity $\Delta I$. We first integrate along elemental tubes of equal current, and then add the results. The power transformed into heat in the elemental tube of current shown in Fig. 6.14 is

$$\begin{aligned} \Delta P_{\text{tube}} &= \int_{\text{tube}} \mathbf{J} \cdot \mathbf{E}\, dv = \int_{\text{tube}} E\, dl\, J\Delta S \doteq \int_1^2 E\, dl\, \Delta I = \Delta I \int_1^2 E\, dl \\ &= (V_1 - V_2)\Delta I, \end{aligned}$$

since the current $\Delta I$ is constant along the tube. Adding the results for all tubes, we have

$$\boxed{P = (V_1 - V_2)I = RI^2 = G(V_1 - V_2)^2} \quad \text{(Joule's law for resistors).} \tag{6.8.2}$$

Computation of the resistance of a two-terminal resistor is in general very difficult. Such a computation requires our knowing the exact distribution of current, which is hard to determine except for some simple geometries. But,

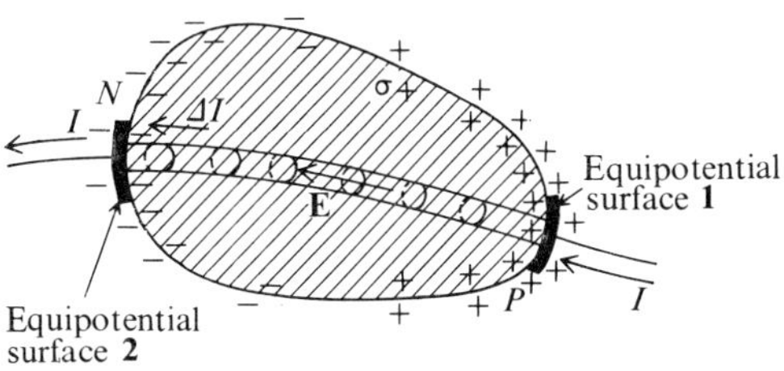


**Fig. 6.14** A generalized two-terminal resistor.

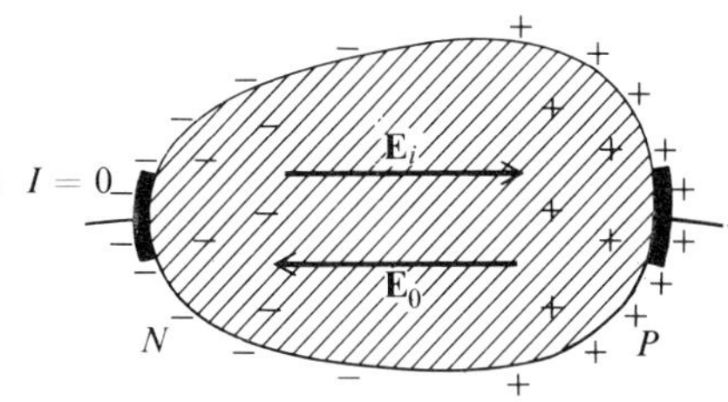


**Fig. 6.15** A generalized open-circuited two-terminal generator.

according to Eq. (6.8.1), the resistance of any resistor can always be measured.

A common type of two-terminal resistor in practice is a thin, straight (or moderately bent) metallic wire. Consider a piece of a thin wire of length $l$, cross-section area $S$, and conductivity $\sigma$, carrying a steady current of intensity $I$. The current density vector is obviously parallel to the wire axis, so that, according to Eq. (6.2.7), $J = I/S$. The electric field intensity is $E = J/\sigma = I/\sigma S$. The potential difference between the ends of the wire is hence

$$V_1 - V_2 = \int_1^2 \mathbf{E} \cdot d\mathbf{l} = I \frac{l}{\sigma S}.$$

The resistance of the piece of wire considered is thus

$$R = l/\sigma S \qquad \text{(thin wire).} \tag{6.8.3}$$

Other simple examples of calculating the resistances of two-terminal resistors are given in Problems 6.G8 to 6.G11, 6.G16, and 6.G17. It should be noted that the concept of resistance between two electrodes in a poor conductor described in Section 6.7.3 is in a sense a special case of the two-terminal resistors explained above.

Consider now the two-terminal open-circuited generator shown in Fig. 6.15. The impressed field $\mathbf{E}_i$ will push the electric charges and cause them to be accumulated as indicated. When equilibrium is attained, the electric field intensity $\mathbf{E}_0$ due to these charges is exactly equal and opposite to $\mathbf{E}_i$. Note that if div $\mathbf{E}_i \neq 0$, then div $\mathbf{E}_0 \neq 0$ as well. The charges will be accumulated not only on the surface, but also inside the generator. According to Eq. (6.4.6), the electromotive force of the generator is equal to the potential difference between the open-circuited electrodes,

$$\boxed{\mathscr{E} = V_P - V_N} \qquad \text{(open-circuited generator).} \tag{6.8.4}$$

Suppose now that the generator generates a certain current, as shown in Fig. 6.16. The electric field of accumulated charges will not compensate for $\mathbf{E}_i$ any longer, since now at all points $(\mathbf{E} + \mathbf{E}_i) = \mathbf{J}/\sigma$, $\sigma$ denoting the conductivity inside

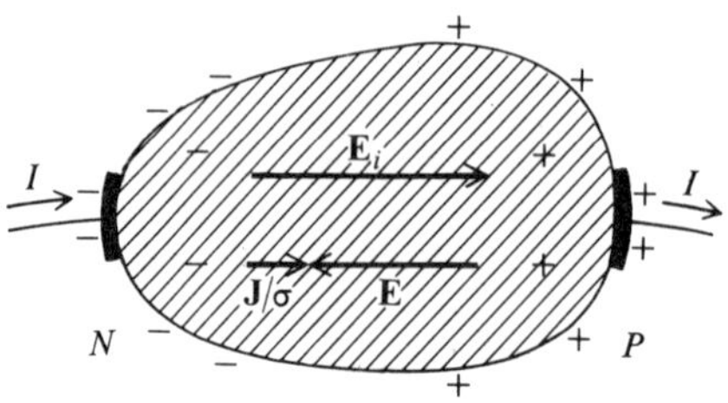


**Fig. 6.16** A generalized generator with current in the direction of its electromotive force.

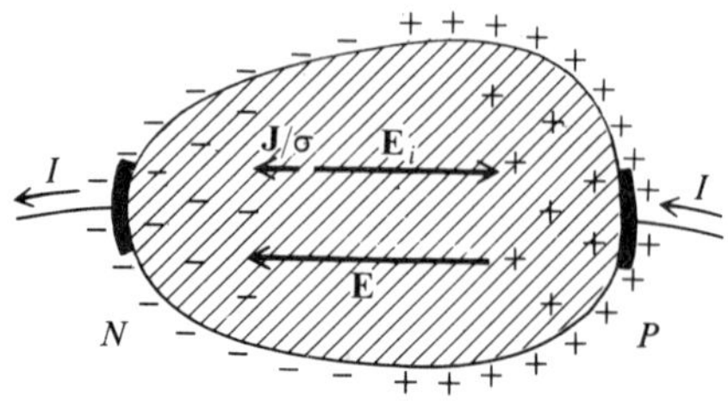


**Fig. 6.17** A generalized generator with current in the direction opposite to its electromotive force.

the generator. *The potential difference between the electrodes will hence be smaller than* $\mathscr{E}$. It is smaller for a potential difference $\Delta(V_P - V_N)$ necessary to maintain current $I$ through the generator regarded as a resistor. This potential difference is proportional to $I$,

$$\Delta(V_P - V_N) = R_g I.$$

$R_g$ is referred to as the *internal resistance of the generator*. The potential difference between the electrodes of a generator when current $I$ flows through it is thus

$$\boxed{V_P - V_N = \mathscr{E} - R_g I} \qquad \text{(current } I \text{ through generator).} \qquad (6.8.5)$$

If a current $I$ is forced through the generator in the direction *opposite* to $\mathbf{E}_i$ (Fig. 6.17), the potential difference between the electrodes will be larger than $\mathscr{E}$:

$$V_P - V_N = \mathscr{E} + R_g I \qquad \text{(current } I \text{ through generator opposite to } \mathbf{E}_i\text{).} \qquad (6.8.6)$$

The power supplied by the generator when the current flows in the direction of $\mathbf{E}_i$ is given by Eq. (6.6.4). If we divide the current field inside the generator in tubes of small constant current intensity $\Delta I$, integrating first along such tubes and then adding the results, Eq. (6.6.4) yields

$$P_g = \mathscr{E} I. \qquad (6.8.7)$$

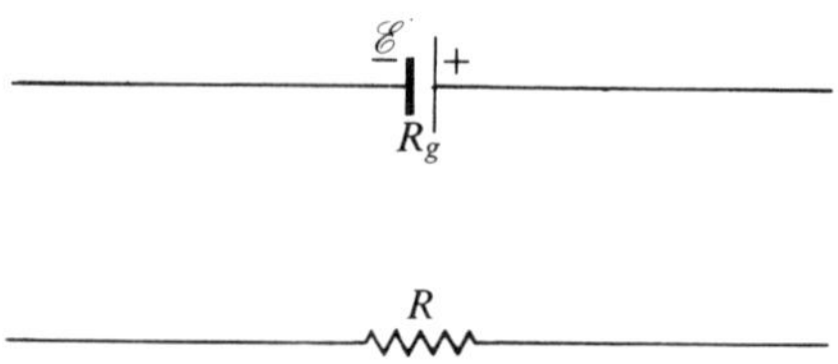


**Fig. 6.18** Usual symbols for a generator of steady electromotive force and for a resistor.

One part of this power, equal to $R_g I^2$, is dissipated in the generator itself. The other part (usually the major part) of $P_g$ is used to push the electric charges inside the generator against the electric forces of the accumulated electric charges. Since a certain amount of work is done *against* the forces of the electric field, its energy should increase. However, the same power is simultaneously dissipated in resistors outside the generator. The propelling force on the charges in the resistors is due to the electric field of the accumulated charges. In this way power is transferred from a generator to other parts of the current field.

Two-terminal generators and resistors are designated by the symbols shown in Fig. 6.18, the longer, thinner electrode being conventionally adopted to represent the positive electrode. The field $\mathbf{E}_i$ inside a generator thus represented is from the shorter to the longer electrode. Plus and minus signs are usually omitted.

*Problems referring to Section 6.8:* 6.T10 to 6.T11, 6.G18 to 6.G24, 6.N10.

## 6.9 DIRECT-CURRENT NETWORKS

**6.9.1 Introduction.** Steady current systems encountered in practice are most often combinations of two-terminal generators and resistors interconnected by thin wires. Theoretically these wires are regarded as if they were made of a perfect conductor. Such systems are referred to as the *electrical networks*. Figure 6.19 is an example of an electrical network, using the symbols from Fig. 6.18.

The small volumes where three or more interconnecting thin conductors meet, such as the small regions 1, 2, and 3 in Fig. 6.19, are referred to as the *nodes* of the network. Series of generators and resistors connected between two nodes are called the *branches* of the network. For example, there are two branches between nodes 1 and 2, that is, the branches $\mathscr{E}_1 R_1$ and $R_2 \mathscr{E}_2$.

To solve a given electrical network essentially means that we must determine the intensities of the currents in all branches, if the values of all the electromotive forces and resistances are given. Some currents may also be given in advance, and

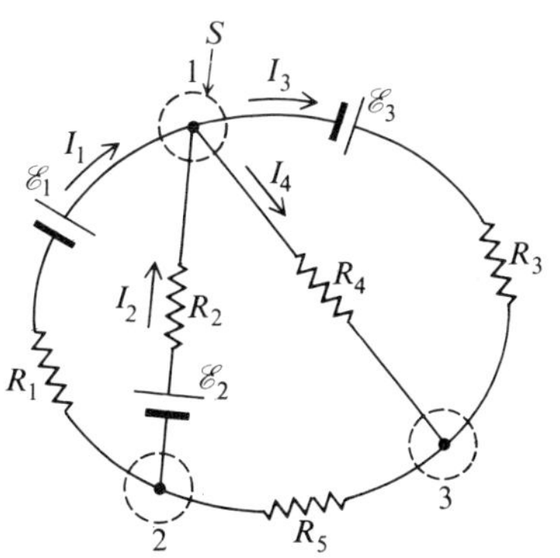


**Fig. 6.19** An electrical network.

either the values of the resistances of some resistors or the values of the electromotive forces of some generators may be regarded as unknowns.

At this point a new science is born, the theory of electrical networks. Except for its two basic laws, which are explained in the following sections, the theory of electrical networks can be developed further by using a formalism in which the electromagnetic theory can hardly be recognized. This is, however, not the subject of the present text, and we shall not deal with it here.

**6.9.2 Kirchhoff's first law.** Kirchhoff's first law is the equation of continuity (6.3.6) specialized for the case of the nodes of an electrical network. If we apply the continuity equation (6.3.6) to a surface enclosing a node of an electrical network, it follows that the sum of all currents leaving the node through the conductors meeting at the node must be zero:

$$\boxed{\sum I = 0} \qquad \text{(Kirchhoff's first law; valid at any node of a network).} \qquad (6.9.1)$$

Kirchhoff's first law is one of the two basic laws for analyzing electrical networks. For example, Eq. (6.9.1) for node 1 of the network shown in Fig. 6.19 has the form

$$-I_1 - I_2 + I_3 + I_4 = 0.$$

If there are $N$ nodes in a network, $(N - 1)$ independent equations for currents in branches can be written for any $(N - 1)$ nodes. The equation for the $N$th node is not independent, since, if Eq. (6.9.1) is satisfied in $(N - 1)$ nodes, it is automatically satisfied in the $N$th node.

**6.9.3 Ohm's law for a simple circuit. Kirchhoff's second law.** Let us consider a simple closed circuit, consisting of a single generator and a single resistor (Fig. 6.20). According to Eqs. (6.8.1) and (6.8.5) we have

$$V_P - V_N = RI \qquad \text{and} \qquad V_P - V_N = \mathscr{E} - R_g I,$$

so that

$$\boxed{\mathscr{E} = (R + R_g)I} \qquad \text{(Ohm's law for a simple circuit).} \qquad (6.9.2)$$

This is *Ohm's law for a simple circuit*.

Let us now consider a closed contour in any network, for example contour $R_1\mathscr{E}_1\mathscr{E}_3R_3R_5R_1$ in Fig. 6.19. Adopting a clockwise direction for the positive direction around the contour, we have, according to Eq. (6.8.5)

$$\begin{aligned} V_1 - V_2 &= \mathscr{E}_1 - R_1 I_1 && \text{(branch 2–1)},\\ V_3 - V_1 &= \mathscr{E}_3 - R_3 I_3 && \text{(branch 1–3)},\\ V_2 - V_3 &= 0 - R_5 I_5 && \text{(branch 3–2)}. \end{aligned}$$

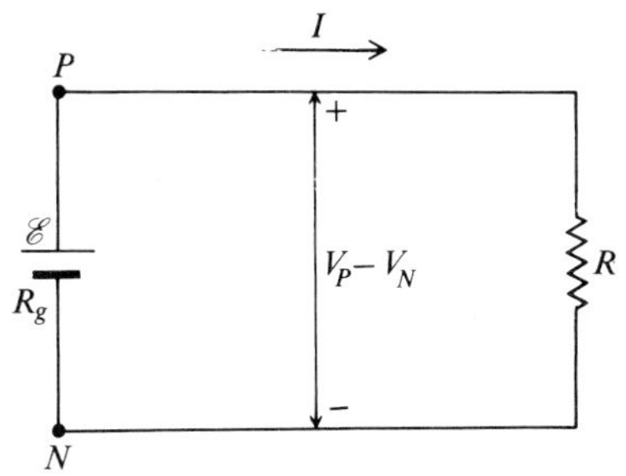


**Fig. 6.20** A simple steady current circuit.

Adding these equations, we obtain

$$0 = \mathscr{E}_1 + \mathscr{E}_3 - (R_1 I_1 + R_3 I_3 + R_5 I_5).$$

It is simple to generalize the above result for any closed contour in any network. Equation

$$\boxed{\Sigma\mathscr{E} - \Sigma RI = 0} \qquad \text{(Kirchhoff's second law; holds for any closed loop in a network)} \tag{6.9.3}$$

must always hold. All electromotive forces and all currents are assumed to be in the adopted positive direction of the contour. This is *Kirchhoff's second law*. Together with the first law given in Eq. (6.9.1), it represents a sufficient tool for analyzing electrical networks with steady currents.

**6.9.4 Series and parallel connections of resistors.** If several, say $n$, resistors are connected end to end, it is said that they are connected in series (Fig. 6.21). A series connection of resistors again has two ends, that is, it is a two-terminal resistor. The equivalent resistance of the resistors connected in series is obtained if we suppose that a current $I$ flows through the connection. The potential difference between the ends of the connection is

$$V = R_1 I + R_2 I + \cdots + R_n I,$$

from which we obtain the equivalent resistance

$$R = V/I = R_1 + R_2 + \cdots + R_n \qquad \text{(series connection).} \tag{6.9.4}$$

If the $n$ resistors are connected as shown in Fig. 6.22, it is said that they are connected in parallel. According to Kirchhoff's second law,

$$R_1 I_1 = R_2 I_2 = \ldots R_n I_n = V,$$

and according to Kirchhoff's first law

$$I_1 + I_2 + \cdots + I_n = I.$$

From the last two equations the equivalent resistance of the parallel connection

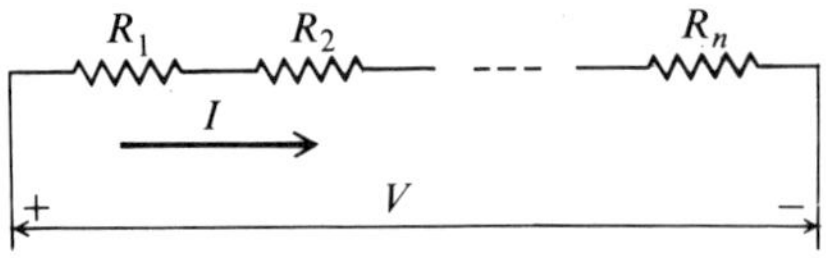


Fig. 6.21 A series connection of resistors.

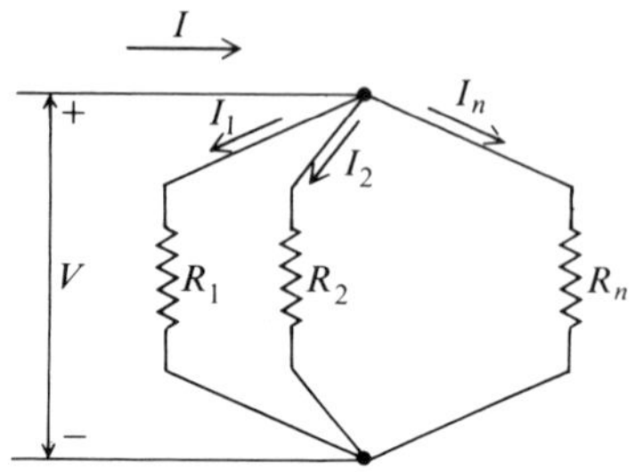


Fig. 6.22 A parallel connection of resistors.

of resistors is determined by the equation

$$1/R = 1/R_1 + 1/R_2 + \cdots + 1/R_n \qquad \text{(parallel connection).} \qquad (6.9.5)$$

## 6.10 RELAXATION TIME

This short section is not strictly related to steady current fields. We shall discuss the question how fast a charge distribution created in a homogeneous conductor disperses under the influence of mutual repulsive forces.

Suppose that, at an instant $t$, the charge density at a point inside a homogeneous conductor of conductivity $\sigma$ and permittivity $\varepsilon$ is $\rho$. The following equations then apply:

$$\operatorname{div} \mathbf{D} = \operatorname{div} \varepsilon\mathbf{E} = \varepsilon \operatorname{div} \mathbf{E} = \rho, \qquad \text{or} \qquad \operatorname{div} \mathbf{E} = \rho/\varepsilon$$
$$\operatorname{div} \mathbf{J} = \operatorname{div} \sigma\mathbf{E} = \sigma \operatorname{div} \mathbf{E} = -\partial\rho/\partial t, \qquad \text{or} \qquad \operatorname{div} \mathbf{E} = -(\partial\rho/\partial t)/\sigma.$$

The charge density $\rho$ as a function of time hence satisfies the equation

$$\frac{\partial \rho}{\partial t} + \frac{\sigma}{\varepsilon}\rho = 0. \qquad (6.10.1)$$

The solution of this equation, with assumed density $\rho_0$ at $t = 0$, is

$$\rho = \rho_0 e^{-\frac{\sigma}{\varepsilon}t}. \qquad (6.10.2)$$

The ratio

$$\varepsilon/\sigma = \tau \qquad (6.10.3)$$

has the dimension of time, and is referred to as the *relaxation time*. It shows how fast the charge density $\rho$ diminishes. The charge density is $e = 2.718$ times larger at the beginning, $t = 0$, than at time $t = \tau$.

For *all* conductors the ralaxation time is extremely small, as we can readily deduce by inspecting Table 6.1. For good insulators it may be of the order of hours and even days, as can be seen from the data in Tables 6.1 and 2.1.

According to Eq. (6.10.2), if, at a certain instant of time and at a certain point, charge density $\rho = 0$, it cannot become different from zero at any later instant. The implicit hypothesis is that the electric field intensity due to electric charges is the only cause of the current flow. Thus, in normal circumstances, accumulations of charges do not exist inside any *homogeneous* conductor or insulator, but, as mentioned in Section 6.7.1, charges can be accumulated in inhomogeneous conductors and on the boundary surface of two different conductors.

*Problems referring to Section 6.10:* 6.T12, 6.N11.

## PROBLEMS

### Test problems

**6.T1.** Prove that the continuity equation (6.3.3) is also valid if the surface $S$ changes its shape in time.

**6.T2.** What is the current intensity through an imaginary spherical surface situated in a cloud of charges of constant density $\rho$, if the radius $a$ of the sphere is imagined to vary in time according to the law: (1) $a(t) = a_0 t$, (2) $a(t) = a_0(t_0 - t)$, and (3) $a(t) = a_0 \sin \omega t$? ($a_0$, $t_0$ and $\omega$ are constants.)

***6.T3.** The charge density in a certain region inside a conductor disperses under the influence of the mutual repulsive forces according to the law $\rho(x, y, z, t) = \rho_0(x, y, z)\, e^{-t/\tau}$. What is the general expression for the current intensity through a closed surface $S$ situated in that region?

**6.T4.** What is the water-flow analogy of a generator of electromotive force?

**6.T5.** What is the work done by a generator of electromotive force when a charge $\Delta Q$ is moved through the generator under the action of the impressed field?

**6.T6.** A steady current of intensity $I$ flows through a wire of conductivity $\sigma$ and cross-section area $S$. A small part of the circuit is made of a thinner wire, of cross-section $S/n$, and of conductivity $\sigma_1$. Determine the power transformed into heat per unit volume in both wires. (This is the principle on which fuses for preventing excessive currents in a system are made.)

**6.T7.** Current flow lines are tangential to a boundary surface of two different conductors. What is the surface density of charges on the boundary surface?

**6.T8.** In an electrolytic trough the electrodes are made of a conductor having ten times larger conductivity than the electrolyte. Discuss the usefulness of such a trough for analyzing electrostatic problems by current-flow analogy.

**6.T9.** From the known expressions for capacitance of an isolated sphere, of a parallel-plate capacitor, of a spherical capacitor, and of a two-wire line, determine the resistance of such systems when the dielectric is not perfect.

**6.T10.** Do the systems form the preceding problem represent two-terminal resistors?

**6.T11.** Two electrodes are immersed in a poorly conducting electrolyte. Is such a system a two-terminal resistor?

***6.T12.** In a small region $v$ inside a piece of a conducting material, at the time $t = 0$ a certain charge distribution exists. According to Eq. (6.10.2) no charges will appear outside $v$ anywhere in the piece of conducting material, since $\rho_0 = 0$ at these points, although the charge inside $v$ decreases. Explain how this is possible, and where the dispersing charges appear.

## General problems

**6.G1.** A dielectric in a parallel-plate capacitor has permittivity $\varepsilon$ and a very small, but non-zero conductivity $\sigma$. The surface of the plates is $S$, their distance $d$. Determine the resistance of the capacitor. If the capacitor is connected to a source of electromotive force $\mathscr{E}$ and a negligible internal resistance, determine the surface density of all free and polarization charges. Show that the electric field intensity inside the capacitor is the sum of the fields due to these charges.

*Solution.* The resistance can be obtained directly from the known value of the capacitance, according to Eq. (6.7.18). We shall derive the expression for the resistance following the procedure suggested in Section 6.7.3. If a current of intensity $I$ is assumed to flow through the capacitor, the current density at all points of the imperfect dielectric is simply $I/S$. Hence the electric field intensity is $E = J/\sigma = I/\sigma S$, so that the potential difference between the electrodes is $Ed = Id/\sigma S$. Thus

$$R = E\,d/I = d/\sigma S. \tag{a}$$

If the potential difference between the electrodes is $\mathscr{E}$, the electric field intensity in the dielectric is $E = \mathscr{E}/d$. The current density $J$ and the electric displacement $D$ are therefore

$$J = \sigma E = \sigma(\mathscr{E}/d), \qquad D = \varepsilon E = \varepsilon(\mathscr{E}/d). \tag{b}$$

According to (b) and Eq. (6.7.12), the surface density of the free charges on the boundary surface of the positive electrode and the dielectric is

$$\rho_s = \varepsilon(\mathscr{E}/d). \tag{c}$$

The surface density of polarization charges,

$$\rho_{sp} = -(\varepsilon - \varepsilon_0)(\mathscr{E}/d). \tag{d}$$

The total surface charge density, which is the source of the electric field inside the dielectric, is thus

$$\rho_{s\ \text{total}} = \rho_s + \rho_{sp} = \varepsilon_0(\mathscr{E}/d).$$

The electric field intensity inside the dielectric is, as we would expect,

$$E = \rho_{s\,\text{total}}/\varepsilon_0 = \mathscr{E}/d.$$

**6.G2.** Repeat the preceding problem for a spherical capacitor of electrode radii $a$ and $b$, and with an imperfect dielectric of permittivity $\varepsilon$ and conductivity $\sigma$.

**6.G3.** The dielectric in a parallel-plate capacitor consists of two slabs of thicknesses $d_1$ and $d_2$, permittivities $\varepsilon_1$ and $\varepsilon_2$, and very small conductivities $\sigma_1$ and $\sigma_2$. Determine the resistance between the electrodes. If the potential difference between the electrodes is $V$, determine the electric field intensity, the electric displacement vector, and the current density in both slabs, as well as the surface density of free and polarization charges on the three boundary surfaces.

*Solution.* Following the procedure outlined in Section 6.7.3, let us suppose that the current through the capacitor is $I$. The current density vector is obviously normal to the plates and the boundary surface of the two slabs. According to the boundary condition given by Eq. (6.7.8a), the current density in the two slabs is the same, and is equal to $J = I/S$.

The electric field intensities $E_1$ and $E_2$ in the two slabs are

$$E_1 = J/\sigma_1 = I/\sigma_1 S, \qquad E_2 = J/\sigma_2 = I/\sigma_2 S. \tag{a}$$

The potential difference between the electrodes when current $I$ flows through the capacitor is

$$V = E_1\, d_1 + E_2\, d_2 = I\left(\frac{d_1}{\sigma_1 S} + \frac{d_2}{\sigma_2 S}\right). \tag{b}$$

Hence the resistance of the capacitor is

$$R = \frac{V}{I} = \frac{d_1}{\sigma_1 S} + \frac{d_2}{\sigma_2 S}. \tag{c}$$

Suppose that the capacitor is connected to a potential difference $V$. According to (b) and (a), the current density and the electric field intensities in the two slabs are

$$J = \frac{I}{S} = \frac{V}{S\left(\dfrac{d_1}{\sigma_1 S} + \dfrac{d_2}{\sigma_2 S}\right)} = \frac{\sigma_1\sigma_2 V}{d_1\sigma_2 + d_2\sigma_1}, \tag{d}$$

$$E_1 = \frac{J}{\sigma_1} = \frac{\sigma_2 V}{d_1\sigma_2 + d_2\sigma_1}, \qquad E_2 = \frac{J}{\sigma_2} = \frac{\sigma_1 V}{d_1\sigma_2 + d_2\sigma_1}. \tag{e}$$

The electric displacement in the two slabs is

$$D_1 = \varepsilon_1 E_1 = \frac{\varepsilon_1\sigma_2 V}{d_1\sigma_2 + d_2\sigma_1}, \qquad D_2 = \varepsilon_2 E_2 = \frac{\varepsilon_2\sigma_1 V}{d_1\sigma_2 + d_2\sigma_1}. \tag{f}$$

The electric displacement vector is *not* the same in the two slabs, as it would be if the two dielectrics were perfect.

The surface density of the free charges on the boundary surface of the two slabs is obtained from (f):

$$\rho_{s12} = D_2 - D_1 = \frac{\varepsilon_2\sigma_1 - \varepsilon_1\sigma_2}{d_1\sigma_2 + d_2\sigma_1} V. \tag{g}$$

Similarly, the surface densities of the free charges on the boundary surfaces of the two slabs and the electrodes are

$$\rho_{s1} = D_1 = \frac{\varepsilon_1\sigma_2 V}{d_1\sigma_2 + d_2\sigma_1}, \qquad \rho_{s2} = -D_2 = -\frac{\varepsilon_2\sigma_1 V}{d_1\sigma_2 + d_2\sigma_1}. \tag{h}$$

It is assumed that the positive electrode is in contact with slab 1.

The surface densities of polarization charges on the three boundary surfaces are

$$\rho_{sp1} = -(\varepsilon_1 - \varepsilon_0)E_1, \qquad \rho_{sp2} = (\varepsilon_2 - \varepsilon_0)E_2, \qquad \rho_{sp12} = (\varepsilon_1 - \varepsilon_0)E_1 - (\varepsilon_2 - \varepsilon_0)E_2.$$

It is left as an exercise for the reader to prove that the electric field intensities in the two slabs, given by (e), are due to all free and polarization charges, the whole system being considered in a vacuum.

**6.G4.** Repeat the preceding problem for a spherical capacitor with an imperfect dielectric, consisting of two concentric layers of permittivities $\varepsilon_1$ and $\varepsilon_2$ and conductivities $\sigma_1$ and $\sigma_2$.

The radii of the electrodes and of the boundary sphere between the two dielectrics are $a$, $b$, and $c$, respectively.

**6.G5.** Repeat Problem 6.G3 for a coaxial cable with an imperfect dielectric consisting of two layers. Use the same symbols as in the preceding problem.

**6.G6.** A hemispherical electrode of very high conductivity is buried in poorly conducting earth, as shown in Fig. 6.G6. Determine the resistance of such a grounding system. If a current $I$ flows from the electrode into the earth, determine the electric field intensity at all points on the earth's surface. Discuss what will happen if a man having noninsulating shoes approaches the grounding electrode.

*Solution.* According to the conclusions of Section 6.7.4, the influence of the boundary surface is equivalent to the influence of an image electrode, if we imagine the whole system to be filled with a medium of conductivity equal to the conductivity $\sigma$ of the earth. The current flow lines hence are radial. For a current intensity $I$ flowing through the hemisphere, the current density at a distance $r$ from the center of the hemisphere is

$$J = I/2\pi r^2. \tag{a}$$

The electric field intensity is thus

$$E = J/\sigma = I/2\pi\sigma r^2. \tag{b}$$

This expression is also valid for any point at the surface of the earth.

The potential of the electrode with respect to the reference point at infinity is

$$V_a = \int_a^\infty E\,dr = \int_a^\infty \frac{I}{2\pi\sigma r^2}\,dr = \frac{I}{2\pi\sigma a}. \tag{c}$$

The resistance between the electrode and very far points hence is

$$R = V_a/I = 1/2\pi\sigma a. \tag{d}$$

The potential at any point on the surface of the earth due to the current flow is

$$V_r = \int_r^\infty E\,dr = I/2\pi\sigma r.$$

The potential difference between two points at a distance $\Delta r = d$ apart is given by

$$\Delta V = \frac{I}{2\pi\sigma r} - \frac{I}{2\pi\sigma(r + d)} = \frac{Id}{2\pi\sigma r(r + d)}.$$

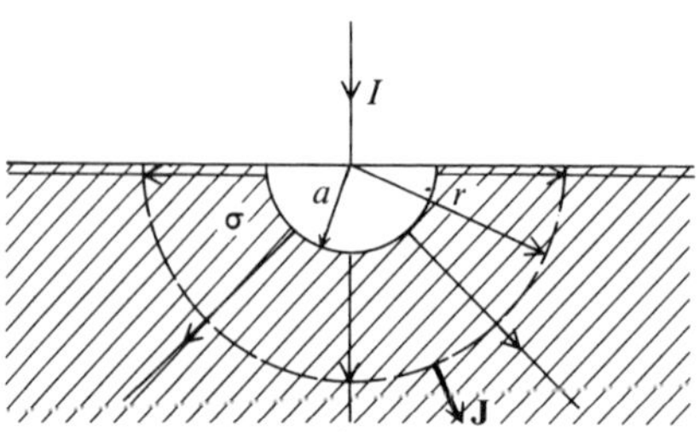

Figure 6.G6

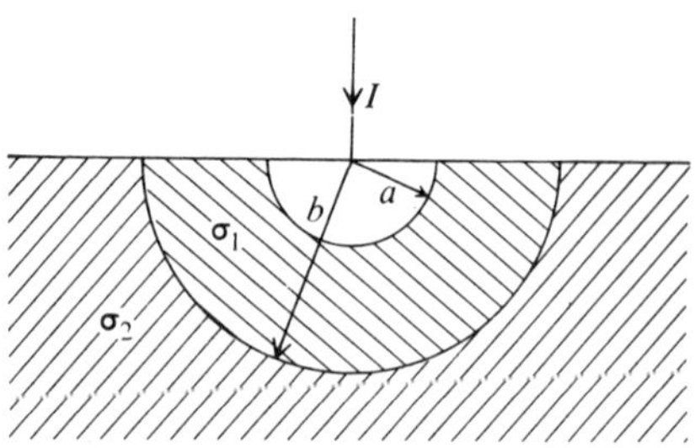

Figure 6.G7

Suppose a man approaches the grounding electrode when a large current flows through it. The potential difference $\Delta V$ between his feet can be very large, even dangerous for his life if his shoes are not made of an insulating material. For example, if $\sigma = 10^{-2}$ S/m, $I = 1000$ A, $d = 0.75$ m, and $r = 1$ m, $\Delta V = 6820$ V!

**6.G7.** Repeat the preceding problem if the earth in the vicinity of the hemisphere is inhomogeneous, of conductivity $\sigma_1$ for $a < r < b$, and of conductivity $\sigma_2$ for $r > b$ (Fig. 6.G7).

*Answer.* For $b = a$, or $b \to \infty$, or $\sigma_1 = \sigma_2$ the result must coincide with that of the preceding problem.

**6.G8.** Two hemispherical electrodes of radius $a$ are buried in the earth of conductivity $\sigma$, as shown in Fig. 6.G8. The distance $d$ between the electrodes is much larger than their radius. Determine the resistance between the two electrodes.

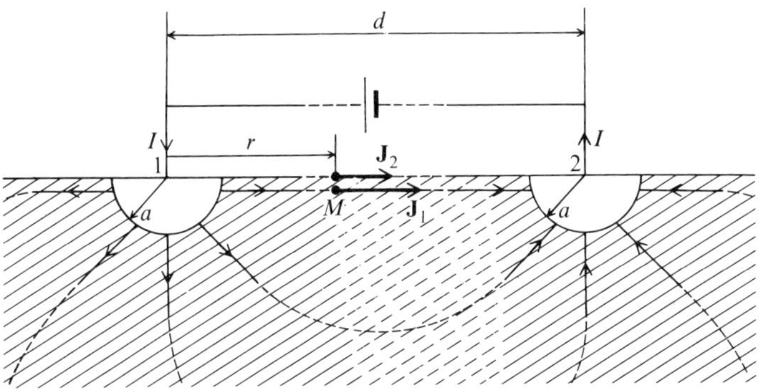


**Figure 6.G8**

*Solution.* Since $d \gg a$, distribution of current in the immediate vicinity of each electrode is very nearly what it would be if it were isolated. At other points, the current density is the vector sum of the current densities due to both electrodes, considered to be isolated. In particular, at points on the surface of the earth, on the line connecting the two electrodes, the two current density vectors are parallel and in the same direction, so that they add algebraically.

Following the procedure for determining the resistance described in Section 6.7.3, suppose that a current of intensity $I$ flows through the system. Current densities at the point $M$, due to currents through the two electrodes, are

$$J_1 = I/2\pi r^2, \qquad J_2 = I/2\pi(d - r)^2.$$

The potential difference between the electrodes is hence

$$V_1 - V_2 = \int_a^{d-a} E\,dr = \int_a^{d-a} \frac{J_1 + J_2}{\sigma}\,dr = \frac{I}{2\pi\sigma}\int_a^{d-a}\left[\frac{1}{r^2} + \frac{1}{(d-r)^2}\right]dr$$

$$= \frac{I}{2\pi\sigma}\left[-\frac{1}{d-a} + \frac{1}{a} + \frac{1}{d-(d-a)} - \frac{1}{d-a}\right] \simeq \frac{I}{\pi\sigma a},$$

since $d \gg a$. The resistance between the electrodes

$$R = (V_1 - V_2)/I = 1/\pi\sigma a.$$

It is twice as large as the resistance of a single electrode, given by (a) of Problem 6.G6, that is, as if the two resistors were connected in series.

**6.G9.** Repeat the preceding problem for the two hemispherical electrodes shown in Fig. 6.G9. Both $a$ and $b$ are much smaller than the distance $d$ between the electrodes.

*Answer.* For $b = a$ or $\sigma_1 = \sigma_2$, the result must be the same as that of the preceding problem.

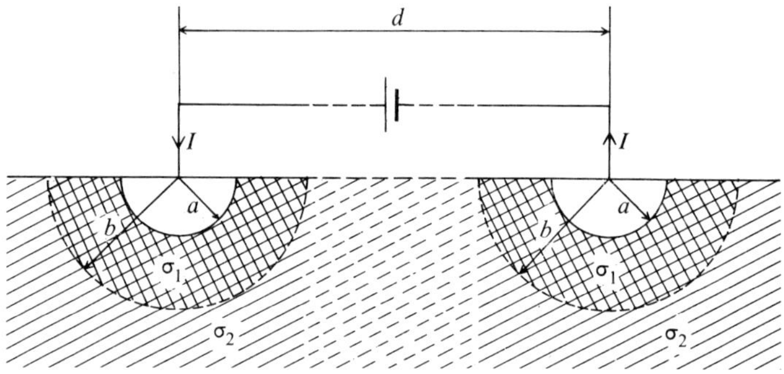


**Figure 6.G9**

**6.G10.** Two very long parallel horizontal conductors of radius $a$ reside on the flat surface of the earth, pressed into the earth for exactly $a$. The conductivity of the earth is $\sigma$, and the distance between the axes of the conductors is $d$ ($d \gg a$). Determine the resistance between the conductors per unit length.

*Answer.* $R_l \simeq (2/\pi\sigma) \ln d/a$.

**6.G11.** Repeat the preceding problem assuming that the earth is not homogeneous, but may be approximated in the vicinity of each conductor by two sections of different conductivities, as shown in Fig. 6.G9. The figure is to be assumed to be a cross-section of the system, with $d$ much larger than both $a$ and $b$.

*Answer.* For $b = a$ or for $\sigma_1 = \sigma_2$ the result must become identical with that in the preceding problem.

**6.G12.** A grounding sphere of radius $a$ is buried at a depth $d$ ($d \gg a$) below the surface of the earth of conductivity $\sigma$. Determine points at the surface of the earth at which the electric field intensity is maximal. Determine the electric field intensity at these points if the intensity of the current through the grounding sphere is $I$.

*Solution.* As explained in Section 6.7.4, the influence of the insulator above the surface of the earth on the current flow is equivalent to an image grounding sphere (Fig. 6.G12). The whole space has to be imagined to be filled with a conductor of conductivity $\sigma$. The current density, and hence also the electric field intensity due to both grounding spheres, are very nearly radial. The total current density and the electric field intensity are obtained as the vector sum of the two radial fields, both of intensity

$$J = J_i = I/4\pi r^2, \qquad E = E_i = I/4\pi\sigma r^2.$$

Subscript $i$ denotes quantities due to the image sphere.

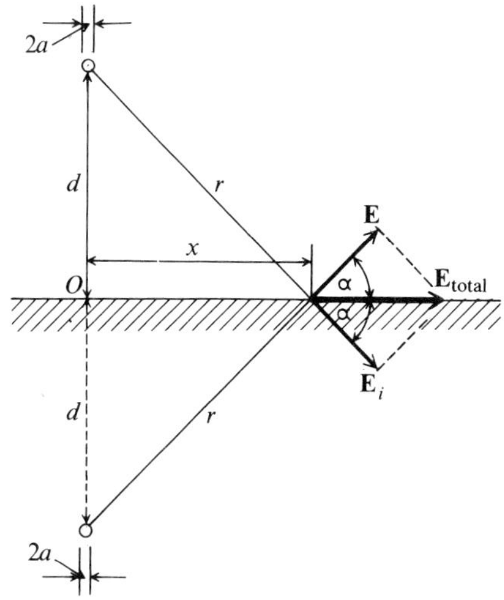


Figure 6.G12

According to Fig. 6.G12, the total electric field intensity at the surface of the earth is given by

$$E_{\text{total}} = (E + E_i)\cos\alpha = 2E\cos\alpha = \frac{I}{2\pi\sigma}\,\frac{x}{(x^2 + d^2)^{3/2}}\,.$$

The maximal value of $E_{\text{total}}$ is obtained for $x$ determined by the equation $\partial E_{\text{total}}/\partial x = 0$. Thus we find that

$$x_{\max} = \frac{d}{\sqrt{2}} \qquad \text{and} \qquad E_{\text{total max}} = \frac{I}{3\sqrt{3}\pi\sigma d^2}\,.$$

**6.G13.** Repeat the preceding problem involving a long thin horizontal conductor buried at a depth $d$ below the surface of the earth, where the current flowing into the earth per unit length of the conductor is $I_l$.

*Answer*. For $I_l = 10$ A/m, $\sigma = 4 \times 10^{-3}$ S/m, and $d = 1$ m, $E_{\text{total wax}} = 400$ V/m.

**6.G14.** Starting from the result of Problem 1.G47, determine the grounding resistance of a very thin circular grounding disk of radius $a$, if the disk: (1) is buried deep in the earth, the conductivity of which is $\sigma$, (2) is half-buried vertically in the earth, and (3) is lying on the surface of the earth. Note that all three cases are approximations of real grounding systems using grounding plates.

**6.G15.** Starting from the result of Problem 1.G48, determine the grounding resistance of a very thin grounding rod of length $L$ and radius $a$, if the rod is: (1) buried deep below the surface of the earth, (2) driven vertically into the earth for its full length, and (3) lying on the surface of the earth and pressed into it for exactly $a$. Note that all three cases (particularly the second) are used as real grounding systems with grounding rods. The conductivity of the earth is $\sigma$.

***6.G16.** Two conducting spheres of radius $a$ are buried at equal depths $d$ below the surface of the earth, at a distance $b$ apart. The conductivity of the earth is $\sigma$, and both $d$ and $b$ are much larger than $a$. Determine an approximate expression for the resistance between the spheres.

*Answer.* For $d \gg b$ the resistance must be twice as large as that of an isolated sphere of radius $a$, given by Eq. (6.7.21).

***6.G17.** Repeat the preceding problem for two very long parallel horizontal rods, assuming that the length $L$ of the rods is much larger than the distance $b$ between them.

*Answer.* For $d \gg b$ the resistance must be equal to the resistance between the conductors of a two-wire line of length $L$ with a lossy dielectric.

**6.G18.** Suppose that the liquid dielectric in Fig. 3.G18 is not perfect, but has a small conductivity $\sigma$, and that the width of the plates is $a$. The mass density of the dielectric is $\rho_m$. Determine the current intensity through the capacitor as a function of the electromotive force $\mathscr{E} = V$ of the generator. Ignore the fringing effect and the internal resistance of the generator.

*Solution.* In Problem 3.G18, we found that, due to electrostatic forces, the level of the liquid between the electrodes would be higher than the level of the liquid outside the electrodes, for

$$h = \frac{(\varepsilon - \varepsilon_0)V^2}{2d^2 \rho_m g}.$$

According to Ohm's law (6.9.2) and the expression (a) of Problem 6.G1 for resistance between electrodes of a parallel-plate capacitor with an imperfect dielectric, if $C$ is the depth of the lower edges of the electrodes, the current through the capacitor is

$$I = \frac{\mathscr{E}}{R} = \frac{\sigma a(c + h)}{d} \mathscr{E} = \sigma \frac{ac}{d} \mathscr{E} + \sigma \frac{ah}{d} \mathscr{E} = \sigma \frac{ac}{d} \mathscr{E} + \sigma \frac{a(\varepsilon - \varepsilon_0)}{2d^3 \rho_m g} \mathscr{E}^3.$$

**6.G19.** Repeat the preceding problem for the case of the coaxial cable in Problem 3.G14, if the conductivity $\sigma$ of the dielectric is nonzero.

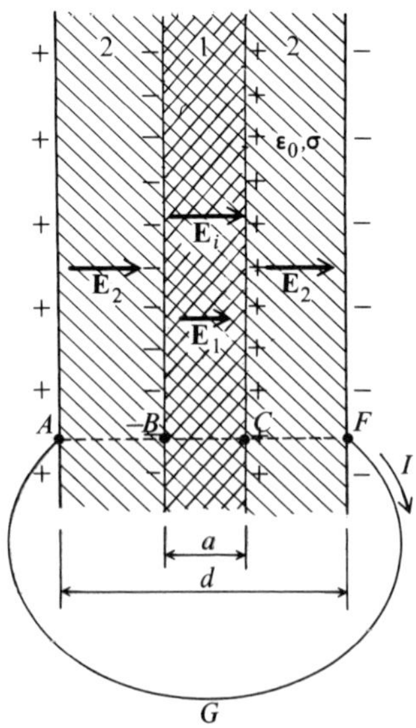


**Figure 6.G20**

**6.G20.** In the doubly crosshatched region of the very large conducting slab of conductivity $\sigma$ and permittivity $\varepsilon_0$ shown in Fig. 6.G20 a uniform impressed field $\mathbf{E}_i$ acts as indicated. The end surfaces of the slab are coated with a conductor of conductivity much larger than $\sigma$, and connected by a wire of negligible resistance. Determine the current density, the electric field intensity, and the charge density at all points of the system. Ignore the fringing effect.

*Solution.* Owing to symmetry, the current flow lines are parallel to $\mathbf{E}_i$ (perpendicular to the end surfaces) at all points. Hence the electric field intensity of the accumulated electric charges will also be normal to the end surfaces. The current density must have the same magnitude at all points. Otherwise a steady process of accumulation of charges would take place, which is impossible. If we denote the electric field intensity in the region of the impressed field by $E_1$, and out of this region by $E_2$, then from Ohm's law in point form and the above conclusion for the current density, it follows that

$$J = \sigma E_2 = \sigma(E_i + E_1), \qquad \text{from which} \qquad E_2 = E_i + E_1. \tag{a}$$

The electric field of accumulated charges is conservative, i.e. Eq. (6.7.1) must be satisfied. Applying Eq. (6.7.1) to the closed contour ABCFGA we have

$$\oint \mathbf{E} \cdot d\mathbf{l} = E_2(d - a) + E_1 a = 0, \qquad \text{from which} \qquad E_1 = -\frac{d - a}{a} E_2. \tag{b}$$

From (a) and (b) we have

$$E_2 = \frac{a}{d} E_i, \qquad E_1 = -\frac{d - a}{d} E_i. \tag{c}$$

The current density in the slab is therefore

$$J = \sigma \frac{a}{d} E_i = \frac{\sigma}{d}(aE_i). \tag{d}$$

This was to be expected, since $aE_i$ is the electromotive force of the slab generator, and $\sigma/d$ represents the conductance of the slab between the end coatings per unit surface.

Charges are accumulated at four surfaces: the two end surfaces, and the two boundary surfaces of the slab generator. According to Eq. (6.7.12), the surface density of charges on the two end surfaces is

$$\rho_{s\,\text{end}} = \varepsilon_0 E_2 = \varepsilon_0 \frac{a}{d} E_i. \tag{e}$$

The charges are the sources of $E_1$ and $E_2$ only, not of $E_i$. According to Eqs. (6.7.10) and (c), the surface density of free charges on the boundary surfaces of the slab generator is

$$\rho_{s\,\text{gen}} = D_2 - D_1 = \varepsilon_0(E_2 - E_1) = \varepsilon_0 \frac{a + d - a}{d} E_i = \varepsilon_0 E_i. \tag{f}$$

The signs of the charges on all surfaces are indicated in the figure.

It is left to the reader to prove that $E_1$ and $E_2$ are indeed due to the charges $\rho_{s\,\text{end}}$ and $\rho_{s\,\text{gen}}$.

**6.G21.** Repeat the preceding problem assuming that the permittivity of the slab is $\varepsilon$ (different from $\varepsilon_0$). Note that in that case polarization charges are also present.

***6.G22.** Inside a cylindrical region of radius $a$ in a very large conducting block of conductivity $\sigma$ and permittivity $\varepsilon_0$ a uniform impressed electric field $\mathbf{E}_i$ acts (Fig. 6.G22a). Determine the

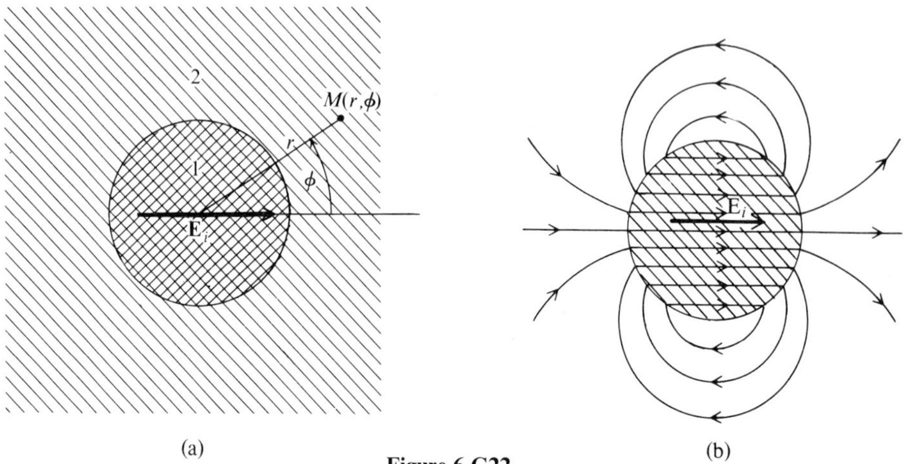


**Figure 6.G22**

electric field intensity, the current density, and the density of the accumulated charges at all points. Is it possible in this case to define a resistance connected to the generator?

*Solution.* The problem is two-dimensional. Let the $z$ axis of a cylindrical coordinate system coincide with the axis of the cylindrical region, and let the direction $\phi = 0$ correspond to the direction of the impressed field. The conducting medium is assumed to be homogeneous; the potential function of the electric field due to accumulated charges must therefore satisfy Laplace's equation in two dimensions. In Section 4.3.3, we found that the general solution of Laplace's equation in such a case is

$$V(r, \phi) = (A_r r^k + B_r r^{-k})(A_\phi \sin k\phi + B_\phi \cos k\phi). \tag{a}$$

($A_r$, $B_r$, $A_\phi$, $B_\phi$, and $k$ are constants.)

In the adopted coordinate system, the potential is an even function of $z$, that is, $V(r, -\phi) = V(r, \phi)$. Outside the generating region, for $r > a$, the potential must decrease with increasing $r$. Inside the region, it must be finite for $r = 0$. If we suppose that $k > 0$, then (a), subject to these restrictions becomes

$$V_1(r, \phi) = Ar^k \cos k\phi \qquad \text{for} \qquad r < a, \tag{b}$$

$$V_2(r, \phi) = Br^{-k} \cos k\phi \qquad \text{for} \qquad r > a. \tag{c}$$

($A$ and $B$ are new constants.)

The potential may be given generally by the series of solutions (b) and (c) with appropriate constants $A$, $B$, and $k$. It will be seen, however, that the boundary conditions in this case can be satisfied with a single term of the type (b) and (c). The constants $A$, $B$, and $k$ must be determined so that the following boundary conditions are satisfied:

$$V_1(a, \phi) = V_2(a, \phi), \qquad \text{whence} \qquad B = Aa^{2k}, \tag{d}$$

and

$$J_{r1}(a, \phi) = J_{r2}(a. \phi), \qquad \text{whence} \qquad \sigma\left(-\frac{\partial V_1}{\partial r} + E_i \cos \phi\right)_{r=a} = -\sigma\left(\frac{\partial V_2}{\partial r}\right)_{r=a}$$

Taking into account (b) and (c), the last condition becomes

$$-Aka^{k-1}\cos k\phi + E_i \cos\phi = Bka^{-k-1}\cos k\phi. \tag{e}$$

This equation has to be satisfied for all $\phi$, which is possible only if $k = 1$. Hence $(-A + E_i) = Ba^{-2}$, from which, using (d), we find

$$A = \tfrac{1}{2}E_i, \qquad B = \frac{a^2}{2}E_i.$$

The potential functions (b) and (c) become

$$V_1(r, \phi) = \tfrac{1}{2}E_i r\cos\phi, \qquad r \le a, \tag{f}$$

$$V_2(r, \phi) = \tfrac{1}{2}E_i a^2 \frac{\cos\phi}{r}, \qquad r \ge a. \tag{g}$$

The electric field intensity in the two regions is

$$E_{1r} = -\frac{\partial V_1}{\partial r} = -\tfrac{1}{2}E_i\cos\phi, \qquad E_{1\phi} = -\frac{1}{r}\frac{\partial V_1}{\partial\phi} = \tfrac{1}{2}E_i\sin\phi, \tag{h}$$

so that

$$\mathbf{E}_1 = -\mathbf{E}_i/2, \tag{i}$$

and

$$E_{2r} = -\frac{\partial V_2}{\partial r} = \tfrac{1}{2}E_i\cos\phi\,\frac{a^2}{r^2}, \qquad E_{2\phi} = -\frac{1}{r}\frac{\partial V_2}{\partial\phi} = \tfrac{1}{2}E_i\sin\phi\,\frac{a^2}{r^2}. \tag{j}$$

The current density inside the generating region is given by

$$J_{1r} = \sigma(E_i\cos\phi + E_{1r}) = \tfrac{1}{2}\sigma E_i\cos\phi, \qquad J_{1\phi} = \sigma(-E_i\sin\phi + E_{1\phi}) = -\tfrac{1}{2}\sigma E_i\sin\phi,$$

so that the current density vector is in the direction of $\mathbf{E}_i$, and

$$\mathbf{J} = \sigma\mathbf{E}_i/2. \tag{k}$$

Outside the generating region

$$J_{2r} = \sigma E_{2r} = \tfrac{1}{2}\sigma E_i\cos\phi\,\frac{a^2}{r^2}, \qquad J_2 = \sigma E_{2\phi} = \tfrac{1}{2}\sigma E_i\sin\phi\,\frac{a^2}{r^2}. \tag{l}$$

Equipotential lines for $r < a$ and $r > a$ are determined by the relations

$$r\cos\phi = C_1 \quad \text{for} \quad r < a, \qquad \cos\phi/r = C_2 \quad \text{for} \quad r > a, \tag{m}$$

where $C_1$ and $C_2$ are constants. The second equation can be rewritten as $r = \cos\phi/C_2$. This represents a family of circles with diameters $1/C_2$, having a common point $r = 0$ and centers on the line $\phi = 0$ and $\phi = \pi$. The lines of the current density vector are orthogonal to these lines. It is not difficult to show that these lines are again circles, having a common point at $r = 0$ and with centers at $\phi = \pm\pi/2$. Figure 6.G22b shows the lines of the current density vector.

On the boundary surface of the generating region, a surface charge is accumulated, of density

$$\rho_s(\phi) = \varepsilon_0 E_{2r}(a, \phi) - \varepsilon_0 E_{1r}(a, \phi),$$

where $E_{1r}$ and $E_{2r}$ represent the electric field intensities due to the *charges*. According to (h) and (j),

$$E_{1r}(a, \phi) = -\tfrac{1}{2}E_i \cos \phi, \qquad E_{2r} = \tfrac{1}{2}E_i \cos \phi.$$

The surface charge density $\rho_s(\phi)$ hence becomes

$$\rho_s(\phi) = \varepsilon_0 E_i \cos \phi. \tag{n}$$

***6.G23.** The systems of the type described in the preceding problem are not encountered in practice. But a thin metal plate with an impressed field in a certain region is used in several devices. Suppose that Fig. 6.G22a represents *such* a system. Discuss whether the results of the preceding problem are also exactly valid in this case.

*Discussion.* Since the current density **J** has no $z$ component, we may imagine that the system of the preceding problem is "sliced" into thin sheets perpendicular to the axis of the system, with very thin insulating foils between them. Nothing will be changed in the field. The boundary conditions on the thin insulating foils are automatically satisfied: the current density vector is tangential to them.

One may now be tempted to conclude that the charge and current distributions will remain *exactly* the same if a single thin sheet, with a coin-like generator in it, is taken away from the system, but such reasoning is incorrect. The electric field intensity in this sheet while in the system was due to charges distributed over the surface of the *infinite* cylinder, according to (n) of the preceding problem. When a single sheet is removed from the system, a circular ring of charges (n) would be the only source of the electric field in the sheet. It is different from the electric field of charges distributed over the infinite cylinder.

From the other side, the potential function has again to satisfy Laplace's equation in two dimensions, and exactly the same boundary conditions as before. The boundary conditions for the current density vector are also the same. According to the uniqueness theorem, the distribution of the potential and the current must be exactly the same as before!

The answer to this seeming paradox is simple. In the case of the sheet, the charges will appear also on the two surfaces of the sheet. They will be distributed so that the electric field intensity in the sheet is the same as before. Determination of the distribution of these charges is, however, far from simple, as it involves solving an integral equation. The unknown, the charge density, appears under the integral sign.

***6.G24.** Determine the power delivered to the system by the generating region in Problem 6.G22, and the power transformed into heat inside and outside the generating region, per unit length of the system.

*Answer.* The calculated power delivered by the generator must be equal to the calculated total power transformed into heat inside and outside the generating region.

## Numerical problems

**6.N1.** Suppose that the number of free electrons in a metal per unit volume is $N = 10^{23}$ per $\text{cm}^3$. Determine the average velocities of the electrons in the metal if the current density is (1) $10^{-2}$ A/mm$^2$, (2) 1 A/mm$^2$, and (3) 10 A/mm$^2$.

**6.N2.** How many electrons per unit time cross a cross-section of a conductor carrying current of intensity $I = 1A$?

**6.N3.** Calculate the electric field intensity in copper, aluminum, see water, distilled water, and glass necessary to produce a current flow of density 1 $\mu$A/mm$^2$. For copper, aluminum, and sea water determine also the electric field intensity necessary to produce a current of density 1 A/mm$^2$ and 10 A/cm$^2$.

**6.N4.** What is the power per unit volume transformed into heat in an aluminum conductor carrying a current of density 5 A/mm$^2$?

**6.N5.** A current of density $J = 10$ A/cm$^2$ flows normal to the boundary surface of the earth and sea water. Assuming that for earth $\sigma_e = 10^{-2}$ S/m and $\varepsilon_e = 10\varepsilon_0$, and for sea water $\sigma_w = 3$ S/m and $\varepsilon_w = 81\varepsilon_0$, determine the surface density of the free and polarization charges on the boundary surface.

**6.N6.** Calculate the conductance per unit length of a coaxial cable with $b/a = e = 2.718\ldots$, when filled with the transformer oil.

**6.N7.** Determine the grounding resistance of a grounding sphere of radius $a = 10$ cm, buried deeply below the surface of the earth, the conductivity of which is $\sigma_e = 10^{-3}$ S/m.

**6.N8.** Determine the grounding resistance of a circular plate, of radius $a = 0.5$ m, immersed deeply in sea water. (See Problem 6.G14.)

**6.N9.** Determine the grounding resistance of a metal rod of length $L = 1$ m, and radius $a = 1$ cm, driven vertically into the earth conductivity of which is $\sigma = 10^{-4}$ S/m. (See Problem 6.G15.)

**6.N10.** Calculate the resistance of a wire of $S = 1$ mm$^2$ cross-section area, and length $l = 1$ m for all metallic conductors of Table 6.1 on p. 203.

**6.N11.** Calculate the relaxation time for copper, distilled water, transformer oil, and mica.

**Suggested reading.** The list progresses towards more advanced texts. For full bibliographical description see Bibliography.

1. SEELY, S., *Introduction to Electromagnetic Fields*, Chapter 1.
2. KRAUS, J. D., *Electromagnetics*, Chapter 3.
3. PLONSEY, R. and COLLIN, R. E., *Principles and Applications of Electromagnetic Fields*, Chapter 5.
4. SMYTHE, W. R., *Static and Dynamic Electricity*, Chapter VI.

CHAPTER 7

# ELECTRIC CURRENT IN A VACUUM

## 7.1 INTRODUCTION

There are two major differences between steady currents in solid and liquid conductors and current flow in a vacuum and in rarefied gases. First, the electric charges in a vacuum or in gases are most often propelled by the electric field of some stationary charges only. The impressed electric field, equivalent to some nonelectric forces, is usually not present. Second, no linear relation between the current density and the electric field intensity, equivalent to Ohm's law, is valid. This conclusion is evident in the case of a vacuum, where the only forces acting on the charges are electric forces and forces of inertia. Consequently, charged particles in an electric field in a vacuum do not follow the lines of the electric field intensity, except in some very special cases. In the case of gases, particularly if rarefied, the paths of accelerated ions between two successive collisions are relatively large, so that they can acquire a considerable kinetic energy. The consequences of this large kinetic energy will not be analyzed in this text. Here it suffices to say that in rarefied gases Ohm's law is also not valid, and that the paths of the electric charges do not coincide generally with the lines of the electric field intensity.

*Problem referring to Section 7.1:* 7.T1.

## 7.2 MOTION OF ELECTRIC CHARGES IN A VACUUM UNDER THE INFLUENCE OF THE ELECTRIC FIELD

Let us consider a point charge $Q$ of mass $m$ moving in a vacuum in an electric field. The force on the charge is $\mathbf{F} = Q\mathbf{E}$. The vectorial differential equation of motion of the particle, valid for small velocities (when relativistic effects can be ignored), is

$$m\frac{d^2\mathbf{r}}{dt^2} = Q\mathbf{E}. \tag{7.2.1}$$

In the scalar form, with respect to a rectangular coordinate system, this becomes

$$m\frac{d^2x}{dt^2} = QE_x, \qquad m\frac{d^2y}{dt^2} = QE_y, \qquad m\frac{d^2z}{dt^2} = QE_z. \tag{7.2.2}$$

Analysis of the motion of the charge implies solving Eqs. (7.2.2), subject to some specified initial conditions. For simple field configurations, this is not a very difficult problem. In addition to the example given below, two further simple examples are presented in Problems 7.G1 and 7.G2. For more complicated field configurations, however, the problem is very difficult, and usually requires either a numerical or an experimental solution. These cases will not be considered in this text.

Very often in practice we require only the final velocity of a charged particle accelerated by the electric field. This problem can be solved easily, provided that the potential difference between the initial and the final points of the particle path is known.

Work done by the electric field in moving the charge for a length $d\mathbf{l}$ is $dW_e = Q\mathbf{E}\cdot d\mathbf{l}$. This work is obviously done for increasing the velocity of the particle, that is, its kinetic energy. Suppose that, at the initial point 0 of the path, the velocity $v_0$ is known. The velocity $v_1$ at any point 1 of the path is then determined by the equation

$$\tfrac{1}{2}mv_1^2 - \tfrac{1}{2}mv_0^2 = Q\int_0^1 \mathbf{E}\cdot d\mathbf{l} = Q(V_0 - V_1). \tag{7.2.3}$$

This is in fact the law of conservation of energy in this case. Thus,

$$v_1 = \sqrt{v_0^2 + \frac{2Q(V_0 - V_1)}{m}}. \tag{7.2.4}$$

In the particular, but common, case when $v_0 = 0$ this becomes

$$v_1 = \sqrt{\frac{2Q(V_0 - V_1)}{m}} \quad \text{(zero initial velocity).} \tag{7.2.5}$$

Equation (7.2.3) and hence also Eqs. (7.2.4) and (7.2.5) are valid for both positive and negative charges. If the charge $Q$ (say, the electron charge) is negative, the charge will be accelerated towards the *increasing* values of the potential, i.e. in the direction opposite to $\mathbf{E}$. The potential difference $(V_0 - V_1)$ will be negative, so that the product $Q(V_0 - V_1)$ will again be positive. For convenience, the charge $Q$ and the potential difference $(V_0 - V_1)$ in Eqs. (7.2.4) and (7.2.5) may hence be considered as absolute values.

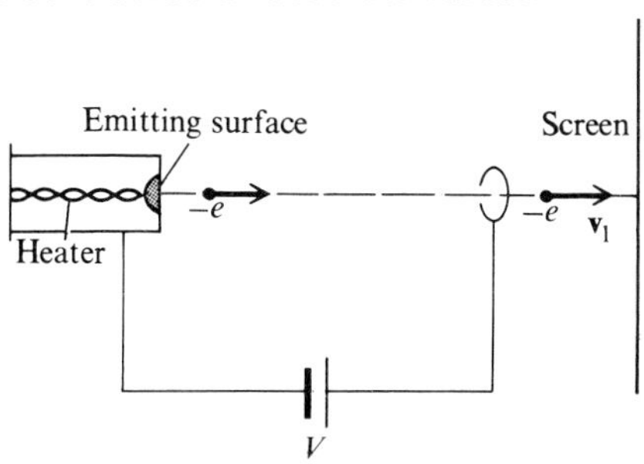


**Fig. 7.1** A basic system for obtaining an electron beam.

As a simple example of the application of Eq. (7.2.5), take the system for accelerating electrons shown in Fig. 7.1. This system is usually referred to as the *electron gun*. The emitting surface shown in Fig. 7.1 is at a very high temperature, and the thermal velocities of the electrons are high, so that their kinetic energies are sufficient to overcome the Coulomb's forces of the positive charges left on the cathode, and to leave the emitting surface. This type of electron emission is known as the *thermionic emission*. The electrons are then accelerated by the electric field towards the ring at a high positive potential $v$, and reaching the ring, they acquire a velocity

$$v_1 = \sqrt{2eV/m}.$$

They continue to travel with approximately this velocity after passing the ring, since the decelerating force in that region is very small.

*Problems referring to Section 7.2:* 7.T2, 7.G1, 7.G2, 7.N1, 7.N2.

## 7.3 STEADY ELECTRIC CURRENT IN A VACUUM. VACUUM DIODE

For obtaining a steady electric current in a vacuum it is necessary to have an emitting surface capable of emitting a large number of electrons per unit time. It has to be large enough to enable the distribution of the moving electrons to be regarded macroscopically as continuous.

In principle, analysis of steady currents in a vacuum follows the pattern sketched in the preceding section. There is only one difference, but an important one. The total electric field which acts on the moving electrons is due to two sources. One part is created by the accumulated charges on the metallic electrodes; the other is due to the charge distributed in space. (These charges are moving, but constantly replacing each other, so that their average density at all points is constant.) This complicates the solution considerably. In fact, a solution of Poisson's equation

$$\text{div grad } V = -\rho/\varepsilon_0 \tag{7.3.1}$$

has to be found which satisfies the prescribed boundary conditions. Such systems are usually referred to as *space-charge systems*.

As an example, let us consider an idealized case of the so-called parallel-plate vacuum diode (Fig. 7.2), and let us suppose that the cathode $C$ emits electrons with zero initial velocities uniformly over its surface. The zero initial velocity implies that there is no current through the diode if the potential difference between the anode $A$ and the cathode $C$ is zero. In reality, electrons always have certain nonzero initial velocities. So, there is a current (albeit a very small one) through the diode, even if the cathode is at a small positive potential with respect to the anode. In addition, we shall ignore the fringing effect due to the finite size of the diode electrodes.

The potential and the charge density in the diode are connected by Poisson's

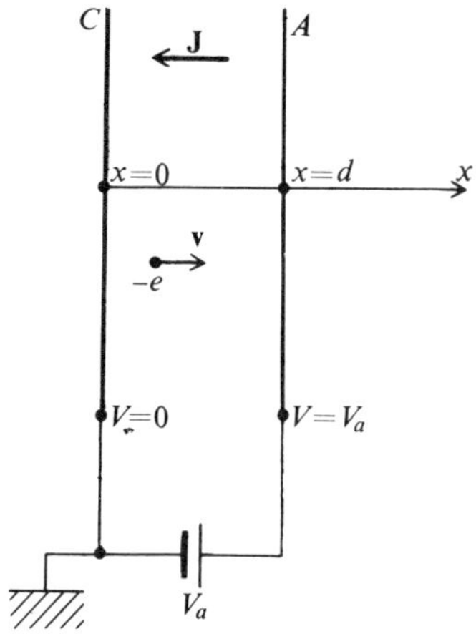


**Fig. 7.2** A parallel-plate vacuum diode.

equation (7.3.1). In this case it becomes

$$d^2V(x)/dx^2 = -\rho(x)/\varepsilon_0. \tag{7.3.2}$$

We assumed that the velocity of the electrons leaving the cathode was zero. According to Eq. (7.2.5) the velocity of the electrons at a distance $x$ from the cathode is hence

$$v(x) = \sqrt{2eV(x)/m}. \tag{7.3.3}$$

$e$ and $m$ denote the charge and the mass of an electron. Now suppose that the current density through the diode is $J$. Since we are considering the stationary state, the current density is the same at all points. Let us denote by $N(x)$ the number of electrons in the diode, per unit volume, at a distance $x$ from the cathode. Then $\rho(x) = -N(x)e$. According to Eqs. (6.2.1) and (7.3.3), we therefore have

$$-\rho(x) = \frac{J}{v(x)} = J\sqrt{\frac{m}{2eV(x)}}. \tag{7.3.4}$$

Observe that the current density vector **J** is opposite to the velocity **v** of the electrons, since they are negative charges (see Fig. 7.2).

Eliminating $\rho(x)$ from Eq. (7.3.2), using Eq. (7.3.4), we are left with

$$\frac{d^2V(x)}{dx^2} = \frac{J}{\varepsilon_0}\sqrt{\frac{m}{2eV(x)}} = AV^{-1/2}(x), \qquad A = \frac{J}{\varepsilon_0}\sqrt{\frac{m}{2e}}. \tag{7.3.5}$$

If we multiply Eq. (7.3.5) by $2(dV/dx)$ and for simplicity write $V$ instead of $V(x)$, this becomes

$$2\frac{d^2V}{dx^2}\frac{dV}{dx} = \frac{d}{dx}\left(\frac{dV}{dx}\right)^2 = 2AV^{-1/2}\frac{dV}{dx} = 4A\frac{d}{dx}(V^{1/2}),$$

from which we derive

$$(dV/dx)^2 = 4AV^{1/2} + C_1 \tag{7.3.6}$$

($C_1$ is a constant). Since the cathode potential is assumed to be zero, $C_1$ equals the square of the electric field intensity at the cathode surface. This electric field intensity is unfortunately a function of several factors and is not known, but we shall assume that it is approximately zero. With assumed $C_1 = 0$, Eq. (7.3.6) becomes

$$V^{-1/4}\, dV = 2\sqrt{A}\, dx,$$

from which, by integrating, we get

$$\tfrac{4}{3}V^{3/4} = 2\sqrt{A}\, x + C_2.$$

($C_2$ also represents a constant.) Since for $x = 0$ we assumed that $V = 0$, $C_2 = 0$. Thus, with the value of $A$ given by Eq. (7.3.5),

$$V = \left(\frac{3x}{2}\right)^{4/3}\left(\frac{J}{\varepsilon_0}\right)^{2/3}\left(\frac{m}{2e}\right)^{1/3}. \tag{7.3.7}$$

To determine the current density $J$ we recall that, for $x = d$, $V = V_a$. From Eq. (7.3.7) we can see that the current density in the diode is

$$\boxed{J = BV_a^{3/2}} \quad \text{(velocity of electrons and field intensity at the cathode surface assumed zero),} \tag{7.3.8}$$

where

$$B = \frac{4\varepsilon_0}{9d^2}\sqrt{\frac{2e}{m}}. \tag{7.3.9}$$

The dependence shown in Eq. (7.3.8) of the current density $J$ on the anode potential $V_a$ in a vacuum diode is referred to as the *three-halves power law*, or the *Langmuir-Child law*. In reality, the exponent of $V_a$ is not exactly 3/2. But the

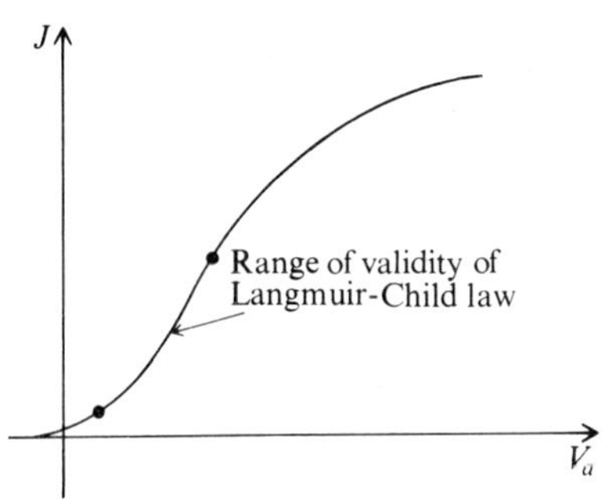


**Fig. 7.3** Measured characteristic of the parallel-plate vacuum diode.

approximation turns out to be a good one even for a cylindrical diode, except for a very small and a very large potential $V_a$ of the anode.

As mentioned above, there is some current through the diode even for negative values of $V_a$, due to nonzero initial velocities of the electrons. For very large values of $V_a$, all electrons emitted by the thermionic emission are immediately sucked off the cathode. The current can therefore increase very little with increasing $V_a$. Figure 7.3 shows the real dependence of the current through the diode on the potential $V_a$ of the anode.

It is important to note that a vacuum diode is a *nonlinear* resistor. The ratio of the potential difference between its terminals and the current density is a function of this potential difference.

*Problems referring to Section 7.3:* 7.T3, T.G3, 7.N3, 7.N4.

## PROBLEMS

### Test problems

**7.T1.** A cloud of identical, charged particles is situated in a vacuum in the gravitational field of the earth. Is there an impressed electric field in addition to the electric field of the charges themselves?

**7.T2.** An electron is emitted parallel to a large uncharged conducting plane. Describe quantitatively the motion of the electron.

**7.T3.** If inside the diode considered in Section 7.3 there is an inert gas (a gas which cannot be ionized) rather than a vacuum, discuss qualitatively whether the three-halves power law will be valid. If not, will the exponent be larger or smaller than 1.5? Discuss what happens with the flow of electrons when the gas pressure inside the diode is increased.

### General problems

**7.G1.** A thin electron beam is formed with some convenient electrode system. The electrons in the beam are accelerated by a potential difference $V_0$. The beam passes between two parallel plates, serving for the electrostatic deflection of the beam, and later falls on the screen $S$ (Fig. 7.G1). Determine the deflection $y_0$ of the beam as a function of the potential difference $V$ between the plates. (This is the method of the electrostatic deflection of the electron beam in cathode-ray tubes.)

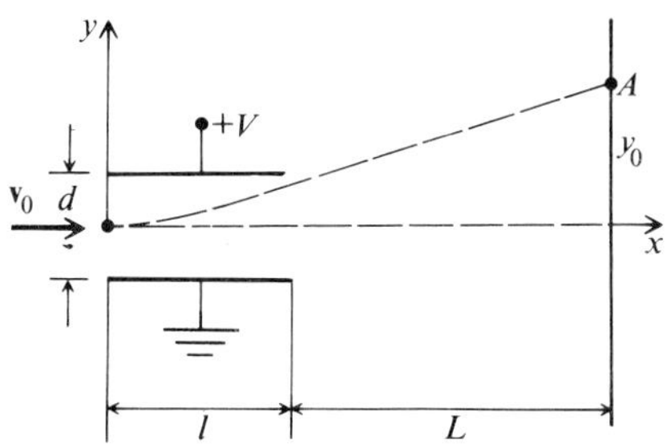


Figure 7.G1

*Solution.* With respect to the coordinate system shown in Fig. 7.G1, the equations of motion of the electrons in the beam are

$$m\frac{d^2x}{dt^2} = 0, \tag{a}$$

$$m\frac{d^2y}{dt^2} = e\,E = e\frac{V}{d}. \tag{b}$$

[$m$ is the mass and $e$ the (absolute) value of the charge of the electron.]

According to (a), the velocity in the $x$ direction is constant. It is equal to the velocity with which the beam enters between the plates:

$$\frac{dx}{dt} = v_x = v_0 = \sqrt{\frac{2eV_0}{m}}. \tag{c}$$

The $x$ coordinate of an electron after a time $t$, measured with respect to the instant when the electron was at the origin, is

$$x = v_0 t. \tag{d}$$

Remembering that for $t = 0$ both $y = 0$ and $dy/dt = 0$, we have, according to (b),

$$\frac{dy}{dt} = \frac{eVt}{md}, \tag{e}$$

$$y = \frac{eVt^2}{2md}. \tag{f}$$

From (d) and (f) it follows that the path of the electrons between the plates is a parabola.

According to (d), the electrons reach the ends of the plates after an interval $t_l = l/v_0$. At this point, deflection $y_l$ and velocity $(dy/dt)_l$ are given by

$$y_l = \frac{eV}{2md}\frac{l^2}{v_0^2}, \qquad \left(\frac{dy}{dt}\right)_l = \frac{eV}{md}\frac{l}{v_0}. \tag{g}$$

The time interval for which an electron travels between the end of the plates and the screen is $t_L = L/v_0$. From (g) it therefore follows that the total deflection $y_0$ of the electrons on the screen is

$$y_0 = y_l + \left(\frac{dy}{dt}\right)_l t_L = \frac{eV}{2md}\frac{l^2}{v_0^2} + \frac{eV}{md}\frac{Ll}{v_0^2} = \frac{Vl}{2V_0 d}\left(L + \frac{l}{2}\right),$$

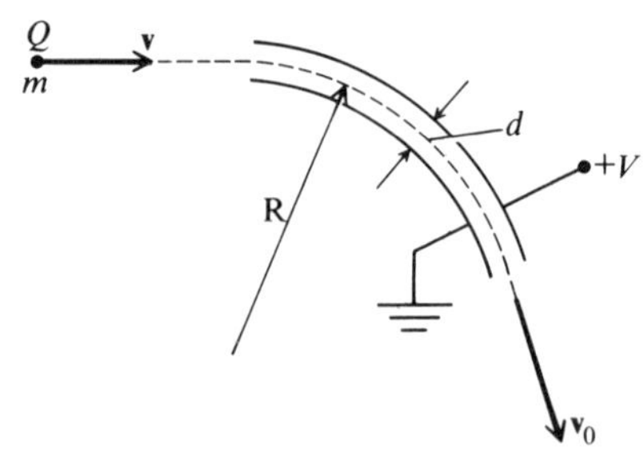


Figure 7.G2

according to (c). Deflection of the beam per unit potential difference between the deflection plates hence is

$$\frac{y_0}{V} = \frac{l}{2V_0 d}\left(L + \frac{l}{2}\right). \tag{h}$$

Ratio (h) is sometimes referred to as the *sensitivity* of the cathode-ray tube.

**7.G2.** A beam of charged particles of positive charge $Q$, mass $m$, and different velocities, enters between two closely spaced curved metal plates. The distance $d$ between the plates is much smaller than radius $R$ of their curvature (see Fig. 7.G2). Determine the velocity $v_0$ of the particles which are deflected by the electric field between the plates so that they leave the plates without hitting any of them.

*Solution.* In the space between the plates, two forces act upon the charged particles. The electric force, $\mathbf{F}_e$, equal in magnitude to $QV/d$, acts towards the grounded plate. The centrifugal force $\mathbf{F}_c$, equal in magnitude to $mv^2/R$, is perpendicular to the trajectory of the charges. The forces are of equal magnitude and in opposite directions. Only those charges will follow a circular arc of radius $R$ which have a velocity $v_0$ determined by

$$mv_0^2/R = Q\,V/d \qquad \text{whence} \qquad v_0 = \sqrt{QRV/md}.$$

**7.G3.** Prove that the total charge in a vacuum diode (including the charge on its electrodes) is zero. Determine the capacitance of the diode.

*Solution.* It is possible to prove that the total charge inside the diode and on its electrodes must be zero in a steady-state case by starting from Gauss' law in integral form. This easy and direct proof is left to the reader. Here we shall show a different method of proof.

Combining Eqs. (7.3.7) to Eq. (7.3.9), we first find that the potential function inside the diode in terms of the potential difference $V_a$ between the electrodes is

$$V(x) = FV_a x^{4/3}. \tag{a}$$

[$F$ is a known constant which can be deduced from Eqs. (7.3.7) to (7.3.9).] From Poisson's equation (7.3.2) it follows that the charge density is given by

$$\rho(x) = -\varepsilon_0 FV_a \frac{d^2}{dx^2}(x^{4/3}) = -\frac{4}{9}\varepsilon_0 FV_a x^{-2/3}.$$

The total charge distributed inside the diode, per unit diode area, is

$$Q_s = \int_0^d \rho(x)\,dx = -\frac{4}{9}\varepsilon_0 FV_a \int_0^d x^{-2/3}\,dx = -\frac{4}{3}\varepsilon_0 FV_a\,d^{1/3}. \tag{b}$$

The electric field intensity on the cathode surface is zero, hence the surface charge density on the cathode is also zero. On the anode surface, from (a),

$$E(d) = -(\partial Vv/\partial x)_{x=d} = -4FV_a d^{1/3}/3. \tag{c}$$

The minus sign indicates that the electric field intensity is directed away from the anode, as it should be. From (c), the positive charge on the anode, per unit area, is

$$\rho_{sA} = \varepsilon_0 |E(d)| = 4\varepsilon_0 FV_a d^{1/3}/3.$$

This is identical with (b), so that the total charge in the diode is zero. The capacitance of the diode per unit area

$$C_s = \rho_{sA}/V_a = 4\varepsilon_0 F d^{1/3}/3.$$

# PART III

# THE STATIC MAGNETIC FIELD

CHAPTER 8

# MAGNETIC FIELD OF STEADY CURRENTS IN THE ABSENCE OF MAGNETIC MATERIALS

## 8.1 INTRODUCTION

The electric force between two small static charged bodies (two point charges) is governed by Coulomb's law given in Eq. (1.2.2). If the two charges are in motion with respect to the observer, the law governing the force between them requires re-examination. The conditions under which Coulomb's law was experimentally proved to be valid are not satisfied any more. We shall show that the correction terms to Coulomb's force are indeed necessary to describe the forces in these cases. We shall see that, for velocities much smaller than the velocity of light, the additional terms are much smaller than Coulomb's term. They are also of quite a different mathematical form.

In the present chapter we shall deal only with the additional forces exerted by a *steady* current system on either a single moving charge or on another steady current system. The *macroscopic* forces in all these cases can be explained by the addition of a single term to the Coulomb force, acting between pairs of elemental electric charges when in motion with respect to the observer. This additional term in the force law is referred to as the *magnetic force*.

The magnetic force exists whenever the charges are in motion with respect to the observer. Therefore, to experience these forces, it is not necessary to move charged macroscopic bodies. For example, steady electric current in metallic conductors represents an organized motion of a very large number of electrons *in otherwise practically uncharged bodies*. Hence it is possible to consider the macroscopic effects of the magnetic forces separately from Coulomb's forces.

There is another important consequence of this possibility. In a current-carrying conductor practically all the conduction electrons are set in organized motion. The magnetic force between two current-carrying conductors is a vector sum of the magnetic forces acting between pairs of practically *all* conduction electrons. We must recall that the electric charges on conductors can reside only on their surfaces. Hence the smallness of the magnetic force when compared with Coulomb's force in the case of two small charged particles is in no way a criterion of the order of magnitude of these forces in *macroscopic* systems. In fact, it will be seen that macroscopic magnetic forces are in practice much larger than the Coulomb's forces.

In this chapter, the magnetic forces and the field from which they may be derived will be considered in a special (theoretically the simplest) case. All substances in the field are supposed to be such that their presence does not influence the field, provided that no macroscopic currents exist inside them. The majority of substances in nature satisfy this requirement very closely. Air, practically all dielectrics, and practically all metals except iron, cobalt, nickel, and some of their alloys and chemical compounds, are not subject to any significant force when placed near a current-carrying system. Other cases will be considered in the next chapter.

## 8.2 THE MAGNETIC FORCE BETWEEN TWO SMALL MOVING CHARGES

**8.2.1 Introduction.** The law governing the force between two small static charged bodies, the Coulomb law, was determined experimentally, measuring the force between two charges directly. Such a procedure is unfortunately not possible for determining the correction terms to the Coulomb's force when the two charges are in motion with respect to the observer. Firstly, for reasonable velocities the correction terms are very small when compared with the Coulomb's force, and are practically indistinguishable from it. Secondly, any experiment with moving charges is obviously much more involved than one with static charges.

It follows that the magnetic force between two moving charges cannot be determined by direct measurement; but it is possible to determine that force by some indirect measurement. For example, it is possible to measure the force between two current-carrying conductors, and reduce it mathematically to elemental forces between pairs of elemental electric charges. Or it is possible to analyze the motion of a single charge in the vicinity of a current-carrying conductor, and again to express the total force on the particle as a sum of the magnetic forces exerted on it by individual electrons moving inside the conductor.

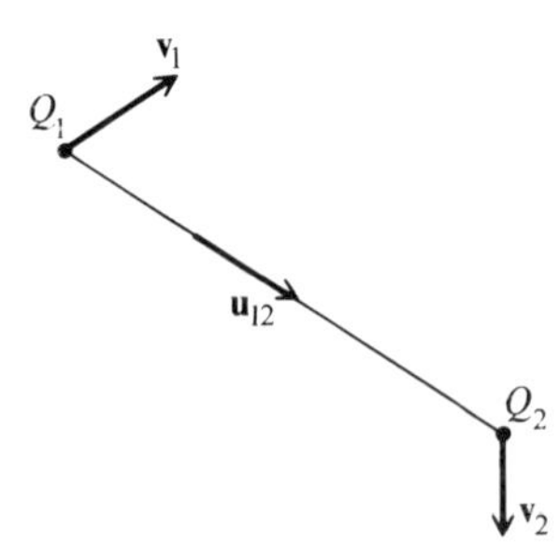


**Fig. 8.1** Two moving point charges.

**8.2.2 The law of magnetic force between two small moving charges.** Consider two point charges $Q_1$ and $Q_2$ moving in a vacuum with velocities $\mathbf{v}_1$ and $\mathbf{v}_2$ with respect to the observer (see Fig. 8.1). Let us suppose that the magnitudes of both $\mathbf{v}_1$ and $\mathbf{v}_2$ are much smaller than the velocity of light. By reference to Fig. 8.1, the *magnetic force* $\mathbf{F}_{m12}$ exerted by the charge $Q_1$ on the charge $Q_2$ is found by indirect experiments, involving *steady* current systems, to be

$$\mathbf{F}_{m12} = K_m \frac{Q_1 Q_2}{r^2} \mathbf{v}_2 \times (\mathbf{v}_1 \times \mathbf{u}_{12}). \tag{8.2.1}$$

The constant $K_m$ depends upon the system of units adopted. In the MKSA system of units $K_m$ is written in the form $K_m = \mu_0/4\pi$. The new constant $\mu_0$ is referred to as *the permeability of the vacuum*, or of free space. The law of the magnetic force hence becomes

$$\boxed{\mathbf{F}_{m12} = \frac{\mu_0}{4\pi} \frac{Q_1 Q_2}{r^2} \mathbf{v}_2 \times (\mathbf{v}_1 \times \mathbf{u}_{12})} \quad \text{(the law of magnetic force between two point charges).} \tag{8.2.2}$$

The parentheses indicate that the vector equal to the cross product $\mathbf{v}_1 \times \mathbf{u}_{12}$ is to be determined first. Only then should $\mathbf{v}_2$ be multiplied vectorially by the vector thus obtained.

The unit of velocity in the MKSA system is meter per second, and one coulomb per second equals one ampere. With this, from Eq. (8.2.2), we see that the unit of permeability is newton per ampere squared. For reasons which will become apparent in Section 11.5.1, the unit *henry per meter* (abbreviation H/m) is used instead. The value of $\mu_0$ is exactly

$$\boxed{\mu_0 = 4\pi \times 10^{-7}\ \text{H/m}} \quad \text{(permeability of a vacuum).} \tag{8.2.3}$$

Air has practically the same permeability.

Comparing Eq. (8.2.2) with Coulomb's law, Eq. (1.2.2), we can see that the product $(\varepsilon_0\mu_0)^{-1}$ has the dimension of velocity squared. Using the approximate value of $\varepsilon_0$ given in Eq. (1.2.3) and the value of $\mu_0$ in Eq. (8.2.3), it is easy to find that $(\varepsilon_0\mu_0)^{-1} \simeq 9 \times 10^{16} = (3 \times 10^8)^2$, so that

$$1/\sqrt{\varepsilon_0\mu_0} \simeq 3 \times 10^8\ \text{m/s} \simeq c_0 \qquad (c_0 = \text{velocity of light in a vacuum}). \tag{8.2.4}$$

If the exact instead of the approximate value of $\varepsilon_0$ is used, the exact value of the velocity of light is obtained. We shall see in Section 15.2 that this coincidence is by no means accidental.

**8.2.3 Notes on the magnetic force between two moving charges.** The first note relates to the magnitude of the magnetic force when compared with the Coulomb force. The magnetic force, given by Eq. (8.2.2), is a function of both the magnitude and the direction of the velocities of the charges. We shall compare the maximal possible magnetic force for given velocities with the Coulomb force. The largest possible magnitude of the cross product is equal to the product of the magnitudes of the two vectors. Hence $\mathbf{F}_{m12}$, given by Eq. (8.2.2), has a magnitude that cannot be larger than

$$|\mathbf{F}_{m12}|_{\max} = \frac{\mu_0}{4\pi}\frac{Q_1 Q_2}{r^2} v_1 v_2. \tag{8.2.5}$$

The ratio of the magnitudes of the maximal possible magnetic force (8.2.5) and the electric force (1.2.2) is

$$|\mathbf{F}_{m12}|_{\max}/|\mathbf{F}_e| = \varepsilon_0\mu_0\, v_1 v_2 = v_1 v_2/c_0^2, \tag{8.2.6}$$

according to Eq. (8.2.4).

The velocities of the electric charges are in practice usually very much smaller than the velocity of light in a vacuum. The magnetic force between two charges is hence negligible when compared with the Coulomb force.

The second note relates to the direction of the magnetic force exerted by the charge $Q_2$ on the charge $Q_1$, which is obtained simply by interchanging subscripts 1 and 2 in Eq. (8.2.2):

$$\mathbf{F}_{m21} = \frac{\mu_0}{4\pi}\frac{Q_2 Q_1}{r^2}\mathbf{v}_1 \times (\mathbf{v}_2 \times \mathbf{u}_{21}). \tag{8.2.7}$$

Although $\mathbf{u}_{21} = -\mathbf{u}_{12}$, in the general case $\mathbf{F}_{m21}$ is *not* equal to $-\mathbf{F}_{m12}$, since generally

$$\mathbf{v}_2 \times (\mathbf{v}_1 \times \mathbf{u}_{12}) \neq -\mathbf{v}_1 \times (\mathbf{v}_2 \times \mathbf{u}_{12}). \tag{8.2.8}$$

For example, if $\mathbf{v}_1$ is parallel to $\mathbf{u}_{12}$, the left-hand side of Eq. (8.2.8) is zero, since $\mathbf{v}_1 \times \mathbf{u}_{12} = 0$, but the right-hand side is not necessarily zero.

Newton's third law is thus *not* satisfied by the magnetic forces between two isolated electric charges. We shall not explain the reasons for this in detail. However, the law of the magnetic force between two charges is one of the basic laws of electromagnetic theory, and a few short qualitative notes at least are necessary.

We must first note that the charges are moving. Their electric field varies in time at every point. There is positive experimental evidence that a varying field must be associated with certain inertial properties. Therefore, a system of two isolated moving charges has to be considered not as a system of two, but rather of three "bodies": the two charges, and the field with its inertial properties. In a three-body system, Newton's third law is *not* valid for forces on only two of the bodies in the system.

Finally, let us consider several charges moving with respect to the observer with velocities much smaller than the velocity of light. The total magnetic force on any charge is obtained as a vector sum of partial magnetic forces, given by Eq. (8.2.2). This *experimental* conclusion is referred to as the *principle of superposition of magnetic forces*.

*Problems referring to Section 8.2:* 8.T1, 8.N1.

## 8.3 THE CONCEPT OF THE MAGNETIC FIELD. MAGNETIC FLUX DENSITY

The law of the magnetic force between two moving charges (8.2.2) can be rewritten as follows:

$$\mathbf{F}_{m12} = Q_2 \mathbf{v}_2 \times \left( \frac{\mu_0}{4\pi} \frac{Q_1 \mathbf{v}_1 \times \mathbf{u}_{12}}{r^2} \right). \tag{8.3.1}$$

At any particular instant of time the expression in the parentheses is not a function of either $Q_2$ or $\mathbf{v}_2$. If, instead of $Q_2$, we imagine at the same point some other charge $Q$, having a velocity $\mathbf{v}$, we can obtain the magnetic force on $Q$ simply by changing $Q_2$ to $Q$ and $\mathbf{v}_2$ to $\mathbf{v}$ in Eq. (8.3.1). The expression in the parentheses in Eq. (8.3.1) can thus be considered as describing a certain physical vector quantity existing around the charge $Q_1$, whenever it moves with respect to the observer. It is characterized by the fact that, if a second charge $Q$ is moving in that region with a velocity $\mathbf{v}$, it is subject to a force given by

$$\mathbf{F}_m = Q \mathbf{v} \times \mathbf{B}, \tag{8.3.2}$$

where

$$\boxed{\mathbf{B} = \frac{\mu_0}{4\pi} \frac{Q_1 \mathbf{v}_1 \times \mathbf{u}_{12}}{r^2}} \quad \text{(definition of the flux density vector produced by a single moving charge).} \tag{8.3.3}$$

The vector function **B** defined by Eq. (8.3.3) is referred to as the *magnetic flux density* produced by the charge $Q_1$. That part of the space where the force $\mathbf{F}_m$ on a moving charge can be described by an expression in the form of Eq. (8.3.2) is referred to as the *magnetic field.* From the definition in Eq. (8.3.2) of the magnetic flux density vector **B**, we can see that it describes the force properties of a magnetic field. In this respect it is in a sense analogous to the electric field intensity in the case of the static electric field.

From Eq. (8.3.2), it is clear that the unit of the magnetic flux density is newton-second per coulomb-meter. A separate name, the *tesla* (abbreviation T) is given to that unit. For reasons which will become apparent in Section 8.6, the unit *weber per square meter* (abbreviation $\mathrm{Wb/m^2}$) is also used.

Now let us suppose that charge $Q$ is moving with a velocity $\mathbf{v}$, and that there are also $n$ charges $Q_1, Q_2, \ldots, Q_n$, moving with velocities $\mathbf{v}_1, \mathbf{v}_2, \ldots, \mathbf{v}_n$. All the velocities are measured with respect to the same coordinate system. According to the principle of superposition of magnetic forces, the total magnetic force on $Q$ due to all charges is

$$\mathbf{F}_m = Q \mathbf{v} \times \left( \frac{\mu_0}{4\pi} \frac{Q_1 \mathbf{v}_1 \times \mathbf{u}_1}{r_1^2} \cdots + \frac{\mu_0}{4\pi} \frac{Q_n \mathbf{v}_n \times \mathbf{u}_n}{r_n^2} \right). \tag{8.3.4}$$

[$\mathbf{u}_1, \mathbf{u}_2, \ldots, \mathbf{u}_n$ are unit vectors directed (at the instant considered) *from* charges $Q_1, \ldots, Q_n$ *towards* the charge $Q$.] From the definition in Eq. (8.3.2), the magnetic flux density vector **B** due to $n$ moving charges is hence

$$\mathbf{B} = \frac{\mu_0}{4\pi} \sum_{k=1}^{n} \frac{Q_k \mathbf{v}_k \times \mathbf{u}_k}{r_k^2}. \tag{8.3.5}$$

Suppose that the magnetic field is due to uncompensated moving charges, or that some additional static charges are also present. In addition to the magnetic force $\mathbf{F}_m$ given by Eq. (8.3.2), a moving charge $Q$ is then also subject to a certain electric force $\mathbf{F}_e = Q\mathbf{E}$ ($\mathbf{E}$ is the electric field intensity at the point occupied by the charge at the instant considered). In that case the total force on the charge is

$$\boxed{\mathbf{F} = \mathbf{F}_e + \mathbf{F}_m = Q(\mathbf{E} + \mathbf{v} \times \mathbf{B})} \qquad \text{(Lorentz force).} \tag{8.3.6}$$

The total force given by Eq. (8.3.6) is referred to as the *Lorentz force.*

In the case of the magnetic field of a single point charge, or a number of point charges, the magnetic flux density **B** created by the charges changes in time at all points. This type of magnetic field is therefore *not* a static field. In this part of the text, however, we shall consider only static magnetic fields, that is, magnetic fields which do not vary in time.

A *macroscopically* static magnetic field is obtained whenever we have a stationary system of electric charges in motion, which means a steady electric current. We saw in Section 6.1 that steady electric current may be produced both in a vacuum and in a conductor. The magnetic field exists, of course, in both cases. In practice, the static magnetic field is always produced by steady electric currents in metallic conductors. But it should be remembered that the magnetic field is present whenever the electric charges are moving with respect to the observer.

*Problems referring to Section 8.3:* 8.T2, 8.N2.

## 8.4 THE MAGNETIC FIELD OF STEADY ELECTRIC CURRENTS. THE BIOT-SAVART LAW

Let us now consider the current-carrying conductor shown in Fig. 8.2. For simplicity it is assumed that the free charges inside the conductor are positive, and the charges at rest negative.

If a test charge $\Delta Q$ resides at a fixed point outside the current-carrying conductor, it will be acted on neither by the electric force nor by the magnetic force. The moving charges inside the conductor are compensated, so that there is no appreciable electric field outside (nor inside) the conductor. Since charge $\Delta Q$ is stationary, according to Eq. (8.3.4) the magnetic force on $\Delta Q$ is also zero. If the charge $\Delta Q$ is set in motion, say with a velocity $\mathbf{v}$, the electric force on it remains zero. But according to Eq. (8.3.4) it will now be acted upon by a magnetic force

$$\mathbf{F}_m = \Delta Q\, \mathbf{v} \times \frac{\mu_0}{4\pi} \sum \frac{Q\, \mathbf{v}_k \times \mathbf{u}_k}{r_k^2}.$$

In this equation $Q$ represents the charge of free charges inside the conductor, and $\mathbf{v}_k$ their average drift velocities (which generally vary at different points). The sum extends to all moving charges inside the conductor. The magnetic flux density at any point is hence given by

$$\mathbf{B} = \frac{\mu_0}{4\pi} \sum \frac{Q\,\mathbf{v}_k \times \mathbf{u}_k}{r_k^2}. \tag{8.4.1}$$

The number of free charges per unit volume of a conductor is enormous. We do not need to consider the current macroscopically as a system of discrete moving charges. Suppose that the number of free charges per unit volume of the conductor is $N$. Then, inside a small volume $\Delta v$, there are $N\Delta v$ charges. They all move with the same velocity $\mathbf{v}$, since $\Delta v$ was supposed to be very small. The sum in Eq. (8.4.1) over the individual charges may hence be replaced by a sum over the volume elements $\Delta v$ (see Fig. 8.2):

$$\mathbf{B} = \frac{\mu_0}{4\pi} \sum_v \frac{N\,Q\,\mathbf{v} \times \mathbf{u}}{r^2} \Delta v. \tag{8.4.2}$$

But according to Eq. (6.2.1) $NQ\mathbf{v}$ represents the current density $\mathbf{J}$ at points inside the volume element $\Delta v$. If $\Delta v$ is assumed to be a "physically small" volume, the sum in Eq. (8.4.2) may be replaced by the volume integral. Hence the magnetic flux density due to the steady current in the conductor is given by

$$\boxed{\mathbf{B} = \frac{\mu_0}{4\pi} \int_v \frac{\mathbf{J} \times \mathbf{u}}{r^2}\,dv} \quad \text{(magnetic flux density produced by macroscopic currents).} \tag{8.4.3}$$

The symbol $\mathbf{u}$ stands for the unit vector directed *from* the volume element $dv$ *towards* the fixed point at which the magnetic flux density is being determined (see Fig. 8.2). If the current distribution inside a conductor is known, we can always calculate the magnetic flux density, at least numerically.

In practice the current is very often flowing through thin conducting wires. In these cases it is practical to use an approximate formula which is readily derived from Eq. (8.4.3). Suppose that the area of the cross-section of the wire is $\Delta S$. Then

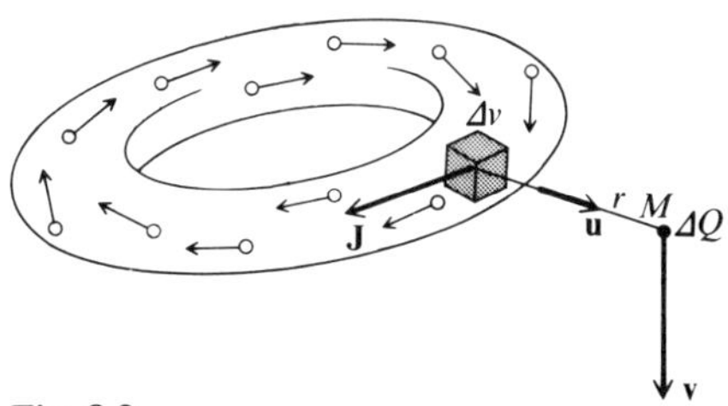


**Fig. 8.2**
A small charge $\Delta Q$ in the field of charges moving inside a conductor.

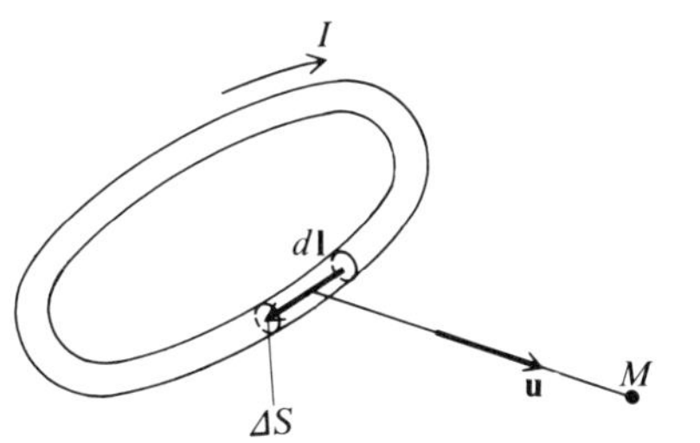


**Fig. 8.3** A quasi-filamentary current-carrying conductor.

we may write $\Delta S\, dl$ in Eq. (8.4.3) instead of $dv$, $dl$ representing an element of the wire. Now, $J\Delta S$ is the intensity of the current in the wire, since $\mathbf{J}$ is normal to $\Delta S$. As the current intensity is not a vector quantity, we cannot transfer the vector sign of $\mathbf{J}$ to $I$. But a vector in the direction of $\mathbf{J}$ must be present in Eq. (8.4.3). It is conventional to orient the contour in the direction of the current density vector, and to consider the length element $d\mathbf{l}$ of the wire as a vector quantity. With these remarks the magnetic flux density vector of a thin wire carrying a current $I$ is (see Fig. 8.3)

$$\boxed{\mathbf{B} = \frac{\mu_0 I}{4\pi} \oint_C \frac{d\mathbf{l} \times \mathbf{u}}{r^2}} \quad \text{(magnetic flux density of current in a thin wire—the Biot-Savart law).} \tag{8.4.4}$$

This important formula is known as the *Biot-Savart law*. It often serves as the starting point in presenting the theory of the static magnetic field.

From Eq. (8.4.4) it follows that the magnetic flux density due to a current element $I d\mathbf{l}$ is given by

$$\boxed{d\mathbf{B} = \frac{\mu_0 I}{4\pi} \frac{d\mathbf{l} \times \mathbf{u}}{r^2}} \quad \text{(flux density of a current element).} \tag{8.4.5}$$

At very high frequencies the current is distributed in a thin sheet on the conductor surface. This effect is known as the *skin effect*, and will be analyzed in Chapter 14. Specialization of the general formula (8.4.3) for that case is straightforward. Let us consider, for this purpose, a very thin sheet of thickness $\Delta d$, carrying a steady current, as shown in Fig. 8.4. The volume element in this case becomes $\Delta d\, dS$. The integration in Eq. (8.4.3) has to be performed over the surface of the sheet:

$$\mathbf{B} = \frac{\mu_0}{4\pi} \int_S \frac{(\mathbf{J}\, \Delta d) \times \mathbf{u}}{r^2}\, dS. \tag{8.4.6}$$

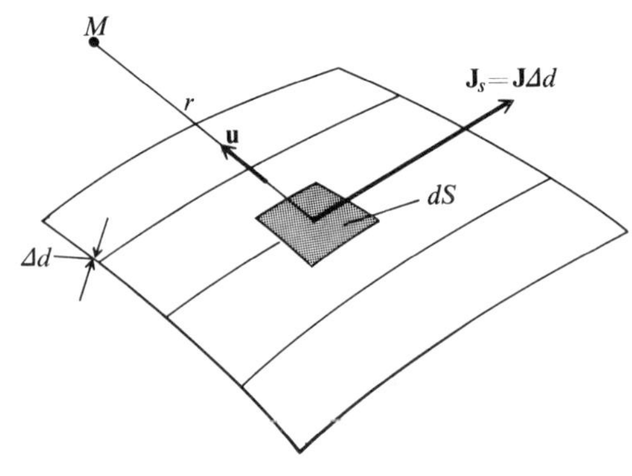


**Fig. 8.4** A current sheet.

The product $\mathbf{J}\Delta d$ represents the current intensity in the sheet per unit length normal to the current flow. It is referred to as the density of the surface current. If we designate the density of the surface current by $\mathbf{J}_s$, Eq. (8.4.6) becomes

$$\mathbf{B} = \frac{\mu_0}{4\pi} \int_S \frac{\mathbf{J}_s \times \mathbf{u}}{r^2} \, dS. \tag{8.4.7}$$

Let us now illustrate the application of the Biot-Savart law for calculating the magnetic flux density vector in some simple cases.

**8.4.1 Flux density at the center of a circular line current.** Consider a very thin circular conductor carrying a steady current of intensity $I$ (see Fig. 8.5). According to Eq. (8.4.5) the magnetic flux density at the center of the contour, due to the current element $d\mathbf{l}$ shown, is

$$d\mathbf{B} = \frac{\mu_0 I}{4\pi} \frac{d\mathbf{l} \times \mathbf{u}}{a^2}. \tag{8.4.8}$$

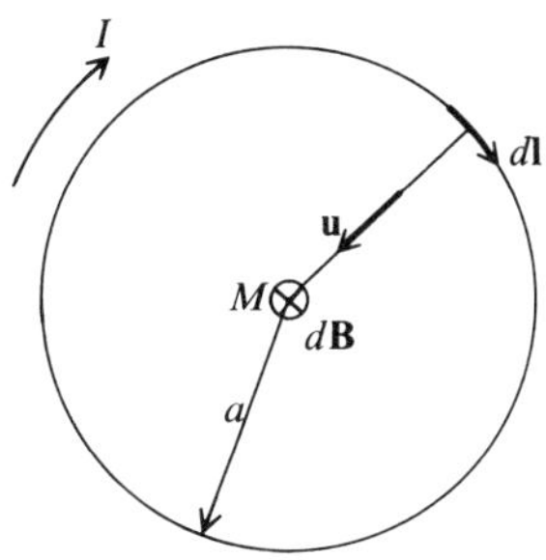


**Fig. 8.5** A circular current-carrying contour.

The product $d\mathbf{l} \times \mathbf{u}$, and hence also the flux density $d\mathbf{B}$ due to the current element $d\mathbf{l}$, is directed away from the observer into the plane of the figure. This is true for the flux density created at $M$ by any other current element of the contour. This direction of $d\mathbf{B}$ is symbolically represented by the sign $\otimes$. If $d\mathbf{B}$ were *towards* the observer, the sign $\odot$ would be used instead. Since the direction of all $d\mathbf{B}$'s is the same, the vector integral (8.4.4) may be calculated as an ordinary integral. The distance of all elements $d\mathbf{l}$ from $M$ is equal to the radius $a$. Vectors $d\mathbf{l}$ and $\mathbf{u}$ are mutually perpendicular. The magnitude of their cross product is simply $dl$. With these remarks, we see that

$$B = \frac{\mu_0 I}{4\pi a^2} \oint_C dl = \frac{\mu_0 I}{4\pi a^2} 2\pi a = \frac{\mu_0 I}{2a}. \tag{8.4.9}$$

**8.4.2 Flux density of a straight current filament.** Consider a thin, very long straight conductor carrying a current of intensity $I$ (see Fig. 8.6). The magnetic field is obviously symmetrical with respect to the conductor axis. The magnetic

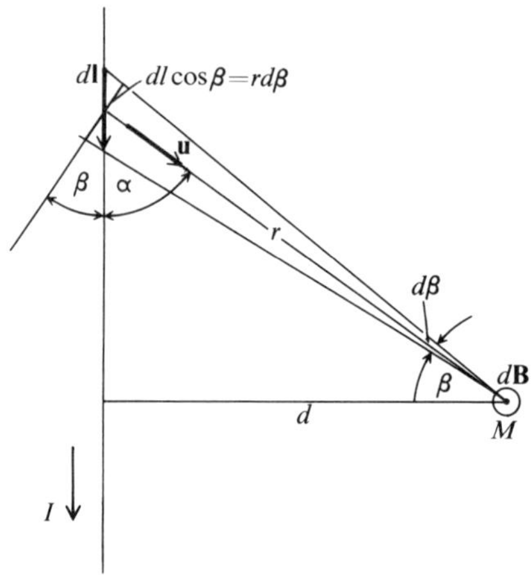


**Fig. 8.6** A very long straight current-carrying conductor.

flux density due to the current element $d\mathbf{l}$ shown, and to all other current elements, is directed from the plane of the drawing towards the observer, as indicated by the sign $\odot$. Hence the vector integral (8.4.4) can again be calculated as an ordinary integral.

According to Eq. (8.4.5) and Fig. 8.6, the magnitude of the flux density vector at $M$ due to the current element $d\mathbf{l}$ shown is

$$dB = \frac{\mu_0}{4\pi}\,\frac{I\,dl\sin\alpha}{r^2} = \frac{\mu_0 I}{4\pi}\,\frac{r\,d\beta}{\cos\beta}\,\frac{\cos\beta}{r^2} = \frac{\mu_0 I}{4\pi d}\cos\beta\,d\beta.$$

The total magnetic flux density at $M$ hence is

$$B = \frac{\mu_0 I}{4\pi d}\int_{-\pi/2}^{\pi/2}\cos\beta\,d\beta = \frac{\mu_0 I}{2\pi d}. \tag{8.4.10}$$

This result was first established experimentally by Biot and Savart. It served as a basis for obtaining the mathematical formula (8.4.4) which is named after them.

**8.4.3 Simplified formula for the flux density of planar currents at points in their plane.** When the magnetic flux density is determined at points of the plane of a planar system of currents, the Biot-Savart law given in Eq. (8.4.4) may be simplified.

Let us consider an arbitrary planar contour carrying current of intensity $I$, and let us determine the magnetic flux density at a point $M$ of the plane of the contour due to the element $d\mathbf{l}$ (see Fig. 8.7). The vector product $d\mathbf{l} \times \mathbf{u}$ is normal to the plane of $d\mathbf{l}$ and $\mathbf{u}$. From Eq. (8.4.5) it follows that the elemental flux density vector at any point of the plane is normal to the plane. The vector integral (8.4.4) may be changed to an ordinary algebraic integral. If we adopt as the reference

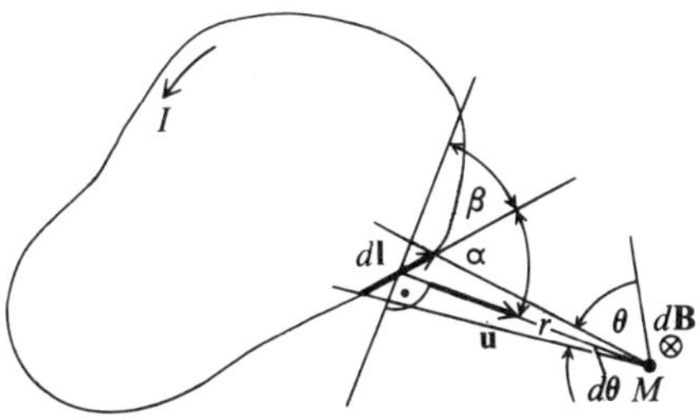


**Fig. 8.7** A planar current-carrying contour.

direction of **B** that shown in Fig. 8.7, then only the magnitude of $d\mathbf{B}$, together with the plus or minus sign, is required. Substituting $dl \sin \alpha$ into Eq. (8.4.5) instead of $d\mathbf{l} \times \mathbf{u}$, we have

$$dB = \frac{\mu_0 I}{4\pi} \frac{dl \sin \alpha}{r^2}. \tag{8.4.11}$$

Now, $dl = r\, d\theta/\cos \beta = r\, d\theta/\sin \alpha$. (Angles $\theta$, $\alpha$, and $\beta$ are shown in Fig. 8.7.) Equation (8.4.11) simplifies therefore to

$$dB = \frac{\mu_0 I}{4\pi} \frac{d\theta}{r}. \tag{8.4.12}$$

The total magnetic flux density due to a closed current-carrying planar contour at points of the plane of the contour is thus given by

$$B = \frac{\mu_0 I}{4\pi} \oint_C \frac{d\theta}{r} \quad \text{($B$ at points of the plane of a planar contour).} \tag{8.4.13}$$

We suggest that, as an exercise, the reader determines the magnetic flux density in the two preceding examples, using this formula.

*Problems referring to Section 8.4:* 8.T3, 8.T4, 8.G1 to 8.G11, 8.N3 to 8.N8.

## 8.5 THE LINES OF THE MAGNETIC FLUX DENSITY VECTOR

The lines of the magnetic flux density vector **B** are defined as lines having the property that vector **B** is tangential to them at all points. If a set of such lines is drawn for a magnetic field, and marked by arrow signs indicating the direction of the magnetic flux density vector, the direction of the vector **B** is known approximately at all points. We shall see in the next section that if we adhere to a certain rule in drawing these lines, their density is a measure of the intensity of vector **B**. The lines of the magnetic flux density vector **B** are very useful for visualizing the magnetic field.

An important example of lines of the magnetic flux density vector are lines of **B** of a single point charge. According to Eq. (8.3.3), vector **B** in that case is perpen-

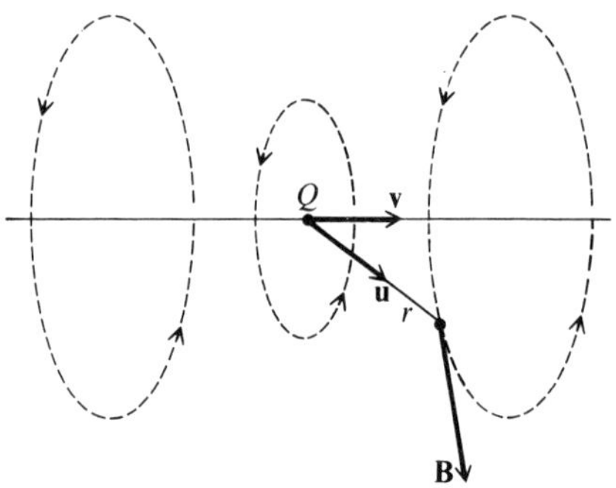


**Fig. 8.8** Lines of vector **B** of a moving point charge.

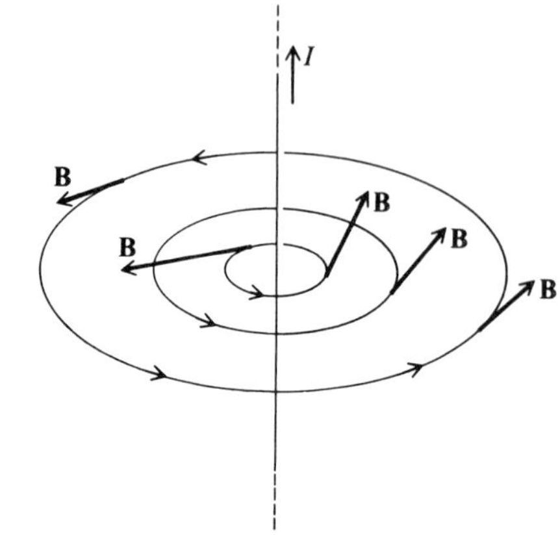


**Fig. 8.9** Lines of vector **B** of a straight current filament.

dicular to the vectors **v** and **u**. *Hence the lines of the magnetic flux density vector of a single charge are circles* with centers on the straight line along which the velocity vector **v** of the charge is directed (see Fig. 8.8).

As another example, we can readily see from Fig. 8.6 and Eq. (8.4.10) that the lines of **B** in the case of a straight current filament are again circles, with their centers at the axis of the filament (see Fig. 8.9). This is, of course, a consequence of the fact that the lines of **B** of the individual charges moving in the filament are circles.

*Problem referring to Section 8.5:* 8.T5.

## 8.6 THE MAGNETIC FLUX AND ITS PROPERTIES

We are by now familiar with the concept of the flux of a vector function. The flux of the magnetic flux density vector **B** through a surface $S$ (see Fig. 8.10) is referred to as the *magnetic flux*. It is usually designated by the symbol $\Phi$:

$$\boxed{\Phi = \int_S \mathbf{B} \cdot d\mathbf{S}} \qquad \text{(definition of the magnetic flux).} \qquad (8.6.1)$$

Magnetic flux is a quantity frequently used in practice. Its unit is given a separate name, the weber (abbreviation Wb). Recalling that the unit of **B** is the tesla (T), we can see that $\text{T} = \text{Wb/m}^2$.

Tubes of magnetic flux are defined as tubular surfaces formed by a bunch of lines of the magnetic flux density vector **B**. By its very definition vector **B** is tangential to a tube of magnetic flux at all points. Hence the magnetic flux through any cross-section of a tube is a constant. It is conventional to divide the whole magnetic field into tubes of *equal* flux, each of which is then represented by a single

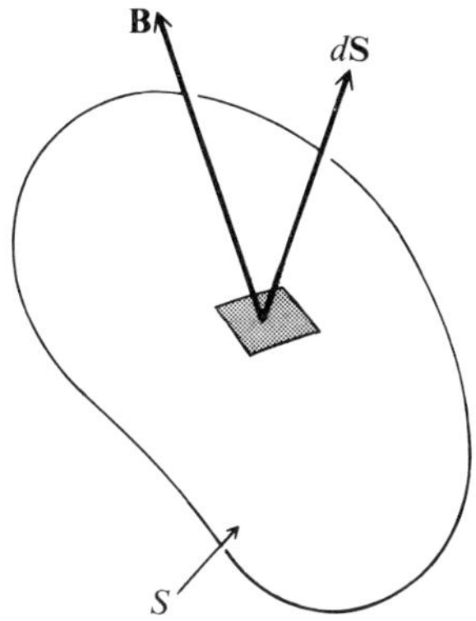


**Fig. 8.10** Illustration for defining magnetic flux.

line of the magnetic flux density vector, say the one along the axis of the tube. The density of the lines is then proportional to the magnitude of vector **B**. We shall always adhere to this convention of drawing the lines of vector **B**. The picture of these lines gives a notion of both the direction and the magnitude of vector **B** at all points.

**8.6.1 The law of conservation of magnetic flux.** Magnetic flux has a very simple but extremely important property: *the magnetic flux through any closed surface is zero*. This is sometimes referred to as the *law of conservation of magnetic flux*, and may be proved as follows.

Any magnetic field is produced by elemental charges in motion. The total flux density vector is obtained as a vector sum of the elemental flux density vectors due to all the elemental charges in a system. According to Eq. (8.6.1), the magnetic flux $\Phi$ through any surface is also equal to the sum of the elemental magnetic fluxes through the surface due to all the elemental moving charges creating the magnetic field. But according to the expression (8.3.3) for the flux density of a single moving charge, the lines of the magnetic flux density vector **B** in that case are closed circles, as shown in Fig. 8.8. If we consider any closed surface situated in the magnetic field of the charge, any line of vector **B** will enter the surface exactly as many times as it leaves it. The lines represent tubes of equal flux. The inward flux is therefore exactly equal to the outward flux, and the total flux through the closed surface is zero.* As a result, magnetic flux through any closed surface produced by any number of moving charges is also zero.

The law of conservation of magnetic flux is expressed by the equation

$$\boxed{\oint_S \mathbf{B} \cdot d\mathbf{S} = 0} \quad \text{(integral form of the law of conservation of magnetic flux).} \qquad (8.6.2)$$

* Strict mathematical proof of this statement is given in Problem 8.G12.

Using the divergence theorem, Eq. (1.11.10), we may write this in the form

$$\int_v \operatorname{div} \mathbf{B}\, dv = 0, \tag{8.6.3}$$

where $v$ represents the volume enclosed by the surface $S$. Since this equation must hold for *any* volume, it follows that

$$\boxed{\operatorname{div} \mathbf{B} = 0} \quad \text{(differential form of the law of conservation of magnetic flux).} \tag{8.6.4}$$

This equation is known as the differential form of the law of conservation of magnetic flux.

The law of conservation of magnetic flux is sometimes also stated in the following two ways. Firstly, it is assumed to represent the fact that the lines of vector **B** always close upon themselves. Secondly, it is also considered as a statement that in nature there are no magnetic charges analogous to the electric charges.

**8.6.2 Magnetic flux through a closed contour.** Let us consider two open surfaces $S_1$ and $S_2$ spanned over a common contour $C$ (see Fig. 8.11), and let the positive direction along the contour $C$ be that shown in the figure. Positive normals to the two surfaces may have either of the two directions. To make the choice of the positive normal spanned over a contour unique, we define it as being determined according to the right-hand screw rule with respect to the positive direction along the contour. *We shall always adhere to this convention.* The positive normals to $S_1$ and $S_2$ according to this convention are shown in the figure.

If $S_1$ and $S_2$ are situated in a magnetic field, then the magnetic flux through them is

$$\Phi_1 = \int_{S_1} \mathbf{B}\cdot d\mathbf{S}_1 = \int_{S_1} \mathbf{B}\cdot \mathbf{u}_{n1}\, dS_1, \qquad \Phi_2 = \int_{S_2} \mathbf{B}\cdot d\mathbf{S}_2 = \int_{S_2} \mathbf{B}\cdot \mathbf{u}_{n2}\, dS_2.$$

The surface $(S_1 + S_2)$ is a closed surface, so that Eq. (8.6.2) applies. But we adopted the convention always to direct the positive normal to a closed surface from the surface outward. Hence, in calculating the flux through the closed sur-

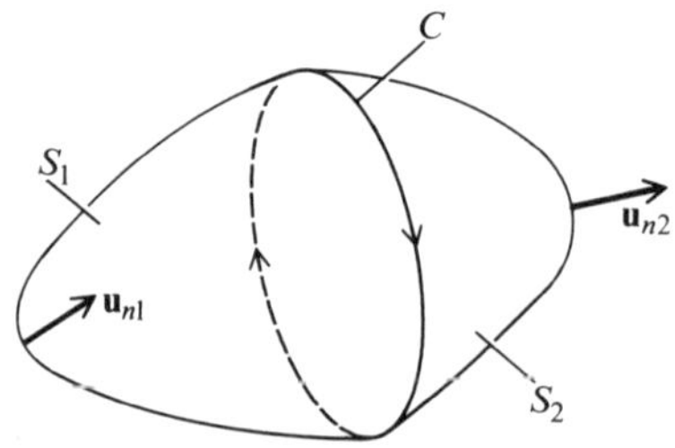


**Fig. 8.11** Two surfaces having a common limiting contour.

face $(S_1 + S_2)$, we have to change $\mathbf{u}_{n1}$ to $-\mathbf{u}_{n1}$. Thus we have

$$\oint_{S_1+S_2} \mathbf{B} \cdot d\mathbf{S} = \int_{S_1} \mathbf{B} \cdot (-\mathbf{u}_{n1})\, dS_1 + \int_{S_2} \mathbf{B} \cdot \mathbf{u}_{n2}\, dS_2 = 0,$$

from which

$$\int_{S_1} \mathbf{B} \cdot d\mathbf{S}_1 = \int_{S_2} \mathbf{B} \cdot d\mathbf{S}_2 \qquad \text{(any two surfaces spanned over a common contour).} \tag{8.6.5}$$

The surfaces $S_1$ and $S_2$ were completely arbitrary; hence it follows that the magnetic flux through *all* surfaces spanned over a common contour are identical. Thus it is not the shape of the surface but *only* the shape of the contour by which it is limited which determines the magnetic flux through the surface. For this reason, reference is often made to the *magnetic flux through a contour* rather than through the surface limited by the contour. We shall frequently use this concept of the magnetic flux through a contour.

In the steady current case, the current density vector satisfies Eq. (6.3.6), analogous to Eq. (8.6.2). Hence we can refer in that case also to the total current through a closed *contour* instead of to the current through the surface limited by the contour.

*Problems referring to Section 8.6:* 8.T6, 8.T7, 8.G12 to 8.G15, 8.N9, 8.N11.

## 8.7 AMPERE'S CIRCUITAL LAW

In addition to the integral relation in Eq. (8.6.2), the magnetic flux density vector $\mathbf{B}$ produced by stationary currents also satisfies another very simple and important relation, known as *Ampère's circuital law*. According to that law, the line integral of the magnetic flux density vector $\mathbf{B}$ along any closed contour is equal to the total current through the contour, times $\mu_0$:

$$\boxed{\oint_C \mathbf{B} \cdot d\mathbf{l} = \mu_0 \int_S \mathbf{J} \cdot d\mathbf{S}} \qquad \text{(Ampère's circuital law, magnetic materials not being present).} \tag{8.7.1}$$

In the case of stationary currents, we can always divide the current flow in elemental current filaments. Thus Ampère's law is proved for all cases if it is proved for a single closed current filament.*

Let us first consider a single charge $Q$ moving with a velocity $\mathbf{v}$. According to Eq. (8.3.3), the line integral of the magnetic flux density vector $\mathbf{B}$ produced by $Q$, along a closed contour $C$ (see Fig. 8.12), is

$$\oint_C \mathbf{B} \cdot d\mathbf{l} = \frac{\mu_0 Q}{4\pi} \oint_C \frac{(\mathbf{v} \times \mathbf{u}) \cdot d\mathbf{l}}{r^2}. \tag{8.7.2}$$

---

* If desired, proof of Ampère's law can be omitted without loss of continuity. In that case move on to Eq. (8.7.7).

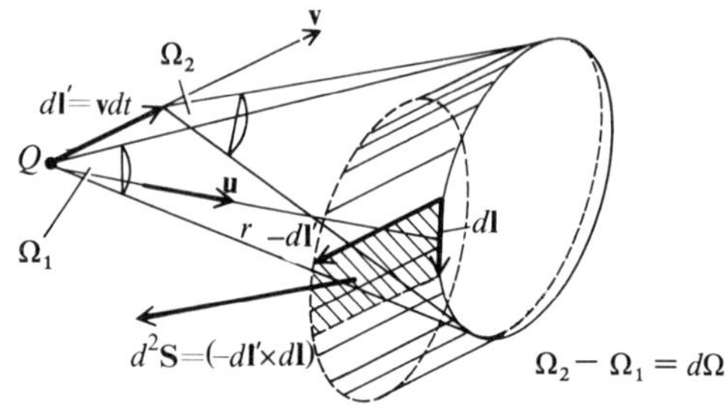


**Fig. 8.12** Construction for determining the line integral of vector **B** due to a single moving point charge.

The line integral has to be evaluated at the instant when the charge $Q$ is at the point shown. Now, if we refer to Fig. 8.12 we see that $\mathbf{v} = d\mathbf{l}'/dt$. The time interval $dt$ may be taken out of the integral. In addition, according to the property of the triple scalar product expressed by Eq. (0.3.15), we have

$$(d\mathbf{l}' \times \mathbf{u}) \cdot d\mathbf{l} = d\mathbf{l} \cdot (d\mathbf{l}' \times \mathbf{u}) = \mathbf{u} \cdot (d\mathbf{l} \times d\mathbf{l}').$$

The line integral in Eq. (8.7.2) of the vector **B** along the contour $C$ thus becomes

$$\oint_C \mathbf{B} \cdot d\mathbf{l} = \frac{\mu_0 Q}{4\pi} \frac{1}{dt} \oint_C \frac{\mathbf{u} \cdot (d\mathbf{l} \times d\mathbf{l}')}{r^2} = \frac{\mu_0 Q}{4\pi} \frac{1}{dt} \oint_C \frac{(-d\mathbf{l}' \times d\mathbf{l}) \cdot \mathbf{u}}{r^2}. \tag{8.7.3}$$

By referring to Fig. 8.12, we see that the cross product $(-d\mathbf{l}' \times d\mathbf{l})$ represents the vector surface element $d^2\mathbf{S}$ of the strip shown crosshatched. This strip is obtained if $Q$ is assumed to be fixed, and the contour moved for $-d\mathbf{l}'$. According to the definition of an elemental solid angle,

$$\frac{(-d\mathbf{l}' \times d\mathbf{l}) \cdot \mathbf{u}}{r^2} = \frac{d^2\mathbf{S} \cdot \mathbf{u}}{r^2} = d^2\Omega \tag{8.7.4}$$

represents the elemental solid angle subtended by the elemental surface $d^2\mathbf{S}$ at the point where the charge $Q$ is situated. Integrating $d^2\Omega$ along the contour $C$ results in the elemental solid angle $d\Omega$ subtended at $Q$ by the whole strip surface. But $d\Omega$ is exactly equal to the difference $(\Omega_2 - \Omega_1)$ of the solid angles subtended by any surface spanned over $C$ at the end and the initial positions of the charge $Q$. Hence

$$\oint_C \mathbf{B} \cdot d\mathbf{l} = \frac{\mu_0 Q}{4\pi} \frac{d\Omega}{dt}. \tag{8.7.5}$$

Consider now a part of a filamentary current $I$ (see Fig. 8.13). A contribution to the line integral in Eq. (8.7.5) of moving charges $dQ$ in an element $dl'$ of the conductor is

$$d \oint_C \mathbf{B} \cdot d\mathbf{l} = \frac{\mu_0 dQ}{4\pi} \frac{d\Omega}{dt} = \frac{\mu_0}{4\pi} \frac{dQ}{dt} d\Omega = \frac{\mu_0}{4\pi} I \, d\Omega.$$

Hence

$$\oint_C \mathbf{B}\cdot d\mathbf{l} = \frac{\mu_0 I}{4\pi}\int_{\Omega_1}^{\Omega_2} d\Omega = \frac{\mu_0 I}{4\pi}(\Omega_2 - \Omega_1). \tag{8.7.6}$$

By definition, $\Omega_1$ and $\Omega_2$ represent solid angles subtended at points $M_1$ and $M_2$ by the contour $C$. Suppose we start at $M_1$. Then the solid angle $\Omega$ in Fig. 8.13 first decreases, and then increases in approaching $M_2$. Let the point move further, along the dashed part of the filament, to coincide with $M_1$. The right-hand side of Eq. (8.7.6) will then become zero, because then $\Omega_2$ will be equal to $\Omega_1$.

Consider now the filament 2 which encircles the contour $C$ of integration. When moving from $M'_1$ to $M'_2$ along the filament, the increment $d\Omega$ of the solid angle is *always* positive. The angle $\Omega_2$ becomes the *outer* solid angle, as indicated at $M'_2$. If $M'_2$ moves further along the dashed line and coincides with $M'_1$, the solid angle $\Omega'_2$ will then be equal to $(\Omega'_1 + 4\pi)$. The difference $\Omega'_2 - \Omega'_1$ is hence equal to $4\pi$, if the filamentary current loop encircles the contour $C$ of integration. Consequently, for a single current filament with a current $I$

$$\oint_C \mathbf{B}\cdot d\mathbf{l} = \begin{cases} \mu_0 I \text{ when } C \text{ encircles current filament } I \\ 0 \quad \text{when } C \text{ does not encircle current filament} \end{cases} \tag{8.7.7}$$

It is very important to note that Eq. (8.7.7) was derived by supposing that the positive directions of the contour $C$ and of the current in the filament are connected according to the right-hand screw rule (see Fig. 8.13).

It is always possible to divide any stationary current system into elemental,

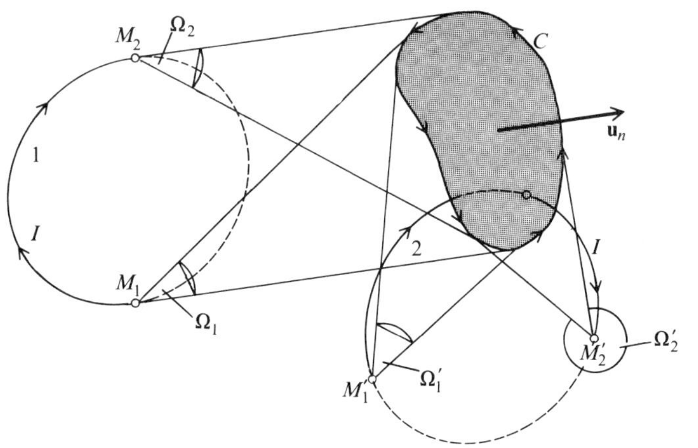


**Fig. 8.13** Construction for determining the line integral of vector **B** due to a current filament.

closed current filaments for which Eq. (8.7.7) applies. Hence we have

$$\oint_C \mathbf{B} \cdot d\mathbf{l} = \mu_0 \int_S \mathbf{J} \cdot d\mathbf{S}, \tag{8.7.8}$$

where $S$ represents any surface limited by the contour $C$. The positive normal to $S$ is determined according to the right-hand screw rule with respect to the positive direction along $C$ (see Fig. 8.14). Ampère's circuital law (8.7.1) is thereby proved. It should be noted that Ampère's law is a direct consequence of Eq. (8.3.3) for the flux density vector of a single moving charge, *and* the solenoidal character of steady current flow. In Section 13.2, Ampère's law for nonstationary electric currents will be derived.

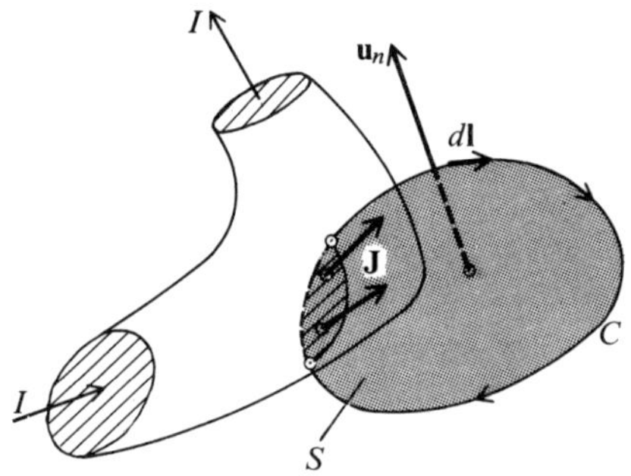


**Fig. 8.14** An illustration of Ampère's circuital law.

Ampère's circuital law (8.7.1) is a very important general law for the static magnetic field in the absence of magnetic materials. Furthermore, by using this law, we may determine the magnetic flux density vector in some special, but important, cases in a simple manner. In this respect it is analogous to Gauss' law in the static electric field. If, by symmetry, the direction of vector **B** is known in advance, Ampère's circuital law (8.7.1) suffices for determining the one unknown quantity, the magnitude of vector **B**. Several such applications of Ampère's law are presented below.

**8.7.1 The magnetic field of a current in a straight circular cylindrical conductor.** Consider a straight conductor of radius $a$, carrying a current of intensity $I$. By symmetry, the lines of the magnetic flux density vector are circles centered at the conductor axis, and the magnitude of vector **B** is a function of the distance $r$ from the conductor axis only. The direction of **B** is determined by the right-hand screw rule. If we apply Ampère's circuital law to a circular contour $C_1$ of radius $r_1 \geq a$ (see Fig. 8.15), since $d\mathbf{l}_1$ and **B** are parallel at all points of the contour of integration and the magnitude of **B** is constant, we have

$$\oint_{C_1} \mathbf{B} \cdot d\mathbf{l}_1 = \oint_{C_1} B\, dl_1 = B \oint_{C_1} dl_1 = B\, 2\pi r_1 = \mu_0 I,$$

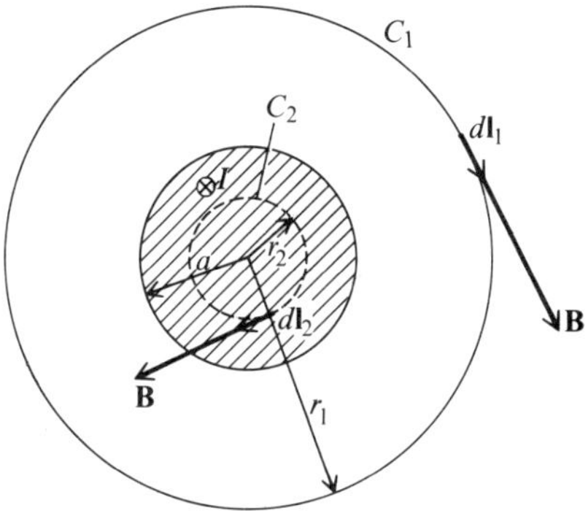


**Fig. 8.15** A straight current-carrying conductor of circular cross-section.

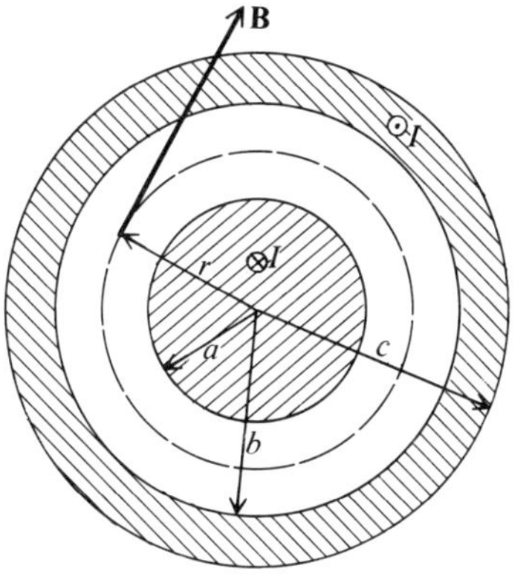


**Fig. 8.16** A coaxial cable.

from which

$$B = \frac{\mu_0 I}{2\pi r_1}, \qquad r_1 \geq a. \tag{8.7.9}$$

Comparing this with Eq. (8.4.10) we see that the magnetic field *outside* a straight conductor of circular cross-section is the same as that of a line current along its axis.

The contour $C_2$ of radius $r_2 < a$ encircles only a part of the total current, of intensity $r_2^2\pi I/a^2\pi$. Thus

$$\oint_{C_2} \mathbf{B}\cdot d\mathbf{l}_2 = \oint_{C_2} B\,dl_2 = B\oint_{C_2} dl_2 = B\,2\pi r_2 = \mu_0\frac{I}{a^2\pi}r_2^2\pi,$$

whence

$$B = \frac{\mu_0 I r_2}{2\pi a^2}, \qquad r_2 \leq a. \tag{8.7.10}$$

The magnetic field of current in a coaxial cable can be determined by exactly the same procedure. The cross-section of the cable is shown in Fig. 8.16. The currents in the inner and outer conductors are of the same intensity $I$, but in opposite directions. The following expressions for the magnitude of the magnetic flux density vector $\mathbf{B}$ are obtained:

$$B = \frac{\mu_0 I}{2\pi a^2}r \qquad \text{for} \qquad r \leq a, \tag{8.7.11}$$

$$B = \frac{\mu_0 I}{2\pi r} \qquad \text{for} \qquad a \leq r \leq b, \tag{8.7.12}$$

$$B = \frac{\mu_0 I(c^2 - r^2)}{2\pi r(c^2 - b^2)} \qquad \text{for} \qquad b \leq r \leq c. \tag{8.7.13}$$

For $r > c$ the flux density $\mathbf{B} = 0$.

**8.7.2 Magnetic field of current in a toroidal coil and in an infinitely long solenoid.** A toroidal coil is obtained if a thin wire is densely wound on an axially sym-

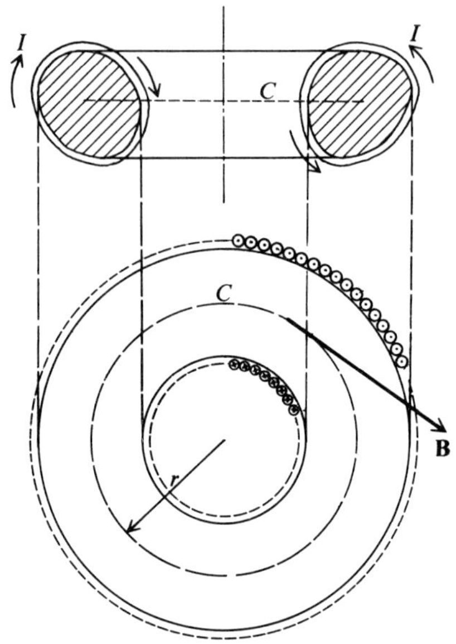


**Fig. 8.17** A densely wound toroidal coil.

metrical toroidal core (see Fig. 8.17). Let us suppose that there are $N$ turns of the wire, with a current of intensity $I$, and let us approximate the coil by a current sheet over the core surface. By symmetry, the lines of the magnetic flux density vector inside the core are circles with their centers at the axis of symmetry, as indicated in Fig. 8.17. The total current through the circles which are outside the core is zero; thus, according to Eq. (8.7.1), **B** is also zero at all points outside the core.

Consider a circular contour $C$ inside the core. The total current through $C$ is exactly $NI$ (see Fig. 8.17). From Eq. (8.7.1) we have

$$\oint_C \mathbf{B}\cdot d\mathbf{l} = 2\pi r\, B = \mu_0 NI, \text{ from which } B = \mu_0 NI/2\pi r \text{ (toroidal coil).} \tag{8.7.14}$$

Suppose that the dimensions of the cross-section of the core are much smaller than its mean radius $r_{\text{mean}}$. The magnitude of the magnetic flux density is then approximately equal at all points inside the core, and equal to

$$B = \mu_0 \frac{N\,I}{2\pi r_{\text{mean}}} \qquad \text{(thin toroidal coil).} \tag{8.7.15}$$

Let $r_{\text{mean}}$ and $N$ be simultaneously increased, but so that the ratio $N_l = N/2\pi r_{\text{mean}}$ (the number of turns per unit length) remains constant. The result is an infinitely long solenoid, having $N_l$ turns per unit length, with a current of intensity $I$. The magnetic flux density inside such a solenoid of *arbitrary* cross-

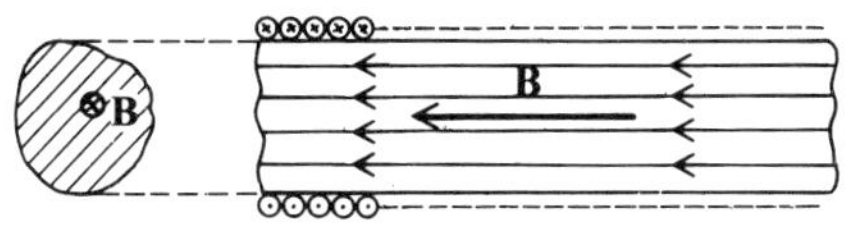


**Fig. 8.18** Section of a very long solenoid.

section is directed along the solenoid axis. Its magnitude inside the solenoid is

$$B = \mu_0 N_l I \qquad \text{(infinitely long solenoid)}, \tag{8.7.16}$$

and is practically zero outside. The directions of the current and of vector **B** are connected by the right-hand screw rule (see Fig. 8.18).

**8.7.3 Magnetic field of a planar current sheet.** Suppose that the density of surface current over a very thin conducting sheet is $\mathbf{J}_s$. By symmetry, the lines of the magnetic flux density vector are parallel to the sheet and normal to $\mathbf{J}_s$ (see Fig. 8.19). The only unknown is the dependence of the magnitude of **B** upon the distance from the sheet.

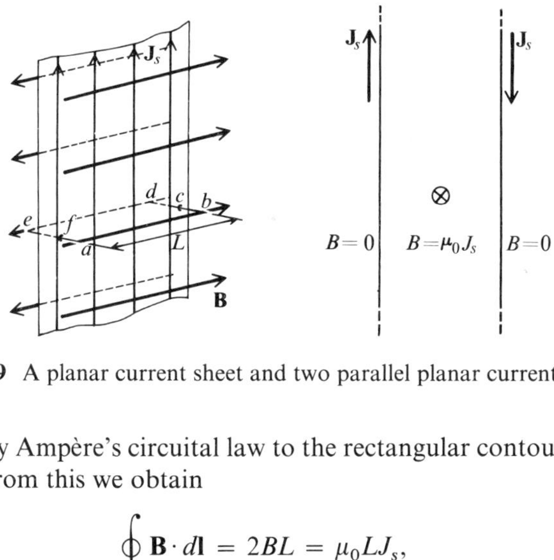


**Fig. 8.19** A planar current sheet and two parallel planar current sheets.

Let us apply Ampère's circuital law to the rectangular contour *abcdefa* shown in Fig. 8.19. From this we obtain

$$\oint \mathbf{B}\cdot d\mathbf{l} = 2BL = \mu_0 L J_s,$$

since the total current encircled by the contour is the current in the part *cf* of the sheet, equal to $LJ_s$. Thus

$$B = \mu_0 J_s/2 \qquad \text{(infinite planar current sheet)}. \tag{8.7.17}$$

Let us next consider two parallel sheets with surface currents of equal densities $J_s$ but in opposite directions. The magnetic field is obtained by using the above result and the principle of superposition. Between the sheets, according to Eq.

(8.7.17) and Fig. 8.19,

$$B = \mu_0 J_s \qquad \text{(between parallel current sheets)}, \tag{8.7.18}$$

and $B = 0$ outside.

*Problems referring to Section 8.7:* 8.T8–8.T11, 8.G16–8.G19, 8.N10–8.N12.

## 8.8 DIFFERENTIAL FORM OF AMPERE'S CIRCUITAL LAW. CURL OF A VECTOR FUNCTION

Ampère's circuital law (8.7.1) in its integral form is an exact relation between the *total* current through a contour and the line integral of the vector **B** along the contour. However, it tells us nothing about the connection between vectors **B** and **J** at a particular point. This information is provided by the differential form of Ampère's circuital law.

Let us apply Ampère's law to a very small planar contour $\Delta C$. In that case Eq. (8.7.1) becomes

$$\oint_{\Delta C} \mathbf{B} \cdot d\mathbf{l} = \mu_0 \mathbf{J} \cdot \Delta \mathbf{S}. \tag{8.8.1}$$

The symbol $\Delta\mathbf{S}$ represents the planar surface element limited by $\Delta C$. If we imagine that $\Delta C$ becomes smaller and smaller, the right-hand side, and hence also the left-hand side, of Eq. (8.8.1) will also become smaller and smaller. In the limit, both are zero. This is not the case, however, if we write $\Delta\mathbf{S}$ in the form $\Delta S\,\mathbf{u}_n$, divide Eq. (8.8.1) by $\Delta S$, and then let $\Delta S$ tend to zero. The limit of the ratio

$$\lim_{\Delta S \to 0} \frac{\oint_{\Delta C} \mathbf{B} \cdot d\mathbf{l}}{\Delta S} = \mu_0 \mathbf{J} \cdot \mathbf{u}_n \tag{8.8.2}$$

is nonzero at all points where **J** is not zero (except if $\mathbf{u}_n$ is taken to be perpendicular to **J**). Equation (8.8.2) represents Ampère's law in differential form. We shall see later that it represents a vectorial partial differential equation.

An expression of the type on the left-hand side of Eq. (8.8.2) is very often encountered in mathematical physics. It depends upon the orientation of $\Delta S$, since the product $\mathbf{J} \cdot \mathbf{u}_n$ varies with the direction of $\mathbf{u}_n$. The left-hand side of Eq. (8.8.2) must therefore be regarded as the projection of some *vector function*, the direction of which coincides with the direction of vector **J**, on the direction of $\mathbf{u}_n$. This vector function is referred to as the *curl of vector* **B**. The left-hand side of Eq. (8.8.2) hence represents the *component of the curl of* **B** *in the direction of* $\mathbf{u}_n$ (which is normal to $\Delta S$).

The curl of a vector function **B** is designated simply as curl **B**. Ampère's circuital law in differential form can hence be written in the form

$$\boxed{\operatorname{curl} \mathbf{B} = \mu_0 \mathbf{J}} \qquad \text{(the differential form of Ampère's circuital law for currents in vacuum)}. \tag{8.8.3}$$

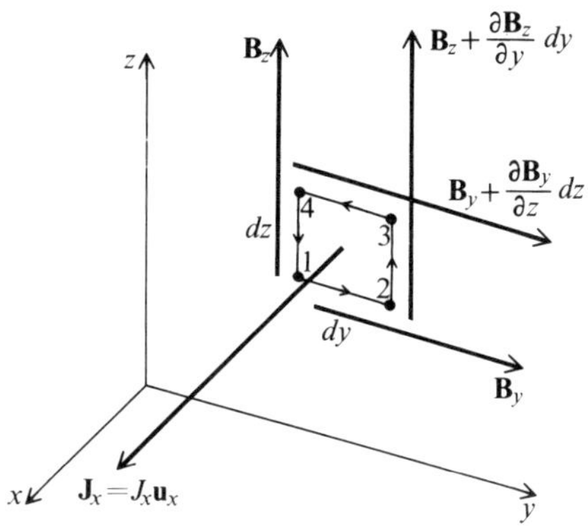


**Fig. 8.20** Construction for determining the curl in a rectangular coordinate system.

Expressions for the three components of curl **B** with respect to any coordinate system may be obtained according to the definition in Eq. (8.8.2) of the component of curl **B** in an arbitrary direction $\mathbf{u}_n$. For convenience this expression can be rewritten as follows:

$$\text{Component of curl } \mathbf{B} \text{ in direction } \mathbf{u}_n = \lim_{\Delta S \to 0} \frac{\oint_{\Delta C} \mathbf{B} \cdot d\mathbf{l}}{\Delta S}, \tag{8.8.4}$$

where $\Delta S$ is perpendicular to $\mathbf{u}_n$.

As an example, let us determine the three components of curl **B** with respect to the rectangular coordinate system. Expressions for curl **B** with respect to cylindrical and spherical coordinate systems are derived in Problems 8.G20 and 8.G21.

To determine the $x$ component, $\text{curl}_x\mathbf{B}$, of curl **B**, let us consider a small rectangle with sides $dy$ and $dz$ (see Fig. 8.20). The surface $dS$ is then normal to the unit vector $\mathbf{u}_x$, as required by Eq. (8.8.4). We have to determine the line integral along the contour with respect to its positive direction. The positive direction along the contour and the unit vector $\mathbf{u}_x$ are connected by the right-hand screw rule. With this in mind, we see that the line integral of **B** around the rectangular contour 12341 is

$$\begin{aligned}\oint_{12341} \mathbf{B} \cdot d\mathbf{l} &= B_y dy + \left(B_z + \frac{\partial B_z}{\partial y} dy\right) dz - \left(B_y + \frac{\partial B_y}{\partial z} dz\right) dy - B_z dz \\ &= \left(\frac{\partial B_z}{\partial y} - \frac{\partial B_y}{\partial z}\right) dy\, dz.\end{aligned} \tag{8.8.5}$$

The elemental surface limited by the contour is $dS = dy\,dz$. From Eqs.

(8.8.4) and (8.8.5) it follows that the $x$ component of curl $\mathbf{B}$ is

$$\operatorname{curl}_x\mathbf{B} = \frac{\oint_{12341} \mathbf{B}\cdot d\mathbf{l}}{dy\,dz} = \frac{\partial B_z}{\partial y} - \frac{\partial B_y}{\partial z}. \tag{8.8.6}$$

The components $\operatorname{curl}_y\mathbf{B}$ and $\operatorname{curl}_z\mathbf{B}$ are obtained in the same way. The result is

$$\operatorname{curl}_y\mathbf{B} = \frac{\partial B_x}{\partial z} - \frac{\partial B_z}{\partial x}, \qquad \operatorname{curl}_z\mathbf{B} = \frac{\partial B_y}{\partial x} - \frac{\partial B_x}{\partial y}. \tag{8.8.7}$$

From Eqs. (8.8.6) and (8.8.7) the curl of the vector function $\mathbf{B}$ with respect to the rectangular coordinate system is given by

$$\operatorname{curl}\mathbf{B} = \left(\frac{\partial B_z}{\partial y} - \frac{\partial B_y}{\partial z}\right)\mathbf{u}_x + \left(\frac{\partial B_x}{\partial z} - \frac{\partial B_z}{\partial x}\right)\mathbf{u}_y + \left(\frac{\partial B_y}{\partial x} - \frac{\partial B_x}{\partial y}\right)\mathbf{u}_z. \tag{8.8.8}$$

Ampère's circuital law in differential form (8.8.3) is thus indeed a vectorial partial differential equation. It represents the following three scalar partial differential equations:

$$\frac{\partial B_z}{\partial y} - \frac{\partial B_y}{\partial z} = \mu_0 J_x, \qquad \frac{\partial B_x}{\partial z} - \frac{\partial B_z}{\partial x} = \mu_0 J_y, \qquad \frac{\partial B_y}{\partial x} - \frac{\partial B_x}{\partial y} = \mu_0 J_z. \tag{8.8.9}$$

From Eq. (8.8.8) we can see that curl $\mathbf{B}$ with respect to the rectangular coordinate system can be represented in the form of a vector determinant:

$$\operatorname{curl}\mathbf{B} = \begin{vmatrix} \mathbf{u}_x & \mathbf{u}_y & \mathbf{u}_z \\ \dfrac{\partial}{\partial x} & \dfrac{\partial}{\partial y} & \dfrac{\partial}{\partial z} \\ B_x & B_y & B_z \end{vmatrix}. \tag{8.8.10}$$

This form of curl $\mathbf{B}$ in a rectangular coordinate system is perhaps easiest to remember.

If we recall the definition in Eq. (1.7.5) of the del operator $\nabla$ and compare Eq. (8.8.10) with Eq. (0.4.13), we can see that, *with respect to the rectangular coordinate system*, curl $\mathbf{B}$ can be written in a symbolic form

$$\operatorname{curl}\mathbf{B} = \nabla\times\mathbf{B} = \left(\frac{\partial}{\partial x}\mathbf{u}_x + \frac{\partial}{\partial y}\mathbf{u}_y + \frac{\partial}{\partial z}\mathbf{u}_z\right) \times (B_x\mathbf{u}_x + B_y\mathbf{u}_y + B_z\mathbf{u}_z). \tag{8.8.11}$$

The differential form of Ampère's circuital law can hence be written in the shorter form

$$\nabla\times\mathbf{B} = \mu_0\,\mathbf{J}. \tag{8.8.12}$$

This is strictly valid in a rectangular coordinate system only. Very often, however, $\nabla\times\mathbf{B}$ is used as an equivalent symbol for curl $\mathbf{B}$ generally.

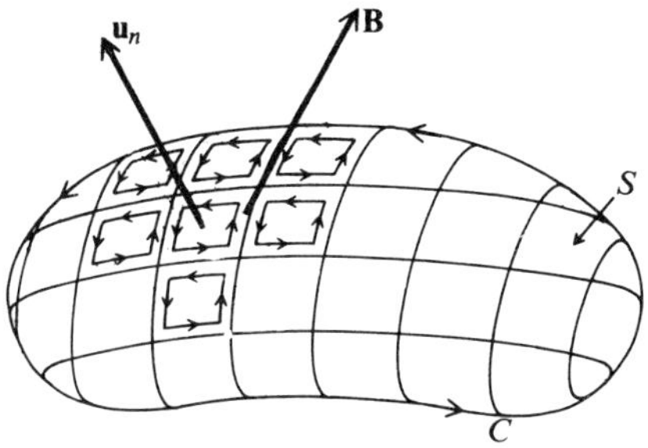


**Fig. 8.21** A surface subdivided into elemental surfaces.

A very important theorem of vector analysis is what is known as *Stokes' theorem.* It can be derived easily from the definition of the curl of a vector function, given by Eq. (8.8.4). Consider an arbitrary surface $S$ limited by a contour $C$, and imagine the surface $S$ to be divided into a large number of small surfaces $dS$ limited by small contours $dC$ (see Fig. 8.21). From Eq. (8.8.2) and (8.8.3) it follows that for any of these small contours we can write

$$\oint_{dC} \mathbf{B} \cdot d\mathbf{l} = \mu_0 \, \mathbf{J} \cdot d\mathbf{S} = \text{curl } \mathbf{B} \cdot d\mathbf{S}. \tag{8.8.13}$$

Note that for pairs of adjacent sides of two small contours the line elements $d\mathbf{l}$ are oppositely directed, and that $\mathbf{B}$ is the same for both. By adding equations of the type of Eq. (8.8.13) for all elemental surfaces we hence obtain the following integral relation

$$\boxed{\oint_C \mathbf{B} \cdot d\mathbf{l} = \int_S \text{curl } \mathbf{B} \cdot d\mathbf{S}} \quad \text{(Stokes' theorem; } \mathbf{B} = \text{any well-behaved vector function).} \tag{8.8.14}$$

This is known as *Stokes' theorem.* We shall refer frequently to this important theorem.

*Problems referring to Section 8.8:* 8.T12, 8.T13, 8.G20, 8.G21.

## 8.9 THE MAGNETIC VECTOR POTENTIAL

**8.9.1 Definition of the magnetic vector potential.** According to Ampère's circuital law (8.7.1), the line integral of the magnetic flux density vector $\mathbf{B}$ around a closed contour is not zero if the contour encircles some current. Therefore, it is not possible to define an auxiliary scalar function completely analogous to the electrostatic potential function $V$, from which vector $\mathbf{B}$ could be obtained as its gradient. (We shall see in the next section that such a function exists nevertheless if certain conditions are met.) But in this case an auxiliary *vector function* can be defined, from which vector $\mathbf{B}$ can always be obtained by differentiation. To understand this, a short mathematical digression is necessary.

The divergence and the curl of a vector function are not defined with respect to any particular coordinate system. Consequently, if any general property of these operations is proved to be valid with respect to any coordinate system, it is valid generally. This important fact also relates to the gradient of a scalar function and to any combination of the three operations.

Consider the divergence of the curl of a vector function $\mathbf{A}(x, y, z)$ with respect to the rectangular coordinate system. According to the expressions in Eq. (1.11.7) for the divergence, and in Eq. (8.8.8) for the curl in a rectangular coordinate system, we have

$$\operatorname{div}(\operatorname{curl}\mathbf{A}) = \frac{\partial}{\partial x}\left(\frac{\partial A_z}{\partial y} - \frac{\partial A_y}{\partial z}\right) + \frac{\partial}{\partial y}\left(\frac{\partial A_x}{\partial z} - \frac{\partial A_z}{\partial x}\right) + \frac{\partial}{\partial z}\left(\frac{\partial A_y}{\partial x} - \frac{\partial A_x}{\partial y}\right) = 0.$$

As we explained above, although we proved it only in the rectangular coordinate system, this relation is valid generally:

$$\boxed{\operatorname{div}(\operatorname{curl}\mathbf{A}) = 0} \qquad (\mathbf{A} = \text{any well-behaved vector function}). \tag{8.9.1}$$

Comparing this with Eq. (8.6.4) we see that vector **B** can always be represented as the curl of an auxiliary vector function, say **A**:

$$\boxed{\mathbf{B} = \operatorname{curl}\mathbf{A}} \qquad (\text{definition of the magnetic vector potential}). \tag{8.9.2}$$

The vector function **A** thus defined is referred to as the *magnetic vector potential*. Note that the magnetic flux density vector **B** is obtained from the magnetic vector potential **A** by differentiation.

The magnetic vector potential **A** is *not* uniquely defined by Eq. (8.9.2). An infinite number of vector functions having the same curl exists, which can be shown as follows. The curl of the gradient of any scalar function $F$ is identically zero:

$$\boxed{\operatorname{curl}(\operatorname{grad} F) = 0} \qquad (F = \text{any well-behaved scalar function}). \tag{8.9.3}$$

The truth of this can be easily checked in the rectangular coordinate system by using Eqs. (1.7.6) and (8.8.8). Suppose that the correct magnetic flux density vector **B** is obtained according to Eq. (8.9.2) from a vector potential **A**. It can also be obtained from a vector potential $\mathbf{A}' = \mathbf{A} + \operatorname{grad} F$, where $F$ is *any* well-behaved scalar function of position. The vector potential **A** is not therefore uniquely defined by Eq. (8.9.2).

**8.9.2 The differential equation for the magnetic vector potential.** The magnetic flux density vector **B** must satisfy Ampère's circuital law given in Eq. (8.8.3). Hence the magnetic vector potential, defined by Eq. (8.9.2), must satisfy the following vectorial differential equation of the second order:

$$\text{curl}\,(\text{curl}\,\mathbf{A}) = \mu_0\,\mathbf{J}. \tag{8.9.4}$$

Let us expand the curl (curl **A**) in the rectangular coordinate system, using the expression in Eq. (8.8.8) for the curl. Consider first the equation corresponding to the $J_x$ component,

$$\text{curl}_x(\text{curl}\,\mathbf{A}) = \mu_0\,J_x.$$

According to Eqs. (8.8.6) and (8.8.7) the left-hand side of this equation can be written as

$$\begin{aligned}\text{curl}_x(\text{curl}\,\mathbf{A}) &= \frac{\partial}{\partial y}\,\text{curl}_z\mathbf{A} - \frac{\partial}{\partial z}\,\text{curl}_y\mathbf{A}\\ &= \frac{\partial}{\partial x}\left(\frac{\partial A_y}{\partial y} + \frac{\partial A_z}{\partial z}\right) - \frac{\partial^2 A_x}{\partial y^2} - \frac{\partial^2 A_x}{\partial z^2}.\end{aligned}$$

Let us add and subtract $\partial^2 A_x/\partial x^2$ from the right-hand side of this equation. Then, according to the expression in Eq. (1.11.7) for the divergence of a vector function and the expression in Eq. (1.12.5) for the Laplacian operator, div grad, we obtain

$$\text{curl}_x(\text{curl}\,\mathbf{A}) = \frac{\partial}{\partial x}\,\text{div}\,\mathbf{A} - \text{div}\,(\text{grad}\,A_x). \tag{8.9.5}$$

In a similar manner we obtain

$$\text{curl}_y(\text{curl}\,\mathbf{A}) = \frac{\partial}{\partial y}\,\text{div}\,\mathbf{A} - \text{div}\,(\text{grad}\,A_y), \tag{8.9.6}$$

$$\text{curl}_z(\text{curl}\,\mathbf{A}) = \frac{\partial}{\partial z}\,\text{div}\,\mathbf{A} - \text{div}\,(\text{grad}\,A_z). \tag{8.9.7}$$

From these equations and the expression in Eq. (1.7.6) for the gradient in a rectangular coordinate system, we have

$$\begin{aligned}\text{curl}\,(\text{curl}\,\mathbf{A}) = \text{grad}\,(\text{div}\,\mathbf{A}) - [\mathbf{u}_x\,\text{div}\,(\text{grad}\,A_x) + \mathbf{u}_y\,\text{div}\,(\text{grad}\,A_y)\\ + \mathbf{u}_z\,\text{div}\,(\text{grad}\,A_z)\,].\end{aligned} \tag{8.9.8}$$

The gradient of a *vector* is not defined. Nevertheless, the following abbreviated notation is usually used:

$$\text{div}\,(\text{grad}\,\mathbf{A}) \equiv \mathbf{u}_x\,\text{div}\,(\text{grad}\,A_x) + \mathbf{u}_y\,\text{div}\,(\text{grad}\,A_y) + \mathbf{u}_z\,\text{div}\,(\text{grad}\,A_z)$$
$$\text{(rectangular coordinate system only).} \tag{8.9.9}$$

With this Eq. (8.9.8) becomes

$$\text{curl}\,(\text{curl}\,\mathbf{A}) = \text{grad}\,(\text{div}\,\mathbf{A}) - \text{div}\,(\text{grad}\,\mathbf{A}). \tag{8.9.10}$$

Although it was derived in a rectangular coordinate system, this equation is valid generally. It should be noted, however, that the simple expression in Eq. (8.9.9) for div (grad **A**) is valid in a rectangular coordinate system only; with respect to any other system the expression for div (grad **A**) has to be determined from its defining formula in Eq. (8.9.10):

$$\text{div}\,(\text{grad}\,\mathbf{A}) \equiv \text{grad}\,(\text{div}\,\mathbf{A}) - \text{curl}\,(\text{curl}\,\mathbf{A}). \tag{8.9.11}$$

According to the Helmholtz theorem,* a vector function is completely specified by its curl and its divergence. The curl of **A** is specified by Eq. (8.9.2); otherwise **A** is an arbitrary vector function. Hence, we may assume the divergence of **A** to be *any* function. Let us define it here so that the simplest possible equation for determining **A** is obtained. From Eqs. (8.9.10) and (8.9.4) we can see that, if

$$\boxed{\text{div}\,\mathbf{A} = 0} \qquad \text{(steady current)}, \tag{8.9.12}$$

the differential equation for **A** is simplified considerably. It then reads simply

$$\boxed{\text{div}\,(\text{grad}\,\mathbf{A}) = -\mu_0\,\mathbf{J}}\,. \tag{8.9.13}$$

The operation implied by div (grad **A**) in a *rectangular coordinate system* is defined by Eq. (8.9.9). In other coordinate systems, div (grad **A**) has to be determined according to Eq. (8.9.11), noting that div **A** = 0.

**8.9.3 Vector potential of a given distribution of currents.** The vectorial partial differential equation (8.9.13) of the second order represents three scalar partial differential equations. According to Eq. (8.9.9), in a rectangular coordinate system it is equivalent to the following three equations:

$$\begin{aligned} \text{div}\,(\text{grad}\,A_x) &= -\mu_0 J_x, \\ \text{div}\,(\text{grad}\,A_y) &= -\mu_0 J_y, \\ \text{div}\,(\text{grad}\,A_z) &= -\mu_0 J_z. \end{aligned} \tag{8.9.14}$$

$A_x$, $A_y$, and $A_z$ are *scalar* functions of position. Therefore, all three equations are of the type of Poisson's equation (1.12.1). The particular solution of Poisson's equation was found to be

$$V = \frac{1}{4\pi\varepsilon_0}\int_v \frac{\rho\,dv}{r}.$$

Hence the particular solutions of Eqs. (8.9.14) are, by analogy,

$$A_x = \frac{\mu_0}{4\pi}\int_v \frac{J_x\,dv}{r}, \quad A_y = \frac{\mu_0}{4\pi}\int_v \frac{J_y\,dv}{r}, \quad A_z = \frac{\mu_0}{4\pi}\int_v \frac{J_z\,dv}{r}. \tag{8.9.15}$$

(*dv* is a volume element in the current-carrying region, and *r* the distance of the point where **A** is being determined from *dv*.)

---

* For the proof of the theorem, see Appendix 2.

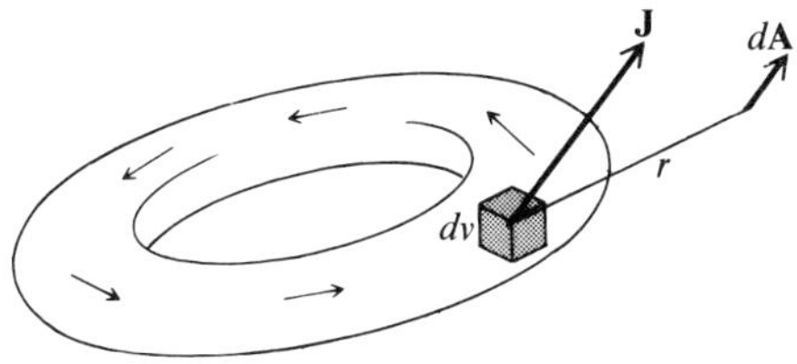


**Fig. 8.22** Vector potential due to a current-carrying conductor.

From Eq. (8.9.15) the total vector potential **A** at a point is given by

$$\mathbf{A} = A_x \mathbf{u}_x + A_y \mathbf{u}_y + A_z \mathbf{u}_z = \frac{\mu_0}{4\pi} \int_v \frac{J_x \mathbf{u}_x + J_y \mathbf{u}_y + J_z \mathbf{u}_z}{r}\, dv,$$

or

$$\boxed{\mathbf{A} = \frac{\mu_0}{4\pi} \int_v \frac{\mathbf{J}\, dv}{r}} \qquad \text{(stationary currents).} \qquad (8.9.16)$$

From this we can see that the elemental vector potential $d\mathbf{A}$ due to a current element $\mathbf{J}\, dv$ is in the same direction as **J** (see Fig. 8.22).

In the case of currents in thin wires, this may be written in an approximate form:

$$\mathbf{A} = \frac{\mu_0}{4\pi} \oint_C \frac{I\, d\mathbf{l}}{r} \qquad \text{(filamentary currents).} \qquad (8.9.17)$$

When current flows through very thin sheets, Eq. (8.9.16) becomes

$$\mathbf{A} = \frac{\mu_0}{4\pi} \int_S \frac{\mathbf{J}_s\, dS}{r} \qquad \text{(current sheet).} \qquad (8.9.18)$$

**8.9.4 The relation between magnetic vector potential and magnetic flux.** There is a very simple relation between the magnetic flux through a closed contour and the line integral of the magnetic vector potential around the contour. According to Eq. (8.9.2) and Stokes' theorem (8.8.14), we have

$$\Phi = \int_S \mathbf{B} \cdot d\mathbf{S} = \int_S \text{curl}\, \mathbf{A} \cdot d\mathbf{S} = \int_C \mathbf{A} \cdot d\mathbf{l}. \qquad (8.9.19)$$

This relation is very useful in calculating the magnetic flux in some special, and important, practical cases.

**8.9.5 Conclusion.** The *electrostatic* scalar potential function is an extremely useful tool for solving many types of electrostatic problems. As an example let us take a system of charged conductors at given potentials. We have seen that it suffices to determine the electrostatic potential functions satisfying these boundary conditions and Laplace's equation. All other quantities of interest, such as the distribution of charges over the bodies, the electric field intensity, etc., may then be determined easily.

In the case of the magnetic field of steady currents, the distribution of the current is always *given*. Consequently, the flux density vector $B$ can be calculated directly from Eq. (8.4.3), with no need to determine the vector potential **A** first. In some cases it is easier to calculate **A** first according to Eq. (8.9.16), and then **B** by using Eq. (8.9.2). But the usefulness of the vector potential **A** in solving problems of the static magnetic field is not comparable with the usefulness of the electric scalar potential $V$ in solving electrostatic problems.

As a theoretical tool, the magnetic vector potential **A** is very useful. The relation in Eq. (8.9.19) between **A** and the magnetic flux serves to determine the self-inductances and the mutual inductances of quasi-filamentary conductors (Section 11.5). We can also easily calculate the energy stored in the magnetic field by using the magnetic vector potential. There are many other cases in which this concept is advantageously used, but perhaps the most important application of the magnetic vector potential is in the case of time-varying fields. We shall see later that additional forces between electric charges in *accelerated* motion can be expressed very concisely in terms of the magnetic vector potential (Section 11.2.1).

*Problems referring to Section 8.9:* 8.T14 to 8.T16, 8.G22 to 8.G28.

## *8.10 THE MAGNETIC SCALAR POTENTIAL

**8.10.1 Definition of the magnetic scalar potential.** According to Ampère's circuital law (8.7.1), the line integral of the magnetic flux density vector **B** around a closed contour is generally not zero. Only when the contour does *not* encircle any current do we have that

$$\oint_C \mathbf{B} \cdot d\mathbf{l} = 0 \qquad (C \text{ encircles no currents}). \tag{8.10.1}$$

The necessary condition for defining an auxiliary scalar potential function analogous to the electrostatic potential is satisfied in this case only.

The electrostatic scalar potential was defined as the line integral of the electric field intensity vector **E** from the point considered (say $M$) to a fixed, reference point $R$,

$$V_M = \int_M^R \mathbf{E} \cdot d\mathbf{l}.$$

Similarly, we define the *magnetic scalar potential* $V_m$ at a point $M$ with respect to a reference point $R$ as

$$V_{mM} = \frac{1}{\mu_0} \int_M^R \mathbf{B} \cdot d\mathbf{l}. \tag{8.10.2}$$

* This section can be omitted without loss of continuity.

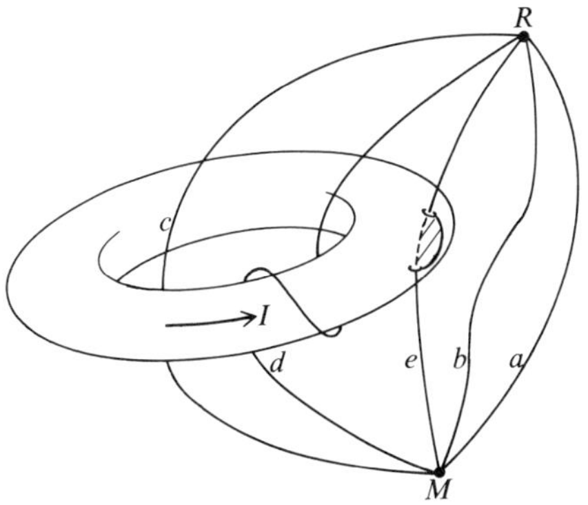


Fig. 8.23 Illustration of multivaluedness of the magnetic scalar potential.

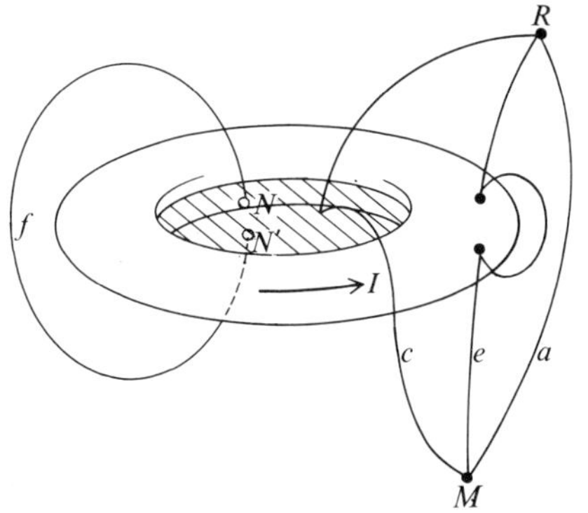


Fig. 8.24 Illustration of how multivaluedness of the magnetic scale potential is avoided.

(The factor $1/\mu_0$ is introduced for convenience.) With no additional restrictions, the magnetic scalar potential $V_m$ as defined by Eq. (8.10.2) is not a single-valued function of position, as we can see from the following reasoning.

Figure 8.23 shows points $M$ and $R$ and the current-carrying conductor creating the magnetic field. The line integral of $\mathbf{B}$ around the closed contour $MaRbM$ is zero, since the contour encircles no current. The magnetic scalar potential defined by Eq. (8.10.2) has therefore the same value along both paths $MaR$ and $MbR$. But this is not the case with paths $McR$, $MdR$, and $MeR$. According to Ampère's circuital law (8.7.1) the value of the integral in Eq. (8.10.2) for path $McR$ differs from that for the path $MaR$ by exactly $\mu_0 I$, for path $MdR$ by $2\mu_0 I$, and for path $MeR$ by a fraction of $\mu_0 I$.

To make the magnetic scalar potential $V_m$ defined by Eq. (8.10.2) a single-valued function of position the following "trick" has been devised. We first imagine arbitrary membranes spanned over all the conductors. Next we consider the conductors and the imaginary membranes as barriers, which cannot be pierced when determining $V_m$ according to Eq. (8.10.2). The paths $McR$, $MdR$, and $MeR$ are not allowed to pierce either the imaginary membrane or the conductor (see Fig. 8.24). The magnetic scalar potential $V_m$ is now a single-valued function of position.

The values of the magnetic scalar potential of points at the conductor surfaces and on the two faces of the membrane are different. In particular, two adjacent points $N$ and $N'$ on the two sides of the membrane (see Fig. 8.24) have potentials $V_m$ differing by exactly $I$. This can be shown as follows. The path $NfN'$ is not closed. However, points $N$ and $N'$ being infinitely close, the line integral of $\mathbf{B}$ from $N'$ to $N$ through the imagined membrane is also infinitely small. The line integral of $\mathbf{B}$ from $N$ over $f$ to $N'$ is thus equal to the line integral around the closed contour, that is to $\mu_0 I$. Therefore

$$V_{mN} - V_{mN'} = \frac{1}{\mu_0}\int_N^R \mathbf{B}\cdot d\mathbf{l} - \frac{1}{\mu_0}\int_{N'}^R \mathbf{B}\cdot d\mathbf{l} = \frac{1}{\mu_0}\int_N^{N'} \mathbf{B}\cdot d\mathbf{l}$$

$$\simeq \frac{1}{\mu_0}\oint_{NbN'N} \mathbf{B}\cdot d\mathbf{l} = I.$$

The magnetic scalar potential of the point $N$ which is on the *positive* side of the membrane is higher than the potential of the point $N'$. The positive face of the membrane is determined by the right-hand screw rule with respect to the direction of the current in the conductor.

**8.10.2 The magnetic scalar potential of a closed current filament.** The lines of the current density vector $\mathbf{J}$ of stationary currents always close upon themselves. Hence we can divide any system of stationary currents into closed elemental current filaments.

Consider the elemental current filament $C$ shown in Fig. 8.25. The difference of the magnetic scalar potential at points $M$ and $M'$ shown is by definition from Eq. (8.10.2):

$$dV_m = V_{mM'} - V_{mM} = \frac{1}{\mu_0}\int_{M'}^{M} \mathbf{B}\cdot d\mathbf{l} = -\frac{1}{\mu_0}\mathbf{B}\cdot d\mathbf{l}', \tag{8.10.3}$$

where $d\mathbf{l}'$ is the position vector of the point $M'$ with respect to the point $M$. It follows that

$$\mathbf{B} = -\mu_0 \operatorname{grad} V_m. \tag{8.10.4}$$

Let us now introduce into Eq. (8.10.3) the value given in Eq. (8.4.4) of the magnetic flux density vector $\mathbf{B}$. Note that $d\mathbf{l}'$ is a constant vector, so that it can be introduced under the integral sign. Using Eq. (0.3.15), we have

$$dV_m = \frac{I}{4\pi}\oint_C \frac{(d\mathbf{l}\times\mathbf{u})\cdot(-d\mathbf{l}')}{r^2} = \frac{I}{4\pi}\oint_C \frac{(-d\mathbf{l}'\times d\mathbf{l})\cdot\mathbf{u}}{r^2}. \tag{8.10.5}$$

The same increment of the magnetic scalar potential would be obtained *at point M* if the conductor $C$ is imagined to be translated for $-d\mathbf{l}'$ (see Fig. 8.25). But, in that case, the cross product $(-d\mathbf{l}' \times d\mathbf{l})$ represents numerically the area of the small crosshatched parallelogram shown in the diagram. The expression under the integral sign then represents the elemental solid angle subtended at $M$ by the small crosshatched surface. The integral is thus equal to the elemental solid angle $d\Omega$ subtended at $M$ by the strip surface obtained by translating the contour $C$ to its position $C'$:

$$dV_m = \frac{I d\Omega}{4\pi}. \tag{8.10.6}$$

It is obvious that $d\Omega$ is the increase in the solid angle subtended at $M$ by the contour $C$ when translated for $-d\mathbf{l}'$. The same increase in the solid angle sub-

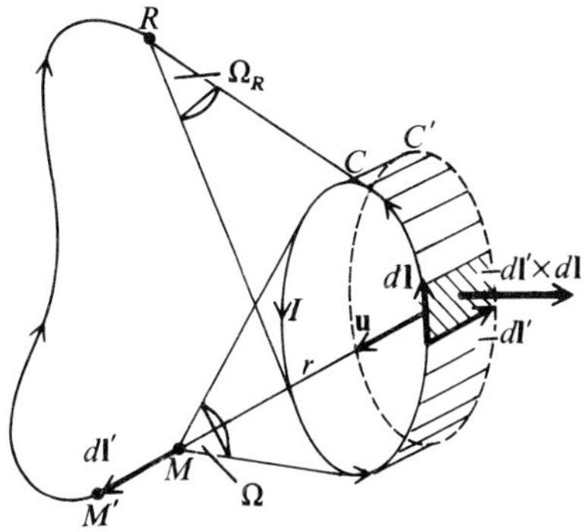


**Fig. 8.25** Construction for determining the magnetic scalar potential of a closed current filament.

tended at $M$ by the contour $C$ is obtained when $M$ is moved for $+d\mathbf{l}'$, to the point $M'$.

From Eqs. (8.10.3) and (8.10.6), the magnetic scalar potential of an arbitrary point, with respect to the reference point $R$ is

$$V_m = \frac{1}{\mu_0}\int_M^R \mathbf{B}\cdot d\mathbf{l}' = -\int_M^R dV_m = -\frac{I}{4\pi}\int_\Omega^{\Omega_R} d\Omega = \frac{I}{4\pi}(\Omega - \Omega_R). \tag{8.10.7}$$

The symbols $\Omega$ and $\Omega_R$ represent solid angles subtended at the point considered and at $R$, respectively, by the *positive* side of a surface limited by the contour, because unit vector $\mathbf{u}$ in Fig. 8.25 is directed *toward M*. If the reference point $R$ is assumed to be at infinity,

$$V_m = \frac{\mathbf{I}}{4\pi}\,\Omega \quad \text{(filamentary current, reference point at infinity).} \tag{8.10.8}$$

Consider now a nonfilamentary closed conductor, such as that represented in Fig. 8.26. To determine the magnetic scalar potential at a point $M$ we must consider the current filaments $dI$, and then integrate the expression in Eq. (8.10.8) over all current filaments:

$$V_m = \frac{1}{4\pi}\int \Omega\, dI. \tag{8.10.9}$$

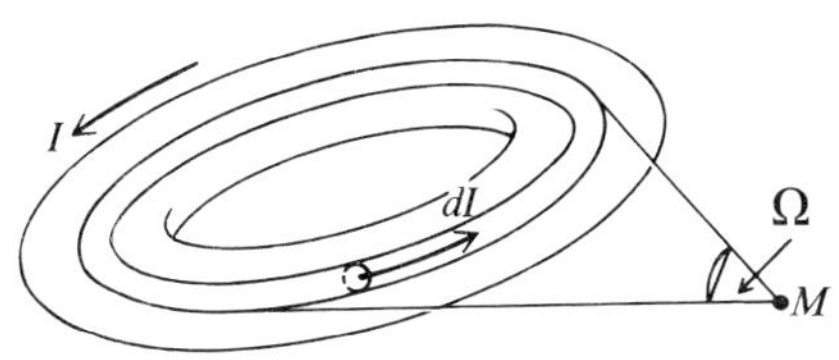


**Fig. 8.26** A current-carrying conductor subdivided into elemental current filaments.

The reader should again note that $\Omega$ represents the solid angle subtended at the point considered by the *positive* side of a surface limited by the elemental current $dI$.

**8.10.3 The magnetic scalar potential at large distances from a closed filamentary current. The concept of the magnetic dipole.** Let us determine the magnetic scalar potential at large distances from a planar filamentary contour of current $I$ and area $S$. In that case Eq. (8.10.8) becomes (see Fig. 8.27)

$$V_m = \frac{I}{4\pi} \frac{S_n}{r^2} = \frac{I}{4\pi} \frac{\mathbf{S} \cdot \mathbf{u}}{r^2}. \tag{8.10.10}$$

The product $I\,\mathbf{S}$ is usually referred to as the *magnetic moment* of the contour $C$, and is designated by $\mathbf{m}$,

$$\mathbf{m} = I\,\mathbf{S}. \tag{8.10.11}$$

The magnetic scalar potential expressed by Eq. (8.10.10) can thus be written in the form

$$V_m = \frac{\mathbf{m} \cdot \mathbf{u}}{4\pi r^2}. \tag{8.10.12}$$

Comparing this with Eq. (2.2.3), we see that the magnetic scalar potential at large distances from a current contour of moment $\mathbf{m}$ is of the same form as the electrostatic potential of a dipole of moment $\mathbf{p}$. Hence, the magnetic flux density vector *at large distances* from a current-carrying contour is also of the same form as the electric field intensity vector *at large distances* from an electric dipole. For this reason, a small current-carrying contour is often referred to as the *magnetic dipole*. Formally, the small current loop can be replaced by two equal magnetic charges, opposite in sign, and constituting a complete (but, in nature, nonexistent) analogue to the electric dipole. Only in this sense is it possible to define the concept of the magnetic charges, analogous to the electric charges. This concept is very useful for the analysis of some problems; but we must remember that no magnetic charges exist in reality.

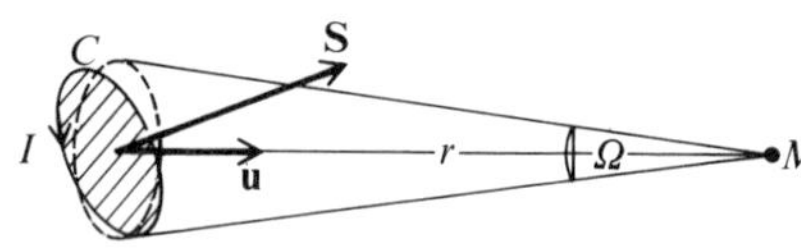


**Fig. 8.27** Construction for determining the magnetic scalar potential at large distances from a planar closed current filament.

**8.10.4 Conclusion** To ensure the uniqueness of the magnetic scalar potential it was essential to introduce certain rather artificial restrictions. The magnetic scalar potential is nevertheless very useful in determining the magnetic field of some current distributions; a few examples of its application are presented in Problems 8.G29 to 8.G31. Its use is even more profitable in analyzing some cases of the magnetic field in the presence of magnetic materials. The method of solution then is the same as in the analogous electrostatic case, and many results can be obtained by simple analogy. This application of the magnetic scalar potential is presented in Section 9.6.

*Problems referring to Section 8.10:* 8.G29 to 8.G31.

## PROBLEMS

### Test problems

**8.T1.** Figure 8.T1 shows four charges $Q_1, \ldots, Q_4$, having at a certain instant the velocities indicated. Determine the magnetic force exerted by all these charges individually on the other three charges. This makes altogether twelve different expressions.

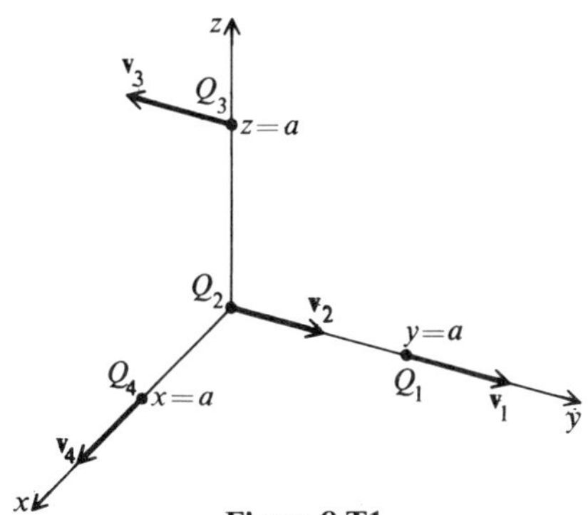


**Figure 8.T1**

**8.T2.** Discuss the possibility of changing the kinetic energy of a charged particle by the action upon it of some appropriate magnetic field only.

**8.T3.** A straight, very long, thin conductor has a charge $Q_l$ per unit length. It also carries a current of intensity $I$. A charge $Q$ is moving with a velocity $v$ parallel to the wire, at a distance $d$ from it, unaffected by the simultaneous action of both electric and magnetic force. Determine the velocity $v$ of the charge, assuming the necessary correct direction of the current and the sign of the charge on the conductor.

**8.T4.** A charge $Q$ is moving along the axis of a circular current-carrying conductor normal to the plane of the contour. Discuss the influence of the magnetic field on the motion of the charge.

**8.T5.** Sketch the lines of the magnetic flux density vector in the case of two long, parallel, straight, thin conductors with equal currents, when the currents are in the same and in the opposite directions.

**8.T6.** A closed surface $S$ encloses a small conducting loop with current $I$. What is the magnetic flux through the surface $S$?

**8.T7.** A conductor carrying a current of intensity $I$ pierces a closed surface $S$. What is the magnetic flux through the surface $S$?

**8.T8.** Does Ampère's circuital law (8.7.1) apply to a closed contour in the magnetic field of a single small charge $Q$, moving with a velocity $\mathbf{v}$?

***8.T9.** Repeat the derivation of Ampère's circuital law assuming the opposite direction for the contour $C$ to that shown in Fig. 8.13.

**8.T10.** In a certain region of space the magnetic flux density vector $\mathbf{B}$ has the same direction at all points, but its magnitude is not constant in the direction perpendicular to its lines. Are there currents in that part of space?

**8.T11.** An infinitely long, straight, cylindrical conductor of rectangular cross-section carries a current of intensity $I$. Is it possible to determine the magnetic flux density inside and outside the conductor by starting from Ampère's circuital law?

**8.T12.** Derive the $y$ and $z$ components of the curl $\mathbf{B}$ given by Eq. (8.8.7).

**8.T13.** Verify that the magnetic flux density outside and inside a straight cylindrical conductor of circular cross-section, carrying a current of intensity $I$, satisfies the differential form of Ampère's circuital law. The expression for the curl in the cylindrical coordinate system is derived in Problem 8.G20.

**8.T14.** Prove Eq. (8.9.3).

**8.T15.** Prove Eqs. (8.9.6) and (8.9.7).

**8.T16.** If the vector potential $\mathbf{A}$ is a constant vector, or equal to $C\mathbf{r}/r^3 = -C \operatorname{grad}(1/r)$, does it define a magnetic field?

## General problems

**8.G1.** Determine the magnetic flux density of a straight current filament of length $l$ and with a current of intensity $I$.

*Solution.* To determine the magnetic flux density at the point $M$ shown in Fig. 8.G1 it is easiest to use the simplified form, given in Eq. (8.4.13), of the Biot-Savart law, valid in the case of planar current filaments:

$$B = \frac{\mu_0 I}{4\pi} \int_{\theta_1}^{\theta_2} \frac{d\theta}{r} = \frac{\mu_0 I}{4\pi} \int_{\theta_1}^{\theta_2} \frac{\sin\theta\, d\theta}{a} = \frac{\mu_0 I}{4\pi a}(\cos\theta_1 - \cos\theta_2). \tag{a}$$

It should be noted that a nonclosed current filament with a stationary current cannot exist. So (a) has no physical meaning unless the current filament is a part of a closed circuit.

**8.G2.** Determine the magnetic flux density at the center of a square filament with sides $a$ and current $I$.

*Answer.* For $I = 1$ A and $a = 10$ cm, $B = 1.13 \times 10^{-5}$ T.

**8.G3.** Determine the magnetic flux density at the point $M$ shown in Fig. 8.G3. The two straight pieces of wire of the circuit are assumed to be very long.

*Answer.* For $I = 2$ A and $a = 5$ cm, $B = 2.06 \times 10^{-5}$ T.

**8.G4.** Determine the magnetic flux density on the axis of a circular current loop of radius $a$ and with current of intensity $I$, at a distance $z$ from the plane of the loop.

*Solution.* According to the Biot-Savart law (8.4.5), the element $d\mathbf{l}$ of the loop shown in

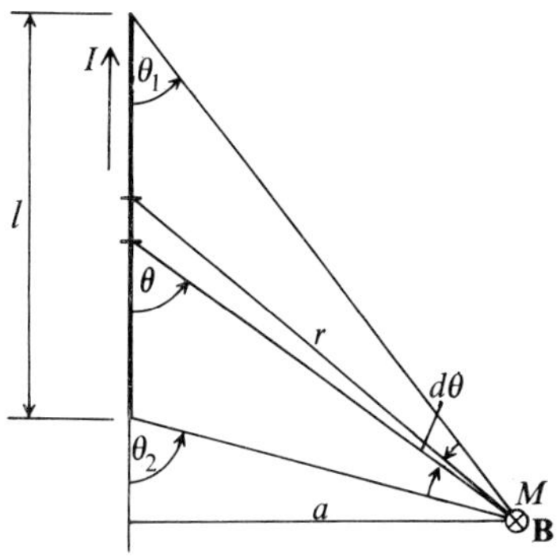


Figure 8.G1

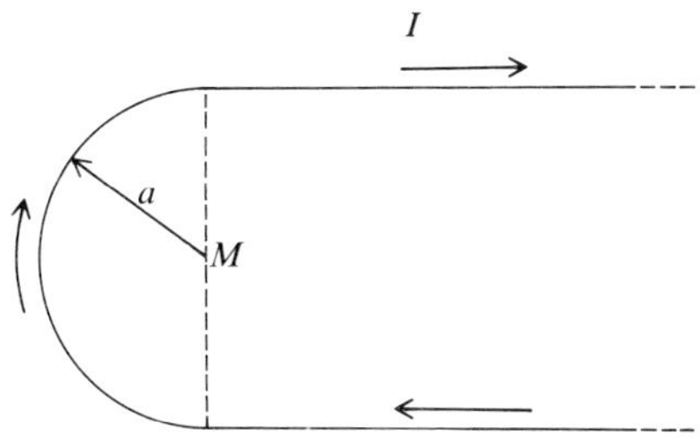


Figure 8.G3

Fig. 8.G4a creates at the point $M$ the magnetic flux density $d\mathbf{B}$ in the direction shown, and of magnitude

$$dB = \frac{\mu_0 I}{4\pi} \frac{dl}{r^2}$$

The loop element $dl'$ symmetric to $dl$ creates the magnetic flux density of the same magnitude and in the direction indicated. The resultant flux density of current elements $dl$ and $dl'$ is hence directed along the $z$ axis. Its magnitude is

$$|d\mathbf{B} + d\mathbf{B}'| = 2\, dB \cos \alpha = 2 \frac{\mu_0 I}{4\pi} \frac{dl}{r^2} \frac{a}{r} = \frac{\mu_0 I}{2\pi} \frac{a\, dl}{(a^2 + z^2)^{3/2}}. \tag{a}$$

The total magnetic flux density is also directed along the $z$ axis. According to (a) its magnitude is

$$B = B_z = \frac{\mu_0 I}{2\pi} \frac{a\, \pi a}{(a^2 + z^2)^{3/2}} = \frac{\mu_0 I a^2}{2(a^2 + z^2)^{3/2}}. \tag{b}$$

For $z = 0$, (b) becomes identical with Eq. (8.4.9), as it should be. In Fig. 8.G4b the lines of the magnetic flux density vector in this case are sketched.

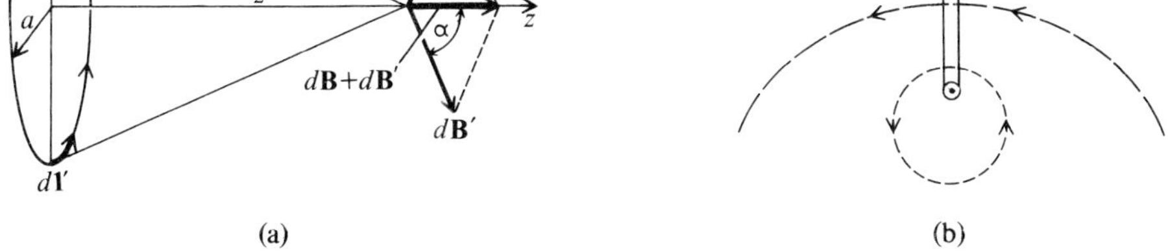


Figure 8.G4

**8.G5.** Using the result of the preceding problem, determine the magnetic flux density on the axis of a solenoidal coil of length $b$ and radius $a$. The coil has a large number $N$ of turns, with a current of intensity $I$.

*Solution.* The number of turns is very large. Hence we can consider the solenoid with its current as a cylindrical current sheet of surface current density $J_s = NI/b$. Consider an elemental ring of width $dz'$ of the current sheet as in Fig. 8.G5a. According to (b) of the preceding problem, the magnetic flux density at a point $M$ on the solenoid axis due to the elemental current ring is

$$dB_z = \frac{\mu_0 J_s\, dz'\, a^2}{2[a^2 + (z - z')^2]^{3/2}} = \frac{\mu_0 J_s\, dz'}{2a} \sin^3\alpha, \tag{a}$$

since $\sin\alpha = a/[a^2 + (z - z')^2]^{1/2}$. Now, according to Fig. 8.G5a,

$$z - z' = \frac{a}{\tan\alpha}, \qquad dz' = \frac{a}{\tan^2\alpha}\frac{1}{\cos^2\alpha}\, d\alpha = \frac{a}{\sin^2\alpha}\, d\alpha,$$

so that (a) becomes

$$dB_z = \frac{\mu_0 J_s}{2} \sin\alpha\, d\alpha. \tag{b}$$

The magnetic flux density at $M$ due to the whole solenoid is hence

$$B_z = \int dB_z = \frac{\mu_0 J_s}{2}\int_{\alpha_1}^{\alpha_2} \sin\alpha\, d\alpha = \frac{\mu_0 J_s}{2}(\cos\alpha_1 - \cos\alpha_2), \tag{c}$$

or, since $J_s = NI/b$,

$$B_z = \frac{\mu_0 NI}{2b}(\cos\alpha_1 - \cos\alpha_2). \tag{d}$$

It is left to the reader to deduce from (c) the magnetic flux density inside a very long (theoretically, infinitely long) solenoid.

Figure 8.G5b shows the lines of the magnetic flux density vector of such a solenoid.

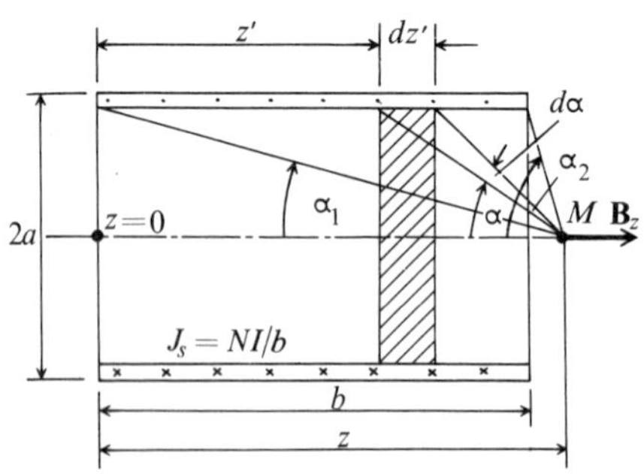


(a)

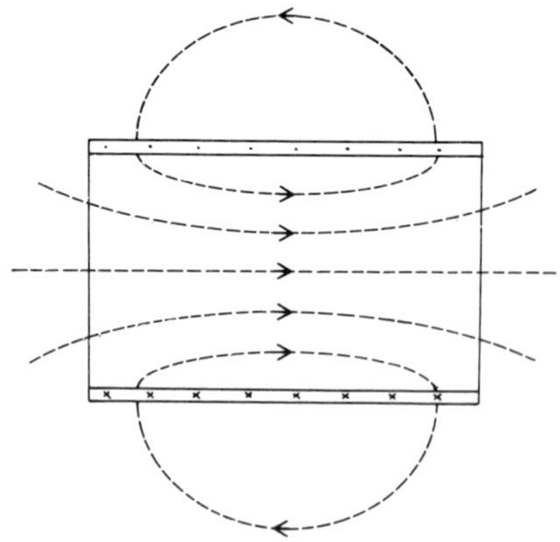

(b)

**Figure 8.G5**

**8.G6.** Two thin parallel coaxial circular coils of radius $a$ and with $N$ turns are placed at a distance $b$. The currents through the two coils are in the same direction (see Fig. 8.G6). Prove that in the vicinity of the point $M$ on the axis, at a distance $b/2$ from both coils, the magnetic field is practically uniform. Sketch the lines of **B**. [*Hint:* it is necessary to prove that the first and second derivatives of $B_z$ at $M$ with respect to $z$ are zero.]

**8.G7.** An insulating cylinder of radius $a$ and length $b$ has a surface charge of density $\rho_s$ (say, created by friction), and rotates with an angular velocity $\omega$. Determine the magnetic flux density on the cylinder axis.

*Answer.* The result is given by (c) of Problem 8.G5, if $J_s$ is changed to $\rho_s\,\omega a$. Why is this?

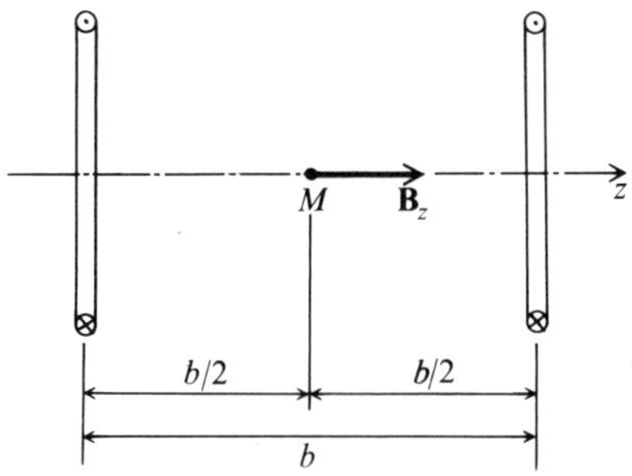


Figure 8.G6

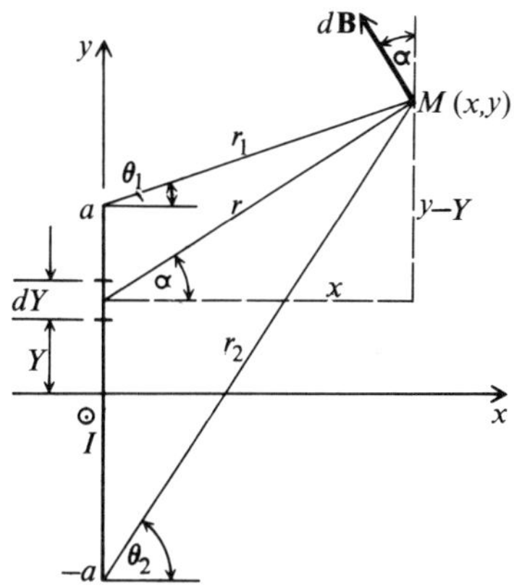


Figure 8.G8

***8.G8.** Determine the magnetic flux density of a current of intensity $I$ through a thin, straight metal strip of width $2a$. What must the result obtained become for the special cases when: (1) we consider the field at large distances from the strip when compared with $a$, and (2) the width of the strip becomes infinite?

*Solution.* If the coordinate system is adopted as shown in Fig. 8.G8, the magnitude $dB$ of the magnetic flux density vector created at a point $M(x, y)$ by the current through an elemental strip of width $dY$ is

$$dB = \frac{\mu_0\, dI}{2\pi r} = \frac{\mu_0 I}{4\pi a}\,\frac{dY}{\sqrt{x^2 + (y - Y)^2}}\,.$$

The projections of $dB$ on the $x$ and $y$ axes are

$$dB_x = -dB \sin\alpha, \qquad dB_y = dB\cos\alpha.$$

(The angle $\alpha$ is shown in Fig. 8.G8.)

Let us determine first the total component $B_x$. Since

$$dB_x = -\frac{\mu_0 I}{4\pi a}\,\frac{dY}{\sqrt{x^2 + (y - Y)^2}}\,\frac{y - Y}{\sqrt{x^2 + (y - Y)^2}}\,,$$

by introducing a new variable $u = (y - Y)$, we have

$$B_x = \frac{\mu_0 I}{4\pi a}\int_{y+a}^{y-a}\frac{u\,du}{x^2 + u^2} = \frac{\mu_0 I}{8\pi a}\ln\frac{x^2 + (y - a)^2}{x^2 + (y + a)^2}. \tag{a}$$

Similarly

$$B_y = \frac{\mu_0 I x}{4\pi a}\int_{-a}^{a}\frac{dY}{x^2 + (y - Y)^2} = \frac{\mu_0 I}{4\pi a}\int_{y-a}^{y+a}\frac{x\,du}{x^2 + u^2} = \frac{\mu_0 I}{4\pi a}\left(\tan^{-1}\frac{y + a}{x} - \tan^{-1}\frac{y - a}{x}\right) \tag{b}$$

Both (a) and (b) can be given a simpler form. Referring to Fig. 8.G8, we see that

$$B_x = \frac{\mu_0 I}{4\pi a}\ln\frac{r_1}{r_2}, \qquad B_y = \frac{\mu_0 I}{4\pi a}(\theta_2 - \theta_1). \tag{c}$$

Suppose that the point $M$ is at a large distance $R$ from the center of the strip. Then, if $\theta = \alpha$ at point $Y = 0$,

$$r_1 \simeq R - a\sin\theta, \qquad r_2 \simeq R + a\sin\theta, \qquad \theta_2 - \theta_1 \simeq 2a\cos\theta/R,$$

so that equations (c) become

$$B_x \simeq \frac{\mu_0 I}{4\pi a}\ln\frac{1 - \delta}{1 + \delta}, \qquad B_y \simeq \frac{\mu_0 I}{2\pi R}\cos\theta, \tag{d}$$

where $\delta = a\sin\theta/R$. The expression for $B_y$ is immediately recognized as the $y$ component of the magnetic flux density due to a straight current filament, as it should be. Since $\delta = a\sin\theta/R$ is much smaller than unity

$$\ln[(1 - \delta)/(1 + \delta)] = \ln(1 - \delta) - \ln(1 + \delta) \simeq (-\delta) - (+\delta) = -2\delta = -2a\sin\theta/R.$$

Hence

$$B_x \simeq -\frac{\mu_0 I}{2\pi R}\sin\theta,$$

as expected.

If the width $a$ of the strips becomes infinitely large but the ratio $I/2a = J_s$ is kept constant, $r_1/r_2$ becomes unity, so that $B_x = 0$. In that case, from Fig. 8.G8 the difference $(\theta_2 - \theta_1)$ becomes $\pi$, so that the second expression in (c) yields $B_y = \mu_0 J_s/2$. The result is identical with Eq. (8.7.17), as expected.

***8.G9.** Determine the magnetic flux density created by a straight, infinitely long strip of width $dl$ with *perpendicular* current of surface density $J_s$, as shown in Fig. 8.G9. Obviously, such a current strip cannot exist; but it may represent part of a complete current system.

*Solution.* According to Eq. (8.4.5), the magnetic flux density created by the element $dl\,dz$ of the strip at the point $M$ shown in Fig. 8.G9 is

$$d^2\mathbf{B} = \frac{\mu_0\,dI\,d\mathbf{l} \times \mathbf{u}}{4\pi r^2}, \tag{a}$$

where $dI = J_s\,dz$. Now, $d\mathbf{l} = dl\,\mathbf{u}_x$, and, from the figure,

$$\mathbf{u} = \cos\alpha\cos\theta\,\mathbf{u}_x + \cos\alpha\sin\theta\,\mathbf{u}_y - \sin\alpha\,\mathbf{u}_z.$$

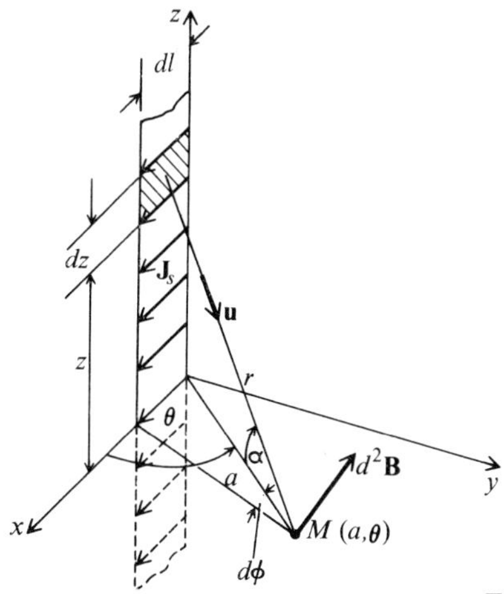


Figure 8.G9

Hence we have

$$d\mathbf{l} \times \mathbf{u} = dl(\cos\alpha \sin\theta\, \mathbf{u}_z + \sin\alpha\, \mathbf{u}_y). \tag{b}$$

According to (a) and (b) the components of the magnetic flux density vector $d^2\mathbf{B}$ are

$$d^2B_x = 0, \qquad d^2B_y = \frac{\mu_0\, dI\, dl}{4\pi r^2} \sin\alpha, \qquad d^2B_z = \frac{\mu_0\, dI\, dl}{4\pi r^2} \cos\alpha \sin\theta.$$

The current element at $-z$ creates at $M$ the $y$ component of the magnetic flux density equal to $d^2B_y(-z) = -d^2B_y(z)$. The total $y$ component of $d\mathbf{B}$ created by the strip is thus zero, so that only the $z$ component exists. Since

$$dI = J_s\, dz, \qquad dz = r\, d\alpha/\cos\alpha \qquad \text{and} \qquad r = a/\cos\alpha,$$

the component $d^2B_z$ may be written in the form

$$d^2B_z = \frac{\mu_0 J_s \sin\theta\, dl}{4\pi a} \cos\alpha\, d\alpha.$$

The total flux density created by the strip is thus

$$dB_z = \frac{\mu_0 J_s \sin\theta\, dl}{4\pi a} \int_{-\pi/2}^{\pi/2} \cos\alpha\, d\alpha = \frac{\mu_0 J_s \sin\theta\, dl}{2\pi a}. \tag{c}$$

The last expression may be simplified. Since $\sin\theta\, dl/a = d\phi$ represents the elemental angle between two planes containing the point $M$ and the two limiting lines of the current strip, (c) becomes

$$dB = dB_z = \frac{\mu_0 J_s}{2\pi}\, d\phi. \tag{d}$$

***8.G10.** Using the formula (d) of the preceding problem, determine the magnetic field: (1) of an infinite cylindrical open current sheet, not necessarily plain, with a current of surface density $J_s$ perpendicular to its generatrices; (2) inside and outside a cylindrical solenoid of arbitrary

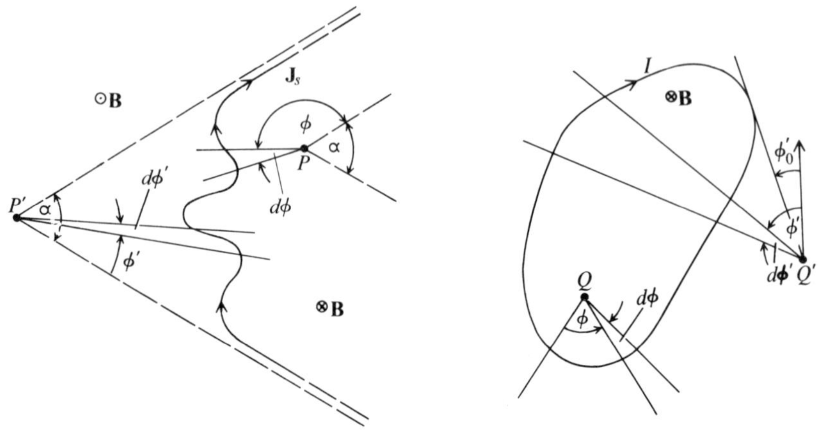


(a) (b)

**Figure 8.G10**

cross-section, having densely wound $N_l$ turns per unit length, with a current of intensity $I$.

*Solution.* (1) Consider a current sheet, the cross-section of which is shown in Fig. 8.G10a. Suppose that the angle between the half-planes from the point $P$, where the flux density is determined, toward the ends of the sheet is $\alpha$. According to (d) of the preceding problem we have

$$B_P = \frac{\mu_0 J_s}{2}\int_0^{2\pi-\alpha} d\phi = \frac{\mu_0 J_s}{2\pi}(2\pi - \alpha),$$

and at the point $P'$

$$B_{P'} = \frac{\mu_0 J_s}{2\pi}\int_0^{\alpha} d\phi' = \frac{\mu_0 J_s}{2\pi}\alpha.$$

In the particular case when $\alpha = \pi$ the flux density is $B_P = B_{P'} = \mu_0 J_s/2$. This is the same as in the case of a planar current sheet (Section 8.7.3), but in our case it need not be planar.

(2) Figure 8.G10b shows the cross-section of a cylindrical solenoid of an arbitrary cross-section. The density $J_s$ of the surface current in (d) of the preceding problem has now to be replaced by $N_l I$. According to the figure, inside the solenoid

$$B_Q = \frac{\mu_0 N_l I}{2\pi}\int_0^{2\pi} d\phi = \mu_0 N_l I,$$

and outside the solenoid

$$B_{Q'} = \frac{\mu_0 N_l I}{2\pi}\int_{\phi_0'}^{\phi_0'} d\phi' = 0.$$

**8.G11.** A thin circular disk of an insulating material is charged over one face with a constant surface charge of density $\rho_s$. The disk rotates about the axis perpendicular to its plane with an angular velocity $\omega$. Determine the magnetic flux density on the disk axis.

*Solution.* Let us divide the disk into elemental rings of radius $R$ and width $dR$ (see Fig. 8.G11).

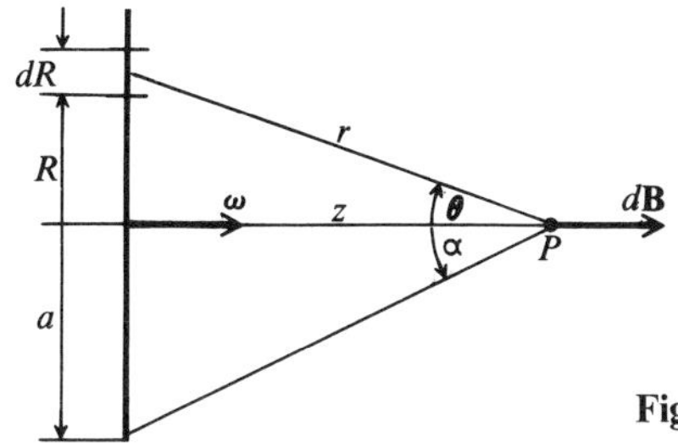


Figure 8.G11

Each of these rings can be treated as a circular current of intensity

$$dI = \rho_s 2\pi R\,dR\,\frac{\omega}{2\pi} = \rho_s \omega R\,dR.$$

According to (b) of Problem 8.G4, the magnetic flux density at a point $P$ on the axis created by the elemental circular current is

$$dB = \frac{\mu_0 \rho_s \omega R\,dR\,R^2}{2(R^2 + z^2)^{3/2}} = \frac{\mu_0 \rho_s \omega R^3\,dR}{2(R^2 + z^2)^{3/2}}. \tag{a}$$

Taking into account that

$$R/(R^2 + z^2)^{1/2} = R/r = \sin\theta, \qquad dR = r\,d\theta/\cos\theta \qquad \text{and} \qquad r = z/\cos\theta,$$

(a) reduces to

$$dB = \frac{\mu_0 \rho_s \omega z}{2}\,\frac{\sin^3\theta}{\cos^2\theta}\,d\theta.$$

The total magnetic flux density is obtained by integrating the above expression from $\theta = 0$ to $\theta = \alpha$:

$$B = \frac{\mu_0 \rho_s \omega z}{2}\int_0^\alpha \frac{\sin^3\theta}{\cos^2\theta}\,d\theta = -\frac{\mu_0 \rho_s \omega z}{2}\int_{\cos\theta=1}^{\cos\alpha} \frac{1-\cos^2\theta}{\cos^2\theta}\,d(\cos\theta)$$

$$= \frac{\mu_0 \rho_s \omega z}{2}\left(\cos\alpha + \frac{1}{\cos\alpha} - 2\right) = \frac{\mu_0 \rho_s \omega}{2}\left(\frac{a^2 + 2z^2}{\sqrt{a^2 + z^2}} - 2z\right). \tag{b}$$

It is left to the reader to check that the magnitude of the magnetic flux density given by (b) is extremely small for reasonable magnitudes of $\rho_s$, $\omega$, and $a$ (*cf.* Problem 8.N8).

***8.G12.** Prove that the divergence of the magnetic flux density vector **B**, as given by the general formula in Eq. (8.4.3), is zero.

*Proof.* According to Eq. (8.4.3)

$$\text{div}\,\mathbf{B} = \frac{\mu_0}{4\pi}\,\text{div}_P \int_v \frac{\mathbf{J}\times\mathbf{u}}{r^2}\,dv.$$

The subscript $P$ of the operator "div" denotes that the operation implied relates to the point $P$ at which the magnetic flux density vector is determined. Since the volume integral refers to the volume occupied by the currents, the two operations are independent, and can hence change places. The unit vector **u** is directed from the volume element $dv$ towards the field point $P$.

Instead of $\mathbf{u}/r^2$ we can write $-\text{grad}_P(1/r)$. In the light of these remarks, we have

$$\text{div}\,\mathbf{B} = -\frac{\mu_0}{4\pi}\int_v \text{div}_P(\mathbf{J} \times \text{grad}_P \frac{1}{r})\, dv. \tag{a}$$

Now, for any two well-behaved vector functions **a** and **b**, formal application of the differential operator $\nabla$ yields

$$\begin{aligned}\text{div}(\mathbf{a} \times \mathbf{b}) &= \nabla\cdot(\mathbf{a} \times \mathbf{b}) = \nabla\cdot(\mathbf{a} \times \overset{*}{\mathbf{b}}) + \nabla\cdot(\overset{*}{\mathbf{a}} \times \mathbf{b}) \\ &= \mathbf{b}\cdot(\nabla \times \mathbf{a}) - \mathbf{a}\cdot(\nabla \times \mathbf{b}) = \mathbf{b}\cdot\text{curl}\,\mathbf{a} - \mathbf{a}\cdot\text{curl}\,\mathbf{b}.\end{aligned}$$

(The asterisk above a function designates that it should be considered as a constant.) So (a) can be written in the form

$$\text{div}\,\mathbf{B} = -\frac{\mu_0}{4\pi}\int_v [\text{grad}_P \frac{1}{r}\cdot \text{curl}_P\,\mathbf{J} - \mathbf{J}\cdot\text{curl}_P(\text{grad}_P \frac{1}{r})]\, dv.$$

The first term in the brackets is zero since the operation (differentiation!) implied relates to the point $P$, and $\mathbf{J}$ is a function of the coordinates of the volume element $dv$. The second term is also zero, since the curl of the gradient of any scalar function is identically zero. Thus $\text{div}\,\mathbf{B} = 0$.

**8.G13.** Determine the magnetic flux through the cross-section of the infinitely long solenoids shown in Fig. 8.G13. The number of turns per unit length of the solenoids is $N_l$, and the current intensity through them is $I$.

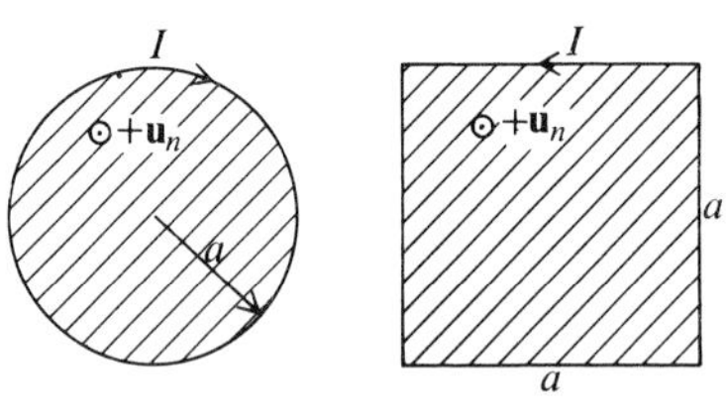


**Figure 8.G13**

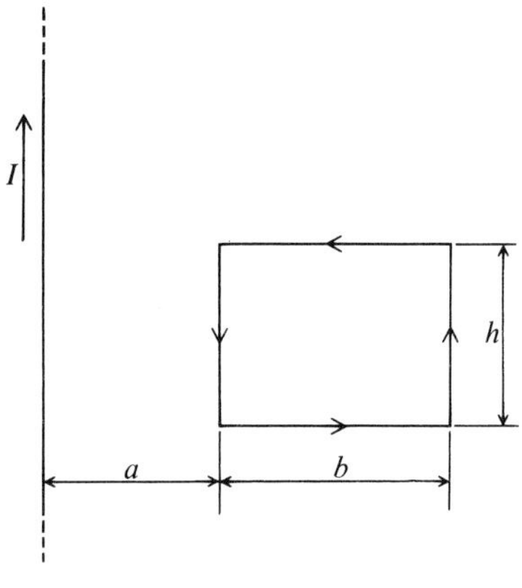


**Figure 8.G14**

**8.G14.** Determine the magnetic flux through the rectangular contour shown in Fig. 8.G14 created by a current of intensity $I$ through the straight, infinitely long, thin conductor.

*Answer*. For $a = b = h = 10$ cm, and $I = 1$ A, the flux $\Phi = -1.39 \times 10^{-8}$ Wb.

**8.G15.** Determine the magnetic flux through a cross-section of the toroidal coils shown in Fig. 8.G15. The coils are densely wound with $N$ turns, with a current of intensity $I$ through them.

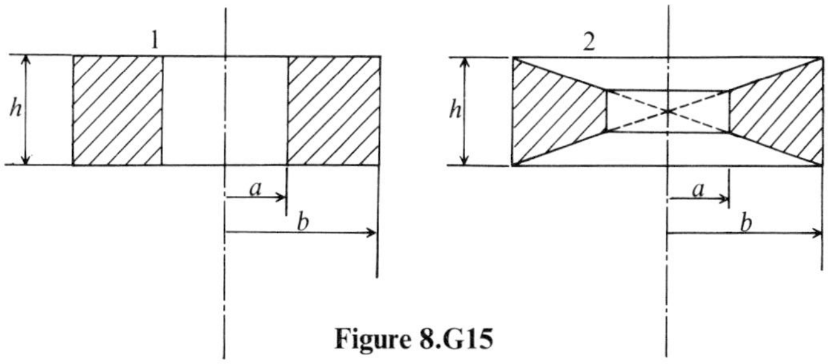

Figure 8.G15

*Answer.* For $I = 1$ A, $N = 1000$, $a = 10$ cm, $b = 15$ cm, and $h = 10$ cm the magnetic flexes are $\Phi_1 = 0.81 \times 10^{-5}$ Wb, and $\Phi_2 = 0.67 \times 10^{-5}$ Wb.

**8.G16.** Determine the magnetic flux density created by a hollow cylindrical conductor with current of intensity $I$. The cross-section of the conductor is shown in Fig. 8.G16.

*Solution.* By symmetry, the lines of the magnetic flux density vector are circles with their centers on the conductor axis. The direction of **B** is determined by the right-hand screw rule with respect to the direction of the current. Applying Ampère's circuital law (8.7.1) first to the circular contour $C_1$, encircling no current, we have

$$\oint_{C_1} \mathbf{B}\cdot d\mathbf{l} = 2\pi r_1 B(r_1) = 0, \qquad \text{whence} \qquad B = 0 \qquad \text{for} \qquad r_1 < b. \tag{a}$$

Applying Ampère's law now to the contour $C_2$, we obtain

$$\oint_{C_2} \mathbf{B}\cdot d\mathbf{l} = 2\pi r_2 B(r_2) = \mu_0 \frac{I}{(c^2 - b^2)\pi}(r_2^2 - b^2)\pi,$$

from which

$$B(r) = \mu_0 \frac{I(r_2^2 - b^2)}{2\pi r_2(c^2 - b^2)}, \qquad b \le r_2 \le c. \tag{b}$$

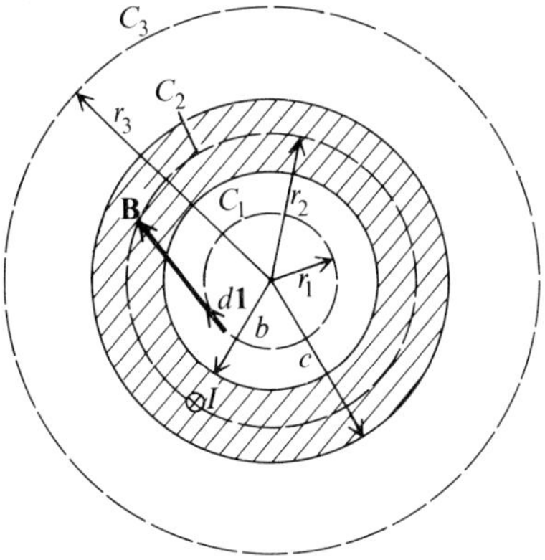

Figure 8.G16

For $r_3 > c$ we find in the same way that

$$B(r_3) = \mu_0 \frac{I}{2\pi r_3}, \qquad r_3 > c.$$

**8.G17.** Determine the magnetic field created by a cylindrical, circular, hollow conductor with a current of density **J**. The circular cylindrical cavity is not coaxial with the conductor (see Fig. 8.G17a).

*Solution.* It may seem that we are confronted with a problem involving very complicated integrations, but this is not so. The problem can be solved easily by using the principle of superposition, as indicated diagrammatically in Fig. 8.G17b. We have only to add vectorially the magnetic flux density of a *solid* conductor, of radius *a* with a current of density **J** and of a *solid* conductor, of radius *b*, replacing the cavity and with a current of density $-\mathbf{J}$. Only the magnetic field inside the cavity will be determined below, using this method. Determination of the field inside and outside the conductor is left as an exercise for the reader.

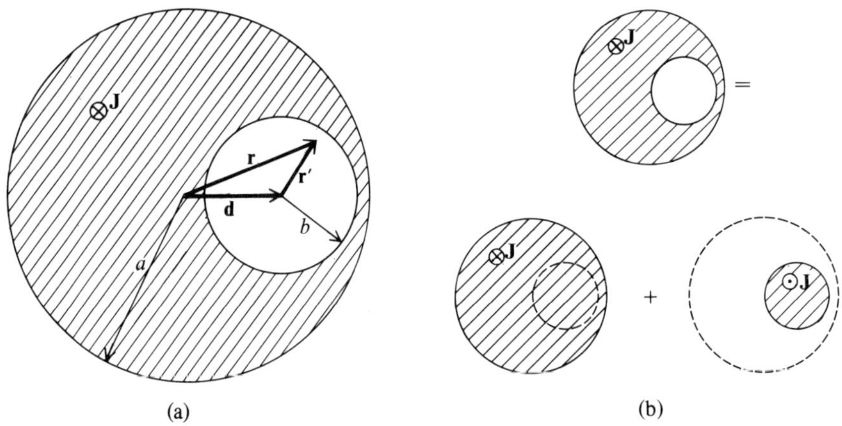


**Figure 8.G17**

It is easy to check that the expression in Eq. (8.7.10) for the magnetic flux density inside a circular, cylindrical conductor with a current of density **J** can be represented in the following vector form:

$$\mathbf{B} = \mu_0 \frac{\mathbf{J} \times \mathbf{r}}{2}.$$

(**r** is the position vector of the point where **B** is determined with respect to the conductor axis.) By using this result and the notation from Fig. 8.G17a, we can easily obtain the magnetic flux density in the cavity:

$$\mathbf{B} = \mu_0 \frac{\mathbf{J} \times \mathbf{r}}{2} - \mu_0 \frac{\mathbf{J} \times \mathbf{r}'}{2} = \mu_0 \frac{\mathbf{J} \times (\mathbf{r} - \mathbf{r}')}{2} = \mu_0 \frac{\mathbf{J} \times \mathbf{d}}{2}.$$

(**d** is the position vector of the cavity axis with respect to the axis of the conductor.) Hence the magnetic field inside the cavity is uniform.

**8.G18.** A very long conducting circular cylinder of radius $a$ has a charge $Q_l$ per unit length, and rotates around its axis with an angular velocity $\omega$. Determine the magnetic field inside and outside the cylinder.

*Solution.* Since the cylinder is conducting, the charge is distributed uniformly over its surface. Its rotation with the cylinder represents a surface current. The density of the surface current is

$$J_s = \frac{Q_l}{2\pi a}\,\omega a = Q_l n, \tag{a}$$

where $n$ is the number of revolutions of the cylinder per unit of time. The problem is hence the same as that dealt with in Section 8.7.2. The result is the same as Eq. (8.7.16), with $J_s$ in (a) replacing $N_l I$ in that equation.

**8.G19.** A very long circular cylinder of radius $a$ is charged throughout its volume with a volume charge of density $\rho$. It rotates around its axis with an angular velocity $\omega$. Determine the magnetic field inside and outside the cylinder.

*Answer.* Outside the cylinder, the magnetic field does not exist. Inside the cylinder, the magnetic flux density vector is axial, and at a distance $r = a/2$ from the axis has the magnitude $B(a/2) = 3\mu_0\omega\rho a^2/8$.

**8.G20.** Starting from the defining expression in Eq. (8.8.4) for components of the curl of a vector function, determine the curl of a vector function with respect to the cylindrical coordinate system.

*Solution.* Let us first determine the $r$ component of the curl of a vector function $\mathbf{A}$. In Fig. 8.G20 is shown the elemental surface perpendicular to the unit vector $\mathbf{u}_r$. According to the definition in Eq. (8.8.4) and to Fig. 8.G20,

$$\begin{aligned}\operatorname{curl}_r\mathbf{A} &= \frac{A_\phi\, r\, d\phi - [A_\phi + (\partial A_\phi/\partial z)\, dz]\, r\, d\phi - A_z\, dz + [A_z + (\partial A_z/\partial\phi)\, d\phi]\, dz}{r\, d\phi\, dz}\\ &= \frac{1}{r}\frac{\partial A_z}{\partial \phi} - \frac{\partial A_\phi}{\partial z}.\end{aligned}$$

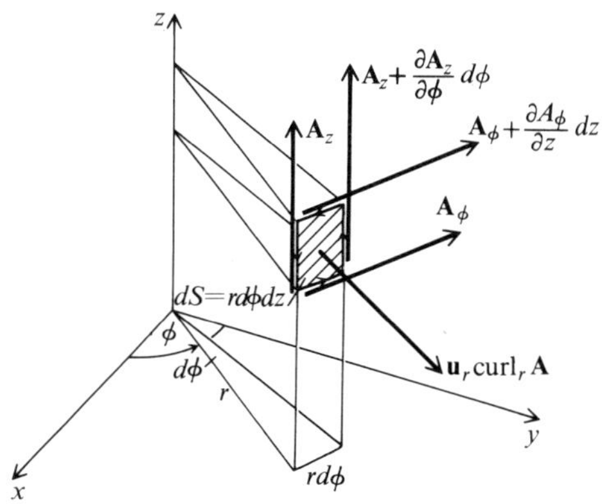


**Figure 8.G20**

We similarly obtain

$$\mathrm{curl}_\phi \mathbf{A} = \frac{A_z\,dz - [A_z + (\partial A_z/\partial r)\,dr]\,dz - A_r\,dr + [A_r + (\partial A_r/\partial z)\,dz]\,dr}{dr\,dz} = \frac{\partial A_r}{\partial z} - \frac{\partial A_z}{\partial r},$$

$$\mathrm{curl}_z \mathbf{A} = \frac{A_r\,dr - [A_r + (\partial A_r/\partial \phi)\,d\phi]\,dr - A_\phi\,r\,d\phi + [A_\phi + (\partial A_\phi/\partial r)\,dr](r + dr)\,d\phi}{r\,dr\,d\phi}$$

$$= \frac{1}{r} A_\phi + \frac{\partial A_\phi}{\partial r} - \frac{1}{r}\frac{\partial A_r}{\partial \phi} = \frac{1}{r}\frac{\partial}{\partial r}(r\,A_\phi) - \frac{1}{r}\frac{\partial A_r}{\partial \phi}.$$

The expression for the curl of a vector function **A** with respect to the cylindrical coordinate system hence has the form

$$\mathrm{curl}\,\mathbf{A} = \left(\frac{1}{r}\frac{\partial A_z}{\partial \phi} - \frac{\partial A_\phi}{\partial z}\right)\mathbf{u}_r + \left(\frac{\partial A_r}{\partial z} - \frac{\partial A_z}{\partial r}\right)\mathbf{u}_\phi + \frac{1}{r}\left[\frac{\partial}{\partial r}(r\,A_\phi) - \frac{\partial A_r}{\partial \phi}\right]\mathbf{u}_z. \qquad \text{(a)}$$

**8.G21.** Prove that the curl of a vector function **A** with respect to the spherical coordinate system is

$$\mathrm{curl}\,\mathbf{A} = \frac{1}{r\sin\theta}\left[\frac{\partial}{\partial\theta}(\sin\theta\,A_\phi) - \frac{\partial A_\theta}{\partial\phi}\right]\mathbf{u}_r + \frac{1}{r}\left[\frac{1}{\sin\theta}\frac{\partial A_r}{\partial\phi} - \frac{\partial}{\partial r}(r\,A_\phi)\right]\mathbf{u}_\theta$$

$$+ \frac{1}{r}\left[\frac{\partial}{\partial r}(r\,A_\theta) - \frac{\partial A_r}{\partial\theta}\right]\mathbf{u}_\phi.$$

**8.G22.** Prove that the magnetic vector potential created by any current distribution of finite density in a finite region of space is also finite.

*Proof.* If the current density is finite at all points and if the current flow is limited to a finite region of space, the integral in Eq. (8.9.16) is obviously finite at all points where $\mathbf{J} = 0$. At points where $\mathbf{J} \neq 0$, however, it may seem that **A** can have an infinitely large value, since $1/r \to \infty$ when we approach the point at which the vector potential is determined.

Consider a point $M$ in the current field, and a very small sphere $S$ of radius $a$ centered at $M$. If the sphere is small enough, **J** can be considered to be constant inside it. The contribution to the vector potential of currents outside the sphere $S$ is finite, as explained above. The part $\mathbf{A}_S$ of the vector potential created at $M$ by currents inside the sphere is, according to Eq. (8.9.16),

$$\mathbf{A}_S = \frac{\mu_0}{4\pi}\int_{v_S}\frac{\mathbf{J}\,dv}{r} = \frac{\mu_0\mathbf{J}}{4\pi}\int_{v_S}\frac{dv}{r} = \frac{\mu_0\mathbf{J}}{4\pi}\int_0^a\frac{4\pi r^2\,dr}{r} = \frac{\mu_0\mathbf{J}a^2}{2}.$$

From this we can see that $\mathbf{A}_S$ is finite however small $a$. The vector potential is hence finite at all points.

**8.G23.** Prove that the vector potential created by surface currents of finite density and distributed in a finite region of space is finite at all points. [*Hint:* consider a point $M$ on a surface with surface currents, and prove that the contribution to **A** at $M$ due to the current inside a small circle centered at $M$ is always finite.]

***8.G24.** Prove that the divergence of the vector potential **A** given by Eq. (8.9.16) is zero, by directly determining its divergence.

*Proof*. According to Eq. (8.9.16)

$$\operatorname{div} \mathbf{A} = \frac{\mu_0}{4\pi} \operatorname{div}_P \int_v \frac{\mathbf{J}\, dv}{vr}.$$

The differentiation implied by the operator "div" refers to the point $P$ where $\mathbf{A}$ is determined. The integral refers to that part of space where currents exist. The two operations are hence independent, and can change places. Thus

$$\operatorname{div} \mathbf{A} = \frac{\mu_0}{4\pi} \int_v \operatorname{div}_P \frac{\mathbf{J}}{r}\, dv. \tag{a}$$

Now,

$$\operatorname{div}_P \frac{\mathbf{J}}{r} = \frac{1}{r} \operatorname{div}_P \mathbf{J} + \mathbf{J} \cdot \operatorname{grad}_P \frac{1}{r} = \mathbf{J} \cdot \operatorname{grad}_P \frac{1}{r}. \tag{b}$$

The expression $\operatorname{div}_P \mathbf{J} = 0$, since $\mathbf{J}$ is a function of the point in the element $dv$, and not of the point $P$. Next, if we denote by $Q$ a point in $dv$, we have

$$\operatorname{grad}_P \frac{1}{r} = -\operatorname{grad}_Q \frac{1}{r}. \tag{c}$$

This is easily understood if we recall that "grad" implies differentiation with respect to the coordinates, and that $r = [(x_P - x_Q)^2 + (y_P - y_Q)^2 + (z_P - z_Q)^2]^{1/2}$. But

$$\operatorname{div}_Q \frac{\mathbf{J}}{r} = \frac{1}{r} \operatorname{div}_Q \mathbf{J} + \mathbf{J} \cdot \operatorname{grad}_Q \frac{1}{r} = \mathbf{J} \cdot \operatorname{grad}_Q \frac{1}{r}. \tag{d}$$

Here $\operatorname{div}_Q \mathbf{J} = 0$, because $\mathbf{J}$ is supposed to represent the current density of a *stationary* current system. Combining (b), (c), and (d) we have

$$\operatorname{div}_P \frac{\mathbf{J}}{r} = -\operatorname{div}_Q \frac{\mathbf{J}}{r} \qquad \text{(the stationary case)}, \tag{e}$$

so that (a) becomes

$$\operatorname{div} \mathbf{A} = -\frac{\mu_0}{4\pi} \int_v \operatorname{div}_Q \frac{\mathbf{J}}{r}\, dv.$$

In this integral the differentiation implied by "$\operatorname{div}_Q$" refers to a point inside $dv$, so that we can apply the divergence theorem (1.11.10). Thus

$$\operatorname{div} \mathbf{A} = -\frac{\mu_0}{4\pi} \oint_S \frac{\mathbf{J} \cdot d\mathbf{S}}{r} = -\frac{\mu_0}{4\pi} \oint_S \frac{J_n\, dS}{r}, \tag{f}$$

$J_n$ denoting the component of $\mathbf{J}$ normal to the surface element $dS$. But the surface $S$ has to enclose *all* currents. Therefore $J_n = 0$, and consequently also $\operatorname{div} \mathbf{A} = 0$.

Suppose now that the system of currents is *not* stationary. In that case, according to Eq. (6.3.5), $\operatorname{div}_Q \mathbf{J} = -\partial\rho/\partial t$ rather than zero. So instead of (e) we have

$$\operatorname{div}_P \frac{\mathbf{J}}{r} = -\operatorname{div}_Q \frac{\mathbf{J}}{r} + \frac{1}{r} \operatorname{div}_Q \mathbf{J} = -\operatorname{div}_Q \frac{\mathbf{J}}{r} - \frac{1}{r} \frac{\partial \rho}{\partial t} \qquad \text{(the nonstationary case)},$$

and (f) becomes

$$\operatorname{div} \mathbf{A} = -\frac{\mu_0}{4\pi} \oint_S \frac{J_n \, dS}{r} - \frac{\mu_0}{4\pi} \int_v \frac{1}{r} \frac{\partial \rho}{\partial t} \, dv. \tag{g}$$

The first integral in (g) vanishes as before, and the second integral can be modified as follows. Since neither $r$ nor $v$ are functions of time

$$\operatorname{div} \mathbf{A} = -\frac{\mu_0}{4\pi} \frac{\partial}{\partial t} \int_v \frac{\rho \, dv}{r} = -\varepsilon_0 \mu_0 \frac{\partial V}{\partial t}. \tag{h}$$

[$V$ is the electric scalar potential defined by Eq. (1.6.12).] Relation (h) is known as the *Lorentz condition* for potentials **A** and $V$ in the case of nonstationary currents. In fact, it represents a disguised form of the continuity equation.

**8.G25.** Starting from the differential equation (8.9.4) for the magnetic vector potential, determine the vector potential of an infinite cylindrical conductor of radius $a$, with a current of density **J**. From the vector potential determine the magnetic flux density.

*Solution.* Let the $z$ axis of a cylindrical coordinate system coincide with the conductor axis. According to Eq. (8.9.16), the magnetic vector potential can have only the $z$ component. So Eq. (8.9.4) becomes

$$\operatorname{curl} \operatorname{curl} (A_z \mathbf{u}_z) = \mu_0 \mathbf{J} = \mu_0 J \mathbf{u}_z.$$

By symmetry, $A_z$ is not a function of the coordinates $\phi$ and $z$. According to the expression (a) of Problem 8.G20 for the curl in the cylindrical coordinate system, the last equation becomes

$$-\frac{1}{r} \frac{d}{dr}\left( r \frac{dA_z}{dr} \right) = \mu_0 J. \tag{a}$$

Consider first the points outside the conductor. There $J = 0$, so that (a) yields

$$\frac{1}{r} \frac{d}{dr}\left( r \frac{dA_z}{dr} \right) = 0, \qquad \text{from which} \qquad A_z = C' \ln r + C'' \qquad \text{for} \qquad r > a. \tag{b}$$

($C'$ and $C''$ are integration constants.) For the purpose of determining the magnetic flux density only, the value of $C''$ is obviously irrelevant.

Inside the conductor, the vector potential is obtained by direct successive integration of (a):

$$r \, dA_z/dr = -\mu_0 J r^2/2 + F', \qquad \text{so that} \qquad A_z = -\mu_0 J r^2/4 + F' \ln r + F'' \qquad r < a. \tag{c}$$

($F'$ and $F''$ denote integration constants.)

According to (b), (c), and the expression for the curl in the cylindrical coordinate system,

$$\mathbf{B} = \left( \frac{\mu_0 J r}{2} - \frac{F'}{r} \right) \mathbf{u}_\phi \qquad \text{for} \qquad r < a, \tag{d}$$

$$\mathbf{B} = -\frac{C'}{r} \mathbf{u}_\phi \qquad \text{for} \qquad r > a. \tag{e}$$

It is left as an exercise for the reader to prove that $F' = 0$, and $C' = -\mu_0 J a^2/2 = -\mu_0 I/2\pi$, $I$ representing the intensity of the current in the conductor. Expressions (b) and (c) for the vector potential thus become

$$A_z = -\frac{\mu_0 I}{2\pi}\ln r + C'' \qquad \text{for} \qquad r > a,$$
$$A_z = -\frac{\mu_0 J r^2}{4} + F'' \qquad \text{for} \qquad r < a. \tag{f}$$

**8.G26.** Determine the vector potential of two parallel infinite straight current filaments with currents of equal intensity $I$ but in opposite directions, (1) using the result of the preceding problem and (2) evaluating the integral in Eq. (8.9.17).

*Solution.* (1) Denote by $r_1$ the distance of the point at which the vector potential is determined from the filament with the current in the positive $z$ direction, and by $r_2$ the distance of that point from the other filament. From (f) of the preceding problem it follows directly that

$$A = A_z = \frac{\mu_0 I}{2\pi}\ln\frac{r_2}{r_1}. \tag{a}$$

(2) To obtain (a) using Eq. (8.9.17), it is necessary to consider first two very long, but *finite*, filaments. Only when the integration is performed can the length of the filaments be assumed to be infinite. The two integrals occurring in the evaluation are divergent when considered isolated. The derivation of (a) following these remarks and using Eq. (8.9.17) is left to the reader.

**8.G27.** Determine the vector potential, and hence the magnetic flux density vector, between and outside two planar parallel infinite current sheets with equal magnitudes $J_s$ of the surface current density, but in opposite directions. The distance between the sheets is $d$.

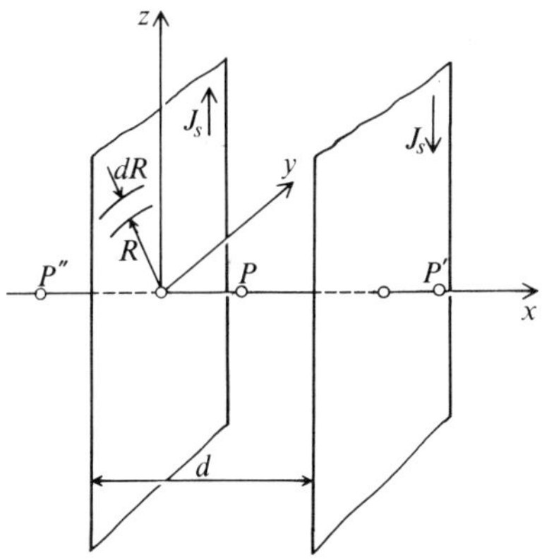


**Figure 8.G27**

*Solution.* Let a rectangular coordinate system be positioned as shown in Fig. 8.G27. According to Eq. (8.9.18), the vector potential is parallel to the $z$ axis. Consider first a point $P$ between the sheets. As the surface element in Eq. (8.9.18) it is most convenient to take an elemental circular ring of area $2\pi R\, dR$ centered at the $x$ axis. All points on that ring are at the same distance from the point $P$. If we simultaneously take into account the vector potential of both sheets, Eq. (8.9.18) yields

$$A = A_z = \frac{\mu_0 J_s}{2}\left[\int_0^\infty \frac{R\,dR}{\sqrt{x^2 + R^2}} - \int_0^\infty \frac{R\,dR}{\sqrt{(d-x)^2 + R^2}}\right], \qquad 0 \le x \le d.$$

Both of the integrals in brackets diverge, but their difference will be shown to be finite. This is why we consider the two integrals simultaneously.

Introducing new variables

$$\sqrt{x^2 + R^2} = u,\ R\,dR = u\,du, \qquad \text{and} \qquad \sqrt{(d-x)^2 + R^2} = v,\ R\,dR = v\,dv,$$

the above expression becomes

$$A = A_z = \frac{\mu_0 J_s}{2} \lim_{r\to\infty}\left(\int_x^r du - \int_{d-x}^r dv\right) = \frac{\mu_0 J_s}{2}(d - 2x), \ \ 0 \le x \le d. \tag{a}$$

The vector potential at points outside the sheets can be determined in the same way. For example, for the point $P'$ everything remains as above, except that $(d - x)$ should be replaced by $(x - d)$. Thus we obtain

$$A = A_z = -\mu_0 J_s d/2 \qquad \text{for} \qquad x > d. \tag{b}$$

In the case of the point $P''$ on the negative $x$ axis we have to put $(d + x)$, instead of $(d - x)$. The result is

$$A = A_z = \mu_0 J_s d/2 \qquad \text{for} \qquad x < 0. \tag{c}$$

The magnetic flux density vector is obtained from the expression

$$\mathbf{B} = \text{curl}\,\mathbf{A} = -\frac{dA_z}{dx}\mathbf{u}_y.$$

($A_z$ is the only component of **A**, and it is a function of $x$ only.) Hence, according to (a), (b), and (c), we have that between the sheets $\mathbf{B} = \mu_0 J_s \mathbf{u}_y$, and that $\mathbf{B} = 0$ outside of them. The result is of course the same as that given in Section 8.7.3.

**8.G28.** Determine the magnetic vector potential inside and outside an infinite circular cylindrical solenoid of radius $a$ having $N_l$ turns per unit length. The intensity of the current through the winding is $I$. Make use of the expressions for the magnetic flux density determined in Section 8.7.2.

*Solution.* By symmetry, the lines of the magnetic vector potential are circles centered at the solenoid axis. Using Eq. (8.7.16) and applying Eq. (8.9.19) to two such circles $C_1$ and $C_2$, one having a radius smaller and the other a radius larger than $a$, we have

$$\oint_{C_1} \mathbf{A}\cdot d\mathbf{l} = A\,2\pi r = \int_{S_1} \mathbf{B}\cdot d\mathbf{S} = \mu_0 N_l\, I\, r^2\pi, \qquad r < a,$$

$$\oint_{C_2} \mathbf{A}\cdot d\mathbf{l} = A\,2\pi r = \int_{S_2} \mathbf{B}\cdot d\mathbf{S} = \mu_0 N_l\, I\, a^2\pi, \qquad r > a.$$

Hence

$$A = \mu_0 N_l\, I\, r/2 \qquad \text{for} \qquad r < a, \tag{a}$$

$$A = \mu_0 a^2\, N_l\, I/2r \qquad \text{for} \qquad r > a. \tag{b}$$

**8.G29.** Determine the magnetic scalar potential on the axis of a circular current filament of radius $a$ and with current $I$. Determine hence the magnetic flux density vector on the axis of the loop.

*Solution.* According to Eq. (8.10.8), the magnetic scalar potential at a point $M$ on the loop axis is (see Fig. 8.G29)

$$V_m = \frac{I}{4\pi}\Omega.$$

($\Omega$ is the solid angle subtended at $M$ by the contour.) By the definition of the solid angle, it is equal to the area cut out of a sphere of unit radius, by the cone from the point considered towards the object. According to Fig. 8.G29, $\Omega$ is given by

$$\Omega = \frac{S}{r^2} = \frac{2\pi r^2(1 - \cos\theta)}{r^2} = 2\pi\left(1 - \frac{z}{\sqrt{a^2 + z^2}}\right). \tag{a}$$

($S$ represents the area of the part of the sphere indicated in dashed line.) Thus

$$V_m = \frac{I}{2}\left(1 - \frac{z}{\sqrt{a^2 + z^2}}\right). \tag{b}$$

The expression for the gradient of a scalar function in a cylindrical coordinate system is derived in Problem 1.G11. Using Eq. (8.10.4) and bearing in mind that by symmetry the magnetic flux density vector has the $z$ component only, we obtain

$$B_z = -\mu_0 \operatorname{grad} V_m = \mu_0 I a^2/2(a^2 + z^2)^{3/2}.$$

This is identical with (b) of Problem 8.G4.

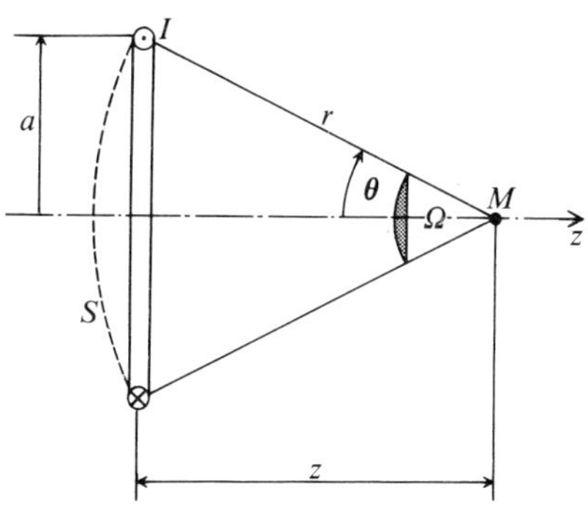


**Figure 8.G29**

**8.G30.** Determine the magnetic scalar potential on the axis of the rotating charged disk of Problem 8.G11. Hence obtain the magnetic flux density on the axis, given by (b) of that problem. [*Hint:* use the formula in Eq. (8.10.9) and (b) of the preceding problem.]

**8.G31.** Determine the magnetic scalar potential on the axis of a cylindrical solenoid of radius $a$ and length $L$, having $N_l$ densely wound turns per unit of length. The intensity of the current through the solenoid is $I$. Hence determine the magnetic flux density vector.

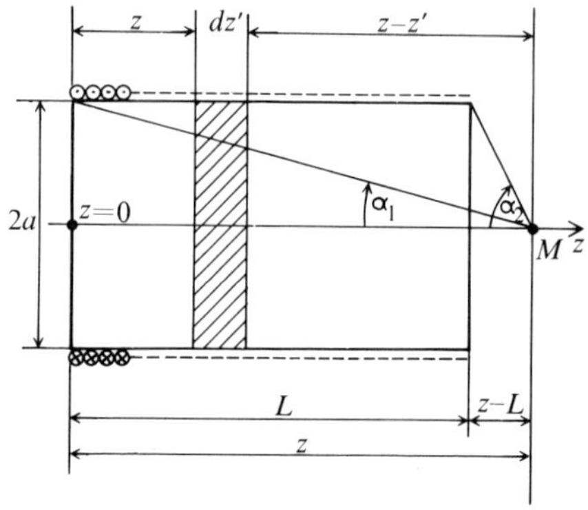


**Figure 8.G31**

*Solution.* By reference to Fig. 8.G31, according to (a) of Problem 8.G29 and Eq. (8.10.9), the magnetic scalar potential at the point $M$ on the solenoid axis is

$$V_m = \frac{1}{4\pi}\int_0^L 2\pi\left[1 - \frac{z - z'}{\sqrt{a^2 + (z - z')^2}}\right] IN_l \, dz'.$$

If we introduce a new variable $(z - z') = t$, $dz' = -dt$, this becomes

$$V_m = \frac{IN_l L}{2} + \frac{IN_l}{2}\int_z^{z-L} \frac{t\,dt}{\sqrt{a^2 + t^2}} = \frac{IN_l L}{2} + \frac{IN_l}{2}\left[\sqrt{a^2 + (z - L)^2} - \sqrt{a^2 + z^2}\right]. \tag{a}$$

By symmetry, on the solenoid axis only the $z$ component of the magnetic flux density vector exists. From (a)

$$B_z = -\mu_0 \operatorname{grad}_z V_m = -\mu_0 \frac{\partial V_m}{\partial z} = \frac{\mu_0 I N_l}{2}\left[\frac{-(z - L)}{\sqrt{a^2 + (z - L)^2}} + \frac{z}{\sqrt{a^2 + z^2}}\right]. \tag{b}$$

From Fig. 8.G31 we readily see that (b) is identical with (d) of Problem 8.G5. The first term in brackets is equal to $-\cos\alpha_2$, and the second term to $+\cos\alpha_1$.

## Numerical problems

**8.N1.** Determine the magnetic and electric forces between two equal small charges $Q = 10^{-10}$ C, at a distance $r = 1$ cm apart, if the velocities of both charges are $v = 10^4$ m/s. The directions of the two velocity vectors are the same, both perpendicular to $r$ at the instant considered.

**8.N2.** Determine the magnetic flux density of an electron moving with a velocity $v = 10^5$ m/s, at a point of the plane of the electron normal to its velocity and at a distance $r = 1$ cm from it. The electron charge is $-1.602 \times 10^{-19}$ C.

**8.N3.** Determine the magnetic flux density at the center of a circular current filament of radius $a = 10$ cm and with a current of intensity $I = 1$ A.

**8.N4.** Determine the magnitude of the magnetic flux density at several points along the axis of the circular contour from the preceding problem, and sketch the dependence of $B_z$ on $z$. (See Problem 8.G4.)

**8.N5.** Determine the magnetic flux density at a distance $r = 10$ cm from an infinite straight filament with a current of intensity $I = 2$ A.

**8.N6.** A solenoid of length $b = 10$ cm and radius $a = 1$ cm has $N_l = 3000$ turns per unit length with a current of intensity $I = 1$ A. Determine the flux density on the axis of the solenoid, at its center and at its ends. (See Problem 8.G5.)

**8.N7.** Determine the magnetic flux density near the strip of Problem 8.G8 for $I = 10$ A, $a = 10$ cm, and $\theta_1 = -\pi/4$, $\theta_2 = \pi/4$.

**8.N8.** Determine the magnetic flux density on the axis of the rotating charged disk of Problem 8.G11, if $\rho_s = 10^{-6}$ C/m$^2$, $\omega = 2\pi \times 100$ rad/s, and $a = z = 10$ cm.

**8.N9.** Determine the magnetic flux through the rectangular contour shown in Fig. 8.G14, if $I = 10$ A, $a = b = 10$ cm, and $h = 5$ cm.

**8.N10.** Determine the magnetic flux density inside a very long cylindrical solenoid with $N_l = 3000$ turns per meter and with a current of intensity $I = 1$ A.

**8.N11.** Determine the magnetic flux through a cross-section of the solenoid from the preceding problem if the area of the cross-section is $S = 10$ cm$^2$.

**8.N12.** Determine the magnetic flux density inside the rotating charged cylinder of Problem 8.G18, if $Q_l = 10^{-9}$ C/m and $n = 6000$ revolutions per minute.

**Suggested reading.** The list progresses towards more advanced texts. For full bibliographical details, see the Bibliography.

1. CARTER, G. W., *The Electromagnetic Field in Its Engineering Aspects*, Chapter 6.
2. SEELY, S., *Introduction to Electromagnetic Fields*, Chapter 6.
3. KRAUS, J. D., *Electromagnetics*, Chapter 4.
4. PLONSEY, R. and R. E. COLLIN, *Principles and Applications of Electromagnetic Fields*, Chapter 6.

CHAPTER 9

# MAGNETIC FIELD OF STEADY CURRENTS IN THE PRESENCE OF MAGNETIC MATERIALS

## 9.1 BEHAVIOR OF MAGNETIC SUBSTANCES IN THE MAGNETIC FIELD

**9.1.1 Introduction.** Atoms of all material bodies consist of a heavy central positive nucleus, and a number of negative electrons revolving round it. This is a very crude, but, for our purpose, an adequate model of the structure of matter. The electrons in their orbits are moving with very high velocities, and from the macroscopic point of view they can be considered as tiny, atomic-scale current loops. In addition, electrons are imagined also to rotate about their axes, which is known as the electron spin. If electrons are considered as small charged balls, their spin, being a motion of electric charge, may again be thought of as equivalent to a tiny closed current.

The foregoing rough model of an atom is sufficient for the analysis of the macroscopic behavior of substances in the magnetic field. In fact, it is quite sufficient to consider atoms as equivalent current loops of atomic dimensions situated in a vacuum. All the other properties of atoms are irrelevant as far as our analysis is concerned.

So, to analyze the influence of substances on the macroscopic magnetic field into which they are introduced, it is necessary for us to analyze first the behavior of a single current loop in a static magnetic field. Due to the extremely small size of equivalent atomic currents, they can be considered as always situated in a uniform magnetic field.

The most important effect for our purpose is the torque by which a uniform magnetic field acts on a current loop. The torque tends to turn the loop so that the magnetic field of the loop increases the external magnetic field. In consequence, the total magnetic field is generally larger than the primary field, a fact of great practical importance.

**9.1.2 Torque on a circular current loop in a uniform magnetic field.** For our purpose it is simplest to imagine atomic currents as elemental circular current loops. Let us consider such a loop of radius $a$ and with a current of intensity $I$, situated in a uniform magnetic field of flux density $\mathbf{B}$. Each moving charge $Q$ inside the contour is acted upon by a force as given by Eq. (8.3.2). Let the number of moving charges per unit volume be $N$, and their average velocity $\mathbf{v}$. The magnetic force on all the charges moving inside an elemental volume $dv$ is then

$$d\mathbf{F} = NQ\mathbf{v} \times \mathbf{B}\, dv.$$

In the case of a thin conductor of cross-section area $\Delta S$

$$NQ\mathbf{v}\,dv = NQ\mathbf{v}\,dl\,\Delta S = NQv\Delta S\,d\mathbf{l} = J\Delta S\,d\mathbf{l} = I\,d\mathbf{l}.$$

So the magnetic force on a small current element $I\,d\mathbf{l}$ ($d\mathbf{l}$ assumed, as usual, to be in the direction of the current) is

$$d\mathbf{F} = I\,d\mathbf{l} \times \mathbf{B}. \tag{9.1.1}$$

Let us first determine the total magnetic force on the contour. Since $\mathbf{B}$ is a constant vector, according to Eq. (9.1.1) the total force is

$$\mathbf{F} = \oint_C I\,d\mathbf{l} \times \mathbf{B} = I\left(\oint_C d\mathbf{l}\right) \times \mathbf{B} = 0. \tag{9.1.2}$$

The result is obviously valid for any current contour in a uniform magnetic field.

We can see from this that no translation of the contour can result. But the elemental forces on the elements of the contour generally tend to rotate it. For convenience, let us resolve vector $\mathbf{B}$ into the component $\mathbf{B}_n$ normal to the surface of the loop (see Fig. 9.1) and the component $\mathbf{B}_p$ parallel to it (see Fig. 9.2). From Fig. 9.1 we see that $\mathbf{B}_n$ tends only to deform the contour (i.e. to increase its radius). But from Fig. 9.2 it is evident that the component $\mathbf{B}_p$ tends to rotate the contour around the axis $OO'$ in the direction indicated. The elemental torque due to the elemental forces $d\mathbf{F}$ and $d\mathbf{F}'$ on elements $d\mathbf{l}$ and $d\mathbf{l}'$ is

$$dT = dF\,a \sin\alpha + dF'\,a \sin\alpha = 2I\,dl\,B_p \sin\alpha\,a \sin\alpha = 2Ia^2B_p \sin^2\alpha\,d\alpha,$$

since $dF = dF' = I\,dl\,B_p \sin\alpha$, and $dl = a\,d\alpha$. The total torque is hence

$$T = \int_0^{\pi} 2Ia^2B_p \sin^2\alpha\,d\alpha = 2Ia^2B_p\frac{\pi}{2} = a^2\pi IB_p. \tag{9.1.3}$$

Now, from Fig. 9.3 $B_p = B \sin\theta$, where $\theta$ is the angle between $\mathbf{B}$ and the

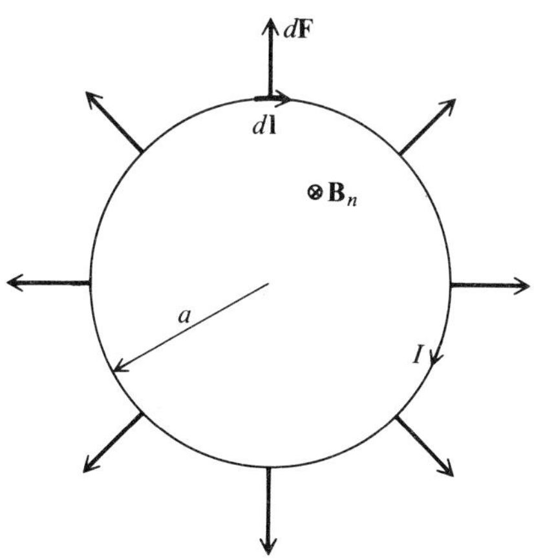


**Fig. 9.1** The total force on a closed current filament situated in a uniform magnetic field is zero.

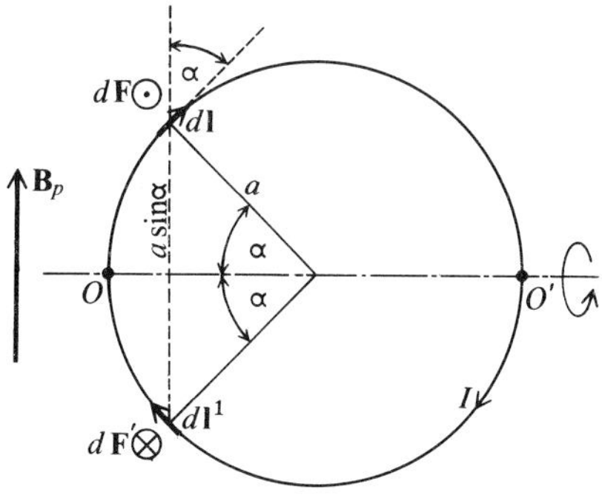


**Fig. 9.2** Construction for calculating the torque of magnetic forces.

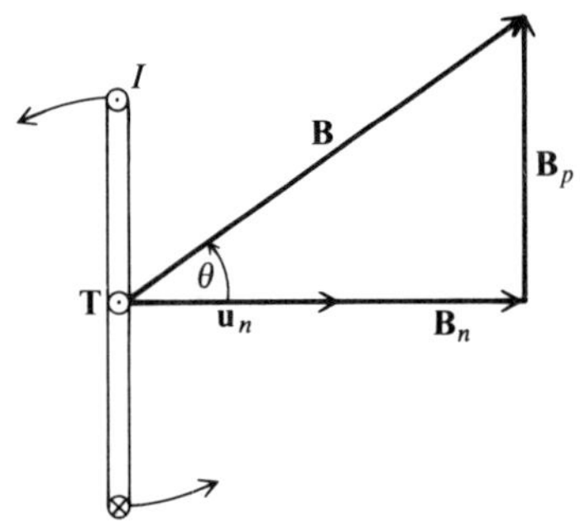


**Fig. 9.3** Torque of magnetic forces tends to align **B** and $\mathbf{u}_n$.

positive normal to the surface of the contour. The torque $T$ given by Eq. (9.1.3) can therefore be represented as a vector

$$\mathbf{T} = a^2\pi I\,\mathbf{u}_n \times \mathbf{B} = I\mathbf{S} \times \mathbf{B}, \tag{9.1.4}$$

where **S** stands for $a^2\pi\mathbf{u}_n$. According to the usual convention, the direction of the torque vector **T** is determined according to the right-hand screw rule with respect to the contour rotation.

The product $I\mathbf{S}$ is known as the *magnetic moment* of the loop. It is usually designated by **m**:

$$\boxed{\mathbf{m} = I\mathbf{S}} \quad \text{(definition of the magnetic moment of a current contour).} \tag{9.1.5}$$

With this definition Eq. (9.1.4) becomes

$$\boxed{\mathbf{T} = \mathbf{m} \times \mathbf{B}} \quad \text{(torque of magnetic forces on a planar closed current filament).} \tag{9.1.6}$$

The torque **T** due to the magnetic forces obviously tends to align vectors **m** and **B**, since it is zero only when they are in the same direction. According to Eq. (9.1.5), this means that it tends to rotate the contour so that the magnetic flux of the external field **B** through the contour is maximal. The magnetic field of the contour then adds to the external field, and the total flux density is increased.

**9.1.3 Qualitative discussion of the behavior of magnetic materials in a magnetic field.** Let us consider a piece of a magnetic material situated in a magnetic field. Before the magnetic field was established, the magnetic moments of the atoms were distributed in space statistically. So, although a microscopic magnetic field existed in the vicinity of each atom, the net macroscopic field was zero.

When the external magnetic field has been turned on, each elemental loop representing an atom is acted on by a torque given by Eq. (9.1.6). This torque tends to align the magnetic moments with the magnetic flux density vector of the external magnetic field. We can imagine the orientation of the elemental current loops as opposed by certain elastic forces, with or without friction. But, even when only partially oriented, the atomic currents will create a secondary macroscopic magnetic field, so that each atomic current is in fact situated in the external field *and* in this secondary macroscopic field of all other elemental currents. Thus, as in the analogous electrostatic case, the equilibrium state is a function of both the geometry of the external field and the geometry and physical properties of the piece of the magnetic material we are considering.

In order to determine the secondary macroscopic field, we must know the distribution of the magnetic moments of the elemental current loops. The secondary field is then found simply to be the field due to a given aggregate of elemental circular currents *considered in a vacuum*. Our problem is thus reduced to determining the flux density vector created by a single circular current loop, as well as that due to an aggregate of such loops.

The reader will probably have realized that the elemental current loop plays a role analogous to that of the electric dipole in the static electric field. A current loop and an electric dipole are quite different physical structures, but we shall see that equations governing distant fields in the two cases are nevertheless very similar. As a consequence, we shall see that a large part of the technique developed for analyzing the electric field is, by a simple analogy, applicable also to the magnetic field.

*Problems referring to Section 9.1:* 9.T1, 9.T2.

## 9.2 MAGNETIC VECTOR POTENTIAL AND THE FIELD OF AN ELEMENTARY CURRENT LOOP AND OF THE AGGREGATE OF THE LOOPS

**9.2.1 The case of a single current loop.** Let us consider a circular current loop of radius $a$, with a current of intensity $I$, and let us determine the magnetic vector potential $\mathbf{A}$ at a distance $r$ from the loop which is much larger than $a$.

According to Eq. (8.9.17), the lines of the magnetic vector potential $\mathbf{A}$ are circles centered at the loop axis. The magnitude of $\mathbf{A}$ along such a circle is constant, which means that $\mathbf{A}$ is a function of the spherical coordinates $r$ and $\theta$ only. Hence the unknown quantity is the magnitude of $\mathbf{A}$ as a function of $r$ and $\theta$. The loop and a line of the magnetic vector potential are sketched in Fig. 9.4.

Figure 9.5 shows the loop on a larger scale. It is assumed that the point $M$ is in the plane $x = 0$, and that $r \gg a$. Since we know in advance that the only existing component of $\mathbf{A}$ at $M$ is $\mathbf{A}_\phi$, only the elemental component $dA_\phi$ due to individual current elements is of interest. According to Eq. (8.9.17), the component $dA_\phi$, due to the element $dl$ shown, is

$$dA_\phi = \frac{\mu_0 I}{4\pi} \frac{dl \sin \phi}{R}. \tag{9.2.1}$$

Now

$$dl = a\, d\phi, \qquad \frac{1}{R} \simeq \frac{1}{r - a \sin \phi \sin \theta} \simeq \frac{1}{r}\left(1 + \frac{a}{r} \sin \phi \sin \theta\right).$$

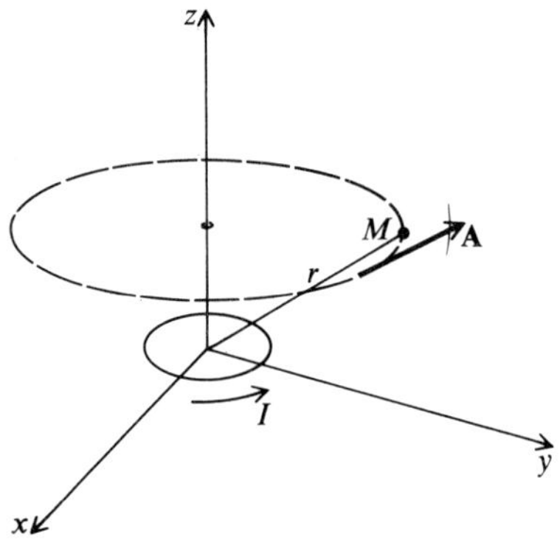


Fig. 9.4 A circular current filament and a line of the magnetic vector potential.

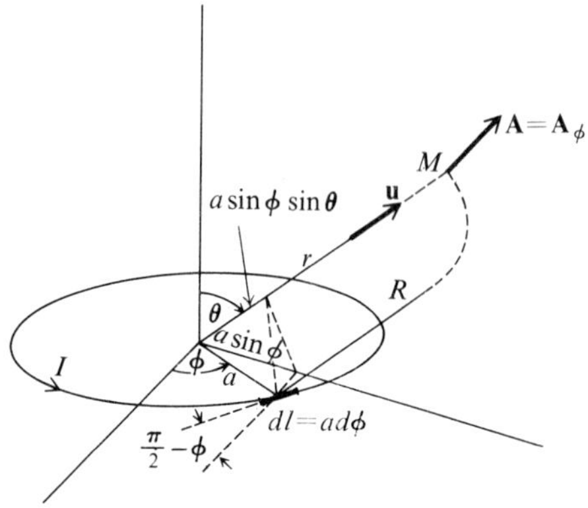


Fig. 9.5 Construction for determining the vector potential of a circular current filament.

So from Eq. (9.2.1) we have

$$A_\phi = \frac{\mu_0 I a}{4\pi r}\int_0^{2\pi}\left(1 + \frac{a}{r}\sin\phi\sin\theta\right)\sin\phi\, d\phi$$

$$= \frac{\mu_0 I a^2 \sin\theta}{4\pi r^2}\int_0^{2\pi}\sin^2\phi\, d\phi = \frac{\mu_0 I a^2 \pi \sin\theta}{4\pi r^2}.$$

According to the definition in Eq. (9.1.5) of the magnetic moment **m** of the loop, this can be written as

$$\mathbf{A} = \frac{\mu_0 \mathbf{m}\times\mathbf{u}}{4\pi r^2}. \tag{9.2.2}$$

The magnetic flux density vector **B** is now obtained by using Eq. (8.9.2):

$$\mathbf{B} = \text{curl}\,\mathbf{A} = \frac{\mu_0}{4\pi}\,\text{curl}\,\frac{\mathbf{m}\times\mathbf{u}}{r^2}. \tag{9.2.3}$$

**9.2.2 The case of the aggregate of loops. Magnetic moment density vector (magnetization vector).** Consider a piece of magnetized matter, and suppose that the magnetic moments **m** of all the elemental current loops equivalent to atomic currents are known. The resulting vector potential at any point is then obtained as a vector sum of the terms in Eq. (9.2.2):

$$\mathbf{A} = \frac{\mu_0}{4\pi}\sum_v \frac{\mathbf{m}_k\times\mathbf{u}_k}{r_k^2}, \tag{9.2.4}$$

$v$ representing the volume of the magnetized piece of matter.

In a macroscopic theory it is unnecessary for us to know the magnetic moments of the individual atoms. Let us suppose that we can determine the vector sum of the elemental magnetic moments inside any "physically small" volume $dv$. Then the density of the magnetic moments is

$$\boxed{\mathbf{M} = \frac{\text{sum of all } \mathbf{m} \text{ inside } dv}{dv}} \quad \text{(definition of the magnetization vector).} \tag{9.2.5}$$

If vector **M** is known at all points, we can determine the vector potential created by small loops inside a volume $v$ as an integral, and not as the sum in Eq. (9.2.4) over all the individual loops. According to Eq. (9.2.5), a small volume $dv$ has a magnetic moment

$$\sum_{dv} \mathbf{m} = \mathbf{M}\, dv,$$

so that Eq. (9.2.4) becomes (see Fig. 9.6)

$$\mathbf{A}_m = \frac{\mu_0}{4\pi} \int_v \frac{\mathbf{M} \times \mathbf{u}}{r^2}\, dv. \tag{9.2.6}$$

(The subscript $m$ is added to **A** to stress that this component of the vector potential is due to magnetized matter.)

The magnetic moment density vector **M**, defined by Eq. (9.2.5), is also known as the *magnetization vector*. For simplicity, we shall more often use this latter name. If the magnetization vector **M** is determined as a function of the coordinates, the vector potential, and hence the magnetic flux density vector **B** can be determined easily, at least theoretically. However, the magnetization vector is a function of both the external magnetic field *and* the state of the induced magnetization at all points, since this produces an additional magnetic field. As in the case of the dielectric substances in the electrostatic field, we are confronted with a strange situation. The total magnetic field at a point inside a piece of magnetized matter is a function of magnetization throughout the volume considered, and the magnetization at a given point is a function of the total magnetic field! Hence analysis of the *transient* process is extremely complicated. However, an equilibrium is always finally attained, and our task is to determine this equilibrium. For that purpose, the dependence of the magnetization vector **M** on the total magnetic flux density vector **B** must be known.

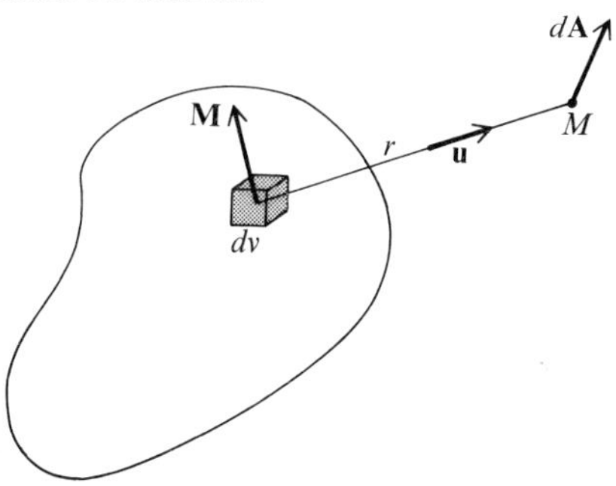


**Fig. 9.6** Magnetic vector potential created by a magnetized body.

For most substances, the magnetization at a given point is a linear function of the total magnetic flux density at that point. This linear relation is usually written in the form

$$\mathbf{M} = \frac{1}{\mu_0} \frac{\chi_m}{1 + \chi_m} \mathbf{B}. \tag{9.2.7}$$

The dimensionless constant $\chi_m$ is known as the *magnetic susceptibility* of the material considered. In Section 9.4, we shall see that for all linear magnetic materials $\chi_m$ is much smaller than unity.

There is an important class of magnetic materials for which $\chi_m$ is much larger than unity. Unfortunately, in all these cases $\chi_m$ is a function of both **B** and the magnetic history of the specimen. This is discussed briefly in Section 9.7. At this point we need to appreciate only that any analysis of magnetic fields in the presence of nonlinear magnetic materials is extremely involved.

*Problems referring to Section 9.2:* 9.T3, 9.T4.

## 9.3 MACROSCOPIC CURRENTS EQUIVALENT TO A MAGNETIZED SUBSTANCE

The magnetic field created by a magnetized substance is due to elemental current loops representing the magnetic properties of individual atoms. These elemental currents, however, can be considered as resulting in certain equivalent macroscopic currents. The magnetic field due to magnetized matter can then be assumed to be due to these macroscopic resultants of microscopic, atomic currents. We shall see that the density of the macroscopic resultant of microscopic currents can be expressed in terms of the magnetization vector **M**.

Consider first a part of the surface of a magnetized piece of matter, and let the elemental currents be perpendicular to the surface, as shown in Fig. 9.7. We can readily see from this that, if the magnetization is uniform, the macroscopic resultant of the elemental currents inside the piece of matter is zero. However, in a thin layer near the surface, uncompensated currents exist, resulting in a thin

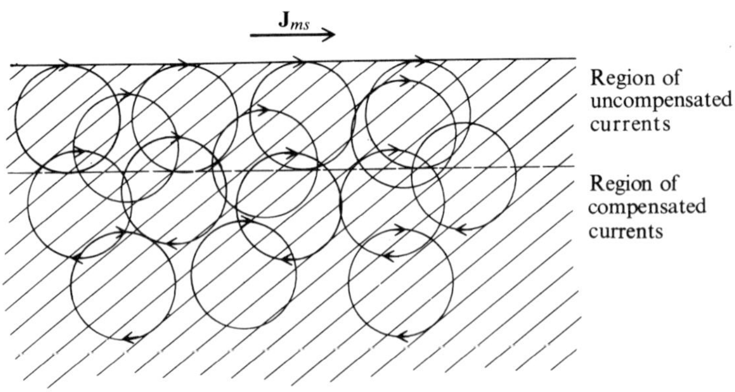


**Fig. 9.7** A region near the surface of a uniformly magnetized body.

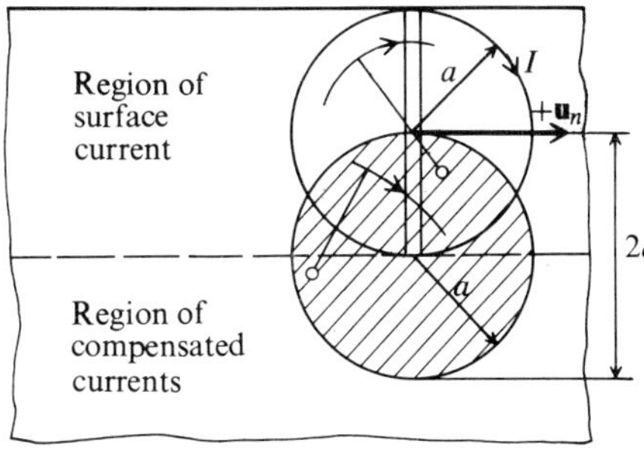


**Fig. 9.8** Construction for determining macroscopic surface current resultant of microscopic currents.

current sheet. The thickness of this current sheet is of the same order as the diameter of the elemental current loops, so it can be regarded as a surface current.

Let us determine the density $J_{ms}$ of the resultant surface current. We must remember that the density of a surface current is defined as the current intensity through a thin strip of unit length, perpendicular to both the surface and the current flow. Figure 9.8 is an enlarged section of Fig. 9.7. Now, let us suppose that the radius of the atomic currents is $a$, and their intensity $I$. To determine the density of the surface current it is necessary to determine the total current through a strip of unit length, the cross-section of which is shown in the double line. Only those loops cross this surface once in its positive direction which are centered in a cylinder of radius $a$ and of unit length. The cross-section of this cylinder is shown crosshatched in Fig. 9.8. The elemental currents centred outside this cylinder might cross the strip, but *twice*, once in the positive and once in the negative direction. So their total contribution is zero.

Suppose there are $N$ elemental loops per unit volume. Then, inside the circular cylinder shown, $Na^2\pi$ loops are centered. The intensity of the current through the strip, equal to the density of the surface current, hence is

$$J_{ms} = INa^2\pi = Nm = M, \tag{9.3.1}$$

by definition of the magnetization vector **M**.

Let us assume now that the angle between the magnetic moments **m** of the elemental loops and the normal to the surface is $\alpha$, as indicated diagrammatically in Fig. 9.9. In that case, the uncompensated loops are centered inside the elliptic cylinder, shown crosshatched, of volume $a^2\pi \sin\alpha$ per unit length. Hence, instead of Eq. (9.3.1), we have

$$J_{ms} = INa^2\pi \sin\alpha = M \sin\alpha. \tag{9.3.2}$$

According to Fig. 9.9, this can be represented in the vector form as

$$\boxed{\mathbf{J}_{ms} = \mathbf{M} \times \mathbf{u}_n} \qquad \text{(surface current resultant of microscopic currents; } \mathbf{u}_n \text{ directed away from the magnetized substance).} \tag{9.3.3}$$

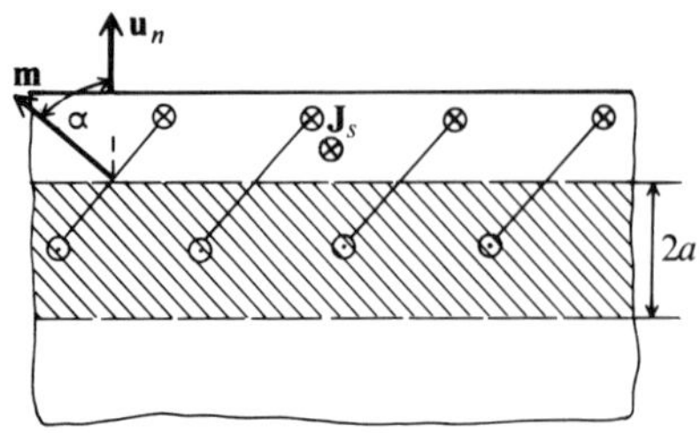


**Fig. 9.9** Construction for determining the macroscopic surface current resultant in the general case.

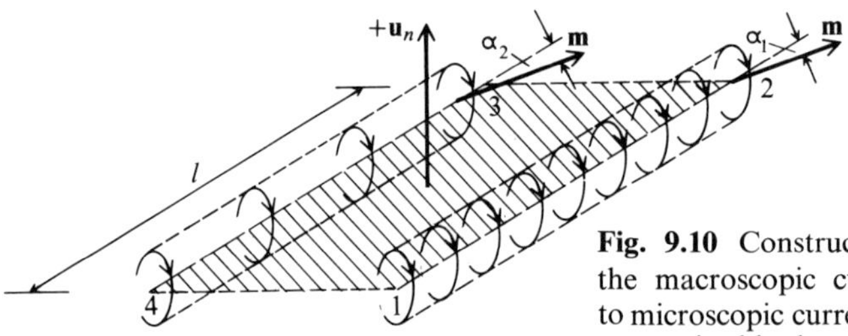


**Fig. 9.10** Construction for determining the macroscopic current density equivalent to microscopic currents inside a nonuniformly magnetized body.

It is obvious that, if the magnetization of a magnetized piece of matter is uniform, the resultant currents are surface currents only. The elemental currents throughout the volume are compensated by the adjacent currents. But if the magnetization is not uniform, the macroscopic resultants of the microscopic currents might appear throughout the magnetized body.

To determine the current density of the macroscopic resultant, let us consider a rectangular contour inside a magnetized body, as shown in Fig. 9.10. The uncompensated current through the surface of the rectangle 12341 is again due to the current loops centered inside a certain region only. It is not difficult to conclude that this region now consists of the two cylinders shown. Their volumes are $a^2\pi l \cos\alpha_1$ and $a^2\pi l \cos\alpha_2$, $\alpha_1$ and $\alpha_2$ representing the angles between the sides of the rectangle and the direction of the magnetic moments along them. Let $N_1$ and $N_2$ represent (different) numbers of loops per unit volume along sides 12 and 34 of the rectangle, and let $M_1$ and $M_2$ denote the magnitudes of the corresponding magnetization vectors. Then the current intensity through the rectangle due to the elemental currents is

$$\begin{aligned} I_{m\,\text{total}} &= IN_1 a^2\pi l \cos\alpha_1 - IN_2 a^2\pi l \cos\alpha_2 \\ &= M_1 l \cos\alpha_1 - M_2 l \cos\alpha_2. \end{aligned} \tag{9.3.4}$$

Using the vector notation, we can write this as

$$I_{m\,\text{total}} = \oint_{12341} \mathbf{M}\cdot d\mathbf{l}.$$

If we denote the area of the rectangle by $\Delta S$, the component of the current density vector normal to the surface is then

$$\mathbf{J}_m \cdot \mathbf{u}_n = \frac{\oint_{12341} \mathbf{M} \cdot d\mathbf{l}}{\Delta S} \tag{9.3.5}$$

According to the definition of the curl in Eq. (8.8.4), for $\Delta S$ sufficiently small (but large enough to enclose a large number of atomic currents) the right-hand side of (9.3.5) precisely represents the component of curl $\mathbf{M}$ in the direction of $\mathbf{u}_n$:

$$\mathbf{J}_m \cdot \mathbf{u}_n = \mathbf{u}_n \cdot \text{curl}\, \mathbf{M}.$$

The required density of the macroscopic resultant of the elemental currents is

$$\boxed{\mathbf{J}_m = \text{curl}\, \mathbf{M}} \quad \text{(density of the macroscopic resultant of the microscopic currents).} \tag{9.3.6}$$

So, if the magnetization vector is known as a function of the coordinates, the resultant macroscopic densities of surface and volume currents can be obtained from Eqs. (9.3.3) and (9.3.6). These currents represent the only source of the secondary magnetic field due to magnetized matter. The total magnetic field can now be analyzed as the field of the original macroscopic currents and these secondary equivalent currents, both existing *in a vacuum*. Analysis of the magnetic fields in the presence of magnetized matter is hence considerably simplified, at least in principle. It is reduced to the analysis of a system of currents situated in a vacuum, which was discussed in the preceding chapter.

*Problems referring to Section 9.3:* 9.T5, 9.T6, 9.G1, 9.G2.

## 9.4 THE GENERALIZED FORM OF AMPERE'S CIRCUITAL LAW. THE MAGNETIC FIELD INTENSITY

According to the conclusions arrived at in the preceding section, the secondary sources of the magnetic field inside magnetized matter can be reduced to a system of macroscopic currents. These currents represent the macroscopic resultants of the elemental, microscopic currents, which in our model are considered as the real sources of the secondary magnetic field. The real currents and these equivalent currents are then considered to exist in a vacuum.

Suppose at a certain point both the real current and the macroscopic resultant of the microscopic currents exist. The density of the total current creating the magnetic field is then their sum. So, according to Eq. (9.3.6), we have that

$$\mathbf{J}_{\text{total}} = \mathbf{J} + \mathbf{J}_m = \mathbf{J} + \text{curl}\, \mathbf{M}. \tag{9.4.1}$$

In this case, Ampère's circuital law (8.7.1) must therefore be modified. $\mathbf{J}_{\text{total}}$ rather than $\mathbf{J}$ only has to be introduced into the surface integral on the right-hand side:

$$\oint_C \mathbf{B} \cdot d\mathbf{l} = \mu_0 \int_S (\mathbf{J} + \text{curl}\, \mathbf{M}) \cdot d\mathbf{S}. \tag{9.4.2}$$

This equation represents the integral form of the generalized Ampère's circuital law. Usually, however, it is modified as follows.

Let us apply Stokes' theorem, Eq. (8.8.14), to the second term on the right-hand side of Eq. (9.4.2). The surface integral of curl **M** can be transformed into a line integral of **M** along the contour $C$ limiting the surface. So Eq. (9.4.2) can be written as

$$\oint_C \left(\frac{\mathbf{B}}{\mu_0} - \mathbf{M}\right) \cdot d\mathbf{l} = \int_S \mathbf{J} \cdot d\mathbf{S}. \tag{9.4.3}$$

The difference of the two vector functions in the parentheses has a very remarkable property. The line integral of this difference along any closed contour is *always* equal to the total *real* current crossing a surface limited by the contour. Because of this simple and extremely useful property, this difference is defined as a new vector function characterizing the magnetic field. It is usually referred to as the *magnetic field intensity* vector, and is designated by **H**:

$$\boxed{\mathbf{H} = \mathbf{B}/\mu_0 - \mathbf{M}} \quad \text{(the generally valid relation, definition of vector } \mathbf{H}\text{).} \tag{9.4.4}$$

So the generalized Ampère's circuital law in the integral form (9.4.3) becomes

$$\boxed{\oint_C \mathbf{H} \cdot d\mathbf{l} = \int_S \mathbf{J} \cdot d\mathbf{S}} \quad \text{(the generalized form of Ampère's circuital law; valid for a static magnetic field only).} \tag{9.4.5a}$$

For some purposes it is more convenient to write it in the following explicit form:

$$\oint_C \mathbf{H} \cdot d\mathbf{l} = \text{sum of all currents through } C. \tag{9.4.5b}$$

It is seen that the unit of the magnetic field intensity is *ampere per meter* (abbreviation A/m).

The differential form of the generalized Ampère's circuital law is obtained readily from the special form of Ampère's law given in Eq. (8.8.3). Replacing **J** in this equation by (**J** + curl **M**) we have

$$\text{curl } \mathbf{B} = \mu_0(\mathbf{J} + \text{curl } \mathbf{M}),$$

or

$$\text{curl } (\mathbf{B}/\mu_0 - \mathbf{M}) = \mathbf{J}. \tag{9.4.6}$$

According to the definition in Eq. (9.4.4) of the magnetic field intensity vector **H**, this becomes

$$\boxed{\text{curl } \mathbf{H} = \mathbf{J}} \quad \text{(the differential form of Ampère's generalized circuital law; valid for a static magnetic field only).} \tag{9.4.7}$$

It should be noted that the generalized form of Ampère's circuital law as defined above is valid for magnetic fields of steady currents only. In Section 13.2 we shall see that in the case of time-varying magnetic fields, created by time-varying currents, an additional term has to be added on the right-hand side.

In the case of linear magnetic materials, the magnetization vector **M** is proportional to the magnetic flux density vector **B**. Hence, according to Eq. (9.4.4), it is also proportional to the magnetic field intensity vector **H**. This linear relation is most often written in the form

$$\mathbf{M} = \chi_m \mathbf{H} \qquad \text{(linear magnetic materials).} \tag{9.4.8}$$

As we noted in Section 9.2.2, the constant $\chi_m$ is known as the magnetic susceptibility of a substance. With Eq. (9.4.8) the relation in Eq. (9.4.4) becomes

$$\mathbf{B} = \mu_0(\mathbf{H} + \mathbf{M}) = \mu_0(1 + \chi_m)\mathbf{H}. \tag{9.4.9}$$

The sum $(1 + \chi_m)$ is denoted by $\mu_r$, and is referred to as the *relative permeability*. The product $\mu_0\mu_r$ is designated by $\mu$, and is known as the *permeability* of the material considered. Thus

$$1 + \chi_m = \mu_r, \qquad \mu_0(1 + \chi_m) = \mu_0\mu_r = \mu, \tag{9.4.10}$$

and Eq. (9.4.9) simplifies to

$$\mathbf{B} = \mu\mathbf{H} \qquad \text{(linear magnetic materials).} \tag{9.4.11}$$

According to Eqs. (9.4.9) and (9.4.11), the magnetization vector **M** inside a linear magnetic material can be expressed by either **B** or **H** and permeability $\mu$:

$$\mathbf{M} = \frac{\mathbf{B}}{\mu_0} - \mathbf{H} = \frac{\mu - \mu_0}{\mu\mu_0}\mathbf{B} = \left(\frac{\mu}{\mu_0} - 1\right)\mathbf{H}. \tag{9.4.12}$$

In Table 9.1 are presented the values of the relative permeability $\mu_r$ and the magnetic susceptibility $\chi_m = \mu_r - 1$ for some linear substances. Representative values of some average permeability of a few nonlinear, highly magnetizable substances are also given for comparison. Note that $\mu_r$ can also be *smaller* than unity, which was not the case with the relative permittivity $\varepsilon_r$. Obviously, substances

**Table 9.1** Relative permeability $\mu_r$ and magnetic susceptibility $\chi_m$ of some substances.

| | $\mu_r$ | $\chi_m$ | | $\mu_r$ | $\chi_m$ |
|---|---|---|---|---|---|
| Bismuth | 0.99983 | $-1.7 \times 10^{-4}$ | Air | 1.00000036 | $3.6 \times 10^{-7}$ |
| Silver | 0.99998 | $-2.6 \times 10^{-5}$ | Aluminum | 1.000021 | $2.1 \times 10^{-5}$ |
| Copper | 0.99999 | $-0.94 \times 10^{-5}$ | Cobalt | 250 | nonlinear |
| Water | 0.99999 | $-0.88 \times 10^{-5}$ | Nickel | 600 | nonlinear |
| Vacuum | 1.00000 | 0.00 | Soft iron | 5000 | nonlinear |

with $\mu_r < 1$ have completely compensated atomic magnetic moments, so they behave as superconductive loops with no currents. A note on the behavior of a superconductive loop in a magnetic field is given in Section 11.4.2.

Substances with $\mu_r < 1$ are known as *diamagnetic*; those having $\mu_r > 1$, but almost equal to unity, are called *paramagnetic* substances. Substances with $\mu_r \gg 1$, which are *always* nonlinear, are referred to as *ferromagnetic*. Note that both diamagnetic and paramagnetic substances have $\mu_r$ differing from unity only slightly. In most practical applications their permeability is assumed to be equal to $\mu_0$. More about the magnetic properties of matter is presented in Section 9.7.

**9.4.1 Lines of the magnetic field intensity vector.** The lines of the magnetic field intensity vector **H** are defined, as usual, as geometrical lines having the property that vector **H** is tangential to them at all points. As a representation of the magnetic field, the lines of vector **H** are at a disadvantage when compared with the lines of vector **B**. Vector **B** is always divergenceless, so that the lines of **B** are continuous, while vector **H** is not divergenceless. Indeed, from Eqs. (8.6.4) and (9.4.4) it follows that $\operatorname{div}\mathbf{H} = -\operatorname{div}\mathbf{M}$. So, in the general case, vector **H** has sources and sinks; that is, there are points at which some of the lines of **H** originate and points at which these lines end. More about this question is presented in Section 9.6.

**9.4.2 The magnetic field in a toroidal coil with a two-layer core.** As an example of the application of the generalized Ampère's law, let us consider a toroidal coil having $N$ turns densely wound on the core shown in Fig. 9.11. Let the current intensity in the coil be $I$. By symmetry, the lines of the magnetic flux density vector **B**, and hence also those of the magnetic field intensity vector **H**, are circles centered at the axis of the toroid. But, since magnetic materials are present, the specialized form of Ampère's circuital law (Eq. 8.7.1) cannot be applied. The generalized form of the law given in Eq. (9.4.5) is valid, however, for static magnetic fields generally. Applying Eq. (9.4.5b) to a circular contour of radius $r$, lying inside the core, we have

$$\oint \mathbf{H}\cdot d\mathbf{l} = H\,2\pi r = NI, \qquad \text{from which} \qquad H = NI/2\pi r, \qquad a < r < c.$$

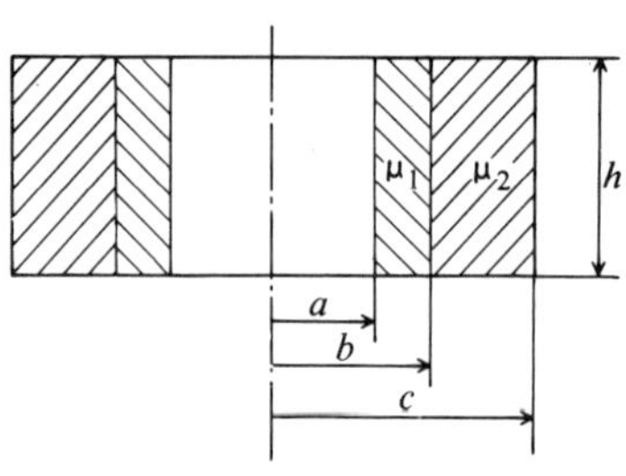


**Fig. 9.11** A toroidal coil with a two-layer core.

According to Eq. (9.4.11), the magnetic flux density has different values in the two layers:

$$B_1 = \mu_1 H = \mu_1(NI/2\pi r), \qquad B_2 = \mu_2 H = \mu_2(NI/2\pi r).$$

**9.4.3 Interpretation of the magnetic field intensity vector H.*** According to Eq. (9.4.4), the magnetic field intensity vector **H** is a difference of two physically different quantities. The magnetization vector **M** is defined as the density of the magnetic moments of elemental, atomic currents at a given point. It is therefore a measure of the local state of the magnetization of matter. The ratio $\mathbf{B}/\mu_0$ is proportional to the magnetic flux density vector at the point considered, due to *all* currents (macroscopic and microscopic). The two quantities might, but equally might not, be connected by some functional relationship. To understand this, let us imagine a magnetized body in which elemental loops are firmly "frozen". No external magnetic field then has a functional relationship with the magnetization. Although this example is an idealized one, it clearly illustrates the idea. Thus we can see that the magnetic field intensity is a hybrid vector function, consisting of two components having a completely different physical meaning. It is defined only as a convenient quantity for analyzing the magnetic field in the presence of magnetic materials.

As an example of the interpretation of the physical meaning of vectors **B** and **H**, let us examine the following simple problem. In vacuum $\mathbf{B} = \mu_0\mathbf{H}$, so the magnetic component of the Lorentz force of Eq. (8.3.6) on a charge moving in a vacuum can also be written as $\mathbf{F} = \mu_0 Q\mathbf{v} \times \mathbf{H}$. We could therefore assume that **H**, and not **B**, is the force vector of the magnetic field. According to the above reasoning this is not correct, but this is not the answer to the question. Only experiment can decide who is right and who wrong. Now, vectors **B** and **H** *inside* a permanent magnet are in essentially opposite directions (see Fig. 9.17). If we imagine a charge is moving inside a permanent magnet so that the interaction between the charge and atoms of the magnet can be neglected, then from our standpoint it should be deflected according to the law of force $\mathbf{F} = Q\mathbf{v} \times \mathbf{B}$. The other approach requires that it be deflected according to the force law $\mathbf{F} = \mu_0 Q\mathbf{v} \times \mathbf{H}$, that is, in the essentially opposite direction. Experiments favor the force law $\mathbf{F} = Q\mathbf{v} \times \mathbf{B}$, as we would expect.

*Problems referring to Section 9.4:* 9.T7 to 9.T9, 9.G3 to 9.G6, 9.N1 to 9.N3.

## 9.5 BOUNDARY CONDITIONS

The meaning of the term "boundary conditions" is the same here as in the electrostatic and steady current cases, i.e. they are relations between vectors **B** and **H** at two adjacent points on the two sides of an interface between two different magnetic substances. It is possible to deduce these relationships by starting from the

---

* This section may be omitted without loss of continuity.

physical model we adopted for analyzing magnetic materials, as well as in a purely formal manner. The latter method is considerably shorter, and will be used here.

The boundary condition for the normal components of the magnetic flux density vector can be obtained as follows. We apply the law of conservation of the magnetic flux to a very thin cylinder (or "pillbox") with its plane bases parallel to the boundary surface (see Fig. 9.12). Since the magnetic flux through the closed surface is zero, it follows that

$$\boxed{B_{1n} = B_{2n}} \quad \text{(boundary condition for the normal components of } \mathbf{B}). \tag{9.5.1}$$

In the case of linear media, $\mathbf{B}_1 = \mu_1 \mathbf{H}_1$, $\mathbf{B}_2 = \mu_2 \mathbf{H}_2$, and this boundary condition becomes

$$\mu_1 H_{1n} = \mu_2 H_{2n} \quad \text{(linear magnetic media)}. \tag{9.5.2}$$

The boundary condition for the tangential components of the magnetic field intensity vector can be derived from the generalized Ampère's circuital law (9.4.5a). Let us assume that there are no real surface currents on the interface. Then, by applying Eq. (9.4.5a) to the narrow rectangular contour shown in Fig. 9.13, we obtain

$$\boxed{H_{1t} = H_{2t}} \quad \text{(boundary condition for the tangential components of } \mathbf{H}\text{, no surface currents on the interface)}. \tag{9.5.3}$$

In the case of linear media, this can be written in the form

$$B_{1t}/\mu_1 = B_{2t}/\mu_2 \quad \text{(linear magnetic media)}. \tag{9.5.4}$$

As an example, let a cavity be made inside a magnetized substance, in the form of a very flat disk perpendicular to the lines of the magnetic flux density vector

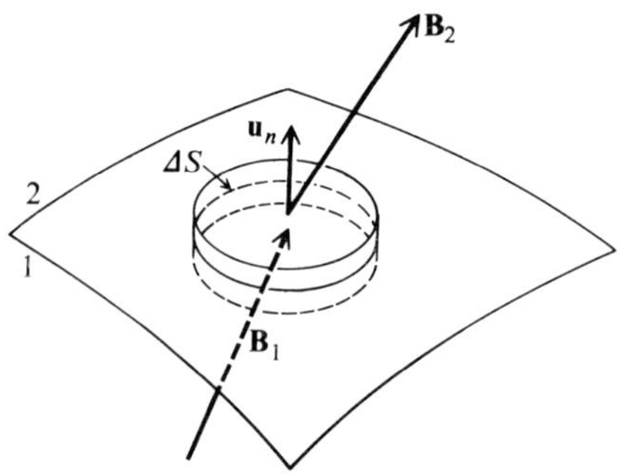


**Fig. 9.12** Gaussian surface for proving that $B_{1n} = B_{2n}$.

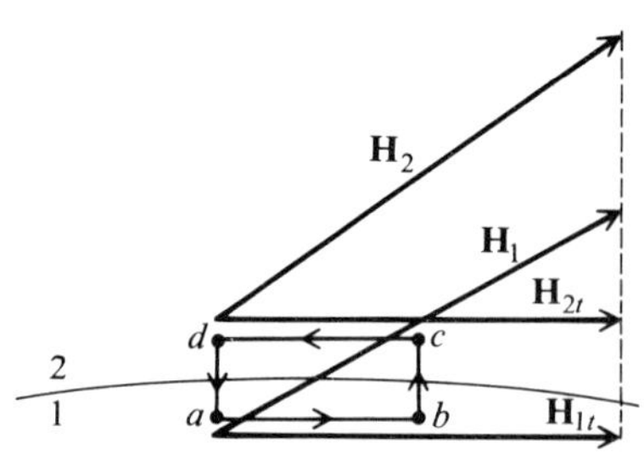


**Fig. 9.13** Elemental contour for proving that $H_{1t} = H_{2t}$ (no surface currents on the interface).

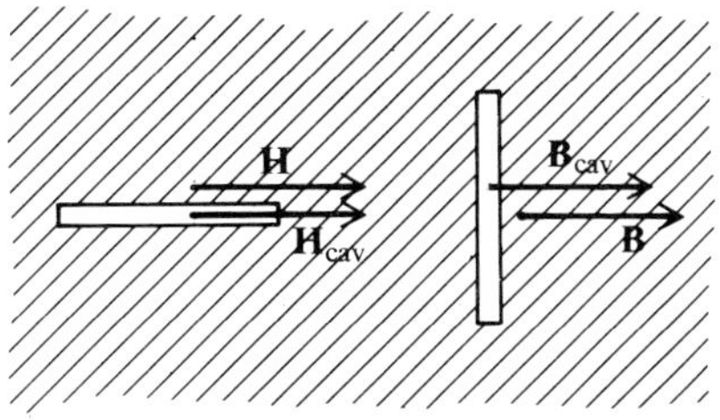


**Fig. 9.14** A needle-like and a disk-like cavity.

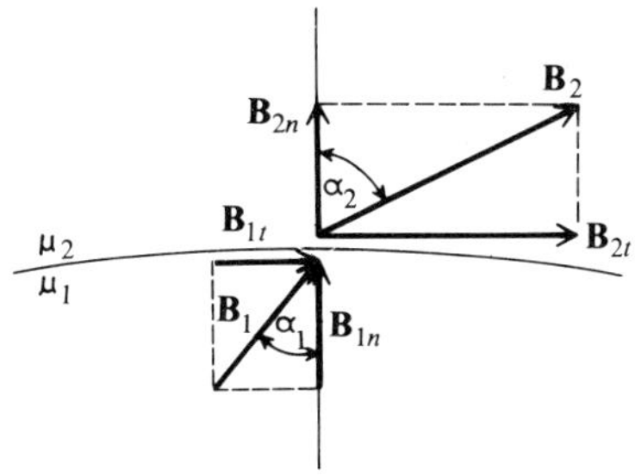


**Fig. 9.15** Construction for determining the law of refraction of magnetic field lines.

(see Fig. 9.14). According to Eq. (9.5.1), the flux density in the cavity will be equal to that in the surrounding magnetized substance. Similarly, according to Eq. (9.5.3), the magnetic field intensity inside the needle-shaped cavity shown in Fig. 9.14 is equal to that in the surrounding medium.

In the case of two linear magnetic media, a simple law of refraction of the lines of vector **B** can be derived. If we denote the angles between vectors $\mathbf{B}_1$ and $\mathbf{B}_2$ and the normal to the surface by $\alpha_1$ and $\alpha_2$, it follows from Fig. 9.15 that

$$\tan \alpha_1 = B_{1t}/B_{1n}, \qquad \tan \alpha_2 = B_{2t}/B_{2n}.$$

So, according to Eqs. (9.5.1) and (9.5.4), we obtain

$$\frac{\tan \alpha_1}{\tan \alpha_2} = \frac{B_{1t}/B_{1n}}{B_{2t}/B_{2n}} = \frac{B_{1t}}{B_{2t}} = \frac{\mu_1}{\mu_2}. \tag{9.5.5}$$

This is known as the *law of refraction* of the magnetic field lines.

Consider the important practical case when medium 1 is air ($\mu_1 \simeq \mu_0$) and medium 2 is some ferromagnetic substance ($\mu_2 \gg \mu_0$). As explained in connection with Table 9.1, ferromagnetic materials are never linear. So $\mu_2$ is not a constant, but it is always much larger than $\mu_0$. In that case

$$\tan \alpha_1/\tan \alpha_2 = \mu_1/\mu_2 \simeq 0 \qquad (1 - \text{air}, 2 - \text{ferromagnetic}). \tag{9.5.6}$$

The lines of the magnetic flux density vector in air are thus practically normal to surfaces of ferromagnetic materials. This important fact is illustrated in Fig. 9.16.

*Problems referring to Section 9.5:* 9.T10, 9.T11, 9.N4, 9.N5.

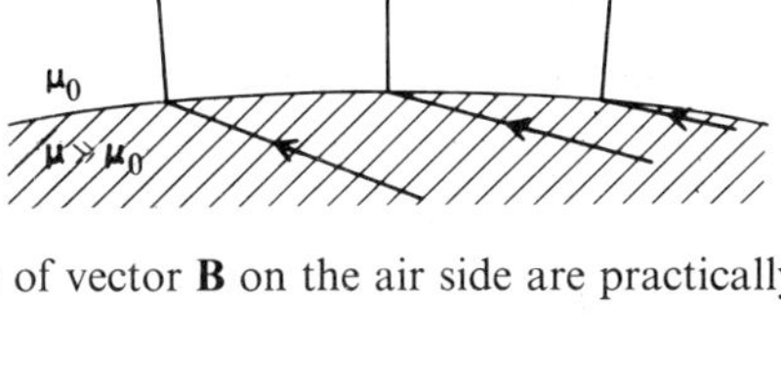


**Fig. 9.16** If $\mu \gg \mu_0$, lines of vector **B** on the air side are practically normal to the boundary surface.

## 9.6 THE MAGNETIC SCALAR POTENTIAL OF A MAGNETIZED SUBSTANCE. CONCEPT OF "MAGNETIC CHARGES"

Consider a permanently magnetized body, that is, a piece of matter inside which elemental currents are oriented even in the absence of an external magnetic field. (How such a permanently magnetized body is obtained is explained in Section 9.7.3.) The magnetic field exists both inside and outside the body. According to Eq. (9.4.2), the magnetic flux density vector **B** has to satisfy the equation

$$\oint_C \mathbf{B} \cdot d\mathbf{l} = \int_S \operatorname{curl} \mathbf{M} \cdot d\mathbf{S}. \tag{9.6.1}$$

Although there are no macroscopic currents, the line integral of vector **B** around a closed contour is not necessarily zero. To make it zero for any closed contour, we can introduce a restriction that the contour is not allowed to pierce the magnetized body. In that case the right-hand side in Eq. (9.6.1) is always zero, and the formal basis for defining a scalar potential function exists. However, the scalar potential thus defined cannot be used for determining the magnetic field inside magnetized bodies.

Consider the generalized form of Ampère's law (9.4.5) in that case. Since no macroscopic currents are assumed to exist, the right-hand side of Eq. (9.4.5) is zero. So

$$\oint_C \mathbf{H} \cdot d\mathbf{l} = 0 \qquad \text{(no macroscopic currents present)} \tag{9.6.2}$$

is valid even if the contour pierces the magnetized body. Thus Eq. (9.6.2) can represent a basis for defining a scalar potential function, from which the magnetic field intensity vector **H** can be derived at points both inside and outside a magnetized body. It is convenient to have as close a mathematical analogy as possible with the electrostatic scalar potential. Results found there then apply directly to the analogous magnetic case. So we can define the magnetic scalar potential $V_m$ by the equation

$$\mathbf{H} = -\operatorname{grad} V_m \qquad \text{(definition of the magnetic scalar potential).} \tag{9.6.3}$$

In a vacuum, the magnetic scalar potential becomes identical with that described in Section 8.10.1. Hence the latter can be considered as a special case of the scalar potential defined by Eq. (9.6.3) when no magnetic materials are present.

The magnetic scalar potential of a magnetized body with a known magnetization vector **M** at all points can be determined by analogy with the electric scalar potential of a polarized dielectric, where the following relations were derived:

Potential of a single dipole (at large distances): $V_{\text{dipole}} = \mathbf{p} \cdot \mathbf{u}/4\pi\varepsilon_0 r^2$.
Potential due to a polarized dielectric:

$$V_p = \frac{1}{4\pi\varepsilon_0} \int_v \frac{-\operatorname{div} \mathbf{P}\, dv}{r} + \frac{1}{4\pi\varepsilon_0} \int_S \frac{\mathbf{P} \cdot \mathbf{u}_n}{r}\, dS.$$

Electric field intensity due to a polarized dielectric: $\mathbf{E} = -\text{grad } V_p$.

In Section 8.10.3 the magnetic scalar potential of a single elemental loop was shown to be of exactly the same form (except for the factor $1/\varepsilon_0$) as the electrostatic potential of an electric dipole given above:

$$V_{m\,\text{loop}} = \mathbf{m} \cdot \mathbf{u}/4\pi r^2.$$

So the potential due to a magnetized body can by simple analogy be written down as:

$$V_m = \frac{1}{4\pi} \int_v \frac{-\text{div } \mathbf{M}}{r} dv + \frac{1}{4\pi} \int_S \frac{\mathbf{M} \cdot \mathbf{u}_n}{r} dS. \tag{9.6.4}$$

In the electrostatic case, $-\text{div } \mathbf{P}$ was shown to be the volume density, and $\mathbf{P} \cdot \mathbf{u}_n$ the surface density, of the polarization charges. In the magnetic case considered, no real magnetic dipoles exist. Consequently, no magnetic charges can be accumulated either inside or on the surface of a magnetized body. But a purely formal *definition* of magnetic charges is possible by analogy; thus, we define the volume density $\rho_m$ of the magnetic charges as

$$\rho_m = -\text{div } \mathbf{M}, \tag{9.6.5}$$

and density $\rho_{ms}$ of the surface magnetic charges as

$$\rho_{ms} = \mathbf{M} \cdot \mathbf{u}_n. \tag{9.6.6}$$

With Eqs. (9.6.5) and (9.6.6), the magnetic scalar potential, Eq. (9.6.4), due to a magnetized body becomes

$$V_m = \frac{1}{4\pi} \int \frac{\rho_m}{r} dv + \frac{1}{4\pi} \int \frac{\rho_{ms}}{r} dS. \tag{9.6.7}$$

This is identical (except that the factor $1/\varepsilon_0$ is missing) with the electrostatic potential of the volume and surface distribution of the electric charges. Now, the magnetic field intensity $\mathbf{H}$ is obtained from $V_m$ in exactly the same way as the electric field intensity from $V$. Hence it follows that the magnetic field intensity produced by magnetized matter has sources and sinks in positive and negative magnetic charges, as defined by Eqs. (9.6.5) and (9.6.6).

The magnetic flux density vector $\mathbf{B}$ is always divergenceless. If we wish to determine the magnetic flux density produced by a magnetized substance, we have to determine the macroscopic resultants of the microscopic currents, according to Eqs. (9.3.3) and (9.3.6). Thus, inside a permanently magnetized body, the magnetic field intensity and the magnetic flux density vectors have different directions. Outside, in a vacuum, they are collinear, since $\mathbf{B} = \mu_0 \mathbf{H}$. To illustrate this conclusion, Fig. 9.17 shows the lines of vectors $\mathbf{M}$, $\mathbf{H}$, and $\mathbf{B}/\mu_0$ in the case of a permanently and uniformly magnetized circular cylinder. Inside the cylinder the vectors $\mathbf{B}/\mu_0$ and $\mathbf{H}$ are in essentially opposite directions. Outside $\mathbf{H} = \mathbf{B}/\mu_0$, and the two sets of lines are identical.

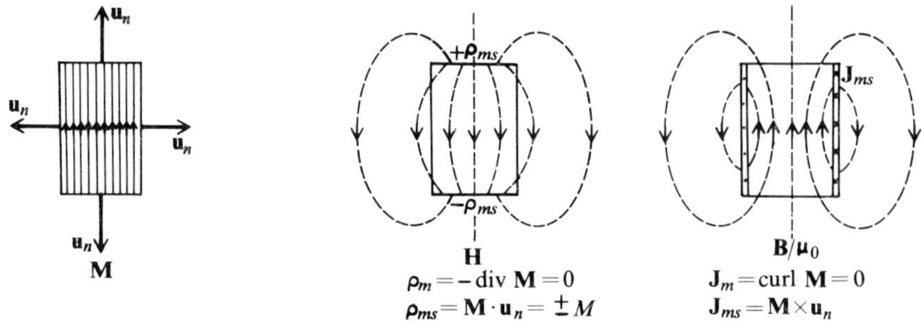


**Fig. 9.17** Lines of vectors **M**, **H**, and $\mathbf{B}/\mu_o$ of a short, uniformly magnetized cylinder.

In the general case of the magnetic field produced by both macroscopic currents and magnetized matter the following general procedure for determining **B** and **H** can be adopted. For determining **B**, it is necessary to determine the macroscopic resultants of the microscopic currents. The magnetic flux density vector at all points is then determined as being created by real currents and these macroscopic resultants, all currents being imagined to exist in a vacuum.

In the case of the magnetic field intensity vector **H**, it is necessary to determine the distribution of the magnetic charges on all magnetized bodies. The magnetic field intensity vector **H** is then determined as partly due to real currents, and partly to these magnetic charges. Both currents and charges are assumed to be situated in a vacuum.

To determine the magnetic scalar potential $V_m$ given by Eq. (9.6.4) we need to know in advance the magnetization vector at all points. This is not usually known, however, since it is a function of the magnetic field, which is precisely what we are trying to determine. When all the media can be considered as linear, and when there are no macroscopic currents in the region, the magnetic scalar potential function $V_m$ satisfies Laplace's equation, and the solving procedure can be borrowed from the electrostatic case.

To prove this, suppose that at all points $\mathbf{B} = \mu\mathbf{H}$ ($\mu$ being a constant), and that there are no macroscopic currents in the region. In that case

$$\text{div } \mathbf{B} = \text{div } \mu\mathbf{H} = \mu \text{ div } \mathbf{H} = 0,$$

which with Eq. (9.6.3) yields

$$\text{div grad } V_m = 0 \qquad \text{(linear, homogeneous media).} \qquad (9.6.8)$$

So solving for $V_m$ reduces to finding a solution of Laplace's equation (9.6.8) which satisfies certain prescribed boundary conditions (for details see Section 10.1.3). According to the uniqueness theorem, which was proved in Section 1.13.2, only one such solution exists and can hence be found in any way. For example, we can obtain the solution simply by analogy from the corresponding electrostatic case, if such a problem was analyzed and solved there.

*Problems referring to Section 9.6:* 9.T12 to 9.T14, 9.C7, 9.G8.

## 9.7 SOME PHYSICAL PROPERTIES OF MAGNETIC MATERIALS

**9.7.1 Introduction.** In Section 9.4 we noted that magnetic materials are divided into three main groups. Materials of the first group are characterized by small *negative* values of the magnetic susceptibility, and are called *diamagnetic* substances. Materials of the second group are characterized by small *positive* values of the magnetic susceptibility, and are known as *paramagnetic* materials. Both diamagnetic and paramagnetic materials are linear, but their permeability is practically equal to that of a vacuum, as we can see from Table 9.1. Hence, for almost all applications, diamagnetic and paramagnetic materials can be considered as having no influence on the magnetic field. They are very seldom used in practice because of their magnetic properties, and therefore no further details of these magnetic properties will be presented here.

Magnetic materials classified in the third group are known as *ferromagnetic* materials. They are iron, cobalt, nickel, and some of their chemical compounds and alloys. Their main characteristic is that they have a very large relative permeability. However, they are all nonlinear, and their magnetic properties are functions of previous magnetization, i.e. their "magnetic history". The application of ferromagnetic materials in electrical engineering is very wide, and a basic knowledge of their most important properties at least is essential. Analysis of these properties is far from simple, and in fact, cannot be performed with the tools of the classical field theory alone. Only a phenomenological description of these properties therefore will be presented here.

**9.7.2 The magnetization curve of ferromagnetic materials.** As mentioned earlier, the magnetic properties of a ferromagnetic material are functions of its magnetic history, that is, of how the specimen attained its magnetization. This is referred to as the *hysteresis* behavior. It will be seen that an infinite number of different histories, resulting in the same magnetization, can be imagined. Consequently, to all these different histories a different subsequent magnetic behavior of the specimen corresponds. Hence, any property of a ferromagnetic material has a meaning only if it is considered together with its complete magnetic history.

It is impossible, however, to characterize a ferromagnetic material by an infinite set of data, corresponding to an infinite number of different histories, so that only a limited idea of the general behavior of such substances can be obtained. A precise mathematical description of the type $\mathbf{B} = \mu\mathbf{H}$, valid for linear magnetic materials, is not possible. This is a very important conclusion, to be kept in mind whenever we are dealing with some new problem involving ferromagnetic materials.

The other important problem in determining the properties of a ferromagnetic material is the following. Because of nonlinearity and hysteresis behavior, no unique results can be obtained by using specimens of the same material but of different shapes. We can understand this by recalling that the total flux density at a given point is due to external macroscopic currents, *and* to microscopic currents throughout the specimen. In the case of ferromagnetic materials the latter usually produce the larger part of the total field. To determine the secondary field for a specimen of an arbitrary shape is not possible. It is a function not only of the external field, but also of the properties of the specimen, which is precisely what has to be determined.

Thus, in determining experimentally the magnetic properties of a ferromagnetic material we are, firstly, compelled to use specimens of some convenient shape, and, secondly, to describe these properties in terms of only a few different magnetization processes.

The most convenient sample for analyzing the magnetic properties of a ferromagnetic material is a *thin* toroidal core, densely and evenly wound with a large number ($N$) of turns (see Fig. 9.18). The magnetization vector in this case is practically uniform over any cross-section of the core. Since the magnetization vector is of a constant magnitude inside the core and is tangential to the core surface, both the volume and the surface density of the magnetic charges, given by Eqs. (9.6.5) and (9.6.6), are zero. So, in this case, the magnetic field intensity vector **H** is produced by a macroscopic current $I$ in the coil only. Using the generalized form of Ampère's circuital law (9.4.5), we find that the magnitude of the magnetic field intensity vector in the core is

$$H = NI/2\pi r. \tag{9.7.1}$$

We shall show in Section 11.4.1 that the total magnetic flux $\Phi$ through a cross-section of the core, rather than the magnetic flux density vector **B**, can be measured. However $\Phi$ and $B$ are connected by a simple relation. If we denote by $S$ the area of the cross-section of the core, then

$$B = \Phi/S. \tag{9.7.2}$$

So the dependence of the magnetic flux density on the magnetic field intensity can be determined point by point and plotted in a convenient diagram. But it should be noted that such a diagram refers to a thin toroidal specimen, and to a given process of magnetization which must be specified.

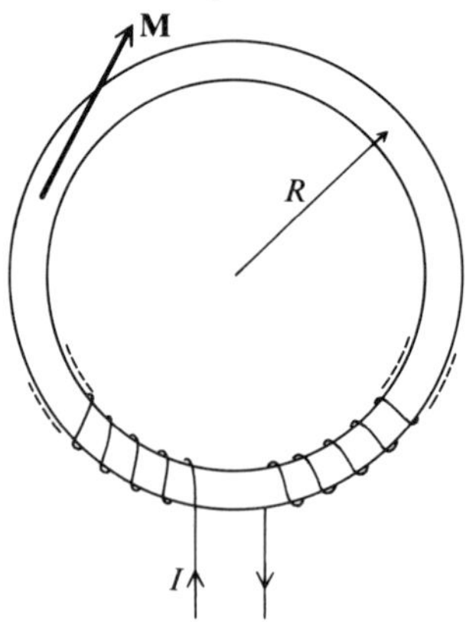


**Fig. 9.18** A thin toroidal coil.

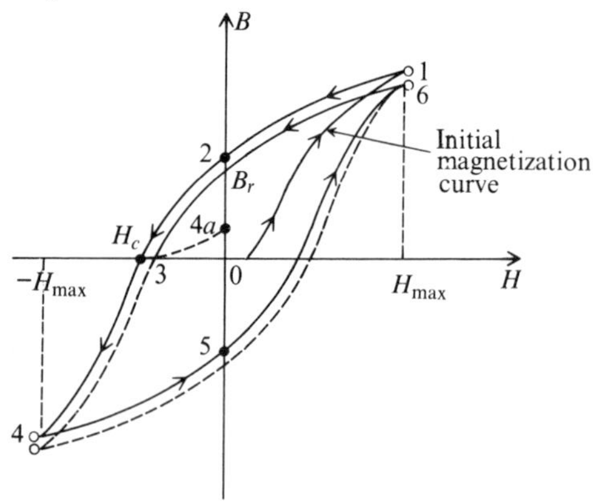


**Fig. 9.19** A typical magnetization process of a ferromagnetic material.

Figure 9.19 shows a typical curve $B(H)$ obtained by using a toroidal specimen. Let us assume that at the beginning the specimen is not magnetized. If the magnetic field intensity is increased (by increasing the current) from zero to a value $H_{max}$, the point in the $B$-$H$ plane will trace the curve 0–1 shown. This curve is known as the *initial magnetization curve*. If $H$ is then decreased, by decreasing the current in the coil, the point will not trace the curve 1–0, but rather the curve 1–2. So, although at point 2 the current in the coil, and hence also the magnetic field intensity in the core, are zero, the magnetic flux density is not zero. This particular value, $B_r$, of $B$ is referred to as the *residual or remanent flux density*. Note that $B_r$ depends upon the value of $H_{max}$ used. The physical meaning of the residual flux density is as follows. Due to certain effects similar to friction, the elemental currents will still be oriented, although there is no external field, and will produce a certain magnetic field inside the core. The remanent flux density explains the existence of permanent magnets.

If the current is reversed, $H$ will be in the opposite direction with respect to its previous direction. This is represented in the figure by the negative values of $H$. The point will trace the curve 2–3. For a certain value $H_c$ of $H$, known as the coercive force or *coercive field intensity*, there will be no flux through the core. Physically, this means that the magnetic flux density due to the macroscopic current in the coil is exactly equal, but opposite, to the magnetic flux density of the elemental loops, which are still oriented so as to create $B$ in the positive direction. To prove this we have simply to turn off the current. The point in the $B$-$H$ plane then moves to a point 4*a* on the $B$ axis. Hence there is still some flux density vector in the assumed positive direction created by the oriented elemental loops. Magnetic materials with large $H_c$ are referred to as magnetically hard, those with small $H_c$ as soft.

If we increase the current still more in the negative direction, the point traces the curve 3–4. If we next decrease the current intensity and reverse its direction again, the point will trace the curve 4–5–6. But the point 6, corresponding to $+H_m$, will *not* coincide with the point 1 on the initial magnetization curve. Only if we repeat such a cycle ten or more times will the curve traced by the point in the $B$-$H$ plane close. The closed curve thus obtained is known as the *hysteresis loop*, a typical example of which is shown in Fig. 9.20. In most of the practical applications of ferromagnetic materials we deal with sinusoidally time-varying macroscopic currents, so that the hysteresis loop is perhaps their most important characteristic.

Hysteresis loops corresponding to different maximum values $H_{max}$ of the magnetic field intensity are of a similar, but not of quite the same, shape. Figure 9.21 shows several hysteresis loops, the largest of which is the largest possible. For still larger values of $H_{max}$, the point in the $B$-$H$ plane traces the same largest loop, and then the straight lines $\mathbf{B} = \mu_0(\mathbf{H} + \mathbf{M}_{max})$, as indicated in Fig. 9.21 by dash-dot lines. In this case it is said that *saturation* is attained. Physically, saturation corresponds to the complete alignment of all the available elemental current loops.

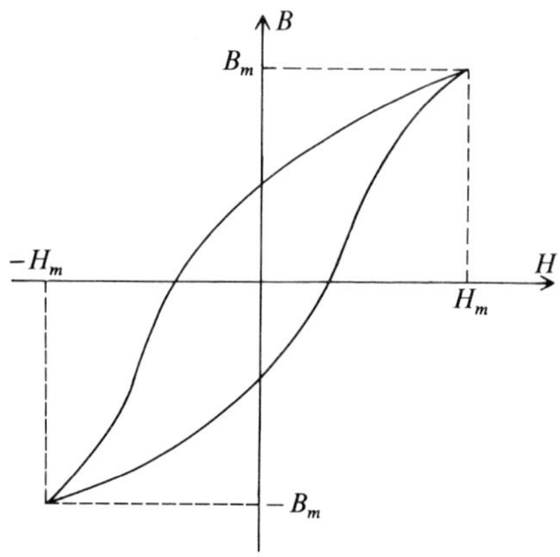


**Fig. 9.20** A hysteresis loop.

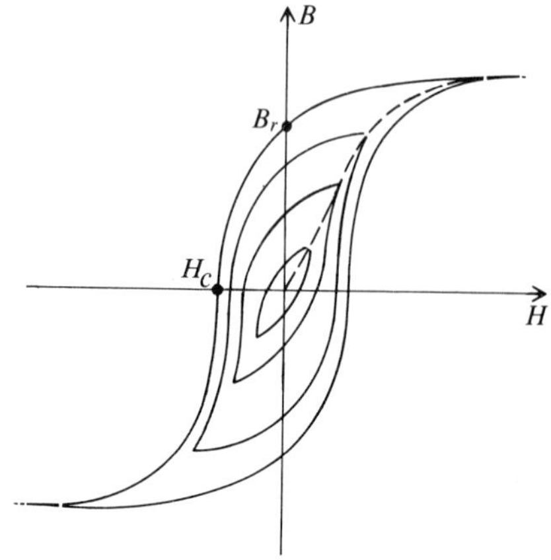


**Fig. 9.21** Normal magnetization curve connects the tips of hysteresis loops corresponding to different $H_{max}$.

The values of the remanent flux density $B_r$ and the coercive force $H_c$ are usually given for that saturation hysteresis loop. The dashed curve shown in Fig. 9.21, which connects the tips of the loops, is known as the *normal magnetization curve*.

We can now understand a frequently used procedure for demagnetizing magnetized pieces of ferromagnetic materials. With the magnetic field varying sinusoidally with time, the body is first alternately magnetized to saturation. The intensity of the external field is then slowly decreased to zero. The hysteresis loops thus become smaller and smaller, and the body is demagnetized.

It is evident that no unique definition of some kind of permeability, defined as the ratio $B/H$, is possible. In fact, the ratio $B/H$ becomes zero, negative, and even infinite, for different points along a hysteresis loop. Nevertheless, several different kinds of permeability, useful for diverse practical applications, are defined below.

First of all, what is called *normal permeability* is the ratio $B/H$ along the normal magnetization curve. Since this is a curve and not a straight line, the normal permeability is not constant.

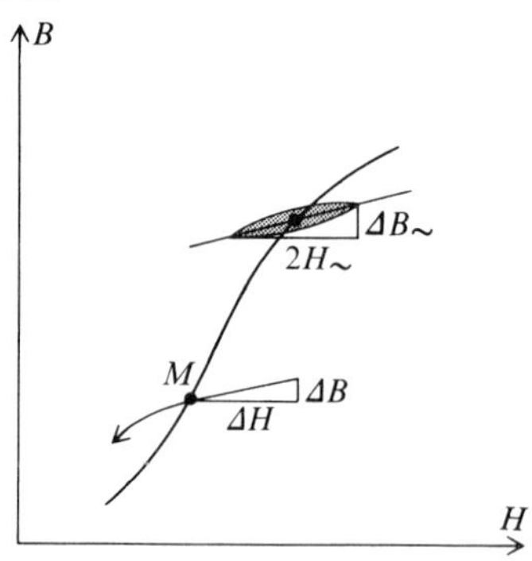


**Fig. 9.22** Illustration of incremental and reversible permeabilities.

Next, the *differential permeability* can be defined as $dB/dH$ along the normal magnetization curve. Differential permeability at the origin is referred to as the *initial permeability*.

There are quite a few practical cases when a small alternating current is superimposed on a large, steady magnetizing current. In these cases, the point in the $B$-$H$ plane describes a small, very flat hysteresis loop, as shown in Fig. 9.22. The ratio $\Delta B_{\sim}/2H_{\sim}$ is referred to as the *incremental permeability*.

Finally, the *reversible permeability* is defined as the slope of the curve obtained when the magnetic field intensity at any point is *reversed* by a very small amount, as indicated in Fig. 9.22 at point $M$. The reversible permeability has a somewhat smaller value than the incremental permeability at the same point.

**9.7.3 A qualitative explanation of the hysteresis properties of ferromagnetic materials.** The hysteresis properties of a ferromagnetic material can be explained by a very simple model. According to this model, a ferromagnetic substance is comprised of small, permanently magnetized *domains*. The domains are permanently magnetized because the magnetic moments of the atoms inside them are aligned. Hence the magnetization of a ferromagnetic substance does not imply the orientation of all the elemental currents separately. Rather, these domains become oriented. They are usually very small, but nevertheless contain a large number of atoms (of the order of $10^{15}$).

Any rotation of domains which results in their alignment with the external magnetizing field is necessarily accompanied by some friction, and this has two important consequences. Firstly, magnetization will *always* lag behind the field producing it. This accounts for the hysteresis properties of ferromagnetic materials, and explains the existence of permanent magnets. Secondly, since any orientation of the domains is accompanied by friction, magnetization of all ferromagnetic materials is necessarily accompanied by the transformation of magnetic energy into heat. These losses are referred to as the *hysteresis losses*, and will be considered in more detail in Section 12.1.4.

The domain model can also be used to help us understand the behavior of another class of magnetic materials, the *ferrimagnetic* materials, in which the domains are of unequal magnitudes. Under the action of the external magnetic field and as the result of mutual interactions, domains with a larger magnetic moment align with the field, and those with a smaller magnetic moment align opposite to the field. The result is a fairly large increase of magnetic flux density, although not so large as in ferromagnetic materials.

An important subclass of ferrimagnetic materials are the *ferrites*, which are characterized by very low conductivity, comparable to that of poor dielectrics. In Chapter 14 we shall show that low conductivity is a necessary requirement for all magnetic materials used at high frequencies. Ferrites are therefore used at high frequencies when a large magnetic flux density in a certain region is desired. Examples of their application are ferrite cores for the antennas built inside radio receivers, and cores for high-frequency transformers.

**9.7.4 Dependence of ferromagnetic properties on temperature.** It was found experimentally that all ferromagnetic substances lose their ferromagnetic properties above a certain temperature. This critical temperature is known as the *Curie temperature*. Above the Curie temperature, ferromagnetic materials behave as ordinary linear paramagnetic materials. Thus, if a permanently magnetized body is heated above the Curie temperature, it loses its magnetization, and, when cooled, becomes completely demagnetized.

The Curie temperatures for most materials exhibiting ferromagnetic properties at room temperature range from a few hundreds to a thousand degrees Celsius. It is worth mentioning that some materials, paramagnetic at room temperature, become ferromagnetic if sufficiently cooled. Obviously, the Curie temperature of these materials is lower than the room temperature.

*Problems referring to Section 9.7:* 9.T15 to 9.T20.

## 9.8 THE MAGNETIC CIRCUIT

The most frequent applications of ferromagnetic materials involve ferromagnetic cores of widely differing shapes. The cores may or may not have air gaps, and they are magnetized by the current in a coil wound on one section of the core. An exact analysis of these important practical cases is impossible for several reasons, the two most outstanding being the usually very complicated boundary of the ferromagnetic body, and insufficient analytical knowledge of the magnetic properties of ferromagnetic materials. Fortunately, in many cases these important problems can be solved approximately, but with sufficient accuracy. The approximate analysis resembles that of steady current circuits.

**9.8.1 The quasi-filamentary linear magnetic circuit.** Consider a coil with $N$ turns and a current of intensity $I$, situated in a linear (not necessarily homogeneous) magnetic material. Let us confine our attention to a tube of small magnetic flux $\Delta\Phi$ (Fig. 9.23), when the following relations hold:

$$\oint_C \mathbf{H}\cdot d\mathbf{l} = NI \qquad \text{(generalized form of Ampère's circuital law)}, \tag{9.8.1}$$

$$\Delta\Phi = B\Delta S \qquad (\Delta S\text{, and hence } B\text{, variable along the tube}), \tag{9.8.2}$$

$$\mathbf{B} = \mu\mathbf{H}, \tag{9.8.3}$$

$$\oint_S \mathbf{B}\cdot d\mathbf{S} = 0 \qquad \text{(any closed surface)}. \tag{9.8.4}$$

A completely analogous set of equations is valid for steady current in a filamentary current tube:

$$\oint_C \mathbf{E}\cdot d\mathbf{l} = \mathscr{E} \qquad (\mathscr{E} - \text{electromotive force acting in the circuit}), \tag{9.8.5}$$

$$\Delta I = J\Delta S \qquad (\Delta S\text{, and hence } J\text{, variable along the tube}), \tag{9.8.6}$$

$$\mathbf{J} = \sigma \mathbf{E}, \tag{9.8.7}$$

$$\oint_S \mathbf{J} \cdot d\mathbf{S} = 0 \qquad \text{(any closed surface).} \tag{9.8.8}$$

Since the two sets of equations are completely analogous, the solutions must be of the same form. In the steady current case, the following relations were derived from Eqs. (9.8.5) to (9.8.8):

$$\Sigma \mathscr{E} - \Sigma RI = 0 \qquad \text{(any closed circuit),} \tag{9.8.9}$$

$$R = \int_C \frac{dl}{\sigma \Delta S} \qquad \text{(any thin current tube),} \tag{9.8.10}$$

$$\Sigma I = 0 \qquad \text{(any node).} \tag{9.8.11}$$

We can readily see that, in the magnetic circuit, the product $NI$ plays the role of the electromotive force and is therefore referred to as the *magnetomotive* force. Permeability $\mu$ corresponds to conductivity $\sigma$. The magnetic flux corresponds to the electric current intensity $I$. Thus we have

$$\Sigma NI - \Sigma R_m \Phi = 0 \qquad \text{(any closed magnetic circuit),} \tag{9.8.12}$$

$$R_m = \int_C \frac{dl}{\mu \Delta S} \qquad \text{or} \qquad R_m = \frac{l}{\mu S} \qquad \text{for } S \text{ constant,} \tag{9.8.13}$$

$$\Sigma \Phi = 0 \qquad \text{(any node of the magnetic circuit).} \tag{9.8.14}$$

$R_m$ is referred to as the *magnetic resistance* or the *reluctance* of the magnetic circuit considered. Equations (9.8.12) and (9.8.14) are known as Kirchhoff's laws for magnetic circuits.

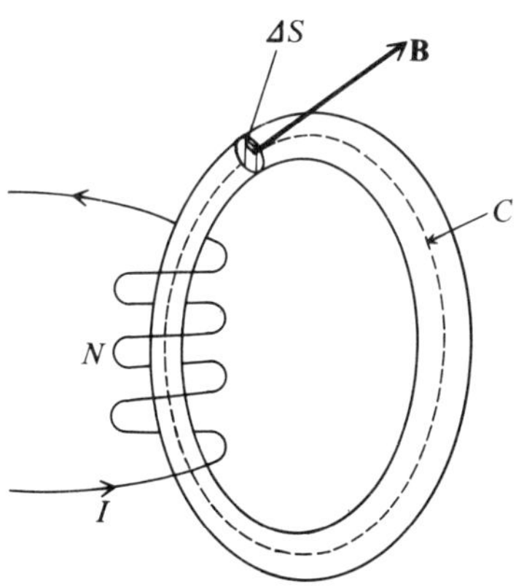


**Fig. 9.23** A tube of small magnetic flux.

It follows that the analysis of quasi-filamentary linear magnetic circuits parallels the analysis of quasi-filamentary steady current circuits. As an example, let us consider a thin toroidal coil of $N$ turns, length $l$, and area of cross-section $S$. Suppose that the permeability of the core is $\mu$, and that a current of intensity $I$

flows through the coil. Using Eqs. (9.8.12) and (9.8.13), we have

$$\Phi = \frac{NI}{R_m} = \frac{NI}{l/\mu S} = \mu \frac{NI}{l} S.$$

This is of course the same result as we obtained when determining $B$ in the coil and then $\Phi$ according to $\Phi = BS$.

So the analysis of quasi-filamentary magnetic circuits is very simple. Unfortunately, magnetic circuits in practice are never filamentary but rather the opposite. The foregoing theory is nevertheless important, since it serves as a basis for the approximate analysis of real, nonfilamentary magnetic circuits.

**9.8.2 The approximate analysis of nonfilamentary linear magnetic circuits.** Consider now a thick U-shaped core of permeability $\mu_1$ much larger than $\mu_0$, closed by a thick bar of permeability $\mu_2$, also much larger than $\mu_0$, as shown in Fig. 9.24. If $\mu \gg \mu_0$ the substances cannot be linear, but we may assume that they can be treated as approximately linear. Let us suppose that $N$ turns, carrying a current of intensity $I$, are wound onto the core. The exact determination of the magnetic field in such a case is impossible, in spite of the assumed linearity. However, two quite general characteristics of the magnetic field inside the core can easily be demonstrated.

Firstly, since $\mu_1 \gg \mu_0$ and $\mu_2 \gg \mu_0$, from Eq. (9.5.4) we can conclude that the tangential component of the magnetic flux density is much larger in the core than outside it. The normal components of **B** being equal, the magnetic flux density inside the core is generally much larger than at any point outside. Hence the magnetic flux can be considered to be approximately restricted to the core, *although this can never exactly be the case*. This represents the first assumption in analyzing the real, nonfilamentary magnetic circuits.

Secondly, we assume that Eqs. (9.8.12) to (9.8.14) are reasonably accurate in this case as well, if an average length $(l_1 + l_2)$ is used instead of the length of the filamentary circuit. The accuracy of this assumption cannot in general be analyzed, but that the error is small we can see from the following example, for which both exact and approximate calculations are possible.

Let us look at the toroidal coil the cross-section of which is shown in Fig. 9.25.

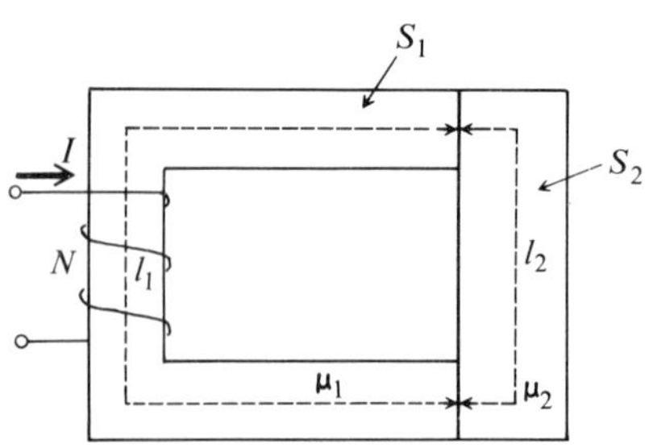


**Fig. 9.24** A real magnetic circuit.

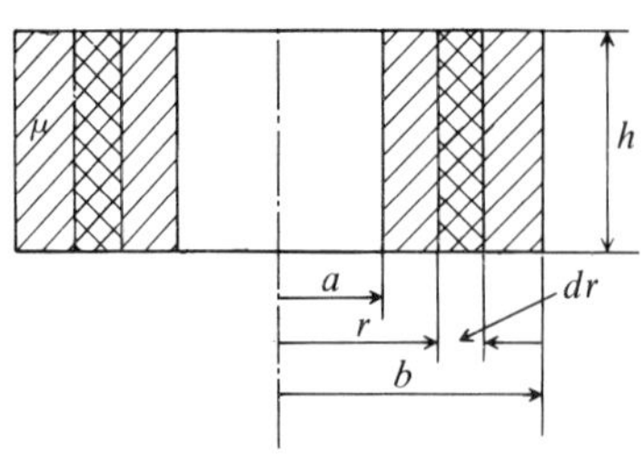


**Fig. 9.25** A toroidal coil.

Suppose that the $N$ turns of the coil carry a current of intensity $I$. Then, according to Ampère's circuital law given in Eq. (9.4.5b), we have

$$H = NI/2\pi r, \qquad B = \mu(NI/2\pi r).$$

The exact value of the magnetic flux through the cross-section of the toroid is hence

$$\Phi_{\text{exact}} = \frac{\mu NIh}{2\pi}\int_a^b \frac{dr}{r} = \frac{\mu NIh}{2\pi}\ln\frac{b}{a}.$$

According to Eqs. (9.8.12) and (9.8.13) and the above approximation,

$$\Phi_{\text{appr}} = \frac{NI}{R_m} = \frac{NI}{\dfrac{\pi(a+b)}{\mu(b-a)h}} = \frac{\mu NIh}{2\pi}\frac{2(b-a)}{b+a}.$$

The relative error is thus

$$\frac{\Phi_{\text{appr}} - \Phi_{\text{exact}}}{\Phi_{\text{exact}}} = \frac{2(b-a)}{(b+a)\ln b/a} - 1 = \frac{2(b/a-1)}{(b/a+1)\ln b/a} - 1,$$

which is very small even for very thick toroids. For example, for $b/a = e = 2.718\ldots$ it is less than $-8$ per cent.

So the magnetic flux in the magnetic circuit represented in Fig. 9.24 can be determined approximately according to the formula

$$\Phi = \frac{NI}{l_1/\mu_1 S_1 + l_2/\mu_2 S_2}.$$

For more complicated magnetic circuits, such as that shown in Fig. 9.26, determination of the magnetic flux through branches of the circuit parallels the process of determining the currents in the analogous steady-current case. The

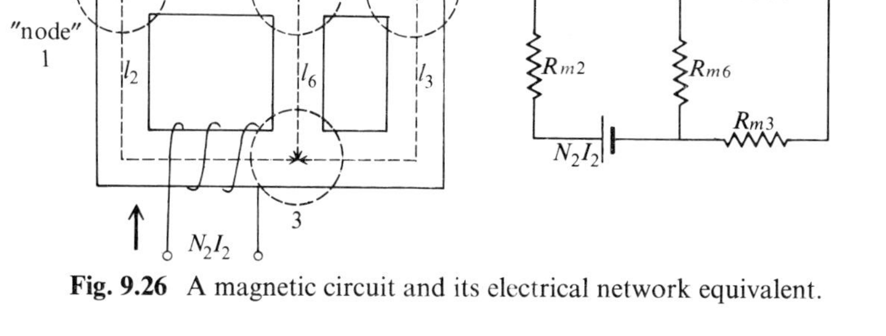


**Fig. 9.26** A magnetic circuit and its electrical network equivalent.

scheme of the magnetic circuit in terms of magnetomotive forces and reluctances is also shown in Fig. 9.26. The narrow air gap is assumed not to introduce any considerable flux leakage. Its reluctance is calculated in exactly the same way as the reluctance of the other parts of the circuit, that is, $R_{m0} = l_0/\mu_0 S_0$. Since $\mu_0$ is always much smaller than $\mu$, even a very narrow air gap has a considerable influence on the distribution of the flux in the circuit, since its reluctance is very high. Finally, the nodes in a magnetic circuit are in fact the surfaces enclosing the whole region where three or more branches of the circuit meet together. The four nodes of the circuit shown are encircled by a dashed line.

**9.8.3 Approximate analysis of nonlinear magnetic circuits,** Real magnetic circuits are always made of ferromagnetic materials, so that they are always nonlinear. Thus, instead of the simple relation $\mathbf{B} = \mu\mathbf{H}$, we have to deal with an experimentally determined magnetization curve. However, as we noted in Section 9.7.1, there is no unique way—indeed no precise way—of describing the properties of a ferromagnetic material. We have instead to adopt the most suitable magnetization curve available. For example, if the material was not magnetized previously, the initial magnetization curve seems to be the best choice. If we are dealing with periodic currents, the normal magnetization curve is obviously a better choice. But whichever is the problem considered, we can use only the following three equations:

$$\oint_C \mathbf{H}\cdot d\mathbf{l} = NI, \qquad \text{or approximately} \qquad \Sigma H_k l_k = NI, \tag{9.8.15}$$

$$\oint_S \mathbf{B}\cdot d\mathbf{S} = 0, \qquad \text{or approximately} \qquad \Sigma \Phi_k = 0, \tag{9.8.16}$$

$$B = B(H) \qquad \text{(suitable experimental magnetization curve).} \tag{9.8.17}$$

Except in the most simple cases, the approximate analysis of nonlinear magnetic circuits is very complicated and will not be described here at any greater length. A few examples of the method are given in Problems 9.G13 to 9.G19.

**9.8.4 The magnetic circuit of a permanent magnet.** Consider a thin ferromagnetic toroid, and suppose that inside it there exists a remanent flux density $B_r$ as the result of a previous magnetization. Let us cut out of the toroid a very thin slice, as shown in Fig. 9.27. Before the slice was cut out, the magnetic flux density existed inside the toroid only, and it was created by the equivalent current sheet on the toroid surface. After the slice has been cut out, the magnetic field also exists outside the toroid. Since the gap is small, the field is practically restricted to the air gap. Inside the toroid, a certain change in magnetization also occurred. It is impossible to analyze the resulting field exactly, but the following approximate analysis will suffice for all practical purposes.

After the slice has been cut out, the magnetic charges appear on the surfaces of the gap, and produce a magnetic field intensity both in the gap and in the

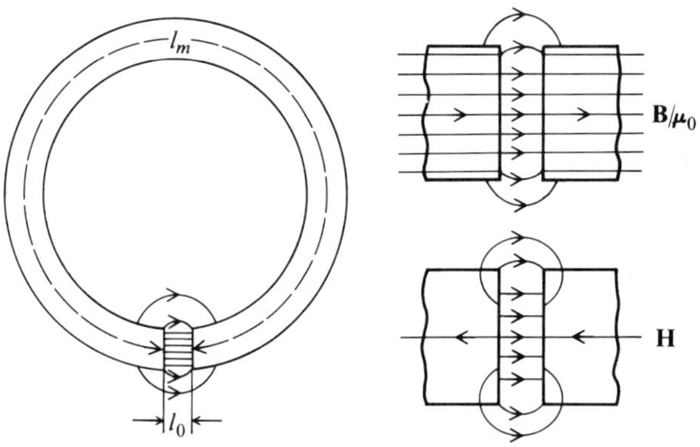


**Fig. 9.27** A thin toroidal permanent magnet and lines of vectors $\mathbf{B}/\mu_o$ and $\mathbf{H}$ in the region of the air gap.

air around the toroid, and in the toroid itself, as shown on the enlarged details of the toroid in Fig. 9.27. We can see that, in the toroid, $B$ and $H$ are in essentially opposite directions.

Since there are no macroscopic currents, Ampère's circuital law in Eq. (9.4.5b) yields

$$\int_{\text{toroid}} \mathbf{H} \cdot d\mathbf{l} + \int_{\text{gap}} \mathbf{H} \cdot d\mathbf{l} = 0. \tag{9.8.18}$$

If we denote the average lengths of the toroid and of the gap by $l_m$ and $l_0$, and the corresponding magnetic field intensities by $\mathbf{H}_m$ and $\mathbf{H}_0$, then Eq. (9.8.18) can be written as

$$\int_{l_m} \mathbf{H}_m \cdot d\mathbf{l}_m = -H_0 l_0. \tag{9.8.19}$$

It should be noted that $\mathbf{H}_m$ is *not* of a constant magnitude inside the magnet, since it is produced only by the magnetic charges located near the gap. Thus Eq. (9.8.19) can be solved only approximately. It is reasonable to replace the integral in Eq. (9.8.19) by $H_{m\,\text{av}} l_m$, $H_{m\,\text{av}}$ representing some average magnetic field intensity inside the toroid. Since $H_0 = B_0/\mu_0 = B_m/\mu_0$ ($B_m \simeq B_0$ according to the law of conservation of magnetic flux), Eq. (9.8.19) can be put in the form

$$B_m = -\frac{\mu_0 l_m}{l_0} H_{m\,\text{av}}. \tag{9.8.20}$$

To determine $H_{m\,\text{av}}$ and $B_m$ we have to use the magnetization curve. In fact, at different points inside the magnet the magnetic field intensity $H_m$ is different. But a reasonable approximation is that it is roughly constant, and equal to $H_{m\,\text{av}}$. Thus the approximate state of magnetization at all points is obtained as the intersection of the magnetization curve and the straight line of Eq. (9.8.20), as illustrated in Fig. 9.28.

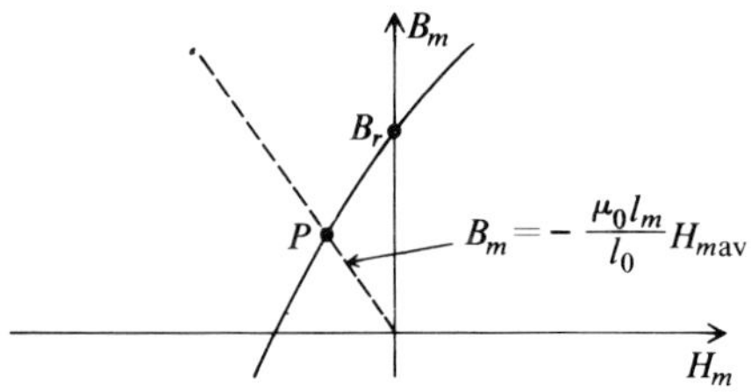


**Fig. 9.28** A demagnetization curve of a ferromagnetic material.

From this simple example we can see that the exact analysis of the magnetic circuit of permanent magnets is practically impossible, so we have to use any reasonable approximate theory. Although we have gathered a certain insight into the magnetic circuits of permanent magnets, there are many other associated problems, which again can be answered only roughly. In Problem 9.G20 we consider an additional detail in connection with permanent magnets. For further details the reader is referred to other texts.*

*Problems referring to Section 9.8:* 9.T21 to 9.T25, 9.G9 to 9.G20, 9.N6 to 9.N8.

## PROBLEMS

### Test problems

**9.T1.** Prove that the expression in Eq. (9.1.6) for the torque of electromagnetic forces is valid also for a rectangular planar loop in a uniform magnetic field. Assume two sides of the loop to be perpendicular to the field lines.

**9.T2.** Dividing an arbitrary planar current loop situated in a uniform magnetic field in small imaginary rectangular loops, and noting that the total torque is the vector sum of partial torques, prove that the expression for the torque on the loop is given by Eq. (9.1.6).

**9.T3.** There are $N$ elemental currents per unit volume, each having a magnetic moment $\mathbf{m}$. What is the magnetization vector $\mathbf{M}$?

**9.T4.** Prove that the unit of the magnetization vector $\mathbf{M}$ is the same as that of $\mathbf{B}/\mu_0$.

**9.T5.** Suppose that all atomic currents contained in the page you are reading can be oriented so that $\mathbf{m}$ is towards you. What is then their macroscopic resultant, and what is (qualitatively) the magnetic field of such a "magnetic leaf"?

***9.T6.** What are the macroscopic resultants of the microscopic currents of a short circular cylindrical piece of magnetized matter with uniform magnetization $\mathbf{M}$ at all points, if: (1) $\mathbf{M}$ is parallel to the cylinder axis, and (2) $\mathbf{M}$ is perpendicular to the axis?

***9.T7.** Sketch roughly the lines of vectors $\mathbf{M}$, $\mathbf{B}$, and $\mathbf{H}$ in the two cases from the preceding problem.

**9.T8.** A thin toroid is uniformly magnetized along its length with a magnetization vector of magnitude $M$. No macroscopic currents are present. Noting that lines of $\mathbf{M}$, $\mathbf{B}$, and $\mathbf{H}$ inside the toroid are circles by symmetry, determine the magnitude of $\mathbf{B}$ and prove that $\mathbf{H} = 0$.

---

* For example, see Attwood, S. S., *Electric and Magnetic Fields*, John Wiley and Sons, Inc., New York, 1958, pp. 381–391.

**9.T9.** If a fast charged elementary particle pierces the toroid of the preceding problem, will it be deflected by the magnetic force or not?

**9.T10.** Using Eq. (8.9.19), prove that the boundary condition for the tangential components of the magnetic vector potential at a boundary surface of two different (not necessarily linear) magnetic materials is $A_{1t} = A_{2t}$.

**9.T11.** Prove that, if a surface current of density $J_s$ exists on a boundary surface, then $H_{t1} - H_{t2} = J_s$. Vectorially $\mathbf{u}_n \times (\mathbf{H}_1 - \mathbf{H}_2) = \mathbf{J}_s$, $\mathbf{u}_n$ being directed from medium 2 to medium 1.

***9.T12.** Determine the distribution of the magnetic charges in the case of magnetized bodies from Problem 9.T6. From this sketch the lines of vectors **H** and **B**.

**9.T13.** Prove that the total magnetic charge on any magnetized body is zero. [*Hint:* use the divergence theorem.]

**9.T14.** Starting from Eq. (9.5.2) derive the boundary condition for derivatives of the magnetic scalar potential in the direction perpendicular to a boundary surface of two different linear magnetic materials.

**9.T15.** Sketch the initial magnetization curve corresponding to a change of $H$ from zero to $-H_{\text{max}}$ in Fig. 9.19.

**9.T16.** Suppose that the magnetization of the toroidal core of Fig. 9.18 corresponds to the point $B_r$ in Fig. 9.19. The coil around the core is removed, and the magnetic flux density decreased to zero by some appropriate *mechanical* or *thermal* treatment. How does the point in the $B$-$H$ plane go to zero?

**9.T17.** Explain why the toroidal core shown in Fig. 9.25 is not acceptable for determining the $B$-$H$ curve using Eqs. (9.7.1) and (9.7.2).

**9.T18.** The initial magnetization curve of a certain ferromagnetic material is determined for a thin toroidal core. Explain the process of determining the magnetix flux through the core of the same material, but of the form shown in Fig. 9.25.

**9.T19.** Make a rough sketch of the curve in the $B$-$H$ plane obtained if $H$ is increased to $H_m$, then decreased to zero, then again increased to $H_m$ and decreased to zero, and so on.

**9.T20.** Is it possible to obtain a higher remanent flux density than that obtained when saturation is attained and then $H$ reduced to zero?

**9.T21.** Suppose that $\mu$ in Fig. 9.26 is not a constant, but rather a function of $H$ (or $B$). Is it admissible to calculate the magnetic fluxes in the branches as the sums of the fluxes produced by the currents in the two coils individually?

**9.T22.** In Section 9.8.2 it was mentioned that it is not possible to have a magnetic circuit with no leakage flux. Explain this statement.

**9.T23.** Is it possible to construct a magnetic circuit closely analogous to a conductor with a steady current if the electric circuit is situated (1) in a vacuum, and (2) in an imperfect dielectric?

**9.T24.** Suppose that the slice which was cut out of the toroid in Fig. 9.27 is replaced, so that a closed toroid is obtained. Will the point $P$ in the $B$-$H$ plane in Fig. 9.28 move back to the point $B_r$?

**9.T25.** One half of the length of a thin toroidal coil is filled with a ferromagnetic material; in the other half is some paramagnetic material. Is it permissible to analyze the problem as a magnetic circuit? Sketch the lines of vectors **B** and **H**.

## General problems

**9.G1.** A ferromagnetic sphere of radius $a$ is uniformly magnetized with a magnetic moment per unit volume **M**. Determine the distribution of the macroscopic resultant of the microscopic currents.

*Solution.* Since **M** is a constant vector, curl $\mathbf{M} = 0$, so that, according to Eq. (9.3.6), there are no macroscopic resultants of the microscopic currents inside the sphere. According to Eq. (9.3.3) and Fig. 9.G1, density of the surface currents is given by

$$J_{ms} = M \sin \theta.$$

The surface currents are circulating currents, as indicated in the figure.

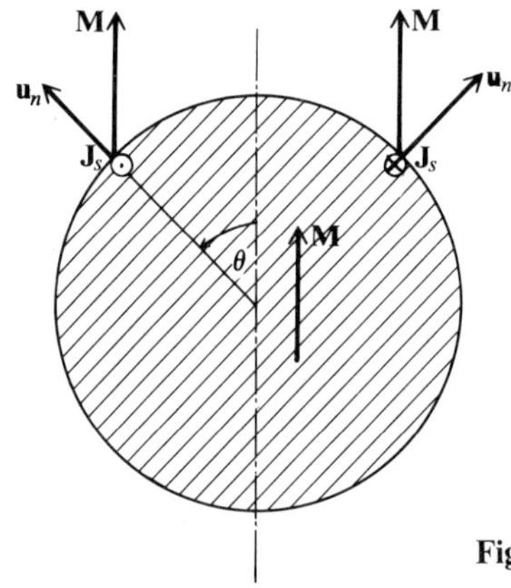


Figure 9.G1

**9.G2.** A parallelepiped of a ferromagnetic material is magnetized so that the magnetization vector **M** is perpendicular to one face of the parallelepiped, and varies linearly from zero to **M** from one face to the other (Fig. 9.G2a). Determine and sketch the macroscopic resultants of the microscopic currents.

*Solution.* According to Eq. (9.3.3), there are no resultant surface currents on faces 5 and 6 because there $\mathbf{M} \times \mathbf{u}_n = 0$, nor on face 3 since there $\mathbf{M} = 0$. But on the other three faces resultant currents exist, with densities of:

$$\text{face 1:} \quad \mathbf{J}_{ms1} = M\mathbf{u}_z \times (-\mathbf{u}_y) = M\mathbf{u}_x,$$

$$\text{face 2:} \quad \mathbf{J}_{ms2} = M\frac{b-y}{b}\mathbf{u}_z \times \mathbf{u}_x = M\frac{b-y}{b}\mathbf{u}_y,$$

$$\text{face 3:} \quad \mathbf{J}_{ms4} = M\frac{b-y}{b}\mathbf{u}_z \times (-\mathbf{u}_x) = -M\frac{b-y}{b}\mathbf{u}_y.$$

According to Eq. (9.3.6), inside the magnetized parallelepiped a resultant volume current exists, of density

$$\mathbf{J}_m = \text{curl}\,\mathbf{M} = \text{curl}\left[M\frac{b-y}{b}\mathbf{u}_z\right] = -\frac{M}{b}\mathbf{u}_x.$$

The distribution of the resultant macroscopic currents is shown diagrammatically in Fig. 9.G2b, in a cross-section of the parallelepiped perpendicular to the $z$ axis. It is left to the

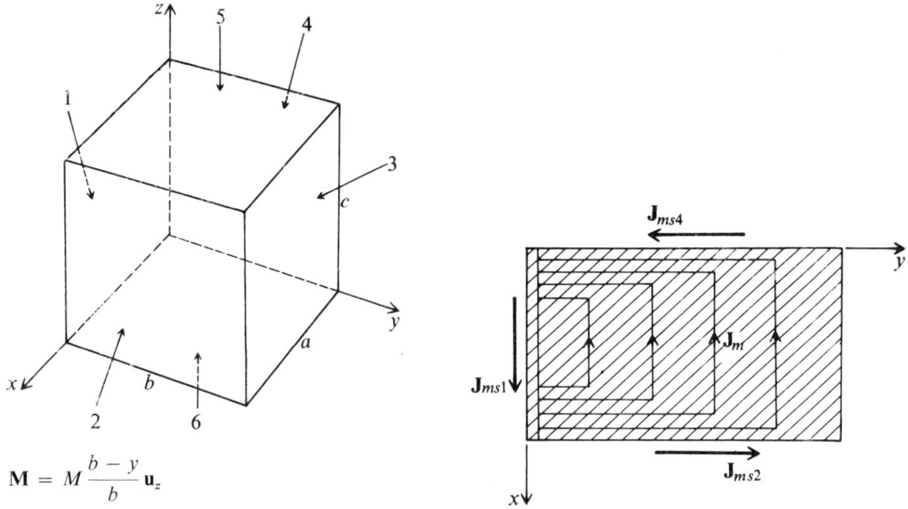


(a) **Figure 9.G2** (b)

reader to sketch the lines of vectors **B** and **H** in a cross-section of the parallelepiped perpendicular to the $x$ axis.

**9.G3.** A thin circular disk of thickness $d$ and radius $a$ ($a \gg d$) is magnetized with a constant magnetization vector **M**, perpendicular to its faces. Determine the magnetic flux density and the magnetic field intensity at points of the disk axis normal to its faces.

*Solution.* It might seem that the most direct solution is to find the magnetic vector potential at a point $P$ at the disk axis (Fig. 9.G3), and then **B** according to $\mathbf{B} = \text{curl}\,\mathbf{A}$. This procedure cannot be applied, however, since the vector potential **A** given by Eq. (9.2.6) is zero at all points of the axis. This does not mean, of course, that the magnetic flux density is also zero. To determine **B** according to $\mathbf{B} = \text{curl}\,\mathbf{A}$, it is necessary to find **A** as a function of *all* coordinates, and not only of $z$.

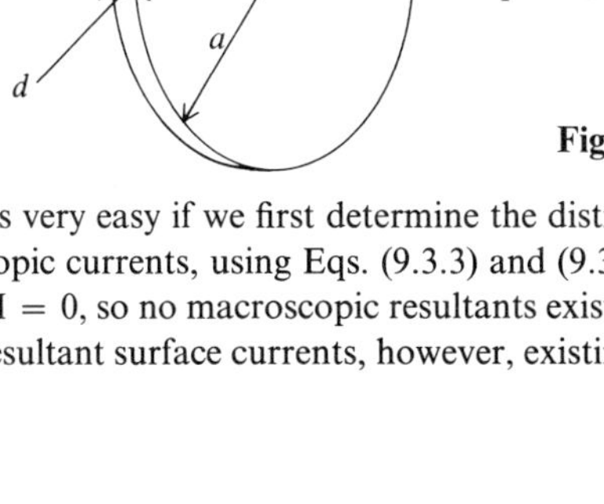


**Figure 9.G3**

The solution becomes very easy if we first determine the distribution of the macroscopic resultants of the microscopic currents, using Eqs. (9.3.3) and (9.3.6). Since **M** is assumed to be constant, $\mathbf{J}_m = \text{curl}\,\mathbf{M} = 0$, so no macroscopic resultants exist inside the disk. According to Eq. (9.3.3), there are resultant surface currents, however, existing along the curved surface, as indicated in the figure. The magnetized disk produces the same magnetic flux density as a current around the disk periphery, of intensity.

$$I = J_{ms}d = Md.$$

The magnetic flux density on the axis of a circular ring was determined in Problem 8.G4. Both inside and outside the magnetized disk, the magnetic flux density on its axis is given by

$$B = B_z(z) = \frac{\mu_0 M\, d\, a^2}{2(a^2 + z^2)^{3/2}}.$$

The magnetic field intensity outside the disk is $\mathbf{H} = \mathbf{B}/\mu_0$, and inside the disk $\mathbf{H} = \mathbf{B}/\mu_0 - \mathbf{M}$.

**9.G4.** A very long conductor of radius $a$ and permeability $\mu$ carries a steady current of intensity $I$. Coaxially with the conductor a tube of the same permeability is placed, with inner radius $b$ and outer radius $c$, as shown in Fig. 9.G4. Determine vectors $\mathbf{H}$, $\mathbf{B}$, $\mathbf{M}$, $\mathbf{J}_m$, and $\mathbf{J}_{ms}$ at all points, given that (1) permeability $\mu$ is constant, and (2) $\mu = \mu_0 H/H_0$, where $H_0$ is a constant.

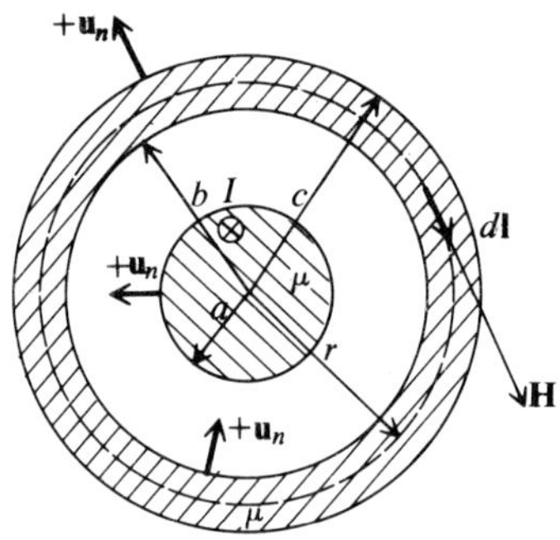


Figure 9.G4

*Solution.* By symmetry, the lines of both **H** and **B** are circles centered at the conductor axis. The magnitudes of these vectors are functions of the distance $r$ from the axis only. Applying the generalized form of Ampère's circuital law (9.4.5a) to a circular contour corresponding to one line of **H**, we have

$$\oint_C \mathbf{H}\cdot d\mathbf{l} = H2\pi r = \begin{cases} Ir^2\pi/a^2\pi & \text{for} \quad r \le a \\ I & \text{for} \quad r \ge a, \end{cases}$$

from which

$$H = Ir/2\pi a^2 \quad \text{for} \quad r \le a, \tag{a}$$

$$H = I/2\pi r \quad \text{for} \quad r \ge a. \tag{b}$$

The magnetic flux density is obtained as $\mathbf{B} = \mu\mathbf{H}$. If $\mu$ is supposed to be constant [case (1)],

$$B = \mu Ir/2\pi a^2 \quad \text{for} \quad r < a, \tag{c}$$

$$B = \mu_0 I/2\pi r \quad \text{for} \quad a < r < b \quad \text{and} \quad r > c, \tag{d}$$

$$B = \mu I/2\pi r \quad \text{for} \quad b < r < c. \tag{e}$$

To obtain the macroscopic resultants of the microscopic currents, it is necessary to determine the magnetization vector **M** at all points. According to Eq. (9.4.12)

$$M = M_\phi = \left(\frac{\mu}{\mu_0} - 1\right)\frac{I}{2\pi a^2} r \qquad \text{for} \qquad r < a,$$

$$M = M_\phi = \left(\frac{\mu}{\mu_0} - 1\right)\frac{I}{2\pi r} \qquad \text{for} \qquad b < r < c.$$

$M$ is zero at all other points. Using the expression for the curl in the cylindrical coordinate system (Problem 8.G20), the density of the equivalent currents in the conductor and the tube are

$$J_m = J_{mz} = \text{curl}_z\,\mathbf{M} = \frac{1}{r}\frac{d}{dr}(rM_\phi) = \left(\frac{\mu}{\mu_0} - 1\right)\frac{I}{a^2\pi} \qquad \text{for} \qquad r < a, \tag{f}$$

$$J_m = 0 \qquad \text{for} \qquad b < r < c.$$

The density of the surface currents is

$$\mathbf{J}_{ms}(a) = \mathbf{M}(a) \times \mathbf{u}_{na}, \qquad \text{or} \qquad J_{ms}(a) = -\left(\frac{\mu}{\mu_0} - 1\right)\frac{I}{2\pi a}, \tag{g}$$

$$\mathbf{J}_{ms}(b) = \mathbf{M}(b) \times \mathbf{u}_{nb}, \qquad \text{or} \qquad J_{ms}(b) = \left(\frac{\mu}{\mu_0} - 1\right)\frac{I}{2\pi b}, \tag{h}$$

$$\mathbf{J}_{ms}(c) = \mathbf{M}(c) \times \mathbf{u}_{nc}, \qquad \text{or} \qquad J_{ms}(c) = -\left(\frac{\mu}{\mu_0} - 1\right)\frac{I}{2\pi c}. \tag{i}$$

It is left as an exercise for the reader to prove that the magnetic flux density given by (c), (d), and (e) is created by current $I$ in the central conductor, and the macroscopic resultants (f) to (i) of the microscopic currents, all the currents being considered to be in a vacuum. Part (2) of the problem is also left to the reader.

**9.G5.** Repeat the preceding problem assuming that the tube represents the outer conductor of a coaxial cable, so that macroscopic currents in the conductor and the tube are of the same intensity $I$, but in opposite directions.

**9.G6.** A straight cylindrical conductor of radius $a$ carries a current of intensity $I$ uniformly distributed over its cross-section. The permeability of the conductor is a function of the distance $r$ from the conductor axis, of the form $\mu(r) = \mu_0(ka - r)/a$, where $k$ is a constant ($k > 2$). Determine the vectors **H**, **B**, and **M** inside the conductor, and prove that **B** is created by the current $I$ and the macroscopic resultants of the microscopic currents, all currents being assumed to exist in a vacuum.

**9.G7.** Solve Problem 9.G3 by first determining the magnetic charges and the magnetic scalar potential. [*Hint:* use Problem 2.G3.]

**9.G8.** A cylindrical magnet of radius $a$ and length $b$ is magnetized uniformly in the direction of its axis. The magnetization of the magnet is $M$. Determine the distribution of the magnetic charges and hence the magnetic scalar potential on the axis. Next find vectors **B** and **H** on the axis, both inside and outside the cylinder. Compare the results with those of Problem 8.G31.

**9.G9.** Determine the magnetic flux in the magnetic circuit shown in Fig. 9.G9, if the current intensities through the two coils are $I_1$ and $I_2$. Note that $I_1$ and $I_2$ produce a magnetic flux of the same direction.

*Answer.* For $I_1 = I_2 = 1$ A, $N_1 = N_2 = 100$ turns, $\mu_1 = 1000\,\mu_0$, $\mu_2 = 5000\,\mu_0$, $l_1 = 50$ cm, $l_2 = 20$ cm, $S_1 = 5$ cm$^2$ and $S_2 = 10$ cm$^2$, the flux $\Phi = 2.42 \times 10^{-4}$ Wb.

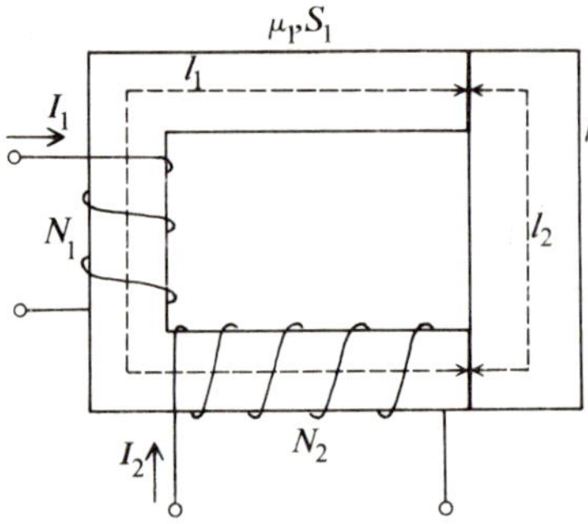


Figure 9.G9

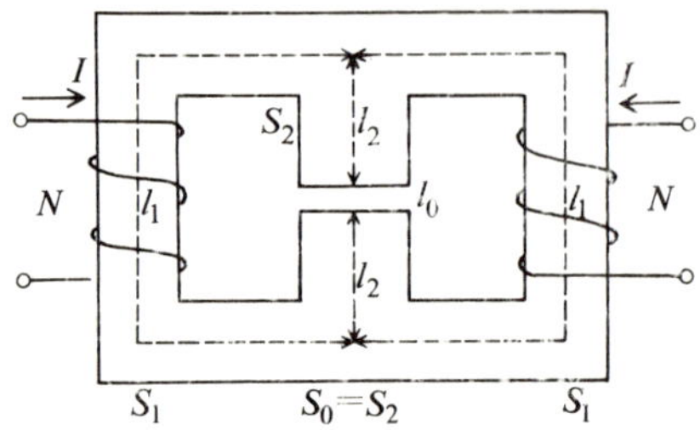


Figure 9.G11

**9.G10.** In the preceding problem let $I_1 = I_2$, and determine currents $I_1 = I_2$ which would produce a required flux $\Phi_{\text{required}}$ in the magnetic circuit.

*Answer.* With the same data as in the answer to the preceding problem, for $\Phi_{\text{required}} = 5 \times 10^{-4}$ Wb, the currents are $I_1 = I_2 = 2.07$ A.

**9.G11.** Determine the magnetic flux density in the air gap of the magnetic circuit shown in Fig. 9.G11. Determine also the flux density if the reluctance (the magnetic resistance) of all the parts of the circuit except of the air gap is ignored.

*Answer.* For $N = 200$, $I = 1$ A, $\mu_r = 5000$, $l_1 = 50$ cm, $l_2 = 10$ cm, $l_0 = 5$ mm, $S_1 = 5$ cm$^2$ and $S_2 = 10$ cm$^2$, the magnetic flux density in the gap $B_0 = 0.0491$ T. If the reluctance of the magnetic part of the circuit is ignored, $B_0 \simeq 0.0503$ T.

**9.G12.** Determine the current intensity $I$ in the preceding problem necessary to produce a given magnetic flux density $B_0$ in the air gap. Determine also $I$ approximately, taking into account the reluctance of the air gap only (i.e., assuming the reluctance of the magnetic part of the circuit to be negligibly small).

*Answer.* For the numerical values (except for $I$) given in the answer to the preceding problem, and $B_0 = 0.100$ T, ignoring the magnetic resistance of iron, $I \simeq 1.99$ A. Without ignoring it, $I \simeq 2.04$ A.

**9.G13.** Figure 9.G13a shows the initial magnetization curve of a ferromagnetic material from which the core shown in Fig. 9.G13b is made. Determine the current intensity $I$ through the coil which is necessary to produce a magnetic flux density in the air gap of $B_0 = 1$ T. The core was not magnetized previously. The average lengths of the core sections and the corresponding cross-section areas are indicated in the figure. Ignore the leakage flux.

*Solution.* A problem of this kind is the most simple problem of nonlinear magnetic circuits. The current intensity producing the desired magnetic flux density in the air gap is determined according to the following procedure.

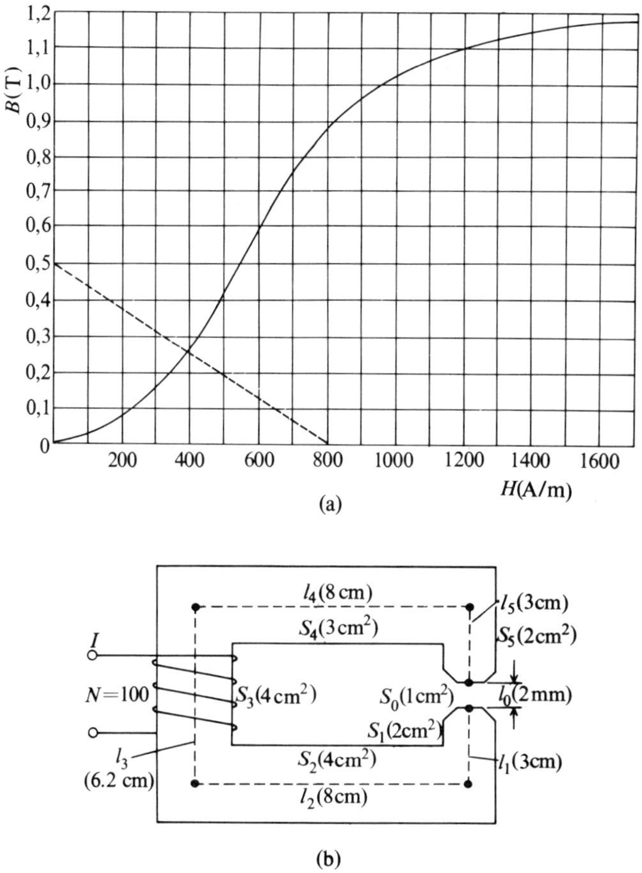


**Figure 9.G13**

(1) From the known $B_0$ in the air gap, the magnetic flux through the gap is $\Phi = B_0 S_0$. Since there is no flux leakage, this is also the flux through the whole circuit.

(2) The magnetic flux density in all parts of the circuit is determined according to $B_k = \Phi/S_k = B_0 S_0/S_k$.

(3) From the magnetization curve, the magnetic field intensities $H_k$ corresponding to the flux densities $B_k$ are next determined. In the air gap $H_0 = B_0/\mu_0$.

(4) The product $H_k l_k$ are calculated for all parts of the circuit.

(5) Using the (approximate) expression in Eq. (9.8.15) of Ampère's circuital law we calculate the required product $NI$:

$$NI = \Sigma H_k l_k.$$

This procedure is simplest if done in tabular form, as shown in the table below. Note that the flux through the circuit is $\Phi = B_0 S_0 = 10^{-4}$ Wb.

| $k$ | $l_k$ (cm) | $S_k$ (cm$^2$) | Material | $B_k$ (T) | $H_k$ (A/m) | $H_k l_k$ (A) |
|---|---|---|---|---|---|---|
| 0 | 0.2 | 1 | Air | 1.00 | $8 \times 10^5$ | 1600 |
| 1 | 3 | 2 | Ferromagnetic | 0.50 | 540 | 16.2 |
| 2 | 8 | 4 | Ferromagnetic | 0.25 | 380 | 30.4 |
| 3 | 6.2 | 4 | Ferromagnetic | 0.25 | 380 | 19.8 |
| 4 | 8 | 3 | Ferromagnetic | 0.33 | 440 | 35.2 |
| 5 | 3 | 2 | Ferromagnetic | 0.50 | 540 | 16.2 |
| | | | | | $NI = \Sigma H_k l_k =$ | 1717.8 |

From the table the required intensity of the current is seen to be

$$I = \Sigma H_k l_k / N = 17.2 \text{ A}.$$

It should be noted that the product $H_0 l_0$ is much larger than all the other products $H_k l_k$ taken together. This is almost always the case. Hence we can obtain a reasonably accurate result by simply ignoring all $H_k l_k$ – s except $H_0 l_0$:

$$(NI)_{\text{appr}} = H_0 l_0 = 1600, \qquad \text{from which} \qquad I_{\text{appr}} = 16 \text{ A}.$$

This approximate procedure is very useful for determining roughly either the product $NI$ for a given flux $\Phi$, or the magnetic flux $\Phi$ corresponding to a given product $NI$ (see Problem 9.G16).

**9.G14.** A toroidal core of cross-section area $S = 2$ cm$^2$ and an average length $l_m = 0.5$ m is made of the ferromagnetic material having the initial magnetization curve shown in Fig. 9.G13a. The core is wound with $N = 200$ turns carrying a current of intensity $I = 2$ A, and has an air gap of length $l_0 = 1$ mm. Determine the magnetic flux density in the air gap. Ignore the flux leakage and assume that the core was not magnetized previously.

*Solution.* Since the areas of the cross-sections of the core and the air gap are equal, it follows that $B_m = B_0$. According to Ampère's circuital law we have the second equation,

$$H_m l_m + H_0 l_0 = NI,$$

or, since $B_m = B_0 = \mu_0 H_0$,

$$H_m l_m + B_m \frac{l_0}{\mu_0} = NI. \tag{a}$$

Now, we do not know the analytical dependence of $B_m$ upon $H_m$. Instead, this dependence is given in the form of the curve shown in Fig. 9.G13a. Thus from Eq. (a) we see that $B_m$ and $H_m$ cannot be determined analytically; a graphical solution is, however, easy. According to Eq. (a), the magnetic flux density inside the core with the air gap is given by

$$B_m = \frac{\mu_0 NI}{l_0} - \frac{\mu_0 l_m}{l_0} H_m. \tag{b}$$

Equation (b) represents an equation of a straight line in the $B$-$H$ plane. $B_m$ and $H_m$, on the other hand, are connected by the magnetization curve, and the point determining the magnitude of the magnetic flux density in the core must lie on both, that is, at their intersection.

With the numerical values given above, Eq. (b) becomes

$$B_m = 0.503 - 0.628 \times 10^{-3} H_m. \tag{c}$$

The straight line (c) is indicated in Fig. 9.G13a by a dashed line. The intersection of this line with the magnetization curve is for $B_m = B_0 = 0.26$ T. Thus is the problem solved.

Consider a core of a variable cross-section, as shown in Fig. 9.G13b. The magnetic flux density in the air gap corresponding to a given current intensity through the coil cannot be determined as above. In this case, it is simplest to draw the curve representing the dependence of $B_0$ on $NI$, assuming a set of values of $B_0$, and determining the corresponding product $NI$ in accordance with the procedure outlined in the preceding problem. We can then determine the magnetic flux density $B_0$ in the air gap for the actual product $NI$ by graphical interpolation.

**9.G15.** Assuming that the initial magnetization curve of the core from the preceding problem can be approximated by the analytical curve

$$B_m = 2H_m/(400 + H_m),$$

determine the flux density $B_0$ in the gap.

*Answer.* $B_0 = 0.433$ T.

**9.G16.** If a current of intensity $I = 5$ A flows through the coil wound on the ferromagnetic core of Problem 9.G13, determine the magnetic flux density $B_0$ in the gap. Ignore the leakage flux.

*Solution.* As explained in Section 9.8.3, the analysis of any nonlinear magnetic circuit has to be based on relations given in Eqs. (9.8.15) to (9.8.17). In the present case these relations result in the following system of equations:

$$(1)\ NI = H_0 l_0 + H_1 l_1 + \cdots + H_5 l_5, \tag{a}$$

$$(2)\ B_k = B_0(S_0/S_k) = \mu_0 H_0(S_0/S_k), \qquad k = 1, 2, \ldots, 5, \tag{b}$$

(3) the magnetization curve given in Fig. 9.G13a. (c)

For determining $H_0$, according to Eq. (a) we have to determine $H_1, H_2, \ldots, H_5$. But from Eqs. (b) and (c) it follows that this cannot be done until $H_0$ is found. A direct solution of the system (a)-(c) is thus not possible.

There are two rather differing ways for approximately solving a problem of this type. The first is to determine the curve $B_0 = B_0(NI)$, by *assuming* a set of values for $B_0$, and determining the product $NI$ required to produce this $B_0$, following the procedure described in Problem 9.G13. The value of $B_0$ corresponding to the given product $NI$ is then determined by graphical interpolation. It is obviously useful to have at least some idea of what the value of $B_0$ will be in order to determine the required part of the $B_0(NI)$ curve most directly.

The other method is essentially an iterative method. We first make as good a guess of $H_0$ as possible. Using Eq. (b), we next calculate the first approximate values of $B_k$, $k = 1, 2, \ldots, 5$. The first approximate values of $H_k$, $k = 1, 2, \ldots, 5$ can be found by using the magnetization curve and the values of $B_k$ thus found. These $H_k$ values are introduced into Eq. (a) to determine a still better value of $H_0$. The process is repeated until the desired accuracy is attained.

For both methods, however, a guess at $H_0$ is essential. It is simplest to assume that the magnetic resistance of all the ferromagnetic parts of the circuit is negligible when compared with the reluctance of the gap. This guess is usually a good one, as we saw in Problems 9.G11 to 9.G13. Thus, in this case, we start with

$$H_{0_{\text{appr}}} = NI/l_0 = 500/(2 \times 10^{-3}) = 250{,}000 \text{ A/m}. \tag{d}$$

In fact, to produce this value of $H_0$ a somewhat larger current is required, since the reluctances of ferromagnetic parts were ignored. Hence, if we are solving the problem by the first (graphical) method, we determine the product $NI$ corresponding to the magnetic flux density in the gap $B_{0_{\text{appr}}} = \mu_0 H_{0_{\text{appr}}}$, and for another two or three somewhat *lower* values of $B_0$. We shall obtain a portion of the curve $B_0(NI)$, and find the correct value of $B_0$ for $NI = 500$ by graphical interpolation.

According to the second (iterative) method, by using $H_{0_{\text{appr}}}$ we can determine (approximate) $B_1, B_2, \ldots, B_5$ from Eqs. (b) and (d), find (approximate) values of $H_1, H_2, \ldots, H_5$ corresponding to them from the magnetization curve (c), and then determine from (a) the first improved value of $H_0$. Usually the first, or at most the second, approximate value will be of sufficient accuracy.

It is left as an exercise for the reader to check that by both methods the same result ($B_0 \simeq 0.27$ T) is obtained.

**9.G17.** Repeat the preceding problem assuming (1) $l_0 = 1$ mm and $l_3 = 6.1$ cm and (2) $l_0 = 5$ mm and $l_3 = 6.5$ cm, all the other data being as given in Fig. 9.G13b. [A method to check the result: the two procedures described in the preceding problem must yield essentially the same result.]

***9.G18.** Figure 9.G18a shows a magnetic circuit made of magnetic material having the magnetization curve represented in Fig. 9.G13a. Determine the magnetic flux through the three branches of the magnetic circuit, if the coil carries a current of intensity $I = 2.5$ A. Ignore the flux leakage.

*Solution.* This problem cannot be solved by the application of Kirchhoff's laws given in Eqs. (9.8.12) and (9.8.14), since $R_m$ is not constant. Only the general laws in Eqs. (9.8.15) to (9.8.17) for solving nonlinear magnetic circuits are applicable.

From Eq. (9.8.16) one independent equation only follows:

$$\Phi_1 - \Phi_2 - \Phi_3 = 0, \qquad \text{or} \qquad B_1 S_1 - B_2 S_2 - B_3 S_3 = 0. \tag{a}$$

The reference directions of the magnetic flux in the three branches are shown in the figure.

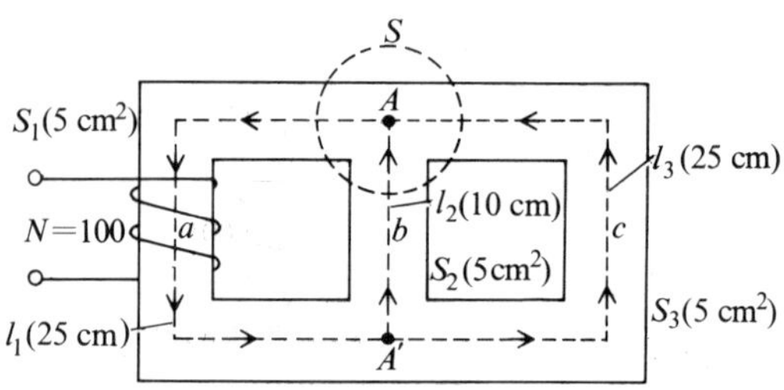


(a)

**Figure 9.G18**

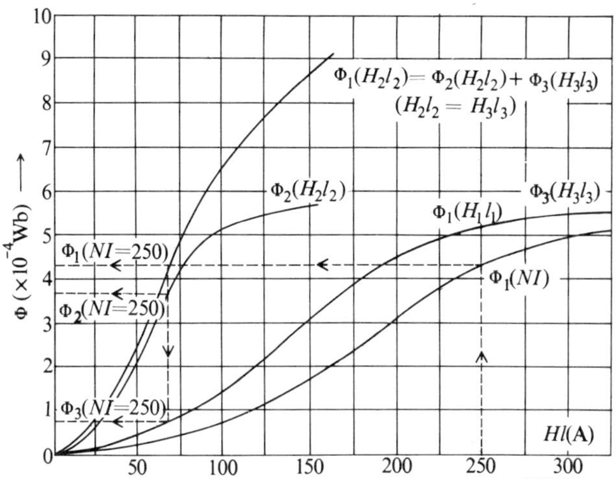


(b)

**Figure 9.G18**

From Eq. (9.8.15) two independent equations follow. For example, applying Eq. (9.8.15) to the closed loops *AcA'bA* and *A'aAbA'* we have

$$H_3 l_3 - H_2 l_2 = 0, \tag{b}$$

$$H_1 l_1 + H_2 l_2 = NI. \tag{c}$$

The solution of Eqs. (a) to (c) subject to the magnetization curve of Fig. 9.G13a can be determined as follows: The magnetization curve gives the dependence of the magnetic flux density $B$ on the magnetic field intensity $H$. If we multiply the abscissas $H$ of the magnetization curve by the length $l_k$ of a branch, and the ordinates $B$ by the cross-section area of the branch, the magnetization curve becomes a curve showing the dependence on the corresponding product $Hl$ of the flux $\Phi$ through the branch. Figure 9.G18b shows such curves, determined from the magnetization curve for branches 1, 2, and 3. For example, the curve $\Phi_2(H_2 l_2)$ is obtained by multiplying the abscissas of the points on the magnetization curve by $l_2 = 0.1$ m, and the ordinates by $S_2 = 5 \times 10^{-4}\,\text{m}^2$. The curves $\Phi_1(H_1 l_1)$ and $\Phi_3(H_3 l_3)$ are identical, since the branches are of the same dimensions. But it should be noted that the curve $\Phi_1(H_1 l_1)$ does *not* represent the total flux in branch 1. It represents the flux that *would* exist in the branch if the magnetic field intensity in it were $H_1$.

According to Eq. (a), the magnetic flux in branch 1 is the sum of the fluxes in branches 2 and 3, $\Phi_1 = \Phi_2 + \Phi_3$. As $H_2 l_2 = H_3 l_3$, according to Eq. (b), pairs of points on the curves $\Phi_2(H_2 l_2)$ and $\Phi_3(H_3 l_3)$ corresponding to the same value of $Hl$ represent the possible pairs of fluxes $\Phi_2$ and $\Phi_3$. Hence the curve $\Phi_1(H_2 l_2)$, obtained by adding these two, as is done in Fig. 9.G18b, represents the dependence of the flux $\Phi_1$ in branch 1 on $H_2 l_2$, viz., $H_3 l_3$ ($H_2 l_2 = H_3 l_3$).

By the help of Eq. (c) we can change this dependence of $\Phi_1$ on $H_2 l_2$ to dependence on $NI$, which is precisely what has to be determined. According to Eq. (c), the flux $\Phi_1$ corresponding to a certain value of $H_2 l_2$ is produced by a magnetomotive force

$$NI = H_1 l_1 + H_2 l_2.$$

Thus $NI$ can be determined by simply adding the abscissas of the curves $\Phi_1(H_1l_1)$ and $\Phi_1(H_2l_2)$ for fixed values of the flux $\Phi_1$. So we add these two curves in the direction of the $Hl$ axis, that is, add the abscissas corresponding to the same ordinate. The curve $\Phi_1(NI)$ shown is obtained in this manner. From this curve we find that, when $NI = 250$, the flux $\Phi_1(250) \simeq 4.25 \times 10^{-4}$ Wb. From the curve $\Phi_1(H_2l_2)$ we next find that $H_2l_2 \simeq 66$ A corresponds to this value of $\Phi_1$. Finally the flux $\Phi_2 \simeq 3.6 \times 10^{-4}$ Wb in branch 2 and (since $H_3l_3 = H_2l_2$) the flux $\Phi_3 \simeq 0.6 \times 10^{-4}$ Wb in branch 3 correspond to $H_2l_2 \simeq 66$ A.

***9.G19.** Repeat the preceding problem given that (1) $I = 2$ A, and (2) $S_2 = 10\ \text{cm}^2$ (instead of $5\ \text{cm}^2$) and $I = 2.5$ A. [*Check:* if all $B$'s and $H$'s are found correctly, Eqs. (a), (b), and (c) from the preceding problem will be satisfied.]

**9.G20.** In Fig. 9.G20a is a toroidal permanent magnet of a ferromagnetic material having the demagnetization curve (a part of the magnetization curve for $H < 0$) shown in Fig. 9.G20b. Find the maximum value of the flux density vector in the air gap, for a given value of the gap volume; that is, change $l_0$ and thereby $B_0$, but simultaneously changing $S_0$ so that the product $l_0S_0 = v_0$ remains constant. Ignore the leakage flux.

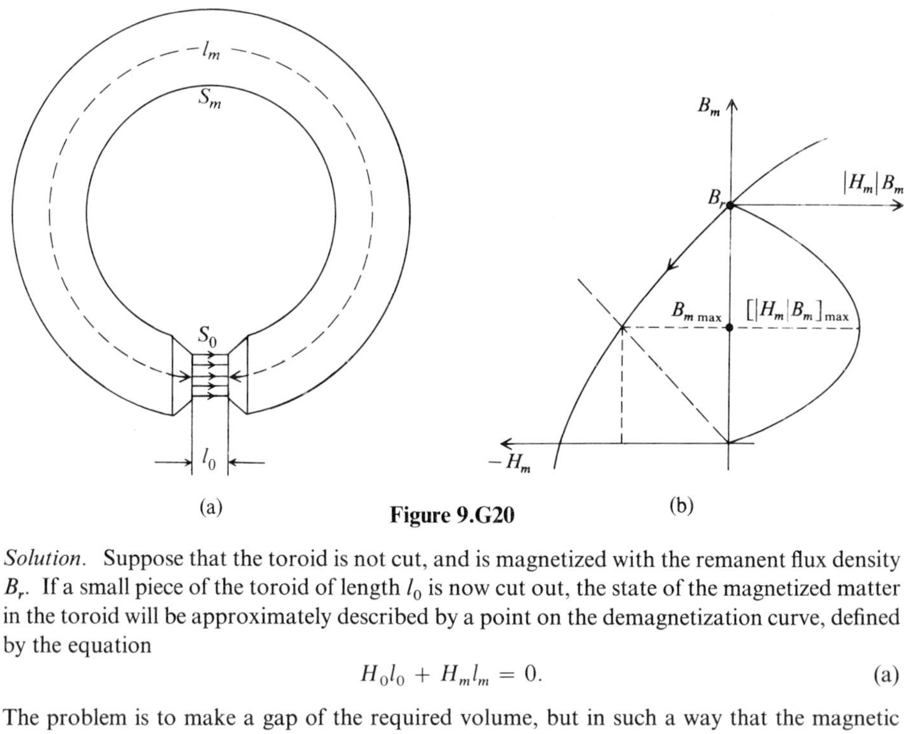


**Figure 9.G20**

*Solution.* Suppose that the toroid is not cut, and is magnetized with the remanent flux density $B_r$. If a small piece of the toroid of length $l_0$ is now cut out, the state of the magnetized matter in the toroid will be approximately described by a point on the demagnetization curve, defined by the equation

$$H_0l_0 + H_ml_m = 0. \tag{a}$$

The problem is to make a gap of the required volume, but in such a way that the magnetic flux density in the gap is the maximum possible. This is obviously an important problem of permanent magnet design. Note that, when the gap is modeled, the volume of the magnet itself is practically unchanged, since $l_0 \ll l_m$.

In addition to Eq. (a), the following two equations are also valid:

$$H_0 = B_0/\mu_0, \qquad B_m = S_0B_0/S_m. \tag{b}$$

Combining Eqs. (a) and (b) we can write the equation

$$B_0^2 = \mu_0 \frac{S_m l_m}{S_0 l_0} B_m |H_m|. \tag{c}$$

This is the equation which answers the problem. Since on the right-hand side of Eq. (c) $\mu_0 S_m l_m / S_0 l_0$ remains very nearly constant while modeling the gap, only the product $B_m|H_m|$ changes. Hence the maximum value of $B_0$, equal to $B_{0\,\max} = S_m\, B_{m\,\max}/S_0$, will be obtained for the maximum value of the product $B_m|H_m|$, as indicated in Fig. 9.G20b.

The product $B_m|H_m|$ thus plays an important role in the practical design of permanent magnets. The best ferromagnetic materials for permanent magnets are those with high values of $(B_m|H_m|)_{\max}$. The expression $B_m|H_m|$ is referred to as the *energy product*.

## Numerical problems

**9.N1.** If the tube enclosing the conductor in Problem 9.G4 has a relative permeability $\mu_r = 1000$, and if $b = 2$ cm, $c = 3$ cm, and $I = 1$ A, determine $J_{ms}$ on the surfaces of the tube.

**9.N2.** A very long cylindrical core is made of soft iron, having a relative permeability $\mu_r = 5000$. The core is densely wound with 1000 turns of thin wire per meter. The coil carries a current of intensity $I = 1$ A. Determine the values of $B$, $H$, and $M$ in the core, and the density of the equivalent surface current on the core surface.

**9.N3.** Repeat the preceding problem for a copper core ($\chi_m \simeq -0.94 \times 10^{-5}$).

**9.N4.** Determine the ratio of the magnitudes of the tangential components of the magnetic flux density vector on the interface between air and a ferromagnetic substance having an average relative permeability $\mu_r = 10{,}000$.

**9.N5.** Suppose that the lines of the magnetic flux density vector on the boundary surface between a ferromagnetic substance with $\mu_r = 7000$ and air make angles of (1) 0°, (2) 30°, (3) 45°, (4) 60°, (5) 85° with the normal to the surface in the ferromagnetic substance. Calculate the corresponding directions of the magnetic flux density in air.

**9.N6.** Determine the exact and approximate values of the reluctance of the toroidal core shown in Fig. 9.25, if $\mu_r = 5000$, $a = 5$ cm, $b = 8$ cm, and $h = 3$ cm.

**9.N7.** Determine the necessary current intensity required to produce a magnetic flux of $10^{-4}$ Wb in the magnetic circuit shown in Fig. 9.24, if $N = 200$, $l_1 = 50$ cm, $l_2 = 15$ cm, $S_1 = S_2 = 10\ \text{cm}^2$, $\mu_{r1} = 1000$, and $\mu_{r2} = 5000$.

**9.N8.** It is found experimentally that the permanent magnet of Fig. 9.27 produces in the air gap a magnetic flux density of $B_0 = 2 \times 10^{-2}$ T. If $l_0 = 5$ mm and $l_m = 50$ cm, determine approximately the surface density of the macroscopic resultants of the microscopic currents on the magnet.

**Suggested reading.** The list progresses from easier to more advanced texts. For full bibliographical details, see the Bibliography.

1. SEELY, S., *Introduction to Electromagnetic Fields*, Chapter 7.
2. REITZ, J. R. and F. J. MILFORD, *Foundations of Electromagnetic Theory*, Chapter 10.
3. SCOTT, W. T., *The Physics of Electricity and Magnetism*, Chapter 8.
4. BOHN, E. V., *Introduction to Electromagnetic Fields and Waves*, Chapter 8.

CHAPTER 10

# METHODS OF SOLVING STATIC MAGNETIC FIELD PROBLEMS

## 10.1 ANALYTICAL METHODS

**10.1.1 Introduction.** In all cases of the static magnetic field, the distribution of the macroscopic currents, that is, of the primary sources of the magnetic field, is given in advance. If no ferromagnetic bodies are present in the field, determination of the magnetic field is reduced to the application of the Biot–Savart law given by Eq. (8.4.3). Determining the value of the integral in the Biot–Savart law can be very complicated but is not difficult otherwise. Several simple examples of determining the magnetic field of a given distribution of currents in a vacuum were presented in Chapter 8. To the same class belong the problems of magnetized bodies of a given magnetization in the absence of macroscopic currents. The macroscopic resultants of the microscopic currents can then be determined, and the problem is reduced to determining the magnetic field of these currents when considered in a vacuum.

If ferromagnetic (magnetically nonlinear) bodies as well as macroscopic currents are present, exact determination of the magnetic field is practically impossible. However, we can gain a certain insight into the magnetic field in these cases by considering ferromagnetic substances as linear, with some (always very high) average value of the relative permeability. We shall adopt this linearity approximation throughout the present chapter, if not explicitly stated otherwise.

The determination of the distribution of the magnetic field, with the permeability assumed to be constant, reduces to determining the flux density and the magnetic field intensity vectors which satisfy certain specified boundary conditions. We considered such boundary-value problems in some detail in connection with the electrostatic field. We shall discover that essentially the same technique applies to many cases in the static magnetic field.

**10.1.2 A method of images.** The idea at the root of applying the method of images to the electrostatic field was to replace the induced charges on the surfaces of the conducting bodies by some simpler system of charges inside them. These imaginary charges have to produce outside the bodies the same electric field as the real induced charges. The imaginary charges are determined correctly if the resultant electric field *outside* the bodies satisfies the required boundary conditions. According to the uniqueness theorem, if this is attained, the solution obtained is the only possible one.

The same idea is the basis of the method of images in the static magnetic field. Take, for example, a straight, thin conductor carrying a current $I$, running in air parallel to the plane surface of a large ferromagnetic slab (Fig. 10.1). According to Eq. (9.5.6), the lines of the magnetic flux density vector *in air* are virtually perpendicular to the boundary surface of the slab. If we remove the ferromagnetic slab, and, symmetrically with respect to the former boundary surface of the slab and air, place a conductor with a current of the same intensity *and the same direction*, as shown in Fig. 10.2, then, by symmetry, the lines of the vector **B** will in this case be perpendicular to the plane of symmetry, that is, to the former interface. The boundary conditions at the interface *from the air side* are hence virtually the same as before. According to the uniqueness theorem, the magnetic field in the upper part of the system shown in Fig. 10.2 is also, for all intents and purposes,

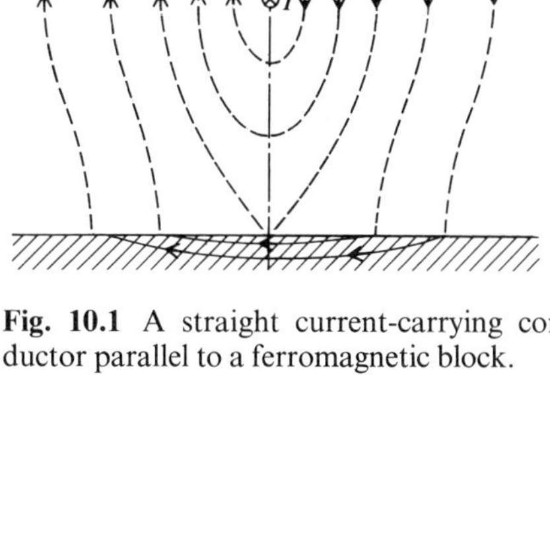


**Fig. 10.1** A straight current-carrying conductor parallel to a ferromagnetic block.

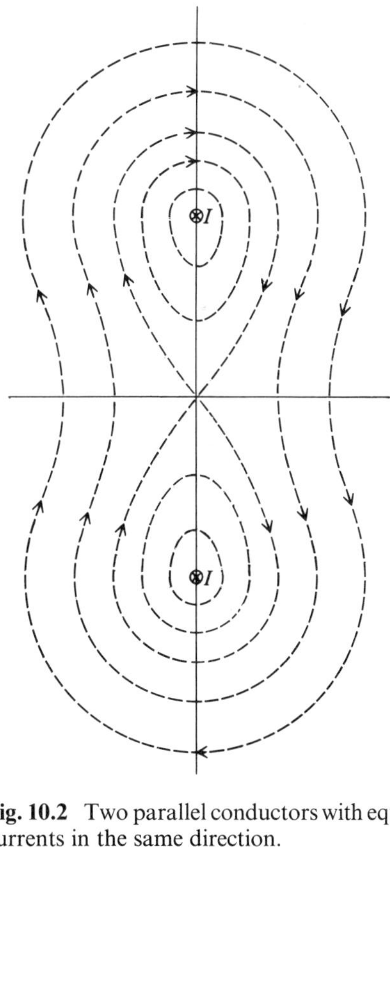


**Fig. 10.2** Two parallel conductors with equal currents in the same direction.

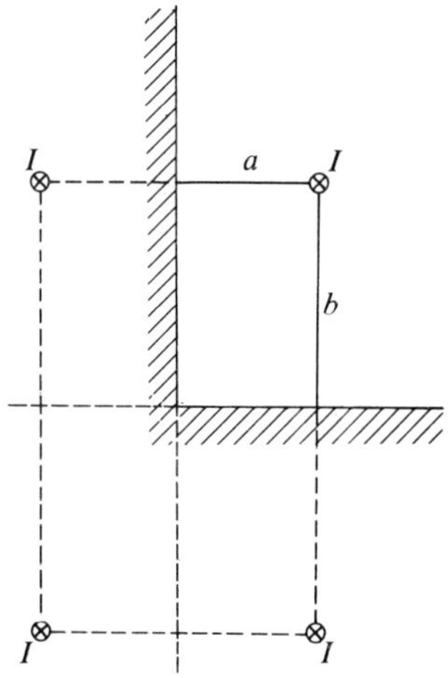


**Fig. 10.3** A straight conductor in a ferromagnetic 90° corner.

equal to that above the ferromagnetic slab in the real case. Thus, the magnetic field of all the microscopic currents in the slab is the same as that of a single current-carrying conductor, the positive image of the real conductor in the surface of the slab. This conclusion is easy to apply to any system of currents in the air. Note that the linearity assumption was not necessary in this case; the only requirement was that the permeability of the slab was much larger than $\mu_0$.

In a similar way, we can conclude that the influence of the microscopic currents in a ferromagnetic 90° corner on the magnetic field *in air* can be reduced to three positive images of the conductor existing in air, as indicated in Fig. 10.3. Several other examples of the application of the method of images in solving static magnetic field problems are given in Problems 10.G1 to 10.G4.

**10.1.3 Direct solving of Laplace's equation for the magnetic scalar potential. Magnetic shielding.** We can solve several important types of problems relating to magnetized bodies in an external magnetic field by determining the solution to Laplace's equation for the magnetic scalar potential. The practical importance of this method cannot be compared, however, with its importance in the electrostatic field.

If we consider electrostatic problems with no free charges present and magnetostatic problems with no macroscopic, real currents present, we see that a complete analogy exists between the two fields. All results, therefore, which we obtained for the electrostatic field are valid for the corresponding magnetostatic problems. It follows, then, that we need not solve problems of the static magnetic field if we solved corresponding electrostatic problems earlier. Since this analogy represents the basis for writing down results directly from those of the analogous electrostatic problem, a survey of the analogous relations is given below.

| *Electrostatic field* (no free charges in the region) | *Magnetostatic field* (no macroscopic currents in the region) |
|---|---|
| $\operatorname{div}\operatorname{grad} V = \nabla^2 V = 0$ | $\operatorname{div}\operatorname{grad} V_m = \nabla^2 V_m = 0$ |
| $\mathbf{E} = -\operatorname{grad} V$ | $\mathbf{H} = -\operatorname{grad} V_m$ |
| $\mathbf{D} = \varepsilon_0\mathbf{E} + \mathbf{P}$ (generally) | $\mathbf{B} = \mu_0(\mathbf{H} + \mathbf{M})$ (generally) |
| $\mathbf{D} = \varepsilon\mathbf{E}$ (linear media) | $\mathbf{B} = \mu\mathbf{H}$ (linear media) |

*Boundary conditions*

| | |
|---|---|
| $E_{1t} = E_{2t}$, or $V_1 = V_2$ | $H_{1t} = H_{2t}$, or $V_{m1} = V_{m2}$ |
| $D_{1n} = D_{2n}$, or $\varepsilon_1 \dfrac{\partial V}{\partial n} = \varepsilon_2 \dfrac{\partial V}{\partial n}$ | $B_{1n} = B_{2n}$, or $\mu_1 \dfrac{\partial V_m}{\partial n} = \mu_2 \dfrac{\partial V_m}{\partial n}$ |

Thus, the solution to a magnetostatic case analogous to an electrostatic case is obtained from the latter by simply changing $V$ to $V_m$, $\mathbf{E}$ to $\mathbf{H}$, $\mathbf{D}$ to $\mathbf{B}$, and $\varepsilon$ to $\mu$.

As the first example of the use of this analogy, let us consider a sphere of radius $a$ and permeability $\mu$, situated in a vacuum in a uniform external magnetic field of intensity $\mathbf{H}_0$. This case is analogous to the case of a dielectric sphere of permittivity $\varepsilon$ situated in a vacuum in a uniform external electric field of intensity $\mathbf{E}_0$. The solution to the latter problem was found at the end of Section 4.3.4. According to Eqs. (4.3.70) to (4.3.72) and the foregoing analogy, the magnetic field inside the sphere in the present case is uniform, of the intensity

$$\mathbf{H}_{\text{in}} = \frac{3\mu_0}{\mu + 2\mu_0}\mathbf{H}_0, \qquad r < a, \tag{10.1.1}$$

and the $r$ and $\theta$ components of the field intensity vector outside the sphere are

$$H_{\text{out}\,r} = \left(1 + 2\frac{\mu - \mu_0}{\mu + 2\mu_0}\frac{a^3}{r^3}\right) H_0 \cos\theta, \qquad r > a, \tag{10.1.2}$$

$$H_{\text{out}\,\theta} = -\left(1 - \frac{\mu - \mu_0}{\mu + 2\mu_0}\frac{a^3}{r^3}\right) H_0 \sin\theta, \qquad r > a. \tag{10.1.3}$$

The magnetic flux density inside the sphere is simply $\mathbf{B}_{\text{in}} = \mu\mathbf{H}_{\text{in}}$; outside the sphere, $\mathbf{B}_{\text{out}} = \mu_0\mathbf{H}_{\text{out}}$. Figure 10.4 shows the lines of vectors $\mathbf{B}/\mu_0$ and $\mathbf{H}$ in this case.*

Because the magnetic field inside the sphere is *uniform*, from Eq. (10.1.1) we can also determine the magnetic field inside a sphere of *nonlinear*, but homogeneous, material. As the external field is uniform, it tends to magnetize uniformly the sphere of whatever material it is made. But a uniformly magnetized sphere again produces *inside* the sphere a uniform field, so that the total field is also

* The rest of this section is devoted to examples, and can be omitted, if desired, without loss of continuity.

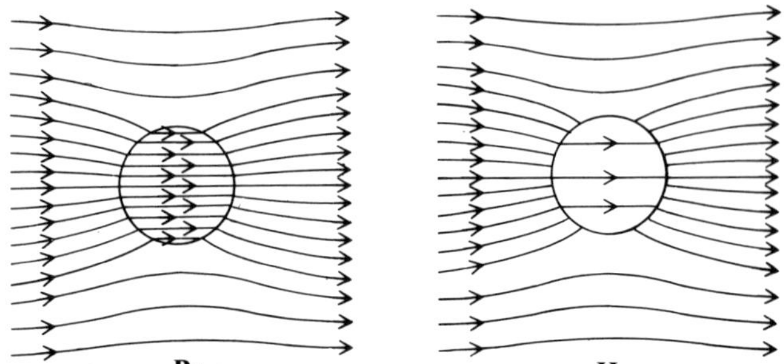


**Fig. 10.4** Lines of vectors $\mathbf{B}/\mu_0$ and $\mathbf{H}$ of a magnetic sphere in an external uniform field.

uniform. During the process of establishing the external field, assumed to be always uniform in increasing from zero to $\mathbf{H}_0$, all points of the sphere had an equal magnetization, increasing from zero to a certain value, the only difference being that it is not now proportional to $\mathbf{H}_0$. From the foregoing reasoning and from Eq. (10.1.1) we can readily determine the magnetic field intensity and the flux density inside a homogeneous ferromagnetic sphere.

For this purpose we first express $\mathbf{H}_{\text{in}}$ in Eq. (10.1.1) as a function of $\mathbf{H}_0$ and the magnetization $\mathbf{M}$ only. Now, $\mathbf{M} = \chi_m \mathbf{H}_{\text{in}} = (\mu/\mu_0 - 1)\mathbf{H}_{\text{in}}$, or

$$\mathbf{M} = \frac{\mu - \mu_0}{\mu_0} \frac{3\mu_0}{\mu + 2\mu_0} \mathbf{H}_0 = \frac{3(\mu - \mu_0)}{\mu + 2\mu_0} \mathbf{H}_0. \tag{10.1.4}$$

Since $\mathbf{B}_{\text{in}} = \mu \mathbf{H}_{\text{in}}$ for a linear medium, according to Eqs. (10.1.1) and (10.1.4)

$$\mathbf{H}_{\text{in}} = \frac{3\mu_0}{\mu + 2\mu_0} \mathbf{H}_0 = \mathbf{H}_0 - \frac{(\mu - \mu_0)}{\mu + 2\mu_0} \mathbf{H}_0 = \mathbf{H}_0 - \frac{\mathbf{M}}{3}. \tag{10.1.5}$$

As explained above, this expression for $\mathbf{H}_{\text{in}}$ is valid also in the present case. The magnetic flux density inside the sphere is thus given by

$$\mathbf{B}_{\text{in}} = \mu_0(\mathbf{H}_{\text{in}} + \mathbf{M}) = \mu_0(\mathbf{H}_0 + \tfrac{2}{3}\mathbf{M}). \tag{10.1.6}$$

In particular, for $\mathbf{H}_0 = 0$ Eqs. (10.1.5) and (10.1.6) represent the magnetic field intensity and the flux density inside a uniformly and permanently magnetized isolated sphere, with magnetization $\mathbf{M}$:

$$\mathbf{H}_{\text{in}} = -\frac{\mathbf{M}}{3}, \qquad \mathbf{B}_{\text{in}} = \mu_0 \frac{2\mathbf{M}}{3}. \tag{10.1.7}$$

The lines of vectors $\mathbf{B}/\mu_0$ and $\mathbf{H}$ of a uniformly magnetized sphere are shown in Fig. 10.5.

In conclusion, let us determine the magnetization, the magnetic flux density, and the magnetic field intensity inside a ferromagnetic sphere of a given magneti-

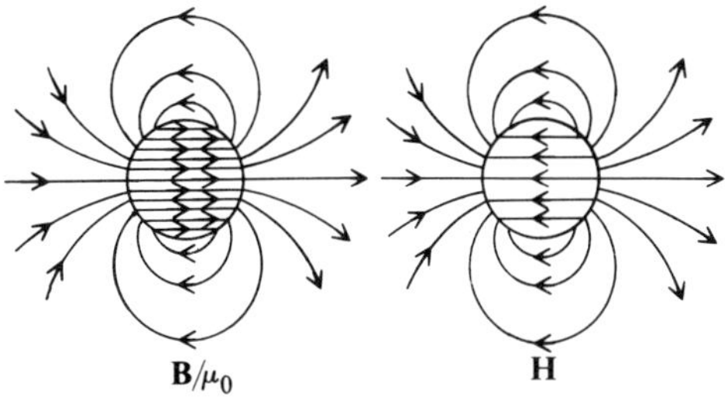


Fig. 10.5 Lines of vectors $\mathbf{B}/\mu_0$ and $\mathbf{H}$ of a uniformly magnetized sphere.

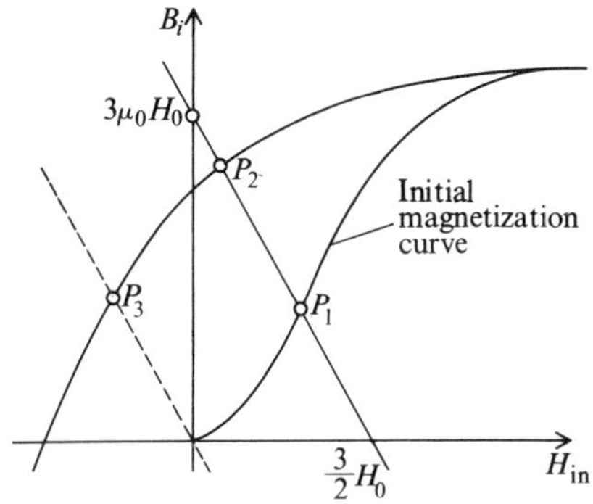


**Fig. 10.6** Construction for determining the working point on the magnetization curve of a ferromagnetic sphere situated in a uniform external field.

zation curve, say that shown in Fig. 10.6. The point describing the state of magnetization must be on the magnetization curve. A single point on the curve refers to all points inside the sphere, since the magnetization is uniform. If we suppose that the external uniform field is $\mathbf{H}_0$, then the point describing the field inside the sphere in the $H_{in}$–$B_{in}$ plane must also be on the straight line obtained by eliminating $\mathbf{M}$ from Eqs. (10.1.5) and (10.1.6). In the scalar form

$$B_{in} = \mu_0[H_0 + 2(H_0 - H_{in})] = \mu_0(3H_0 - 2H_{in}). \tag{10.1.8}$$

This straight line is also shown in Fig. 10.6, from which we can see that two solutions are possible, corresponding to the two different magnetization processes. If the field intensity $H_0$ was established from zero to the value $H_0$ without reversals, the state of magnetization is described by the point $P_1$. If, however, the external magnetic field intensity was increased until saturation was attained, and then reduced to the same value $H_0$, the point $P_2$ is the correct one. In both cases the magnetization of the sphere could easily be found by using the general relation $\mathbf{M} = \mathbf{B}_{in}/\mu_0 - \mathbf{H}_{in}$.

If the external field is first increased until saturation is attained and then switched off, the working point is the point $P_3$, on the magnetization curve and the straight line $B_{in} = -2\mu_0 H_{in}$.

In the same way it is possible to solve all other problems analogous to electrostatic problems of which we know the solution. This is not, however, of much practical importance, because only rarely can cases of static and quasi-static magnetic fields be approximated with simple geometrical forms. In addition, the magnetic materials used never have a constant permeability. For this reason, we shall not continue with other examples of this kind here, especially as several other cases are considered in Problems 10.G5 to 10.G11.

Another important practical problem of the static magnetic field relates to magnetic shielding. A region in space can be partially shielded from magnetic fields by enclosing it within a layer of some ferromagnetic substance. With a hypothetical material, which had infinitely large $\mu$, the shielding would be perfect, analogous to electrostatic shielding with conducting shells. But permeability is always finite; magnetic shielding is therefore imperfect. The simple example given below is intended to shed some light on this problem.

Let us take a spherical shell of outer radius $a$, inner radius $b$, and permeability $\mu$, situated in an external magnetic field of intensity $\mathbf{H}_0$, and from this data let us determine the magnetic field intensity $H_{\text{cav}}$ in the sphere cavity. Figure 10.7 shows the hollow sphere, with the axis of a spherical coordinate system assumed to be parallel to the external field $\mathbf{H}_0$. There are no macroscopic currents in the region considered, and the sphere is assumed to be homogeneous. The magnetic scalar potential function $V_m$ satisfies Laplace's equation (9.6.8). To solve for the magnetic field at all points, we must find the solution of Laplace's equation which satisfies the boundary conditions in this case.

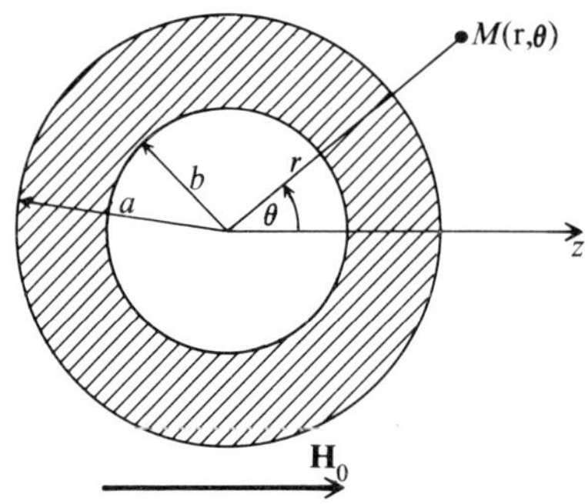


**Fig. 10.7** A hollow sphere in a uniform external magnetic field.

By symmetry, the potential $V_m$ is a function of coordinates $r$ and $\theta$ only. It can be represented in the form of Eq. (4.3.53),

$$V_{m(n)}(r, \theta) = [A_n r^n + B_n r^{-(n+1)}]\Theta_n(\theta), \tag{10.1.9}$$

where $\Theta_n(\theta)$ is a solution of the Legendre equation (4.3.54).

Now, at large distances from the sphere, the potential $V_m$ must reduce to the potential $V_{m0}$ of the external field $H_0$. With respect to the reference plane $\theta = \pi/2$, $V_{m0}$ is given by

$$V_{m0} = -H_0 r \cos\theta. \tag{10.1.10}$$

So, for points outside the sphere, we have to put $A_1 = -H_0$ and $A_n = 0$ for

$n \neq 1$. Thus, according to Eq. (10.1.9), outside the sphere

$$V_{m\,\text{out}} = (-H_0 r + B_1 r^{-2}) \cos\theta + \sum_{n\neq 1} B_n r^{-n}\Theta_n(\theta), \tag{10.1.11}$$

since $\Theta_1(\theta) = \cos\theta$.

The potential inside the sphere wall for $r = a$ must be equal to that given by Eq. (10.1.11) for $r = a$, for *all* $\theta$. Hence, inside the wall the potential must be of the form

$$V_{m\,\text{in}} = (C_1 r + C_2 r^{-2}) \cos\theta + \sum_{n\neq 1} C_n r^{-n}\Theta_n(\theta), \tag{10.1.12}$$

where $C_1, C_2, \ldots$ are constants.

The potential $V_{m\,\text{cav}}$ in the cavity cannot be infinite for $r = 0$, so that all $B_n$ in Eq. (10.1.9) must be zero. In addition, for $r = b$ it must have the value in Eq. (10.1.12) for $r = b$, for *all* $\theta$. So inside the cavity

$$V_{m\,\text{cav}} = Dr\cos\theta, \tag{10.1.13}$$

where $D$ is a constant. The magnetic field inside the cavity is thus uniform. Comparing Eq. (10.1.13) with Eq. (10.1.12), we conclude that except $C_1$ and $C_2$ all $C_n$–s must be zero. Otherwise, the condition $V_{m\,\text{cav}}(b) = V_{m\,\text{in}}(b)$ could not be satisfied for all $\theta$. In a similar way, we can find that in Eq. (10.1.11) all $B_n$–s must be zero for $n \neq 1$.

Assuming that $D = -H_{\text{cav}}$, let us determine $H_{\text{cav}}$ as a function of $H_0$. The following boundary conditions can be used to this purpose:

$$V_{m\,\text{out}}(a) = V_{m\,\text{in}}(a), \qquad \text{whence} \qquad -H_0 + B_1 a^{-3} = C_1 + C_2 a^{-3}; \tag{10.1.14}$$

$$V_{m\,\text{in}}(b) = V_{m\,\text{cav}}(b), \qquad \text{whence} \qquad C_1 + C_2 b^{-3} = -H_{\text{cav}}; \tag{10.1.15}$$

$$\mu_0 \frac{\partial V_{m\,\text{out}}}{\partial r} = \mu \frac{\partial V_{m\,\text{in}}}{\partial r} \qquad \text{for} \qquad r = a, \qquad \text{whence}$$

$$\mu_0(H_0 + 2B_1 a^{-3}) = \mu(-C_1 + 2C_2 a^{-3}); \tag{10.1.16}$$

$$\mu \frac{\partial V_{m\,\text{in}}}{\partial r} = \mu_0 \frac{\partial V_{m\,\text{cav}}}{\partial r} \qquad \text{for} \qquad r = b, \qquad \text{whence}$$

$$\mu_0 H_{\text{cav}} = \mu(-C_1 + 2C_2 b^{-3}). \tag{10.1.17}$$

From these equations we can obtain the following expression for $H_{\text{cav}}$:

$$H_{\text{cav}} = \frac{9H_0}{2\dfrac{\mu}{\mu_0} + 5 + 2\dfrac{\mu_0}{\mu} - 2\dfrac{(\mu - \mu_0)^2}{\mu_0\mu}\left(\dfrac{b}{a}\right)^3}. \tag{10.1.18}$$

In all cases of magnetic shielding $\mu \gg \mu_0$. The general expression in Eq. (10.1.18) then simplies to

$$H_{\text{cav}} \simeq \frac{9}{2\mu_r}\frac{1}{1 - (b/a)^3} H_0, \tag{10.1.19}$$

where $\mu_r = \mu/\mu_0$. (It should be noted that the above analysis is valid only if $\mu$ is constant.)

Some idea of the order of magnitude of the effect of magnetic shielding is useful. In the table below we give the ratios $H_{\text{cav}}/H_0$ for $\mu_r = 400$ (corresponding roughly to the permeability of soft iron for small fields), and $a = 10$ cm. The difference $d = a - b$ denotes the thickness of the sphere wall.

| $d$ (mm) | 2 | 4 | 10 | 20 |
|---|---|---|---|---|
| $H_{\text{cav}}/H_0$ | 0.188 | 0.098 | 0.041 | 0.023 |

These values correspond to the *static* field $H_0$. If the field varies in time, the shielding effect is more pronounced, due to eddy currents induced in the sphere wall.

*Problems referring to Section 10.1:* 10.T1 to 10.T6, 10.G1 to 10.G11, 10.N1 to 10.N5.

## 10.2 APPROXIMATE METHODS

**10.2.1 Introductory remarks.** The exact, or practically exact, methods described in the preceding sections are extremely useful for determining the magnetic field in some important cases. Unfortunately, they are all restricted to relatively simple geometrical structures. In practice, we most often encounter quite irregular structures, or, at least, structures not solvable by any exact method. Some approximate methods for their solution are therefore of considerable practical importance.

We dealt with one such type of approximate method in analyzing magnetic circuits in Sections 9.8.2 to 9.8.4. But many problems cannot be attacked by this method, and we have to develop still further methods, three of which we shall describe in some detail.

**10.2.2 The graphical method.** As in the case of the electrostatic field, the applicability of the graphical method is restricted to two types of magnetic fields: two-dimensional fields and axially symmetrical fields. In addition, we must assume that the medium is magnetically linear.

Consider a two-dimensional magnetic field with three different regions: (1) a region with $\mu \simeq \mu_0$ (air); (2) a region with $\mu \gg \mu_0$ (an idealized ferromagnetic substance having a constant permeability); and (3) a current-carrying region having $\mu \simeq \mu_0$, which is almost always the case.

Let us first consider region 1 (air). Figure 10.8 shows the cross-section of a part of that region. Let the differences $\Delta V_m$ between the two adjacent surfaces of equal magnetic potential, shown in thin lines, be very small. The magnitude of the magnetic field intensity vector **H** is then very nearly given by

$$H \simeq \Delta V_m / \Delta a_{Vm}. \tag{10.2.1}$$

If we denote by $\Delta\Phi_l$ the magnetic flux (per unit length along the two-dimensional system considered) represented by each line of the magnetic flux density, then,

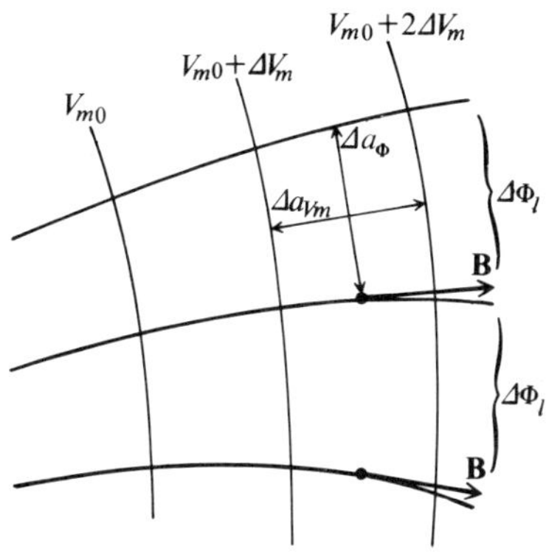


**Fig. 10.8** A two-dimensional static magnetic field.

according to Eq. (10.2.1), the approximate value of $\Delta\Phi_l$ is

$$\Delta\Phi_l \simeq B\Delta a_\Phi \simeq \mu_0 \frac{\Delta V_m}{\Delta a_{Vm}} \Delta a_\Phi, \tag{10.2.2}$$

from which

$$\frac{\Delta a_\Phi}{\Delta a_{Vm}} \simeq \frac{\Delta\Phi_l}{\mu_0 \Delta V_m}. \tag{10.2.3}$$

Thus, in the region where $\mu \simeq \mu_0$, the ratio $\Delta a_\Phi/\Delta a_{Vm}$ is very nearly constant throughout the cross-section of the system, provided that $\Delta V_m$ and $\Delta\Phi_l$ are sufficiently small. Since the lines of the magnetic flux density vector are perpendicular to the equipotential lines, the map obtained is in the form of very nearly similar curvilinear rectangles. If $\Delta\Phi_l$ and $\Delta V_m$ are chosen so that $\Delta\Phi_l/\mu_0\Delta V_m = 1$, the map of these curvilinear rectangles becomes a map of very nearly curvilinear squares.

Next, we consider region 2, that is, the region of high (but, by hypothesis, constant) permeability. The above reasoning remains valid, because the magnetic scalar potential can be defined for that region as well.

At the boundary surface of regions 1 and 2, the lines of the magnetic flux density vector refract according to the law given by Eq. (9.5.6), so the lines are virtually normal to the surface on the air side, but not normal on the ferromagnetic side. Hence, *from the air side only*, the surface of region 2 (the region of high permeability) is practically an equipotential surface. This fact is of great help in starting a plot of curvilinear squares in air when ferromagnetic boundaries are present. As an example, Fig. 10.9 is a rough sketch of the magnetic field lines and equipotential lines for a simple two-dimensional system.

In the case of region 3 (the current-carrying region) the magnetic scalar potential cannot be defined, as was explained in Section 9.6. However, the lines

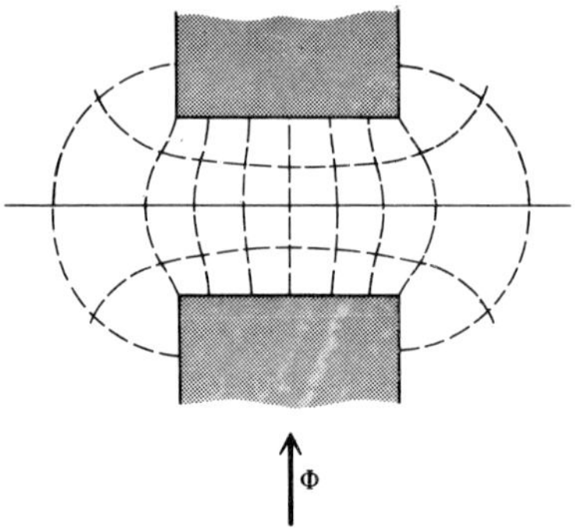


**Fig. 10.9** An example of the map of curvilinear squares.

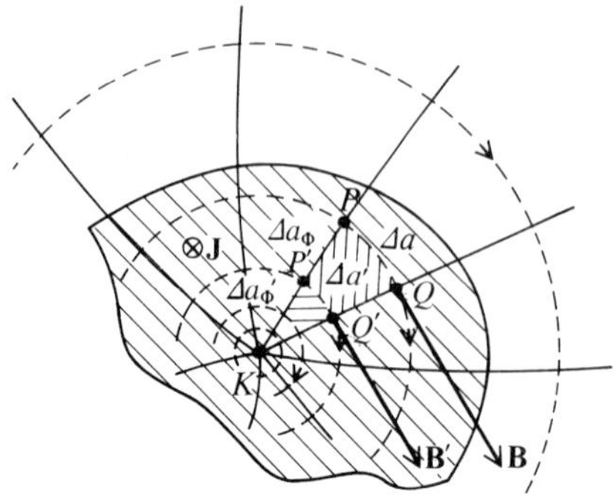


**Fig. 10.10** Lines of vector **B** and lines of no work inside a conductor.

of the magnetic flux density vector are *not* refracted at the surface of the conductor. This can be proved as follows. As no *surface* currents exist, $H_{1t} = H_{2t}$ or, the permeability of the conductor being equal to that of air, $B_{1t} = B_{2t}$. In addition, $B_{1n} = B_{2n}$, so that neither the lines of the magnetic flux density vector nor the lines *continuing* on the lines of equal magnetic potential (but not representing them any more, since $V_m$ cannot be defined inside a current-carrying conductor) are refracted. So two mutually orthogonal sets of lines can also be drawn inside the conductor, as indicated in Fig. 10.10.

Successive closed lines of the magnetic flux density vector inside the conductor will necessarily shrink around a certain point $K$. This point is usually referred to as the *core* or *kernel*, and represents the trace of the line along which $B = 0$. The lines continuing on the lines of equal magnetic scalar potential inside the conductor, which are referred to as the *lines of no work*, will consequently all meet together at the kernel, as shown in Fig. 10.10. So, if the position of the kernel is known, the general behavior of the plot inside the conductor is known as well.

Unfortunately, it is only possible to determine the position of the kernel in advance in certain special cases. But let us assume that the position of the kernel has been determined, and that the current density is equal at all points inside the conductor. Noting that **B** is perpendicular to the lines of no work, let us apply Ampère's circuital law to the curvilinear triangles PKQ and P′KQ′. We then obtain the following approximate relations:

$$B\Delta a = SJ, \qquad B'\Delta a' = S'J, \tag{10.2.4}$$

where $S$ and $S'$ stand for the areas of the two curvilinear triangles. Let the lines of **B** represent tubes of equal magnetic flux per unit length along the system; then, in addition,

$$B\Delta a_{\Phi} = B'\Delta a_{\Phi}'. \tag{10.2.5}$$

Comparing Eqs. (10.2.4) and (10.2.5),

$$\frac{\Delta a}{\Delta a_\Phi} = \frac{\Delta a'}{\Delta a_\Phi'}\frac{S}{S'}, \quad \text{or} \quad \frac{\Delta a'}{\Delta a_\Phi'} = \frac{S'}{S}\frac{\Delta a}{\Delta a_\Phi}. \tag{10.2.6}$$

Inside the current-carrying conductor the curvilinear rectangles are not similar, and the ratio $\Delta a'/\Delta a_\Phi'$ of their sides becomes smaller and smaller as the kernel is approached. Although much more difficult than sketching curvilinear squares in the current-free region, we can make a sketch of the magnetic field lines in a current-carrying region. Such a sketch is very useful for determining approximately the magnetic field of complex two-dimensional systems.

Once the field plot has been constructed, it is easy to determine approximately the magnetic flux density at any point. First note that an approximate form of Ampère's circuital law applicable to a field plot is

$$\Sigma B\,\Delta a_{Vm} \simeq \mu_0 I, \tag{10.2.7}$$

where the sum refers to a closed contour encircling a total current $I$. From the other side, according to Eq. (10.2.3) and the convention that curvilinear squares are obtained everywhere outside the current-carrying region, we have that $\Delta a_{Vm} \simeq \Delta a_\Phi$. Now, the product $B\Delta a_\Phi$ has the same value for all points, since each line represents a tube of equal magnetic flux. Thus, if $N_V$ is the number of equipotential lines, Eq. (10.2.7) yields

$$N_V B\,\Delta a_{Vm} \simeq \mu_0 I, \quad \text{so that} \quad B \simeq \mu_0 I/(N_V\,\Delta a_{Vm}). \tag{10.2.8}$$

From the known value of $B$ we can determine the magnetic flux per unit length of the system through any surface. We multiply $B$, as given by Eq. (10.2.8), by $\Delta a_\Phi$, and sum the expressions thus obtained over the desired number $N_\Phi$ of the tubes of flux:

$$\Phi_l \simeq N_\Phi B\,\Delta a_\Phi \simeq \mu_0 I N_\Phi/N_V, \tag{10.2.9}$$

since

$$\Delta a_{Vm} \simeq \Delta a_\Phi.$$

As an example, let us consider the coaxial cable shown in Fig. 10.11, with $b/a \simeq 3.3$. From the rough plot shown, according to Eq. (10.2.9), the magnetic flux between the conductors, through the strip $P_1P_2$ of unit length, is

$$\Phi_{lP_1P_2} \simeq \mu_0 I\tfrac{3}{16} = 0.187\,\mu_0 I.$$

The exact formula yields

$$\Phi_{lP_1P_2} = \frac{\mu_0 I}{2\pi}\int_a^b \frac{dr}{r} = \frac{\mu_0 I}{2\pi}\ln\frac{b}{a} \simeq 0.190\,\mu_0 I.$$

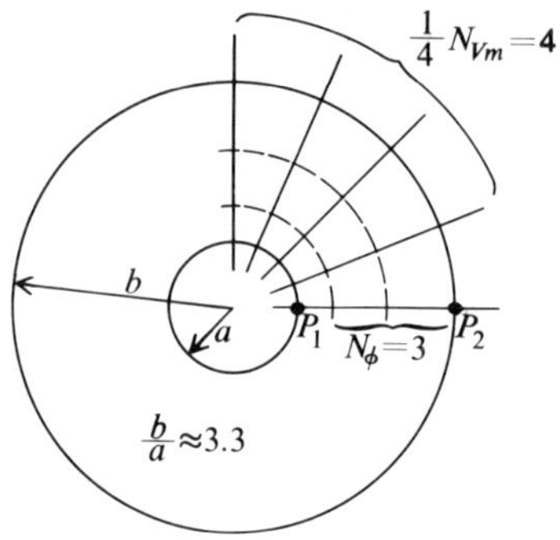


**Fig. 10.11** A coaxial cable as a two-dimensional static magnetic field problem.

In spite of the rather crude plot, the difference between the approximate and the exact values is very small.

As an example of plotting the magnetic field of current in a conductor of an irregular cross-section, Fig. 10.12 gives a rough sketch of curvilinear squares and rectangles for a solid cylindrical conductor of rectangular cross-section. In another example, the same conductor is assumed to reside on a ferromagnetic surface (see Fig. 10.13). A few other examples are given in Problems 10.G12 to 10.G15, but for further details and examples the reader is referred to other sources.*

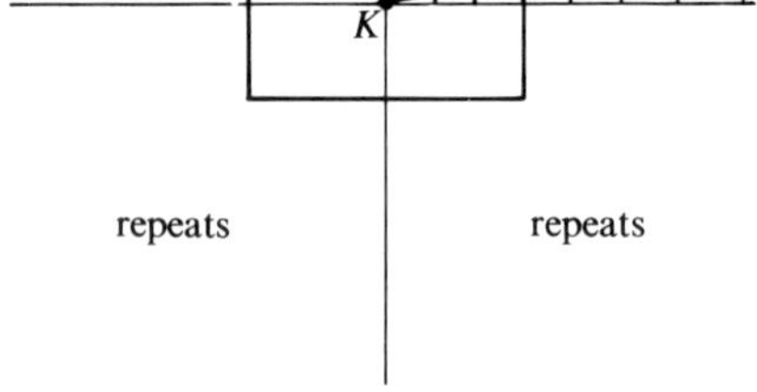


**Fig. 10.12** A rectangular current-carrying conductor.

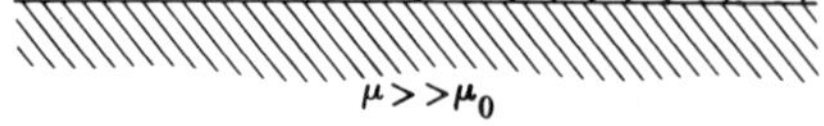


**Fig. 10.13** A rectangular current-carrying conductor residing on a ferromagnetic block.

* For example, Bewley, L. V., *Two-Dimensional Fields in Electrical Engineering*, Macmillan, New York, 1948, pp. 167–181, and Attwood, S. S., *op. cit.*, pp. 392–407.

**10.2.3 The numerical solution.** The approximate numerical solution of magnetostatic problems is used mainly for solving two-dimensional and axially symmetrical systems. For determining the magnetic field in the current-free region, Laplace's equation for $V_m$ is replaced, as in the electrostatic case, by a system of linear equations (difference equations). Determining the magnetic scalar potential in air between ferromagnetic bodies is completely analogous to that of determining the electric scalar potential in air between conducting charged bodies. The results derived there can be used in this case by direct analogy, and we shall not repeat the procedure here.

**10.2.4 Experimental methods.** In practice it is very often necessary to determine the magnetic field in a certain region of space only (for example, in air gaps of electrical motors and generators). In addition, the geometry of the system might be such that it is not possible to determine the field by any of the approximate methods mentioned. In such cases it is possible to use some experimental methods, three of which are described below.

A frequently used system for determining the magnetic flux density vector at a discrete set of points is by what is called *search coil.* The search coil is a very small coil when compared with the radius of curvature of the field lines of the field which is being examined. By a convenient method, it is possible to measure the magnetic flux of the external field through the coil, and hence to calculate the flux density. The complete experimental procedure cannot be explained, however, until the electromagnetic induction phenomena are considered (Section 11.4.1). But it is obvious that the particular direction of the search coil at a given point for which the magnetic flux through the coil is maximal has to be determined. The direction perpendicular to the coil in that case corresponds to the direction of the flux density vector.

Another method for determining the magnetic flux density at a given point is based on what is known as the *Hall effect.* To demonstrate this let us consider a thin strip carrying a direct current of density $\mathbf{J}$ and situated in a uniform magnetic field. For simplicity, let us assume that the magnetic flux density is perpendicular to the strip (see Fig. 10.14), and that the free charges $Q$, constituting the current in the strip, are positive. Since they are moving in the magnetic field, they are acted on by the force $\mathbf{F} = Q\mathbf{v} \times \mathbf{B}$, $\mathbf{v}$ designating the average charge velocity. Thus an "impressed" force acts on the moving charges, and they will move in the direction of the force. This will continue, however, only until the impressed force is compensated by the electrostatic force, due to the accumulated charges on the strip edges. If a sensitive instrument is connected as shown in Fig. 10.14, we can measure the electromotive force $\mathscr{E}$ between the edges of the strip. Now, $\mathscr{E}$ is proportional to the magnitude of the magnetic flux density vector $\mathbf{B}$, and conversely. Noting that $J = NQv$ ($N$ designating the number of free charges per unit volume of the conductor), we have

$$\mathscr{E} = avB = aJB/NQ, \qquad \text{or} \qquad B = \alpha\mathscr{E}, \quad \alpha = NQ/aJ.$$

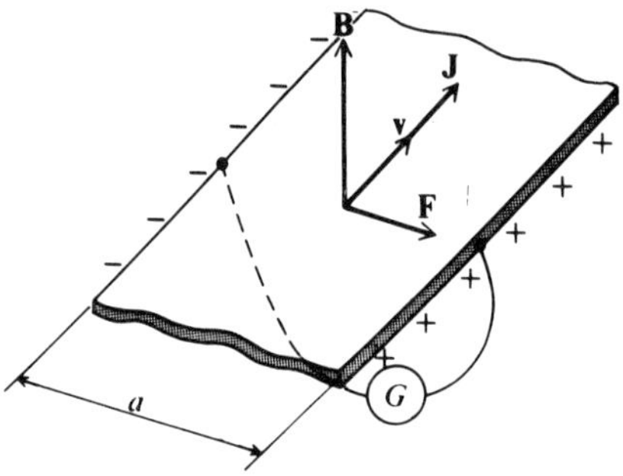


Fig. 10.14 Illustration of the Hall effect.

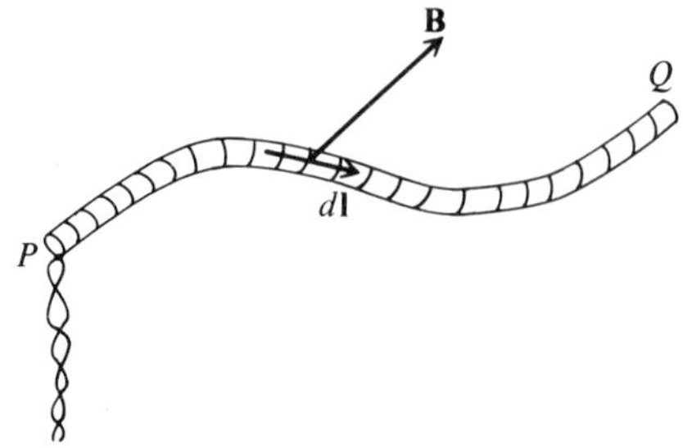


Fig. 10.15 A flexible coil for measuring the line integral of vectors **B** and **H**.

The constant $1/NQ$ is a characteristic of each conductor and is referred to as the *Hall coefficient*. It is interesting that neither the sign nor the magnitude of the Hall coefficient obtained experimentally always agree with the predictions of the classical theory. They agree, however, with the predictions of the quantum theory.

Finally, it is also possible to measure the line integral of either **B** or **H**. For this purpose, a thin nonferromagnetic strip (say, made of rubber) is wound densely along its length with $N_l$ turns per unit length. The two ends of such a flexible coil are placed at points $P$ and $Q$, between which the line integral has to be measured (see Fig. 10.15). If the cross-section area of the strip is $S$, the magnetic flux through the $N_l\,dl$ turns in the length element $d\mathbf{l}$ of the strip shown is

$$d\Phi = N_l\,dl\,\mathbf{S}\cdot\mathbf{B} = N_l S\mathbf{B}\cdot d\mathbf{l}.$$

Thus the total flux through the whole coil is

$$\Phi = \int d\Phi = N_l S\int_P^Q \mathbf{B}\cdot d\mathbf{l}.$$

The magnetic flux through the coil can be determined in exactly the same way as through the search coil, which will be explained in Section 11.4.1. The line integral of **B** along any open or closed contour can be measured in this manner. In particular, if points $P$ and $Q$ are allowed to coincide, the method can be used to prove experimentally Ampère's circuital law.

*Problems referring to Section 10.2:* 10.T7 to 10.T11, 10.G12 to 10.G15, 10.N6.

## PROBLEMS

### Test problems

**10.T1.** A coaxial cable with currents $I$ and $-I$ runs in air parallel to a plane ferromagnetic block. Is it necessary in this case to use the method of images for determining the magnetic field?

**10.T2.** A plane circular contour with current $I$ resides in air on the plane surface of a ferromagnetic block. What is roughly the magnetic flux through the contour as compared with the flux through it when isolated in space?

**10.T3.** Discuss whether the method of images described in Section 10.1.2 remains valid if the permeability of the block is high, but is a function of the magnetizing field.

**10.T4.** Suppose that a solution for the electric scalar potential is found for a dielectric cube situated in air, in an external uniform electric field. Is it possible to use this result for obtaining the magnetic scalar potential of a magnetic cube in a uniform external magnetic field, if (1) the permeability of the cube is high, but constant, and (2) the cube is made of a real ferromagnetic material which exhibits hysteresis?

**10.T5.** A solid sphere of a real ferromagnetic material (exhibiting hysteresis properties) is introduced into a uniform magnetic field through a nonuniform fringe field. Is the result obtained in Section 10.1.3 applicable in this case?

***10.T6.** It was shown in Section 10.1.3 that the magnetic field intensity $H$ inside a magnetic sphere situated in a uniform external magnetic field of intensity $H_0$ is smaller than $H_0$. What is the physical meaning of this result?

**10.T7.** Determine the positions of the kernels in two straight cylindrical conductors of circular cross-sections if they are touching, and if the currents in them are (1) in the same, and (2) in opposite directions.

**10.T8.** A cylindrical circular conductor carries a steady current and resides on the surface of a plane ferromagnetic block. Determine the position of the kernel.

**10.T9.** A cylindrical conductor resides in air parallel to a ferromagnetic block. In which direction is the kernel inside the conductor shifted under the influence of the secondary field due to the block?

**10.T10.** Is it possible for the kernel to be situated *outside* a current-carrying region?

**10.T11.** In Fig. 19.14 the free charges are assumed to be positive. Discuss the situation if the free charges are supposed to be negative electrons, which corresponds to the classical theory.

## General problems

**10.G1.** If a thin conductor, carrying a current of intensity $I$, runs parallel to two ferromagnetic blocks, as shown in Fig. 10.G1, determine the magnetic field in the air between the two blocks.

*Solution.* The problem can be solved by the method of magnetic images. We first determine the magnetic image of the current $I$ in, say, the lower block. This image is numbered 1. Now,

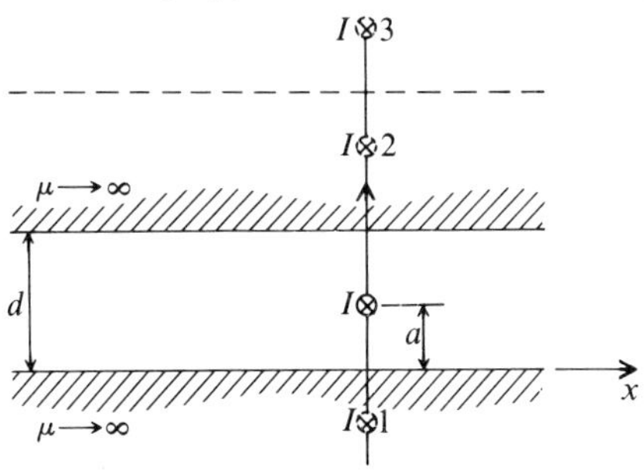


**Figure 10.G1**

the boundary conditions on the surface of the upper block are not satisfied. To satisfy them also, it is necessary to place magnetic images 2 and 3 of the current $I$ and its image 1 with respect to the upper block, as shown in the figure. This disturbs the boundary conditions on the lower block, so that we have to add images 4 and 5 (not shown) of 2 and 3 with respect to the lower block. We can see that this process continues indefinitely. Determination of the analytical expression for the $x$ and $y$ components of the resultant magnetic flux density at a certain point in the air layer is left to the reader.

**10.G2.** Repeat the preceding problem for the conductors of the cross-sections shown in Fig. 10.G2. Determine also the magnetic flux density inside the rectangular conductor.

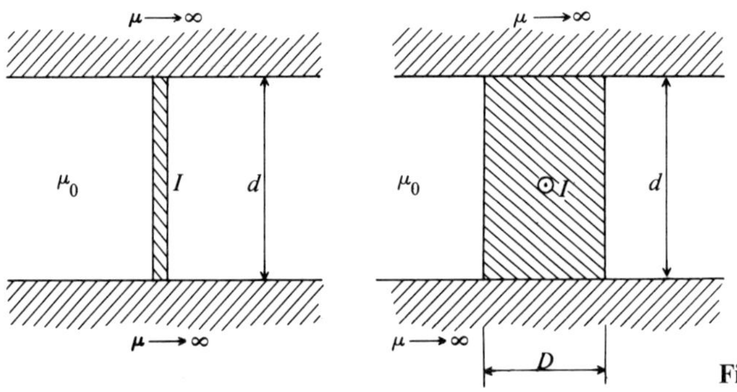


**Figure 10.G2**

***10.G3.** Two very large magnetic blocks of permeabilities $\mu_1$ and $\mu_2$ are divided by a plane surface, as shown in Fig. 10.G3. In medium 1, of permeability $\mu_1$, a thin straight conductor with a current of intensity $I$ runs parallel to the boundary surface. Prove:

(1) That the influence of medium 2 on the magnetic field in medium 1 can be reduced to the field of a straight current filament of intensity $\alpha I$. The filament is at the place of the image of current $I$ in the boundary surface. The whole space is assumed to be filled with a medium of permeability $\mu_1$.

(2) That the influence of medium 1 on the magnetic field produced in medium 2 by the current $I$ can be reduced to an additional current $\beta I$ in the conductor. The whole space is assumed to be filled with a medium of permeability $\mu_2$.

Determine the coefficients $\alpha$ and $\beta$.

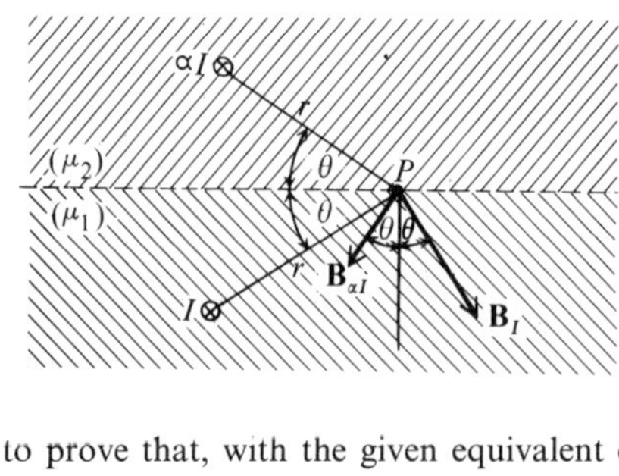


**Figure 10.G3**

*Solution.* It is necessary to prove that, with the given equivalent currents, it is possible to satisfy the necessary boundary conditions. If this is attained, then, according to the uniqueness theorem, the solution thus found is the only possible one.

Let us first determine the magnetic field at a point $P$ of the boundary surface, assuming that the whole space has permeability $\mu_1$. Assume that two conductors with currents exist—the real conductor, and a conductor at the place of its image and with current $\alpha I$, as required in statement (1) and shown in Fig. 10.G3. The flux density produced at $P$ by the currents $I$ and $\alpha I$ is respectively

$$B_I = \mu_1 \frac{I}{2\pi r}, \qquad B_{\alpha I} = \mu_1 \frac{\alpha I}{2\pi r}.$$

The directions of the two vectors are indicated in the figure.

The normal component of the resultant flux density vector is hence

$$B_{In} = B_I \cos\theta + B_{\alpha I}\cos\theta = \frac{\mu_1 I}{2\pi r}(1+\alpha)\cos\theta. \tag{a}$$

The tangential component of the resultant magnetic field intensity is similarly

$$H_{1t} = \frac{B_I}{\mu_1}\sin\theta - \frac{B_{\alpha I}}{\mu_1}\sin\theta = \frac{I}{2\pi r}(1-\alpha)\sin\theta. \tag{b}$$

According to statement (1), the components (a) and (b) have to be equal to the real components at $P$, in medium 1.

Let us now determine the magnetic field created at $P$ by a current $(1+\beta)I$ at the place of the real current and assume that the whole space has permeability $\mu_2$, as required by statement (2). The magnetic flux density at $P$ due to this current is

$$B_{(1+\beta)I} = \mu_2 \frac{(1+\beta)I}{2\pi r}.$$

The normal component of the magnetic flux density vector at $P$ is

$$B_{2n} = B_{(1+\beta)I}\cos\theta = \mu_2 \frac{(1+\beta)I}{2\pi r}\cos\theta, \tag{c}$$

and the tangential component of the magnetic field intensity vector

$$H_{2t} = \frac{B_{(1+\beta)I}}{\mu_2}\sin\theta = \frac{(1+\beta)I}{2\pi r}\sin\theta. \tag{d}$$

According to statement (2), the components (c) and (d) have to be equal to the real components at $P$, in medium 2.

Let us try to determine the coefficients $\alpha$ and $\beta$ so that the real boundary conditions, $B_{1n} = B_{2n}$ and $H_{1t} = H_{2t}$, are satisfied. According to Eqs. (a) to (d), these conditions result in the following two equations:

$$\mu_1(1+\alpha) = \mu_2(1+\beta), \qquad \text{and} \qquad 1-\alpha = 1+\beta,$$

from which

$$\alpha = (\mu_2 - \mu_1)/(\mu_1 + \mu_2), \qquad \beta = (\mu_1 - \mu_2)/(\mu_1 + \mu_2). \tag{e}$$

It is hence possible to determine the magnetic field in both media as described in the text of the problem.

***10.G4.** Analyze the special cases of the preceding problem when: (1) $\mu_2 \gg \mu_1$, and (2) $\mu_1 \gg \mu_2$.

*Solution.* (1) According to Eq. (e) of the preceding problem, for $\mu_2 \gg \mu_1$ we obtain that $\alpha \simeq 1$ and $\beta \simeq -1$. Hence, the magnetic field in medium 1 is identical with the field of the current $I$ and its positive image $\alpha I = I$ existing in a homogeneous medium of permeability $\mu_1$. This is in fact the image method described in Section 10.1.2. In medium 2, the field is practically nonexistent, since it is equal to the field of current $(1 + \beta)I = (1 - 1)I = 0$. The lines of the magnetic flux density vector thus enter the ferromagnetic substance virtually at right angles, and close upon themselves through a thin layer near the surface of the ferromagnetic block.

(2) For $\mu \gg \mu_2$, Eq. (e) of the preceding problems yields $\alpha = -1$ and $\beta = 1$. This means that the magnetic field in medium 1 is the same as the field of the current $I$ and its negative image $-I$, situated in a homogeneous medium of permeability $\mu_1$. In medium 2, the magnetic field is the same as that produced by a current $(1 + \beta)I = 2I$ at the place of the real current $I$, situated in a medium of permeability $\mu_2$. The lines of the magnetic flux density vector in medium 2 are thus parts of circles centered at the conductor axis. Note that this is an example in which the magnetic field lines are *not* perpendicular to the surface of very high permeability (see Fig. 10.G4). This is not in contradiction to the boundary condition of Eq. (9.5.6). In this case $\alpha_2 = \pi/2$, so that $\alpha_1$ can be *any* angle, and the condition is still satisfied.

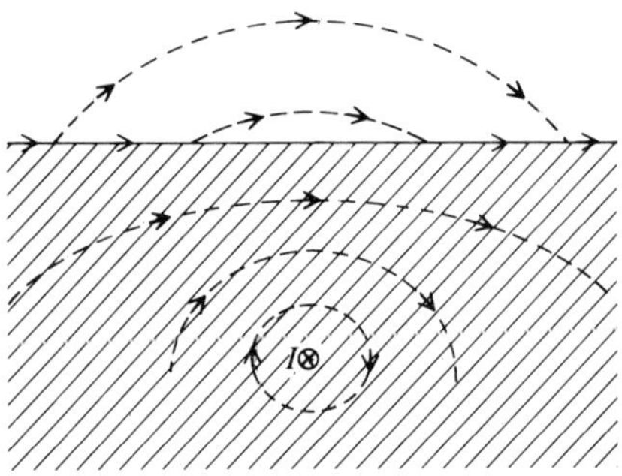


**Figure 10.G4**

***10.G5.** Derive the expressions in Eqs. (10.1.1) to (10.1.3) for the magnetic field intensity inside and outside a sphere of radius $a$ and permeability $\mu$, situated in a uniform external magnetic field of intensity $H_0$, starting from Laplace's equation for the magnetic scalar potential $V_m$. In other words, do *not* use the result found for the analogous electrostatic case, but follow the same procedure.

**10.G6.** A sphere of a high permeability $\mu$ is situated in an external magnetic field of intensity $H_0$. A small cavity is made inside this sphere, far from its surface. Determine the approximate value of the magnetic field intensity inside the cavity, if the cavity is (1) spherical, (2) needle-shaped, in the direction of $H_0$, and (3) disk-shaped, perpendicular to $H_0$. From your result work out what is the maximal theoretical possibility of the magnetic shielding with a cavity inside a sphere of high permeability.

*Solution.* The cavity is small and its presence cannot appreciably alter the magnetization of the sphere, except in the immediate vicinity of the cavity. So the spherical cavity can be considered simply as a sphere of permeability $\mu_0$, situated in an external uniform magnetic field in a medium of permeability $\mu$. This field in which the cavity is situated is, of course, the uniform field $H_{\text{in}}$ inside the solid sphere, residing in air in a uniform external field of intensity $H_0$. The intensity of that field is given by Eq. (10.1.1):

$$H_{\text{in}} = \frac{3\mu_0}{\mu + 2\mu_0} H_0.$$

The field intensity in the spherical cavity is obtained by using the same formula, but changing $H_0$ to $H_{\text{in}}$ and interchanging $\mu$ and $\mu_0$:

$$H_{\text{cav}} = \frac{3\mu}{\mu_0 + 2\mu} H_{\text{in}} = \frac{9\mu_0\mu}{(\mu + 2\mu_0)(\mu_0 + 2\mu)} H_0 \qquad \text{(spherical cavity).} \tag{a}$$

For $\mu \gg \mu_0$ this equation becomes

$$H_{\text{cav}} \simeq \frac{9\mu_0}{2\mu} H_0 = \frac{9}{2\mu_r} H_0 \qquad \text{(spherical cavity).} \tag{b}$$

The field inside the cavity is hence very much smaller than the external field. For example, if $\mu_r = 5000$, $H_{\text{cav}} \simeq 9 \times 10^{-4} H_0$.

If we consider the needle-shaped cavity in the direction of the field $\mathbf{H}_0$, and hence also of $\mathbf{H}_{\text{in}}$, then, in this case, according to the boundary condition for the tangential components of the magnetic field intensity

$$H_{\text{cav}} = H_{\text{in}} = \frac{3\mu_0}{\mu + 2\mu_0} H_0 \simeq \frac{3}{\mu_r} H_0 \qquad \text{(needle-shaped cavity).} \tag{c}$$

Thus, the needle-shaped cavity gives approximately fifty per cent better protection than the spherical cavity. It is, however, essential that the cavity should have the proper orientation.

When the disk-shaped cavity is perpendicular to $\mathbf{H}_0$ (and hence also to $\mathbf{H}_{\text{in}}$), according to the boundary conditions we have that

$$H_{\text{cav}} = \frac{\mu}{\mu_0} H_{\text{in}} = \frac{3\mu}{\mu + 2\mu_0} H_0 \simeq 3H_0 \qquad \text{(disk-shaped cavity),}$$

so that no shielding effect exists.

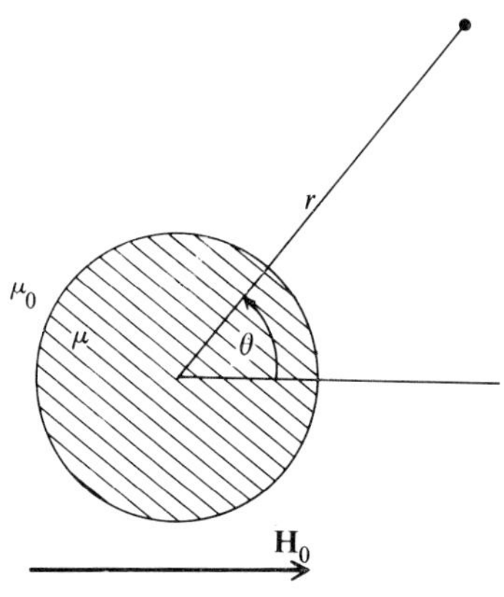


**Figure 10.G7**

***10.G7.** A very long solid cylinder of radius $a$ and permeability $\mu$ is situated in a uniform external magnetic field of intensity $H_0$. The vector $\mathbf{H}_0$ is perpendicular to the cylinder axis (see Fig. 10.G7). Determine the magnetic scalar potential and hence the vectors **B** and **H** at all points.

*Answer.* $V_{m\,\text{in}} = -\dfrac{2\mu_0}{\mu + \mu_0} H_0 r \cos\theta$, $V_{m\,\text{out}} = \left(\dfrac{\mu - \mu_0}{\mu + \mu_0}\dfrac{a^2}{r} - r\right) H_0 \cos\theta$.

***10.G8.** In the cylinder in the preceding problem a small coaxial cylindrical cavity is made, of radius $b$ which is much smaller than $a$. Determine the magnetic field intensity in the cavity if $\mu \gg \mu_0$.

*Answer.* For $\mu_r = 1000$, $H_{\text{cav}} = 0.004\,H_0$.

***10.G9.** A very long cylinder of radius $a$ is uniformly magnetized with magnetization **M**, perpendicular to its axis. Determine the vectors **B** and **H** at all points.

*Solution.* According to Eqs. (9.3.3) and (9.3.6), the macroscopic resultants of the microscopic currents exist only on the surface of the cylinder, and their density is $J_s = |\mathbf{M} \times \mathbf{u}_n| = M \sin\theta$ (see Fig. 10.G9a). Exactly the same distribution of surface current is obtained if two cylinders of radius $a$, carrying currents of the same density $J$, but in opposite directions, are parallel and almost coincidental, having axes at a very small distance $b$ (see Fig. 10.G9b). The surface current density in that case is given by

$$J_s = Jad\theta b \sin\theta/(ad\theta) = bJ \sin\theta.$$

The magnetic field both inside and outside the cylinder can hence be determined in the same way as the field of the two parallel cylinders, with $bJ = M$. Determination of the magnetic field of the magnetized cylinder, using this reasoning, is left to the reader. [*Hint:* use vector forms for the magnetic field inside and outside a current-carrying conductor of circular cross-section. See Problem 8.G17.]

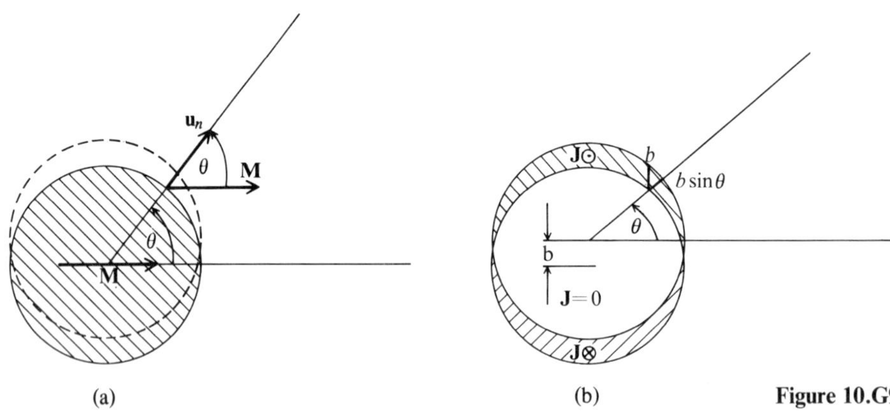


**Figure 10.G9**

***10.G10.** A hollow cylinder, of inner radius $b$, outer radius $a$, and permeability $\mu$, is situated in an external uniform magnetic field of intensity $H_0$. Determine the magnetic field intensity inside the cavity.

*Solution.* Since it is assumed that no macroscopic currents are present, the magnetic scalar potential $V_m$ is in this case defined at all points. Let the $z$ axis of the cylindrical coordinate system coincide with the cylinder axis, and the polar axis be in the direction of the external field $H_0$. The potential is then not a function of the coordinate $z$. We found the general solution of Laplace's equation in this particular case of symmetry in Section 4.3.3. According to Eqs. (4.3.22), (4.3.29), and (4.3.32), it is in the form of a sum of terms

$$V_{m(k)}(r, \theta) = (A_{1k} \sin k\theta + A_{2k} \cos k\theta)(B_{1k} r^k + B_{2k} r^{-k}), \tag{a}$$

where $k$ is an arbitrary number. By symmetry, the potential is an even function of $\theta$, so that $A_{1k} = 0$. If we introduce the new constants $C_{1k}$ and $C_{2k}$, the magnetic scalar potential outside the cylinder must be in the form

$$V_m = \Sigma\,(C_{1k} r^k + C_{2k} r^{-k}) \cos k\theta, \qquad k > 0. \tag{b}$$

At large distances from the cylinder, the potential $V_m$ must approach the potential $V_{m0} = -H_0 r \cos\theta$ of the external magnetic field. This is possible if of all the $C_{1k} r^k$ terms the term $C_{11} r$ only is nonzero. Thus we have

$$V_{m\,\text{out}} = (-H_0 r + C_{21} r^{-1}) \cos\theta + \sum_{k \neq 1} C_{2k} r^{-k} \cos k\theta, \qquad r \geq a. \tag{c}$$

Inside the cylinder, on its outer surface, the potential $V_{m\,\text{in}}$ must be equal to $V_{m\,\text{out}}$ for *all* $\theta$. So inside the cylinder

$$V_{m\,\text{in}} = (D_{11} r + D_{21} r^{-1}) \cos\theta + \sum_{k \neq 1} D_{2k} r^{-k} \cos k\theta, \qquad b < r < a. \tag{d}$$

The potential at the axis of the cylinder, in the cavity, must be finite. For $r = b$ it must be equal to $V_{m\,\text{in}}$ for *all* $\theta$. Hence it follows that the potential inside the cavity is given by

$$V_{m\,\text{cav}} = -H_{\text{cav}} r \cos\theta, \tag{e}$$

and that all $C_{2k}-$s and $D_{2k}-$s in the sum terms of Eqs. (c) and (d) must be zero. In Eq. (e) $-H_{\text{cav}}$ stands for an as yet undetermined constant. The magnetic field inside the cavity is uniform.

$V_{m\,\text{out}}$, $V_{m\,\text{in}}$, and $V_{m\,\text{cav}}$ have to satisfy the boundary conditions on the outer and inner surfaces of the cylinder, analogous to Eqs. (10.1.14) to (10.1.17). Four linear equations in $C_{21}$, $D_{11}$, $D_{21}$, and $H_{\text{cav}}$ are obtained, from which $H_{\text{cav}}$ can be determined. The result is

$$H_{\text{cav}} = \frac{4\mu/\mu_0}{(\mu/\mu_0 + 1)^2 - (\mu/\mu_0 - 1)^2 (b/a)^2} H_0. \tag{f}$$

If $\mu \gg \mu_0$, Eq. (f) can be written in a simpler, approximate form:

$$H_{\text{cav}} \simeq \frac{4}{\mu_r} \frac{1}{1 - (b/a)^2} H_0, \qquad \mu_r \gg 1. \tag{g}$$

For the numerical example, let us assume that $\mu_r = 400$. Then for $a = 10$ cm, the following values of $H_{\text{cav}}/H_0$ are obtained for different values $d = a - b$ of the cylinder thickness:

| $d$ (mm) | 2 | 4 | 10 | 20 |
|---|---|---|---|---|
| $H_{\text{cav}}/H_0$ | 0.252 | 0.128 | 0.053 | 0.028 |

***10.G11.** In the cavity of the cylinder in the preceding problem, another hollow cylinder, of outer radius $c$ and inner radius $d$, is coaxially placed. Determine the magnetic field intensity inside the cavity of the inner cylinder.

*Answer.* The magnetic field inside the cavity of the inner cylinder is uniform, in the direction of $H_0$ and, for $\mu \gg \mu_0$, approximately of intensity

$$H_{\text{cav}} = \frac{16H_0}{(\mu/\mu_0)^2(1 - b^2/a^2)(1 - c^2/b^2)(1 - d^2/c^2) + 4(\mu/\mu_0)(1 - b^2d^2/a^2c^2)}.$$

For example, if $\mu_r = 400$, $a = 10$ cm, $b = 9.8$ cm, $c = 5$ cm, and $d = 4.8$ cm, then $H_{\text{cav}} \simeq 0.0294\, H_0$. Comparing this result with that for $d = 4$ mm from the preceding problem, we conclude that with less ferromagnetic material approximately four times better shielding effect is obtained. It is always better to use several thin sheets as a shield rather than one thick sheet.

**10.G12.** Sketch the plot of the magnetic field produced by two very thin parallel strips carrying equal currents in opposite directions, assuming that the distance $d$ between the strips is equal to their width. From this plot determine the magnetic flux density at the center between the strips, and compare the result with the exact value, which can be calculated from the result of Problem 8.G8.

**10.G13.** Two cylindrical conductors of equal radii are touching. Sketch the plot of the magnetic field if the currents in the two conductors are of equal magnitude, and (1) in the same, and (2) in opposite directions.

**10.G14.** Sketch the magnetic field plot for the two-dimensional systems shown in Fig. 10.G14a and 10.G14b.

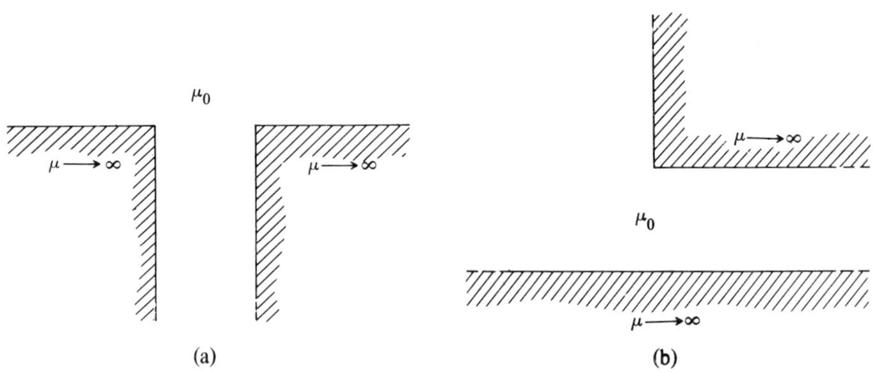


**Figure 10.G14**

**10.G15.** A cylindrical conductor of radius $a = 1$ cm carries a steady current of intensity $I = 100$ A, and runs parallel to a ferromagnetic block of very high permeability. The axis of the conductor is at a distance $h = 2a$ from the block. Sketch the plot of the magnetic field, and hence determine the magnitude of the flux density vector at the surface of the conductor, at the point closest to the block. Check the result by determining this flux density exactly.

## Numerical problems

**10.N1.** A ferromagnetic sphere of permeability $\mu_r = 500$ is introduced into a uniform magnetic field of intensity $H_0 = 10$ A/m. Determine the magnetization of the sphere, and the density of the equivalent surface currents and magnetic charges.

**10.N2.** Determine the thickness of the wall of a ferromagnetic hollow sphere, of permeability $\mu_r = 5000$ and outer radius $a = 20$ cm, so that the magnetic field inside the cavity is 1000 times smaller than the external uniform field.

**10.N3.** Repeat the preceding problem for a cylinder of the same outer radius and the same permeability (see Problem 10.G10).

***10.N4.** Compute the ratio $H_{cav}/H_0$ for fixed $a$, $c$, $d$, and $\mu$ in Problem 10.G11, for different values of $b$. Plot this dependence and discuss the influence of the thickness of the outer shell.

***10.N5.** Repeat the preceding problem for $c$ variable only, $a$, $b$, $d$, and $\mu$ being as before.

**10.N6.** Compute the electromotive force of a copper Hall element of width $a = 1$ cm, situated in a magnetic field of flux density $B = 0.01$ T, if the density of the current through the element is $J = 1$ A/mm$^2$. [The number of free electrons in copper is approximately $N = 8.4 \times 10^{22}$ per cubic centimeter (assuming one free electron per atom).]

**Suggested reading.** The list progresses towards more advanced texts. For full bibliographical details, see the Bibliography.

1. BEWLEY, L. V., *Two-Dimensional Fields in Electrical Engineering*, Chapters 5 and 6.

2. WEBER, E., *Electromagnetic Fields, Theory and Applications*, Chapters 5, 6, 7 and 8.

# PART IV

# THE QUASI-STATIONARY ELECTROMAGNETIC FIELD

CHAPTER 11

# ELECTROMAGNETIC INDUCTION

## 11.1 THE TOTAL FORCE BETWEEN TWO SMALL MOVING CHARGES

**11.1.1 Introduction.** In Sections 8.1 to 8.3 we noted that the total force between two small *moving* electric charges is not given by Coulomb's law alone. Certain corrective terms have to be added to take account of the additional force. We noted also that, for reasonable velocities of the charges, the corrective terms are very small when compared with Coulomb's force, and, consequently, that it is not possible to determine them experimentally in a direct way.

So, in dealing with these additional forces in the case of steady current systems, an additional term was determined not by direct measurements on the two moving charges, but in an indirect way. The total force on a large number of charges in a known state of *steady* motion was measured, and reduced *mathematically* to contributions of pairs of individual moving charges. The additional term to Coulomb's force

$$\mathbf{F}_{m12} = \mu_0 \frac{Q_1 Q_2}{4\pi} \frac{\mathbf{v}_2 \times (\mathbf{v}_1 \times \mathbf{u}_{12})}{r^2} \tag{11.1.1}$$

was thus introduced, and is referred to as the magnetic force. The symbols $\mathbf{v}_1$ and $\mathbf{v}_2$ represent the velocities of the charges $Q_1$ and $Q_2$, and $\mathbf{F}_{m12}$ the additional force by which the moving charge $Q_1$ acts on the moving charge $Q_2$. This additional term is sufficient for describing the forces in all cases between *steady* electric current systems. But this is *not* the total additional force between two charges in an arbitrary state of motion, not even when two isolated charges are moving with uniform velocities. Such a simple system is not a steady-state system, since the distance between the two charges and the observer varies in time. So, the steady-state additional force term, the magnetic force, is not necessarily the only corrective term to the Coulomb force. To obtain the total force between two charges in slow motion, forces between *nonsteady* current systems have to be investigated. This could be done by experiments with the nonsteady motion of a large number of charges, that is, by considering the mutual effects of nonsteady current systems with both the geometry and the current density varying with time. The most simple experiments are those with varying currents and charges in stationary conductors.

In Chapter 13 we shall see that if currents and charges are time-varying, the fields which they produce do not vary in synchronism with them. Instead, changes of the fields are always retarded with respect to changes of currents and charges. The more rapid the changes and the larger the distances from their sources, the larger is the retardation. If currents and charges vary sufficiently slowly with time and are distributed in a sufficiently small region of space, the retardation effect in determining their interactions can be ignored.

Let us suppose that the currents and charges do indeed vary so slowly with time that the retardation can be ignored. Then both the electric and the magnetic field due to them are practically the same at any instant as if they were created by the currents and charges existing at that instant, and assumed to be stationary. We can observe additional effects between nonsteady current systems by comparing the expected effects of steady currents and static electric charges with the real existing effects in the nonsteady case. Systems satisfying the foregoing requirement are referred to as *quasi-stationary systems* (that is, systems which are almost stationary), and the field by which the forces between the electric charges can be described is known as the *quasi-stationary electromagnetic field.* In this part of the text that is the only field we shall consider. We shall analyze the general electromagnetic field in the fifth and final part of the text.

**11.1.2 The force between two slow small charges in an arbitrary state of motion.** The forces acting between *all* systems of quasi-stationary macroscopic electric currents and charges can be explained if a *single* term is added to the Coulomb force and the magnetic force between two electric charges. This term is of the form

$$\mathbf{F}'_{e12} = -\mu_0 \frac{Q_1 Q_2}{4\pi} \frac{\partial}{\partial t}\left(\frac{\mathbf{v}_1}{r}\right)_{\mathbf{v}_2=0}. \tag{11.1.2}$$

$\mathbf{F}'_{e12}$ represents the force exerted by a charge $Q_1$ on a charge $Q_2$. The latter is assumed to be stationary at the particular instant considered, whence the subscript $\mathbf{v}_2 = 0$ (see Fig. 11.1). The additional force $\mathbf{F}'_{e21}$ exerted by charge $Q_2$ on charge $Q_1$ is obtained from Eq. (11.1.2) by interchanging subscripts 1 and 2.

Hence, from *macroscopic* experiments with quasi-stationary currents and clusters of elemental charges, we deduce that the total force exerted by a small

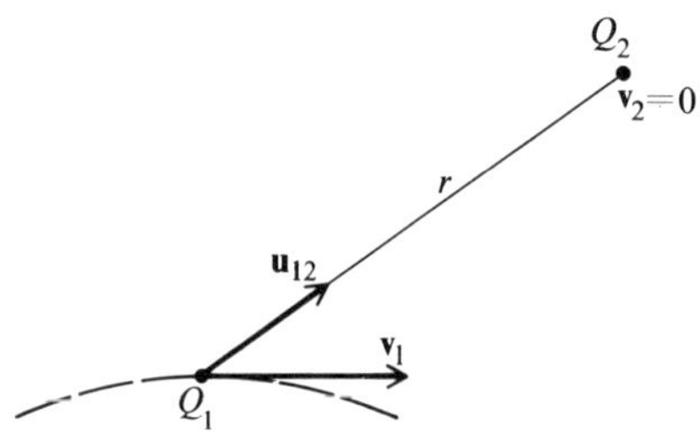


**Fig. 11.1** An illustration for the third term in the law of force between two moving charges.

charge $Q_1$ on a charge $Q_2$ is

$$\mathbf{F}_{12} = \frac{1}{4\pi\varepsilon_0}\frac{Q_1 Q_2}{r^2}\mathbf{u}_{12} + \frac{\mu_0}{4\pi}\frac{Q_1 Q_2}{r^2}\mathbf{v}_2 \times (\mathbf{v}_1 \times \mathbf{u}_{12}) - \frac{\mu_0}{4\pi} Q_1 Q_2 \frac{\partial}{\partial t}\left(\frac{\mathbf{v}_1}{r}\right)_{\mathbf{v}_2 = 0}$$

(the total force exerted by charge $Q_1$ on charge $Q_2$, as deduced from macroscopic experiments). (11.1.3)

The force $\mathbf{F}_{21}$ exerted by the charge $Q_2$ on charge $Q_1$ is obtained from this equation by interchanging subscripts 1 and 2.

We can see that in the general case $\mathbf{F}_{12}$ is *not* equal to $\mathbf{F}_{21}$. A qualitative explanation of this phenomenon is the same as that in the case of the magnetic force, that is, the system of two moving charges has to be considered as a *three*-body problem, the third "body" being the electric and magnetic fields, which vary in time. Newton's third law cannot be expected to hold for the forces between only *two* bodies of such a system.

**11.1.3 The generalized concept of the electric field. The physical meaning of the electromagnetic field.** The expression in Eq. (11.1.3) for the total force exerted by a charge $Q_1$ on a charge $Q_2$ can be *conveniently* split into two parts. One part depends explicitly upon the velocity $\mathbf{v}_2$ of charge $Q_2$, the other is not a function of $\mathbf{v}_2$:

$$\mathbf{F}_{12} = Q_2\left[\frac{1}{4\pi\varepsilon_0}\frac{Q_1}{r^2}\mathbf{u}_{12} - \frac{\mu_0 Q_1}{4\pi}\frac{\partial}{\partial t}\left(\frac{\mathbf{v}_1}{r}\right)_{\mathbf{v}_2=0}\right] + Q_2\mathbf{v}_2 \times \left[\frac{\mu_0}{4\pi}\frac{Q_1}{r^2}(\mathbf{v}_1 \times \mathbf{u}_{12})\right] \tag{11.1.4}$$

We recognize the first term in the first brackets as the Coulomb electric field intensity produced by the charge $Q_1$. The only difference when compared with the static electric field intensity is that it varies with time, since $r$ varies with time. The force on $Q_2$, due to the second term in the first brackets, is also obtained by multiplying the charge $Q_2$ by a given vector function. By analogy, it is natural to refer to it as an additional electric field intensity created by a *moving* charge. The *total electric field intensity* of a point charge $Q$ at a fixed point (with respect to an observer) can be defined as

$$\mathbf{E} = \frac{1}{4\pi\varepsilon_0}\frac{Q}{r^2}\mathbf{u} - \frac{\mu_0 Q}{4\pi}\frac{\partial}{\partial t}\left(\frac{\mathbf{v}}{r}\right)$$

(definition of the total electric field intensity of a moving charge). (11.1.5)

($\mathbf{v}$ is the instantaneous velocity of the charge, and the unit vector $\mathbf{u}$ is directed conventionally *away* from the charge.)

The third term in Eq. (11.1.4) is the familiar expression for the force on a charge $Q_2$ moving in the magnetic field of charge $Q_1$. The magnetic flux density vector $\mathbf{B}$ created by a point charge $Q$, moving with instantaneous velocity $\mathbf{v}$, is given by the known expression

$$\mathbf{B} = \frac{\mu_0}{4\pi} \frac{Q\mathbf{v} \times \mathbf{u}}{r^2}. \tag{11.1.6}$$

[$\mathbf{v}$, $\mathbf{u}$, and $r$ have the same meaning as in Eq. (11.1.5).] With Eqs. (11.1.5) and (11.1.6) the total force exerted on a charge $Q_2$ by a charge $Q$ can be written *formally* in the same form as the Lorentz force in Eq. (8.3.6):

$$\mathbf{F}_{\text{on } Q_2} = Q_2\mathbf{E} + Q_2\mathbf{v}_2 \times \mathbf{B}. \tag{11.1.7}$$

In this equation, however, $\mathbf{E}$ stands for the *generalized* electric field intensity defined by Eq. (11.1.5), and hereafter, in this book, the symbol $\mathbf{E}$ will *always* be used with that meaning. For static charges, it is identical with the electric field intensity defined earlier. Thus, in this generalized sense, if a stationary charge in a region of space is acted on by a force $\mathbf{F} = Q\mathbf{E}$, we say that an electric field exists in this region. $\mathbf{E}$ is not, however, necessarily a vector function of the space coordinates only, and is not necessarily conservative.

A very important additional conclusion can be drawn from the foregoing considerations. The force on a charge $Q_2$ given by either Eq. (11.1.3) or Eq. (11.1.7) is a physical *entity*. Only for convenience is it split in Eq. (11.1.7) into two distinct parts, derivable from the two vector functions of the coordinates and of time, referred to as the electric and magnetic fields. But it should be clearly understood that the *electric and magnetic fields are two parts of a unique physical phenomenon*. By a simple argument we can always trace the simultaneous existence of both of them in *all* cases, although one of them might be almost completely suppressed in a certain region of space.

For example, in the case of an electrostatic field it might be stated that no magnetic field is present at all. This is, however, basically incorrect. There are no static charges in nature, if only we go to dimensions small enough. Electrons revolve around nuclei, and the electrons in metallic conductors are in perpetual, unorganized, Brownian-like motion. Thus the magnetic field exists even in the electrostatic case, but its *macroscopic resultant* is negligible.

Similarly, a permanent magnet may be considered as the source of a pure magnetic field, but this again is true only for macroscopic resultants. A microscopic electric field exists inside and in the immediate vicinity of every atom, and the macroscopic resultants of this field again are negligible, but not nonexistent.

This unique physical field, which we split for convenience into two parts, and refer to as the generalized electric and the magnetic fields, is known as the *electromagnetic field. The electromagnetic field is the only really existing field created by electric charges*, for, as explained, the electric and magnetic fields are only names for two extreme cases of this unique electromagnetic field. But strictly there is no physical phenomenon in which one of these two parts is present by itself.

*Problems referring to Section 11.1:* 11.T1 to 11.T4.

## 11.2 ELECTROMAGNETIC INDUCTION IN THE CASE OF A STATIONARY CLOSED LOOP

**11.2.1 The electric field of a quasi-stationary current system.** Let us now consider a quasi-stationary current system. By definition, the currents of such a system vary so slowly that the magnetic field due to them is virtually equal to that produced by steady currents which, at that precise instant, have the same distribution as varying currents. This can include slowly moving, current-carrying conductors, or stationary conductors with slowly varying currents, or both. We shall see in Section 13.4.3 that, in sinusoidal steady-state cases, circuits of laboratory dimensions can usually be considered as quasi-stationary up to frequencies of several million cycles per second.

At every point of such a current system the current density $\mathbf{J}$ is a function of time. If we denote by $Q$ the charge of the free charge carriers, by $\mathbf{v}$ their average velocity, and by $N$ the number of free charge carriers per unit volume, we find that, according to Eq. (11.1.5) and the principle of superposition, the electric field intensity (in the generalized sense) created by these moving charges is given by

$$\mathbf{E} = -\frac{\mu_0}{4\pi} \sum_v Q \frac{\partial}{\partial t}\left(\frac{\mathbf{v}}{r}\right) + \text{a possible Coulomb term.} \tag{11.2.1}$$

The sum refers to the volume $v$ occupied by the varying currents. Macroscopically, this sum can be represented as an integral:

$$\mathbf{E} = -\frac{\mu_0}{4\pi} \int_v \frac{\partial}{\partial t}\left(\frac{NQ\mathbf{v}}{r}\right) dv + \text{a possible Coulomb term.} \tag{11.2.2}$$

(Since $N$ and $Q$ are constants, they can be introduced under the sign of partial differentiation with respect to time.) But $NQ\mathbf{v} = \mathbf{J}$, the current density, so that Eq. (11.2.2) becomes

$$\mathbf{E} = -\frac{\mu_0}{4\pi} \int_v \frac{\partial}{\partial t}\left(\frac{\mathbf{J}}{r}\right) dv + \text{a possible Coulomb term.} \tag{11.2.3}$$

The volume $v$ is assumed to enclose all the moving charges, and it can be imagined to be very large, and constant in time. The partial derivative sign can then be taken outside the integral:

$$\mathbf{E} = -\frac{\partial}{\partial t}\left(\frac{\mu_0}{4\pi} \int_v \frac{\mathbf{J}\, dv}{r}\right) + \text{a possible Coulomb term.} \tag{11.2.4}$$

But, according to Eq. (8.9.16), the integral in parentheses is exactly the magnetic vector potential $\mathbf{A}$ (at the instant of time considered) created by all the currents of the system. Thus we have finally

$$\boxed{\mathbf{E} = -\frac{\partial \mathbf{A}}{\partial t} + \text{a possible Coulomb term}}\;. \tag{11.2.5}$$

**11.2.2 Faraday's law of electromagnetic induction for a stationary closed loop.** Let us now consider a closed conducting loop situated in this electric field. Every free charge inside the loop is acted on by the electric field defined by Eq. (11.2.5). The line integral of the possible Coulomb term around a closed circuit is always zero, as we proved in Section 1.6. The first term, however, is not of the Coulomb type, and acts as an impressed field along the conductor. The electromotive force induced in the whole closed stationary loop is given by

$$\mathscr{E} = \oint_C \mathbf{E} \cdot d\mathbf{l} = -\oint_C \frac{\partial \mathbf{A}}{\partial t} \cdot d\mathbf{l} = -\frac{\partial}{\partial t} \oint_C \mathbf{A} \cdot d\mathbf{l}. \tag{11.2.6}$$

According to Eq. (8.9.19), the line integral of $\mathbf{A}$ is equal to the magnetic flux through the closed contour. The electromotive force induced in any *stationary* closed contour by any system of varying currents, or by steady currents in moving conductors, or by both, is hence simply equal to the negative rate of change of the magnetic flux through the contour:

$$\boxed{\oint_C \mathbf{E} \cdot d\mathbf{l} = -\frac{\partial \Phi}{\partial t} = -\int_S \frac{\partial \mathbf{B}}{\partial t} \cdot d\mathbf{S}} \qquad \text{(stationary loop).} \tag{11.2.7}$$

This remarkably simple result is of great practical importance, and is referred to as *Faraday's law of electromagnetic induction* for a stationary loop. For it was Faraday who, in 1831, discovered experimentally that, in a stationary closed conducting loop, current is induced whenever the magnetic flux through the loop varies.

As the first example of the application of this law, let us consider a permanent magnet approaching a stationary loop, as shown in Fig. 11.2. The permanent magnet is equivalent to a system of macroscopic currents, and since it is moving, the magnetic flux created by these currents through the loop varies in time. According to the assumed positive direction of the contour, $\partial\Phi/\partial t > 0$, so that the electromotive force induced in the loop is in the direction indicated. This electromotive force creates the current in the loop, producing a magnetic field, which is shown in the figure in dashed lines. As a result, the change of the magnetic flux, caused initially by the magnet motion, is reduced. So the induced current tends to *lessen* the change of the magnetic flux through the contour. This qualitative rule for determining the induced current in a stationary loop is true for any direction of the motion of the magnet, and is known as the *Lenz law*.

For the second example, let us consider a very long cylindrical solenoid of radius $a$, tightly wound by $N_l$ turns per unit length, and carrying a varying current $i = I_m \cos \omega t$. Suppose that a single turn of radius $b$ ($b > a$) encircles the solenoid, as shown in Fig. 11.3. The accelerated charges which constitute the varying solenoid current create a varying electric field (in the generalized sense) at all points. By symmetry, the lines of the vector potential $\mathbf{A}$ are circles centered at the solenoid axis. According to Eqs. (8.9.19) and (8.7.16), the instantaneous value of the vector

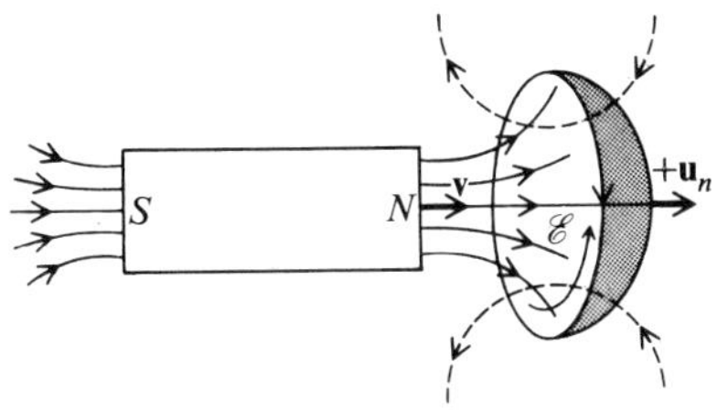


Fig. 11.2 An illustration of the Lentz law.

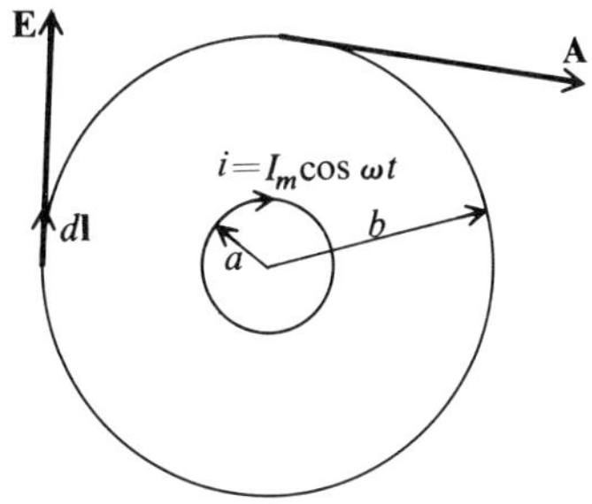


Fig. 11.3 Electric field of a solenoid with varying current.

potential at a distance $r$ from the solenoid axis ($r > a$) is

$$A = \mu_0 N_l i a^2 \pi / 2\pi r = \mu_0 a^2 N_l I_m \cos \omega t / 2r. \tag{11.2.8}$$

So, from Eq. (11.2.5), the instantaneous value of the induced electric field is

$$E = -\frac{\partial A}{\partial t} = -\mu_0 \frac{a^2 N_l}{2r} \frac{\partial i}{\partial t} = \mu_0 \frac{a^2 N_l I_m \omega \sin \omega t}{2r}. \tag{11.2.9}$$

The lines of vector **E** are also circles centered at the solenoid axis. The induced electromotive force in the circular loop shown in the figure is hence

$$\oint_C \mathbf{E} \cdot d\mathbf{l} = \oint_C E\, dl = E(b) \oint_C dl = E(b) 2\pi b = \mu_0 a^2 \pi\, N_l I_m \omega \sin \omega t. \tag{11.2.10}$$

This result can be obtained directly from Eqs. (11.2.7) and (8.7.16). However, a much clearer physical picture of the process of induction is created in this way.

The system we have described represents an idealized electric transformer. In real transformers, a ferromagnetic core with two coils, primary and secondary, is used, as Fig. 11.4 shows. Varying current in the primary coil creates in the core a varying magnetization, which is equivalent to the varying macroscopic currents, represented in the figure by dashed lines. All these varying currents create a varying electric field in the surrounding space, and an electromotive force is thus

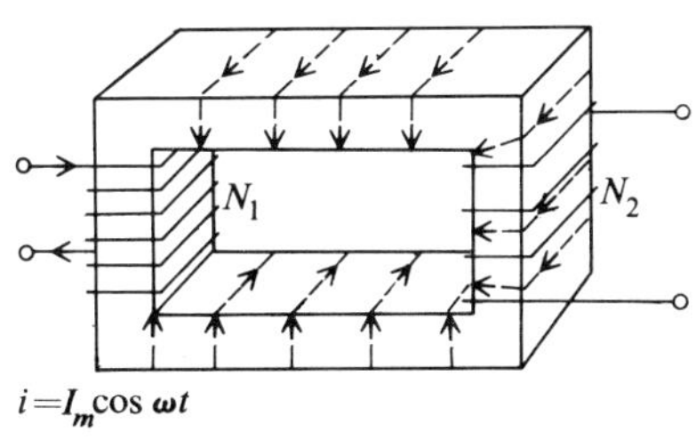


Fig. 11.4 A transformer.

induced in the secondary coil. If the core has a very high permeability, the magnetic flux $\Phi$ is practically localized in the core. So Eq. (11.2.7) can be used for determining very accurately the electromotive force induced in the secondary coil, with $\Phi$ representing the flux through the core. Note that *all* the turns of the secondary coil are situated in this varying electric field; consequently, if $N_2$ is the number of secondary turns, the electromotive force induced will be very nearly $N_2$ times larger than that induced in a single turn.

The electric transformer is a device used frequently in practice. The electromotive force induced by the varying currents in stationary conductors is therefore often referred to as the *transformer electromotive force.*

**11.2.3 Maxwell's generalization of Faraday's law.** As we noted above, Faraday discovered the law of electromagnetic induction, as shown in Eq. (11.2.7), experimentally, by measuring the current induced in a closed *conducting* loop. He therefore postulated the law as valid for any *conducting* closed loop. However, from the foregoing derivation of the law of electromagnetic induction, it is obvious that no conducting closed loop is implied in Eq. (11.2.7). The generalized electric field intensity **E**, created by a moving electric charge, is of such a mathematical form that Eq. (11.2.7) is satisfied for any closed contour in space. If this closed contour coincides with a conducting closed contour, the secondary, easily detectable effect of an induced current appears. But, if the conducting contour is removed, the electric field with the general property described by Eq. (11.2.7) still exists.

This fact was first postulated by James Clerk Maxwell, the founder of electromagnetic theory in its present form. The statement that Eq. (11.2.7) is valid for any closed contour, not necessarily a conducting one, is referred to as the *generalized law of electromagnetic induction.*

Let us now consider the law of electromagnetic induction from the standpoint of the formal, but extremely useful, splitting of the unique electromagnetic field into two separate fields, the electric and the magnetic. The relation in Eq. (11.2.7), which the generalized electric field intensity **E** satisfies in all circumstances, can then be interpreted as a mathematical expression stating that a varying magnetic field is always accompanied by a varying electric field. But the electric and magnetic fields are two inseparable parts of the unique electromagnetic field, so it might be inferred that the inverse formal effect must also exist, that is, that a varying electric field must produce a varying magnetic field. In Chapter 13 we shall see that this follows directly from the general force law of Eq. (11.1.3) and the definitions of the generalized electric and magnetic fields.

We shall refer to Eq. (11.2.7) as the *Faraday-Maxwell law* in integral form. The differential form of this law is obtained by expressing first of all the left-hand side of Eq. (11.2.7) as a surface integral of curl **E**,

$$\oint_C \mathbf{E}\cdot d\mathbf{l} = \int_S \operatorname{curl}\mathbf{E}\cdot d\mathbf{S} = -\int_S \frac{\partial \mathbf{B}}{\partial t}\cdot d\mathbf{S} \qquad \text{(the integral form of the Faraday-Maxwell law).} \tag{11.2.11}$$

Now, the contour $C$ is quite arbitrary, and $S$ represents any surface limited by it. From the integral equation it therefore follows that the generalized electric field intensity and the magnetic flux density are connected by the differential equation

$$\boxed{\operatorname{curl} \mathbf{E} = -\frac{\partial \mathbf{B}}{\partial t}} \quad \text{(the differential form of the Faraday-Maxwell law).} \tag{11.2.12}$$

Both forms of the Faraday-Maxwell law are derived for quasi-stationary electromagnetic fields. In Chapter 13, however, we shall show that it is a generally valid law, as it describes correctly the relation between the electric and magnetic parts of the unique electromagnetic field in all cases of macroscopic electromagnetic fields.

*Problems referring to Section 11.2:* 11.T5 to 11.T9, 11.G1, 11.N1.

## 11.3 THE ELECTROMOTIVE FORCE INDUCED IN CONDUCTORS MOVING IN A STATIC MAGNETIC FIELD

Let us consider a piece of conducting wire moving with a velocity $\mathbf{v}$ in a static, but otherwise arbitrary magnetic field (see Fig. 11.5). All the elemental charges inside the wire are acted on by an electromagnetic force $\mathbf{F} = Q\mathbf{v} \times \mathbf{B}$, since they, together with the wire, are moving in the magnetic field. Freely movable electrons inside the metallic wire will be pushed towards one end of the wire, making the other end positively charged, as indicated in the figure. In the static and quasi-static case, the forces on the free charges cannot exist inside a conductor. The distribution of the charges along the conductor will be exactly such that the electric field created by these charges is of equal magnitude, but of opposite direction to the product $\mathbf{v} \times \mathbf{B}$. It can be said that an external electric field acts along the conductor, producing a total electromotive force

$$\boxed{\mathscr{E} = \int_1^2 (\mathbf{v} \times \mathbf{B}) \cdot d\mathbf{l}} \quad \text{(a moving piece of wire).} \tag{11.3.1}$$

Since the conducting path in this case is not closed, no current through the wire will result.

Now let us move on to consider a closed conducting filament moving in a static magnetic field (see Fig. 11.6). In this case, the electromotive force around the contour is given by

$$\mathscr{E} = \oint_C (\mathbf{v} \times \mathbf{B}) \cdot d\mathbf{l} = \oint_C (d\mathbf{l} \times \mathbf{v}) \cdot \mathbf{B}. \tag{11.3.2}$$

This expression can be transformed as follows. First, if we denote by $d\mathbf{l}'$ the distance covered by an element $d\mathbf{l}$ in a time interval $dt$, we have that $\mathbf{v} = d\mathbf{l}'/dt$. Now,

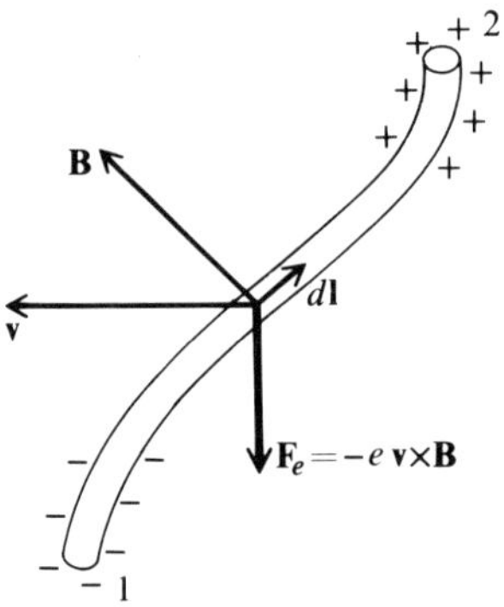


Fig. 11.5 A piece of wire moving in a magnetic field.

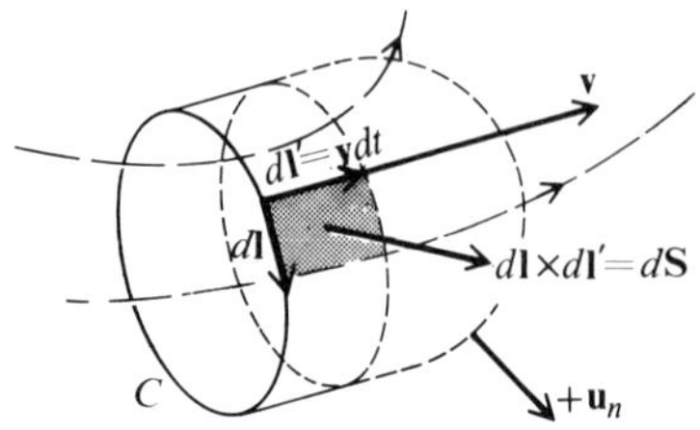


**Fig. 11.6** Construction for deriving the law of electromagnetic induction for moving contours.

$dt$ is a constant and can be taken outside the integral, so Eq. (11.3.2) becomes (see Fig. 11.6):

$$\mathscr{E} = \frac{1}{dt} \oint_C \mathbf{B} \cdot (d\mathbf{l} \times d\mathbf{l}') = \frac{1}{dt} \int_{dS} \mathbf{B} \cdot d\mathbf{S}. \qquad (11.3.3)$$

The integral on the right-hand side represents the magnetic flux through the strip surface which is swept by the contour $C$ in the time interval $dt$. But Fig. 11.6 tells us that we may regard this strip surface as part of a surface limited by the contour $C$ at the initial instant of time. After the time interval $dt$, the magnetic flux through $C$ will hence be *smaller* than before, by exactly the amount of the flux through the swept strip. If we denote by $\Phi$ the magnetic flux at the initial instant, then the *increase* in magnetic flux through the contour $C$ in the interval $dt$ is

$$d\Phi = -\oint_C \mathbf{B} \cdot (d\mathbf{l} \times d\mathbf{l}') = -\int_{dS} \mathbf{B} \cdot d\mathbf{S},$$

so that Eq. (11.3.3) finally becomes

$$\boxed{\mathscr{E} = -\frac{d\Phi}{dt}} \qquad \text{(a moving closed contour).} \qquad (11.3.4)$$

The change of flux through the contour is assumed to be due to the motion of the contour only, the magnetic field being constant in time. The electromotive force given in Eq. (11.3.4) is frequently referred to as the *motional electromotive force.*

The expression in Eq. (11.3.4) for the electromotive force induced in a closed contour due to the contour motion in a static magnetic field is equivalent to Eq. (11.3.2). However, if we deal with a moving piece of wire which is not closed Eq. (11.3.4) cannot be used directly, and it is better and more natural to use Eq. (11.3.1).

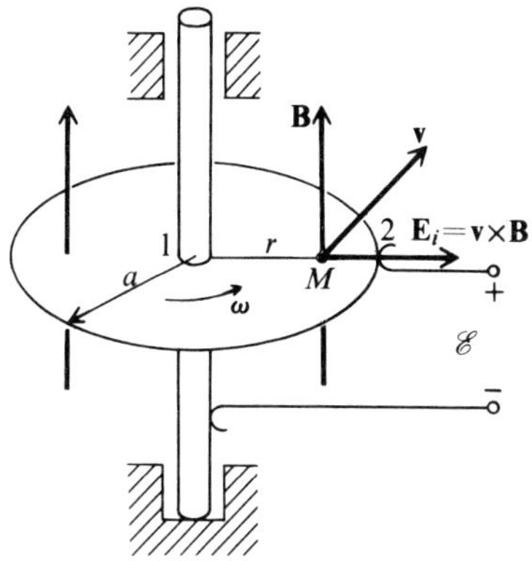


**Fig. 11.7** Faraday's disk generator.

**11.3.1 The Faraday disk generator.** The Faraday disk generator is sketched in Fig. 11.7. A metal disk of radius $a$ rotates with a uniform angular velocity $\omega$, in a uniform magnetic field of flux density $\mathbf{B}$, perpendicular to the plane of the disk. Sliding contacts are provided at the axis and periphery of the disk. Electrons rotating with the disk are moving in the magnetic field, and so an impressed electric field exists in the disk. By reference to Fig. 11.7 and using Eq. (11.3.1), we can easily determine the magnitude of the electromotive force of such a generator:

$$\mathscr{E} = \int_1^2 (\mathbf{v} \times \mathbf{B}) \cdot d\mathbf{l} = \int_0^a \omega r B \, dr = \frac{\omega a^2 B}{2}. \tag{11.3.5}$$

The direction of $\mathscr{E}$ is indicated in the figure.

**11.3.2 A simple generator of alternating electromotive force.** Consider a single rectangular loop with sides $a$ and $b$, rotating in a uniform magnetic field of flux density $\mathbf{B}$ perpendicular to the axis of rotation (see Fig. 11.8). The two ends of the loop are connected to two rings provided with sliding contacts. If the instant corresponding to the situation shown is taken as $t = 0$, the magnetic flux through

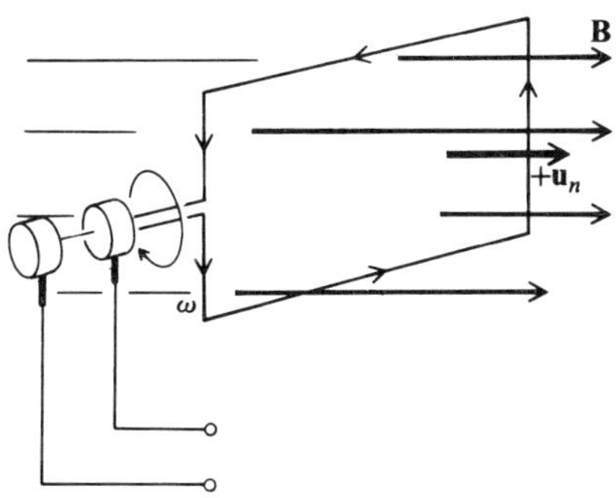


**Fig. 11.8** A simple generator of sinusoidal electromotive force.

the loop is $\Phi = abB \cos \omega t$. According to Eq. (11.3.4), the electromotive force of such a generator is

$$\mathscr{E} = -d\Phi/dt = abB\omega \sin \omega t. \tag{11.3.6}$$

So it varies in time according to the simple harmonic law.

The same result can also be obtained by starting from Eq. (11.3.2); this is left as an exercise for the reader.

*Problems referring to Section 11.3:* 11.T10 to 11.T12, 11.G2, 11.N2.

## 11.4 THE GENERAL LAW OF ELECTROMAGNETIC INDUCTION (FARADAY'S LAW)

We now consider a *closed* filamentary contour moved and deformed in an arbitrary manner in a *time-varying* magnetic field. This time-varying magnetic field is accompanied by a time-varying electric field at all points, and hence also along the contour. In addition, parts of the contour are moving in the magnetic field, and an electromagnetic force on the charges is also present. This can be represented by the equivalent electric field intensity $\mathbf{v} \times \mathbf{B}$. The total electromotive force induced in the contour in this case is given by the sum of Eqs. (11.2.6) and (11.3.2), viz. Eq. (11.3.4):

$$\mathscr{E} = -\oint_C \frac{\partial \mathbf{A}}{\partial t}\cdot d\mathbf{l} + \oint_C (\mathbf{v} \times \mathbf{B})\cdot d\mathbf{l} = -\int_S \frac{\partial \mathbf{B}}{\partial t}\cdot d\mathbf{S} + \oint_C (\mathbf{v} \times \mathbf{B})\cdot d\mathbf{l}. \tag{11.4.1}$$

Now, we have seen that both terms on the right-hand side can be written in the same mathematical form, as negative rates of change of the magnetic flux through the contour. They are due to two different causes. The first is due to a change of flux through the contour as a result of the time-varying magnetic field; the second is a consequence of the change of flux due to the motion and deformation of the contour in the magnetic field considered to be static. So Eq. (11.4.1) can be written as

$$\mathscr{E} = \left(-\frac{\partial \Phi}{\partial t}\right)_{\text{varying field}} + \left(-\frac{\partial \Phi}{\partial t}\right)_{\text{motion in the field}}, \tag{11.4.2}$$

or in a more compact form:

$$\boxed{\mathscr{E} = -\frac{d\Phi}{dt}} \quad \text{(Faraday's law of electromagnetic induction).} \tag{11.4.3}$$

The change of flux through the contour can be due to one or both of the causes explicitly explained in Eq. (11.4.2). Equation (11.4.3) represents *Faraday's law of electromagnetic induction in the general form.* In applying Eq. (11.4.3) we should always remember that it represents a short mathematical form of the clumsy but complete and unambiguous expression for the induced electromotive force given in Eq. (11.4.2). Otherwise, by formally interpreting Eq. (11.4.3), we can come in some instances to erroneous conclusions. Several such examples are described in Problems 11.G5 to 11.G10.

**11.4.1 The measurement of flux density by a search coil.** Let us now consider a small flat coil of area $S$, made of $N$ turns of very thin wire, and situated in a static magnetic field, assuming that the dimensions of the coil are much smaller than the curvature of the field lines. The ends of the coil are tightly twisted together, and connected to a device, called the ballistic galvanometer, which is used for measuring the charge that flows through the coil (see Fig. 11.9). Let the magnetic flux density at the point where the coil is placed have a magnitude $B$, and let the vector $\mathbf{B}$ and the adopted positive normal to the coil surface make an angle $\theta$. Then the total magnetic flux through the coil is $\Phi = NBS\cos\theta$.

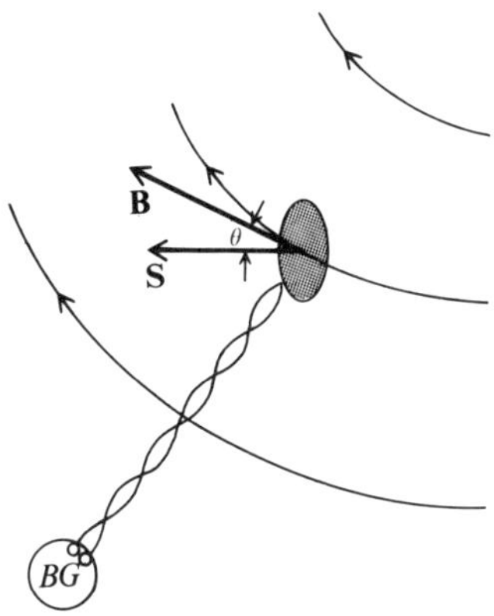


**Fig. 11.9** A search coil.

Suppose now that we either take the coil briskly out of the field, or extinguish the field by switching off the current producing it. During the transient period, the electromotive force will act in the coil, its magnitude and direction in both cases being given by Eq. (11.4.3). If we denote by $R$ the total resistance of the search coil, the leads, and the ballistic galvanometer, and by the $i$ the instantaneous intensity of the current (in the assumed positive direction), then the following equation is valid for any instant during the transient period:

$$\mathscr{E} = -d\Phi/dt = Ri. \tag{11.4.4}$$

If $dQ$ is the charge crossing any cross-section of the circuit in a time interval $dt$, then, by definition, the current intensity is $i = dQ/dt$. From Eq. (11.4.4) it follows that the total charge that will pass through the circuit when the flux through the coil is reduced from $\Phi = NBS\cos\theta$ to zero, in *any* way, is given by

$$Q = \int_0^\infty i\,dt = -\frac{1}{R}\int_0^\infty \frac{d\Phi}{dt}\,dt = -\frac{1}{R}\int_\Phi^0 d\Phi = \frac{\Phi}{R} = \frac{NBS\cos\theta}{R}. \tag{11.4.5}$$

Hence the component of the magnetic flux density vector in the direction of the positive normal to the coil at the initial instant (i.e., when the coil was in the field) is

$$B_n = B\cos\theta = QR/NS. \tag{11.4.6}$$

The maximum flux through the coil, and hence also, according to Eq. (11.4.5), the maximum charge that could be measured by the ballistic galvanometer in the case considered, is obtained when **B** is perpendicular to the surface of the search coil. When we have determined experimentally both this particular initial position of the search coil and the corresponding maximal value of the charge $Q$, we know both the direction and the magnitude of the flux density vector **B**. Theoretically, the process of reducing the flux $\Phi$ to zero can be as long as we wish. Ballistic galvanometers, however, can measure the charge provided that it flows through it in a sufficiently short period of time. The reduction of the flux to zero has, therefore, to be performed as quickly as possible.

**11.4.2 The superconductive loop in a magnetic field.** Let us now consider a closed thin loop made of some perfect conducting material. (In practice, conductors which are virtually perfect are obtained when certain metals are cooled to temperatures of only a few degrees Kelvin.) Suppose that in this loop there exists a current of intensity $I_0$, producing a flux $\Phi_0$ through it. If we introduce the loop into a magnetic field, an electromotive force will be induced in it during the transient process. This will produce an additional current in the loop, resulting in some additional flux through it. Let us denote by $\Phi_e$ the external flux through the loop, and by $\Phi_s$ the flux due to the current in it. Note that both $\Phi_e$ and $\Phi_s$ vary in time while the loop is being introduced into the field. Since the resistance of the loop is zero, it must, according to Faraday's law given in Eq. (11.4.3), be

$$-d\Phi_e/dt - d\Phi_s/dt = 0,$$

from which

$$\Phi_e + \Phi_s = \text{constant}. \tag{11.4.7}$$

Now, when the loop was out of the field, the flux through it was $\Phi_0$. So

$$\Phi_e + \Phi_s = \Phi_0 \tag{11.4.8}$$

at *any* instant. Hence it is impossible to change the magnetic flux through a superconductive loop. Any external flux is immediately compensated by the secondary flux due to the current induced in the loop.

*Problems referring to Section 11.4:* 11.G3 to 11.G20.

## 11.5 MUTUAL INDUCTANCE AND SELF-INDUCTANCE OF QUASI-FILAMENTARY CLOSED LOOPS

**11.5.1 Mutual inductance between two quasi-filamentary loops.** Let us consider two closed stationary contours $C_1$ and $C_2$, made of a very thin conducting wire (quasi-filamentary loops). Suppose that they are situated in a linear magnetic medium, and that there exists in $C_1$ a varying current $i_1(t)$ (see Fig. 11.10). The

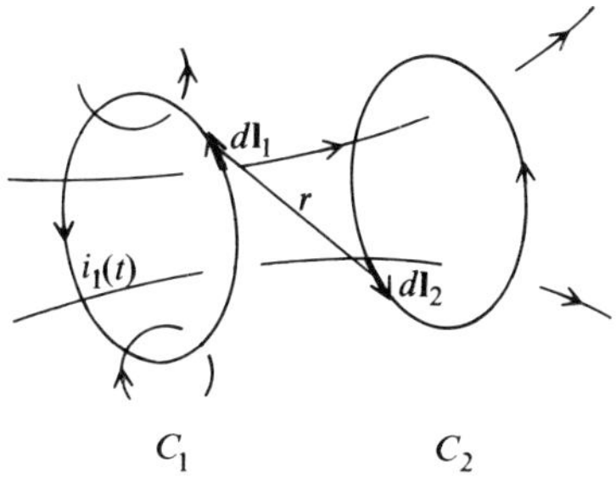


**Fig. 11.10** Two quasi-filamentary magnetically coupled circuits.

accelerated charges in contour $C_1$ constituting the varying current $i_1(t)$ produce a varying electromagnetic field in the vicinity of $C_1$, and also at points in $C_2$. The electric part of this field acts on the charges in $C_2$, resulting in an electromotive force acting along $C_2$. We saw in Section 11.2 that this electromotive force could be calculated either by finding the intensity of this induced electric field along $C_2$, or by determining the negative rate of change of the magnetic flux through the contour. The second method was shown in fact to follow from and to be *mathematically* equivalent to the first method. Unfortunately, it conceals to a certain extent the true physical nature of the induced electromotive force. However, it is very often convenient to say simply that the induced electromotive force is produced by the varying magnetic flux through the contour, and we shall frequently use this short reference to the electromagnetic induction phenomena. Now, knowing the varying magnetic flux through $C_2$, we can readily determine the electromotive force in $C_2$ which it produces. It is consequently clear that determining the magnetic flux through a contour due to varying currents in other contours is an important problem of electromagnetic induction.

The present part of the text deals with quasi-stationary electromagnetic fields, that is, with fields in which all the currents are assumed to vary so slowly that the magnetic field created by them is, for all intents and purposes, equal to that produced by *steady* currents, having magnitudes equal to those of the varying currents at a given moment of time. The determination of the magnetic field of the quasi-stationary current in $C_1$ therefore reduces to determining the magnetic field of a *steady* current in $C_1$, and then multiplying this field by the time factor implicit in $i_1(t)$. So the magnetic flux produced through a contour by a *steady* current in another contour is sufficient for determining the electromotive force induced in the contour when the current is quasi-stationary.

Let us assume that the magnitude of a *steady* current in $C_1$ is $I_1$. According to the Biot-Savart law given in Eq. (8.4.4) the magnitude of the flux density vector at all points is proportional to $I_1$. The magnetic flux $\Phi_{12}$ created by the current $I_1$ through the contour $C_2$ is therefore also proportional to $I_1$:

$$\boxed{\Phi_{12} = L_{12} I_1} \quad \text{(the flux definition of mutual inductance between filamentary loops).} \tag{11.5.1}$$

The symbol $L_{12}$ designates a constant depending upon the geometry of the system and the magnetic properties of the surrounding medium (which must be linear!), but *not* on $I_1$, and is referred to as the *mutual inductance* between loops $C_1$ and $C_2$. From the foregoing reasoning and the definition in Eq. (11.5.1) of the mutual inductance, it follows that, if the mutual inductance $L_{12}$ is determined in an appropriate manner, the electromotive force induced in loop 2 is simply

$$\mathscr{E}_{12} = -L_{12}\frac{di_1}{dt} \qquad \text{(the dynamic definition of mutual inductance).} \qquad (11.5.2)$$

According to Eq. (11.5.1) the unit of mutual inductance is *weber per ampere*. Because of the extensive use of the concept of inductance in practice, this unit is given a separate name, the *henry* (abbreviation H), which is equal to Wb/A. As this turns out to be a very large unit for many practical applications, submultiples of the henry, the millihenry (abbreviation mH; equal to $10^{-3}$ H) and the microhenry ($\mu$H; equal to $10^{-6}$ H) are frequently used instead.

From Eq. (11.5.1) it is clear that mutual inductance can be positive or negative, depending upon the arbitrary choice of the positive direction of loops $C_1$ and $C_2$. If the positive current in $C_1$ produces a positive flux through $C_2$, the mutual inductance is positive; if it produces a negative flux through $C_2$, the mutual inductance is negative. So, the sign of $L_{12}$ is determined by the geometry of the circuit *and* by the *arbitrarily* assigned positive directions along *both* loops. We shall not bother therefore about the sign of $L_{12}$ unless we expressly state otherwise.

The defining equation (11.5.1) of mutual inductance enables us in some instances to make a direct determination of $L_{12}$, according to the following pattern:

(1) Assume current $I_1$ in loop $C_1$, and determine (if feasible) the magnetic flux density due to this current at all points.

(2) Determine the magnetic flux $\Phi_{12}$ created by $I_1$ through $C_2$.

(3) Determine the mutual inductance as $L_{12} = \Phi_{12}/I_1$.

Before illustrating this procedure by a few examples, let us determine the general expression for the mutual inductance of two quasi-filamentary loops in which only the geometry of the system is involved. Note that the magnetic flux through $C_2$ can be calculated as a line integral of the vector potential $\mathbf{A}_1$ created by the current $I_1$:

$$\Phi_{12} = \oint_{C_2} \mathbf{A}_1 \cdot d\mathbf{l}_2. \qquad (11.5.3)$$

Now, according to Eq. (8.9.17), the vector potential $\mathbf{A}_1$ at a point on the line element $d\mathbf{l}_2$ is given by

$$\mathbf{A}_1 = \frac{\mu_0 I_1}{4\pi}\oint_{C_1}\frac{d\mathbf{l}_1}{r}. \qquad (11.5.4)$$

If we combine Eq. (11.5.1) with Eqs. (11.5.3) and (11.5.4), we find that the mutual inductance between loops $C_1$ and $C_2$ is

$$L_{12} = \frac{\mu_0}{4\pi} \oint_{C_1} \oint_{C_2} \frac{d\mathbf{l}_1 \cdot d\mathbf{l}_2}{r}. \tag{11.5.5}$$

This equation is known as the *Neumann formula* for the mutual inductance between two quasi-filamentary loops. We can see that interchanging subscripts 1 and 2 does not change the value of the integral, which means that the mutual inductance $L_{21}$ between loops 2 and 1 is exactly equal to the mutual inductance $L_{12}$ between loops 1 and 2:

$$\boxed{L_{21} = L_{12} = \Phi_{12}/I_1 = \Phi_{21}/I_2} \quad \text{(the flux definition of mutual inductance).} \tag{11.5.6}$$

This is a very important practical conclusion, for in some cases it is very difficult to evaluate the double integral in Eq. (11.5.5), or the flux $\Phi_{12}$, but very easy to evaluate the flux $\Phi_{21}$.

For example, let us consider a toroidal coil tightly wound with $N_2$ turns of thin wire, having the cross-section shown in Fig. 11.11, encircled with one turn of thin wire of arbitrary shape. Obviously, Neumann's formula, Eq. (11.5.5), is of no use in this case. Also, if we denote by $C_1$ the wire encircling the toroid and by $C_2$ the toroid itself, it is practically impossible to determine the magnetic flux $\Phi_{12}$ created by a current in $C_1$ through the toroidal coil $C_2$. It is, however, very simple to determine the flux $\Phi_{21}$ created by the current $I_2$ in the toroidal coil through the loop $C_1$. The magnetic field exists only inside the toroid, and the lines of the magnetic flux density vector of the current in the toroidal coil are circles centered on the toroid axis. Using Ampère's circuital law, we can easily find the magnitude of $\mathbf{B}$ to be $\mu_0 N_2 I_2/2\pi r$. Hence the magnetic flux through any cross-section of the toroid is

$$\Phi_{\text{toroid}} = \mu_0 \frac{N_2 h I_2}{2\pi} \int_a^b \frac{dr}{r} = \mu_0 \frac{N_2 h I_2}{2\pi} \ln \frac{b}{a}. \tag{11.5.7}$$

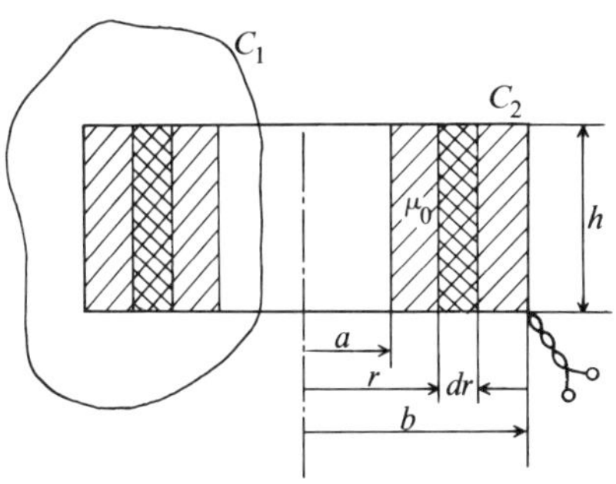


**Fig. 11.11** A toroidal coil encircled by a conductor of an arbitrary shape.

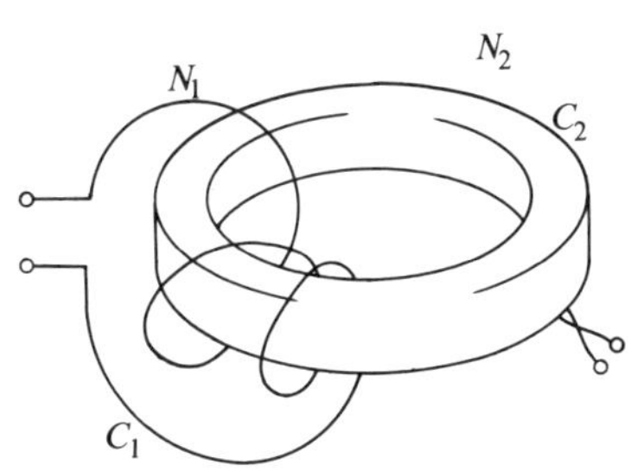


**Fig. 11.12** A toroidal coil and an arbitrary secondary circuit.

But this is also exactly the flux $\Phi_{21}$. Thus the mutual inductance is

$$L_{12} = L_{21} = \frac{\Phi_{21}}{I_2} = \mu_0 \frac{N_2 h}{2\pi} \ln \frac{b}{a}, \tag{11.5.8}$$

for *any* shape of loop $C_1$.

Now let us suppose that loop $C_1$ has not only one but, say, $N_1$ turns, as sketched in Fig. 11.12. It seems difficult to determine the flux through such a complicated surface, but it is easy to conclude that in this case $\Phi_{21}$ is exactly $N_1$ times larger than for a single turn. We can visualize the problem as follows. Consider first a simple case of one turn formed as shown in the first sketch of Fig. 11.13. The positive side of the surface is towards the observer, which is indicated by the plus sign and by crosshatch from the left upwards to the right. If we now rotate the right-hand part of the surface as shown in the middle of Fig. 11.13, its negative side is turned towards the observer, and the total flux through the loop is zero. But if we now "press" away from the observer the half which we rotated and ultimately obtain a loop of the shape shown at the right of Fig. 11.13, the flux through the loop is approximately equal to that at the beginning of the process, i.e. to $2\Phi_0$. Using this general pattern, it is not difficult for us to visualize in Fig. 11.12 a membrane, like, say, that made from soapy water, spanned over the twisted contour of loop $C_2$, and to imagine the positive side of the surface, for example, to be white and the negative side black. This membrane is pierced by the toroid exactly $N_1$ times, and so the flux through $N_1$ turns in this case is exactly $N_1$ times larger than through a single turn. Thus the mutual inductance between the two coils shown in Fig. 11.12 is $N_1$ times larger than that given in Eq. (11.5.8).

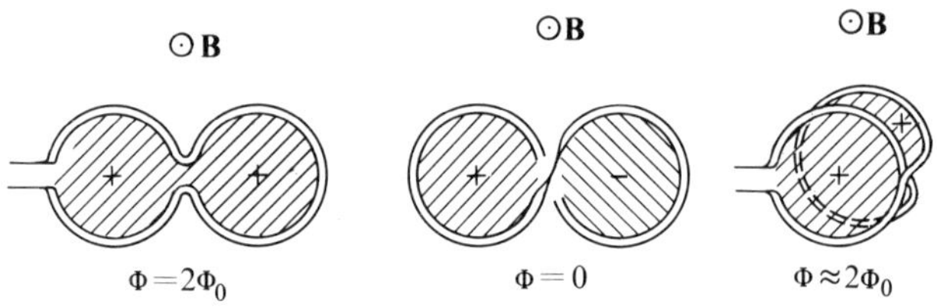


**Fig. 11.13** Determination of magnetic flux through two turns of wire.

Finally, let us determine the mutual inductance per unit length between two parallel two-wire lines, the geometry of which is shown in Fig. 11.14. The two wires of line 1 are designated by 1 and 1′, and the wires of line 2 by 2 and 2′. Let us suppose that a current of intensity $I_1$ flows through line 1 in the direction indicated, and take as the reference direction along line 2 the one shown in the figure. The magnetic flux through line 2 is produced by currents $I_1$ in both conductors 1 and 1′. Figure 11.14 shows the lines of vector $\mathbf{B}_1$ created by the current in conductor 1. The flux through line 2 can be calculated through any surface spanned between conductors 2 and 2′. The most simple surface for determining the flux is that

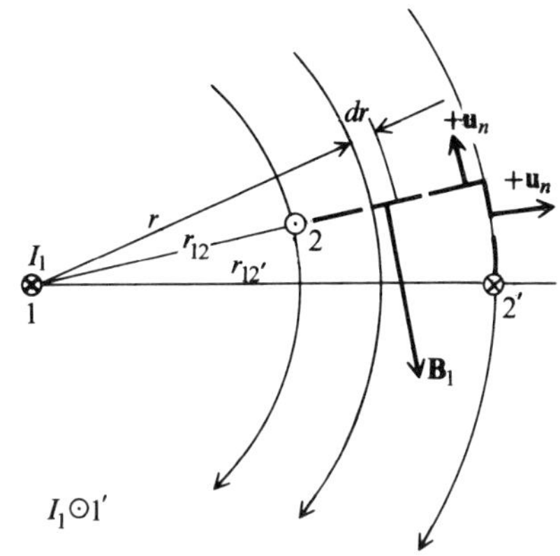


**Fig. 11.14** Two parallel two-wire lines.

shown in the figure in dashed line. Vector $\mathbf{B}_1$ is perpendicular to the flat part of that surface and tangential to its curved part. Since $B_1 = \mu_0 I_1 / 2\pi r$ in the direction indicated, the partial magnetic flux per unit length of line 2 produced by the current in conductor 1 is

$$\Phi_{1l} = -\frac{\mu_0 I_1}{2\pi} \int_{r_{12}}^{r_{12'}} \frac{dr}{r} = -\frac{\mu_0 I_1}{2\pi} \ln \frac{r_{12'}}{r_{12}} .$$

By analogy the magnetic flux per unit length of line 2 produced by the current in conductor 1′ is

$$\Phi_{1'l} = \frac{\mu_0 I_1}{2\pi} \ln \frac{r_{1'2'}}{r_{1'2}} .$$

Hence the total magnetic flux per unit length of line 2 produced by the current $I_1$ in line 1 is

$$\Phi_{12l} = \Phi_{1l} + \Phi_{1'l} = \frac{\mu_0 I_1}{2\pi} \ln \frac{r_{12} r_{1'2'}}{r_{12'} r_{1'2}} .$$

The mutual inductance per unit length of the two lines is

$$L_{12l} = \frac{\Phi_{12l}}{I_1} = \frac{\mu_0}{2\pi} \ln \frac{r_{12} r_{1'2'}}{r_{12'} r_{1'2}} . \tag{11.5.9}$$

**11.5.2 Self-inductance of a quasi-filamentary loop. The coefficient of coupling between two loops.** Let us now consider an isolated loop made of a very thin wire (a quasi-filamentary loop), and let a time-varying current be maintained in it by an appropriate generator of electromotive force. The accelerated free charges along the loop create at all points a varying electric field, which also exists along the loop. Thus, due to the mutual interaction of the accelerated charges in the loop *itself*, an additional electric field acts on all the charges along the loop. The resultant current is produced by both the external (impressed) electric field *and* the secondary electric field created by the accelerated charges. This phenomenon is known as *self-induction*. It is obvious that self-induction is present in any loop with a varying current.

We can represent the overall action of the self-induced electric field along the loop in terms of an equivalent electromotive force. We can easily conclude that the reasoning of Sections 11.2.1 and 11.2.2 is also valid in this case, so the resultant electromotive force induced in the loop by the varying current in the loop itself is given by Eqs. (11.2.6) and (11.2.7). In this case, the magnetic flux $\Phi_{11}$ through the loop is produced by the current in the loop itself. By exactly the same reasoning as in the preceding section we can conclude that the magnetic flux $\Phi_{11}$ through any surface spanned over the loop is proportional to the current in the contour at any instant of time. Thus, it can be taken to be a steady current $I_1$:

$$\boxed{\Phi_{11} = L_{11} I_1} \qquad \text{(the flux definition of self-inductance).} \qquad (11.5.10)$$

The electromotive force induced in the loop by the current in the loop itself can be written in the form

$$\mathscr{E}_{11} = -L_{11}\frac{di_1}{dt} \qquad \text{(the dynamic definition of self-inductance).} \qquad (11.5.11)$$

The coefficient $L_{11}$ is known as the *self-inductance* of the loop, and the unit of self-inductance is the same as that of mutual inductance, i.e. the *henry*. The procedure for determining $L_{11}$ from its defining formula, given in Eq. (11.5.10), follows an analogous pattern to that described for evaluating $L_{12}$ from Eq. (11.5.1). However, Neumann's formula, which is given in Eq. (11.5.5), can be applied in this case only after some necessary precautions are undertaken, as shown below.

Let us first analyze the connection between the self-inductance $L_{11}$ and $L_{22}$ of two loops and their mutual inductance $L_{12}$. We note that the magnetic flux created by the current in a contour is always larger through that contour than through any other contour, since at least some leakage flux exists. Thus $\Phi_{11} \geq \Phi_{12}$ and $\Phi_{22} \geq \Phi_{21}$. Multiplying these two inequalities together, and dividing the inequality thus obtained by $I_1 I_2$, we have

$$\frac{\Phi_{11}}{I_1}\frac{\Phi_{22}}{I_2} = L_{11}L_{22} \geq \frac{\Phi_{12}}{I_1}\frac{\Phi_{21}}{I_2} = L_{12}^2. \qquad (11.5.12)$$

Thus quite generally

$$L_{12} = k\sqrt{L_{11}L_{22}}, \qquad 0 \leq k \leq 1. \qquad (11.5.13)$$

The constant $k$ is known as the *coefficient of coupling*. It is a function of the geometry of the two circuits.

We will now consider the closed quasi-filamentary loop shown in Fig. 11.15 (for clarity the loop is shown to be relatively thick). To obtain the magnetic flux through the loop created by the current in it let us consider the two loops in

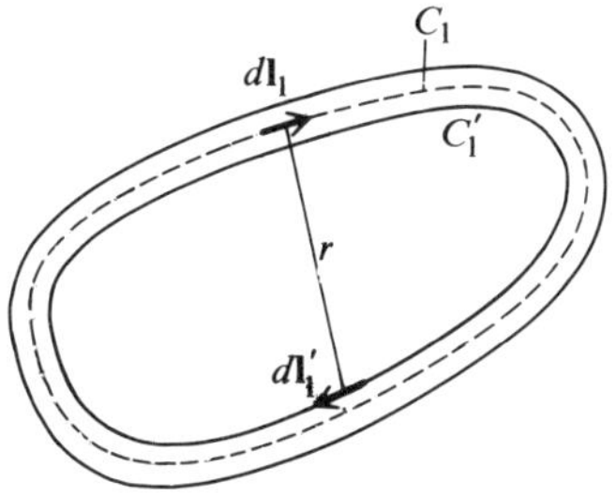


**Fig. 11.15** A quasi-filamentary conductor.

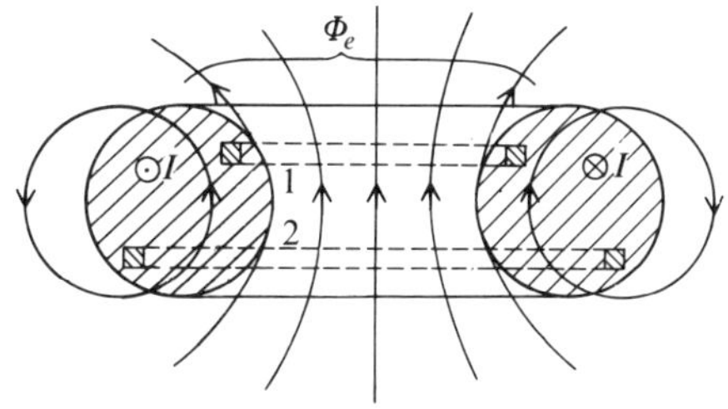


**Fig. 11.16** A thick current-carrying conductor and its magnetic field.

Fig. 11.10 as coming closer and closer and finally coinciding. The corresponding expression in Eq. (11.5.5) for $L_{12}$ then becomes exactly $L_{11}$. However, it can be shown that the integral in that case is an improper one, that is, has an infinitely large value, since $r$ can be zero. Observe, however, that $r$ can be zero because of the assumed filamentary current, which in reality does not exist. A reasonable practical approximation is to consider the current to be concentrated along the loop axis, i.e. along loop $C_1$ in Fig. 11.15, and then to determine the magnetic flux created by this filamentary current through any closed filamentary loop on the conductor surface, say the loop $C_1'$ shown in the figure. If the conductor is very thin this is a very good approximation. Thus Neumann's formula from (11.5.5) for approximately calculating the self-inductance of a quasi-filamentary loop is

$$L_{11} = \frac{\mu_0}{4\pi} \oint_{C_1} \oint_{C_1'} \frac{d\mathbf{l}_1 \cdot d\mathbf{l}_1'}{r}. \tag{11.5.14}$$

If the conductor is very thin and of circular cross-section, the self-inductance obtained according to Eq. (11.5.14) gives quite accurately the ratio between the magnetic flux *external* to the conductor and the current through it. For this reason it is referred to as the *external inductance* of the loop. The *internal inductance* of a loop is associated with the magnetic field inside the conductor constituting the loop, and cannot be defined as any simple ratio of some magnetic flux and the current through the conductor, since the concept of the magnetic flux through a nonfilamentary contour has no meaning. For example, in the case shown in Fig. 11.16 the magnetic flux denoted by $\Phi_e$ is common for all current filaments in the thick conductor. But the magnetic fluxes through two representative filaments 1 and 2 are obviously different. In the case of a varying current through the thick conductor this amounts to different electromotive forces induced along different current filaments. So a definition of internal inductance of a thick conductor is not possible within the framework of the electromagnetic induction phenomena. In Section 12.2 we shall see that a more general definition of mutual inductance and self-inductance, based on energy considerations, is possible, enabling us also to make a simple definition of internal inductance.

In considering thin loops at low frequencies as two-terminal networks we must know their total self-inductance. This is equal to the external plus the internal inductance. The internal inductance of a conductor, of *circular cross-section* of length $l$ and permeability $\mu$, is

$$L_{\text{internal}} = \frac{\mu}{8\pi} l \qquad \text{(the circular conductor, direct current case).} \tag{11.5.15}$$

It is not a function of the conductor radius. This simple formula, which will be derived in Section 12.2.1, is very important, since conductors of circular, or nearly circular, cross-section are those most often encountered in practice.

As an example of calculating self-inductance starting from the definition in Eq. (11.5.10), let us determine the self-inductance of the toroidal coil shown in Fig. 11.11. The magnetic flux through a *cross-section* of the toroid core is given by Eq. (11.5.7). The magnetic flux through $N_2$ turns on the core is $N_2$ times larger. Thus we discover that the self-inductance of the toroid is

$$L_{\text{toroid}} = \frac{N_2 \Phi_{\text{toroid}}}{I_2} = \frac{\mu_0 N_2^2 h}{2\pi} \ln \frac{b}{a}. \tag{11.5.16}$$

As an example of the application of Neumann's formula, Eq. (11.5.14), for determining the external inductance of a quasi-filamentary conductor, let us determine the self-inductance of the square loop of sides $b$, made of a wire of radius $a$ ($b \gg a$), shown in Fig. 11.17. By symmetry, it is sufficient to calculate the integral in Eq. (11.5.14) for that part of $C'$ from point $P_1$ to point $P_2$. The total double integral is simply four times larger. In addition, the elements $d\mathbf{l}$ along sections $P_2P_3$ and $P_4P_1$ of $C$ are perpendicular to elements $d\mathbf{l}'$ along $P_1P_2$, so that only the two integrals along $C$ have to be evaluated, from $P_1$ to $P_2$ and from $P_3$ to $P_4$. Thus,

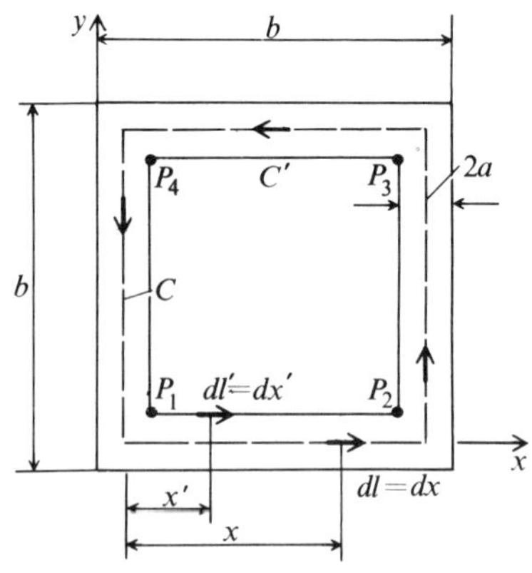


**Fig. 11.17** A square loop.

in this case, Eq. (11.5.14) becomes very approximately

$$L_{11} = 4\frac{\mu_0}{4\pi}\left\{\int_0^b\int_0^b \frac{dx\,dx'}{\sqrt{a^2 + (x - x')^2}} - \int_0^b\int_0^b \frac{dx\,dx'}{\sqrt{b^2 + (x - x')^2}}\right\}. \qquad (11.5.17)$$

The negative sign of the second integral is obtained because $d\mathbf{l}$ along $P_3P_4$ and $d\mathbf{l}'$ along $P_1P_2$ are oppositely directed. The two integrals are of the same type, and can be evaluated as follows. Integration of the first integral with respect to $x$ yields

$$\int_0^b \frac{dx}{\sqrt{(x - x')^2 + a^2}} = \int_{-x'}^{b-x'} \frac{d(x - x')}{\sqrt{(x - x')^2 + a^2}}$$

$$= \ln[(b - x') + \sqrt{(b - x')^2 + a^2}\,] - \ln(-x' + \sqrt{x'^2 + a^2}). \qquad (11.5.18)$$

Integration with respect to $x'$ can now be performed noting that, if $f(x')$ denotes an arbitrary function of $x'$, then integration by parts yields

$$\int \ln[f(x')]\,dx' = x' \ln f(x') - \int x' \frac{1}{f(x')}\frac{df(x')}{dx'}\,dx'. \qquad (11.5.19)$$

The final result for the total self-inductance of the square loop is

$$L_{\text{total}} = \frac{2\mu_0 b}{\pi}\left[\ln\frac{2b}{a(1 + \sqrt{2})} + \sqrt{2} - 2\right] + \frac{\mu b}{2\pi}, \qquad (11.5.20)$$

the second term representing the internal inductance [see Eq. (11.5.15)] of the loop wire.

*Problems referring to Section 11.5:* 11.T13 to 11.T16, 11.G21 to 11.G40, 11.N3 to 11.N7.

## PROBLEMS

### Test problems

**11.T1.** A charge $Q_1$ approaches a stationary charge $Q_2$ along a straight line which passes through $Q_2$. The velocity $v_1$ of $Q_1$ is approximately constant. Determine the additional electric force on $Q_2$ due to the motion of $Q_1$.

**11.T2.** The charge $Q_1$ from the preceding problem moves along a straight line, and $Q_2$ is at a distance $d$ from that line. Determine the additional electric force on the charge $Q_2$.

**11.T3.** A charge $Q$ moves around a circle of radius $a$ with a constant angular velocity $\omega$. Discuss qualitatively the magnitude and direction of the additional electric field, due to the motion of $Q$, at various points in the vicinity of the circle. Determine the additional electric field intensity vector at the center of the circle.

**11.T4.** A charge $Q$ is forced to move with a velocity $\mathbf{v} = v_0 \sin \omega t\, \mathbf{u}_x$. Discuss qualitatively the electric field due to the charge motion.

***11.T5.** A circular metal ring carries a varying current, which creates a varying electric field. If the ring is set in oscillatory motion about the axis normal to its plane, will this electric field be changed?

**11.T6.** A device for accelerating electric charges known as the *betatron* (see Fig. 11.T6) consists of a powerful electromagnet and an evacuated tube which is bent into a circle. Elementary charged particles are accelerated in the tube during the period of time when the magnetic flux through the core is forced to rise approximately as a linear function of time. What is the physical mechanism of accelerating the charges by using the betatron?

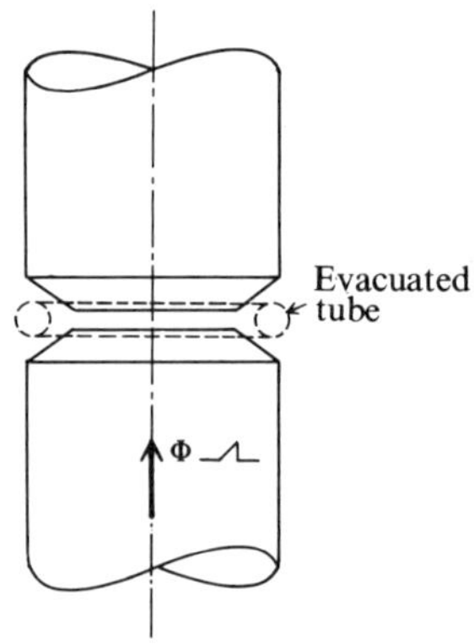


**Figure 11.T6**

**11.T7.** A solid conducting body resides near a conducting loop carrying a varying current. Are there any forces acting on the free charges inside the body?

**11.T8.** A conducting circular ring made of insulated wire resides on the surface of a plane conducting sheet. If the ring carries a varying current, discuss what happens in the sheet. Is the power required to drive the current through the ring different when the ring resides on the sheet from that when it is isolated in space?

**11.T9.** By reference to Eq. (11.2.4), is it physically admissible to speak about electromotive force induced in a *section* of a closed circuit by a varying current in a *section* of another closed circuit?

***11.T10.** A vertical conducting sheet is permitted to fall under the action of gravity between the poles of a powerful permanent magnet. Is the motion of the sheet affected by the presence of the magnet?

***11.T11.** When no load is connected to the elementary generator shown in Fig. 11.8, the torque opposing the rotation of the loop is virtually only that due to friction. Discuss what happens when a resistor is connected between the terminals of the generator.

**11.T12.** An elastic metallic circular loop carrying a steady current $I$ is periodically deformed to a flat ellipse and then released to retain its original circular shape. An electromotive force will be induced in the loop. Explain the origin of this electromotive force.

**11.T13.** With reference to Problem 11.T9 and Eq. (11.5.2), is it physically feasible to define the mutual inductance between two pieces of wire, being two sections of two different circuits, or even of a single closed circuit?

**11.T14.** With reference to Eqs. (11.2.4) and (11.5.11), is it physically feasible to define the self-inductance of a *section* of a thin closed loop?

**11.T15.** To obtain a resistive wire with the smallest self-inductance possible, the wire is sharply bent in the middle and the two mutually insulated halves are pressed tightly together, as shown in Fig. 11.T15. Explain why the self-inductance of such a wire is very small, in terms of Eq. (11.2.5) and in a formal manner, using the definition of self-inductance in Eq. (11.5.10).

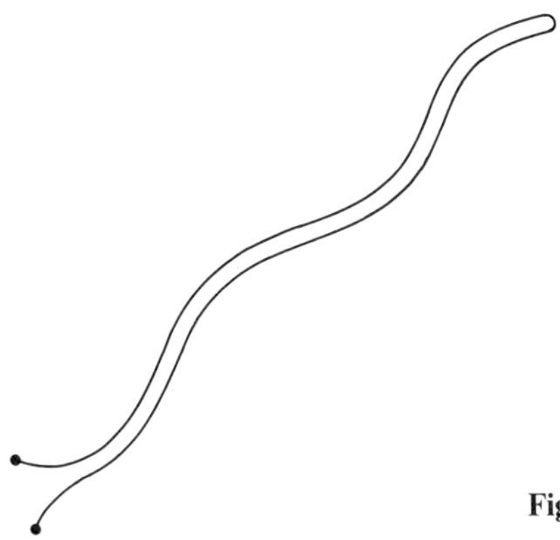

**Figure 11.T15**

**11.T16.** A thin coil is made of $N$ turns of very thin wire pressed tightly together. Explain why the self-inductance of such a coil is approximately $N^2$ times larger than that of a single turn, using (1) the relation in Eq. (11.2.5) and the dynamic definition of self-inductance in Eq. (11.5.11), and (2) the static definition of the self-inductance in Eq. (11.5.10).

## General problems

**11.G1.** A small charge $Q$ is set in forced oscillatory motion in the vicinity of a square conducting loop, as shown in Fig. 11.G1. The instantaneous position of the charge is given by $x = 0$, $y = y_0 \sin \omega t$, $y_0$ and $\omega$ being constants. The maximal displacement $y_0$ of the charge is much smaller than the side $a$ of the loop. Determine the electromotive force induced in the loop.

*Solution.* The electromotive force induced in the loop can be obtained by integrating the electric field intensity vector due to the charge motion around the loop. The electric field intensity due to the motion of charge $Q$ is

$$\mathbf{E} = -\frac{\mu_0}{4\pi} Q \frac{\partial}{\partial t}\left(\frac{\mathbf{v}}{r}\right) \simeq -\frac{\mu_0 Q}{4\pi r}\frac{d\mathbf{v}}{dt}, \tag{a}$$

since the distance $r$ of the charge from all points of the loop is approximately constant (because $y_0 \ll a$). Now,

$$\mathbf{v} = \frac{dy}{dt}\mathbf{u}_y = y_0\, \omega \cos \omega t\, \mathbf{u}_y,$$

so that Eq. (a) becomes

$$\mathbf{E} = E\,\mathbf{u}_y = \frac{\mu_0 Q \omega^2 y_0}{4\pi r} \sin \omega t\, \mathbf{u}_y. \tag{b}$$

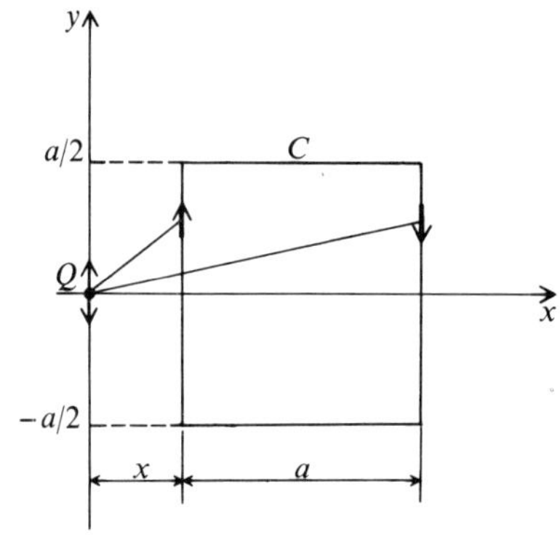


**Figure 11.G1**

The electromotive force induced in the loop is obtained as the line integral of **E** along the loop. Adopting the direction indicated in the figure for the positive direction along the loop, we have

$$\oint_C \mathbf{E}\cdot d\mathbf{l} = \int_{-a/2}^{a/2} E\,\mathbf{u}_y\cdot dy\,\mathbf{u}_y + \int_x^{a+x} E\,\mathbf{u}_y\cdot dx\,\mathbf{u}_x + \int_{a/2}^{-a/2} E\,\mathbf{u}_y\cdot(-dy)(-\mathbf{u}_y) + \int_{a+x}^{x} E\,\mathbf{u}_y\cdot(-dx)(-\mathbf{u}_x).$$

The second and the fourth integrals are zero, since $\mathbf{u}_y\cdot\mathbf{u}_x = 0$. Thus the electromotive force induced in the loop (note that $\mathbf{u}_y\cdot\mathbf{u}_y = 1$) is

$$\int_C \mathbf{E}\cdot d\mathbf{l} = \frac{\mu_0 Q\omega^2 y_0 \sin\omega t}{4\pi}\left[\int_{-a/2}^{a/2}\frac{dy}{\sqrt{x^2+y^2}} + \int_{a/2}^{-a/2}\frac{dy}{\sqrt{(x+a)^2+y^2}}\right]$$

$$= \frac{\mu_0 Q\omega^2 y_0 \sin\omega t}{4\pi}\left[\ln\frac{\sqrt{x^2+(a/2)^2}+a/2}{\sqrt{x^2+(a/2)^2}-a/2} - \ln\frac{\sqrt{(x+a)^2+(a/2)^2}+a/2}{\sqrt{(x+a)^2+(a/2)^2}-a/2}\right].$$

It is left as an exercise for the reader to prove that the same result is obtained by starting from Faraday's law [Eq. (11.4.3)] and the expression in Eq. (8.3.3) for the magnetic flux density of a single moving charge. It should be noted, however, that Faraday's law gives only the *total* electromotive force induced in a contour, while the above procedure also gives an idea of how it is distributed along the contour.

**11.G2.** If a rectangular conducting loop of sides $a$ and $b$ recedes with a constant velocity **v** from a straight conductor carrying a steady current $I$, as shown in Fig. 11.G2, determine the electromotive force induced in the loop as a function of time. Assume that, at time $t = 0$, the distance of the left side of the loop from the conductor is $r = r_0$.

*Solution.* The electromotive force can be determined by starting from either Eq. (11.3.2) or its equivalent form, Eq. (11.3.4). Let us determine the electromotive force by starting from the latter.

The magnetic flux through the rectangular contour when it is at a distance $r$ from the conductor is

$$\Phi = \frac{\mu_0 Ib}{2\pi} \int_r^{r+a} \frac{dr}{r} = \frac{\mu_0 Ib}{2\pi} \ln \frac{r+a}{r}.$$

From this Faraday's law gives

$$\mathscr{E} = -\frac{d\Phi}{dt} = -\frac{\mu_0 Ib}{2\pi} \frac{d}{dt} \ln \frac{r+a}{r} = \frac{\mu_0 Iba}{2\pi r(r+a)} \frac{dr}{dt} = \frac{\mu_0 Iba}{2\pi r(r+a)} v.$$

Use was made of the relation $r = r_0 + vt$. It is left to the reader to prove that the same result is obtained by starting from Eq. (11.3.2).

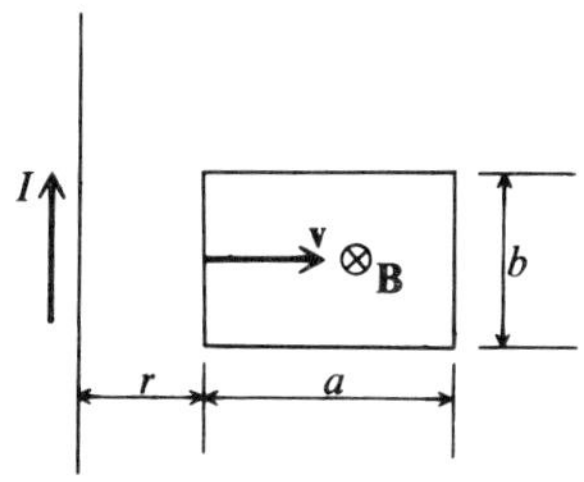


Figure 11.G2

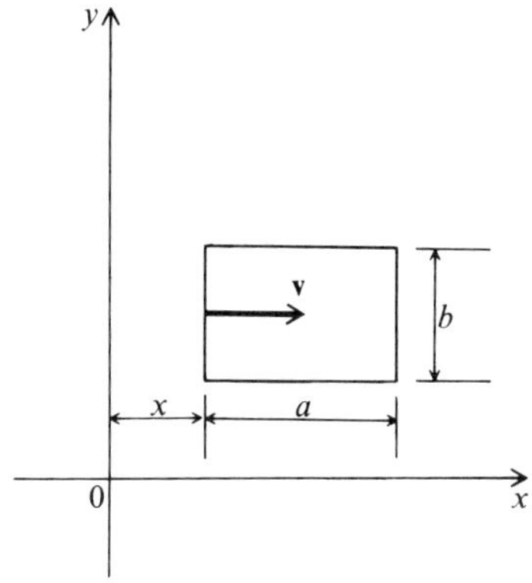


Figure 11.G3

**11.G3.** A plane rectangular loop of sides $a$ and $b$ moves in its plane with a uniform velocity $\mathbf{v}$. The magnetic flux density vector $\mathbf{B}$ is perpendicular to the plane of the loop, and is a function of both the position in the plane and of time. In the plane of the coordinate system shown in Fig. 11.G3, the flux density varies according to the law $B(x, t) = B_0 \cos \omega t \cos kx$, where $B_0$, $\omega$, and $k$ are constants. Determine the dependence on time of the electromotive force induced in the loop. Assume that, at $t = 0$, $x = 0$ also.

*Solution.* The electromotive force induced in the loop is given by the general form of Faraday's law of electromagnetic induction, Eq. (11.4.3). In this case it becomes

$$\mathscr{E} = -\frac{d\Phi}{dt} = -\left(\frac{\partial\Phi}{\partial t} + \frac{\partial\Phi}{\partial x}\frac{dx}{dt}\right) = -\left(\frac{\partial\Phi}{\partial t} + \frac{\partial\Phi}{\partial x} v\right). \tag{a}$$

Now, the magnetic flux through the contour at an instant $t$ is given by

$$\Phi(x, t) = \int_x^{x+a} B(x, t)\, b\, dx = \int_x^{x+a} B_0 b \cos \omega t \cos kx\, dx$$

$$= \frac{B_0 b \cos \omega t}{k} [\sin k(x + a) - \sin kx].$$

So

$$\frac{\partial\Phi}{\partial t} = -\frac{B_0 b\omega \sin \omega t}{k} [\sin k(x + a) - \sin kx], \tag{b}$$

$$\frac{\partial\Phi}{\partial x} = B_0 b \cos \omega t \,[\cos k(x + a) - \cos kx]. \tag{c}$$

According to Eqs. (a) to (c), the total electromotive force induced in the loop is

$$\mathscr{E} = \frac{B_0 b \omega}{k} \sin \omega t \,[\sin k(x + a) - \sin kx] + B_0 b v \cos \omega t \,[\cos kx - \cos k(x + a)].$$

The first component is due to the magnetic field varying with time (the "transformer" component). The second is the consequence of the motion of the loop (the "motional" electromotive force).

**11.G4.** In a uniform time-varying magnetic field of flux density $B(t)$, a circular loop of radius $a$ rotates with an angular velocity $\omega$ around its diameter perpendicular to the lines of vector $\mathbf{B}$. Determine the electromotive force induced in the loop.

**11.G5.** Wire is being wound from a drum $D_1$ onto a drum $D$, at a rate of $N'$ turns per unit time (see Fig. 11.G5). The end of the wire on drum $D$ is fastened to the ring $R$, which has the sliding contact $F$. Between the contact $F$ and another contact $G$ a voltmeter is connected. Through the drum $D$ there is a constant flux $\Phi$, as indicated. What is the electromotive force measured by the voltmeter?

*Answer.* The magnetic flux through the closed loop $FGF$ varies with time according to the law

$$d\Phi_{\text{total}}/dt = d(N\Phi)/dt = \Phi\, dN/dt = \Phi N'.$$

We may perhaps draw the conclusion that an electromotive force is induced in the loop, according to Faraday's law in Eq. (11.4.3):

$$\mathscr{E} = -d\Phi_{\text{total}}/dt = -\Phi N',$$

but such a conclusion is incorrect. We recall that according to Eq. (11.4.2) the electromotive force can be induced in only two ways: by moving sections of the contour in the magnetic field, or by varying the magnetic field in time. This case involves neither a time-varying magnetic field nor the motion of any loop section in the magnetic field, so the induced electromotive force is zero.

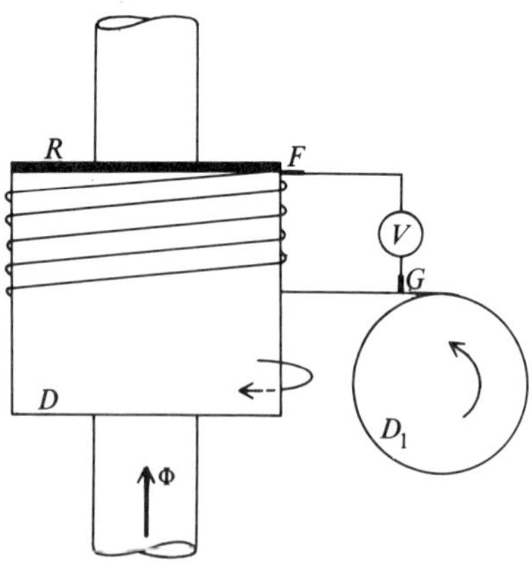


Figure 11.G5

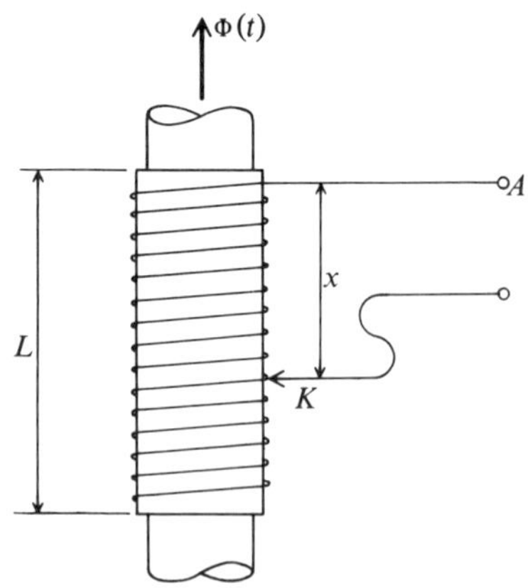


Figure 11.G6

**11.G6.** Around a ferromagnetic core with time-varying magnetic flux $\Phi(t) = \Phi_m \cos \omega t$ a cylindrical, tightly wound coil is placed, as shown in Fig. 11.G6. The length of the coil is $L$, and the number of turns of the coil is $N$. If a sliding contact $K$ moves along the coil according to the law $x = L(1 + \cos \omega_1 t)/2$, what is dependence or time of the electromotive force between contacts $A$ and $K$?

*Solution.* According to Faraday's law of electromagnetic induction given in Eq. (11.4.3), an electromotive force is induced in each turn of the coil, the value of which is

$$\mathscr{E}_1 = -d\Phi/dt = \Phi_m \omega \sin \omega t.$$

Since between contacts $A$ and $K$ there are at the same instant

$$n = Nx/L = N(1 + \cos \omega_1 t)/2$$

turns, the electromotive force induced in all of them is

$$\mathscr{E} = n\mathscr{E}_1 = \frac{\Phi_m \omega N}{2} \sin \omega t \, (1 + \cos \omega_1 t).$$

Note that in this case Faraday's law *cannot* be written in the form

$$\mathscr{E} = -\frac{d\Phi_{\text{total}}}{dt} = -\frac{d(n\Phi)}{dt} = -n\frac{d\Phi}{dt} - \Phi\frac{dn}{dt}.$$

The second term on the right represents the electromotive force that would be induced as the result of an increase in the number of turns. By the same arguments as in the preceding problem, it is clear that such an electromotive force does not exist.

**11.G7.** A circular metal ring of radius $a$ carries a steady current $I$, and rotates about the axis perpendicular to its plane with a constant angular velocity $\omega$. A small charge $Q$ is fastened to the ring and rotating with it, as shown in Fig. 11.G7. Is there a force acting on the charge $Q$, and is there a force on the stationary charge $Q_1$?

*Solution.* When the ring is stationary, there is some average drift velocity $\mathbf{v}_e$ of the electrons, the motion of which makes the current $I$. In that case the positive charges inside the ring are stationary. If the number of both positive and negative charges per unit volume of the ring is $N$, the current density inside the ring is

$$\mathbf{J} = (-e)N\mathbf{v}_e.$$

Let us now imagine the ring rotating with an angular velocity $\boldsymbol{\omega}$. In that case the velocity of the positive charges is no longer zero, but instead is equal to $\boldsymbol{\omega} \times \mathbf{a}$. The velocity of the negative charges (the electrons) is also changed, and is equal to $(\mathbf{v}_e + \boldsymbol{\omega} \times \mathbf{a})$. Hence the resulting current density when the ring rotates is

$$\mathbf{J}' = (-e)\,N\,(\mathbf{v}_e + \boldsymbol{\omega} \times \mathbf{a}) + (+e)\,N\boldsymbol{\omega} \times \mathbf{a} = (-e)\,N\mathbf{v}_e = \mathbf{J}.$$

Thus we see that the current density is not changed by rotating the ring. (This conclusion is obviously valid for any slow motion of any current-carrying conductor.) The magnetic field in the vicinity of the ring is consequently also unchanged. The charge $Q$ moves in the magnetic field which is exactly the same as if the ring were stationary. Accordingly it will be acted upon by the force in the direction indicated in the figure.

Following the same line of reasoning, we can easily see that the force on the stationary charge $Q_1$ is zero, even if the ring rotates with a variable angular velocity.

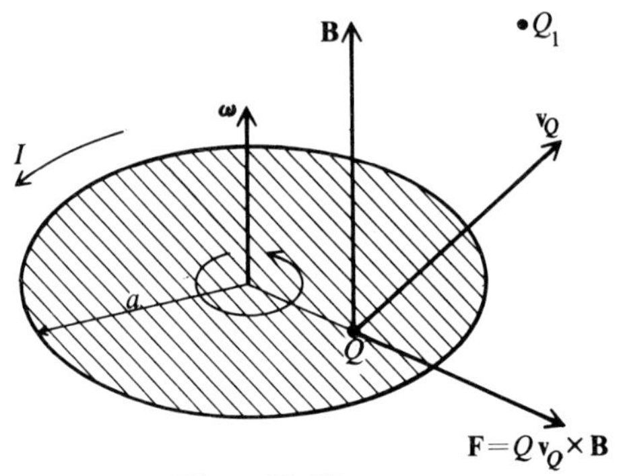


Figure 11.G7

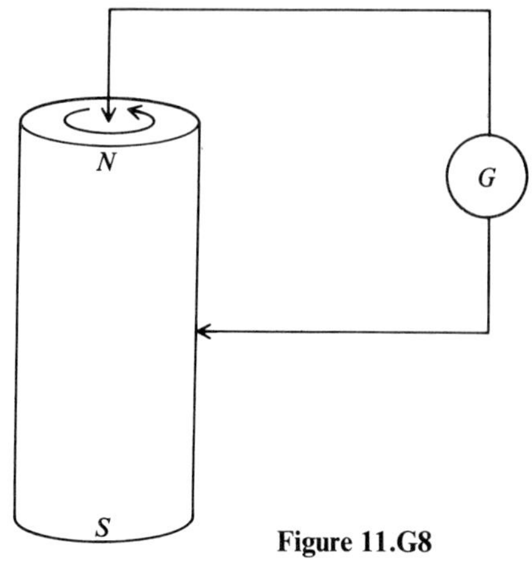


Figure 11.G8

**11.G8.** A cylindrical conducting magnet of circular cross-section rotates about its axis with a uniform angular velocity. A galvanometer $G$ is connected to the equator of the magnet and to the center of one of its bases by means of sliding contacts, as shown in Fig. 11.G8. If such an experiment is performed, the galvanometer indicates a certain current through the circuit (Faraday, 1832). Where is the electromotive force induced: in the stationary conductors connecting the sliding contacts with the galvanometer, or in the magnet itself?

*Answer*. A complete symmetry of the magnetic field is assumed with respect to the magnet's axis of rotation. The conductors connecting the galvanometer with the sliding contacts are hence situated in a time-constant magnetic field, irrespective of whether the magnet rotates or not. These conductors, in addition, do not move with respect to the observer, and consequently no electromotive force can be induced in them. The electromotive force is induced in the magnet, however strange such an assertion may seem at first glance.

The reasoning from which this conclusion follows is analogous to that in the preceding problem. The density of the macroscopic resultants of the microscopic currents is unaffected by rotation, for exactly the same reasons as before. Therefore, at *all* points in space, both inside and outside the magnet, the magnetic field is exactly the same as if the magnet did not rotate. But the electrons inside the magnet are moving in this magnetic field, and are consequently acted upon by the electromagnetic force, which causes the current to flow through the closed loop magnet-galvanometer.

Until recently the Faraday problem was often put in the following form: Do the magnetic field lines rotate with the magnet or not? This corresponds to Faraday's ideas that the lines of the magnetic flux density vector are physically real objects. If they do rotate, the electromotive force is induced in the stationary leads to the galvanometer, since they are cut by the rotating lines. If the lines do not rotate, the electromotive force is induced in the magnet itself, which rotates with respect to these stationary lines. Nevertheless, the question here asked is not physically correct. The lines of any vector field, and hence also of the magnetic field, are only hypothetical geometrical lines to which no physical reality can be ascribed.

**11.G9.** Coaxially with the magnet from the preceding problem a hollow metallic nonferromagnetic cylinder, enclosing the magnet, is added. The sliding contacts are placed on the cylinder in the same way as they were previously placed on the magnet. The cylinder rotates with a certain constant angular velocity. How does the electromotive force induced in the galvanometer circuit depend upon the angular velocity of the *magnet*, which can rotate independently inside the cylinder?

**11.G10.** A ferromagnetic toroid with no air gap is magnetized so that no magnetic field exists outside it. The toroid is encircled by an elastic metal loop, as in Fig. 11.G10a. The loop is now taken out of the toroid in such a way that, during the process, the loop is always electrically closed through the conducting material of the toroid, as in Figs. 11.G10b and 11.G10c. The magnetic flux through the contour was obviously changed from a value $\Phi$, the flux through the toroid, to zero. A formal application of Faraday's law leads to the conclusion that a certain charge will flow through the circuit during the process; but in this case it is not possible to detect any current (Herring, 1908). Explain the negative result of the experiment.

*Explanation.* Let us analyze Herring's experiment from the standpoint of two observers, one stationary with respect to the toroid and one stationary with respect to the galvanometer. It will be seen that the two observers arrive at the same conclusion by different lines of reasoning.

For the first observer Eq. (11.4.2) directly explains the negative result of the experiment. In his coordinate system the magnetic field is constant at all points, and, since it is localized inside the toroid only, the elastic loop does not move in the magnetic field while being taken out of the toroid. Thus, no electromotive force can be induced.

The second observer sees the result quite differently. In the coordinate system of this observer the magnetic field is *not* constant in time. For example, at a point which is immediately outside the toroid, in the direction of its motion, the magnetic flux density suddenly rises from zero to the value of the (remanent) flux density inside the toroid when it crosses that point. But, in addition, there is also the motional component of the electromotive force, since the conducting substance of the magnet moves in the magnetic field of the magnet. (For an explanation of this seeming paradox see Problems 11.G7 and 11.G8.) Now let us consider

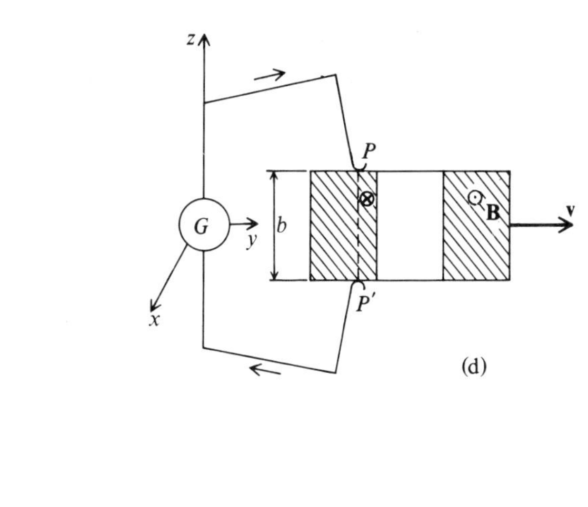


**Figure 11.G10**

an electrically closed contour through the substance of the magnet, for example, the contour $GP\,P'G$ in Fig. 11.G10d. The induced electromotive force along it according to Eq. (11.4.1) is

$$\mathscr{E} = -\int_S \frac{\partial \mathbf{B}}{\partial t}\cdot d\mathbf{S} + \oint_C (\mathbf{v} \times \mathbf{B})\cdot d\mathbf{l}, \tag{a}$$

$S$ denoting the surface limited by $C$ at the instant considered, $d\mathbf{l}$ an element of the contour, and $\mathbf{v}$ the (instantaneous) velocity of that element.

Due to the motion of section $P'P$ of the contour in the magnetic field, an electromotive force is induced, and its value in the adopted positive direction of the contour is

$$\mathscr{E}' = \oint_C (\mathbf{v} \times \mathbf{B})\cdot d\mathbf{l} = \int_{P'}^{P} (\mathbf{v} \times \mathbf{B})\cdot d\mathbf{l} = -bvB, \tag{b}$$

$B$ being the magnitude of the flux density vector along $P'P$.

The increment of the magnetic flux through $C$, regarded as stationary, and with respect to the adopted positive direction along the contour, is

$$d\Phi = -Bbv\,dt.$$

So the electromotive force induced in the contour due to the change of the magnetic flux through the contour (assumed to be stationary) is

$$\mathscr{E}'' = -\int_S \frac{\partial \mathbf{B}}{\partial t}\cdot d\mathbf{S} = -\frac{\partial \Phi}{\partial t} = \frac{Bbv\,dt}{dt} = Bbv. \tag{c}$$

According to Eqs. (b) and (c), the total electromotive force (a) induced in the contour is zero at any instant. Hence no current through the contour can result.

**11.G11.** A narrow slot is cut out of the toroid from the preceding problem, and the loop is taken out of the toroid through that slot. What is the value of the electromotive force induced in the loop from the standpoint of (1) an observer who is stationary with respect to the magnet, and (2) one who is stationary with respect to the loop?

**11.G12.** Discuss the possibility of constructing a generator of electromotive force constant in time operating on the basis of electromagnetic induction.

*Discussion.* Let us denote by $\mathscr{E}$ the desired *constant* electromotive force. According to Faraday's law in Eq. (11.4.3), such an electromotive force can be obtained provided that the magnetic flux through the circuit of the generator satisfies the differential equation

$$d\Phi/dt = -\mathscr{E} = \text{constant}, \qquad \text{or} \qquad \Phi = \Phi_0 - \mathscr{E}t,$$

$\Phi_0$ standing for the magnetic flux at time $t = 0$. Thus, a generator of a constant electromotive force, based on the principle of electromagnetic induction, can be made, provided that we can produce a magnetic flux which increases linearly in time (i.e. decreases with respect to the positive direction of the contour) during an arbitrary period. Since this is not possible, such a generator cannot be constructed.

It is interesting to mention that many attempts have been made to construct such a generator; indeed, several patents were applied for. But, from the above simple reasoning, we can see that all of them were, from the outset, doomed to failure.

**11.G13.** One side of a rectangular loop of wire is partially shielded from an external magnetic field perpendicular to the plane of the loop by means of a hollow ferromagnetic cylinder, as shown in Fig. 11.G13. Thus the sides *ad* and *bc* are obviously situated in a magnetic field of different magnitudes. If the loop, together with its shielding cylinder, is moving with a constant velocity **v** in a uniform external magnetic field of flux density **B**, as shown in the figure, will the galvanometer $G$ detect a current through the circuit? If the answer is "yes", such a device could serve for measuring the relative velocity of vehicles with respect to the terrestrial magnetic field.

*Solution.* It is true that the *total* magnetic field is smaller along the shielded side of the loop; but the magnitude of this total magnetic field is the sum of the external field and the field produced by the microscopic currents in the ferromagnetic shield. The loop is *not* moving with respect to the latter, so that no electromotive force can result from that component. Consequently, in the sides *ad* and *bc* electromotive forces of equal magnitudes will be induced, but in opposite directions along the loop, whether the shield is present or not. The only result will be a very small accumulation of charges along the upper and lower sides of the loop, as shown. They are, however, too small to be measured.

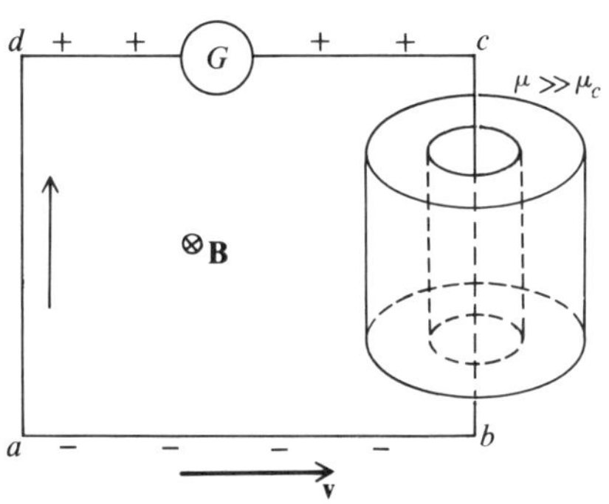


Figure 11.G13

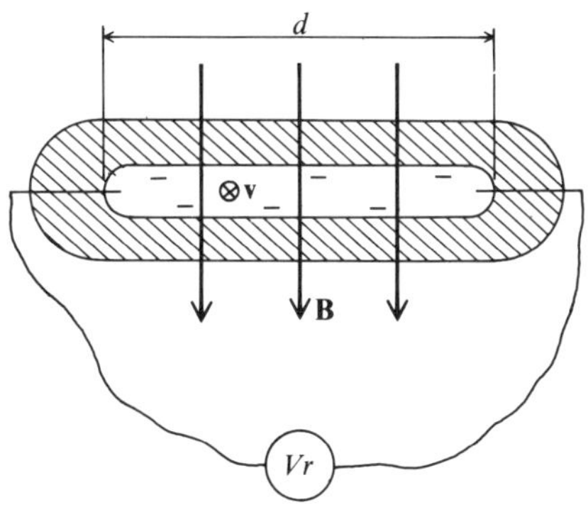


Figure 11.G14

**11.G14.** A fluid of a small, but nonzero conductivity flows through a tube of an insulating material with an unknown velocity $v$. The velocity is very nearly constant over the cross-section of the tube. To measure the velocity of the fluid, a part of the tube is subjected to a uniform magnetic field of flux density $B$. Two small electrodes which are in contact with the fluid are installed through the tube walls, as shown in Fig. 11.G14. A voltmeter $Vr$, of a very high internal resistance, detects the induced electromotive force $\mathscr{E}_0$. What is the velocity of the fluid?

*Answer.* For $\mathscr{E}_0 = 1$ mV, $d = 5$ cm, and $B = 0.1$ T, the velocity of the fluid is $v = 0.2$ m/s.

**11.G15.** On a thin toroidal ferromagnetic core of mean radius $a = 10$ cm, $N = 1000$ turns of thin wire are densely wound. Figure 11.G15a shows the idealized hysteresis loop of the ferromagnetic core corresponding to the saturation attained. A sinusoidal current of an amplitude $I_m = 200$ mA flows through the coil. As a secondary coil, several turns of wire are wound over the primary coil. Sketch the approximate form of the electromotive force induced in the secondary coil, during one cycle of the sinusoidal current in the primary coil.

*Solution.* Only the qualitative shape of the electromotive force induced in the secondary circuit is required. It is therefore sufficient to draw the curve showing the change of the flux

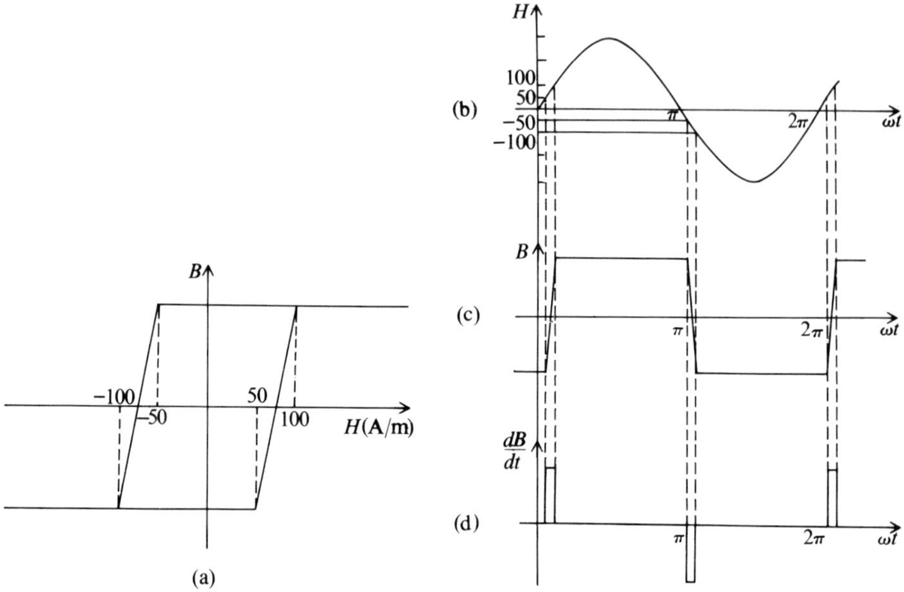


**Figure 11.G15**

density $B$ with time, and therefrom its time derivative $dB/dt$. According to Faraday's law, the electromotive force induced in the secondary circuit is proportional to $-dB/dt$.

The magnetic field intensity in the toroidal core produced by the sinusoidal current in the primary coil is

$$H = Ni/2\pi a = 200 \sin \omega t/(2\pi \cdot 0.1) \simeq 300 \sin \omega t.$$

The amplitude of $H$ (300 A/m) is consequently larger than that at which saturation of the core occurs (100 A/m). Thus the section of the sinusoid of the magnetic field intensity $H$ in Fig. 11.G15b, in the interval $50 \leq H \leq 100$, can be approximately represented by a straight line.

Bearing this in mind and taking into account the magnetization curve we can see from Fig. 11.G15c the change of the flux density $B$, and from Fig. 11.G15d the dependence of $dB/dt$ on time. The latter curve, as we have seen, represents a qualitative form of the electromotive force induced in the secondary coil.

**11.G16.** A very long, tightly wound solenoidal coil of radius $a$ has $N_l$ turns per unit length, carrying a current $i = I_m \cos \omega t$. The medium inside and outside the solenoid is air. Two coaxial secondary coils with $N_1$ and $N_2$ turns have respective radii $a_1 > a$ (the secondary coil encircles the solenoid), and $a_2 < a$ (the secondary coil is inside the solenoid). What is the electromotive force induced in both cases, and what is the electric field intensity existing along the two coils?

**11.G17.** A rectangular secondary loop with sides $a$ and $b$ is placed symmetrically with respect to a very long cylindrical solenoid of radius $a$, as shown in Fig. 11.G17. What are the partial electromotive forces induced in sides $a$ and $b$ of the rectangular loop? The magnetic flux through the solenoid varies following simple harmonic law, $\Phi = \Phi_m \cos \omega t$, and $a/b = \sqrt{3}$. [*Hint:* rather than calculate the line integral of the electric field intensity along sides $a$ and $b$, apply Faraday's law to contours $AA'OA$ and $AA''OA$.]

*Answer.* $\mathscr{E}_a = \Phi_m \omega \sin \omega t/6$; $\mathscr{E}_b = \Phi_m \omega \sin \omega t/3$.

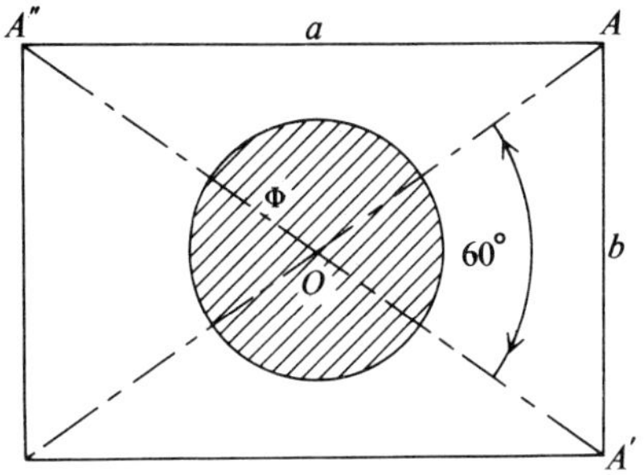


**Figure 11.G17**

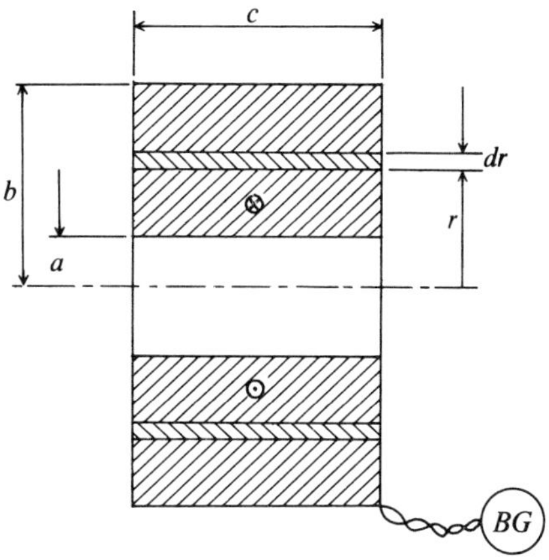


**Figure 11.G18**

**11.G18.** A search coil with a large number of turns $N$, and the cross-section shown in Fig. 11.G18, is situated in a magnetic field of unknown flux density **B**. It is known that its axis coincides with the direction of **B**. The ballistic galvanometer $BG$ connected to the coil indicates a positive current when the current in the coil is in the direction shown in the figure. The total resistance of the search coil, the leads, and the galvanometer is $R$. When the magnetic field is abruptly switched off, the ballistic galvanometer measured a quantity of charge $-Q$. Find the direction and magnitude of the magnetic flux density vector **B**.

*Solution.* Since the charge $Q$ flowed through the circuit in its *negative* direction, we know from Eq. (11.4.6) that the magnetic flux density vector **B** is directed from left to right in the figure. In order to determine the magnitude of **B**, we must first find the total magnetic flux through the coil. Note that the individual turns have different radii. Equation (11.4.6) is derived, however, assuming that all $N$ turns have equal radii.

The total magnetic flux through the coil shown is calculated in the following manner. Through a layer of turns of thickness $dr$, the magnetic flux is

$$d\Phi = \left[\frac{N}{(b-a)c}\, c\, dr\right] r^2 \pi B = \frac{N}{b-a}\pi B r^2\, dr.$$

(The expression in brackets denotes the total number of turns in the elemental layer considered.) So the total flux through all turns of the coil is

$$\Phi = \frac{N\pi B}{b-a}\int_a^b r^2\, dr = \frac{N\pi B}{3(b-a)}(b^3 - a^3) = \frac{N\pi B}{3}(b^2 + ab + a^2).$$

Had all turns the same radius and hence the same area, say $S$, flux through the coil would be simply $NSB$. Therefore, the expression

$$S_{\text{eff}} = \pi(b^2 + ab + a^2)/3$$

is referred to as the effective or average cross-section of the search coil shown in Fig. 11.G18. Thus $\Phi = NS_{\text{eff}}B$, and the magnitude of $B$ according to Eq. (11.4.6) is

$$B = \Phi/NS_{\text{eff}} = RQ/NS_{\text{eff}}.$$

***11.G19.** Repeat the preceding problem for the search coil the cross-section of which is shown in Fig. 11.G19.

*Answer.* For $Q = 10^{-5}$ C, $R = 20\ \Omega$, $a = 0.5$ cm, and $N = 200$, the magnetic flux density $B \simeq 0.034$ T.

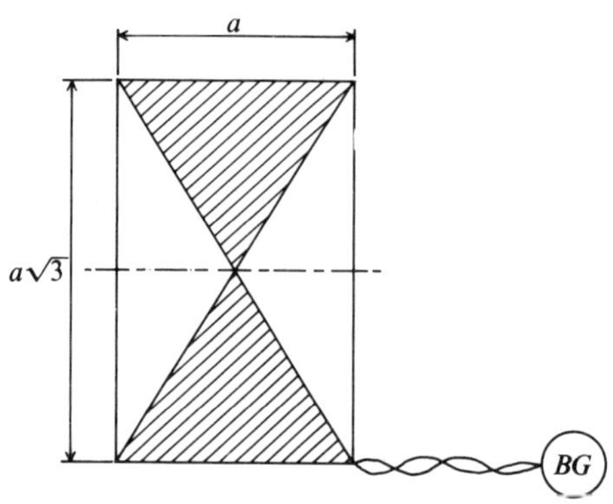


Figure 11.G19

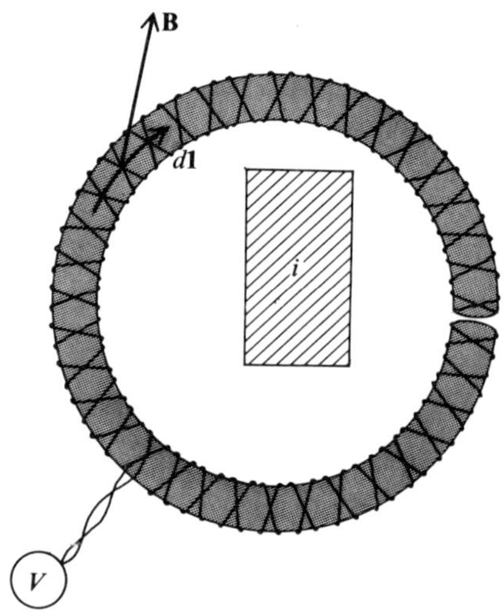


Figure 11.G20

**11.G20.** A conductor of an arbitrary cross-section carries a sinusoidal current of frequency $f$ and an unknown amplitude $I_m$. The conductor is encircled by a flexible, thin, nonferromagnetic strip of the area of the cross-section $S$, densely wound by $N_l$ turns per unit length (see Fig. 11.G20). The measured amplitude of the electromotive force induced between the ends of the strip coil is $\mathscr{E}_m$. What is the amplitude of the current flowing through the encircled conductor?

*Solution.* Consider an element $dl$ of the strip. The magnetic flux through turns on that element is

$$d\Phi = N_l\, dl\, \mathbf{B}\cdot\mathbf{S} = \mu_0 N_l S \mathbf{H}\cdot d\mathbf{l}.$$

So the flux through the whole strip coil

$$\Phi = \mu_0 N_l S \oint_C \mathbf{H}\cdot d\mathbf{l} = \mu_0 N_l S i,$$

according to Ampère's generalized circuital law in Eq. (9.4.5a), because the current $i$ is encircled by the strip coil. The electromotive force induced in the coil is thus

$$\mathscr{E} = -\frac{d\Phi}{dt} = -\mu_0 N_l S \frac{d}{dt}[I_m \cos(2\pi f t)] = \mu_0 N_l S 2\pi f I_m \sin 2\pi f t.$$

Since the measured amplitude of $\mathscr{E}$ is $\mathscr{E}_m$, amplitude of the current is

$$I_m = \frac{\mathscr{E}_m}{2\pi f \mu_0 N_l S}.$$

**11.G21.** What is the mutual inductance between a thin, infinitely long, straight conductor and the loops 2 and 3 in Fig. 11.G21? The medium is air.

*Answer.* $$L_{12} = \frac{\mu_0 h}{2\pi} \ln \frac{a+b}{a}, \qquad L_{13} = \mu_0(b - \sqrt{b^2 - a^2}).$$

The determination of $L_{13}$ involves a rather difficult integral.

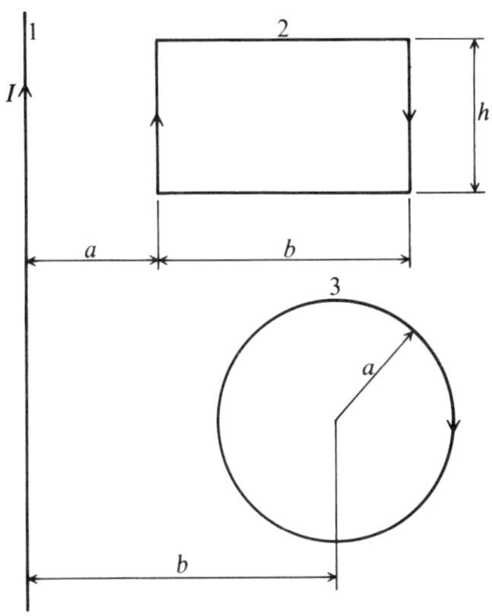


Figure 11.G21

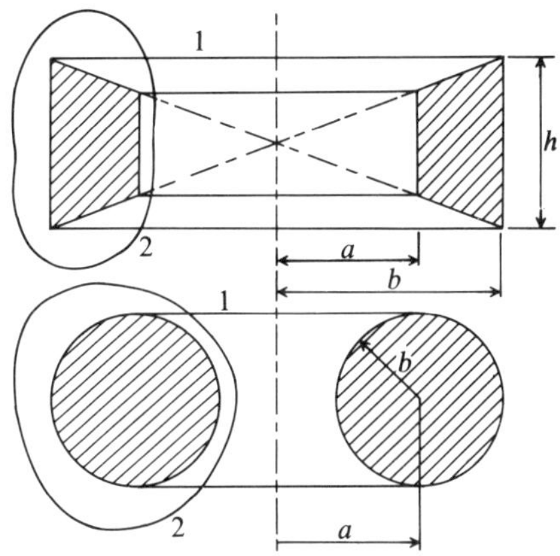


Figure 11.G22

**11.G22.** What is the mutual inductance between the toroidal coils with $N$ densely wound turns on cores of permeability $\mu$ in Fig. 11.G22 and an arbitrary closed loop encircling each toroid?

*Answer.* For the core of trapezoidal cross-section $L_{12} = \mu h N(b - a)/2\pi b$; for that with a circular cross-section $L_{12} = \mu N(a - \sqrt{a^2 - b^2})$. The latter involves evaluating a fairly uncommon integral.

**11.G23.** A conductor carrying a sinusoidal current of frequency $f$ and an unknown amplitude $I_m$ passes through a toroidal coil of $N$ turns tightly wound on a core of the cross-section shown in Fig. 11.G23 and permeability $\mu$. If a voltmeter connected to the coil indicates an amplitude $\mathscr{E}_m$ of the sinusoidal electromotive force induced in the coil, what is the value of $I_m$?

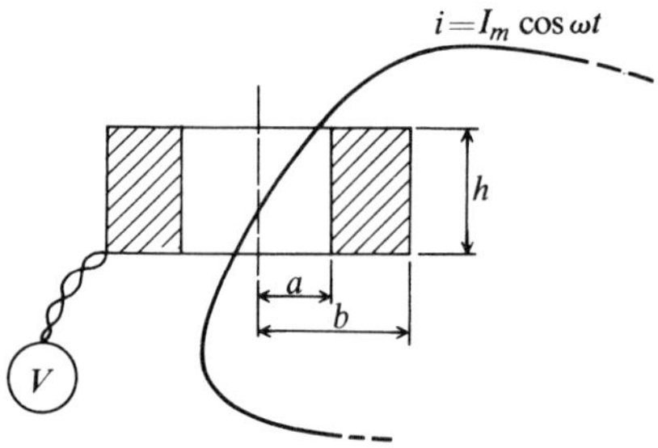


**Figure 11.G23**

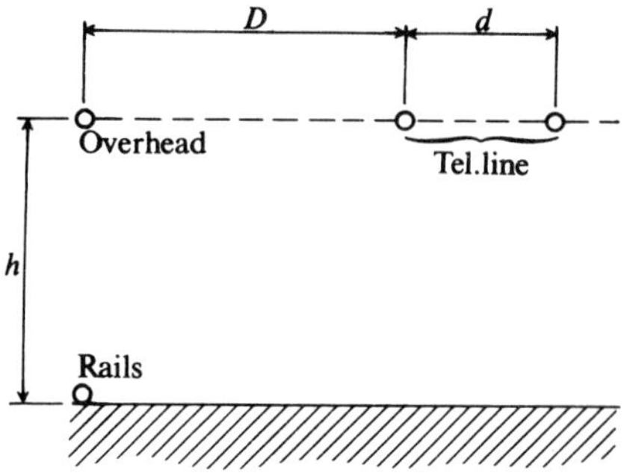


**Figure 11.G24**

*Answer.* For $\mathscr{E}_m = 1$ V, $\mu_r = 100$, $N = 500$, $h = 2$ cm, $f = 60$ Hz, and $b/a = e = 2.718$, $I_m \simeq 13.3$ A.

**11.G24.** Running parallel to a long, straight section of railway driven by alternating current is a telephone two-wire line. The rails and the overhead conductor can be considered approximately as a two-wire line with conductors at a distance $h$. The amplitude of the current through the overhead conductor and the rails is $I_m$, and its frequency is $f$. The cross-section of the system is shown in Fig. 11.G24. What is the amplitude of the electromotive force induced in a section of the telephone line of length $b$?

*Answer.* For $f = 60$ Hz, $I_m = 100$ A, $b = 10$ km, $h = 10$ m, $D = 10$ m, and $d = 25$ cm, the amplitude of the induced electromotive force $\mathscr{E}_m \simeq 0.92$ V.

**11.G25.** A two-wire telephone line runs parallel to a power two-wire line for a distance $D$. The cross-section of the two lines is shown in Fig. 11.G25. What is the amplitude of the electromotive force induced in the telephone line when the power line carries an alternating current of amplitude $I_m$ and frequency $f$?

*Answer.* For $D = 1$ km, $a = 1$ m, $b = 25$ cm, $d = 5$ m, $I_m = 100$ A, and $f = 60$ Hz, the amplitude of the induced electromotive force $\mathscr{E}_m \simeq 0.076$ V.

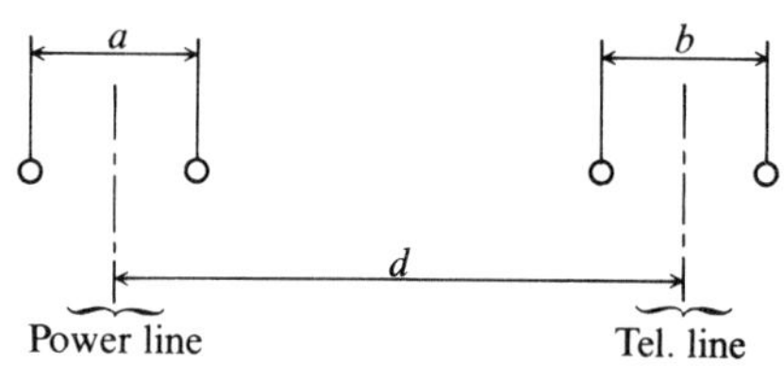


**Figure 11.G25**

**11.G26.** Determine approximately the mutual inductance between two thin, coaxial circular rings in air, of radii $a$ and $b$, if the distance $d$ between their planes is much larger than both $a$ and $b$. [*Hint:* use the result of Problem 8.G4.]

*Answer.* For $a = b = 10$ cm, and $d = 1$ m, $L_{12} \simeq 2 \times 10^{-10}$ H.

**11.G27.** Determine approximately the mutual inductance between two coplanar and concentric circular rings in air, of radii $a$ and $b$, if $b \gg a$.

*Answer*. For $b = 10$ cm, and $a = 1$ cm, $L_{12} \simeq 2 \times 10^{-10}$ H.

***11.G28.** Prove that the mutual inductance of two closed loops $C'$ and $C''$ situated in air can be written in the form

$$L_{12} = \frac{\mu_0}{4\pi} \sum_{p=1}^{m} \sum_{q=1}^{n} \int_{l'_p} \int_{l''_q} \frac{d\mathbf{l}' \cdot d\mathbf{l}''}{r}, \tag{a}$$

where $d\mathbf{l}'$ is an element of $C'$, $d\mathbf{l}''$ an element of $C''$, and $r$ the distance between the two elements. $C'$ is divided into $m$ sections, $l'_1, l'_2, \ldots, l'_p, \ldots, l'_m$, and $C''$ into $n$ sections, $l''_1, l''_2, \ldots, l''_q, \ldots, l''_n$, as shown in Fig. 11.G28.

*Proof*. The proof of Eq. (a) follows directly from Neumann's formula in Eq. (11.5.5) for mutual inductance. The sum of the integrals along $l'_p$, $p = 1, 2, \ldots, m$, represents the integral along $C'$, and similarly for $C''$. In a dynamic case, the mutual inductance can be defined also between two *sections* of two closed circuits, or even for two sections of a single closed circuit. Indeed, Eq. (11.5.2) does not imply that $\mathscr{E}$ is determined along a closed circuit, due to the current in a closed circuit (although in the quasi-stationary case this is always so). Hence the expression

$$L_{pq} = \frac{\mu_0}{4\pi} \int_{l'_p} \int_{l''_q} \frac{d\mathbf{l}' \cdot d\mathbf{l}''}{r} \tag{b}$$

can be referred to as the mutual inductance between sections $l'_p$ and $l''_q$. With Eq. (b) the mutual inductance of Eq. (a) becomes

$$L_{12} = \sum_{p=1}^{m} \sum_{q=1}^{n} L_{pq}. \tag{c}$$

Thus, according to this method the two contours are first divided into convenient sections. The mutual inductances $L_{pq}$ between all the pairs of these sections are calculated according to Eq. (b). The total mutual inductance is determined finally according to Eq. (c). We shall refer to this method as the *method of sections*, although it is essentially only a variation of Neumann's formula.

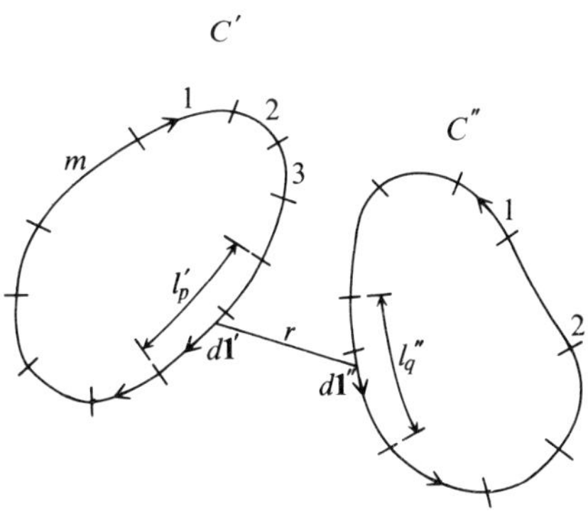


Figure 11.G28

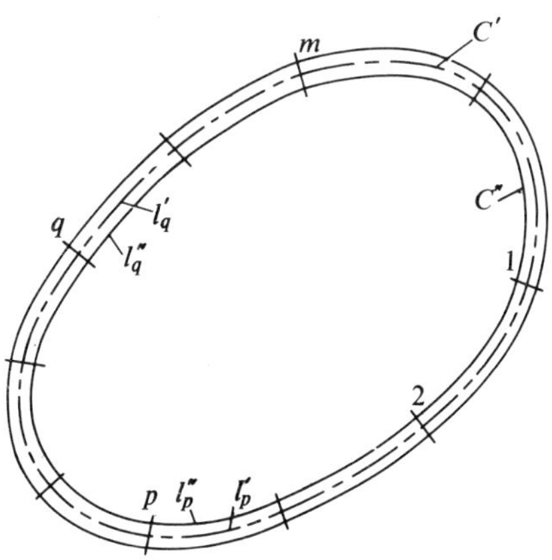


Figure 11.G29

***11.G29.** Adapt the method of sections, described in the preceding problem, for calculating the self-inductance of closed loops made of thin wires of circular cross-section.

*Solution.* According to the approximate Eq. (11.5.14) for the external inductance of a thin closed wire, we must determine the mutual inductance between two filamentary loops—the loop corresponding to the axis of the contour, and a closed loop along the conductor surface. Consequently, according to Eq. (a) of the preceding problem and adding the expression (11.5.15) for internal inductance, the total self-inductance of such a loop can be written in the form

$$L = \frac{\mu_0}{4\pi} \sum_{p=1}^{m} \sum_{q=1}^{m} \int_{l'_p} \int_{l''_q} \frac{d\mathbf{l}' \cdot d\mathbf{l}''}{r} + \frac{\mu_0 l}{8\pi}. \tag{a}$$

Sections $l'_p$ are parts of the contour coinciding with the conductor axis, and sections $l''_q$ are parts of an arbitrary contour along the surface of the conductor. In Eq. (a) it is assumed that the conductor, and hence the two contours, are divided into $m$ sections, denoted by subscripts $1, 2, \ldots p, \ldots q, \ldots, m$ (see Fig. 11.G29).

The terms with $p = q$ refer to the same section, so the expression

$$L_{ep} = \frac{\mu_0}{4\pi} \int_{l'_p} \int_{l''_p} \frac{d\mathbf{l}' \cdot d\mathbf{l}''}{r} \tag{b}$$

can be referred to as the external inductance of section $p$. The terms with $p \neq q$ are the mutual inductances $L_{pq}$ between sections $p$ and $q$.

The total self-inductance can hence be written in the form

$$L = \sum_{p=1}^{m} L_{ep} + \sum_{p=1}^{m} \sum_{q=1}^{m} L_{pq} + \frac{\mu_0 l}{8\pi}, \qquad p \neq q. \tag{c}$$

***11.G30.** Adapt the method of sections for determining the mutual inductance and self-inductance of thin closed conductors when the conductors are in the form of a closed broken line.

*Solution.* In the case of two straight sections, the mutual inductance between them, given by Eq. (b) of Problem 11.G28, can be written in a simpler form

$$L_{pq} = \frac{\mu_0 \cos\theta_{pq}}{4\pi} \int_{l'_p} \int_{l''_q} \frac{dl' \cdot dl''}{r}, \tag{a}$$

since the angle $\theta_{pq}$ is the same for all the pairs of the elements $d\mathbf{l}'$ and $d\mathbf{l}''$.

Similarly, the external self-inductance of a straight section, according to Eq. (b) of the preceding problem, becomes

$$L_{ep} = \frac{\mu_0}{4\pi} \int_{l'_p} \int_{l''_q} \frac{dl' \cdot dl''}{r}, \tag{b}$$

since $d\mathbf{l}'$ and $d\mathbf{l}''$ are in the same direction.

***11.G31.** Find the mutual inductance between two parallel thin conductors of length $b$, and at a distance $d$. Assume the same positive directions along the conductors.

*Answer.* $$L_{12} = \frac{\mu_0 b}{2\pi}\left\{\ln[(b/d) + \sqrt{(b/d)^2 + 1}] - \sqrt{(d/b)^2 + 1} + d/b\right\}.$$

***11.G32.** Find the external self-inductance of a thin straight conductor of length $b$ and radius $a$.

*Answer.* $$L_e = \frac{\mu_0 b}{2\pi}\left\{\ln[(b/a) + \sqrt{(b/a)^2 + 1}] - \sqrt{(a/b)^2 + 1} + a/b\right\}.$$

and $$L_e \simeq \frac{\mu_0 b}{2\pi}\left(\ln\frac{b}{a} - 0.3\right) \quad \text{for} \quad b \gg a.$$

***11.G33.** Determine the self-inductance of a section, of length $b$, of a two-wire line, if the distance between the conductors is $d$ ($d \ll b$), and their radius $a$ ($d \gg a$).

*Solution.* In this case we can use Eq. (c) of Problem 11.G29 for the self-inductance of a conductor divided into sections, which, in the case of two conductors of a two-wire line, becomes

$$L = 2L_e + 2L_{12} + L_i. \tag{a}$$

$L_e$ stands for the external inductance of each of the two conductors, $L_{12}$ for the mutual inductance between them, and $L_i$ for the total internal inductance of the line, $L_i = 2\mu b/8\pi$.

According to the approximate expression for $L_e$ given in the answer to the preceding problem

$$2L_e \simeq \frac{\mu_0 b}{\pi}\left(\ln\frac{b}{a} - 0.3\right). \tag{b}$$

According to the answer to Problem 11.G31, for the case when $b \gg d$,

$$2L_{12} \simeq -\frac{\mu_0 b}{\pi}\left(\ln\frac{2b}{d} - 1\right) \simeq -\frac{\mu_0 b}{\pi}\left(\ln\frac{b}{d} - 0.3\right),$$

so that, according to Eqs. (a) and (b),

$$L = \frac{\mu_0 b}{2\pi}\left(\ln\frac{b}{a} - \ln\frac{b}{d}\right) + 2\frac{\mu b}{8\pi} = \frac{\mu_0 b}{\pi}\left(\ln\frac{d}{a} + \frac{\mu_r}{4}\right). \tag{c}$$

***11.G34.** Find the self-inductance of a rectangular loop of sides $a$ and $b$, if the radius of the conductor is $R$.

*Answer.* $$L = \frac{\mu_0}{\pi}\left[a\ln\frac{2ab}{R(a+d)} + b\ln\frac{2ab}{R(b+d)} - 2(a+b-d)\right] + \frac{\mu_0}{4\pi}(a+b),$$

where $d = (a^2 + b^2)^{1/2}$.

***11.G35.** Determine the mutual inductance between two parallel square loops of equal sides $a$, at a distance $b$ from one another, coinciding with two opposite sides of a rectangular parallelepiped.

*Answer.*

$$L_{12} = \frac{2\mu_0}{\pi}\left[b - 2\sqrt{a^2 + b^2} + \sqrt{2a^2 + b^2} + a\ln\frac{a + \sqrt{a^2 + b^2}}{b} - a\ln\frac{a + \sqrt{2a^2 + b^2}}{\sqrt{a^2 + b^2}}\right].$$

**11.G36.** Starting from the flux definition in Eq. (11.5.10) of self-inductance, determine the self-inductance per unit length of a very long cylindrical solenoid, of cross-section area $S$, tightly wound by $N_l$ turns per unit length.

*Answer.* For air, $N_l = 1000$ turns per meter, and $S = 5\ \text{cm}^2$, the self-inductance per unit length $L_l = 0.628$ mH/m.

**11.G37.** On the toroidal core shown in Fig. 11.11 two coils having respectively $N_1$ and $N_2$ turns of thin wire are tightly wound, one over the other. Find the self-inductance of both coils, the mutual inductance of the coils, and the coefficient of coupling.

**11.G38.** What is the self-inductance of the toroidal coils shown in Fig. 11.G22, if they have $N$ tightly wound turns of thin wire?

*Answer.* For the trapezoidal toroid $L = \mu h N^2(b - a)/2\pi a$; for the circular toroid $L = \mu N^2(a - \sqrt{a^2 - b^2})$. A rather complicated integral is involved in the second case.

**11.G39.** Determine the external and the total self-inductance per unit length of a two-wire line, starting from the flux definition of the self-inductance. The radius of the wires is $a$, the distance between them $d$ ($d \gg a$).

*Solution.* Figure 11.G39 shows a section of the line. The external flux entering Eq. (11.5.10) is the flux through the crosshatched surface in the figure. This flux is produced by the flux density vectors $\mathbf{B}_1$ and $\mathbf{B}_2$ due to the currents in both conductors 1 and 2. The direction of $B_1$ and $B_2$ being the same, the magnitude of the total flux density at a distance $r$ from conductor 1, in the plane of the conductors, is

$$B = B_1 + B_2 = \frac{\mu_0 I}{2\pi r} + \frac{\mu_0 I}{2\pi(d - r)}. \qquad \text{(a)}$$

The simplest way to find the external flux per unit length of the line is first to determine the flux through the strip shown in Fig. 11.G39 double crosshatched, and then to sum (integrate) all these elemental fluxes. Thus we have that the external flux per unit length is

$$\Phi_{el} = \int_a^{d-a} B\, dr = \frac{\mu_0 I}{2\pi}\int_a^{d-a}\left(\frac{1}{r} + \frac{1}{d - r}\right) dr$$

$$= \frac{\mu_0 I}{2\pi}\left[\ln\frac{d - a}{a} + \ln\frac{d - (d - a)}{d - a}\right] = \frac{\mu_0 I}{\pi}\ln\frac{d - a}{a} \simeq \frac{\mu_0 I}{\pi}\ln\frac{d}{a},$$

since $d \gg a$. Hence the external inductance of a two-wire line per unit length is

$$L_{el} = \frac{\Phi_{el}}{I} = \frac{\mu_0}{\pi}\ln\frac{d}{a}.$$

The total inductance per unit length

$$L_l = L_{el} + L_{il} = \frac{\mu_0}{\pi}\ln\frac{d}{a} + \frac{\mu}{4\pi}. \qquad \text{(b)}$$

**11.G40.** The inner conductor, of radius $a$, of a coaxial line is solid, and the outer conductor is in the form of a very thin tube, of radius $b$. The dielectric is air, and the permeability of the inner conductor is very nearly $\mu_0$. Determine the external inductance and the total inductance of the line per unit length.

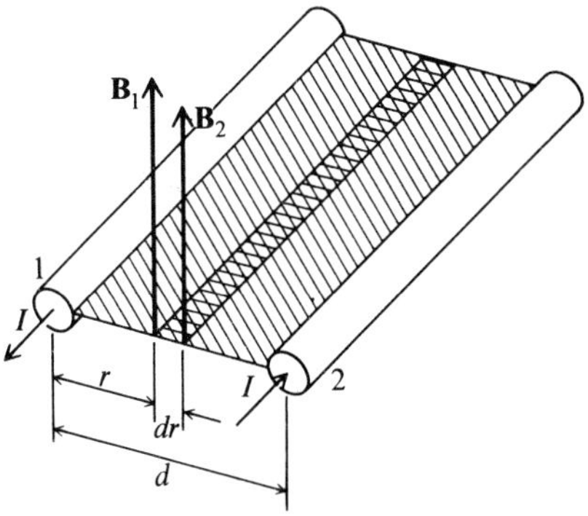


Figure 11.G39

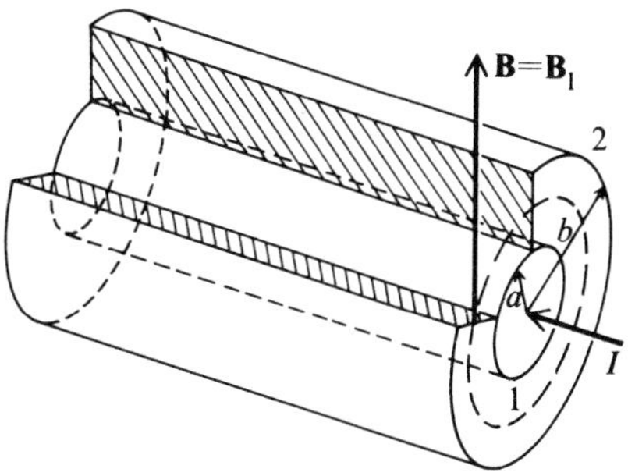


Figure 11.G40

*Solution.* By symmetry, the lines of vector **B** are circles. From Ampère's circuital law in integral form, we readily deduce that there is no magnetic field outside the cable, and that, between the two conductors, $B = \mu_0 I/2\pi r$. To determine the external flux per unit length of the line we must integrate over the unit length of the strip surface shown crosshatched in Fig. 11.G40. We can thus easily find that

$$L_{el} = \frac{\mu_0}{2\pi} \ln \frac{b}{a}, \tag{a}$$

so that the total inductance per unit length is

$$L_l = L_{el} + L_{il} = \frac{\mu_0}{2\pi}\left(\ln \frac{b}{a} + \frac{1}{4}\right). \tag{b}$$

This formula is valid only if thickness of the outer conductor is negligible, and if the current in the inner conductor is distributed uniformly over its cross-section (direct and slowly varying currents).

## Numerical problems

**11.N1.** A very long solenoid, of radius $a = 1$ cm, tightly wound with 1000 turns per meter carries a sinusoidal current $i(t) = \sin(2\pi\, 60t)$ A. Determine the amplitude of the electric field intensity inside and outside the coil.

**11.N2.** What is the electromotive force of a Faraday's disk generator, of disk radius $a = 5$ cm, with $n = 10{,}000$ revolutions per minute, in a uniform magnetic field of flux density $B = 0.1$ T?

**11.N3.** What is the self-inductance of the toroid shown in Fig. 11.11, if $N = 200$, $h = 2$ cm, $b = 3$ cm, and $a = 2$ cm?

***11.N4.** Find the self-inductance of a square loop, of sides $b = 10$ cm, made of a copper wire of radius $a = 1$ mm.

**11.N5.** Determine the self-inductance per unit length of a two-wire line in air, with copper conductors of radius $a = 1.5$ mm, and the distance between them $d = 30$ cm. (See Problem 11.G39.)

**11.N6.** Determine the self-inductance of the trapezoidal toroidal coil shown in Fig. 11.G22 with air core, if $N = 500$, $b = 5$ cm, $a = 2$ cm, and $h = 3$ cm, starting from the exact formula, and assuming that the average flux density in the toroid is that corresponding to the mean radius.

**11.N7.** The radii of the conductors of a coaxial line are $a = 1$ mm and $b = 3$ mm. The dielectric has permability equal to $\mu_0$. Find the external self-inductance of the line per unit length, and the total self-inductance, if the inner conductor is solid, and the outer in the form of a tube of very small thickness. (See Problem 11.G40.)

**Suggested reading.** The list progresses towards more advanced texts. For full bibliographical description see Bibliography.

1. CARTER, G. W., *The Electromagnetic Field in Its Engineering Aspects*, Chapter 8.
2. BEWLEY, L. W., *Flux Linkages and Electromagnetic Induction*, complete book.
3. SMYTHE, W. R., *Static and Dynamic Electricity*, Chapter VIII.

CHAPTER 12

# ENERGY AND FORCES IN THE STATIC AND QUASI-STATIC MAGNETIC FIELD

## 12.1 ENERGY RELATIONS IN QUASI-STATIONARY ELECTROMAGNETIC FIELD

**12.1.1 Introduction.** To establish any static system it is necessary to have a transient, nonstationary process until the final, static state is attained. For example, in the case of the static electric field, the transient process implies a non-electrostatic process of bringing the charges to their desired positions or onto conducting bodies. In the case of the static magnetic field, this transient process starts at the instant when the currents are switched on, or (theoretically) when the current systems begin to be brought from infinity; it ends when all the currents of the system can be considered as steady.

During this transient process varying currents will induce certain electromotive forces in all contours, although at the end of the process no induced electromotive forces exist. In computing the energy required to establish the desired distribution of currents the existence of these induced electromotive forces must be taken into account. This is precisely the reason why it was not possible to analyze the energy relations in a static magnetic field before the phenomenon of electromagnetic induction was explained. In this section, the energy relations for an arbitrary quasi-stationary system will be analyzed, and the case of the static magnetic field then deduced as a special case.

**12.1.2 The energy required to establish a magnetic field.** Let us consider an arbitrary quasi-stationary current system, produced by a time-varying impressed electric field $\mathbf{E}_i$. The time-varying currents produced by this field have a density $\mathbf{J}$ which is given by

$$\mathbf{J} = \sigma(\mathbf{E}_i + \mathbf{E}_{\text{total}}) = \sigma(\mathbf{E}_i - \partial \mathbf{A}/\partial t - \text{grad } V). \tag{12.1.1}$$

The free charges in the system are propelled at all points by both $\mathbf{E}_i$ and $\mathbf{E}_{\text{total}}$. $\mathbf{E}_{\text{total}}$ is due to the quasi-stationary charges on the conductor surfaces (the term $-\text{grad } V$), and to the accelerated charges (varying currents) inside the conductors (the term $-\partial\mathbf{A}/\partial t$).

Our aim is to obtain an energy relation enabling us to determine the power required to maintain the time-varying currents, and hence also the time-varying (but, by hypothesis, quasi-stationary) magnetic field. Let us multiply Eq. (12.1.1) by $\mathbf{J}\, dv/\sigma$, where $dv$ represents a volume element in which the current density is $\mathbf{J}$ and the conductivity $\sigma$. Integrating the equation thus obtained over the whole region $v$ occupied by the currents we have

$$\int_v \mathbf{E}_i \cdot \mathbf{J}\, dv = \int_v \frac{J^2}{\sigma}\, dv + \int_v \mathbf{J} \cdot \frac{\partial \mathbf{A}}{\partial t}\, dv + \int_v \mathbf{J} \cdot \operatorname{grad} V\, dv. \tag{12.1.2}$$

The term on the left-hand side of this equation represents the power of all the "external" (nonelectric) forces acting on the free charges in the whole system at the instant considered. The terms on the right-hand side describe how this power is used up.

The first term on the right represents the power transformed over the whole system into heat (Joule's heat). The third term represents that part of the power of the external forces which is used against the forces of the electric field created by the quasi-stationary distribution of the charges. The second term similarly represents that part of the power of the external forces which is used against the forces of the (quasi-stationary) electric field $-\partial \mathbf{A}/\partial t$ created by all the varying currents of the system at the instant considered. Only this latter term is associated with the energy necessary to establish some distribution of currents, that is, a certain magnetic field. If we denote by $P_m$, the (instantaneous) power necessary to maintain a given quasi-stationary magnetic field we have

$$\boxed{P_m = \int_v \mathbf{J} \cdot \frac{\partial \mathbf{A}}{\partial t}\, dv} \qquad \text{(valid generally).} \tag{12.1.3}$$

Note that no restrictions are imposed on this relation except that the field must be quasi-stationary. For example, region $v$ may contain nonlinear, ferromagnetic materials, or a part of $P_m$ may be used to perform mechanical work in moving some bodies of the system by means of the magnetic forces.

If no nonlinear magnetic materials are present and if all the conductors are stationary, Eq. (12.1.3) under certain restrictions can be simplified. Let the field be quasi-stationary, and let the current density at all points vary with time following the same law, say, that described by a dimensionless function of time $f(t)$. Then we can write $\mathbf{J} = \mathbf{J}_0 f(t)$, where $\mathbf{J}_0$ is a function of position only. In the case of a magnetically linear medium, the secondary sources of the magnetic field (the macroscopic resultants of the microscopic currents) will vary in time following the same law. All these currents are to be assumed to exist in a vacuum. From Eq. (8.9.16) it follows that the vector potential $\mathbf{A}$ must also be of the form $\mathbf{A} = \mathbf{A}_0 f(t)$, $\mathbf{A}_0$ being a function of position only. Thus

$$\mathbf{J} \cdot \frac{\partial \mathbf{A}}{\partial t} = \mathbf{J}_0 \cdot \mathbf{A}_0 f(t) \frac{df(t)}{dt} = \tfrac{1}{2} \mathbf{J}_0 \cdot \mathbf{A}_0 \frac{d}{dt}[f^2(t)] = \frac{1}{2} \frac{\partial}{\partial t}(\mathbf{J} \cdot \mathbf{A}),$$

and Eq. (12.1.3) becomes

$$P_m = \frac{\partial}{\partial t}\int_v \tfrac{1}{2}\mathbf{J}\cdot\mathbf{A}\,dv \qquad \text{(linear, stationary media).} \tag{12.1.4}$$

Let us determine the total work done by the external forces in establishing the field existing at an instant $t$. We multiply Eq. (12.1.3) i.e. (12.1.4) by $dt$, and integrate with respect to time from 0 to $t$ (it is assumed that, at $t = 0$, there was no field). If we designate the work done by the external forces by $W_m$, we have

$$W_m = \int_0^t\int_v \mathbf{J}\cdot\frac{\partial\mathbf{A}}{\partial t}\,dt\,dv = \int_0\int_v^{\mathbf{A}} \mathbf{J}\cdot d\mathbf{A}\,dv \qquad \text{(valid generally),} \tag{12.1.5}$$

$$W_m = \int_0^t d\int_v \tfrac{1}{2}\mathbf{J}\cdot\mathbf{A}\,dv = \tfrac{1}{2}\int_v \mathbf{J}\cdot\mathbf{A}\,dv \qquad \text{(linear, stationary media).} \tag{12.1.6}$$

Let us suppose that the currents are gradually decreased, that is, that the magnetic field is gradually switched off. Then $d\mathbf{A}$ in Eq. (12.1.5) will be *negative*, so that the work done *by the external forces* in reducing the currents to zero will also be negative. This means that a certain amount of the energy of the external sources used to build up the field is returned to them. If the medium is magnetically nonlinear, one part only of this energy is returned to the sources. In that case, the building up and diminishing of the field do *not* follow the same process (in opposite directions). A certain part of that energy is irreversibly transformed into heat in the ferromagnetic material. In the case of *linear* materials, however, the demagnetization process is the complete reverse of the magnetization process, so the total energy used to establish the field is returned to the sources. What happens next to that energy depends upon the nature of the sources. Either it will be transformed into the same type of energy (e.g. chemical energy) as that used by the sources (the reversible sources), or part of that energy, or the whole of it will be transformed into heat inside the sources (the irreversible sources). Consequently, in the case of *linear* magnetic materials, *and only in that case*, it is possible to say that $W_m$ as defined by Eq. (12.1.6) is the energy stored in the system of currents, or more simply the *magnetic energy* of the system. So the concept of magnetic energy, defined as the energy stored in a system of currents and completely available at any instant of time (this is referred to as *free* energy) is possible only for linear media.

In the light of these remarks, we see that Eq. (12.1.6) represents magnetic energy stored in any linear system, although it was derived under certain restrictions (i.e. that **J** and **A** vary in time according to the same law). The energy of a linear system cannot be a function of the process by which the final state of the system is attained.

**12.1.3 The distribution of energy in the magnetic field.** Equation (12.1.5) for the energy required to establish a magnetic field implies integration over the volume of the conductors in which the currents exist. This does not mean, however, that work was done only in creating the currents, and that the equation shows how much energy is absorbed by each elemental volume of the current field. If material bodies are present, a certain amount of work must be done in magnetizing all the elemental volumes of these bodies, against the "elastic" and "frictional" forces tending to prevent the alignment of the magnetic moments of small magnetized domains inside the bodies. Therefore, a certain part at least of the total work in Eq. (12.1.5) has to be done at points inside the magnetic materials. To determine the work done per unit volume *throughout the field*, let us transform Eq. (12.1.5) for the total work done as follows.

The integration in Eq. (12.1.5) refers to the volume occupied by currents, but it can also be assumed to refer to the whole space. The contribution of the integration over the space external to the conductors is zero, since at all points of that region $\mathbf{J} = 0$. Thus, instead of Eq. (12.1.5), we can write

$$W_m = \int_0^{\mathbf{A}} \int_{\text{all space}} \mathbf{J} \cdot d\mathbf{A}\, dv = \int_0^{\mathbf{A}} \int_{\text{all space}} \operatorname{curl} \mathbf{H} \cdot d\mathbf{A}\, dv,$$

since $\mathbf{J} = \operatorname{curl} \mathbf{H}$. The dot product $\operatorname{curl} \mathbf{H} \cdot d\mathbf{A}$ can be expanded as follows. Let $\mathbf{a}$ and $\mathbf{b}$ be two arbitrary well-behaved vector functions of position. Then, according to Eq. (0.3.15), a formal application of the (differential!) del operator yields

$$\begin{aligned}\nabla \cdot (\mathbf{a} \times \mathbf{b}) = \nabla \cdot (\mathbf{a} \times \overset{*}{\mathbf{b}}) + \nabla \cdot (\overset{*}{\mathbf{a}} \times \mathbf{b}) &= \mathbf{b} \cdot (\nabla \times \mathbf{a}) + \mathbf{a} \cdot (\mathbf{b} \times \nabla) \\ &= \mathbf{b} \cdot (\nabla \times \mathbf{a}) - \mathbf{a} \cdot (\nabla \times \mathbf{b}).\end{aligned}$$

The asterisk above a vector function indicates that it has to be considered as a constant in applying the del operator. Thus, quite generally,

$$\boxed{\operatorname{div} (\mathbf{a} \times \mathbf{b}) = \mathbf{b} \cdot \operatorname{curl} \mathbf{a} - \mathbf{a} \cdot \operatorname{curl} \mathbf{b}}\,. \tag{12.1.7}$$

Let $\mathbf{a} = \mathbf{H}$ and $\mathbf{b} = d\mathbf{A}$; then the work $W_m$ can be transformed to

$$W_m = \int_0^{\mathbf{A}} \int_{\text{all space}} \mathbf{H} \cdot \operatorname{curl} d\mathbf{A}\, dv + \int_0^{\mathbf{A}} \int_{\text{all space}} \operatorname{div} (\mathbf{H} \times d\mathbf{A})\, dv.$$

If we transform the last integral into the surface integral over an infinitely large surface, it vanishes, because, at large distances from the system, $\mathbf{H}$ is proportional to $1/r^2$, $\mathbf{A}$ to $1/r$, and the area only to $r^2$. Also, $\operatorname{curl} d\mathbf{A} = d(\operatorname{curl} \mathbf{A}) = d\mathbf{B}$. Hence the work done by the external forces in establishing a magnetic field can be represented as

$$\boxed{W_m = \int_v \left( \int_0^{\mathbf{B}} \mathbf{H} \cdot d\mathbf{B} \right) dv} \qquad \text{(valid generally)}, \tag{12.1.8}$$

where $v$ is the total volume occupied by the *field*. As with Eq. (12.1.5), this expression is also valid for an arbitrary (but quasi-stationary) magnetic field.

In the case of linear media, $\mathbf{B} = \mu\mathbf{H}$, $d\mathbf{B} = \mu\, d\mathbf{H}$, so that Eq. (12.1.8) becomes

$$W_m = \int_v \int_0^{\mathbf{H}} \mathbf{H} \cdot \mu\, d\mathbf{H}\, dv = \int_v \tfrac{1}{2}\mu H^2\, dv = \int_v \tfrac{1}{2}\mathbf{B} \cdot \mathbf{H}\, dv \qquad \text{(linear media).} \tag{12.1.9}$$

According to Eq. (12.1.8), the correct expression for the *total* work done in establishing a magnetic field is obtained if it is assumed that a work

$$\boxed{\frac{dW_m}{dv} = \int_0^{\mathbf{B}} \mathbf{H} \cdot d\mathbf{B}} \qquad \text{(valid generally)} \tag{12.1.10}$$

has to be done per unit volume *throughout the region occupied by the field.* In the case of linear magnetic materials

$$\frac{dW_m}{dv} = \tfrac{1}{2}\mu H^2 = \tfrac{1}{2}\mathbf{B} \cdot \mathbf{H} = \frac{1}{2}\frac{B^2}{\mu} \qquad \text{(linear media).} \tag{12.1.11}$$

As we have noted, in the case of magnetic materials we can expect that a certain work has to be done at all points where the magnetization is established. The mathematical equivalence of Eqs. (12.1.5) and (12.1.8), i.e. of Eqs. (12.1.6) and (12.1.9), requires, however, that *every* elemental volume in the field be assumed to require some energy for its magnetization. This is true even when the magnetic field exists in a vacuum. As with the distribution of energy in the static electric field, it is not possible to check experimentally whether this hypothesis is correct or not: we cannot cut out a piece of the magnetic field and measure the work done in establishing the field in that particular volume. All the remarks we have made in connection with the density of the energy of the static electric field may at this point be profitably read again and applied to the question whether or not energy is really located at every point of the magnetic field.

In the case of a *static* magnetic field, the above formulas are directly applicable if the variable limits of integration **A** and **B** in Eqs. (12.1.5) and (12.1.8) are considered as the final values of **A** and **B**. In the case of *linear* media, it is possible to speak about the energy located in the field, and to assume that it is distributed throughout the field with the density given by Eq. (12.1.11). The symbols **B** and **H** now stand for the flux density and the magnetic field intensity in the static magnetic field.

As a simple example, let us consider a thin toroidal coil of circular cross-section, with $N$ tightly wound turns, made of a linear magnetic material of permeability $\mu$. Suppose that the mean radius of the core is $R$, the area of its cross-section $S$, and that a current of intensity $I$ flows through the coil. As shown in Section 8.7.2, Eq. (8.7.15), the magnetic flux density in the coil is $B = \mu NI/2\pi R$, so, according to Eq. (12.1.10), the density of the energy located in the magnetic field in the toroid is

$$\frac{dW_m}{dv} = \tfrac{1}{2}\mu \frac{N^2 I^2}{(2\pi R)^2}.$$

Since $dW_m/dv$ is the same at all points, the total energy located in the core is

$$W_m = \frac{dW_m}{dv} S 2\pi R = \tfrac{1}{2}\mu S \frac{N^2 I^2}{2\pi R}. \tag{12.1.12}$$

We can obtain the same result by using Eq. (12.1.6). Integration has to be performed along the coil wire only. If we denote by $dl$ an elemental length of the wire and by $\Delta S$ its cross-section, then

$$W_m = \tfrac{1}{2}\int_v \mathbf{J}\cdot\mathbf{A}\,dl\,\Delta S = \tfrac{1}{2}\int_{\text{coil}} J\,\Delta S\,\mathbf{A}\cdot d\mathbf{l} = \tfrac{1}{2}I\int_{\text{coil}} \mathbf{A}\cdot d\mathbf{l} = \tfrac{1}{2}IN\Phi_{\text{core}},$$

since the total flux through the coil is $N$ times the flux through a cross-section of the core. If we replace $\Phi_{\text{core}}$ in the last expression by $BS = \mu SNI/2\pi R$, we get the result in Eq. (12.1.12) immediately.

**12.1.4 Hysteresis losses.** Let us now consider a quasi-stationary system of currents varying according to the simple harmonic (sinusoidal) law. The magnetization at all points will then also be periodic, but not harmonic, unless the medium is linear. If the medium is assumed to exhibit hysteresis properties, let us consider a point at which the magnetic field intensity varies from $H_m$ to $-H_m$. The hysteresis loop along which the working point in the $B$–$H$ plane moves is diagrammatically represented in Fig. 12.1. Let us determine the work done per unit volume at the point considered, during one cycle of the periodic change of the field.

Since the medium is nonlinear, we must use the general Eq. (12.1.10). Now, $\mathbf{H}\cdot d\mathbf{B}$ in that expression represents the area of a small rectangle of sides $H$ and

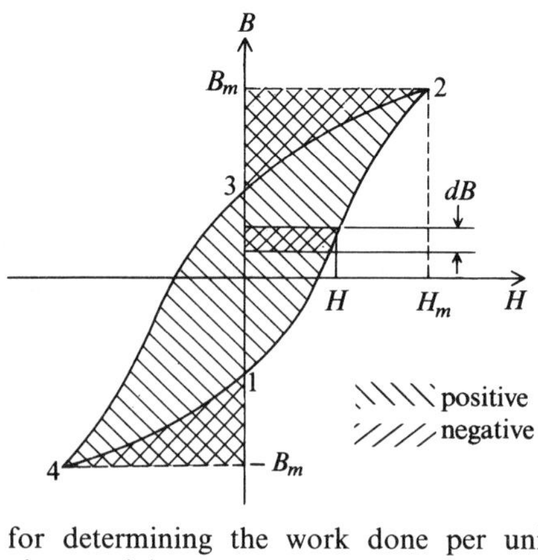


**Fig. 12.1** Construction for determining the work done per unit volume per cycle in magnetizing a ferromagnetic material.

$dB$ in the $B$–$H$ diagram. Figure 12.1 shows one such rectangle. In the present case integration in Eq. (12.1.10) implies the summation of the areas of all these small rectangles which are obtained when the working point moves around the hysteresis loop 12341.

Starting from point 1, we note that, from point 1 to point 2, both $H$ and the increment $dB$ of $B$ are positive. The work done in changing the state of the magnetization from point 1 to point 2 is hence also positive, and is given by the area between part 12 of the hysteresis loop and the $B$ axis. From point 2 to point 3, the magnetic field intensity $H$ is again positive, but the increment $dB$ of $B$ is negative (from point 2 to point 3 the flux density $B$ decreases). The work done in changing the state of the magnetization from point 2 to point 3 is hence also *negative*. This means that the energy per unit volume (represented in Fig. 12.1 by the double crosshatched surface $23B_m2$) is returned back to its sources. From point 3 to point 4, both $H$ and $dB$ are negative, so that the work done is positive. From point 4 back to point 1, $H$ is negative and $dB$ positive, so that again the energy per unit volume, represented by the area $41(-B_m)4$ is negative, i.e. it returns back to its sources.

The total work done per unit volume during *one cycle* of the periodic field is therefore represented by the area of the hysteresis loop. It is transformed inside the magnetic material into heat, and represents unavoidable losses of energy associated with the periodic magnetization of any ferromagnetic substance. Such losses are referred to as the *hysteresis losses*; physically, they represent the energy transformed into heat in periodically orienting magnetized domains, between which a kind of friction exists.

**12.1.5 The magnetic energy of a system of quasi-filamentary current loops.** A system of currents often encountered in practice is that made up of a number of quasi-filamentary current loops situated in a *linear* medium. The energy stored in the magnetic field of such a current system is given by the general formula in Eq. (12.1.6). It can also be expressed in terms of the currents in the loops and the coefficient of inductance as follows.

If the loops are quasi-filamentary, the volume element $dv$ in Eq. (12.1.6) can be represented as the product $S_k dl_k$ of the cross-section $S_k$ of the $k$th loop and the length element $dl_k$ of that loop. Also, $\mathbf{J}\cdot\mathbf{A}\,dl_k\,S_k = JS_k\mathbf{A}\cdot d\mathbf{l}_k = i_k\mathbf{A}\cdot d\mathbf{l}_k$, where $i_k$ is a (generally time-varying) current in loop $k$, which is constant along the loop. In the light of these remarks, the integral in Eq. (12.1.6) can be represented as a sum over all $n$ loops $C_1, C_2, \ldots, C_n$ of the system:

$$W_m = \tfrac{1}{2}\int_v \mathbf{J}\cdot\mathbf{A}\,dv = \tfrac{1}{2}\sum_{k=1}^{n}\oint_{C_k} i_k\mathbf{A}\cdot d\mathbf{l}_k = \tfrac{1}{2}\sum_{k=1}^{n} i_k\oint_{C_k}\mathbf{A}\cdot d\mathbf{l}_k$$

$$= \tfrac{1}{2}\sum_{k=1}^{n} i_k\Phi_k. \tag{12.1.13}$$

In the case of a linear medium, the flux $\Phi_k$ through loop $k$ is equal to the sum of the fluxes $\Phi_{1k}, \Phi_{2k}, \ldots, \Phi_{kk}, \ldots, \Phi_{nk}$ through the loop created by the currents $i_1, i_2, \ldots, i_k, \ldots, i_n$ separately. According to Eq. (11.5.1), the flux $\Phi_k$ through the loop $k$ can be represented in the form

$$\begin{aligned}\Phi_k &= \Phi_{1k} + \Phi_{2k} + \cdots + \Phi_{kk} + \cdots + \Phi_{nk} \\ &= L_{1k}i_1 + L_{2k}i_2 + \cdots + L_{kk}i_k + \cdots + L_{nk}i_n \\ &= \sum_{j=1}^{n} L_{jk}i_j,\end{aligned}$$

so that Eq. (12.1.13) becomes

$$W_m = \tfrac{1}{2}\sum_{j=1}^{n}\sum_{k=1}^{n} L_{jk}i_ji_k \qquad \text{(the magnetic energy of } n \text{ quasi-filamentary current loops).} \tag{12.1.14}$$

In the case of a single loop, Eqs. (12.1.13) and (12.1.14) simplify to

$$W_m = \tfrac{1}{2}\Phi i = \tfrac{1}{2}Li^2 \qquad \text{(an isolated quasi-filamentary loop).} \tag{12.1.15}$$

As an example, let us consider the thin toroidal coil from the end of Section 12.1.3. According to the definition of self-inductance in Eq. (11.5.1) and Eq. (8.7.15) for the flux density in the core, the self-inductance of such a coil is

$$L = \frac{NBS}{I} = \mu\frac{N^2S}{2\pi R}.$$

The magnetic energy stored in the coil if current $i$ flows through it is hence

$$W_m = \frac{1}{2}\frac{\mu N^2 S i^2}{2\pi R},$$

which is the same as Eq. (12.1.12).

*Problems referring to Section 12.1:* 12.T1 to 12.T10, 12.G1 to 12.G8, 12.N1 to 12.N4.

## 12.2 THE GENERALIZED CONCEPT OF INDUCTANCE

**12.2.1 The definition of self-inductance in terms of magnetic energy.** Let us consider a closed conductor of an arbitrary shape, carrying a *quasi-stationary* current of (varying) intensity $i$, and let us suppose that the medium is linear. The magnetic flux density $B$ and the magnetic field intensity $H$ are then both proportional to current $i$, which is valid at *all* points, both inside and outside the conductor. According to Eq. (12.1.9), the magnetic energy of the system can therefore be written in the form

$$W_m = \text{constant} \cdot i^2. \tag{12.2.1}$$

Comparing this equation with Eq. (12.1.15), we can see that it is possible to define the self-inductance of a nonfilamentary current loop in a *linear* medium by the equation

$$\boxed{L = \frac{2W_m}{i^2}} \qquad \text{(the energy definition of self-inductance).} \qquad (12.2.2)$$

This definition of self-inductance may be referred to as the energy definition.

Now, the magnetic energy $W_m$ can be represented as the sum of the energy $W_{mi}$ stored inside the volume $v_i$ of the conductor, and the energy $W_{me}$ stored in the volume $v_e$ outside the conductor,

$$W_m = W_{mi} + W_{me}. \qquad (12.2.3)$$

Introducing this expression for $W_m$ into Eq. (12.2.2), we can represent self-inductance as a sum of two terms

$$L = L_i + L_e, \qquad (12.2.4)$$

where, according to Eq. (12.1.9),

$$L_e = \frac{2W_{me}}{i^2} = \frac{1}{i^2}\int_{v_e} \mathbf{B}\cdot\mathbf{H}\,dv \qquad \text{(the energy definition of external inductance),} \qquad (12.2.5)$$

$$L_i = \frac{2W_{mi}}{i^2} = \frac{1}{i^2}\int_{v_i} \mathbf{B}\cdot\mathbf{H}\,dv \qquad \text{(the energy definition of internal inductance).} \qquad (12.2.6)$$

The part $L_e$ of the self-inductance $L$ is associated with the energy of the magnetic field external to the conductor. It is therefore referred to as the *external inductance*. Similarly, the part $L_i$ of $L$ is associated with the energy of the magnetic field inside the conductor, and is known as the *internal inductance*.

With conductors of an arbitrary shape, both the external and internal inductance are very difficult to determine. Fortunately, conductors are most often of circular cross-section, for which the internal inductance can be easily derived.

Let us consider a straight cylindrical conductor of radius $a$, carrying a steady current of intensity $I$. The magnetic flux density inside the conductor is given by Eq. (8.7.10):

$$B = \frac{\mu I}{2\pi a^2} r, \qquad r < a, \qquad (12.2.7)$$

where $r$ is the distance from the conductor axis. In evaluating the integral in Eq. (12.2.6) for a length $l$ of the conductor, it is simplest to use as the elemental volume $dv$ a thin cylindrical shell of radius $r$ and thickness $dr$, $dv$ being equal to $2\pi rl\,dr$.

With Eq. (12.2.7), $L_i$ given by Eq. (12.2.6) becomes

$$L_i = \frac{1}{I^2}\int_0^a \frac{\mu I^2 r^2}{4\pi^2 a^4} 2\pi r l \, dr = \frac{\mu l}{8\pi}.$$

The internal inductance of a cylindrical conductor of circular cross-section is thus not a function of the conductor radius, but we must note that this is true for steady currents only (the uniform current distribution over the cross-section of the conductor). The internal inductance per unit length is

$$\boxed{L_{il} = \frac{\mu}{8\pi}} \quad \text{(cylindrical conductor of circular cross-section; steady current).} \tag{12.2.8}$$

**12.2.2 Definition of self-inductance in terms of flux linkages.*** As explained in Section 11.5.2, we cannot define the magnetic flux through a nonfilamentary contour. Consequently, a definition of self-inductance in the form of $L = \Phi/I$ is impossible for thick conductors. However, by an appropriate definition of the magnetic flux through a nonfilamentary conductor we may also *formally* use this relation in that case.

Let us *define* the magnetic flux through any current-carrying conductor as

$$\Phi = 2W_m/i. \tag{12.2.9}$$

We can see that Eq. (12.2.2) then becomes simply $L = \Phi/i$. Using Eq. (12.1.6), we can give the flux thus defined an interesting interpretation. First we obtain

$$\Phi = \frac{1}{i}\int_{v_i} \mathbf{J}\cdot\mathbf{A}\, dv. \tag{12.2.10}$$

The integration refers only to the volume of the $v_i$ conductor, since $\mathbf{J} = 0$ outside, Let us divide the volume $v_i$ into elemental tubes of current of the (variable) cross-section area $dS$, with a constant current $di_{\text{tube}}$. The volume element $dv$ can be represented as $dS\, dl$, where $dl$ is the elemental length of a current tube. Then $\mathbf{J}\, dv = \mathbf{J}\, dS\, dl = di_{\text{tube}}\, d\mathbf{l}$, so that Eq. (12.2.10) becomes

$$\Phi = \frac{1}{i}\int_{\text{all tubes}} \oint_{\text{tube}} di_{\text{tube}}\, \mathbf{A}\cdot d\mathbf{l} = \int_{\text{all tubes}} \frac{di_{\text{tube}}}{i} \oint_{\text{tube}} \mathbf{A}\cdot d\mathbf{l}$$

$$= \int_{\text{all tubes}} \Phi_{\text{tube}} \frac{di_{\text{tube}}}{i}. \tag{12.2.11}$$

The expression $\Phi_{\text{tube}}$ stands for the magnetic flux through the current tube under consideration, and it varies from one tube of current to another. However, a certain flux is common to all the tubes. If we denote this flux by $\Phi_e$, we can write $\Phi_{\text{tube}} = \Phi_e + \Delta\Phi_{\text{tube}}$, where $\Delta\Phi_{\text{tube}}$ is the flux linking the tube in addition to $\Phi_e$

* This section may be omitted without loss of continuity.

(see Fig. 12.2). Accordingly, Eq. (12.2.11) can be written as

$$\Phi = \frac{\Phi_e}{i}\int_{\text{all tubes}} di_{\text{tube}} + \int_{\text{all tubes}} \Delta\Phi_{\text{tube}}\frac{di_{\text{tube}}}{i} = \Phi_e + \int_{\text{all tubes}} \Delta\Phi_{\text{tube}}\frac{di_{\text{tube}}}{i}. \quad (12.2.12)$$

The flux $\Phi_e$ is external to the conductor, since it is common to all the tubes of the current. It is proportional to the total current $i$ in the conductor:

$$\Phi_e = L_{e\Phi} i \qquad \text{(the flux definition of external inductance).} \quad (12.2.13)$$

The coefficient $L_{e\Phi}$ is referred to as the external inductance of the conductor defined by the external flux. The last integral on the right-hand side of Eq. (12.2.12) is similarly referred to as the internal flux $\Phi_i$. The internal inductance $L_{i\Phi}$, associated with the internal flux $\Phi_i$, is defined as

$$L_{i\Phi} = \frac{\Phi_i}{i} = \frac{1}{i}\int_{\text{all tubes}} \Delta\Phi_{\text{tube}}\frac{di_{\text{tube}}}{i} \qquad \begin{array}{l}\text{(the flux definition of}\\ \text{internal inductance).}\end{array} \quad (12.2.14)$$

The energy definition from Eq. (12.2.5) of $L_e$ implies a determination of the energy external *to the conductor*. For example, in Fig. 12.2 the energy stored both in the shaded domain and in part of the unshaded domains enters into Eq. (12.2.5). But, according to Eq. (12.2.13), $L_{e\Phi}$ is implicitly defined by the energy stored in all the *tubes* which do not pierce the conductor. This corresponds to the shaded region only in Fig. 12.2. Therefore, the external and internal inductances obtained by the two methods will generally differ somewhat, though their sum is equal in both cases. In the important practical case of a cylindrical conductor of circular cross-section, the two internal inductances must be equal. In that case, integration in both Eq. (12.2.6) and Eq. (12.2.14) refers to the same region of space, since the lines of vector **B** are circles centered at the conductor axis. Let us divide the

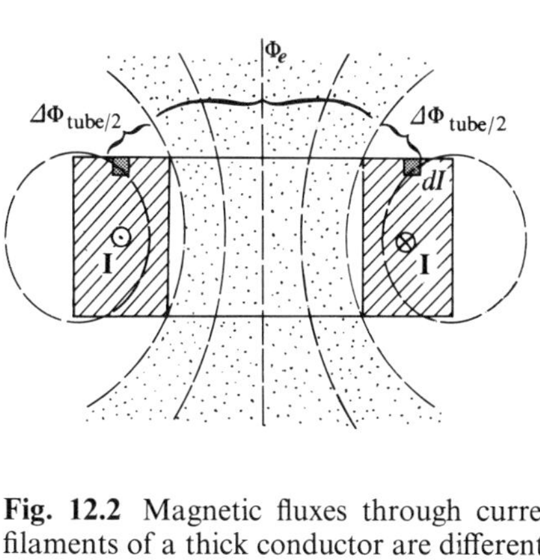


**Fig. 12.2** Magnetic fluxes through current filaments of a thick conductor are different.

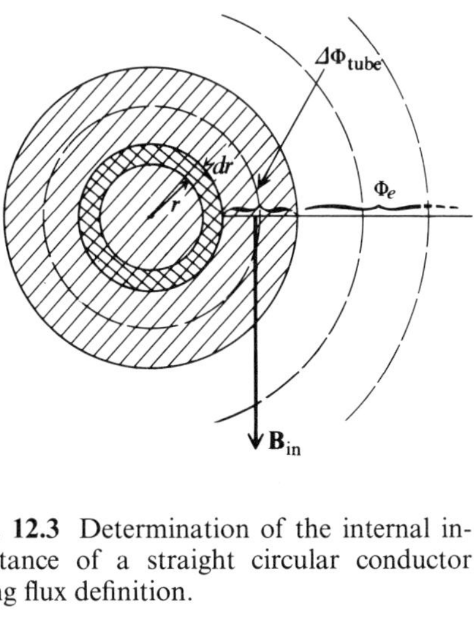


**Fig. 12.3** Determination of the internal inductance of a straight circular conductor using flux definition.

current in the conductor into cylindrical tubes of radius $r$ and thickness $dr$. According to the flux definition in Eq. (12.2.14) and that for the flux density inside the conductor in Eq. (12.2.7), with reference to Fig. 12.3 the internal inductance $L_{i\Phi l}$ per unit length is

$$L_{i\Phi l} = \frac{1}{I}\int_0^a \frac{I2\pi r\,dr}{a^2\pi I}\int_r^a \frac{\mu I}{2\pi a^2}\,r\,dr = \frac{\mu}{a^4\pi}\int_0^a r\,dr\left(\frac{a^2}{2}-\frac{r^2}{2}\right) = \frac{\mu}{8\pi},$$

which is the same as Eq. (12.2.8).

*Problems referring to Section 12.2:* 12.T11, 12.N5.

## 12.3 FORCES BETWEEN STATIONARY CURRENT SYSTEMS

**12.3.1 The Ampere-Laplace law.** Let us consider two current-carrying conductors with steady currents, the charges moving inside which act on each other by the forces given by Eq. (8.2.2). As a result there is a resultant force on the two conductors, which is equal to the sum of the forces between all pairs of elemental charges. In determining the macroscopic forces between the two current systems we do not, of course, add all these elemental forces together. If we know the average velocity of the charges and their density at all points, to determine the total force we consider it as an (infinite) sum of the elemental forces acting between pairs of physically small *volumes* with moving charges inside them. This is a typical macroscopic approach.

Let us now consider two elemental volumes, $dv_1$ and $dv_2$, inside a stationary (or quasi-stationary) current field (Fig. 12.4). The two elemental volumes might represent two elements of the same current-carrying conductor, or of two conductors, between which the force is to be determined. Suppose that the charge of the free charge carriers is $Q$, that their average velocities in $dv_1$ and $dv_2$ are $\mathbf{v}_1$ and $\mathbf{v}_2$, and that in $dv_1$ and $dv_2$ there are $N_1$ and $N_2$ charges per unit volume, respectively. If we denote by $r$ the distance between $dv_1$ and $dv_2$, then the magnetic force on all the moving charges in $dv_2$, exerted by all the moving charges in $dv_1$, is, according to Eq. (8.2.2),

$$\begin{aligned} d\mathbf{F}_{m12} &= \frac{\mu_0}{4\pi}\frac{N_2Q\mathbf{v}_2 \times (N_1Q\mathbf{v}_1 \times \mathbf{u}_{12})}{r^2}\,dv_1\,dv_2 \\ &= \frac{\mu_0}{4\pi}\frac{\mathbf{J}_2 \times (\mathbf{J}_1 \times \mathbf{u}_{12})}{r^2}\,dv_1\,dv_2. \end{aligned} \tag{12.3.1}$$

Let $dv_1$ be an element of a macroscopic volume $v_1$, and $dv_2$ an element of a macroscopic volume $v_2$. The total force by which the currents in volume $v_1$ act upon those in volume $v_2$ is then

$$\mathbf{F}_{m12} = \frac{\mu_0}{4\pi}\int_{v_1}\int_{v_2}\frac{\mathbf{J}_2 \times (\mathbf{J}_1 \times \mathbf{u}_{12})}{r^2}\,dv_1\,dv_2 = \int_{v_2}\mathbf{J}_2 \times \mathbf{B}_1\,dv_2, \tag{12.3.2}$$

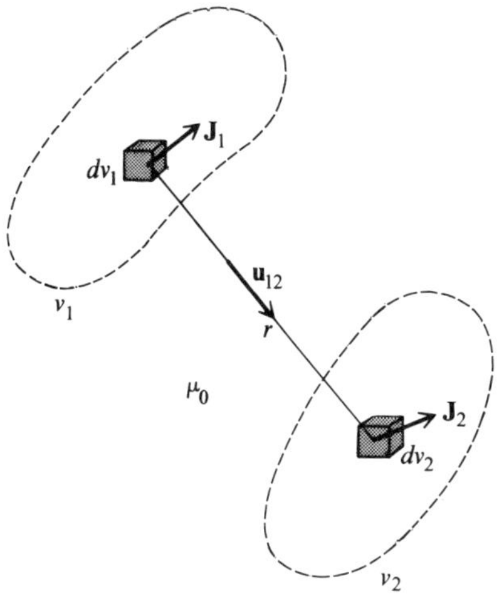


**Fig. 12.4** Two current-carrying regions.

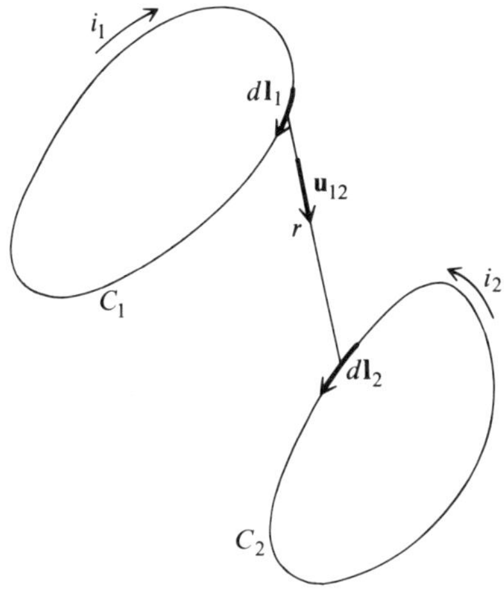


**Fig. 12.5** Two closed current filaments.

since according to Eq. (8.4.3) the integral over $v_1$ represents the flux density vector $\mathbf{B}_1$, created by the currents in volume $v_1$ at points in $v_2$. This is the most general formula for the magnetic force acting between two current-carrying regions.

Let us consider the special, but nevertheless practically important, case of two quasi-filamentary current-carrying contours with currents $i_1$ and $i_2$. The products $\mathbf{J}_1\, dv_1$ and $\mathbf{J}_2\, dv_2$ can be then written in the form $\mathbf{J}_1\, dv_1 = \mathbf{J}_1\, \Delta S_1\, dl_1 = i_1\, d\mathbf{l}_1$, and the second similarly as $\mathbf{J}_2\, dv_2 = i_2\, d\mathbf{l}_2$. So, for two such closed contours $C_1$ and $C_2$, Eq. (12.3.2) becomes

$$\mathbf{F}_{m12} = \frac{\mu_0 i_1 i_2}{4\pi} \oint_{C_1} \oint_{C_2} \frac{d\mathbf{l}_2 \times (d\mathbf{l}_1 \times \mathbf{u}_{12})}{r^2} \qquad \text{(Ampère-Laplace law)} \qquad (12.3.3)$$

(see Fig. 12.5). This expression for the force between two quasi-filamentary contours is usually referred to as the Ampère-Laplace law.

Now, according to the Biot-Savart law given in Eq. (8.4.4), the expression

$$\mathbf{B}_{12} = \frac{\mu_0 i_1}{4\pi} \oint_{C_1} \frac{d\mathbf{l}_1 \times \mathbf{u}_{12}}{r^2}$$

represents the magnetic flux density created by the current in contour $C_1$ at the point where element $d\mathbf{l}_2$ of contour $C_2$ is situated. Hence Eq. (12.3.3) can also be written in the general form (dropping all subscripts)

$$\mathbf{F}_m = \oint_C i\, d\mathbf{l} \times \mathbf{B}. \qquad (12.3.4)$$

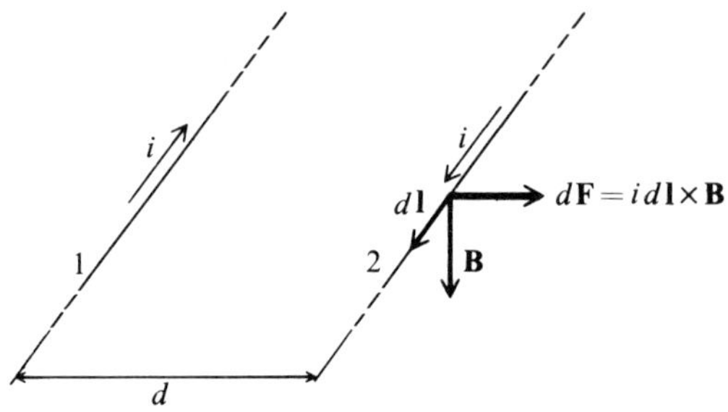


**Fig. 12.6** Force on a conductor of a two-wire line.

If a current element $i\,d\mathbf{l}$ is situated in a magnetic field of flux density $\mathbf{B}$, the elemental force on the current element is given by

$$d\mathbf{F}_m = i\,d\mathbf{l} \times \mathbf{B} \qquad \text{(the magnetic force on a current element).} \qquad (12.3.5)$$

As an example, let us determine the force (per unit length) acting between the conductors of a two-wire line, with current $i$ and the distance between the conductors $d$ (see Fig. 12.6). The magnetic flux density created by the current $i$ in conductor 1 at points of conductor 2 is $B = \mu_0 i/2\pi d$. Its direction is shown in the figure. The force per unit length on conductor 2 is repellent. According to Fig. 12.6 and Eq. (12.3.5) its intensity is

$$F_l = iB = \mu_0 \frac{i^2}{2\pi d}. \qquad (12.3.6)$$

The elemental forces acting on a body generally produce a resultant force and a resultant torque. If the distribution of the forces acting on the body is known, the resultant force and the resultant torque on the body can always be determined.

**12.3.2 The general method of determining the forces in the static and the quasi-static magnetic field.** If only the *total* forces or torques acting either on the current-carrying conductors or on the magnetized bodies are required, a simple general method, based on the principle of virtual displacements, can be applied. As explained in Section 3.5, the principle of virtual displacements is based on the law of conservation of energy.

All current-carrying conductors and magnetized bodies act on each other by means of magnetic forces and torques. A system of currents and magnetized bodies could be held stationary only if some mechanical forces, exactly equal but opposite to the magnetic forces, prevent the latter from moving the bodies. Let us *imagine*, however, that the magnetic forces moved a certain *rigid* body for an infinitesimal distance $\delta x$ against these external, mechanical forces, and that we are able to determine the work $\delta W$ done by the magnetic forces during this displacement. Let $F_x$ be the component of the magnetic force in the direction of the imagined displacement $\delta x$. Then $\delta W = F_x\,\delta x$, from which

$$F_x = \delta W/\delta x \qquad (\delta W = \text{the work done by the magnetic forces}). \qquad (12.3.7)$$

Consequently, if we can determine $\delta W$ as a function of $x$, we can also determine the component $F_x$ of the force on the body.

Determination of the imaginary ("virtual") work done by the magnetic forces during the imaginary ("virtual") displacement of a *rigid* body in the case of *linear* media can be reduced to determining the accompanying change of energy stored in the magnetic field. This can be done in the following two special cases: (1) when all the currents of the system are assumed to be kept constant, and (2) when the magnetic flux linkages through all the closed current loops are assumed to stay constant. The second case corresponds to keeping the vector potential **A** constant at all points inside the conductors (the flux linkages are equal to the line integral of the magnetic vector potential around the closed contours). It is the simpler of the two cases and will be considered first.

Let the vector potential at all points be kept constant during the virtual displacement. Then, according to Eq. (12.1.5), the work done by the *external forces* on the system is zero (since **A** is constant, $d\mathbf{A} = 0$, and the integral refers to the somewhat shifted, but *fixed* volumes of the conductors). This means that the energy required for the work $\delta W$ done by the magnetic forces is supplied *by the system*. Assume that no other energy transformation took place during the displacement (for example, no transformation of energy into heat in ferromagnetic materials). This condition is always fulfilled in the case of linear magnetic materials. According to the law of conservation of energy, the work $\delta W$ done by the magnetic forces must be equal to the *decrease* in energy stored in the system. Now the energy $W_m$ stored in the system is *less* after the virtual displacement, and its increment (the difference between the energies before and after the displacement) is therefore *negative*. Thus $\delta W$ (the *positive* work done by the magnetic forces) is equal to $-\delta W_m$, and Eq. (12.3.7) becomes

$$\boxed{F_x = -\frac{\partial W_m}{\partial x}} \qquad (\mathbf{A}, \text{ i.e. the flux linkages, kept constant}). \qquad (12.3.8)$$

If the torque on a body due to the magnetic forces is to be determined rather than the total force, the same procedure applies. The virtual work $\delta W$ done by the component $T_z$ of the total torque **T** in rotating the body through a virtual angle of $\delta\phi$ about, say, the $z$ axis of rotation is $\delta W = T_z\delta\phi$. Therefore

$$T_z = -\partial W_m/\partial\phi \qquad (\mathbf{A}, \text{ i.e. flux linkages, kept constant}). \qquad (12.3.9)$$

Consider now the first case, when all the currents (that is, **J** at all points) are kept constant during the displacement. According to Eq. (12.1.5), the work done in this case by the external sources of energy is *not* zero, but

$$\delta W_{\text{ext. sources}} = \int_0^{\mathbf{A}+d\mathbf{A}} \int_v \mathbf{J} \cdot d\mathbf{A}\, dv - \int_{v0}^{\mathbf{A}} \int_v \mathbf{J} \cdot d\mathbf{A}\, dv$$

$$= \int_{\mathbf{A}}^{\mathbf{A}+d\mathbf{A}} \int_v \mathbf{J} \cdot d\mathbf{A}\, dv = \int_v \mathbf{J} \cdot d\mathbf{A}\, dv. \qquad (12.3.10)$$

In the case of *linear* media, the energy *stored* in the field is given by Eq. (12.1.6). Let **J** be constant at all points of the *fixed* volumes of the conductors, and let **A** be changed during the virtual displacement to ($\mathbf{A} + d\mathbf{A}$). The difference in the energy stored in the field after and before the displacement is then

$$\delta W_m = \tfrac{1}{2} \int_v \mathbf{J} \cdot (\mathbf{A} + d\mathbf{A})\, dv - \tfrac{1}{2} \int_v \mathbf{J} \cdot \mathbf{A}\, dv = \tfrac{1}{2} \int_v \mathbf{J} \cdot d\mathbf{A}\, dv. \qquad (12.3.11)$$

This is exactly one-half of the elemental work in Eq. (12.3.10) done by the external sources. According to the law of conservation of energy, the other half of the work in Eq. (12.3.10) done by the external sources must be equal to the work $\delta W = F_x\, \delta x$ done by the magnetic forces during the displacement. Hence $\delta W = \delta W_{\text{ext. sources}}/2 = \delta W_m$, so that Eq. (12.3.7) becomes

$$\boxed{F_x = +\frac{\partial W_m}{\partial x}} \qquad (\mathbf{J}\text{, i.e. currents, kept constant}). \qquad (12.3.12)$$

If instead of the total force we want to find the component $T_z$ of the total torque, it is equal to

$$T_z = +\partial W_m / \partial \phi \qquad (\mathbf{J}\text{, i.e. currents, kept constant}). \qquad (12.3.13)$$

From the above expressions it is possible to understand some general properties of magnetic forces and torques. Let us consider, for instance, a system of quasi-filamentary current loops. Their magnetic energy is given by Eq. (12.1.13). Suppose that the currents (i.e. current density **J**) in all the conductors are kept constant. According to Eqs. (12.3.12) and (12.3.13), if the *magnetic* forces and torques are allowed to deform the system, the work $\delta W$ of these forces, and hence also $\delta W_m$, will be *positive*. It therefore follows from Eq. (12.1.13) that the general tendency of the flux through all the contours is to increase, which means that the magnetic forces and torques tend to move or deform them in such a way as to *increase* the magnetic flux through the individual contours. For example, a planar current loop situated in a uniform external magnetic field is acted on by a torque which tends to align its magnetic moment with the direction of the flux density **B** of the external field. Or, the forces acting on the elements of an isolated flexible current contour tend to shape the loop in the form of a circle. (This statement can easily be proved by using a light metal chain on a flat horizontal surface.)

If we consider Eq. (12.1.14) rather than Eq. (12.1.13), for magnetic energy, we may conclude that the magnetic forces and torques are such as to increase the self-inductance and mutual inductance of the system (if the mutual inductance is negative, the forces tend to make it *less* negative).

As an example of calculating the magnetic forces by using the method of virtual displacements, let us determine the force per unit length acting between the conductors of a two-wire line. Let the radius of the conductors be $a$, the distance between their axes $d$ ($d \gg a$), and let the conductors carry a current of intensity $i$. According to Eq. (12.1.15) and Eq. (b) of Problem 11.G39, the magnetic energy per unit length of the line is

$$W_{ml} = \frac{\mu_0}{2\pi}\left(\ln\frac{d}{a} + \frac{\mu_r}{4}\right) i^2.$$

The force per unit length $F_{dl}$ which tends to *increase* the coordinate $d$ (the distance between the conductors) is hence

$$F_{dl} = \frac{\partial W_m}{\partial d} = \frac{\mu_0}{2\pi}\frac{1}{d/a}\frac{1}{a}i^2 = \frac{\mu_0 i^2}{2\pi d}.$$

This is the same as Eq. (12.3.6), which we obtained by using the Ampère-Laplace law given in Eq. (12.3.3). The same result is obtained also by using Eq. (12.3.8). The working out is left as an exercise for the reader.

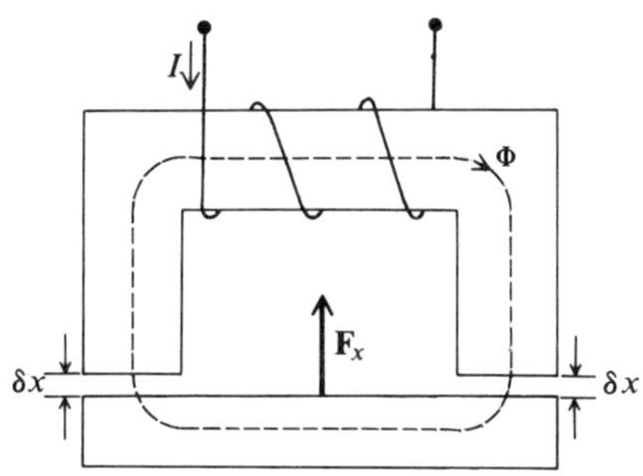


**Fig. 12.7** An electromagnet.

An important application of Eq. (12.3.8) is in determining the lifting force of an electromagnetic. Figure 12.7 shows an electromagnetic with a small air gap of thickness $\delta x$ between its upper and lower sections. Let the cross-section area of the gap be $S$, and imagine the magnetic force to pull the lower section for $\delta x$, but at the same time let the current be changed so that the magnetic flux $\Phi$ in the magnetic circuit remains constant. The flux density $B = \Phi/S$ also remains constant at all points. Consequently, the change in the magnetic energy of the system is only the change of energy stored in the air gap, which (taking into account both gaps) is equal to

$$\delta W_m = -2\frac{1}{2}\frac{B^2}{\mu_0}S\,\delta x = -\frac{\Phi^2}{\mu_0 S}\delta x.$$

According to Eq. (12.3.8) the force $F_x$ acting on the lower section of the electromagnet is in the direction shown, and its intensity is

$$F_x = -\partial W_m/\partial x = \Phi^2/\mu_0 S. \tag{12.3.14}$$

**12.3.3 Pressure on the interface between two different magnetic media.*** The general method for determining magnetic forces can serve as a basis for determining the pressure on the interface between two different (linear) magnetic materials. The derivations parallel those in Section 3.6. This method does not locate the forces inside magnetized bodies, but gives only the resulting pressures, as in the case of the static electric field. The method is based on the hypothesis that magnetic energy is distributed throughout the magnetic field, with the density given by Eq. (12.1.11).

Derivation of the total pressure acting on the interface between two media of permeabilities $\mu_1$ and $\mu_2$ parallels that in Section 3.6.2. We note first that, since no change in energy takes place if the interface is imagined to be displaced in the direction parallel to itself, it follows that no tangential forces act on the boundary surface.

The normal component of force (per unit area) acting on the interface depends upon the direction of **B** and **H** in both media. All other cases can be deduced if we analyze just the following two: (1) when **B** and **H** are perpendicular, and (2) when they are parallel to the interface. In the first example, during a virtual displacement of the interface, **B** and hence also flux $\Phi$ remain constant, so that Eq. (12.3.8) applies. In the second example **H** (i.e. the currents) remains constant, and Eq. (12.3.12) applies. A procedure analogous to that of Section 3.6.2 leads to the following general result:

$$\boxed{p = \frac{dF_n}{dS} = \tfrac{1}{2}(\mu_2 - \mu_1)\left(H_t^2 + \frac{B^2}{\mu_1\mu_2}\right)} \quad \text{(directed from 2 to 1).} \tag{12.3.15}$$

For example, the lifting force of an electromagnet [Eq. (12.3.14)] per unit area of the cross-section of the air gap is obtained easily from Eq. (12.3.15). Note that in air practically only $B_n$ exists, and that the permeability of iron is much larger than that of air.

*Problems referring to Section 12.3:* 12.T12 to 12.T14, 12.G9 to 12.G21, 12.N6 to 12.N9.

## 12.4 THE MOTION OF A CHARGED PARTICLE IN MAGNETIC AND ELECTRIC FIELDS

**12.4.1 The motion of a charged particle in a static magnetic field.** If we take a charged particle of mass $m$ and charge $Q$, moving in a vacuum in a magnetic field of flux density **B**, the magnetic force on the charge is then given by the Lorentz

* This section may be omitted without loss of continuity.

force in Eq. (8.3.6), for $\mathbf{E} = 0$:

$$\mathbf{F}_m = Q\mathbf{v} \times \mathbf{B}, \tag{12.4.1}$$

where $v$ is the instantaneous velocity of the particle. The equation of the motion of the particle is thus

$$m\frac{d\mathbf{v}}{dt} = Q\mathbf{v} \times \mathbf{B}. \tag{12.4.2}$$

Since $\mathbf{v} \times \mathbf{B}$ is perpendicular to the instantaneous velocity $\mathbf{v}$ of the particle, the magnitude of $\mathbf{v}$, that is, the kinetic energy of the particle, remains constant.

Suppose $\mathbf{B}$ is known as a function of the coordinates with respect to a convenient coordinate system. Simultaneous solution of the three scalar equations represented generally by Eq. (12.4.2), subject to an appropriate set of initial conditions, yields three components of the particle's velocity $\mathbf{v}$. It is well to note, however, that the solving of Eq. (12.4.2) might be extremely involved. Let us therefore consider here only the case where the magnetic field is static and uniform, of flux density $\mathbf{B}$. If we first assume that the initial velocity of the particle is parallel to $\mathbf{B}$, then $\mathbf{F}_m = 0$, and the motion of the particle is unaffected by the magnetic field. It will move with its initial velocity along the lines of vector $\mathbf{B}$.

If the initial velocity is perpendicular to $\mathbf{B}$, it follows from Eq. (12.4.1) that the magnitude of $\mathbf{F}_m$ is simply $QvB$, and that $\mathbf{F}_m$ tends to bend the trajectory of the particle (see Fig. 12.8). Let the radius of the curvature of the trajectory be $R$. The centrifugal force $\mathbf{F}_{\text{centrifugal}}$ balances exactly the magnetic force $\mathbf{F}_m$. Since the magnitude of the centrifugal force is $mv^2/R$, we obtain an equation

$$mv^2/R = QvB,$$

from which

$$R = mv/QB. \tag{12.4.3}$$

All the quantities on the right-hand side being constant, the radius $R$ of the curvature is also constant. That is, the trajectory is a circle of radius $R$.

Let the angle between the particle's velocity $\mathbf{v}$ and the flux density vector $\mathbf{B}$ be $\alpha$. The component of the velocity $\mathbf{v}$ parallel to $\mathbf{B}$ is $v_B = v \cos \alpha$, and the component in the direction perpendicular to $\mathbf{B}$ is $v_{\perp B} = v \sin \alpha$. The trajectory of the

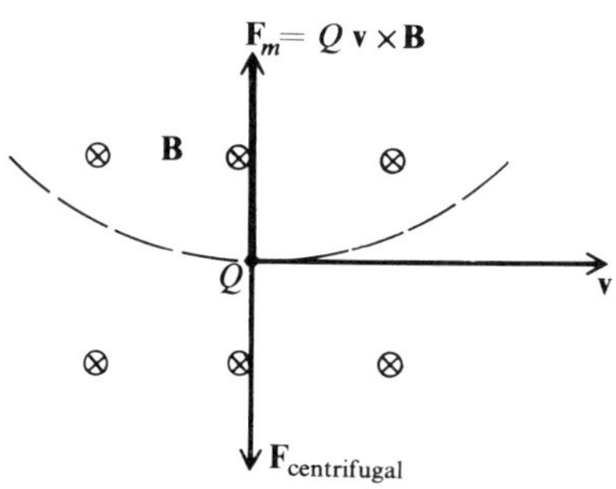


**Fig. 12.8** A charged particle moving in a uniform magnetic field.

particle is a helix, of radius

$$R_\alpha = mv_{\perp B}/QB = mv \sin \alpha/QB, \tag{12.4.4}$$

and of pitch

$$k = \frac{2\pi R_\alpha}{v_{\perp B}} v_B = \frac{2\pi mv \cos \alpha}{QB}. \tag{12.4.5}$$

The axis of the helix coincides with the direction of **B**. If the lines of **B** are curved, with a curvature radius much larger than $R_\alpha$, the helix will practically follow the lines of **B**. For example, charged particles entering the magnetic field of the earth from outer space spiral about the lines of the terrestrial magnetic field, and drift towards either the north or south pole. The interaction between these particles and the molecules of the air in the upper layers of the atmosphere is believed to be the cause of the aurora.

**12.4.2 The motion of a charged particle under the simultaneous action of the magnetic and electric fields.** If a charged particle is moving in a region where both magnetic and electric fields exist, the total force on the particle is the complete Lorentz force in Eq. (8.3.6). The vectorial equation of motion is

$$m\frac{d\mathbf{v}}{dt} = Q\mathbf{E} + Q\mathbf{v} \times \mathbf{B}. \tag{12.4.6}$$

As an example of determining the motion of a charged particle in the static electric and magnetic fields, consider a parallel-plate diode with a static magnetic field of flux density **B** parallel to the plates (see Fig. 12.9). We shall neglect the influence of the space charge in the diode. Let us determine the trajectory of an electron (of charge $-e$ and mass $m$), leaving the negative plate (cathode $C$), with a zero velocity, from the point which coincides with the origin of the Cartesian coordinate system shown. The trajectory of the electron being in the $y0z$ plane, Eq. (12.4.6) gives the following two equations of motion:

$$m\frac{dv_y}{dt} = ev_zB, \tag{12.4.7}$$

$$m\frac{dv_z}{dt} = eE - ev_yB. \tag{12.4.8}$$

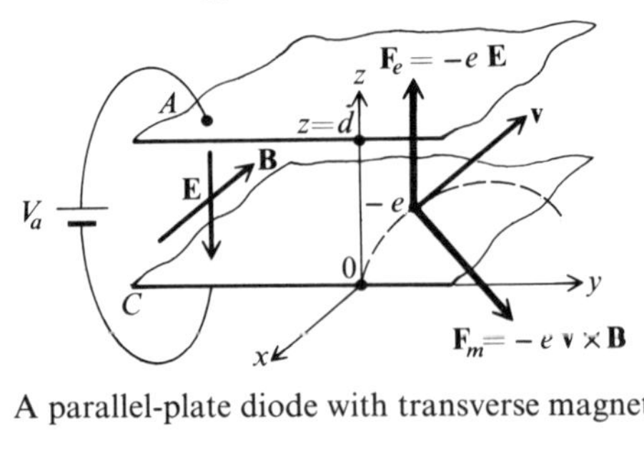


**Fig. 12.9** A parallel-plate diode with transverse magnetic field.

Since $v_z = dz/dt$, Eq. (12.4.7) can be integrated once, and $v_y = ezB/m$ thus found substituted into Eq. (12.4.8). From these operations we obtain

$$d^2z/dt^2 + \omega^2 z = A, \tag{12.4.9}$$

where

$$\omega^2 = e^2B^2/m^2 \qquad \text{and} \qquad A = eE/m. \tag{12.4.10}$$

The solution of Eq. (12.4.9) is

$$z = C_1 \sin \omega t + C_2 \cos \omega t + A/\omega^2, \tag{12.4.11}$$

$C_1$ and $C_2$ designating constants to be determined from the assumed initial conditions. Since when $t = 0$ both $z = 0$ and $dz/dt = 0$, we have

$$z = A(1 - \cos \omega t)/\omega^2. \tag{12.4.12}$$

The other equation of motion is obtained by combining Eq. (12.4.12) with Eq. (12.4.7):

$$y = A(\omega t - \sin \omega t)/\omega^2. \tag{12.4.13}$$

Equations (12.4.12) and (12.4.13) represent the parametric equations of a cycloid.

The largest distance of the electron from the cathode $C$ is $z_{max} = 2A/\omega^2$. The current through the diode exists therefore only if $2A/\omega^2 > d$. According to Eq. (12.4.10), this means that if, for a given $E$, the flux density is such that

$$B > \sqrt{2Em/ed} \qquad (E = V_a/d), \tag{12.4.14}$$

there will be (theoretically) no current through the diode. If $B$ is smaller than the value given by Eq. (12.4.14), all the electrons emitted will reach the anode. This effect is known as the *magnetron effect*.

*Problems referring to Section 12.4:* 12.G21 to 12.G26.

## PROBLEMS

### Test problems

**12.T1.** The current through a thin loop 1 is increased from zero to a constant value $I$. A thin resistive loop 2 has no generators in it, but is in the magnetic field of the current in loop 1. Both loops are made of a magnetically linear material. Are the power $P_m$ of the external sources and the final value $W_m$ of the energy *stored* in the system affected by the presence of loop 2?

**12.T2.** Repeat the preceding problem with loop 2 open-circuited.

**12.T3.** In the vicinity of loop 1 in Problem 12.T1 a body of a linear magnetic material is situated. Is there some energy associated with the magnetization of the body?

***12.T4.** Equation (12.1.6) was derived by assuming that the currents were increased inside *stationary* conductors. Using the law of conservation of energy as an argument, prove that this expression must be valid for the magnetic energy of the system considered, irrespective of the process by which the current system is established.

**12.T5.** Using a sound physical argument, explain why the work given in Eq. (12.1.5) done by external forces in establishing a given *static* magnetic field is a function of the process by which the system of currents is established when ferromagnetic materials are present in the field.

**12.T6.** Is it possible to *derive* Eq. (12.1.8) from Eq. (12.1.9)?

***12.T7.** If the hysteresis loop shown in Fig. 12.1 corresponds to the sinusoidal variation of $H$, will the hysteresis losses be equal to the crosshatched area if $H$ varies as a triangular function of time, that is, varies *linearly* from $-H_m$ to $H_m$, then back to $-H_m$, and so on.

**12.T8.** If the frequency of the alternating current producing a magnetic field is $f$ (cycles per second), what is the power per unit volume necessary to maintain the periodic field in a piece of ferromagnetic substance?

**12.T9.** Will the magnetic energy of a system of fixed quasi-filamentary current loops be changed if a closed conducting loop with *no* current is introduced into the system?

**12.T10.** Is it possible to determine theoretically the self-inductances and mutual inductances in a system of current loops by starting from Eq. (12.1.14), if $W_m$ is known?

***12.T11.** Which internal inductance is larger in the general case: that defined by Eq. (12.2.6), or that defined by Eq. (12.2.14)?

**12.T12.** Direct current due to a lightning stroke on a three-phase line propagates along the three conductors. Will the force repel or attract the conductors?

**12.T13.** If a lightning stroke hits a transformer, in some cases it may be noticed that the transformer is swollen. Explain how this happens.

***12.T14.** A short solenoid carries a *varying* current. Will there be a force on a closed conducting ring which is brought near one of its ends? If you think that it will be, will it be attractive or repellent? [*Hint:* note that induced current in the ring is such as to *decrease* the external flux "causing" it.]

## General problems

**12.G1.** The self-inductance of the toroidal coil shown in Fig. 11.11 is given by Eq. (11.5.16). Prove that the same expression for the self-inductance is obtained by using Eq. (12.1.15), with the value of $W_m$ given by Eq. (12.1.9).

**12.G2.** On a thin ferromagnetic toroidal core, of cross-section area $S = 1\ \text{cm}^2$, $N = 1000$ turns of thin wire are tightly wound. The mean radius of the core is $R = 16$ cm. It may be assumed that the magnetic field is uniform over the cross-section of the toroid. The idealized curve of the initial magnetization is shown in Fig. 12.G2. Determine the work $W_m$ done in establishing the magnetic field inside the toroid, if the intensity of the current through the coil is $I = 2$ A.

*Answer.* $W_m \simeq 0.05$ J.

**12.G3.** Repeat the preceding problem if the idealized initial magnetization curve is as shown in Fig. 12.G3.

*Answer.* $W_m \simeq 0.1$ J.

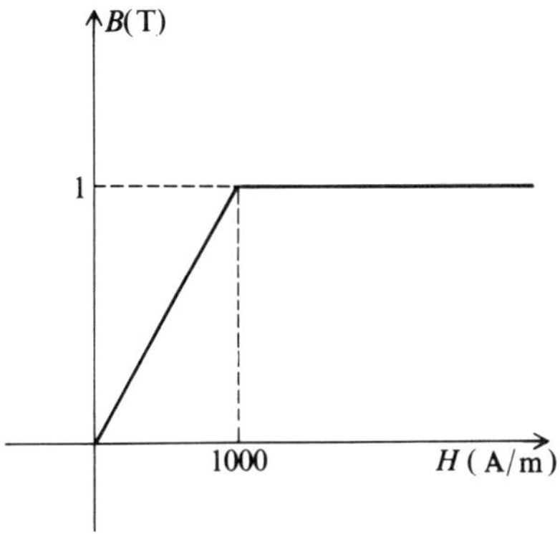


Figure 12.G2

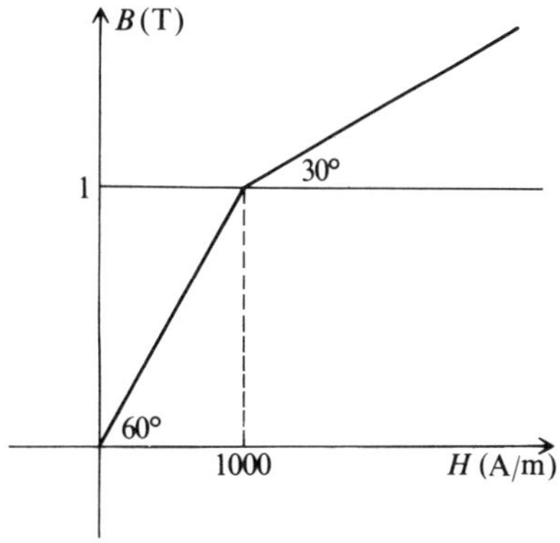


Figure 12.G3

**12.G4.** On the toroidal ferromagnetic core shown in Fig. 12.G4, $N = 628$ turns of thin wire are tightly wound. The intensity of the direct current through the coil is $I = 2$ A, and the idealized initial magnetization curve of the core is as shown in Fig. 12.G2. Determine the work done in establishing the magnetic field in the core, if $a = 5$ cm, $b = 15$ cm, and $h = 10$ cm.

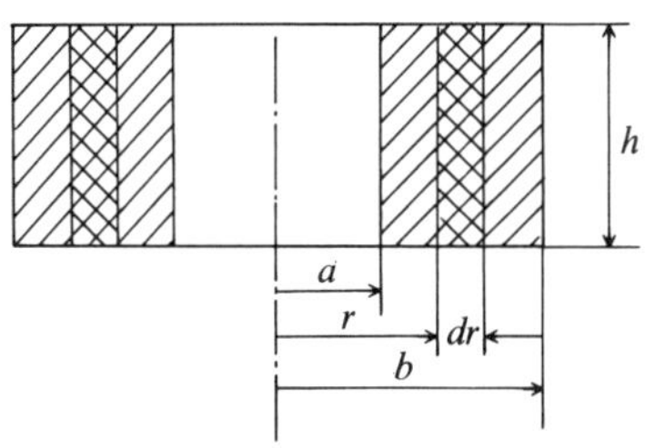


Figure 12.G4

*Solution.* Since the toroid is not thin, we cannot assume that the magnetic field is uniform over its cross-section. Consequently, the connection between the flux density $B$ and the field intensity $H$ in the general case is determined by different points on the magnetization curve for different points inside the core. Similarly, the work done per unit volume in magnetizing the core is different for different points.

For the numerical data given above, the magnetic field intensity inside the core, as a function of the distance $r$ of the point considered from the core axis, is given by

$$H = \frac{NI}{2\pi r} = \frac{200}{r} \text{ A/m}.$$

Hence the minimum field in the core, for $r = b = 15$ cm, is $H_{\min} = 200/0.15 \simeq 1330$ A/m, and saturation is reached at all points inside the core. So, at all points, the work done per unit volume of the core is

$$\frac{dW_m}{dv} = \int_0^B H\,dB = \tfrac{1}{2}1 \times 1000 = 500 \text{ J/m}^3.$$

Hence

$$W_m = \frac{dW_m}{dv} h\pi(b^2 - a^2) \simeq 3.14 \text{ J}.$$

**12.G5.** Repeat the preceding problem for intensities of current through the coil $I_1 = 0.5$ A and $I_2 = 1$ A.

*Answer.* $W_{m1} \simeq 0.86$ J; $W_{m2} \simeq 2.45$ J.

**12.G6.** The initial magnetization curve of a ferromagnetic substance can be approximated by $B = B_0H/(H_0 + H)$, where $B_0$ and $H_0$ are constants. Determine the work done per unit volume in magnetizing this material from zero to the magnetic field intensity $H$.

*Answer.* For $B_0 = 1$ T and $H = H_0 = 1000$ A/m, $dW_m/dv \simeq 193$ J/m$^3$.

**12.G7.** Determine approximately the work done in establishing in the coil of Problem 12.G2 a current of intensity $I = 1$ A, if the initial magnetization curve is that shown in Fig. 9.G13a. [*Hint:* the integral in Eq. (12.1.10) has to be determined graphically.]

*Answer.* $W_m \simeq 0.056$ J.

**12.G8.** On the toroidal core shown in Fig. 12.G4, $N$ turns of thin wire are tightly wound. The current through the coil is sinusoidal, of amplitude $I_m$ and frequency $f$. The idealized shape of the hysteresis loops of the ferromagnetic core is shown in Fig. 12.G8. Determine the power of the hysteresis losses in the core, assuming that the distribution of the magnetic field in the core is the same as in the case of direct current. [It will be shown in Section 14.2 that this assumption is justified only if either frequency is very low (of the order of a few cycles per second), or if the core is made of thin, mutually insulated sheets, or of small mutually insulated ferromagnetic particles.]

*Solution.* The energy lost per unit volume of the core and for one cycle is numerically equal to the area of the hysteresis loop corresponding to the point considered. In the present case, the hysteresis loops are different for different points, since the amplitude of the magnetic field intensity is $H_m = NI_m/2\pi r$. Consequently, we have to determine the elemental losses in the thin circular rings of radius $r$ and thickness $dr$ ($a < r < b$), the total losses being given as their sum (integral).

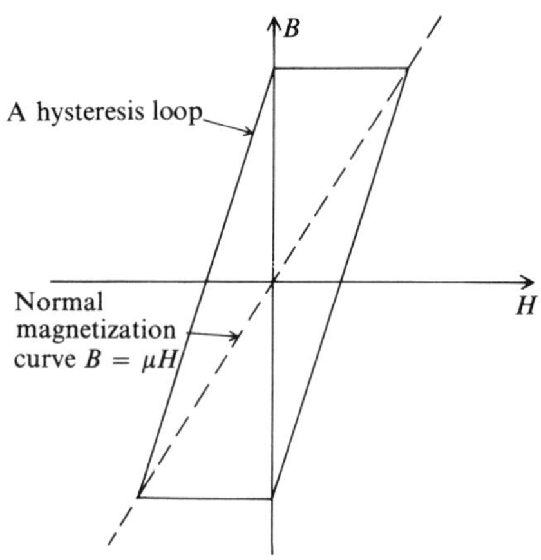


**Figure 12.G8**

Losses in one such ring for one cycle are numerically equal to the area of the idealized hysteresis loop shown in Fig. 12.G8 multiplied by the volume $2\pi rh\,dr$ of the ring:

$$dW_m = 2B_m H_m 2\pi rh\,dr = 4\mu H_m^2 \pi rh\,dr = \frac{\mu N^2 I_m^2 h}{\pi r}\,dr.$$

Losses in the whole core per cycle are

$$W_m = \int_a^b \frac{\mu N^2 I_m^2 h}{\pi r}\,dr = \frac{\mu N^2 I_m^2 h}{\pi}\ln\frac{b}{a}.$$

The power of the hysteresis losses (losses per unit time, that is, in one second) is obtained by multiplying $W_m$ by the number $f$ of cycles per second, $P_{\text{hyst. losses}} = fW_m$.

**12.G9.** One conductor of a two-conductor line is in the form of one half of a thin circular cylinder. The other conductor is a thin wire running along the axis of the first (see Fig. 12.G9). If a current of intensity $I$ flows through the two conductors in opposite directions, determine the force per unit length on the conductors.

*Answer.* The force is repellent, of intensity $F_l = \mu_0 I^2/\pi^2 a$.

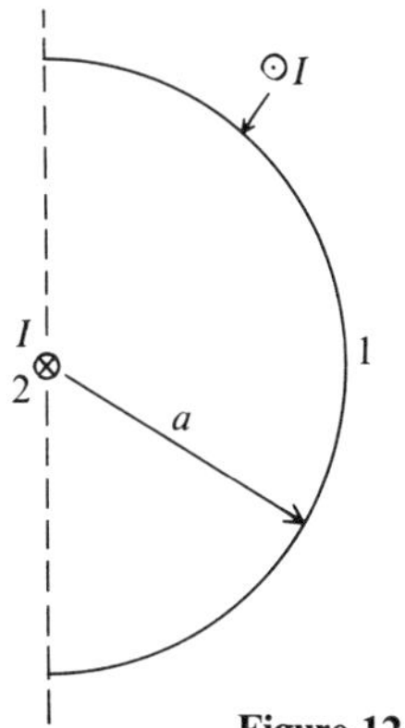


Figure 12.G9

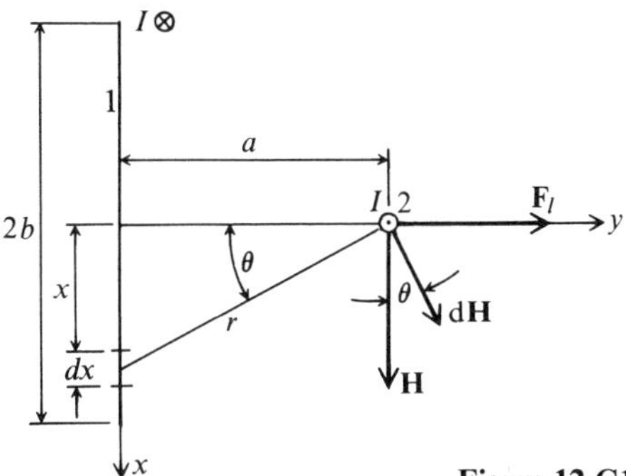


Figure 12.G10

**12.G10.** A thin conductor 2 runs parallel to a thin metal strip 1 (see Fig. 12.G10). Both $a$ and $b$ are much larger than the thickness of the strip. Determine the force per unit length on the two conductors for a current of intensity $I$ flowing through them (in opposite directions).

*Solution.* Consider an elemental strip, of width $dx$, of the strip conductor 1. The current $dI = I\,dx/2b$ through the elemental strip creates at points of conductor 2 a magnetic field intensity $dH$, having the direction shown in the figure, and a magnitude of

$$dH = \frac{dI}{2\pi r} = \frac{I\,dx}{4\pi br} = \frac{I\,dx}{4\pi b\sqrt{a^2 + x^2}}.$$

By symmetry, the resultant magnetic field intensity $H$, created by the current in the whole conductor 1 at points on conductor 2, has only the $x$ component. The magnitude of the resultant field intensity is

$$H = \int dH \cos\theta = \int_{-b}^{b} \frac{I\,dx}{4\pi b\sqrt{a^2 + x^2}}\,\frac{a}{\sqrt{a^2 + x^2}} = \frac{I}{2\pi b}\tan^{-1}\frac{b}{a}.$$

The force per unit length on conductor 2 is

$$F_l = F_{ly} = I\mu_0 H = \mu_0 \frac{I^2}{2\pi b} \tan^{-1}\frac{b}{a}.$$

**12.G11.** Determine the force per unit length on the conductors of the line the cross-section of which is shown in Fig. 12.G11. The current in conductors has an intensity $I$, and medium is air.

*Answer.* For $I = 10$ A and $b = 2$ cm, $F_l \simeq 0.222$ N/m (repulsive).

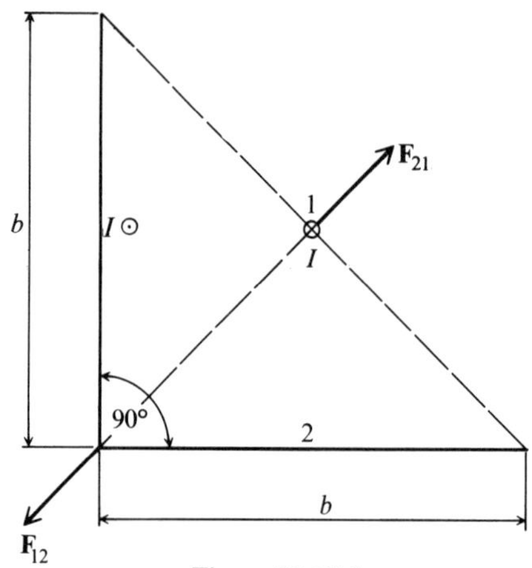


**Figure 12.G11**

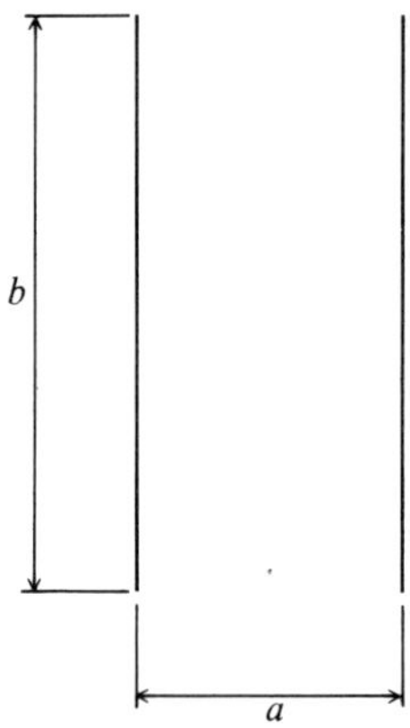


**Figure 12.G12**

***12.G12.** Determine the force per unit length on the conductors of the strip line shown in Fig. 12.G12. The conductors carry a current of intensity $I$ in opposite directions.

*Answer.* The force is repulsive, of magnitude per unit length

$$F_l = \frac{\mu_0 I^2}{2\pi b^2}\left(2b \tan^{-1}\frac{b}{a} - a \ln \frac{a^2 + b^2}{a^2}\right).$$

**12.G13.** A two-wire line with conductors of circular cross-section, of radius $a$ and the distance between their axes $d$, is short-circuited by a straight conducting bar, as shown in Fig. 12.G13. If current $I$ flows through the line, what is the force on the bar? (Assume that the section of the line to the left of the bar is very long, and that the medium is air.)

*Solution.* Both conductors create at points on the bar a flux density equal to one half of the flux density created by an infinitely long conductor. The total flux density due to the current in the two conductors at points on the element $dx$ of the bar is hence

$$B = \frac{\mu_0 I}{4\pi x} + \frac{\mu_0 I}{4\pi(d - x)}.$$

The elemental force on the element $dx$ is

$$dF = I\,dx\,B = \frac{\mu_0 I^2}{4\pi}\left(\frac{dx}{x} + \frac{dx}{d - x}\right),$$

so that the total force on the bar

$$F = \frac{\mu_0 I^2}{4\pi}\int_a^{d-a}\left(\frac{dx}{x} + \frac{dx}{d-x}\right) = \frac{\mu_0 I^2}{2\pi}\ln\frac{d-a}{a}.$$

In the case of large currents this force can have a rather large value (see Problem 12.N8). Thus, if for example the bar represents the lever of a switch for large currents, pivoted at one end, a sudden current surge (due, say, to a short circuit) can push the lever so hard as to open the switch.

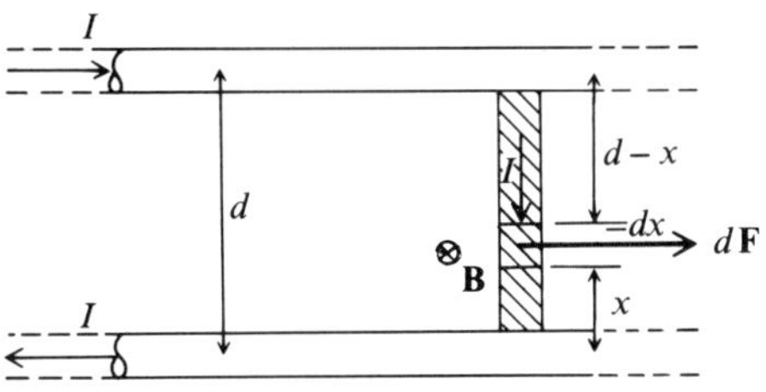


Figure 12.G13

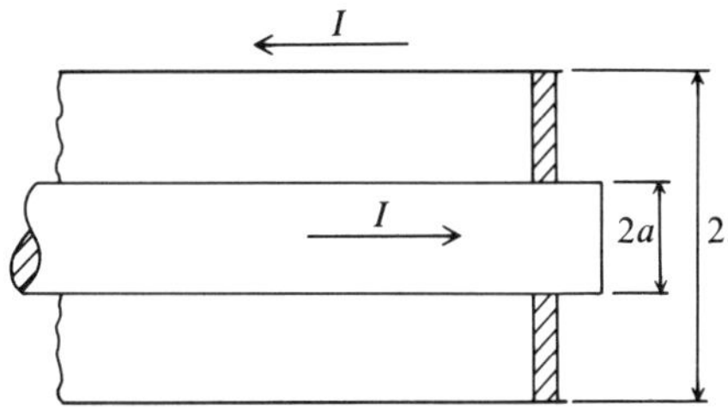


Figure 12.G14

**12.G14.** A long coaxial cable with dielectric air is short-circuited at its end by a thin conducting plate, as shown in Fig. 12.G14. Determine the force on the end plate corresponding to a current of intensity $I$ through the cable by using both Eqs. (12.3.5) and (12.3.12).

*Answer.* For $b/a = e$ and $I = 10$ A, the force $F = 10^{-5}\ N$.

**12.G15.** Determine approximately the force between two parallel coaxial circular loops with currents $I_1$ and $I_2$, if the radii of the loops are $a$ and $b$, respectively, with $a \gg b$, and the distance between their centers is $z$.

*Solution.* Since $b \ll a$, we can consider the magnetic flux produced by current $I_1$ through loop 2 to be equal to the product of the flux density $B_1$ on the axis of the loops, and the area $b^2\pi$ of loop 2. The flux density $B_1$ was found in Problem 8.G4 to be

$$B_1 = B_{1z} = \frac{\mu_0 I_1 a^2}{2(a^2+z^2)^{3/2}}.$$

Since $\Phi_{12} = B_1 b^2\pi$, it follows that the mutual inductance between the two loops is

$$L_{12} = \frac{\Phi_{12}}{I_1} = \frac{\mu_0 a^2 b^2 \pi}{2(a^2+z^2)^{3/2}}.$$

To determine the force acting on loop 2 (or on loop 1), let us apply Eqs. (12.3.12) and (12.1.14) to this case. Since during the virtual displacement along the $z$ axis only the mutual inductance $L_{12}$ varies, we have

$$F_z = \frac{\partial W_m}{\partial z} = I_1 I_2 \frac{\partial L_{12}}{\partial z} = -\frac{3\mu_0 I_1 I_2 a^2 b^2 \pi z}{2(a^2+z^2)^{5/2}}.$$

The minus sign indicates that the force is attractive, which is obvious since we assumed that $\Phi_{12} > 0$.

**12.G16.** Determine the force on loops 2 and 3 of Problem 11.G21 due to current $I$, if the currents through loops 2 and 3 are $I_2$ and $I_3$, respectively.

*Answer.* $F_{12} = \dfrac{\mu_0 h I I_2 b}{2\pi a(a + b)}$, $\quad F_{13} = \mu_0 I I_3 \left( \dfrac{b}{\sqrt{b^2 - a^2}} - 1 \right)$.

Both forces are attractive.

***12.G17.** A U-shaped glass tube is filled with a paramagnetic liquid of unknown susceptibility $\chi_m$. A part of the tube inside the dashed square (see Fig. 12.G17) is subjected to a uniform magnetic field of intensity $H$. Under the influence of the magnetic forces a difference $h$ of the levels of the liquid in the two sections of the tube is obtained. Given that the mass density of the liquid is $\rho_m$ and the medium above the liquid is air, determine $\chi_m$.

*Solution.* Since only the tangential component of $H$ exists, according to Eq. (12.3.15) the upward pressure on the surface of the liquid in the magnetic field is

$$p = \tfrac{1}{2}(\mu - \mu_0)H^2.$$

Hence, if $S$ is the area of the tube cross-section, the force pulling the liquid upwards is

$$F_m = \tfrac{1}{2}(\mu - \mu_0)H^2 S.$$

In a state of equilibrium this force is compensated by the weight $\rho_m g S h$ of the liquid above the lower level. Thus we obtain that

$$\mu - \mu_0 = \mu_0(\mu_r - 1) = \mu_0 \chi_m = 2\rho_m h g / H^2,$$

from which

$$\chi_m = \frac{2\rho_m g h}{\mu_0 H^2},$$

$g$ being the strength of the earth's gravitational field, $g = 9.81\ \text{m/s}^2$. For example, if $\rho_m \simeq 10^3\ \text{kg/m}^3$, $B = \mu_0 H = 1\ \text{T}$, and $h = 4\ \text{mm}$, $\chi_m \simeq 10^{-4}$.

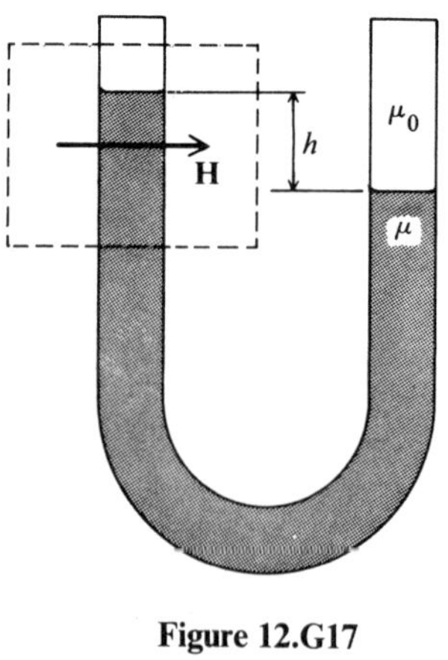


Figure 12.G17

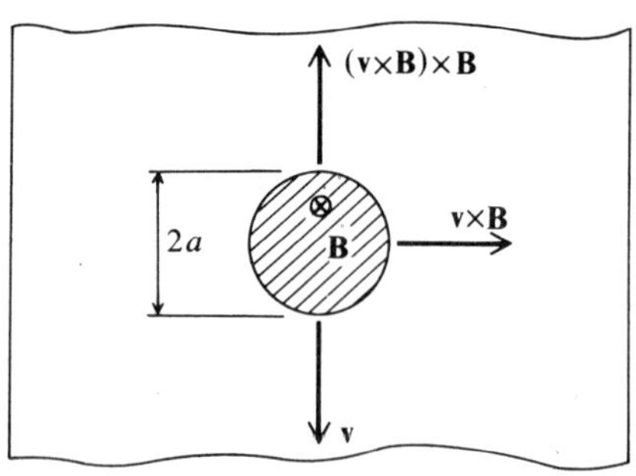


Figure 12.G18

**12.G18.** A metal strip, of conductivity $\sigma$ and of small thickness $b$, moves with a uniform velocity $v$ between the round poles of a permanent magnet. The radius of the magnet poles is $a$, and the width of the strip is much larger than $a$ (see Fig. 12.G18). The flux density $B$ is very nearly constant over the circle shown crosshatched, and practically zero outside it. Determine the induced currents in the strip, and the force on the strip due to the action of the magnetic field of the magnet on these currents. Ignore the effect of the finite dimensions of the strip, and the magnetic field of the induced currents.

*Solution.* Due to the motion of the strip, the free charges in the circular region of the strip (shown crosshatched) are acted upon by the magnetic force. This force can be represented by an equivalent impressed electric field of intensity $\mathbf{E}_i = \mathbf{v} \times \mathbf{B}$. The distribution of the current in the strip produced by such a generating region was determined in Problems 6.G22 and 6.G23. We found there that the current density in the circular region where the impressed field exists is constant, and given by $\mathbf{J} = \sigma\mathbf{E}_i/2 = \sigma\mathbf{v} \times \mathbf{B}/2$. According to Eq. (12.3.2) the magnetic force per unit volume acting on these currents is

$$\frac{d\mathbf{F}}{dv} = \mathbf{J} \times \mathbf{B} = \frac{\sigma}{2}(\mathbf{v} \times \mathbf{B}) \times \mathbf{B}.$$

The force $d\mathbf{F}/dv$ per unit volume acting on the strip in the crosshatched region is constant. It is directed in the direction opposite to that of $\mathbf{v}$, that is, it *opposes* the motion of the strip. The total force is in the same direction, of intensity

$$F = a^2\pi b\frac{dF}{dv} = \tfrac{1}{2}\sigma a^2\pi bvB^2. \qquad \text{(a)}$$

**12.G19.** A thin metal plate is falling between the poles of a permanent magnet under the action of the gravitational field. The pole radius is $a$, and the flux density in the gap is $B$ (see Fig. 12.G19). Determine approximately the velocity of the plate, if its thickness is $b$, its surface $S$, conductivity $\sigma$, and its mass density $\rho_m$.

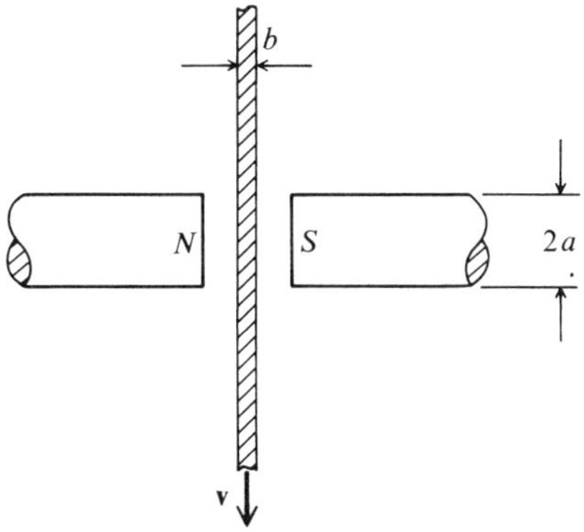


Figure 12.G19

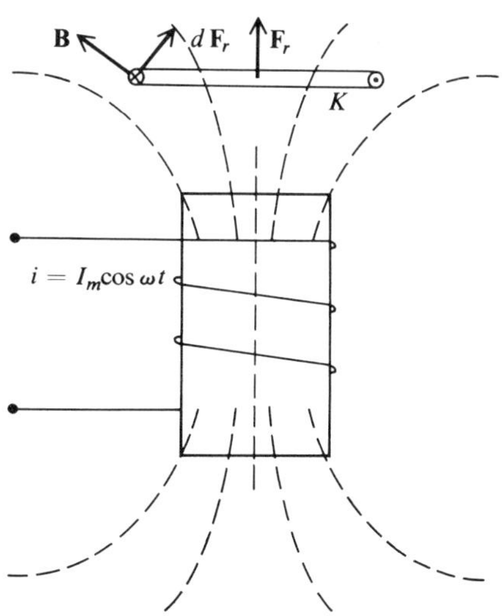


Figure 12.G20

*Answer.* For $a = 2$ cm, $b = 0.5$ mm, $S = 100$ cm$^2$, $B = 1$ T, $\sigma = 36 \times 10^6$ S/m (aluminum), and $\rho_m = 2.7$ gr/cm$^3$ (aluminum), the velocity of the plate is $v \simeq 1.17$ cm/s.

**12.G20.** A metal ring $K$, of negligible resistance, is placed above a short cylindrical electromagnet, as shown in Fig. 12.G20. Determine qualitatively the dependence of the total force on the ring on time, if the current through the electromagnet coil is of the form $i = I_m \cos \omega t$.

*Solution.* The ring is situated in a time-varying magnetic field. Ignoring any possible nonlinearity of the core, we know that the field is proportional to the current producing it, i.e. $\mathbf{B} = \mathbf{B}_m \cos \omega t$. ($\mathbf{B}_m$ is, of course, different at different points.) According to Faraday's law, an electromotive force will be induced in the ring, which will vary in time according to the law

$$\mathscr{E}(t) = -\frac{d\Phi}{dt} = \Phi_m \omega \sin \omega t = \mathscr{E}_m \sin \omega t.$$

Since the resistance of the ring can be ignored, this electromotive force is balanced by the induced electromotive force due to the varying current $i_r$ in the ring. If $L$ is the self-inductance of the ring we have

$$L\frac{di_r}{dt} = \mathscr{E}(t) = \mathscr{E}_m \sin \omega t,$$

whence the current $i_r$ in the ring is of the form $i_r = I_{rm} \cos \omega t$.

The elements of the ring are acted upon by the magnetic force, in the direction shown in the figure. Due to the nonuniformity of the magnetic field, these elemental forces have an axial resultant $F_r$, which is of the form $F_r = k i_r B = F_{rm} \cos^2 \omega t$, where $k$ is a constant. The mean value of the force $F_r$ in one cycle is

$$F_{r\,\text{mean}} = \frac{1}{T}\int_0^T F_{rm} \cos^2 \omega t \, dt = \tfrac{1}{2} F_{rm},$$

where $T$ is the period, $1/f = 2\pi/\omega$, of the harmonic current. The force tends to throw the ring up and out of the field.

**12.G21.** An electron beam of velocity $v$ leaves the system, by which it was accelerated, at a point $P$. The velocities of the electrons in the beam are very nearly of the same magnitude, but they are not all directed along the axis of the beam. The velocity vectors of the most diverging electrons make a (very small) angle $\alpha$ with the beam axis (see Fig. 12.G21). What are the direction and magnitude for the magnetic flux density vector $\mathbf{B}$ of a uniform magnetic

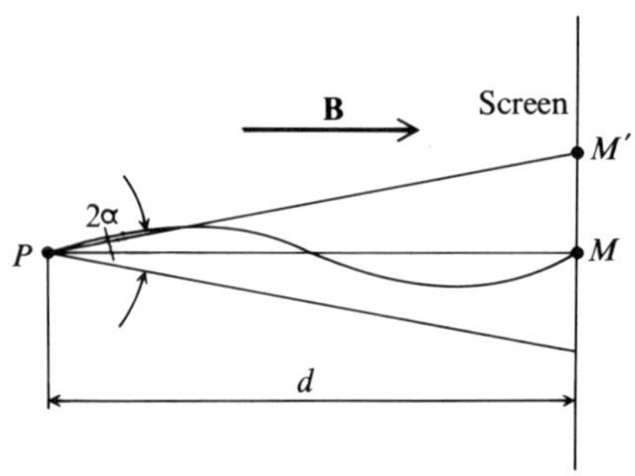


**Figure 12.G21**

field which would concentrate the electron beam practically at the point $M$ of the screen shown in the figure? Ignore the interactions between the electrons of the beam.

*Solution.* According to Eq. (12.4.5), if the magnetic flux density vector $\mathbf{B}$ is parallel to the axis of the beam and its intensity is

$$B = \frac{2\pi mv \cos\alpha}{ed} \simeq \frac{2\pi mv}{ed}, \tag{a}$$

the most divergent electrons will move along the helixes of pitch $d$. So, instead of hitting the screen at point $M'$, they will hit it approximately at $M$. Although with the flux density of the magnitude given by Eq. (a) not all electrons will hit exactly the same point, the effect of such *magnetic focusing* of the beam is very satisfactory. To demonstrate this, let us replace $B$ in Eq. (12.4.5) by the value in Eq. (a). We then discover that the pitches of the helixes of the various electrons of the beam are given by

$$k(\alpha) = d\cos\alpha \simeq d(1 - \alpha^2/2),$$

since $\alpha$ is very small. The largest difference of the helix pitches is

$$\Delta k = k(\alpha) - k(0) = -\alpha^2 d/2.$$

For example, for $\alpha = 3.6° = 2\pi/100$ radians, $\Delta k \simeq 2 \times 10^{-3}d$. All the electrons of the beam hit practically the same point of the screen.

In practice, magnetic focusing is usually not effected by a uniform magnetic field. The magnetic field of a coil coaxial with the beam axis is used instead. The principle of focusing the beam remains the same, but the designing of such systems is mainly experimental.

**12.G22.** An electron beam, accelerated by traversing a potential difference $V$, passes through a region of a practically uniform magnetic field of flux density $B$, as shown in Fig. 12.G22. What is the deflection angle $\alpha$ of the beam?

*Solution.* Under the action of the magnetic field, the beam follows a circular path, of radius $R = mv/eB$ [see Eq. (12.4.3)]. From the figure it is seen that $R = d/\sin\alpha$. Thus

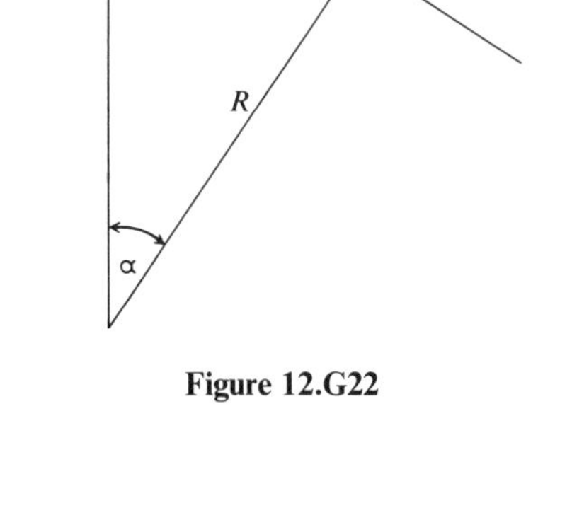

Figure 12.G22

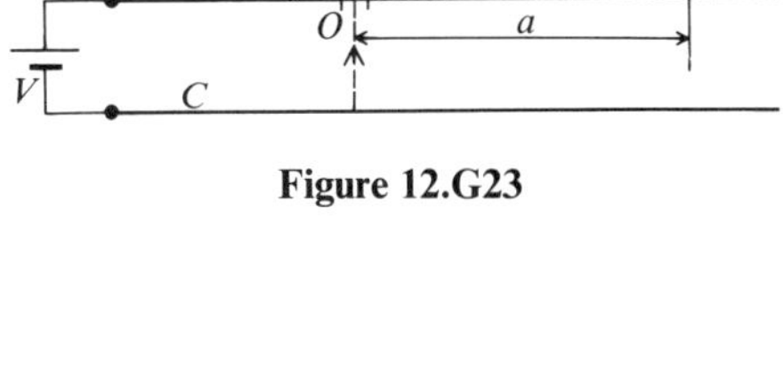

Figure 12.G23

$$\alpha = \sin^{-1}\frac{d}{R} = \sin^{-1}\frac{eBd}{mv} = \sin^{-1}\left(Bd\sqrt{\frac{e}{2mV}}\right),$$

according to Eq. (7.2.5).

**12.G23.** An electron leaves the cathode $C$ with a zero initial velocity, and is accelerated towards the anode $A$ by the electric field existing between the electrodes (see Fig. 12.G23). The potential difference between the anode and the cathode is $V$. The electron next passes through a small hole $O$ in the anode and enters a region with a uniform magnetic field of flux density $B$. It is deflected by the magnetic field, and hits the anode $A$ at a distance $a$ from $O$. Determine the ratio $e/m$ of the electron charge and mass.

*Answer.* $e/m = 8V/a^2B^2$.

**12.G24.** In the vacuum above an uncharged horizontal conducting sheet, a magnetic field of flux density $B$ exists, with its lines parallel to the sheet. If a point charge $Q$ moves with a uniform velocity parallel to the sheet and perpendicular to the lines of $B$, at a distance $d$ from the sheet, what is the magnitude $v$ of the charge velocity? What is the direction of $\mathbf{B}$ in this case? [*Hint:* the charge is acted upon by the magnetic force and by the electric force of the charges it induces on the conducting sheet. The latter can be reduced to the force on the charge due to its "image".]

*Answer.* $v = Q/16\pi\varepsilon_0 h^2 B$.

**12.G25.** A beam of identically charged particles of different velocities passes through a region in which a uniform electric field of intensity $E$ and a uniform magnetic field of flux density $B$ simultaneously exist. The directions of vectors $\mathbf{E}$ and $\mathbf{B}$ are shown in Fig. 12.G25. What is the velocity $v$ of the particles traveling through this region along straight lines, and passing through the hole $O$ in the screen $S$?

*Answer.* $v = E/B$.

**12.G26.** A system for accelerating the charged particles, known as the *cyclotron*, consists of two halves of a very flat, hollow metal cylinder. The two halves are very close but not in contact, and are connected to a source of sinusoidal potential difference of amplitude $V_m$.

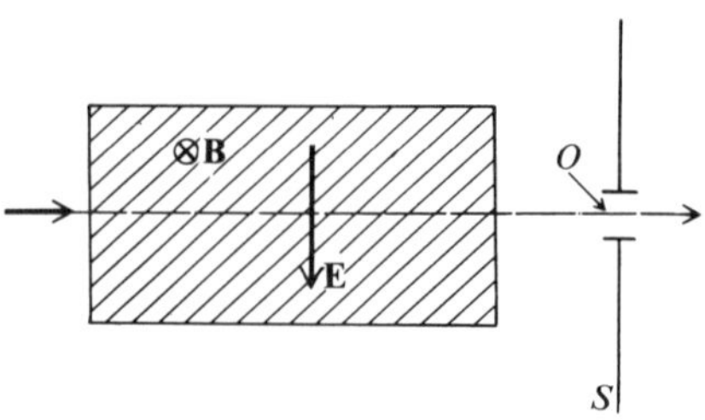


Figure 12.G25

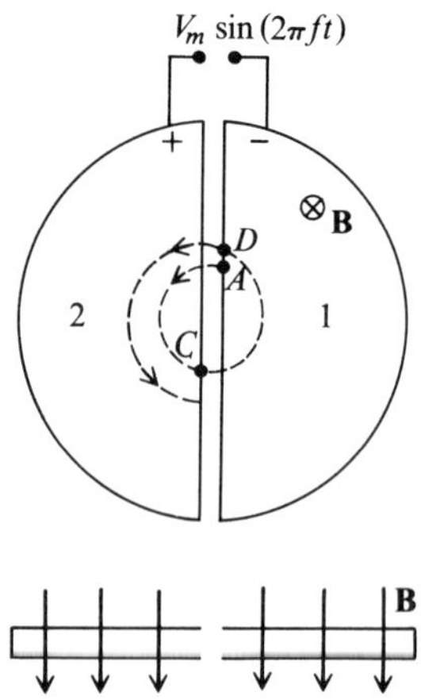


Figure 12.G26

The whole system is situated in a vacuum in a uniform magnetic field, of flux density **B**, perpendicular to the cylinder bases (see Fig. 12.G26). Determine the frequency $f$ of the sinusoidal potential difference, so that a particle, which leaves point $A$ with zero initial velocity and proceeds towards the other half of the cylinder, is subsequently always *accelerated* by the electric field. If the radius of the cylinder is $R$, what is the maximal velocity of the particles that can be obtained by such an accelerator? Assume that the velocities of the particles are much smaller than the velocity of light.

*Solution.* Suppose that, at the instant when the particle is emitted from point $A$, the potential difference between the halves is $V_m$, and that it remains approximately $V_m$ during the passage of the particle from one half into the other. Then the particle acquires a velocity

$$v_1 = \sqrt{2QV_m/m}. \tag{a}$$

In the cavity of half 2 practically only a magnetic field exists, so that the particle moves along a circular path, the radius of which is

$$R_1 = mv_1/QB. \tag{b}$$

To have an accelerating electric field again when the particle reaches point $C$, period $T = 1/f$ of the sinusoidal potential difference must be equal to the double time required for the particle to travel along the semicircle $AC$. Since the velocity $v$ of the particle is constant, we have that

$$T/2 = \pi R_1/v_1 = \pi m/QB.$$

Hence the required period of the potential difference is

$$T = 2\pi m/QB. \tag{c}$$

In this case, the velocity $v_2$ of the particle after reaching the point $D$ in half 1 is determined by the equation

$$\tfrac{1}{2}mv_2^2 - \tfrac{1}{2}mv_1^2 = QV_m,$$

from which, according to Eq. (a),

$$v_2 = \sqrt{2QV_m/m + v_1^2} = v_1\sqrt{2}. \tag{d}$$

Following the same line of reasoning, we can easily conclude that the trajectory of the particle is a sort of spiral. The velocity of the particle along the $n$th semicircle of the spiral is $v_n = v_1\sqrt{n}$, and the radius of the $n$th semicircle is

$$R_n = mv_n/QB = mv_1\sqrt{n}/QB.$$

The velocity corresponding to $R_n = R$ is hence $v_{max} = QRB/m$. It is interesting that the final velocity $v_{max}$ of the particle is not a function of $V_m$, but only of $B$ and $R$. If $V_m$ is smaller the spiral will be more dense, but the final velocity is the same.

In the case of the velocities of particles comparable with the velocity of the light, mass of particles is a function of their velocity (the relativistic increase of mass with velocity). The particles cannot follow changes of the potential difference if period $T$ is constant.

# PART V

# MAXWELL'S EQUATIONS AND THEIR APPLICATION

CHAPTER 13

# MAXWELL'S EQUATIONS

## 13.1 INTRODUCTION

In the preceding chapters we have dealt with special cases of the electromagnetic field. These are the electric field due to static charges, the magnetic field due to steady currents, and the electromagnetic field due to quasi-stationary currents and charges. The basis from which all these concepts were derived was the law of force between two small charges slowly moving in an arbitrary manner with respect to the observer. The method of analysis was typically inductive; that is, starting from the force law and a model of the atomic structure of matter, we defined the concept of the electric and magnetic fields and investigated the properties of these vector fields. As a result, we have gained a deep understanding of the basic relations between the electric and magnetic fields. We solved many related engineering problems with elegance and simplicity. In consequence, it would be hard to overemphasize the importance of the preceding chapters.

At this point, however, our knowledge is sufficient for us to consider the most general types of electromagnetic fields. We shall see that, with a minor addition to the relations derived for the special fields, a consistent system of generally valid equations is obtained. These equations are known as *Maxwell's equations*, and comprise a system of partial differential equations which electric and magnetic components of the unique electromagnetic field must in all cases satisfy. Relations derived earlier can be considered as special cases of these general equations.

Maxwell's equations predict also the existence of some qualitatively new types of electromagnetic fields. Thus Maxwell anticipated the possibility of the existence of varying electromagnetic fields independent of the currents and charges which produced them some twenty years before Hertz proved their existence experimentally. The propagation of energy in the form of time-varying electromagnetic fields along transmission lines and through hollow conducting tubes (wave-guides), as well as systems for producing free electromagnetic waves (antennas) cannot be analyzed without a knowledge of Maxwell's equations.

This chapter is devoted exclusively to deriving, in a sense, these powerful equations and analyzing some of their general important consequences. In the following chapters the analysis will start from Maxwell's equations—that is, the method of analysis will no longer be inductive, but rather deductive. Special cases will be analyzed by means of these generally valid equations.

The importance of understanding the meaning and power of Maxwell's equations cannot be sufficiently stressed. The reader is advised to take the utmost care in reading—and rereading—this chapter.

## 13.2 AMPERE'S CIRCUITAL LAW FOR NONSTATIONARY CURRENTS

In Section 8.7 we saw that Ampère's circuital law, as given in Eq. (8.7.1), for stationary currents is a consequence of Eq. (8.3.3) for the magnetic flux density of a single point charge and of the solenoidal character of the stationary current field, which is, of the continuity equation (6.3.6) for steady currents.

Ampère's circuital law has a different form in the case of nonstationary currents. It again follows from Eq. (8.3.3) and the general continuity equation (6.3.3), but its usual form is obtained by making use also of Gauss' law, as given in Eq. (1.8.5).*

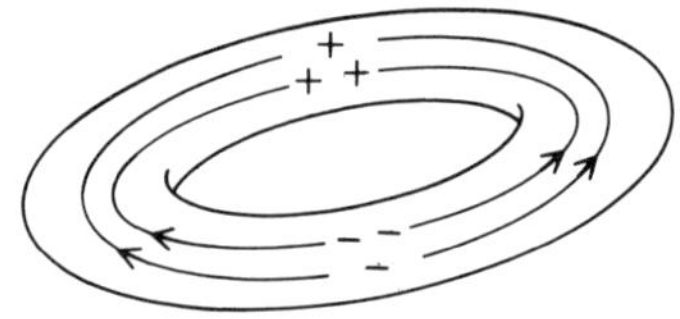

**Fig. 13.1** A conductor with nonstationary current.

Consider a nonstationary current system at a certain instant of time (Fig. 13.1). Let us divide the current field into very thin, almost filamentary tubes, to the surfaces of which the current density vector $\mathbf{J}$ is tangential at all points. In the *stationary* case such tubes close onto themselves, and the current intensity through any cross-section of a tube is the same. In general, this is *not* so in the nonstationary case, because time-varying charge distribution can then exist along the length of the tubes. Since vector $\mathbf{J}$ is tangential to such a tube, no charges can leave or enter it, which means that we can always imagine the charge distribution along the tube to be divided into pairs of equal and opposite time-varying charges, between which a time-varying current exists. The total nonstationary current at a cross-section of the tube is given by the sum of partial currents between all such time-varying associated charge pairs situated on *opposite* sides of the cross-section being considered. These partial currents are connected with the charges on which they end by the continuity equation. We can always represent a nonstationary current system as a bunch of such open, overlapping, filamentary tubes of partial currents. Thus, an *open* current filament, with charges of equal magnitude and of opposite signs at its ends, may be regarded as a building block of any nonstationary current system.

* If desired, the proof of Ampère's circuital law can be omitted. In that case Eq. (13.2.7) is to be regarded as proved and the rest of the section following it should be read carefully.

Let us consider therefore an open quasi-filamentary current situated in a vacuum, and suppose that the instantaneous value of the (time-varying) current in the filament is $i$, and the corresponding (time-varying) charges at its ends are $Q$ and $-Q$ (see Fig. 13.2). The intensity $i$ of the current is of course constant along the filament at any instant; consequently, the line integral of the magnetic field intensity vector $\mathbf{B}$ along a nearby closed contour $C$ is given by Eq. (8.7.6):

$$\oint_C \mathbf{B}\cdot d\mathbf{l} = \frac{\mu_0 i}{4\pi}(\Omega_2 - \Omega_1), \tag{13.2.1}$$

where $\Omega_1$ and $\Omega_2$ are solid angles subtended at starting point 1 and end point 2 of the filament by the contour $C$. The only difference between Eq. (13.2.1) and Eq. (8.7.6) is that $i$ stands for the *instantaneous* intensity of current, whereas $I$ in Eq. (8.7.6) denotes a constant current in *part* of a closed current filament. The reader is recommended to reread the derivation of Eq. (8.7.6) before proceeding further.

The right-hand side of Eq. (13.2.1) can be transformed as follows. Denote the charge $-Q$ at 1 by $Q_1$, and the charge $+Q$ at 2 by $Q_2$. According to the continuity equation and by reference to Fig. 13.2, $i = dQ_2/dt$ if expressed in terms of the charge at 2, and $i = -dQ_1/dt$ if expressed in terms of the charge at 1. The solid angles $\Omega_1$ and $\Omega_2$ in Eq. (13.2.1) are constant. So we can write

$$\oint_C \mathbf{B}\cdot d\mathbf{l} = \frac{\mu_0}{4\pi}\left[\frac{d(Q_2\Omega_2)}{dt} + \frac{d(Q_1\Omega_1)}{dt}\right] = \frac{\mu_0}{4\pi}\frac{d}{dt}(Q_2\Omega_2 + Q_1\Omega_1). \tag{13.2.2}$$

We turn our attention now to two surfaces, $S_1$ and $S_2$, limited by the contour $C$ (see Fig. 13.3). Suppose $S_1$ is not pierced by the filament, but $S_2$ is. Looking first at surface $S_1$, we find that, according to Eq. (1.8.3), $Q_2\Omega_2$ equals the flux of the electric field intensity vector $\mathbf{E}$ through $S_1$, multiplied by $4\pi\varepsilon_0$, created by charge $Q_2 = +Q$. Similarly, $Q_1\Omega_1$ equals $4\pi\varepsilon_0$ multiplied by the flux of $\mathbf{E}$ through the same surface created by charge $Q_1 = -Q$. The sum in the parentheses on the

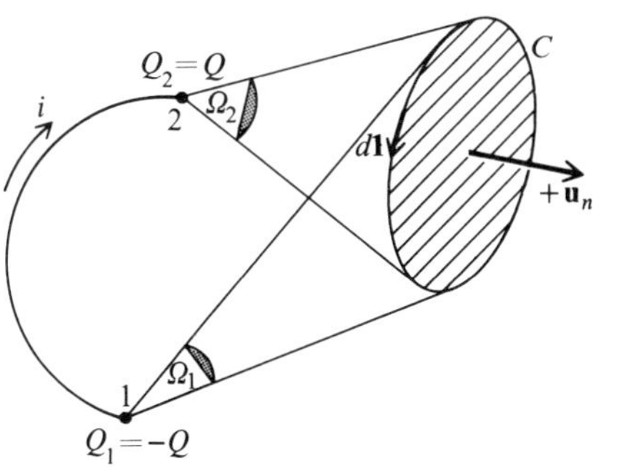


**Fig. 13.2** A current filament with equal charges of opposite signs at its ends.

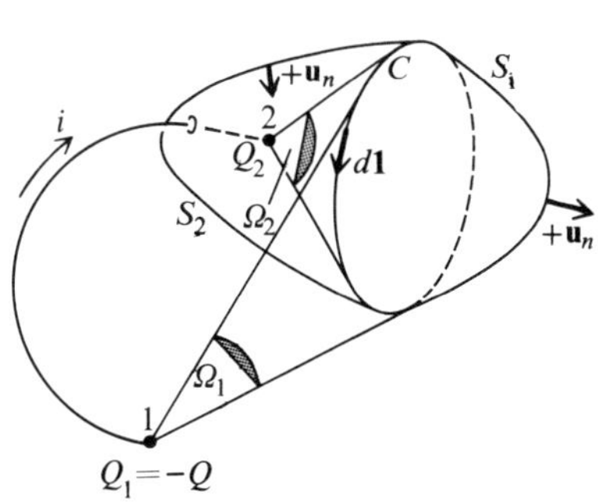


**Fig. 13.3** An open current filament piercing one of the surfaces limited by a closed contour.

right-hand side of Eq. (13.2.2) hence equals the *total* flux of **E** through $S_1$, multiplied by $4\pi\varepsilon_0$:

$$\oint_C \mathbf{B}\cdot d\mathbf{l} = \mu_0\varepsilon_0 \frac{d}{dt}\int_{S_1} \mathbf{E}\cdot d\mathbf{S} \qquad \text{(no conduction current through } S_1\text{).} \qquad (13.2.3)$$

With $S_2$ on the other hand it is not so. Indeed, $\Omega_2$ is *not* the solid angle subtended at 2 by the negative side of $S_2$, and the solid angle subtended at 2 by the *positive* side of $S_2$ is $(4\pi - \Omega_2)$. The flux of **E** through $S_2$ created by the charge $Q_2$ at 2 is hence $-Q_2(4\pi - \Omega_2)/4\pi\varepsilon_0$. To obtain the right-hand side of Eq. (13.2.2) in the form of a surface integral, as in Eq. (13.2.3), we first add and substract $4\pi Q_2$ inside the parentheses of Eq. (13.2.2):

$$\oint_C \mathbf{B}\cdot d\mathbf{l} = \frac{\mu_0}{4\pi}\frac{d}{dt}[4\pi Q_2 - (4\pi - \Omega_2)Q_2 + \Omega_1 Q_1].$$

The last two terms in the brackets represent again the total flux of **E** through $S_2$, multiplied by $4\pi\varepsilon_0$; the time derivative of the first term equals $4\pi i$. Thus we finally have

$$\oint_C \mathbf{B}\cdot d\mathbf{l} = \mu_0 i + \mu_0\varepsilon_0 \frac{d}{dt}\int_{S_2} \mathbf{E}\cdot d\mathbf{S} \qquad \text{(conduction current } i \text{ through } S_2\text{).} \qquad (13.2.4)$$

The right-hand sides in Eqs. (13.2.3) and (13.2.4) are equal, although their form is different. We can prove this by the following reasoning. Surface $(S_1 + S_2)$ is a closed surface. From Gauss' law we thus have

$$Q_2 = \varepsilon_0 \oint_{S_1+S_2} \mathbf{E}\cdot d\mathbf{S} = \varepsilon_0 \int_{S_1} \mathbf{E}\cdot d\mathbf{S} - \varepsilon_0 \int_{S_2} \mathbf{E}\cdot d\mathbf{S}. \qquad (13.2.5)$$

The positive normal to $S_2$ in this equation is $-\mathbf{u}_n$, rather than $+\mathbf{u}_n$ (that is, opposite to $+\mathbf{u}_n$), since $S_2$ is regarded as a part of a *closed* surface. Hence the negative sign in front of the last integral. Now, according to Fig. 13.3 and the continuity equation, $i = dQ_2/dt$. Bearing this in mind and differentiating Eq. (13.2.5) with respect to time, we can prove that the right-hand sides of Eqs. (13.2.3) and (13.2.4) are equal.

We move on to consider an arbitrary nonstationary current system in a vacuum. As we have seen, this can always be regarded as a system of open current filaments with equal, but opposite, charges at their ends. From Eq. (13.2.4) we learn that Ampère's circuital law in this case is (see Fig. 13.4)

$$\oint_C \mathbf{B}\cdot d\mathbf{l} = \mu_0 \sum_S i + \varepsilon_0\mu_0 \frac{d}{dt}\int_S \mathbf{E}\cdot d\mathbf{S}. \qquad (13.2.6)$$

Now, the total current is equal to the flux of the current density vector **J** through the surface $S$. Also, since the surface $S$ is stationary, the time derivative in the second term acts on **E** only, and can be introduced under the integral sign. Thus

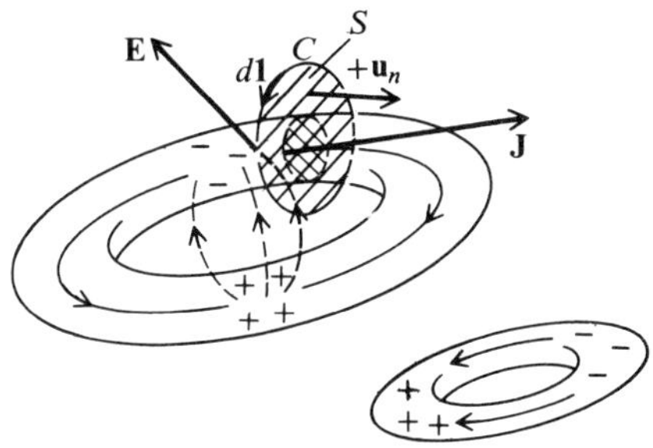


**Fig. 13.4** A nonstationary current system in a vacuum.

Eq. (13.2.6) becomes

$$\boxed{\oint_C \mathbf{B}\cdot d\mathbf{l} = \mu_0 \int_S \left(\mathbf{J} + \varepsilon_0 \frac{\partial \mathbf{E}}{\partial t}\right)\cdot d\mathbf{S}} \qquad \text{(dielectric and magnetic materials not present).} \qquad (13.2.7)$$

This equation is extremely important. It shows that the term $\varepsilon_0 \partial \mathbf{E}/\partial t$ may be considered formally to produce the magnetic flux density just as the real current density $\mathbf{J}$ does. Following Maxwell, this is referred to as the *displacement current density in free space*. This name is adopted generally, but it may be somewhat misleading. From the foregoing it is evident that the term $\varepsilon_0 \partial \mathbf{E}/\partial t$ is only *equivalent* in producing the magnetic field to the real current density $\mathbf{J}$. This was shown to be simply a consequence of the particular expression for the magnetic flux density of a single moving charge.

Using Stokes' theorem [Eq. (8.8.14)] we can easily write the differential form of Eq. (13.2.7) as

$$\text{curl}\,\mathbf{B} = \mu_0 \left(\mathbf{J} + \varepsilon_0 \frac{\partial \mathbf{E}}{\partial t}\right) \qquad \text{(dielectric and magnetic materials not present).} \qquad (13.2.8)$$

We now consider the case when dielectric and magnetic materials *are* present. In the case of dielectrics, the varying polarization at a point represents the real motion of charged particles at that point, that is, the electric current. The density of this *polarization current* is obtained as follows. It was shown in Section 2.3 that the polarization vector $\mathbf{P}$ represents the charge that, in the process of polarization, crosses a surface normal to $\mathbf{P}$, per unit area. By the definition of the current density, the *polarization current density* is hence $\partial \mathbf{P}/\partial t$, which is a real current density, and has to be added to $\mathbf{J}$ in Eqs. (12.3.7) and (12.3.8). The possible varying polarization charges at the ends of the tubes of varying dipoles produce an electric field which has simply to be added to $\mathbf{E}$ in these equations.

With magnetic materials the reasoning is still simpler. In Section 9.3, we saw that the density of the macroscopic resultants of the microscopic currents inside a magnetized body is equal to curl $\mathbf{M}$, the vector $\mathbf{M}$ representing the magnetization at the point considered. If the magnetization varies with time, this current density varies also with time, and it has to be added to the real current density whenever present.

A quite general expression of Ampère's circuital law is thus

$$\oint_C \mathbf{B}\cdot d\mathbf{l} = \mu_0 \int_S \left(\mathbf{J} + \text{curl}\,\mathbf{M} + \frac{\partial \mathbf{P}}{\partial t} + \varepsilon_0 \frac{\partial \mathbf{E}}{\partial t}\right)\cdot d\mathbf{S}, \tag{13.2.9}$$

or, in differential form,

$$\text{curl}\,\mathbf{B} = \mu_0 \left(\mathbf{J} + \text{curl}\,\mathbf{M} + \frac{\partial \mathbf{P}}{\partial t} + \varepsilon_0 \frac{\partial \mathbf{E}}{\partial t}\right). \tag{13.2.10}$$

The last equation simplifies if we recall that $\mathbf{B}/\mu_0 - \mathbf{M} = \mathbf{H}$ and $\varepsilon_0\mathbf{E} + \mathbf{P} = \mathbf{D}$. Thus

$$\boxed{\text{curl}\,\mathbf{H} = \mathbf{J} + \frac{\partial \mathbf{D}}{\partial t}} \quad \text{(the generalized form of Ampère's circuital law in differential form).} \tag{13.2.11}$$

Integrating over a surface $S$, bounded by a contour $C$, and applying Stokes' theorem, we have

$$\boxed{\oint_C \mathbf{H}\cdot d\mathbf{l} = \int_S \left(\mathbf{J} + \frac{\partial \mathbf{D}}{\partial t}\right)\cdot d\mathbf{S}} \quad \text{(the generalized form of Ampère's circuital law in integral form).} \tag{13.2.12}$$

These equations are very important. They represent the last of the four general and always valid equations of the electromagnetic field.

*Problems referring to Section 13.2:* 13.G1 to 13.G3, 13.N1.

## 13.3 GENERAL EQUATIONS OF THE ELECTROMAGNETIC FIELD (MAXWELL'S EQUATIONS)

This section is perhaps the most important in the whole book. The mathematical relations which electric and magnetic parts of the unique electromagnetic field satisfy in special cases will be shown to represent a consistent system of equations in *all* cases.

The most general differential equations from all the special cases of fields considered are given below for convenience, together with an indication of where they were derived.

$$\text{div}\,\mathbf{D} = \rho \quad \text{(Gauss' law, the static electric field),} \tag{13.3.1}$$

$$\text{div}\,\mathbf{J} + \partial\rho/\partial t = 0 \quad \text{(the law of conservation of electric charge, valid generally),} \tag{13.3.2}$$

$$\text{div}\,\mathbf{B} = 0 \quad \text{(the law of conservation of magnetic flux, the static magnetic field),} \tag{13.3.3}$$

$\text{curl}\,\mathbf{E} = -\partial\mathbf{B}/\partial t$ (Maxwell-Faraday law, the quasi stationary electromagnetic field), (13.3.4)

$\text{curl}\,\mathbf{H} = \mathbf{J} + \partial\mathbf{D}/\partial t$ (the generalized Ampère's circuital law, the nonstationary electromagnetic field). (13.3.5)

The generality of Eq. (13.3.2) cannot be questioned. Let us analyze the meaning of the other four.

Equation (13.3.1) states that the Coulomb-type of sources of vector **D** are only free charges, which we have seen to be true for the static electric field. However, we cannot imagine that any new type of sources of **D** can appear in the case of arbitrary time-varying fields, since the charges cannot be created in any way. We therefore *postulate* that Eq. (13.3.1) is a generally valid law.

Equation (13.3.3) states that in nature there are no sources of **B** analogous to the electric charges, that is, that no magnetic charges exist. No experimental evidence ever indicated the opposite, so Eq. (13.3.3) also represents a generally valid relation.

The Maxwell-Faraday law was derived by starting from the law appertaining to the force between two charges in slow motion. However, no velocity of charges appears as a parameter in Eq. (13.3.4). Therefore, we can *postulate* that its validity is not restricted to quasi-stationary electromagnetic fields. A similar argument and the same conclusion is valid in the case of the generalized Ampère's circuital law, Eq. (13.3.5).

Although the foregoing generalizations are *logical*, they *cannot* be proved in a formal manner. We have to consider them as postulates, to be justified, in the case of arbitrary electromagnetic fields, by experimental evidence. These equations have been known for over a century, yet there are no indications that they might not be valid in any cases of macroscopic electromagnetic fields. Consequently, we consider them as being always true. They are referred to as *Maxwell's equations*, after the British physicist who first postulated them, in the early sixties of the last century. The order of writing these equations is irrelevant, since each is equally important. For convenience, however, we shall group them in pairs of "curl" and "divergence" equations:

$$\boxed{\text{curl}\,\mathbf{E} = -\frac{\partial\mathbf{B}}{\partial t}} \quad \text{(valid generally)}, \tag{13.3.6}$$

$$\boxed{\text{curl}\,\mathbf{H} = \mathbf{J} + \frac{\partial\mathbf{D}}{\partial t}} \quad \text{(valid generally)}, \tag{13.3.7}$$

$$\boxed{\text{div}\,\mathbf{D} = \rho} \quad \text{(valid generally)}, \tag{13.3.8}$$

$$\boxed{\operatorname{div} \mathbf{B} = 0} \qquad \text{(valid generally).} \tag{13.3.9}$$

To complete the set of Maxwell's equations, we add to them the relations between vectors **E**, **P**, **D**, **J**, and $\mathbf{E}_i$ (the impressed electric field), and between vectors **B**, **M**, and **H**:

$$\boxed{\mathbf{D} = \varepsilon_0 \mathbf{E} + \mathbf{P}} \quad \text{(valid generally)}, \qquad \boxed{\mathbf{D} = \varepsilon \mathbf{E}} \quad \text{(linear media)}, \tag{13.3.10}$$

$$\boxed{\mathbf{B} = \mu_0(\mathbf{H} + \mathbf{M})} \quad \text{(valid generally)}, \qquad \boxed{\mathbf{B} = \mu \mathbf{H}} \quad \text{(linear media)}, \tag{13.3.11}$$

$$\boxed{\mathbf{J} = \mathbf{J}(\mathbf{E}, \mathbf{E}_i)} \quad \text{(general case)}, \qquad \boxed{\mathbf{J} = \sigma(\mathbf{E} + \mathbf{E}_i)} \quad \text{(linear media).} \tag{13.3.12}$$

The continuity equation is not usually regarded as one of Maxwell's equations, but we add it here for completeness:

$$\boxed{\operatorname{div} \mathbf{J} + \frac{\partial \rho}{\partial t} = 0} \qquad \text{(the continuity equation).} \tag{13.3.13}$$

The differential form of Maxwell's equations (13.3.6) to (13.3.9) is their most important one. The integral form is also useful, for example, in deriving boundary conditions. The integral form of Eq. (13.3.6) and (13.3.7) is obtained by multiplying (dot product) the equations by an element $d\mathbf{S}$ of any closed surface $S$, integrating over the surface $S$, and applying Stokes' theorem to the left-hand side. Similarly, the integral form of Eqs. (13.3.8) and (13.3.9) is obtained by multiplying them by a volume element $dv$, integrating any volume $v$, and applying the divergence theorem to the left-hand side. Thus we obtain

$$\boxed{\oint_C \mathbf{E} \cdot d\mathbf{l} = -\int_S \frac{\partial \mathbf{B}}{\partial t} \cdot d\mathbf{S}} \qquad \text{(valid generally),} \tag{13.3.14}$$

$$\boxed{\oint_C \mathbf{H} \cdot d\mathbf{l} = \int_S \left(\mathbf{J} + \frac{\partial \mathbf{D}}{\partial t}\right) \cdot d\mathbf{S}} \qquad \text{(valid generally),} \tag{13.3.15}$$

$$\boxed{\oint_S \mathbf{D}\cdot d\mathbf{S} = \int_v \rho\, dv} \qquad \text{(valid generally)}, \tag{13.3.16}$$

$$\boxed{\oint_S \mathbf{B}\cdot d\mathbf{S} = 0} \qquad \text{(valid generally)}. \tag{13.3.17}$$

The importance of Maxwell's equations in the theory of electromagnetic fields is tremendous. They represent mathematical laws satisfied by the electric and magnetic components of all real electromagnetic fields. From now on they will always be the starting point of our analysis. It cannot be overemphasized that the reader is supposed to learn and understand them before proceeding even to the next section. Their special forms, valid for fields with certain restrictions, can serve as good practice for memorizing their general form, and simultaneously for understanding them better. For that purpose many simple, but helpful test problems are provided at the end of the chapter.

**13.3.1 The complex form of Maxwell's equations.** The primary sources of any electromagnetic field are currents and free charges. In practice these most often vary in time following closely simple harmonic (sinusoidal) law. Provided that the medium is *linear*, all the field vectors will also vary according to the sinusoidal law.

Let us suppose the frequency of the current producing an electromagnetic field is $f$. The product $2\pi f = \omega$ is usually referred to as the angular frequency. The field vectors will vary at all points with the same frequency. Thus we can write, for example,

$$\mathbf{E}(x, y, z, t) = \mathbf{E}(x, y, z) \cos\left[\omega t + \theta_E(x, y, z)\right], \tag{13.3.18}$$

$$\mathbf{B}(x, y, z, t) = \mathbf{B}(x, y, z) \cos\left[\omega t + \theta_B(x, y, z)\right]. \tag{13.3.19}$$

Vectors $\mathbf{E}$ and $\mathbf{B}$ are functions of the space coordinates and of time. The set $(x, y, z)$ of the coordinates is written for simplicity, and it should be regarded only as indication of dependence upon the space coordinates, representing any possible set of coordinates. We shall use this designation frequently.

The phases $\theta_E$ and $\theta_B$ of the two vectors are also functions of the coordinates. The field vectors are not necessarily of the same phase at all points. The knowledge that all field vectors vary according to the sinusoidal law is not sufficient for determining even $\theta_E$ and $\theta_B$, and it might seem that no considerable simplification can result if the field vectors are assumed to be in the form of Eqs. (13.3.18) and (13.3.19).

The formal analysis of time-harmonic fields, however, can be greatly simplified by means of a mathematical trick, very frequently used, and certainly one of the cleverest ever devised in electrical engineering. The essence of the trick is described below.

According to the well-known mathematical identity, for any $x$

$$\cos x = (e^{jx} + e^{-jx})/2, \tag{13.3.20}$$

where $j$ stands for the imaginary unit, $j \equiv \sqrt{-1}$. Consequently, instead of the cosine in, say, Eq. (13.3.18), we can write

$$\mathbf{E}(x, y, z, t) = \frac{1}{2}\,\mathbf{E}(x, y, z)\left\{e^{j[\omega t + \theta_E(x, y, z)]} + e^{-j[\omega t + \theta_E(x, y, z)]}\right\}. \tag{13.3.21}$$

Analogous expressions can be written for all other field vectors and the charge density $\rho$. Let us substitute **E** and **B**, of the form in Eq. (13.3.21), into Maxwell's equation (13.3.6). Noting that

$$e^{\pm j(\omega t + \theta)} = e^{\pm j\omega t} e^{\pm j\theta}$$

and that $e^{\pm j\theta}$ only is a function of the space coordinates, we obtain

$$\begin{aligned} &e^{j\omega t}\,\text{curl}\,[\mathbf{E}(x, y, z)\,e^{j\theta_E(x, y, z)}] + e^{-j\omega t}\,\text{curl}\,[\mathbf{E}(x, y, z)\,e^{-j\theta_E(x, y, z)}] \\ &\qquad = -j\omega\mathbf{B}(x, y, z)\,e^{j\theta_B(x, y, z)}\,e^{j\omega t} + j\omega\mathbf{B}(x, y, z)\,e^{-j\theta_B(x, y, z)}\,e^{-j\omega t}. \end{aligned}$$

This equation must be satisfied for any $t$, which is possible only if the factors of $e^{j\omega t}$ and of $e^{-j\omega t}$ on both sides are equal:

$$\text{curl}\,[\mathbf{E}(x, y, z)\,e^{j\theta_E(x, y, z)}] = -j\omega[\mathbf{B}(x, y, z)\,e^{j\theta_B(x, y, z)}], \tag{13.3.22}$$

$$\text{curl}\,[\mathbf{E}(x, y, z)\,e^{-j\theta_E(x, y, z)}] = j\omega[\mathbf{B}(x, y, z)\,e^{-j\theta_B(x, y, z)}]. \tag{13.3.23}$$

Now, we *know* that $\mathbf{E}(x, y, z, t)$ and $\mathbf{B}(x, y, z, t)$ are of the form in Eqs. (13.3.18) and (13.3.19). In these equations, the unknowns are the magnitudes $\mathbf{E}(x, y, z)$ and $\mathbf{B}(x, y, z)$, and the phase angles $\theta_E(x, y, z)$ and $\theta_B(x, y, z)$ of the two vectors as functions of the coordinates. *Both* the magnitudes and phase angles of each vector function can immediately be deduced if *either* of the two types of *complex* expressions in the brackets of the last two equations is found. Consequently, when the field vectors vary with time according to the simple harmonic law, *each* of the above equations is equivalent to Eq. (13.3.6). It is customary to use the first, i.e. Eq. (13.3.22).

To simplify the notation, the complex functions in the brackets are designated by a single symbol. To stress that these are *complex* vectors, an additional sign is often used, such as $\mathbf{E}_c$, $\hat{\mathbf{E}}$, $\bar{\mathbf{E}}$ or $\underline{\mathbf{E}}$. In the present text, complex quantities will not usually be specially designated. When necessary, the subscript "c" ("complex") will be used, as in $\mathbf{E}_c$, $\mathbf{B}_c$, $\rho_c$, etc. If there is no possibility of confusion, the simple symbols **E**, **B**, etc. will be used instead. We can always clearly recognize equations with **E**, **B**, etc., which represent complex functions by the presence of $\omega$. In addition, the use of complex functions will be called to the reader's attention whenever necessary.

In the light of these remarks Maxwell's equations (13.1.6) to (13.3.9), and the equation of continuity (13.3.13) in the case of sinusoidally time-varying fields become

$$\boxed{\operatorname{curl} \mathbf{E}_c = -j\omega \mathbf{B}_c} \qquad \text{(sinusoidal excitation, linear media),} \qquad (13.3.24)$$

$$\boxed{\operatorname{curl} \mathbf{H}_c = \mathbf{J}_c + j\omega \mathbf{D}_c} \qquad \text{(sinusoidal excitation, linear media),} \qquad (13.3.25)$$

$$\boxed{\operatorname{div} \mathbf{D}_c = \rho_c} \qquad \text{(sinusoidal excitation, linear media),} \qquad (13.3.26)$$

$$\boxed{\operatorname{div} \mathbf{B}_c = 0} \qquad \text{(sinusoidal excitation, linear media),} \qquad (13.3.27)$$

$$\boxed{\operatorname{div} \mathbf{J}_c + j\omega\rho_c = 0} \qquad \text{(sinusoidal current).} \qquad (13.3.28)$$

In these equations $\mathbf{E}_c$ stands for

$$\mathbf{E}_c = \mathbf{E}(x, y, z)\, e^{j\theta_E(x, y, z)}, \qquad (13.3.29)$$

and similarly for other field vectors and charge density $\rho_c$.

For practical applications, the set of Eqs. (13.3.24) to (13.3.27) is perhaps even more important than their general form [Eqs. (13.3.6) to (13.3.9)], not only because sinusoidal time-varying fields are most often used in practice. Any time-varying field can be represented in the form of a series (Fourier series) or integral (Fourier integral) of sinusoidal terms. The reader is advised to study and learn the complex form of Maxwell's equations with the same thoroughness as their general form.

If complex field vectors are determined, we can easily find their instantaneous real value. For example, if we know the complex electric field intensity $\mathbf{E}_c(x, y, z)$, we can always write it in the form of Eq. (13.3.29), where $\mathbf{E}(x, y, z)$ denotes a *real* function. Now,

$$e^{j\omega t} e^{j\theta_E(x, y, z)} = e^{j[\omega t + \theta_E(x, y, z)]}$$
$$= \cos[\omega t + \theta_E(x, y, z)] + j \sin[\omega t + \theta_E(x, y, z)]. \qquad (13.3.30)$$

From these two equations and Eq. (13.3.18) we can see that the correct value of $\mathbf{E}(x, y, z, t)$ is obtained by simply multiplying the right-hand side of Eq. (13.3.29) by $e^{j\omega t}$, and taking the *real* part of the expression thus obtained:

$$\mathbf{E}(x, y, z, t) = \operatorname{Re}\{\mathbf{E}_c(x, y, z)\, e^{j\omega t}\} = \mathbf{E}(x, y, z) \cos[\omega t + \theta_E(x, y, z)]. \quad (13.3.31)$$

(*Re* is an indication that the *real* part of the expression in braces only is considered. It stands for "the real part of".)

*Problems referring to Section 13.3:* 13.T1 to 13.T19, 13.G4 to 13.G5.

## 13.4 SOME IMPORTANT CONSEQUENCES OF MAXWELL'S EQUATIONS

**13.4.1 General boundary conditions.** As earlier, the term "boundary conditions" refers to relations between normal and tangential components of the field vectors on the two sides of an interface between two different media. The general boundary conditions can be derived in exactly the same way as in the case of special boundary conditions. In fact, the *form* of the boundary conditions is exactly the same as before, although the meaning is, of course, different. Now they are applicable to all cases of electromagnetic fields. The reader is advised to reread Sections 2.6 and 9.5, since only a condensed version will be presented here.

The tangential components of the electric field intensity vector on an interface of two different media are equal (see Fig. 13.5):

$$\boxed{E_{1t} = E_{2t}} \qquad \text{(valid generally).} \tag{13.4.1}$$

This boundary condition is obtained by applying Eq. (13.3.14) to a narrow rectangular contour, as shown, and noting that the magnetic flux through the contour vanishes. In the case of two linear dielectrics of permittivities $\varepsilon_1$ and $\varepsilon_2$ this condition can also be written as

$$D_{1t}/\varepsilon_1 = D_{2t}/\varepsilon_2 \qquad \text{(linear dielectrics).} \tag{13.4.2}$$

The tangential components of the magnetic field intensity vector **H** are also equal (see Fig. 13.5), provided that there are no surface currents on the interface:

$$\boxed{H_{1t} = H_{2t}} \qquad \text{(no surface currents on the interface, otherwise valid generally).} \tag{13.4.3}$$

This boundary condition is obtained by applying Eq. (13.3.15) to the thin rectangular contour shown, noting that, if there are no surface currents, the flux of

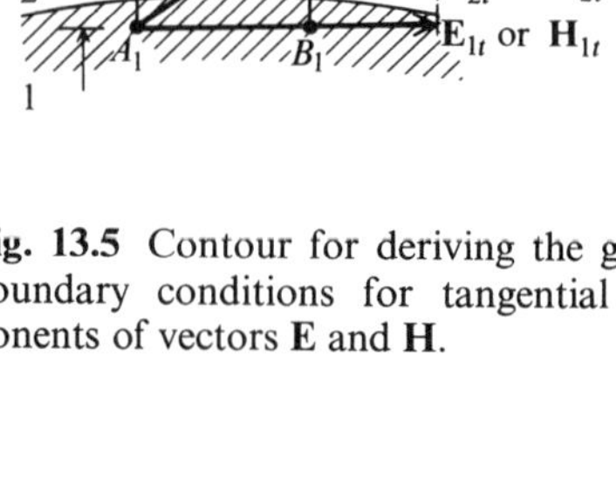


Fig. 13.5 Contour for deriving the general boundary conditions for tangential components of vectors **E** and **H**.

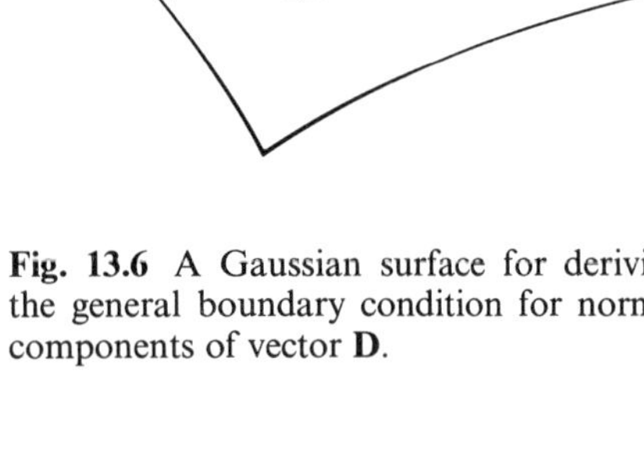


Fig. 13.6 A Gaussian surface for deriving the general boundary condition for normal components of vector **D**.

$(\mathbf{J} + \partial\mathbf{D}/\partial t)$ through the contour vanishes. If the two media are linear, Eq. (13.4.3) can be written in the form

$$B_{1t}/\mu_1 = B_{2t}/\mu_2 \qquad \text{(linear media).} \tag{13.4.4}$$

The normal components of the electric displacement vector **D** on the interface with surface charges of density $\rho_s$ satisfy the following condition (Fig. 13.6):

$$\boxed{D_{2n} - D_{1n} = \rho_s} \qquad \text{(valid generally).} \tag{13.4.5}$$

This condition is obtained by applying Eq. (13.3.16) to the thin "pillbox" shown in the figure.

Finally, by applying Eq. (13.3.17) to such a "pillbox", we obtain that the normal components of the magnetic flux density vector **B** are equal:

$$\boxed{B_{1n} = B_{2n}} \qquad \text{(valid generally).} \tag{13.4.6}$$

An important case of interface is that between a dielectric and a "perfect" conductor. Perfect conductors do not, in fact, exist, but whenever we are dealing with the best available conductors, such as silver, copper, and aluminium, we can in many cases treat them as though they were perfect. No electric field can exist inside a perfect conductor, for its presence would produce currents of infinite density, which is impossible. Nor can a *varying* magnetic field exist inside a perfect conductor; according to Faraday-Maxwell's law [Eq. (13.3.6)] it would produce a varying electric field in the conductor which, as explained, cannot be. Thus, in the case of a perfect conductor (say, medium 1), the boundary conditions of Eqs. (13.4.1), (13.4.5), and (13.4.6) become

$$\boxed{E_t = 0} \tag{13.4.7}$$

$$\boxed{D_n = \rho_s} \qquad \text{(interface of a dielectric and a perfect conductor).} \tag{13.4.8}$$

$$\boxed{B_n = 0} \tag{13.4.9}$$

Practically, the only interface with surface currents is the surface of the perfect conductor. Let us suppose that the density of the surface current is $\mathbf{J}_s$, and note that there is no field inside the conductor. With this, the application of Eq. (13.3.15) to the thin rectangular contour shown in Fig. 13.7 yields

$$H_t\,\Delta l = J_s\,\Delta l, \qquad \text{or} \qquad \boxed{H_t = J_s} \qquad \text{(perfect conductor).} \tag{13.4.10}$$

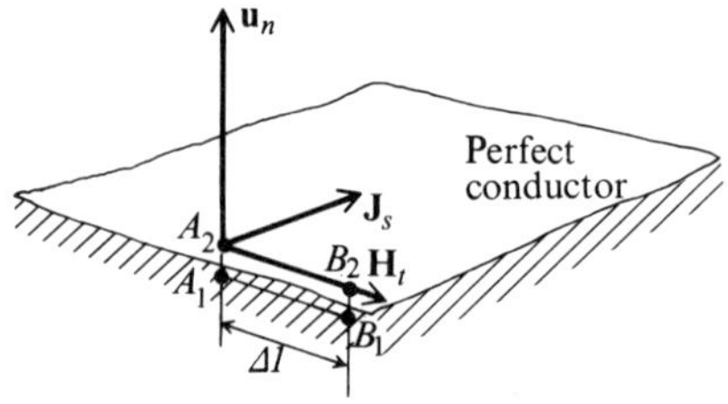


**Fig. 13.7** Surface of a perfect conductor.

It is often useful to write this condition in the vector form, showing also the relation between the directions of the two vectors. If we orient the positive normal away from the perfect conductor, by reference to Fig. 13.7 we have

$$\boxed{\mathbf{H} = \mathbf{J}_s \times \mathbf{u}_n} \quad \text{or} \quad \boxed{\mathbf{J}_s = \mathbf{u}_n \times \mathbf{H}} \quad \text{(perfect conductor).} \tag{13.4.11}$$

**13.4.2 Poynting's theorem.** Poynting's theorem is the expression of the law of conservation of energy as applied to electromagnetic fields.

To derive Poynting's theorem, we multiply Eq. (13.3.6) scalarly by $\mathbf{H}$ and Eq. (13.3.7) by $-\mathbf{E}$. Adding the two equations thus obtained, we have

$$\mathbf{H}\cdot\operatorname{curl}\mathbf{E} - \mathbf{E}\cdot\operatorname{curl}\mathbf{H} = -\mathbf{H}\cdot\frac{\partial\mathbf{B}}{\partial t} - \mathbf{E}\cdot\mathbf{J} - \mathbf{E}\cdot\frac{\partial\mathbf{D}}{\partial t}. \tag{13.4.12}$$

According to Eq. (12.1.7), the left-hand side simply equals div $(\mathbf{E} \times \mathbf{H})$. If we suppose that the medium is linear, then

$$\mathbf{H}\cdot\frac{\partial\mathbf{B}}{\partial t} = \mu\mathbf{H}\cdot\frac{\partial\mathbf{H}}{\partial t} = \frac{\partial}{\partial t}\left(\frac{1}{2}\mu\mathbf{H}\cdot\mathbf{H}\right) = \frac{\partial}{\partial t}\left(\frac{1}{2}\mu H^2\right),$$

$$\mathbf{E}\cdot\frac{\partial\mathbf{D}}{\partial t} = \varepsilon\mathbf{E}\cdot\frac{\partial\mathbf{E}}{\partial t} = \frac{\partial}{\partial t}\left(\frac{1}{2}\varepsilon E^2\right).$$

Now, for linear media, $\mathbf{J} = \sigma(\mathbf{E} + \mathbf{E}_i)$ so that $\mathbf{E} = \mathbf{J}/\sigma - \mathbf{E}_i$. With that in mind, we find that after simple manipulations Eq. (13.4.12) becomes

$$\mathbf{E}_i\cdot\mathbf{J} = \frac{J^2}{\sigma} + \frac{\partial}{\partial t}\left(\frac{1}{2}\varepsilon E^2 + \frac{1}{2}\mu H^2\right) + \operatorname{div}(\mathbf{E}\times\mathbf{H}). \tag{13.4.13}$$

Let us multiply this equation by a volume element $dv$ and integrate over an arbitrary volume $v$ of the field. Making use of the divergence theorem in the last term, we have

$$\int_v \mathbf{E}_i\cdot\mathbf{J}\,dv = \int_v \frac{J^2}{\sigma}\,dv + \frac{\partial}{\partial t}\int_v\left(\frac{1}{2}\varepsilon E^2 + \frac{1}{2}\mu H^2\right)dv + \oint_S (\mathbf{E}\times\mathbf{H})\cdot d\mathbf{S}. \tag{13.4.14}$$

This equation is an energy equation. The left-hand side represents the power of all the sources inside the volume $v$; the terms on the right-hand side show how this power is used. One part of it (the first term) is transformed into heat; the other (the second term) is used to increase the energy located in both the electric and magnetic fields inside $v$. The third term is of a new form, not encountered before, and it may seem at first that the physical meaning of this term cannot be explained in any simple manner, since all the possibilities of using the power of the generators inside $v$ are comprized in the first two terms. But it is not so. The generators inside $v$ can also supply power to the field regions outside $v$. A third term on the right-hand side of Eq. (13.4.14) is indispensable if account is taken of this possibility. We thus interpret the third term as follows:

$$\boxed{\oint_S (\mathbf{E} \times \mathbf{H}) \cdot d\mathbf{S} = \text{power transferred through } S \text{ to region outside } v.} \qquad (13.4.15)$$

This is known as *Poynting's theorem*.

According to this theorem, the term $|\mathbf{E} \times \mathbf{H}|$ can be interpreted as the power transferred at a point by the electromagnetic field per unit surface normal to vector $\mathbf{E} \times \mathbf{H}$. This vector is referred to as the *Poynting vector*, and is usually designated by $\mathscr{P}$ (not to be confused with the polarization vector $\mathbf{P}$):

$$\boxed{\mathscr{P} = \mathbf{E} \times \mathbf{H}} \qquad \text{(definition of the Poynting vector).} \qquad (13.4.16)$$

The Poynting vector $\mathscr{P}$ may be interpreted to indicate the direction and rate of power flow at *all* points. This does *not* follow, however, from Eq. (13.4.15), since only the flux of $\mathscr{P}$ through a *closed* surface is defined. The Poynting vector is nevertheless usually interpreted as mentioned. We shall also adhere to this interpretation of vector $\mathscr{P} = \mathbf{E} \times \mathbf{H}$. In the MKSA system of units the Poynting vector is given in watts per square meter ($W/m^2$).

As the derivation of Poynting's theorem is based on Maxwell's equations and the law of conservation of energy, it must give a correct result in all cases.

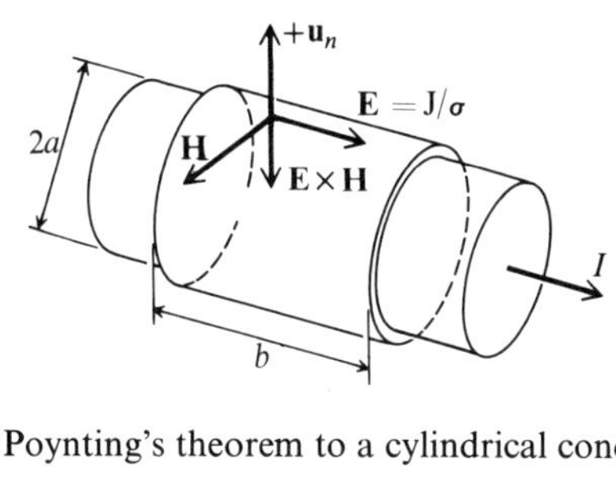


**Fig. 13.8** Application of Poynting's theorem to a cylindrical conductor with direct current.

As an example, let us take first a straight conductor of circular cross-section, of radius $a$ and conductivity $\sigma$, and suppose that the intensity of the steady current in the conductor is $I$. If we apply Poynting's theorem to the surface of the cylinder shown in Fig. 13.8, we note that the Poynting vector $\mathscr{P}$ is normal to the curved surface of the cylinder, and parallel to its bases. On the surface of the conductor $H(a) = I/2\pi a$, and $E(a) = J/\sigma = I/a^2\sigma\pi$, in the directions indicated. Noting that the vector $\mathbf{E} \times \mathbf{H}$ is in the direction opposite to $\mathbf{u}_n$ we have

$$\oint_S (\mathbf{E} \times \mathbf{H}) \cdot d\mathbf{S} = -E(a)\, H(a)\, 2\pi ab = -\frac{b}{\sigma a^2 \pi} I^2.$$

The minus sign suggests that the power is absorbed by the volume enclosed by $S$, which is, of course, the case. The power transformed into heat in the cylinder shown we recognize to be the right-hand side of this equation, since $b/\sigma a^2\pi$ is precisely the resistance of the length $b$ of the conductor.

As a second example let us consider the strip line shown in Fig. 13.9. Let us suppose that a surface $S$ encloses the load $R$, and cuts the line at right angles. Ignoring the fringing effect, we find that the electric and magnetic fields exist only between the strips. There $E = \mathscr{E}/d$ and $H = I/b$, so, according to the directions of the vectors indicated in the figure, we have

$$\oint_S (\mathbf{E} \times \mathbf{H}) \cdot d\mathbf{S} = -EHbd = -\mathscr{E}I.$$

It is obvious that the resistor $R$ *absorbs* power $\mathscr{E}I$. If we apply Poynting's theorem to a surface $S'$ (indicated in dashed line) enclosing the *generator*, we obtain the result that the power $\mathscr{E}I$ is *leaving* this surface, which is again obvious.

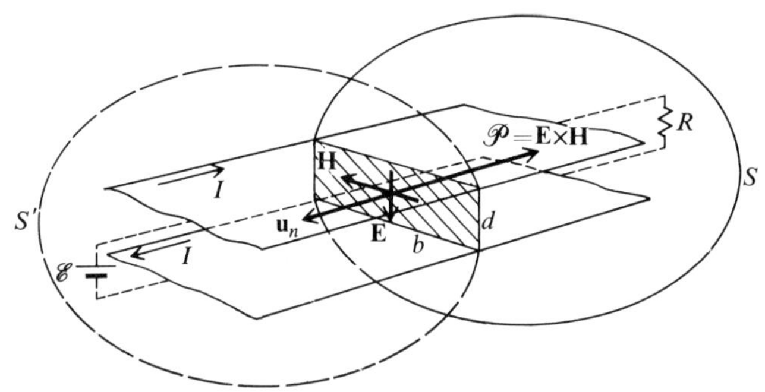


**Fig. 13.9** Application of Poynting's theorem to a strip line.

**13.4.3 Retarded potentials.** The electric scalar potential $V$ was a very convenient auxiliary function in analyzing the static and quasi-static electric fields. Equally so was the magnetic vector potential $\mathbf{A}$ in analyzing the static and quasi-static magnetic fields. Both potential functions are very useful also in the case of arbitrary electromagnetic fields. Maxwell's equations, however, require a different interpretation of these potentials.

Suppose we wish to determine the auxiliary functions $V$ and $\mathbf{A}$ so that

$$\mathbf{E} = -\text{grad } V - \partial\mathbf{A}/\partial t \qquad \text{(as earlier)}, \tag{13.4.17}$$

and

$$\mathbf{B} = \text{curl } \mathbf{A} \qquad \text{(as earlier)}. \tag{13.4.18}$$

If we substitute these expressions into Maxwell's equations (13.3.6) and (13.3.9), we see that both equations are satisfied. The other two, Eqs. (13.3.7) and (13.3.8), will serve to determine $V$ and $\mathbf{A}$ in terms of the source functions $\rho$ and $\mathbf{J}$.

Let us assume that the medium is linear, of permittivity $\varepsilon$ and permeability $\mu$. Then, substituting Eqs. (13.4.17) and (13.4.18) into Eqs. (13.3.7) and (13.3.8), we obtain

$$\text{curl (curl } \mathbf{A}) = \mu\mathbf{J} - \varepsilon\mu \frac{\partial}{\partial t}(\text{grad } V) - \varepsilon\mu \frac{\partial^2 \mathbf{A}}{\partial t^2}, \tag{13.4.19}$$

$$\text{div (grad } V) = -\frac{\rho}{\varepsilon} - \text{div} \frac{\partial \mathbf{A}}{\partial t}. \tag{13.4.20}$$

Now, according to (8.9.10)

$$\text{curl (curl } \mathbf{A}) = \text{grad (div } \mathbf{A}) - \text{div (grad } \mathbf{A}).$$

[The meaning of the term div (grad $\mathbf{A}$) is explained in Section 8.9.] With this, Eq. (13.4.19) becomes

$$\text{div (grad } \mathbf{A}) = -\mu\mathbf{J} + \varepsilon\mu \text{ grad} \frac{\partial V}{\partial t} + \varepsilon\mu \frac{\partial^2 \mathbf{A}}{\partial t^2} + \text{grad (div } \mathbf{A}). \tag{13.4.21}$$

Both $\mathbf{A}$ and $V$ are supposed only to satisfy Eqs. (13.4.17) and (13.4.18); otherwise they are arbitrary functions. Since curl $\mathbf{A}$ is specified, div $\mathbf{A}$ may be adopted to represent any function, but it is wise to choose it so that the foregoing differential equations have the simplest forms possible. This is achieved if

$$\boxed{\text{div } \mathbf{A} = -\varepsilon\mu \frac{\partial V}{\partial t}} \qquad \text{(the Lorentz condition for potentials)}. \tag{13.4.22}$$

By use of the Lorentz condition, Eqs. (13.4.20) and (13.4.21) reduce to

$$\text{div (grad } V) - \varepsilon\mu \frac{\partial^2 V}{\partial t^2} = -\frac{\rho}{\varepsilon}, \tag{13.4.23}$$

$$\text{div (grad } \mathbf{A}) - \varepsilon\mu \frac{\partial^2 \mathbf{A}}{\partial t^2} = -\mu\mathbf{J}. \tag{13.4.24}$$

In the rectangular coordinate system, the equation for **A** is identical in form to Eq. (13.4.23) for all three components:

$$\text{div}\,(\text{grad}\,A_x) - \varepsilon\mu\,\frac{\partial^2 A_x}{\partial t^2} = -\mu J_x,$$

and similarly for $A_y$ and $A_z$.

We may now consider Eq. (13.4.23) as typical for determining both potentials. All cases can be deduced if the solutions of these equations are found for a varying point charge and a varying current elment. In the case of a point charge, the solution of Eq. (13.4.23) must be a function only of the distance $r$ from the charge and time. The Laplacian div (grad $V$) in that case is

$$\text{div}\,(\text{grad}\,V) = \frac{1}{r^2}\,\frac{\partial}{\partial r}\left(r^2\,\frac{\partial V}{\partial r}\right) = \frac{\partial^2 V}{\partial r^2} + \frac{2}{r}\,\frac{\partial V}{\partial r}$$

(see Problem 1.G37). Thus, in the case of a varying point charge for all points out of the charge, Eq. (13.4.23) becomes

$$\frac{\partial^2 V}{\partial r^2} + \frac{2}{r}\,\frac{\partial V}{\partial r} - \varepsilon\mu\,\frac{\partial^2 V}{\partial t^2} = 0. \qquad (13.4.25)$$

This equation can be simplified by introducing a new variable $W$, defined by $V(r, t) = W(r, t)/r$. We then obtain

$$\frac{\partial^2 W}{\partial r^2} - \varepsilon\mu\,\frac{\partial^2 W}{\partial t^2} = 0. \qquad (13.4.26)$$

This is known as the *wave equation* (the justification for its name will be given in Chapter 15). By substitution it is not difficult to show that the solution of the wave equation is *any* function of variables $(t - r/c)$ and $(t + r/c)$, where $c = (\varepsilon\mu)^{-1/2}$. For physical reasons, which will become apparent in Chapter 15, only the solution in the form $W(t - r/c)$ is possible in this case. Thus we have

$$V(r, t) = \frac{W(t - r/c)}{r} \qquad (W = \text{an arbitrary function}). \qquad (13.4.27)$$

This result has a very interesting and important meaning. The potential $V$ at a distance $r$ from the charge does not vary exactly as the charge does. More precisely, at an instant $t$ the potential at a distance $r$ is a function of the charge that existed *at the instant* (t − r/c), that is, a time interval $\Delta t = r/c$ elapses before an observer at a distance $r$ from the charge is able to notice any change occurring in the charge. This potential is therefore referred to as the *retarded potential.*

To determine the function $W(t - r/c)$ more precisely, let us consider a point very close to the charge. In this instance, the retardation may be ignored. If the charge varies according to the law $Q(t)$, the potential is the same as in the static case, with $Q(t)$ replacing the constant charge $Q$:

$$V(r, t) = \frac{Q(t)}{4\pi\varepsilon r} \quad \text{(close to the charge).} \tag{13.4.28}$$

If we compare the last two equations, we see that $W(t - r/c) = Q(t - r/c)/4\pi\varepsilon$. Hence the resulting potential created by a varying point charge is

$$V(r, t) = \frac{Q(t - r/c)}{4\pi\varepsilon r}, \qquad c = (\varepsilon\mu)^{-1/2}. \tag{13.4.29}$$

The retarded potential at a point due to a cloud of charges of density $\rho(t)$ is given by

$$\boxed{V(t) = \frac{1}{4\pi\varepsilon}\int_v \frac{\rho(t - r/c)}{r}\,dv, \qquad c = (\varepsilon\mu)^{-1/2}} \quad \text{(the retarded scalar potential).} \tag{13.4.30}$$

(See Fig. 13.10.)

Each rectangular component of the magnetic vector potential is of this form. The total retarded magnetic vector potential is thus (see Fig. 13.10)

$$\boxed{\mathbf{A}(t) = \frac{\mu}{4\pi}\int_v \frac{\mathbf{J}(t - r/c)}{r}\,dv, \qquad c = (\varepsilon\mu)^{-1/2}} \quad \text{(the retarded vector potential).} \tag{13.4.31}$$

The retarded potentials shed a new light on many aspects of electromagnetic theory. We now see that electric and magnetic fields in the case of varying charges and currents need some time to change at points distant from the sources. The quasi-stationary electromagnetic field, where we ignored the retardation, is only a limiting case. The approximation that fields vary in synchronism with charges and currents is permissible only if the requirement that the retardation can be ignored is met. For example, ordinary electrical networks meet this requirement at relatively low frequencies, and can be considered as quasi-stationary systems, but, at very high frequencies, this is not generally so.

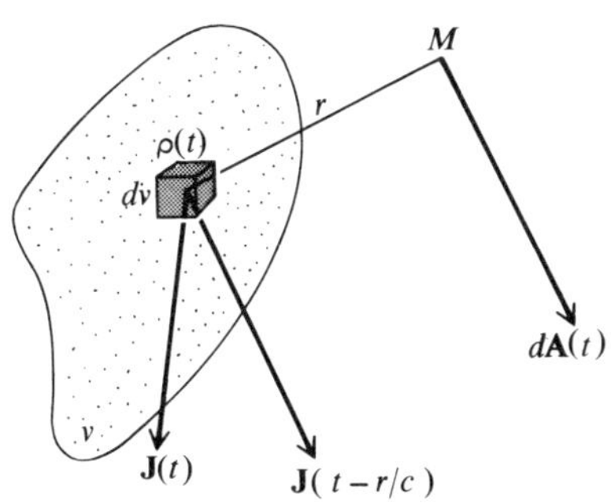


**Fig. 13.10** An illustration of the retarded vector potential.

A very important special case of retarded potentials is when the charge and current density vary with time according to the simple sinusoidal law. In that case the useful formalism with complex field vectors can be used. The retarded potentials, corresponding to the charges and currents of the form $\rho e^{j\omega t}$ and $\mathbf{J}e^{j\omega t}$, are of the form $Ve^{j\omega t}$ and $\mathbf{A}e^{j\omega t}$. Since $e^{j\omega t}$ cancels out, we are left with

$$V = \frac{1}{4\pi\varepsilon}\int_v \frac{\rho e^{-j\omega r/c}}{r}\,dv \qquad \text{(the complex form of the retarded scalar potential)}, \tag{13.4.32}$$

$$\mathbf{A} = \frac{\mu}{4\pi}\int_v \frac{\mathbf{J}e^{-j\omega r/c}}{r}\,dv \qquad \text{(the complex form of the retarded magnetic vector potential)}. \tag{13.4.33}$$

In these formulae $V$, $\rho$, $\mathbf{A}$, and $\mathbf{J}$ are, of course, complex quantities.

If $\omega r/c \ll 1$, the retardation can be ignored, since the formulae then reduce to those valid for the quasi-stationary case. Hence a criterion of quasi-stationarity for sinusoidally time-varying fields is possible. If we denote by $d$ the largest dimension of the system, $d$ must satisfy the inequality

$$d \ll c/\omega = 1/\omega\sqrt{\varepsilon\mu} \qquad \text{(the condition of quasi-stationarity)}. \tag{13.4.34}$$

**13.4.4 The generalized definition of conductors and insulators.** If we take a linear medium, of permittivity $\varepsilon$ and conductivity $\sigma$, then $\mathbf{J} = \sigma\mathbf{E}$ and $\mathbf{D} = \varepsilon\mathbf{E}$, and Maxwell's equation (13.3.25) in complex form becomes

$$\text{curl}\,\mathbf{H} = (\sigma + j\omega\varepsilon)\,\mathbf{E}. \tag{13.4.35}$$

If the medium is a perfect insulator $\sigma = 0$, and so only the second term in the parentheses is present. If the medium is a conductor, both terms are generally needed. Now, if

$$\sigma \gg \omega\varepsilon \qquad \text{(the definition of a good conductor)} \tag{13.4.36}$$

the term $j\omega\varepsilon$ can be ignored. The conduction current density is much larger than the "displacement" current density, in which case we may say that the medium is a good conductor. (Note that at some higher frequency $\omega\varepsilon$ can be comparable with $\sigma$.) The criterion, however, that a substance is a good conductor cannot be universal; frequency is equally as important as permittivity and conductivity.

Similarly, a good insulator is defined by the inequality

$$\sigma \ll \omega\varepsilon \qquad \text{(the definition of a good insulator)}. \tag{13.4.37}$$

**13.4.5 The superposition principle.** Maxwell's equations (13.3.6) to (13.3.9) are *linear* partial differential equations, that is, no terms in the form of the *product* of the two field vectors are present. The superposition principle is the consequence of this linearity.

Let us suppose that $\mathbf{E}_1$, $\mathbf{B}_1$, $\mathbf{H}_1$, $\mathbf{J}_1$, $\mathbf{D}_1$, and $\rho_1$ are functions of position and time which satisfy Maxwell's equations. Let another set $\mathbf{E}_2$, $\mathbf{B}_2$, $\mathbf{H}_2$, $\mathbf{J}_2$, $\mathbf{D}_2$, and $\rho_2$ also satisfy these equations. Then, since for example

$$\text{curl } \mathbf{E}_1 + \text{curl } \mathbf{E}_2 = \text{curl } (\mathbf{E}_1 + \mathbf{E}_2),$$

and

$$\partial \mathbf{B}_1/\partial t + \partial \mathbf{B}_2/\partial t = \partial(\mathbf{B}_1 + \mathbf{B}_2)/\partial t,$$

it follows that the sum of the two satisfies Maxwell's equations as well.

In general, any sum of the solutions of Maxwell's equations is again their solution. This statement is known as the *superposition principle.* It is extremely useful in many cases. Combined with the uniqueness theorem for solutions of Maxwell's equations, it justifies any series or sum solution.

**13.4.6 The uniqueness of the solution.** Only one solution of Maxwell's equations satisfies any set of prescribed boundary conditions. Thus, if in *any way* a set of field functions is found which satisfy simultaneously Maxwell's equations and the prescribed boundary conditions, this set is the only one possible.

We shall not prove this important theorem, for the proof does not add anything to the clarity of its meaning. But it is necessary to remember and to understand its statement.

*Problems referring to Section 13.4:* 13.T20 to 13.T28, 13.G6 to 13.G11, 13.N2.

## PROBLEMS

### Test problems

**13.T1.** Write the full set of Maxwell's equations for the special case of a static electric field, assuming that the dielectric is linear, but inhomogeneous.

**13.T2.** Write the full set of Maxwell's equations for the special case of a static electric field created by the charges on a set of conducting bodies situated in a vacuum.

**13.T3.** Write the full set of Maxwell's equations for the special case of a steady current flow in a homogeneous conductor of conductivity $\sigma$, with no impressed electric field.

**13.T4.** Write the full set of Maxwell's equations for the special case of a steady current flow in an inhomogeneous poor dielectric, with impressed electric field $\mathbf{E}_i$ present.

**13.T5.** Write the full set of Maxwell's equations for the case of a static magnetic field in a linear medium of permeability $\mu$, produced by a steady current flow.

**13.T6.** Write the full set of Maxwell's equations for the case of a static magnetic field produced by a permanent magnet of magnetization $\mathbf{M}$ (a function of position).

**13.T7.** Write the full set of Maxwell's equations for the case of a static magnetic field produced by both steady currents and magnetized matter, if the medium is not linear.

**13.T8.** Write the full set of Maxwell's equations for the special case of a quasi-stationary electromagnetic field, produced by quasi-stationary currents in nonferromagnetic conductors.

**13.T9.** Write Maxwell's equations for an arbitrary electromagnetic field in a vacuum, no free charges being present.

**13.T10.** Write Maxwell's equations for an arbitrary electromagnetic field in a homogeneous perfect dielectric of permittivity $\varepsilon$ and permeability $\mu$.

**13.T11.** Write the full set of Maxwell's equations *in scalar form* in the rectangular coordinate system. Note that *eight* scalar simultaneous partial differential equations result. Write these equations nearly, and attach them to page 455.

**13.T12.** Write the full set of Maxwell's equations *in scalar form* in the cylindrical coordinate system. Attach the sheet with the equations to page 455.

**13.T13.** Write the full set of Maxwell's equations *in scalar form* in the spherical coordinate system. Attach the sheet with the equations to page 455.

**13.T14.** Write Maxwell's equations in scalar form for the particular case of an electromagnetic field in a vacuum ($\mathbf{J} = 0$, $\rho = 0$), if the field vectors are only functions of the Cartesian coordinate $z$ and time $t$.

**13.T15.** Write Maxwell's equations in scalar form for the particular case of an axially symmetrical system with dependence of the field vectors only on the cylindrical coordinate $r$ and time $t$. Assume that $\mathbf{J} = J_z\mathbf{u}_z$ and $\rho = 0$.

**13.T16.** Repeat the preceding problem for $\mathbf{B} = B_z\mathbf{u}_z$.

**13.T17.** Write Maxwell's equations in complex form for an arbitrary electromagnetic field in a very good conductor, of conductivity $\sigma$ and permeability $\mu$.

**13.T18.** Write Maxwell's equations in complex form for a quasi-stationary electromagnetic field.

**13.T19.** Write Maxwell's equations in complex form for an arbitrary electromagnetic field in a perfect dielectric of permittivity $\varepsilon$ and permeability $\mu$, no free charges being present.

**13.T20.** The current intensity through a resistor of resistance $R$ is $I$. What is the flux of the Poynting vector through any closed surface enclosing the resistor? [*Hint:* specialize Eq. (13.4.14) for this problem.]

**13.T21.** A capacitor, of capacitance $C$, is charged with a charge $Q$. What is the flux of the Poynting vector through any surface enclosing the capacitor if the charge $Q$ (1) is constant, and (2) varies according to the law $Q = Q_m \cos \omega t$?

**13.T22.** A coil, of inductance $L$, carries a current $i(t)$. What is the flux of the Poynting vector through any surface enclosing the coil?

**13.T23.** A direct-current generator of electromotive force $\mathscr{E}$ is open-circuited. What is the flux of the Poynting vector through any surface enclosing the generator?

**13.T24.** Repeat the preceding problem if a current $i(t)$ flows through the generator, and if the internal resistance of the generator is $R_g$.

**13.T25.** The largest dimension of a coil at a very high frequency is of the order $(\omega\sqrt{\varepsilon\mu})^{-1}$. Is it possible at such high frequencies to define the inductance of the coil in the same way as in a quasi-stationary case?

**13.T26.** The length of a very long power transmission line, operating at a frequency $f$, is equal to 0.5 of $(\omega\sqrt{\varepsilon\mu})^{-1}$. Is this a quasi-stationary system?

**13.T27.** A parallel-plate capacitor has plates of linear dimensions comparable with $(\omega\sqrt{\varepsilon\mu})^{-1}$, where $\omega$ is the operating angular frequency. Is it possible to determine the capacitance of such a capacitor in the same way as in the static and quasi-static case?

**13.T28.** An electrical network operates at a very high frequency $f$. The largest linear dimension of the network is $2(\omega\sqrt{\varepsilon\mu})^{-1}$. Are Kirchhoff's laws applicable in this case for analyzing the network?

## General problems

**13.G1.** Prove that the flux of the sum $(\mathbf{J} + \partial\mathbf{D}/\partial t)$ through any closed surface is zero.

*Proof.* According to Eq. (13.2.12), the flux of $(\mathbf{J} + \partial\mathbf{D}/\partial t)$ is equal through all surfaces limited by a common contour. If we apply Eq. (13.2.12) to the two surfaces $S_1$ and $S_2$ shown in Fig. 13.G1, we have

$$\oint_C \mathbf{H}\cdot d\mathbf{l} = \int_{S_1}\left(\mathbf{J} + \frac{\partial\mathbf{D}}{\partial t}\right)\cdot d\mathbf{S} = \int_{S_2}\left(\mathbf{J} + \frac{\partial\mathbf{D}}{\partial t}\right)\cdot d\mathbf{S}. \tag{a}$$

Surfaces $S_1$ and $S_2$ form a closed surface. According to the adopted convention, the positive normal to a closed surface is always directed outward. Hence, we have to change $d\mathbf{S}$ to $-d\mathbf{S}$ in the last integral, if it is considered with respect to the outward positive normal. So we obtain

$$\oint_{S_1+S_2}\left(\mathbf{J} + \frac{\partial\mathbf{D}}{\partial t}\right)\cdot d\mathbf{S} = 0. \tag{b}$$

The sum $(\mathbf{J} + \partial\mathbf{D}/\partial t)$ is frequently referred to as the *total current density*, but this term is misleading, since $\varepsilon_0\,\partial\mathbf{E}/\partial t$ does *not* represent a real current. It follows from Eq. (b) that the lines of this total current always close upon themselves.

Equation (b) follows also from the continuity equation and from the generalized form of Gauss' law,

$$\operatorname{div}\mathbf{J} + \partial\rho/\partial t = 0 \qquad \text{and} \qquad \operatorname{div}\mathbf{D} = \rho,$$

from which

$$\operatorname{div}(\mathbf{J} + \partial\mathbf{D}/\partial t) = 0. \tag{c}$$

This is the differential form of Eq. (b).

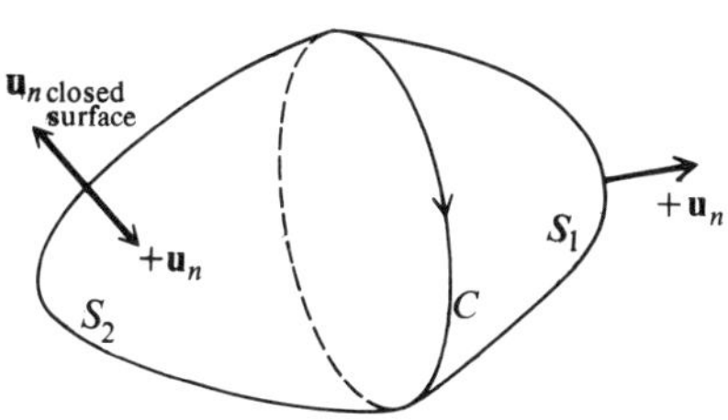


Figure 13.G1

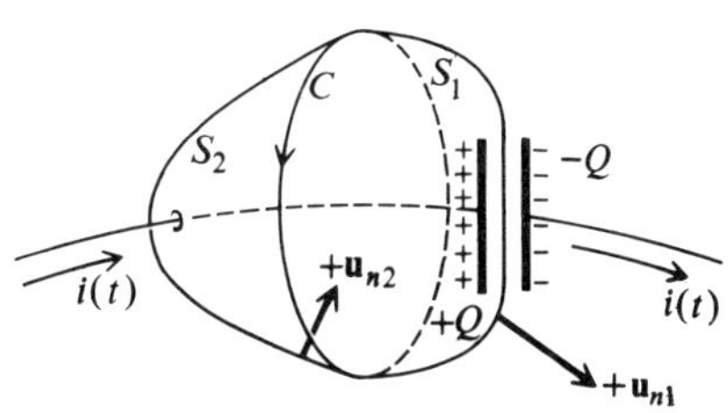


Figure 13.G2

**13.G2.** A current $i(t) = I_m \cos \omega t$ flows through the leads of a parallel-plate capacitor, the area of the plates being $S$ and the distance between them $d$. Prove that the displacement current through the capacitor is exactly $I_m \cos \omega t$. Ignore the fringing effect.

*Proof.* The proof follows directly from the generally valid equation (a) of the preceding problem and Fig. 13.G2. Note that the positive normals to $S_1$ and $S_2$ are determined by the right-hand screw rule with respect to the indicated positive direction around $C$. Through $S_2$ the flux of $(\mathbf{J} + \partial\mathbf{D}/\partial t)$ is simply $i(t)$, since there $\mathbf{D} \simeq 0$. $\mathbf{J}$ is zero at all points of $S_1$, but not $\partial\mathbf{D}/\partial t$. The proof is simple in the light of these facts.

In this case, proof is also possible without the use of the general conclusion, since the electric field in the capacitor is known. The charge on the capacitor plates is given by

$$Q = \int i \, dt = \int I_m \cos \omega t \, dt = \frac{I_m}{\omega} \sin \omega t.$$

Hence the electric field intensity in the capacitor is

$$E = \frac{Q}{\varepsilon S} = \frac{I_m}{\varepsilon \omega S} \sin \omega t,$$

and the displacement current through the capacitor is

$$S \frac{\partial D}{\partial t} = S\varepsilon \frac{\partial E}{\partial t} = I_m \cos \omega t.$$

**13.G3.** A spherically symmetrical charge distribution disperses under the influence of mutually repulsive forces. Suppose that the charge density $\rho(r, t)$, as a function of the distance $r$ from the center of symmetry and of time, is known. Prove that the total current density at any point is zero.

*Solution.* Consider a sphere $S$ of radius $R$. The total convection current through $S$ (that is, the real current) is

$$i_c(R, t) = -\frac{\partial Q(R, t)}{\partial t} = -\frac{\partial}{\partial t} \int_0^R \rho(r, t)\, 4\pi r^2 \, dr,$$

so that the density of this current

$$J_c(R, t) = \frac{i_c}{4\pi R^2} = -\frac{1}{R^2} \int_0^R \frac{\partial \rho(r, t)}{\partial t} r^2 \, dr.$$

The density $J_d$ of the displacement current is given by

$$J_d = \frac{\partial D}{\partial t} = \frac{\partial}{\partial t} \frac{Q(R, t)}{4\pi R^2} = \frac{1}{R^2} \int_0^R \frac{\partial \rho(r, t)}{\partial t} r^2 dr.$$

This is one of the rare examples in which the macroscopic magnetic field is not created by the electric current. The magnetic field is zero since it is produced by *both* the real and the displacement current densities, and the sum of the two is zero at all points.

**13.G4.** Prove that the two divergence Maxwell's equations can be deduced from the two curl equations and the continuity equation.

*Proof.* The divergence of a curl of any vector function is zero [see Eq. (8.9.1)]. Taking the divergence of both sides of the two Maxwell's curl equations, we obtain

$$-\text{div}\,(\partial \mathbf{B}/\partial t) = -\partial(\text{div}\,\mathbf{B})/\partial t = 0, \tag{a}$$

$$\text{div}\,\mathbf{J} + \text{div}\,(\partial \mathbf{D}/\partial t) = \text{div}\,\mathbf{J} + \partial(\text{div}\,\mathbf{D})/\partial t = 0. \tag{b}$$

According to Eq. (a), div **B** cannot be a function of time at any point. Now, either **B** has changed in the past or will change in the future, and even be zero, since physical systems which remain the same forever do not exist. So div **B** cannot be a function of time, and it follows that it must be zero always if it was zero at any time.

With the continuity equation, Eq. (b) can be modified to read

$$\partial(\text{div}\,\mathbf{D} - \rho)/\partial t = 0. \tag{c}$$

By exactly the same argument as given above, we conclude that div **D** must be equal to $\rho$, since at least once there were no fields or charges at the point considered.

***13.G5.** Small-scale models are used often in engineering practice; in particular, they are also used in electrical engineering. Starting from Maxwell's equations, derive the necessary conditions for an electromagnetic field in a small-scale model to be similar to the field in a real, $n$ times larger, object. (These conditions are frequently referred to as the conditions of the *electrodynamic similitude*.)

*Solution.* Suppose we made a small-scale model *geometrically* similar to a real system. An example of a full-scale system and a small-scale model is shown in Fig. 13.G5. The parameters $\varepsilon$, $\mu$, and $\sigma$ of the full-scale system are *known* functions of the coordinates. Frequency $f = \omega/2\pi$ of the generators in the full-scale system is also known. It is necessary to determine the parameters $\varepsilon'$, $\mu'$, and $\sigma'$ as functions of the coordinates in the small-scale system, and the corresponding frequency $f' = \omega'/2\pi$ of generators so that the electromagnetic fields in the two cases are similar.

The two Maxwell's curl equations in the complex form for the full-scale system have the form

$$\text{curl}\,\mathbf{E} = -j\omega\mu\mathbf{H}, \tag{a}$$

$$\text{curl}\,\mathbf{H} = \sigma\mathbf{E} + j\omega\varepsilon\mathbf{E}. \tag{b}$$

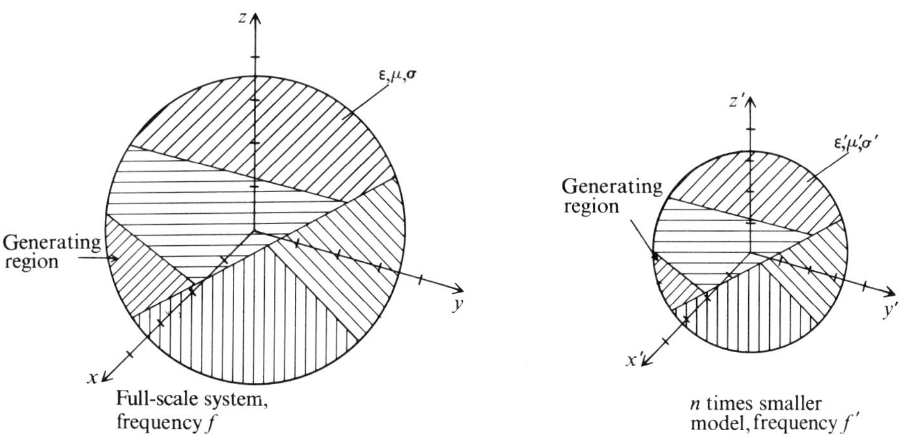


Figure 13.G5

In the small-scale system all the lengths are $n$ times smaller. In order that the curl at the respective points of the model be the same as in the real system, the length coordinates, with respect to which differentiation is performed, must be $n$ times as small. For example, $\partial E'_x/\partial y' = \partial E'_x/\partial(y/n) = n\,\partial E'_x/\partial y$, which means that the two Maxwell's equations for the model must have the form

$$n \operatorname{curl} \mathbf{E}' = -j\omega'\mu'\mathbf{H}', \tag{c}$$

$$n \operatorname{curl} \mathbf{H}' = \sigma'\mathbf{E}' + j\omega'\varepsilon'\mathbf{E}'. \tag{d}$$

The differentiations implied by the curl in these equations is performed with respect to coordinates of the same "size" as in (a) and (b). The boundary conditions in the two cases are identical. Hence the solutions will also be equal, provided that pairs (a), (b) and (c), (d) of the basic equations are equal. This is the case if

$$\omega'\mu' = n\omega\mu, \qquad \sigma' = n\sigma, \qquad \text{and} \qquad \omega'\varepsilon' = n\omega\varepsilon. \tag{e}$$

In practice, permittivity $\varepsilon$ and permeability $\mu$ cannot have any desired value. Models are therefore usually made of the same dielectric, i.e. $\varepsilon' = \varepsilon$ and $\mu' = \mu$. In that case, *both* the first and the third equations are satisfied if $\omega' = n\omega$. This means that the *frequency of the generators in an* n *times smaller model must be* n *times as large.*

The second of conditions in (e) cannot be satisfied in the general case. For example, if the conductors in the full-scale system are made of silver, it is impossible to fulfill this condition. At room temperature there is no better conductor than silver. Practical restrictions therefore make it impossible to obtain the exact electrodynamic similitude in the small-scale model. Fortunately, this obstacle is important only in special cases, where conductivity is a decisive parameter. Whenever this is not the case, the impossibility of fulfilling the condition $\sigma' = n\sigma$ is not important. For obtaining a practically similar system, it is sufficient to use an $n$ times higher frequency in the $n$ times smaller model.

**13.G6.** A lossless coaxial cable, of conductor radii $a$ and $b$, carries a steady current of intensity $I$. The potential difference between the cable conductors is $V$. Prove that the flux of the Poynting vector through any cross-section of the cable is $VI$, using the known expressions for vectors **E** and **H** in the cable. Sketch the dependence of the magnitude of the Poynting vector on the distance $r$ $(a < r < b)$ of a point from the cable axis.

**13.G7.** Suppose that, instead of the instantaneous values of the electric and magnetic field intensities, their *complex* amplitudes are known. How is it possible to determine the *average* value of the Poynting vector in this case?

*Solution.* Suppose that the complex field vectors $\mathbf{E}_c$ and $\mathbf{H}_c$ are given by their real and imaginary parts. That is,

$$\mathbf{E}_c = \mathbf{E}_r + j\mathbf{E}_i, \qquad \mathbf{H}_c = \mathbf{H}_r + j\mathbf{H}_i. \tag{a}$$

The real sinusoidally time-varying field vectors **E** and **H** are, according to the pattern of Eq. (13.3.31),

$$\begin{aligned}\mathbf{E} &= \operatorname{Re}\{\mathbf{E}_c e^{j\omega t}\} = \operatorname{Re}\{(\mathbf{E}_r + j\mathbf{E}_i)(\cos\omega t + j\sin\omega t)\}\\ &= \operatorname{Re}\{(\mathbf{E}_r\cos\omega t - \mathbf{E}_i\sin\omega t) + j(\mathbf{E}_r\sin\omega t + \mathbf{E}_i\cos\omega t)\}\\ &= \mathbf{E}_r\cos\omega t - \mathbf{E}_i\sin\omega t,\end{aligned} \tag{b}$$

and similarly

$$\mathbf{H} = \mathbf{H}_r\cos\omega t - \mathbf{H}_i\sin\omega t. \tag{c}$$

The *instantaneous* value of the Poynting vector is hence

$$\mathscr{P} = \mathbf{E} \times \mathbf{H} = \mathbf{E}_r \times \mathbf{H}_r \cos^2\omega t + \mathbf{E}_i \times \mathbf{H}_i \sin^2\omega t - \tfrac{1}{2}(\mathbf{E}_r \times \mathbf{H}_i + \mathbf{E}_i \times \mathbf{H}_r) \sin 2\omega t, \tag{d}$$

since $\sin \omega t \cos \omega t = \sin 2\omega t/2$. The *average* value of the Poynting vector in one cycle is obtained if Eq. (d) is multiplied by $dt$, integrated from $t = 0$ to $t = T = 2\pi/\omega$, and the result divided by $T$. The average value of $\cos^2\omega t$ and $\sin^2\omega t$ is 1/2, and the average value of $\sin 2\omega t$ is zero. Hence

$$\mathscr{P}_{\text{av}} = (\mathbf{E} \times \mathbf{H})_{\text{av}} = \tfrac{1}{2}(\mathbf{E}_r \times \mathbf{H}_r + \mathbf{E}_i \times \mathbf{H}_i). \tag{e}$$

Now, according to Eq. (a) the average value given in Eq. (e) of the Poynting vector can also be obtained directly from the complex field vectors $\mathbf{E}_c$ and $\mathbf{H}_c$ as follows:

$$\mathscr{P}_{\text{av}} = \tfrac{1}{2}\text{Re}\{\mathbf{E}_c \times \mathbf{H}_c^*\} = \tfrac{1}{2}\text{Re}\{(\mathbf{E}_r + j\mathbf{E}_i)(\mathbf{H}_r - j\mathbf{H}_i)\} = \tfrac{1}{2}(\mathbf{E}_r \times \mathbf{H}_r + \mathbf{E}_i \times \mathbf{H}_i). \tag{f}$$

The asterisk as superscript in $\mathbf{H}_c^*$ denotes the complex conjugate to $\mathbf{H}_c$. That is, $\mathbf{H}_c^* = \mathbf{H}_r \mp j\mathbf{H}_i$ if $\mathbf{H}_c = \mathbf{H}_r \pm j\mathbf{H}_i$.

Therefore, for determining the *average* value of the Poynting vector at a point it is sufficient to find, from the complex field vectors $\mathbf{E}_c$ and $\mathbf{H}_c$, the product

$$\boxed{\mathscr{P}_c = \tfrac{1}{2}\mathbf{E}_c \times \mathbf{H}_c^*} \qquad \text{(definition of the complex Poynting vector).} \tag{g}$$

Its *real* part is exactly equal to the *average* value of the Poynting vector. $\mathscr{P}_c$ thus defined is referred to as the *complex Poynting vector*. Note that this conclusion cannot be deduced from Eq. (a) without knowing Eq. (e). Another very frequent error is to forget that $\mathbf{H}_c^*$, rather than $\mathbf{H}_c$, enters this expression.

**13.G8.** Derive the equation analogous to Eq. (13.4.14), in the case of sinusoidally time-varying fields, in complex notation.

*Solution.* Note that in the complex form of Eq. (13.4.14) the complex Poynting vector defined in the preceding problem must enter. Let us therefore take Eq. (13.3.24) and the *complex conjugate* of Eq. (13.3.25) as our starting point:

$$\text{curl}\,\mathbf{E}_c = -j\omega\mu\mathbf{H}_c, \qquad \text{curl}\,\mathbf{H}_c^* = \mathbf{J}_c^* - j\omega\varepsilon\mathbf{E}_c^*. \tag{a}$$

Multiplying scalarly the first equation by $\mathbf{H}_c^*$, the second by $\mathbf{E}_c$, and subtracting the results we have

$$\mathbf{H}_c^* \cdot \text{curl}\,\mathbf{E}_c - \mathbf{E}_c \cdot \text{curl}\,\mathbf{H}_c^* = \text{div}\,(\mathbf{E}_c \times \mathbf{H}_c^*) = -j\omega(\mu\mathbf{H}_c \cdot \mathbf{H}_c^* - \varepsilon\mathbf{E}_c \cdot \mathbf{E}_c^*) - \mathbf{E}_c \cdot \mathbf{J}_c^*. \tag{b}$$

Now, $\mathbf{H}_c \cdot \mathbf{H}_c^* = H^2$, $\mathbf{E}_c \cdot \mathbf{E}_c^* = E^2$, $H$ and $E$ representing the real maximal values (amplitudes) of the two sinusoidally time-varying vectors at a point. Also, from $\mathbf{J}_c = \sigma(\mathbf{E}_c + \mathbf{E}_{ic})$ we have that $\mathbf{E}_c = \mathbf{J}_c/\sigma - \mathbf{E}_{ic}$. In the light of these remarks, and after dividing it by 2, (b) becomes

$$\text{div}\left(\frac{1}{2}\mathbf{E}_c \times \mathbf{H}_c^*\right) = -2j\omega\left(\frac{\mu H^2}{4} - \frac{\varepsilon E^2}{4}\right) - \frac{J^2}{2\sigma} + \frac{1}{2}\mathbf{E}_{ic} \cdot \mathbf{J}_c^*. \tag{c}$$

The first term on the right is multiplied and divided by two for convenience. It will be shown that $\mu H^2/4$, where $H$ is the *amplitude* of $\mathbf{H}$, represents the average density of the magnetic energy, and similarly for $\varepsilon E^2/4$. Multiplying Eq. (c) by the volume element $dv$, integrating over any volume $v$, and transforming the volume integral on the left-hand side into a surface integral, using the divergence theorem, we finally obtain

$$\int_v \frac{1}{2}\mathbf{E}_{ic}\cdot\mathbf{J}_c^*\,dv = \int_v \frac{J^2}{2\sigma}\,dv + 2j\omega\int_v\left(\frac{\mu H^2}{4} - \frac{\varepsilon E^2}{4}\right)dv + \oint_S \frac{1}{2}(\mathbf{E}_c\times\mathbf{H}_c^*)\cdot d\mathbf{S}. \tag{d}$$

This is the required expression corresponding to Eq. (13.4.14) in the case of complex notation.

This complex equation stands for the following two real equations:

$$\int_v \operatorname{Re}\left\{\frac{1}{2}\mathbf{E}_{ic}\cdot\mathbf{J}_c^*\right\}dv = \int_v \frac{J^2}{2\sigma}\,dv + \oint_S \operatorname{Re}\left\{\frac{1}{2}(\mathbf{E}_c\times\mathbf{H}_c^*)\right\}\cdot d\mathbf{S}, \tag{e}$$

$$\int_v \operatorname{Im}\left\{\frac{1}{2}\mathbf{E}_{ic}\cdot\mathbf{J}_c^*\right\}dv = 2\omega\int_v\left(\frac{\mu H^2}{4} - \frac{\varepsilon E^2}{4}\right)dv + \oint_S \operatorname{Im}\left\{\frac{1}{2}(\mathbf{E}_c\times\mathbf{H}_c^*)\right\}\cdot d\mathbf{S}. \tag{f}$$

A simple physical interpretation of these two equations is possible.

Suppose that the instantaneous value of, say, $\mathbf{J}$ is given by $J\cos(\omega t + \theta_J)\,\mathbf{u}_J$. The average value of the *square* of $\mathbf{J}$ in one cycle is (remember that $T = 2\pi/\omega$)

$$(\mathbf{J}^2)_{\text{av}} = \frac{1}{T}\int_0^T J^2\cos^2(\omega t + \theta_J)\,dt = J^2/2. \tag{g}$$

Consequently, the first integral on the right-hand side of Eq. (e) represents the *average* power in $v$ transformed into heat. According to Eq. (f) of the preceding problem, the second integral on the right-hand side of Eq. (e) is the average power flowing out of $S$. Since the average power delivered to the sinusoidally time-varying electric and magnetic fields inside $v$ is zero, the left-hand side of Eq. (e) represents the average power of all the generators inside $v$. This power is often referred to as the active power of the generators,

$$\boxed{P_{g\,\text{active}} = \int_v \operatorname{Re}\left\{\frac{1}{2}\mathbf{E}_{ic}\cdot\mathbf{J}_c^*\right\}dv} \quad \text{(definition of the active power of a generator).} \tag{h}$$

The first term on the right-hand side of Eq. (f) represents the *difference* of the *average* magnetic and electric energies stored in the electromagnetic field inside $v$. The medium is assumed to be linear, so that this energy travels back and forth from the generator to the field. The other two terms in Eq. (f) consequently have the following meanings. The left-hand side represents the total average power of the generators inside $v$ which they periodically *exchange* with the system. This power is termed the *reactive power*:

$$\boxed{P_{g\,\text{reactive}} = \int_v \operatorname{Im}\left\{\frac{1}{2}\mathbf{E}_{ic}\cdot\mathbf{J}_c^*\right\}dv} \quad \text{(definition of the reactive power of a generator).} \tag{i}$$

By virtue of the law of conservation of energy, the second term on the right of Eq. (f) represents the average value of the power crossing $S$ and being later completely returned to $v$.

If the surface integral in Eq. (e) is nonzero, it is often said that the outside region is an *active load* for the generators inside $v$. Similarly, if the surface integral in Eq. (f) is nonzero, it is said that the outside region is a *reactive load* for the generators in $v$. Generally they are both nonzero, and the outside region is both an active and a reactive load for the generators inside $v$.

**13.G9.** One end of the lossless strip-line of Fig. 13.9 is connected to a generator of sinusoidal electromotive force, of angular frequency $\omega$. A capacitor, of capacitance $C$, is connected to the other end. The frequency is low enough for the system to be regarded as quasi-stationary. Ignoring the edge effect, apply Eqs. (e) and (f) of the preceding problem to the volumes enclosed by both surfaces $S$ and $S'$.

*Solution.* The charge on the capacitor plates $Q(t) = C\mathscr{E}(t)$. In complex notation $Q = C\mathscr{E}$ (for simplicity we omit the subscript $c$). The current intensity through the line $i(t) = dQ(t)/dt$. In complex notation this is $I = j\omega Q$. The *complex* amplitudes of the **E** and **H** fields between the strips are

$$E = \mathscr{E}/d, \qquad H = I/b = j\omega Q/b = j\omega C\mathscr{E}/b.$$

The flux of the complex Poynting vector through the crosshatched area of the surface $S$ is hence

$$\oint_S \frac{1}{2}(\mathbf{E} \times \mathbf{H}^*)\cdot d\mathbf{S} = -\frac{1}{2}EH^*bd = -\frac{1}{2}\frac{\mathscr{E}}{d}\frac{-j\omega C\mathscr{E}}{b}bd = 2j\omega\frac{C\mathscr{E}^2}{4}.$$

The expression $C\mathscr{E}^2/4$ is precisely the average energy stored in the electric field of the capacitor. (Explain why $C\mathscr{E}^2/4$, and not $C\mathscr{E}^2/2$, as in the static case!)

The flux of the complex Poynting vector through $S'$ has the same magnitude, but is of the opposite sign. Thus a capacitor appears as a *negative* reactive load to a generator when complex notation is used.

**13.G10.** Repeat the preceding problem, given that the line is terminated by a coil of inductance $L$ and negligible resistance.

*Solution.* In this case, $\mathscr{E}(t) = L\,di(t)/dt$, which in complex notation becomes $\mathscr{E} = j\omega LI$. Hence the complex amplitudes of the electric and magnetic fields between the strips are

$$E = \mathscr{E}/d, \qquad H = I/b = \mathscr{E}/j\omega Lb,$$

so that

$$\oint_S \frac{1}{2}(\mathbf{E} \times \mathbf{H}^*)\cdot d\mathbf{S} = -\frac{1}{2}EH^*bd = -\frac{1}{2}\frac{\mathscr{E}}{d}\frac{\mathscr{E}}{-j\omega Lb}bd = -2j\omega\frac{LI^2}{4}.$$

$LI^2/4$ is precisely the average energy stored in the magnetic field of the coil. (Explain why $LI^2/4$, and not $LI^2/2$, as in the static case!)

The flux of the complex Poynting vector through the surface $S'$ enclosing the generator is simply of the opposite sign. A coil appears therefore as a *positive* reactive load to a generator when complex notation is used.

**13.G11.** Repeat the two preceding problems, given that the line is a lossless coaxial line of conductor radii $a$ and $b$.

## Numerical problems

**13.N1.** Compare the amplitudes of vectors **J** and $\partial\mathbf{D}/\partial t$ in the conductors of Table 6.1, for frequencies 60 Hz, $10^4$ Hz, $10^8$ Hz, and $10^{10}$ Hz. Assume $\varepsilon = \varepsilon_0$ for all metallic conductors. For sea water assume $\varepsilon_r = 81$ and $\sigma = 4$ S/m, and for earth $\varepsilon_r = 10$ and $\sigma = 10^{-2}$ S/m.

**13.N2.** Suppose a system is regarded as approximately quasi-stationary if its largest dimension $d$ satisfies the inequality $d\omega\sqrt{\varepsilon\mu} \leqslant 0.1$. Determine the largest value of $d$ thus defined for the electrodynamic systems in a vacuum if the frequency of the generators is 60 Hz, $10^4$ Hz, and $10^{10}$ Hz.

**Suggested reading.** The list progresses towards more advanced texts. For full bibliographical description see Bibliography.

1. JORDAN, E. C., and K. G. BALMAIN, *Electromagnetic Waves and Radiating Systems*, Chapters 4 and 6.
2. RAMO, S., J. R. WHINNERY, and T. VAN DUZER, *Fields and Waves in Communication Electronics*, Chapter 4.

CHAPTER 14

# SKIN EFFECT AND CIRCUIT CONCEPTS

## 14.1 INTRODUCTION

In this chapter Maxwell's equations are used for analyzing two practical problems, the first of which is the distribution of time-varying current and magnetic flux density in conductors.

We have seen in Chapter 6 that a steady current is distributed throughout a conductor through which it flows. If the conductor is cylindrical and of a uniform cross-section, the distribution of steady current is also uniform, but this is *not* so if the current is time-varying. We shall see that time-varying currents have a tendency to concentrate near the surfaces of conductors. If the frequency is very high, the current flows through a very thin layer near the conductor surfaces, practically over the surfaces themselves. This is known as the *skin effect*, a term which we use when referring to the nonuniform distribution of time-varying currents in general. Skin effect is of a considerable practical importance: for example, we do not use a thick solid conductor if time-varying currents are used. A hollow conductor of some reasonable thickness may well be equally good. At very high frequencies, silver-coated conductors are used to minimize losses. Almost all the current then flows through the layer of that best available conductor.

A time-varying magnetic field has the same property: the magnetic flux density of time-varying magnetic fields in conducting materials is greatest near their surfaces. Several practical consequences result from this fact. For example, solid ferromagnetic cores are not used for alternating current electrical motors, generators, and transformers. Instead, laminated cores made of thin, mutually insulated sheets are used. At very high frequencies even these are inadequate, and cores are made of pressed ferromagnetic powder with mutually insulated particles.

The cause of both phenomena is the same. A time-varying magnetic field "produces" a time-varying electric field, which in turn creates secondary time-varying currents and a secondary magnetic field. The total electric and magnetic fields are the sums of the two components. Thus, the skin effect is essentially an electromagnetic induction phenomenon.

The second problem analyzed in this chapter is that of electrical networks with time-varying currents. We shall see that the conventional network theory is only an approximation. This approximation is extremely good in most cases, but an engineer is expected to understand the limitations implicit in circuit theory.

## 14.2 SKIN EFFECT IN A STRIP CONDUCTOR

Let us consider a very long straight strip conductor of width $a$ and thickness $d$ (Fig. 14.1). Let the conductivity of the strip be $\sigma$, and its permeability $\mu$. We shall suppose that $a \gg d$, so that the effect of the finite width of the strip can be ignored. Assume that a sinusoidal current $i(t) = I_m \cos \omega t$ flows through the strip, the reference direction of the current being the direction of the $z$ axis (see Fig. 14.1). Our task is to determine the distribution of the current as a function of the frequency $f = \omega/2\pi$, $\sigma$, and $\mu$.

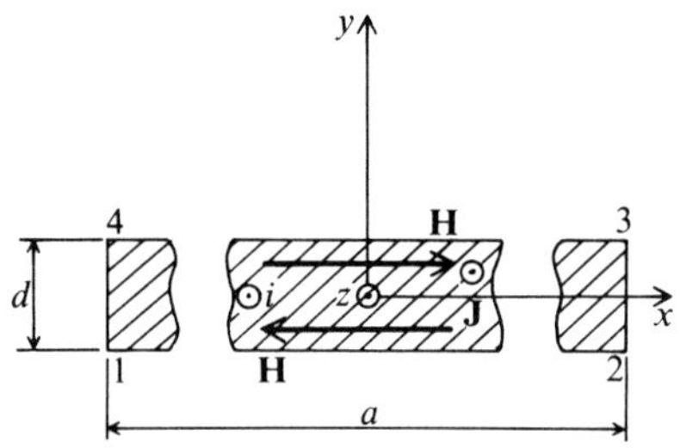


**Fig. 14.1** A strip conductor.

In all good conductors, the density $\partial\mathbf{D}/\partial t$ of the displacement current is much smaller than the density $\mathbf{J}$ of the true current (see Problem 13.N1). The Maxwell's curl equations (13.3.6) and (13.3.7), omitting the term $\partial\mathbf{D}/\partial t$, are

$$\text{curl } \mathbf{E} = -\partial\mathbf{B}/\partial t, \qquad \text{curl } \mathbf{H} = \mathbf{J}. \tag{14.2.1}$$

Now, $\mathbf{B} = \mu\mathbf{H}$ and $\mathbf{J} = \sigma\mathbf{E}$. Since we are considering the sinusoidally time-varying current and fields, the complex forms of Eqs. (14.2.1) will be used. So we have

$$\text{curl } \mathbf{J}_c = -j\omega\mu\sigma\mathbf{H}_c, \qquad \text{curl } \mathbf{H}_c = \mathbf{J}_c. \tag{14.2.2}$$

(The subscript $c$, meaning complex, is added here for clarity, but it will be omitted hereafter.) The current density is in the $z$ direction, $\mathbf{J} = J_z\mathbf{u}_z$. The magnetic field intensity $\mathbf{H}$ is consequently parallel to the $x$ axis, $\mathbf{H} = H_x\mathbf{u}_x$. If we ignore the edge effect, both $J_z$ and $H_x$ are functions of the coordinate $y$ only. Taking into account Eq. (8.8.8) for curl in a rectangular coordinate system, Eqs. (14.2.2) become

$$\partial J_z/\partial y = -j\omega\mu\sigma\, H_x, \qquad -\partial H_x/\partial y = J_z. \tag{14.2.3}$$

Differentiating the first equation with respect to $y$, and using the value of $\partial H_x/\partial y$ from the second equation, we have

$$\frac{d^2 J_z}{dy^2} = j\omega\mu\sigma\, J_z. \tag{14.2.4}$$

The partial derivative signs are replaced by ordinary derivative signs since the complex amplitude $J_z$ of the current density is a function of the coordinate $y$ only. If we put

$$j\omega\mu\sigma = K^2, \tag{14.2.5}$$

Eq. (14.2.4) becomes

$$\frac{d^2 J_z}{dy^2} = K^2 J_z. \tag{14.2.6}$$

Its solution is of the form

$$J_z(y) = J_1 e^{Ky} + J_2 e^{-Ky}. \tag{14.2.7}$$

$J_1$ and $J_2$ are integration constants which can be determined as follows. First, in the coordinate system adopted in Fig. 14.1, $J_z(y)$ is an even function of $y$, that is, $J_z(-y) = J_z(y)$. This is possible only if $J_1 = J_2$. So

$$J_z(y) = J_1(e^{Ky} + e^{-Ky}). \tag{14.2.8}$$

From this we see that $2J_1 = J_z(0)$. Hence

$$J_z(y) = \tfrac{1}{2}J_z(0)\,(e^{Ky} + e^{-Ky}). \tag{14.2.9}$$

The complex constant $K$ is given by Eq. (14.2.5). Since $j = e^{j\pi/2}$,

$$\sqrt{j} = j^{1/2} = e^{j\pi/4} = \cos(\pi/4) + j\sin(\pi/4) = (1 + j)/\sqrt{2}. \tag{14.2.10}$$

The complex constant $K$ can be rewritten as

$$K = (1 + j)\sqrt{\omega\mu\sigma/2} = (1 + j)\,k, \qquad \text{where} \qquad k = \sqrt{\omega\mu\sigma/2}. \tag{14.2.11}$$

If we recall that $e^{\pm jky} = \cos ky \pm j \sin ky$, Eq. (14.2.9) becomes

$$\begin{aligned} J_z(ky) &= \tfrac{1}{2}\, J_z(0)\,(e^{ky} e^{jky} + e^{-ky} e^{-jky}) \\ &= \tfrac{1}{2}\, J_z(0)\,[\cos ky\,(e^{ky} + e^{-ky}) + j \sin ky\,(e^{ky} - e^{-ky})]. \end{aligned} \tag{14.2.12}$$

By definition of the hyperbolic sine and cosine functions,

$$e^{ky} + e^{-ky} = 2\cosh ky, \qquad e^{ky} - e^{-ky} = 2\sinh ky. \tag{14.2.13}$$

So we have

$$J_z(ky) = J_z(0)\,(\cos ky \cosh ky + j \sin ky \sinh ky). \tag{14.2.14}$$

This is the complex amplitude of the current density expressed in terms of the current density $J_z(0)$ at the middle of the strip. Assuming $J_z(0)$ to be real, the magnitude $|J_z(ky)|$ of the current density is

$$|J_z(ky)| = J_z(0)\sqrt{\cos^2 ky \cosh^2 ky + \sin^2 ky \sinh^2 ky}.$$

Recalling that

$$\cos^2 ky = 1 - \sin^2 ky, \qquad \text{and} \qquad \sinh^2 ky = \cosh^2 ky - 1, \tag{14.2.15}$$

we can write this in a shorter form:

$$|J_z(ky)| = J_z(0)\sqrt{\cosh^2 ky - \sin^2 ky}. \tag{14.2.16}$$

Figure 14.2 shows the ratio $|J_z(ky)|/J_z(0)$. We can see that, if $ky_{\max} = kd/2$ is much larger than unity, practically all the current flows near to the two strip surfaces. If $ky_{\max} = kd/2$ is much smaller than unity, the current is distributed practically uniformly over the cross-section of the strip. The resistance of the strip is then at its smallest. As a consequence of the skin effect, *the resistance of conductors with varying currents is always larger than with direct current*. In addition, the internal inductance is now *smaller*.

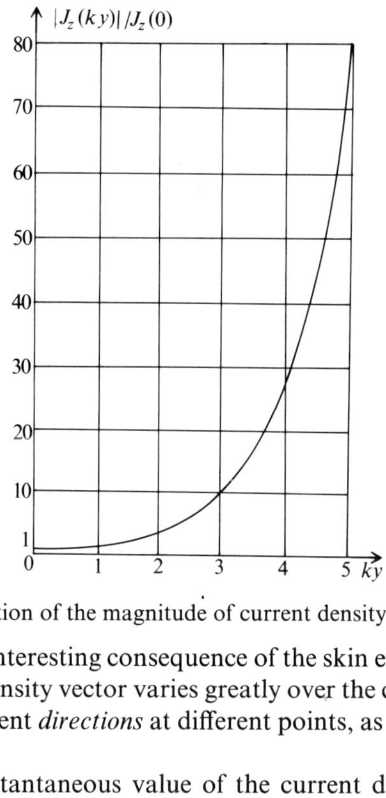


**Fig. 14.2** Distribution of the magnitude of current density in a strip conductor.

There is another interesting consequence of the skin effect. The instantaneous value of the current density vector varies greatly over the cross-section of the strip. It can even be in different *directions* at different points, as the following paragraph shows.

To obtain the instantaneous value of the current density from its complex magnitude, as given in Eq. (14.2.14), we follow the pattern indicated in Eq. (13.3.31). Assuming that $J_z(0)$ is real, we have

$$\begin{aligned} J_z(ky, t) &= \mathrm{Re}\{J_z(ky)\, e^{j\omega t}\} \\ &= J_z(0)\, \mathrm{Re}\{(\cos ky \cosh ky + j \sin ky \sinh ky)(\cos \omega t + j \sin \omega t)\} \\ &= J_z(0)\,(\cos ky \cosh ky \cos \omega t - \sin ky \sinh ky \sin \omega t). \end{aligned} \tag{14.2.17}$$

For example, at $t = 0$ this becomes

$$J_z(ky, 0) = J_z(0) \cos ky \cosh ky.$$

The ratio $J_z(ky, 0)/J_z(0)$ is shown in Fig. 14.3. At points where this ratio is negative, the direction of the current flow is in the direction opposite to the $z$ axis. This picture of course varies in time, but we can see that the distribution of the current is qualitatively different from what we perhaps expected. The current density in the strip is not only of a different magnitude, but also in different directions at different distances from the strip surfaces.

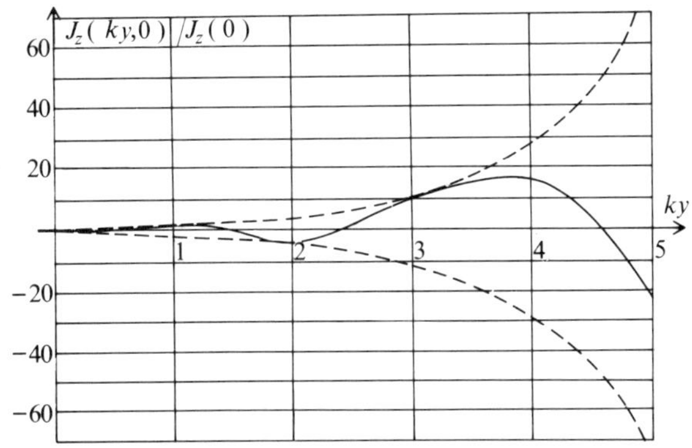


**Fig. 14.3** Instantaneous values of current density in a strip conductor.

Let us now consider the case when the skin effect is very pronounced, that is, when $ky_{\max} = kd/2 \gg 1$. Then, in Eq. (14.2.12), we can ignore $e^{-ky}$ with respect to $e^{ky}$ at all points where $ky \gg 1$. In fact, even when $ky = 3$, $e^{ky}/e^{-ky} = e^{2ky} = e^6 \simeq 400$! In this case Eq. (14.2.12) becomes

$$J_z(ky) \simeq \tfrac{1}{2} J_z(0)\, e^{ky} (\cos ky + j \sin ky) = \tfrac{1}{2} J_z(0)\, e^{ky} e^{jky} \qquad \text{(valid for } ky > 3). \tag{14.2.18}$$

Since $J_z(0)$ is very small, let us replace it by the magnitude $J_z(kd/2)$ of the current density at the strip surfaces,

$$J_z(kd/2) = \tfrac{1}{2} J_z(0)\, e^{kd/2} e^{jkd/2}.$$

Equation (14.2.18) can be thus written in the form

$$J_z(ky) \simeq J_z(kd/2)\, e^{k(y-d/2)} e^{jk(y-d/2)} \qquad \text{(valid for } ky > 3). \tag{14.2.19}$$

If we suppose that $y = (d/2 - 1/k)$, we see that the magnitude of the current density at the distance $1/k$ below the strip surface is equal to $1/e = 0.368$ times that on the surface. This is a convenient way of showing how pronounced the skin effect is. The ratio $1/k$ is usually designated by $\delta$, and is referred to as the *skin depth*:

$$\delta = \frac{1}{k} = \sqrt{\frac{2}{\omega\mu\sigma}} \qquad \text{(the skin depth).} \tag{14.2.20}$$

It is logical to expect this to be valid for any cross-section of the conductor if the skin effect is pronounced, and this is indeed the case. So Table 14.1, which shows the skin depths for several common conductors, is valid for all cases where the skin effect is pronounced, and not only for the strip conductor.

From this table we see that the skin effect for silver, copper, and aluminum is not very pronounced at 60 Hz; but iron conductors are not suitable at even such a low frequency.

**Table 14.1.** Skin depth $\delta$ (in mm) for some common conductors at different frequencies

| | $f$ = 60 Hz | 500 Hz | $10^4$ Hz | $10^6$ Hz |
|---|---|---|---|---|
| Silver | 8.48 | 2.94 | 0.66 | 0.066 |
| Copper | 8.63 | 2.99 | 0.68 | 0.068 |
| Aluminum | 10.1 | 3.81 | 0.85 | 0.085 |
| Iron ($\mu_r \simeq 1000$) | 0.65 | 0.22 | 0.05 | 0.005 |

*Problems referring to Section 14.2:* 14.T1 to 14.T9, 14.G1 to 14.G4, 14.N1 to 14.N4.

## 14.3 THE DISTRIBUTION OF A TIME-VARYING MAGNETIC FIELD IN IRON SHEETS*

Cores of alternating current electric machinery are always made of thin iron sheets, and the magnetic flux density is parallel to these sheets (see Fig. 14.4). If it were normal, a large flux would be encircled by imaginary contours lying in the sheets, and large electromotive forces would result, producing large secondary, "eddy" currents. This would have two equally bad effects. Firstly, the losses in the cores would be large. Secondly, the skin effect would be very pronounced, so that one part of the core only would be used for the magnetic circuit.

Let us consider a thin sheet, of conductivity $\sigma$, permeability $\mu$, and thickness $d$, and suppose that the amplitude of the flux density at the surface of the sheet is $B_0$, and that the frequency of the time-varying field is $f = \omega/2\pi$. Let the reference direction of $B_0$ coincide with the direction of the $x$ axis (see Fig. 14.5). The direction of the induced current density will then be in the $z$ direction, as indicated in dashed lines. Both $B_x$ and $J_z$ in the sheet are functions of coordinate $y$ only, if the edge effect is ignored. With these assumptions, Maxwell's equations specialized

* This section may be omitted without loss of continuity.

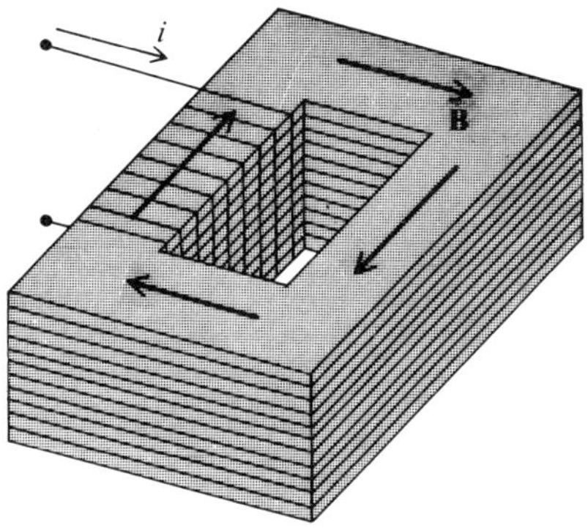


Fig. 14.4 Sheeted ferromagnetic core of a transformer.

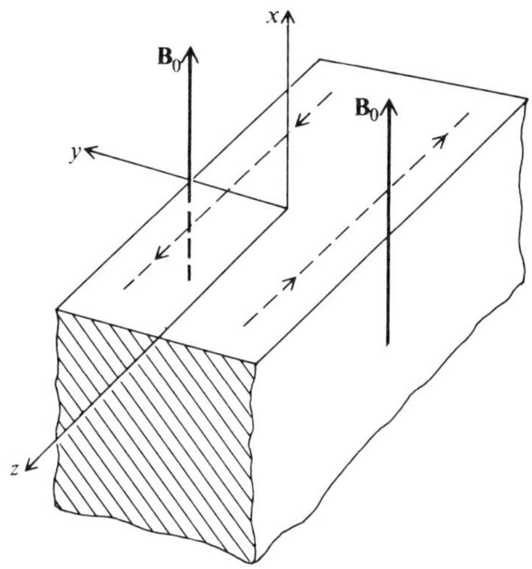


Fig. 14.5 Part of a thin iron sheet.

for this case are given by Eq. (14.2.3). Eliminating $J_x$ from these equations, we obtain

$$d^2B_x/dy^2 = K^2B_x, \tag{14.3.1}$$

where $K$ is given by Eq. (14.2.5) or Eq. (14.2.11). Since $B_x$ is an even function of $y$, the solution is analogous to Eq. (14.2.8):

$$B_x(y) = B_1(e^{Ky} + e^{-Ky}). \tag{14.3.2}$$

Now, $B_x(\pm d/2) = B_0$, so that we can rewrite the last equation as

$$B_x(y) = B_0 \frac{e^{Ky} + e^{-Ky}}{e^{Kd/2} + e^{-Kd/2}}. \tag{14.3.3}$$

Replacing $K$ in this equation by $(1 + j)k$ and transforming the sums in the numerator and denominator as in Eqs. (14.2.12) to (14.2.14), we obtain

$$B_x(ky) = B_0 \frac{\cos ky \cosh ky + j \sin ky \sinh ky}{\cos (kd/2) \cosh (kd/2) + j \sin (kd/2) \sinh (kd/2)}. \tag{14.3.4}$$

Hence the magnitude $|B_x(ky)|$ of the complex amplitude $B_x(ky)$ is

$$|B_x(ky)| = B_0 \sqrt{\frac{\cos^2 ky \cosh^2 ky + \sin^2 ky \sinh^2 ky}{\cos^2(kd/2) \cosh^2(kd/2) + \sin^2(kd/2) \sinh^2(kd/2)}}.$$

Using Eq. (14.2.15), we can write this in a shorter form,

$$|B_x(ky)| = B_0\sqrt{\frac{\cosh^2 ky - \sin^2 ky}{\cosh^2(kd/2) - \sin^2(kd/2)}}\,. \tag{14.3.5}$$

Figure 14.6 shows the ratio $|B_x(ky)|/B_0$ for several values of the parameter $kd$. The ratio $|B_x(ky)|/|B_x(0)|$ is represented by the same curve as $|J_z(ky)|/J_z(0)$ in Fig. 14.2.

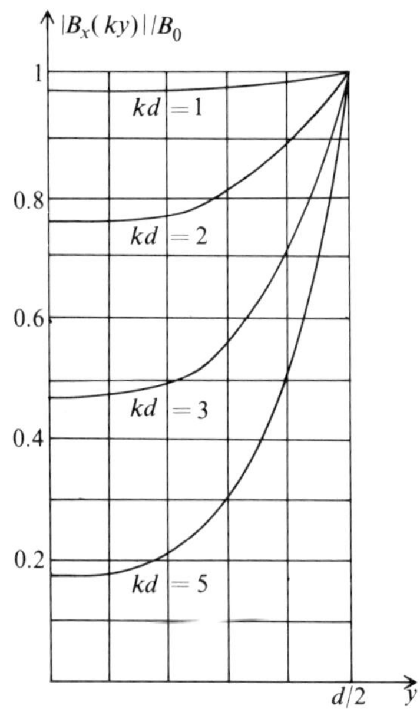


**Fig. 14.6** Distribution of magnetic flux density in a thin iron sheet.

Obviously, the iron sheets must be made very thin to minimize the skin effect and to make use of the whole thickness of the sheet for the magnetic circuit. Since $k$ is also proportional to $\sqrt{\sigma}$, silicon is added to the iron to reduce the conductivity. With these precautions, $kd \ll 1$, and the magnetic flux density is distributed very nearly uniformly over the cross-section of the sheet. Induced currents, causing some losses, nevertheless exist. Let us determine these losses in the case where $kd \ll 1$.

According to the second part of Eq. (14.2.3) and to Eq. (14.3.3), the complex amplitude of the induced current density in the sheet is

$$J_z(y) = -\frac{1}{\mu}\frac{dB_x}{dy} = -B_0\frac{K}{\mu}\frac{e^{Ky} - e^{-Ky}}{e^{Kd/2} + e^{-Kd/2}}\,. \tag{14.3.6}$$

Since $|Ky| \ll 1$ and $|Kd| \ll 1$,

$$e^{Ky} - e^{-Ky} = \left[1 + Ky + \frac{(Ky)^2}{2!} + \cdots\right] - \left[1 - Ky + \frac{(Ky)^2}{2!} - \cdots\right] \simeq 2Ky,$$

$$e^{Kd/2} + e^{-Kd/2} = \left[1 + (Kd/2) + \frac{(Kd/2)^2}{2!} + \cdots\right] + \left[1 - (Kd/2) + \frac{(Kd/2)^2}{2!} - \cdots\right] \simeq 2.$$

So Eq. (14.3.6) becomes

$$J_z(y) \simeq -\frac{1}{\mu} B_0 K^2 y, \qquad (kd \ll 1). \tag{14.3.7}$$

This is the approximate value of the complex *amplitude* of the current density in the sheet. Since the current is sinusoidally time-varying, the *average* power transformed into heat per unit volume is

$$\frac{dP}{dv} \simeq \frac{1}{T}\int_0^T \frac{|J_z(y)|^2 \cos^2 \omega t}{\sigma}\, dt = \frac{1}{2}\frac{|J_z(y)|^2}{\sigma}.$$

From this, the average value of the power transformed into heat in a piece of the sheet, of area $S$, is

$$P \simeq \int_{-d/2}^{d/2} \frac{1}{2}\frac{|J_z(y)|^2}{\sigma} S\, dy = \frac{B_0^2 S |K|^4}{2\sigma\mu^2}\int_{-d/2}^{d/2} y^2\, dy$$

$$= \frac{B_0^2 S|K|^4}{2\sigma\mu^2}\frac{1}{3}\left[(d/2)^3 - (-d/2)^3\right] = \frac{B_0^2 S |K|^4 d^3}{24\sigma\mu^2}.$$

Now, $Sd$ is the volume $v$ of the sheet considered. Substituting into this result the value of $|K|^4$ according to Eq. (14.2.5), we finally obtain

$$\frac{P}{v} \simeq \frac{1}{24} B_0^2 \omega^2 \sigma\, d^2 \qquad (d\sqrt{\omega\mu\sigma} \ll 1). \tag{14.3.8}$$

Losses due to eddy currents in thin sheets are proportional to the *square* of the frequency and of the sheet thickness.

*Problem referring to Section 14.3:* 14.N5.

## 14.4 SKIN EFFECT IN A ROUND WIRE*

Wires of circular cross-section are perhaps the type of conductor most often used. A time-varying current is not distributed uniformly over the cross-section of such wires, that is, skin effect is present (as in the case of any cross-section).

---

* This section may be omitted without loss of continuity.

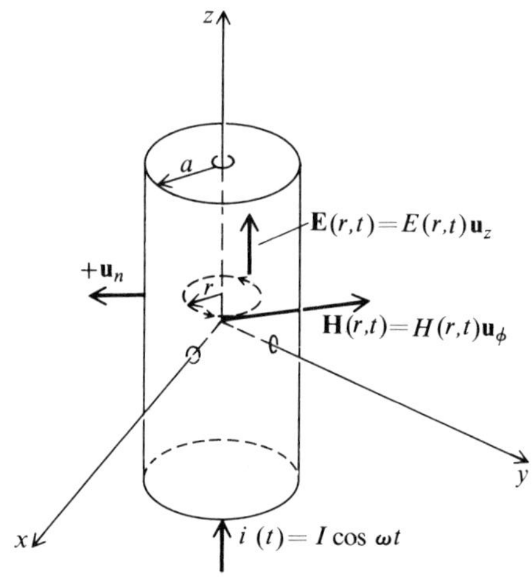


**Fig. 14.7** A cylindrical conductor of circular cross-section.

Let us consider a straight wire of radius $a$, conductivity $\sigma$, and permeability $\mu$, and suppose that the current through the wire is sinusoidal, of amplitude $I$ and frequency $f = \omega/2\pi$. It is natural, in analyzing this problem, to use a cylindrical coordinate system, with the $z$ axis coinciding with the conductor axis (see Fig. 14.7). By symmetry, the electric field intensity has only the $z$ component, and the magnetic field intensity only the $\phi$ component. Both are functions of time and the distance $r$ from the conductor axis, but not of $z$ and $\phi$. Let us write the two Maxwell's curl equations for this case. The expression for curl in a cylindrical coordinate system was derived in Problem 8.G20. The two curl equations in complex form, Eqs. (13.3.24) and (13.3.25), are

$$-\frac{dE}{dr} = -j\omega\mu H, \qquad \frac{1}{r}\frac{d}{dr}(rH) = J. \tag{14.4.1}$$

In the second equation, the density of the displacement current was neglected with respect to the true current density $J$. Now, $J = \sigma E$. Multiplying the first of these equations by $\sigma$ and then substituting the value of $H$ expressed in terms of $dJ/dr$ into the second equation, we have

$$\frac{1}{r}\frac{d}{dr}\left(r\frac{dJ}{dr}\right) = j\omega\mu\sigma J. \tag{14.4.2}$$

After we have carried out the indicated differentiation, this becomes

$$\frac{d^2J}{dr^2} + \frac{1}{r}\frac{dJ}{dr} - j\omega\mu\sigma J = 0. \tag{14.4.3}$$

Introducing a new variable

$$u = r\sqrt{-j\omega\mu\sigma} = \sqrt{-j}\, k'r, \qquad k' = \sqrt{\omega\mu\sigma} \tag{14.4.4}$$

we obtain

$$\frac{d^2J}{du^2} + \frac{1}{u}\frac{dJ}{du} + J = 0. \tag{14.4.5}$$

This equation is known as Bessel's equation, which is an equation of the second order and has two solutions. One solution is known to be infinite for $u = 0$. Since $J$ is finite for $r = 0$, this solution is not applicable in our case. The solution which is finite at $r = 0$ can be always represented in the form of an infinite series

$$J(u) = a_0 + a_1u + a_2u^2 + a_3u^3 + a_4u^4 + \cdots \tag{14.4.6}$$

Coefficients $a_0, a_1, \ldots,$ have to be determined so that this series satisfies Bessel's equation (14.4.5). Substituting this series into Bessel's equation we have

$$2a_2 + 2\cdot 3a_3u + 3\cdot 4a_4u^2 + \cdots + a_1u^{-1} + 2a_2 + 3a_3u + 4a_4u^2 + \cdots$$
$$+ a_0 + a_1u + a_2u^2 + a_3u^3 + a_4u^4 + \cdots = 0,$$

or

$$a_1u^{-1} + (2\cdot 2a_2 + a_0) + (3\cdot 3a_3 + a_1)\,u + (4\cdot 4a_4 + a_2)\,u^2 + \cdots = 0. \tag{14.4.7}$$

This equation has to be satisfied for *any* $u$, which is possible only if each of the coefficients is zero. So we obtain

$$a_1 = 0, \qquad a_3 = 0, \qquad a_5 = 0, \qquad \ldots, \tag{14.4.8}$$

$$a_2 = -\frac{a_0}{2^2}, \qquad a_4 = -\frac{a_2}{4^2} = \frac{a_0}{2^2\cdot 4^2}, \qquad a_6 = -\frac{a_0}{2^2\cdot 4^2\cdot 6^2}, \qquad \ldots \tag{14.4.9}$$

Hence the series solution of Bessel's equation (14.4.5) is

$$J(u) = a_0\left(1 - \frac{u^2}{2^2} + \frac{u^4}{2^2\cdot 4^2} - \frac{u^6}{2^2\cdot 4^2\cdot 6^2} + \frac{u^8}{2^2\cdot 4^2\cdot 6^2\cdot 8^2} - \cdots\right). \tag{14.4.10}$$

The solution of Bessel's equation (14.4.5) in the form of the series in parentheses in Eq. (14.4.10) is known as the Bessel function of the first kind of zeroth order, and is usually designated by $J_0(u)$. Thus we can write that $J(u) = a_0J_0(u)$.

Now, $u$ is a complex variable, as we saw from Eq. (14.4.4), the right-hand side of which has both real and imaginary parts. Therefore $J_0(u)$ is also complex. The constant $a_0$ is obviously the current density along the axis. Designating it by $J(0)$, and noting that

$$(\sqrt{-j})^2 = -j, \qquad (\sqrt{-j})^4 = -1, \qquad (\sqrt{-j})^6 = j, \qquad (\sqrt{-j})^8 = +1, \ldots$$

we have

$$J(k'r) = J(0)\{\mathrm{Re}[J_0(\sqrt{-j}\,k'r)] + j\,\mathrm{Im}[J_0(\sqrt{-j}\,k'r)]\}$$

$$= J(0)\left\{\left[1 - \frac{(k'r)^4}{2^2 \cdot 4^2} + \frac{(k'r)^8}{2^2 \cdot 4^2 \cdot 6^2 \cdot 8^2} - \cdots\right] + j\left[\frac{(k'r)^2}{2^2} - \frac{(k'r)^6}{2^2 \cdot 4^2 \cdot 6^2} + \cdots\right]\right\}.$$

The series in the first brackets is usually designated by *ber*, standing for "Bessel function, real part of", so

$$\mathrm{Re}\{J_0(\sqrt{-j}\,k'r)\} = ber\,k'r = 1 - \frac{(k'r)^4}{2^2 \cdot 4^2} + \frac{(k'r)^8}{2^2 \cdot 4^2 \cdot 6^2 \cdot 8^2} - \cdots. \tag{14.4.11}$$

Similarly, the series in the second brackets is designated by *bei*, meaning "Bessel function, imaginary part of", so

$$\mathrm{Im}\{J_0(\sqrt{-j}\,k'r)\} = bei\,k'r = \frac{(k'r)^2}{2^2} - \frac{(k'r)^6}{2^2 \cdot 4^2 \cdot 6^2} + \cdots. \tag{14.4.12}$$

From this we have

$$J(k'r) = J(0)\,(ber\,k'r + j\,bei\,k'r). \tag{14.4.13}$$

Values of functions $ber\,k'r$ and $bei\,k'r$ are given in extensive tables.* They are shown graphically in Fig. 14.8, together with $(ber^2k'r + bei^2k'r)^{1/2}$, representing the ratio $|J(k'r)|/J(0)$. We can see that, for $k'r \leq 1$, the skin effect is not very pronounced. But if $k'r \gg 1$, the current density at the axis is much smaller than near the surface. So, since $k' = \sqrt{\omega\mu\sigma}$, the skin effect is more pronounced for higher frequencies and for larger permeabilities and conductivities of the conductor.

According to the pattern in Eq. (13.3.31), the instantaneous value of the current density is

$$J(k'r, t) = \mathrm{Re}\{J(k'r)\,e^{j\omega t}\} = J(0)\,(ber\,k'r \cos\omega t - bei\,k'r \sin\omega t).$$

The characteristic shown in Fig. 14.3 is present here also. For example, at $t = 0$ the distribution of the current is represented by the $ber\,k'r$ function shown in Fig. 14.8. If we suppose that $k'a = 6$, then, at $t = 0$, the current density vector is in the $z$ direction up to $k'r \simeq 2.7$, but from $k'r \simeq 2.7$ to $k'r = k'a \simeq 6$ it is in the opposite direction.

The above results use as a phase reference the current density $J(0, t)$ at the conductor axis. They give the correct distribution of the current inside the conductor, but in practice the total current $i(t)$, rather than $J(0, t)$, is known. To express Eq. (14.4.13) in terms of the amplitude $I$ of the total current through the conductor we proceed as follows.

According to Ampère's circuital law we know that

$$2\pi a H(a, t) = i(t),$$

* For example, Jahnke, E. and F. Emde, *Tables of Functions*, Dover Publications, New York, 1945.

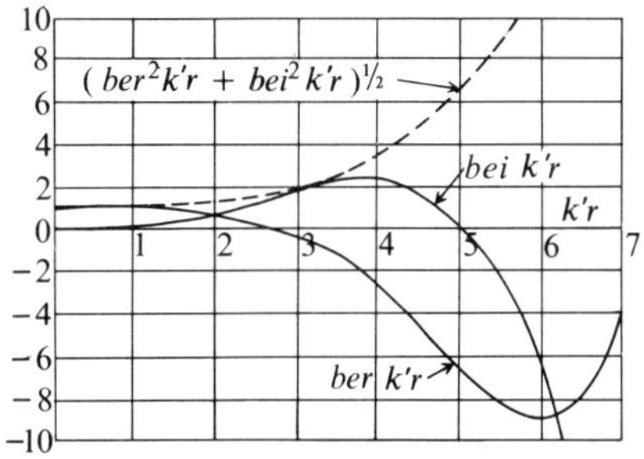


Fig. 14.8 Graph of the functions *ber k′r* and *bei k′r*.

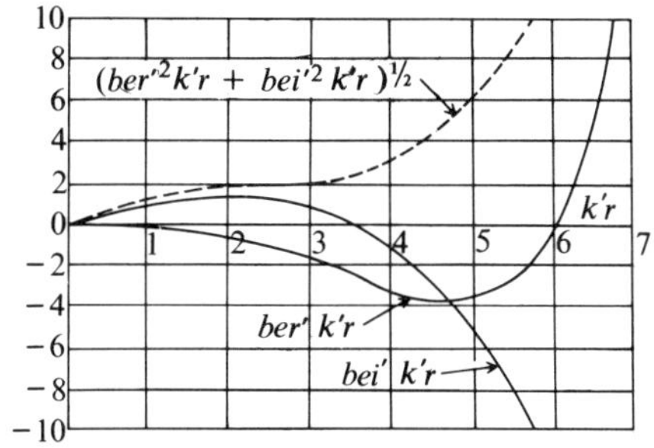


Fig. 14.9 Graph of the functions *ber′ k′r* and *bei′ k′r*.

which in complex notation becomes

$$2\pi a H(k'a) = I. \tag{14.4.14}$$

According to the first part of Eq. (14.4.1) and Eq. (14.4.13)

$$\begin{aligned} H(k'r) &= \frac{1}{j\omega\mu\sigma}\frac{dJ(k'r)}{dr} = \frac{k'}{j\omega\mu\sigma}\frac{dJ(k'r)}{d\,k'r} \\ &= \frac{k'}{j\omega\mu\sigma} J(0)\,(ber'\,k'r + j\,bei'\,k'r). \end{aligned} \tag{14.4.15}$$

*ber′ k′r* and *bei′ k′r* denote the first derivatives of the series shown in Eqs. (14.4.11) and (14.4.12) with respect to the variable $k'r$:

$$ber'\,k'r = \frac{d\,ber\,k'r}{d\,k'r} = -4\frac{(k'r)^3}{2^2\cdot 4^2} + 8\frac{(k'r)^7}{2^2\cdot 4^2\cdot 6^2\cdot 8^2} - \cdots, \tag{14.4.16}$$

$$bei'\,k'r = \frac{d\,bei\,k'r}{d\,k'r} = 2\frac{(k'r)}{2^2} - 6\frac{(k'r)^5}{2^2\cdot 4^2\cdot 6^2} + \cdots. \tag{14.4.17}$$

These functions represent the distribution of the magnetic field inside the conductor; they are shown in Fig. 14.9, together with $(ber'^2k'r + bei'^2k'r)^{1/2}$, which is proportional to the amplitude of the magnetic field intensity at a point.

Combining Eqs. (14.4.14) and (14.4.15), and noting that $k' = \sqrt{\omega\mu\sigma}$, we have

$$J(0) = \frac{jIk'}{2\pi a(ber'\,k'a + j\,bei'\,k'a)}.$$

From this, Eqs. (14.4.13) and (14.4.15) become

$$J(k'r) = \frac{jIk'}{2\pi a}\frac{ber\,k'r + j\,bei\,k'r}{ber'\,k'a + j\,bei'\,k'a}, \tag{14.4.18}$$

$$H(k'r) = \frac{I}{2\pi a} \frac{ber'\, k'r + j\, bei'\, k'r}{ber'\, k'a + j\, bei'\, k'a}. \tag{14.4.19}$$

*Problems referring to Section 14.4:* 14.T9 to 14.T10, 14.G5 to 14.G8, 14.N6 to 14.N7.

### 14.5 THE PROXIMITY EFFECT*

The term "skin effect" is used in referring to the nonuniform distribution of the current in an isolated conductor. If another conductor is present, varying current through it causes additional change in the current distribution in the first conductor. This latter phenomenon is referred to as the *proximity effect*. Essentially, both are electromagnetic induction phenomena, and the distinction between them is, though clear, rather artificial. As an example of the proximity effect, consider a coaxial cable with its currents in opposite directions. The proximity effects will cause a larger current density at the points of the two conductors which are nearest. This is illustrated in Fig. 14.10.

Analysis of the proximity effect obviously cannot be done separately from the skin effect: they are two faces of the same phenomenon. Furthermore, analysis of

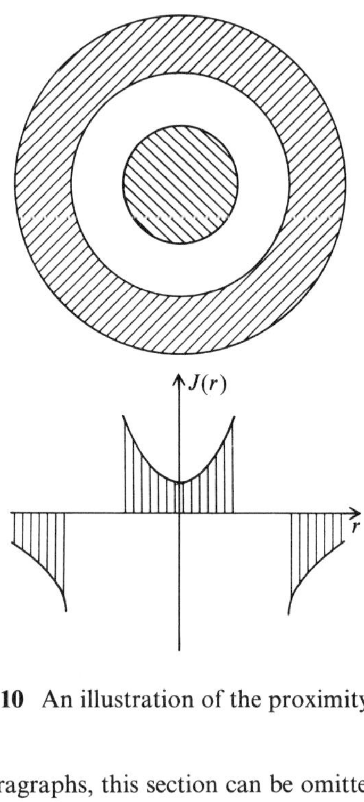


**Fig. 14.10** An illustration of the proximity effect.

* Except for the first two paragraphs, this section can be omitted without loss of continuity.

the combined proximity and skin effects is usually very involved. For example, the exact solution of this combined effect is possible for the coaxial cable, but it is extremely difficult even for a two-wire line with round conductors. Fortunately, the influence of the proximity effect is usually much smaller than that of the skin effect.

To gain some insight into this combined effect, let us consider two conducting sheets of conductivity $\sigma$, permeability $\mu$, and thickness $d$. Let us suppose that the currents in the two sheets are equal but in opposite directions, that their frequency is $f = \omega/2\pi$, and that the sheets are very large, so that the effects of their finite size can be ignored.

The sheet shown in Fig. 14.11 has its current in the positive $z$ direction. By symmetry, **J** and **H** are functions only of the coordinate $y$. Since the coordinate axes are oriented as in Fig. 14.1, the differential equations corresponding to Maxwell's curl equations are given by Eq. (14.2.3). The solution for the current density $J_z(y)$ is the same as for Eq. (14.2.7):

$$J_z(y) = J_1 e^{Ky} + J_2 e^{-Ky}. \tag{14.5.1}$$

To determine the constants $J_1$ and $J_2$, we note that, due to the presence of the other sheet, with equal current in the opposite direction, $H_x(d)$ must be zero. Since from Eq. (14.2.3)

$$H_x(d) = -\frac{1}{j\omega\mu\sigma}(J_1 K e^{Kd} - J_2 K e^{-Kd}),$$

the condition $H_x(d) = 0$ requires that $J_2 = J_1 e^{2Kd}$. Hence Eq. (14.5.1) becomes

$$J_z(y) = J_1[e^{Ky} + e^{K(2d-y)}] = J_1 e^{Kd}[e^{K(d-y)} + e^{-K(d-y)}].$$

Now, for $y = 0$

$$J_z(0) = J_1 e^{Kd}(e^{Kd} + e^{-Kd}),$$

so we can write

$$J_z(y) = J_z(0)\frac{e^{K(d-y)} + e^{-K(d-y)}}{e^{Kd} + e^{-Kd}}.$$

The numerator and denominator on the right-hand side can be transformed by following the procedure described by Eqs. (14.2.10) to (14.2.13). The result is

$$J_z(ky) = J_z(0)\frac{\cos k(d-y)\cosh k(d-y) + j\sin k(d-y)\sinh k(d-y)}{\cos kd\cosh kd + j\sin kd\sinh kd}. \tag{14.5.2}$$

By using the same transformations as those used to obtain Eq. (14.2.16) we can write the ratio $|J_z(ky)|/|J_z(0)|$ in the form

$$\frac{|J_z(ky)|}{|J_z(0)|} = \sqrt{\frac{\cosh^2 k(d-y) - \sin^2 k(d-y)}{\cosh^2 kd - \sin^2 kd}}. \tag{14.5.3}$$

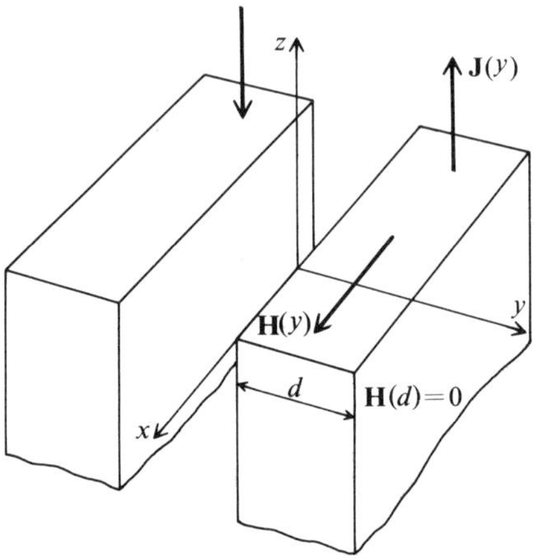


Fig. 14.11 Two parallel-strip conductors with equal currents in opposite directions.

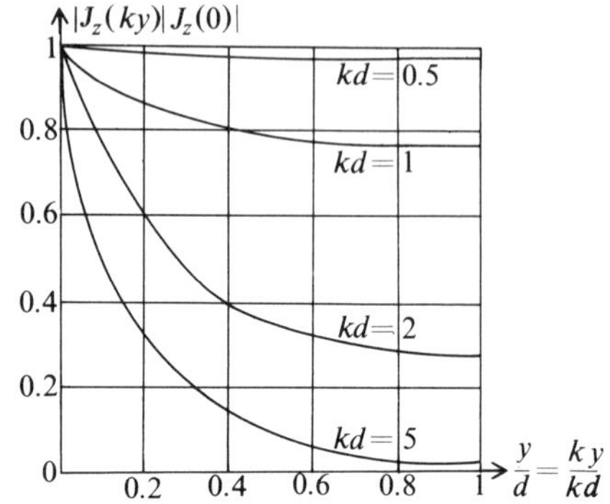


Fig. 14.12 Distribution of current in strips of Fig. 14.11.

Figure 14.12 shows this ratio for several values of $kd$. Note that if the other sheet were not present, the curves would by symmetrical with respect to $y = d/2$, being minimal at that point.

In this idealized case of two theoretically infinite parallel current sheets with opposite currents, the proximity effect is not a function of the distance between the sheets. This is because the magnetic field of an infinite current sheet is of the same magnitude at all distances from it. For sheets of finite width this would not be the case. Qualitatively the same conclusion would be attained, with the addition of the fact that the proximity effect is more pronounced if the sheets are closer. Generally speaking, the proximity effect results in an increase in the resistance of the conductors and a decrease in the internal inductance when compared with those valid for a steady current.

*Problems referring to Section 14.5:* 14.T9, 14.T11, 14.G9.

## 14.6 CIRCUIT CONCEPTS FROM THE FIELD THEORY STANDPOINT

**14.6.1 Introduction.** The four commonest elements in circuit theory are two-terminal generators, resistors, capacitors, and inductors (coils). Any number of these elements, connected by conducting wires, makes an electrical network. Figure 14.13 is an example of such a network, but it is a very unconventional representation. Instead of using the usual symbols for the elements, it represents them just as they might be in practice. The interconnecting wires and their junctions, too, are represented in a way resembling their real shape. *This is what a real circuit is like.*

From the field theory standpoint such a network is an extremely complicated

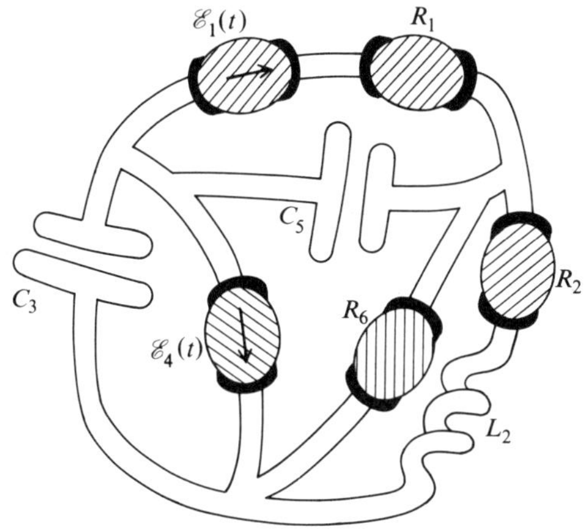


**Fig. 14.13** An electric circuit.

system. The boundaries are quite irregular, and the medium is inhomogeneous. We have learned that even much simpler boundaries (in fact, all but some of the very simplest forms) were too complicated for an analytical solution. In other words, we can never expect to find an *exact* solution to even the simplest real network.

Yet, millions of such networks were solved in the past by using the simple methods of circuit theory, and the solutions thus found were the basis of the designs for many of the devices we encounter every day. The agreement between practice and the predictions of this simple theory has been excellent. That circuit theory is an approximate theory is patently obvious, but the accuracy of its results in most instances is so high that no better solution is required.

However, as with any approximate theory, circuit theory is valid only in certain circumstances, and a knowledge of the range of validity of the simple circuit theory is important for any engineer who has to deal with it. Obviously, to understand its limitations we cannot start from circuit theory itself; we have to start from the general theory to which it is an approximation, that is, from the theory of electromagnetic fields.

Essentially, circuit theory is based on the following two assumptions:

(1) No charges are accumulated in the regions of the nodes of a circuit and on the interconnecting conductors. If this condition is met, Kirchhoff's first law in Eq. (6.9.1) is also valid for time-varying currents:

$$\sum_{\text{any node}} i_k(t) = 0. \tag{14.6.1a}$$

(2) A unique value of the line integral of the *total* electric field intensity can

be associated with any pair of nodes or any pair of terminals of two-terminal elements from which the network is built. If this condition is met, Kirchhoff's second law [Eq. (6.9.3)] is also valid, only in a somewhat different form:

$$\sum \mathscr{E}(t) - \sum [V_k(t) - V_{k'}(t)] = 0 \qquad \text{(any closed loop).} \qquad (14.6.1b)$$

The difference $[V_k(t) - V_{k'}(t)]$ stands for the potential difference between the terminals of two-terminal elements along a given closed contour.

We shall now consider in what circumstances these basic assumptions are valid.

**14.6.2 Restriction on the electric field of the circuit and on circuit dimensions.** The first condition, that no charges are present on the interconnecting conductors, implies that the electric field is localized inside the elements themselves. Therefore, the electric displacement vector is also negligibly small everywhere except inside the elements themselves. The displacement current density, $\partial \mathbf{D}/\partial t$, is hence also negligible. Now, in Section 13.4.3 it was shown that precisely the presence of $\partial \mathbf{D}/\partial t$ in Maxwell's equation (13.3.7) results in retardation of potentials. Consequently, condition (1) implies that the *retardation can be ignored.* In the sinusoidally time-varying case of frequency $f = \omega/2\pi$, according to Eq. (13.4.34), this imposes the following condition on the dimensions of the circuit:

$$\boxed{d = \text{largest dimension of the circuit} \ll 1/\omega\sqrt{\varepsilon\mu}}\,. \qquad (14.6.2)$$

For example, let us consider a circuit of the dimensions of a radio receiver, say $d = 0.5$ m. Then, for all frequencies much smaller than $10^8/2\pi d \simeq 32 \times 10^6$ Hz, this can be analyzed by using standard circuit equations. The circuits for very high frequencies must be much smaller if circuit theory is to be applied to their analysis.

**14.6.3 Definition of voltage between two points. Restrictions on the magnetic field of the circuit.** In the case of time-varying fields, the total electric field intensity is given by Eq. (13.4.17):

$$\mathbf{E} = -\text{grad } V - \partial \mathbf{A}/\partial t. \qquad (14.6.3)$$

Let us now consider two points 1 and 2 in a time-varying electromagnetic field (see Fig. 14.14). The line integral of the *total* electric field intensity *from* point 1 *to* point 2 is referred to as the *voltage* between points 1 and 2. It will be designated by $V_{12}$:

$$\boxed{V_{12} = \int_1^2 \mathbf{E}\cdot d\mathbf{l} = -\int_1^2 \left(\text{grad } V + \frac{\partial \mathbf{A}}{\partial t}\right)\cdot d\mathbf{l}} \qquad \begin{array}{l}\text{(definition of}\\ \text{voltage between}\\ \text{points 1 and 2).}\end{array} \quad (14.6.4)$$

The first term on the right-hand side is precisely the potential difference be-

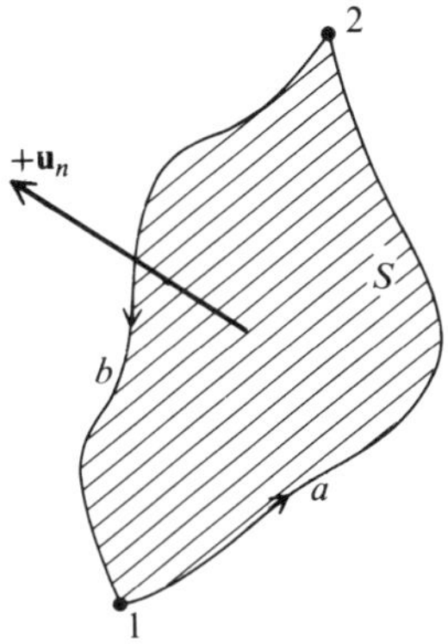


**Fig. 14.14** Two points in a time-varying electromagnetic field.

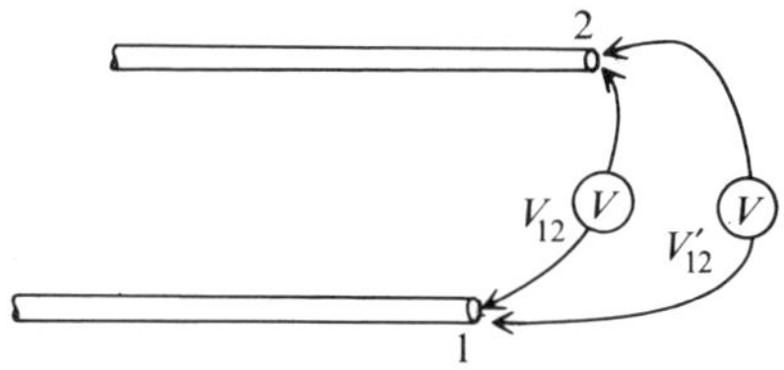


**Fig. 14.15** The readings of the two voltmeters shown are generally different.

tween the two points. So

$$V_{12} = (V_1 - V_2) - \frac{\partial}{\partial t}\int_1^2 \mathbf{A}\cdot d\mathbf{l}. \tag{14.6.5}$$

Hence the concepts of "voltage between two points" and "potential difference between two points" are *not* equal. Only in the static case do they become identical.

The potential difference $(V_1 - V_2)$ is a function of points 1 and 2 only. But the second term on the right is a function of the particular path between points 1 and 2. This is easily seen if two different paths are considered. For example, in Fig. 14.14

$$\underbrace{\int_1^2 \mathbf{A}\cdot d\mathbf{l}}_{\text{along } a} + \underbrace{\int_2^1 \mathbf{A}\cdot d\mathbf{l}}_{\text{along } b} = \int_S \mathbf{B}\cdot d\mathbf{S} = \Phi,$$

so that

$$\underbrace{\int_1^2 \mathbf{A}\cdot d\mathbf{l}}_{\text{along } a} = \underbrace{\int_1^2 \mathbf{A}\cdot d\mathbf{l}}_{\text{along } b} + \Phi. \tag{14.6.6}$$

Thus

$$V_{12\text{ along } a} = V_{12\text{ along } b} - \frac{\partial\Phi}{\partial t}, \tag{14.6.7}$$

where $\Phi$ is the magnetic flux through the contour $1a2b1$ shown. In practice, this implies that, in measuring the voltage between two points with a voltmeter, we generally obtain different readings for different shapes of the voltmeter leads (Fig. 14.15). This is a very important general conclusion for time-varying electromagnetic fields.

Now, basic assumption (2) of circuit theory implies that the voltage between two points is unique. That is, that the reading of a voltmeter connected between any two points of a network does not depend on the shape of the voltmeter leads. According to Eq. (14.6.7), this implies that the magnetic field created by the currents in the circuit can be ignored everywhere except inside the elements of the circuit themselves.

**14.6.4 Basic circuit elements: resistors, capacitors, and inductors.** In principle, the requirements for circuits explained above impose one general restriction on all circuit elements: the electromagnetic field of the charges and currents inside the elements must be of negligible magnitude outside them. To write circuit equations we need one further piece of data: the relation between the potential difference between the terminals of an element and the current through it. Solving for currents and voltages in the circuit, then, implies solving the equations corresponding to Kirchhoff's laws.

Although any kind of element which satisfies these restrictions may be connected in a circuit, those most often used are of four main types. The first is the *resistor*. By resistors we mean here two-terminal elements inside which only a negligible amount of electromagnetic energy can be stored. They are regarded as pure lossy elements. The kind most often used are linear resistors, in which the potential difference between their terminals is proportional to the current intensity through the resistor:

$$V_R(t) = Ri_R(t). \tag{14.6.8}$$

$R$ is the resistance of the resistor; the symbol for resistors is shown in Fig. 14.16.

The second type is the *capacitor*. By a capacitor we mean a lossless element, inside which no appreciable magnetic field exists. The electric field, however, is large and the energy associated with a capacitor is practically the energy of this electric field. Using the definition $i(t) = dQ(t)/dt$ for the current through a capacitor (see Fig. 14.16), we find that the current-voltage relationship is

$$V_C(t) = \frac{Q(t)}{C} = \frac{1}{C}\int i_C(t)\,dt, \qquad \text{or} \qquad i_C(t) = C\frac{dV_C(t)}{dt}. \tag{14.6.9}$$

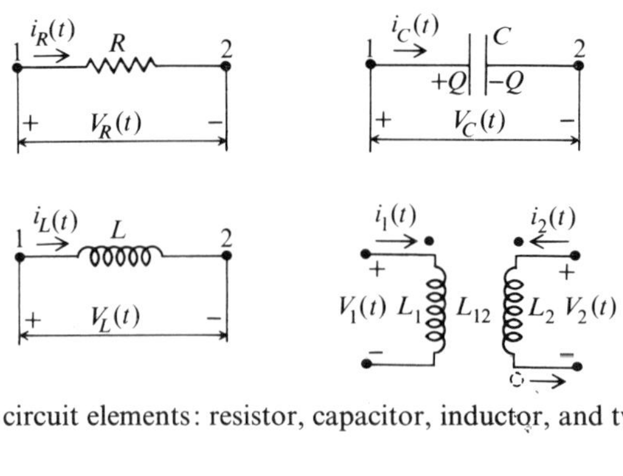


**Fig. 14.16** Basic circuit elements: resistor, capacitor, inductor, and two coupled coils.

The third type is the *inductor*; it is assumed that it is lossless, that the magnetic field is limited to the region of the coil itself, and that the electric field can be ignored. The current-voltage relationship for an inductor is given by

$$V_L(t) = \frac{d\Phi}{dt} = L\frac{di_L(t)}{dt}, \qquad \text{or} \qquad i_L(t) = \frac{1}{L}\int V_L(t)\,dt \tag{14.6.10}$$

(see Fig. 14.16).

Finally, the fourth type is a pair of coupled coils. It is assumed that both coils are lossless, that the electric field inside them can be ignored, and that the magnetic field by which the coils are coupled exists practically only inside such a *four-terminal* network. Two equations are necessary to describe its current-voltage relationship:

$$V_1(t) = L_1\frac{di_1}{dt} \pm L_{12}\frac{di_2}{dt}, \qquad V_2(t) = L_2\frac{di_2}{dt} \pm L_{12}\frac{di_1}{dt}. \tag{14.6.11}$$

The sign of the second terms depends upon the actual direction of the windings. In diagrams, the convention is to denote this by two dots, as indicated in Fig. 14.16. If the dots are positioned in the same way with respect to the directions of the two currents, a positive sign is understood. (This is the case shown.) If the dot on the right coil were in its lower part, as indicated by a small dashed circle, a negative sign would be implied.

If resistors, capacitors, and inductors only are present in a circuit, Kirchhoff's second law [Eq. (14.6.1b)] takes the form

$$\sum\mathscr{E}(t) - \sum\left[Ri(t) + L\frac{di(t)}{dt} + \frac{1}{C}\int i(t)\,dt\right] = 0 \qquad \text{(any closed contour).} \tag{14.6.12}$$

**14.6.5 Circuits with sinusoidal time-varying currents and voltages.** In the case of circuits with linear elements and sinusoidal time-varying currents, the use of complex notation is a great convenience. Equations (14.6.8) to (14.6.10) then read

$$V_R = RI_R, \qquad V_L = j\omega LI_L, \qquad V_C = \frac{1}{j\omega C}I_C. \tag{14.6.13}$$

Of course, the $V$'s and $I$'s now represent *complex* amplitudes.

All three equations are of the same type, $V = ZI$. Consequently, we can formally consider all three elements in the same way, with appropriate $Z$ only. $Z$ is referred to as the *impedance* of an element. If we characterize all the elements by their appropriate complex impedances, the further procedure is completely analogous to the analysis of steady current networks. Impedances replace resistances, and complex amplitudes of currents and voltages are dealt with instead of steady current intensities and potential differences; but the procedure is otherwise the same.

For example, the impedance of a series $RLC$ circuit at an angular frequency $\omega$ is

$$Z_{\text{series}\,RLC} = Z_R + Z_L + Z_C = R + j\omega L + 1/j\omega C = R + j(\omega L - 1/\omega C). \tag{14.6.14}$$

The impedance of any combination of $R$, $L$, and $C$ elements can be determined in the same way.

**14.6.6 Conclusion.** Circuit theory is an approximate theory for analyzing electrical networks, but the degree of approximation is usually extremely high. The following limitations are implicitly assumed: (1) The dimensions of a network are very small when compared with $(\omega\sqrt{\varepsilon\mu})^{-1}$; (2) The electric and magnetic fields are for all practical purposes localized inside the circuit elements.

In practice this is not always so. Nevertheless, by an ingenious treatment and a complete understanding of its limitations, circuit theory can be used beyond the limits imposed by these restrictions. One example is the transmission lines, which we shall study in a later chapter. Similarly, "stray" capacitance is a concept by which a field phenomenon is approximated by the use of circuit concepts. An approximation of this sort is, of course, not so good or so simple as in the case of "concentrated" elements, but it is often much easier than the real electromagnetic field problem, and is used extensively.

The foregoing is a short account of the connections between field theory and circuit theory, which are normally sufficient for an understanding of the limitations implicit in circuit theory. The further development of circuit theory has practically no connection with the electromagnetic field theory. It is the subject of many specialized texts, and will not be discussed any further here.

*Problems referring to Section 14.6:* 14.T12 to 14.T15.

## PROBLEMS

### Test problems

**14.T1.** Suppose that the origin in Fig. 14.1 is moved downward for $\Delta y = d/2$, to coincide with the lower face of the strip. What are the constants $J_1$ and $J_2$ in that case?

**14.T2.** Sketch the functions $J_z(ky, \pi/2\omega)/J_z(0)$ and $J_z(ky, \pi/\omega)/J_z(0)$ [see Eq. (14.2.17)].

**14.T3.** Starting from Eq. (14.2.17), write the integral representing the instantaneous value of the total current $i(t)$ in the strip shown in Fig. 14.1. (Ignore the fringing effect and do *not* evaluate the integral.)

**14.T4.** Starting from Eq. (14.2.16), write the integral representing the *average* power $P_l$ transformed into heat per unit length of the strip (along the $z$ axis) shown in Fig. 14.1. (Ignore the fringing effect and do *not* evaluate the integral.)

**14.T5.** From the results of the two preceding problems, write the expression from which the resistance $R_l$ per unit length of the strip can be determined. The resistance $R_l$ is defined by $P_l = R_l I^2/2$, where $I$ is the *amplitude* of the current $i(t)$ in the strip.

***14.T6.** From Eqs. (14.2.3) and (14.2.9), determine the complex amplitude of the magnetic field intensity in the strip, and then reduce it to a form analogous to Eq. (14.2.14), using Eqs. (14.2.10) to (14.2.13).

***14.T7.** From the result of the preceding problem, deduce the instantaneous value $H_x(ky, t)$ of the magnetic field intensity. Using Ampère's circuital law, determine from this the instantaneous value of the total current through the strip.

***14.T8.** A special transmission line for large alternating currents is designed in the form of two thick parallel bus-bars. To determine its losses experimentally on a model, bus-bars of $n$ times smaller sides than the real ones are used, with current of $n$ times larger frequency. Using Eqs. (14.2.11) and (14.2.14) discuss the validity of the results thus obtained. Will the losses in the real system be larger or smaller than those predicted by the measurements on the model?

**14.T9.** What is the instantaneous value of the current density given by Eqs. (14.2.19), (14.4.13), (14.4.18), and (14.5.2)?

***14.T10.** Starting from Eq. (14.4.18), describe the procedure for determining the resistance per unit length of a round wire at high frequencies.

***14.T11.** Starting from Eq. (14.5.2), describe the procedure for determining the resistance per unit length of a strip line of width $a$. (The strips are so close that the edge-effect can be ignored.)

***14.T12.** Figure 14.T12 shows two closely spaced resistors in a network. A stray electric field is present, as indicated by dashed lines. Discuss whether this stray electric field can be taken into account *exactly* by means of the two capacitors shown, which are of appropriately chosen capacitances.

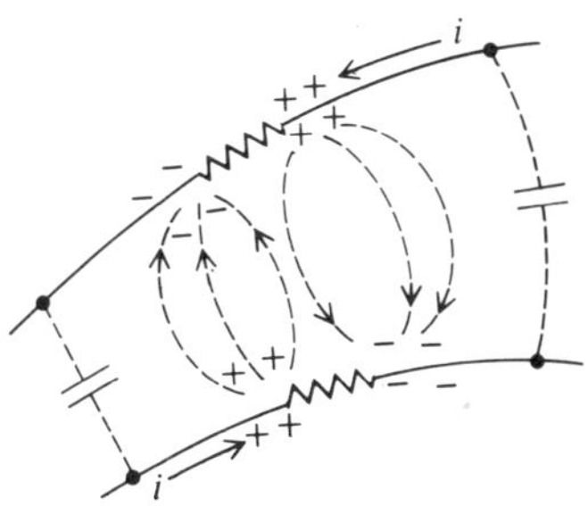


**Figure 14.T12**

**14.T13.** Figure 14.T13 shows a strip transmission line with an alternating current, of surface density $J_s(t) = J_{sm} \cos \omega t$. The potential difference between the strips is $V_1(t) - V_2(t) = V_m \cos \omega t$. Determine the voltage between points $A$ and $B$ along paths $AaB$ and $AbB$.

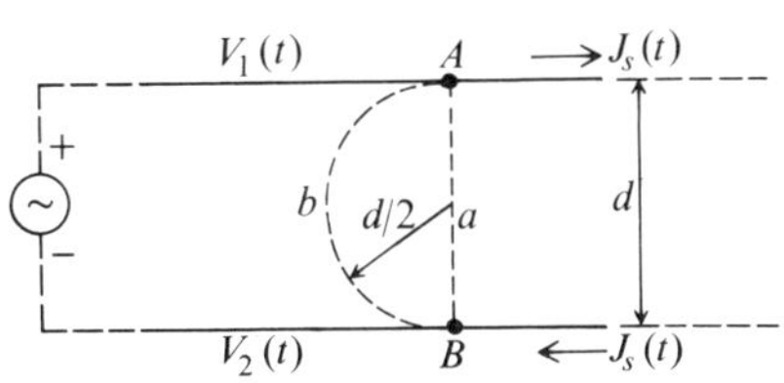


**Figure 14.T13**

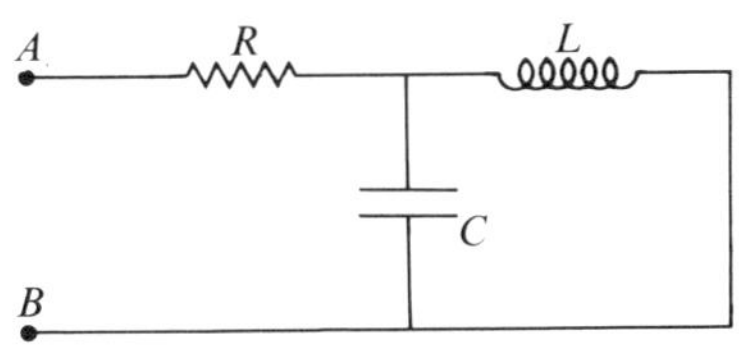


**Figure 14.T14**

**14.T14.** Determine the equivalent impedance of a coil of inductance $L$ and a resistor of resistance $R$ connected in parallel, if the angular frequency is $\omega$.

**14.T15.** Determine the equivalent impedance between points $A$ and $B$ in Figure 14.T15. The angular frequency is $\omega$.

## General problems

**14.G1.** The complex amplitude of the total current in the strip from Section 14.2 is $I$. Express the constant $J_z(0)$ in terms of $I$.

*Solution.* The most direct way to express $J_z(0)$ in terms of $I$ is to find the magnetic field intensity $H_x(d/2) = -H_x(-d/2)$ on the surfaces of the strip, and then to apply Ampère's circuital law to the rectangular contour 12341 shown in Fig. 14.1.

According to Eqs. (14.2.3), (14.2.5), and (14.2.9),

$$H_x(y) = -\frac{1}{j\omega\mu\sigma}\frac{dJ_z(y)}{dy} = -\frac{1}{2K}J_z(0)\,(e^{Ky} - e^{-Ky}).$$

Applying Ampère's circuital law to the contour mentioned, we find that $-2aH_x(d/2) = I$. Hence

$$J_z(0) = \frac{KI}{a(e^{Kd/2} - e^{-Kd/2})}.$$

Using Eqs. (14.2.10) to (14.2.13) this can be transformed to

$$J_z(0) = \frac{KI}{2a(\cos kd/2 \sinh kd/2 + j \sin kd/2 \cosh kd/2)}. \tag{a}$$

**14.G2.** Determine the distribution of the current in the two-layer strip shown in Fig. 14.G2. (The angular frequency of the current through the strip is $\omega$, and the effects of the finite width of the strip can be ignored.)

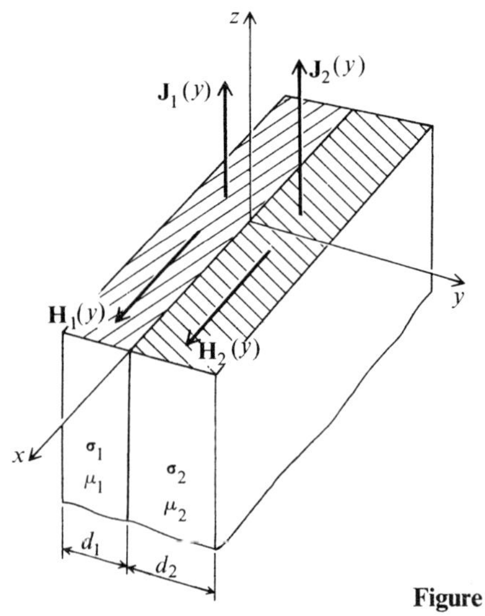


Figure 14.G2

*Solution.* The directions of the coordinate axes are the same as in Fig. 14.1; equations of the type of Eq. (14.2.3) therefore apply for both layers. The complex amplitudes of the current density in the two layers are given by Eq. (14.2.7):

$$J_{z1}(y) = J_1' e^{K_1 y} + J_2' e^{-K_1 y}, \quad -d_1 < y < 0, \tag{a}$$

$$J_{z2}(y) = J_1'' e^{K_2 y} + J_2'' e^{-K_2 y}, \quad 0 < y < d_2, \tag{b}$$

where

$$K_1 = \sqrt{j\omega\mu_1\sigma_1} = (1 + j)\sqrt{\frac{\omega\mu_1\sigma_1}{2}} = (1 + j)k_1,\ K_2 = \sqrt{j\omega\mu_2\sigma_2} = (1 + j)k_2. \tag{c}$$

According to Eq. (14.2.3), and Eqs. (a) and (b) of this problem, the magnetic field intensity in the two layers is

$$H_{x1}(y) = -\frac{1}{K_1^2}\frac{dJ_{z1}}{dy} = -\frac{1}{K_1}(J_1' e^{K_1 y} - J_2' e^{-K_1 y}), \tag{d}$$

$$H_{x2}(y) = -\frac{1}{K_2^2}\frac{dJ_{z2}}{dy} = -\frac{1}{K_2}(J_1'' e^{K_2 y} - J_2'' e^{-K_2 y}). \tag{e}$$

The four unknown constants $J_1'$, $J_2'$, $J_1''$ and $J_2''$ can be determined by using the following boundary conditions:

$$J_{z1}(0)/\sigma_1 = J_{z2}(0)/\sigma_2, \quad \text{from which} \quad J_1' + J_2' = \frac{\sigma_1}{\sigma_2}(J_1'' + J_2''). \tag{f}$$

$$H_{x1}(0) = H_{x2}(0), \quad \text{from which} \quad J_1' - J_2' = \frac{K_1}{K_2}(J_1'' - J_2''), \tag{g}$$

$$-H_{x1}(-d_1) = H_{x2}(d_2), \quad \text{from which}$$

$$J_1' e^{-K_1 d_1} - J_2' e^{K_1 d_1} = -\frac{K_1}{K_2}(J_1'' e^{K_2 d_2} - J_2'' e^{-K_2 d_2}). \tag{h}$$

Finally, if the complex amplitude of the *total* current through the strip per unit width is $I_l$, we have, according to Ampère's circuital law and Eq. (h)

$$2H_{x2}(d_2) = I_l. \tag{i}$$

The determination of the constants and of the current distribution is left to the reader. As a check, for $d_1 = d_2 = d/2$, $\sigma_1 = \sigma_2$ and $\mu_1 = \mu_2$, the final result must be the same as that derived in Section 14.2. Note how much more intricate this case of a two-layer sheet is.

***14.G3.** Determine the resistance and internal inductance per unit length of the strip in Section 14.2, using the result found there and Poynting's theorem in complex form (see Problem 13.G8).

*Solution.* The energy *definitions* of resistance $R$ and internal inductance $L_i$ are

$$P_{\text{av}} = R\frac{|I_c|^2}{2}, \quad \text{from which} \quad R = \frac{2P_{\text{av}}}{|I_c|^2}, \tag{a}$$

$$W_{mi\ \text{av}} = \frac{1}{2}L_i\frac{|I_c|^2}{2}, \quad \text{from which} \quad L_i = \frac{4W_{mi\ \text{av}}}{|I_c|^2}. \tag{b}$$

In these equations $P_{av}$ is the *average* power transformed into heat in the resistor, $I_c$ is the complex *amplitude* of the current, and $W_{mi\,av}$ is the *average* value of the magnetic energy stored *inside* the element considered.

Inside the strip $\mathbf{E}_{ic} = 0$, and $\mu H^2/4 \gg \varepsilon E^2/4$. Combining Eqs. (a) and (b) with Eqs. (e) and (f) of Problem 13.G8 we find therefore that

$$R = -\frac{1}{|I_c|^2}\,\mathrm{Re}\left\{\oint_S (\mathbf{E}\times\mathbf{H}^*)\cdot d\mathbf{S}\right\}, \tag{c}$$

$$L_i = -\frac{1}{\omega|I_c|^2}\,\mathrm{Im}\left\{\oint_S (\mathbf{E}\times\mathbf{H}^*)\cdot d\mathbf{S}\right\}. \tag{d}$$

These formulae are valid generally. By definition, the surface element $d\mathbf{S}$ is directed *away* from the surface $S$. In the special case considered the closed surface $S$ is the surface of the unit length of the strip. We therefore have to determine $\mathbf{E}$ and $\mathbf{H}$ on the strip surface.

Combining Eq. (14.2.14) with Eq. (a) of Problem 14.G1 we obtain the following value of the complex amplitude $E_z(kd/2) = J_z(kd/2)/\sigma$:

$$E_z(kd/2) = \frac{KI}{2a\sigma}\,\frac{\cos kd/2\cosh kd/2 + j\sin kd/2\sinh kd/2}{\cos kd/2\sinh kd/2 + j\sin kd/2\cosh kd/2}. \tag{e}$$

According to Ampère's circuital law (see Fig. 14.1)

$$H_x(d/2) = -I/2a, \qquad \text{from which} \qquad H_x^*(d/2) = -I^*/2a. \tag{f}$$

From Fig. 14.1 we see that both $\mathbf{E}(d/2) \times \mathbf{H}^*(d/2)$ and $\mathbf{E}(-d/2) \times \mathbf{H}^*(-d/2)$ are directed away from the strip. Since they are constant on the surface of the strip and since $II^* = |I|^2$, we obtain from Eqs. (c) to (f)

$$R_l = \mathrm{Re}\left\{\frac{K}{2a\sigma}\,\frac{\cos kd/2\cosh kd/2 + j\sin kd/2\sinh kd/2}{\cos kd/2\sinh kd/2 + j\sin kd/2\cosh kd/2}\right\}, \tag{g}$$

$$L_{l\,\mathrm{int}} = \frac{1}{\omega}\,\mathrm{Im}\left\{\frac{K}{2a\sigma}\,\frac{\cos kd/2\cosh kd/2 + j\sin kd/2\sinh kd/2}{\cos kd/2\sinh kd/2 + j\sin kd/2\cosh kd/2}\right\}. \tag{h}$$

For $kd/2 \ll 1$ Eqs. (g) and (h) must approach those valid for steady current. This check is left as an exercise for the reader.

**14.G4.** A straight insulated conducting strip, of thickness $d$ and width $a$ ($a \gg d$), is situated in a homogeneous conducting medium of permeability $\mu$ and conductivity $\sigma$. The amplitude of the current through the strip is $I$, and its angular frequency $\omega$. Ignoring the finite width of the strip, determine the electric and magnetic fields in the surrounding conducting medium, and the average losses in the medium per unit length of the strip, using Eqs. (c) and (d) of Problem 14.G3.

*Solution.* Figure 14.G4 shows the strip and the coordinate system in relation to which the problem will be analyzed. The directions of the axes are the same as in Fig. 14.1. Equation (14.2.7) is consequently valid also for the current density in the surrounding conducting medium. We consider points $y > 0$ only, since the field is symmetrical with respect to the strip. Then $J_1 = 0$ (the current density must approach zero for $y \to \infty$). Hence

$$J_z(y) = J_z(0)e^{-Ky}, \qquad K = (1 + j)k = (1 + j)\sqrt{\omega\mu\sigma/2}\,. \tag{a}$$

The current density $J_z(0)$ can be expressed in terms of the current intensity $I$ in the strip as follows. Applying Ampère's circuital law to the closed contour 12341, we have $2aH_x(0) = -I$, from which $H_x(0) = -I/2a$. According to Eq. (14.2.3)

$$H_x(y) = -\frac{1}{K^2}\frac{dJ_z(y)}{dy} = \frac{1}{K}J_z(0)e^{-Ky}, \qquad H_x(0) = \frac{1}{K}J_z(0), \tag{b}$$

so that $J_z(0) = -KI/2a$. With this Eqs. (a) and (b) become

$$J_z(y) = -\frac{KI}{2a}e^{-Ky}, \qquad H_x(y) = -\frac{I}{2a}e^{-Ky},$$

or

$$J_z(y) = -\frac{(1+j)kI}{2a}e^{-ky}e^{-jky}, \qquad H_x = -\frac{I}{2a}e^{-ky}e^{-jky}. \tag{c}$$

To determine the power transformed into heat in the surrounding medium we need the electric and magnetic field intensities on the two faces of the strip. According to Eq. (c) they are

$$E_{\text{surface}} = J_z(0)/\sigma = -(1+j)kI/2a\sigma, \qquad H_{\text{surface}} = H_x(0) = -I/2a. \tag{d}$$

We can now determine the external resistance and inductance by which the surrounding medium loads the conductor per unit length. Noting that $\mathbf{E}_{\text{surface}} \times \mathbf{H}^*_{\text{surface}}$ is directed *into* the medium, while the positive normal is directed away from it (see Fig. 14.G4), we find, from Eq. (d), and from Eqs. (c) and (d) of Problem 14.G3, that

$$R_{l\,\text{ext}} = 2\frac{k}{4a^2\sigma}a = \frac{k}{2a\sigma}, \qquad L_{l\,\text{ext}} = 2\frac{k}{\omega 4a^2\sigma}a = \frac{k}{2a\sigma\omega}. \tag{e}$$

If we use Ampère's circuital law it is not difficult to prove that the amplitude of the *total* current induced in the medium is exactly $-I$.

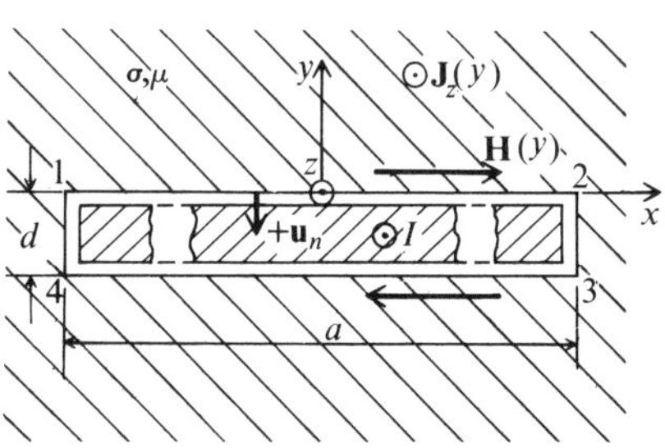


Figure 14.G4

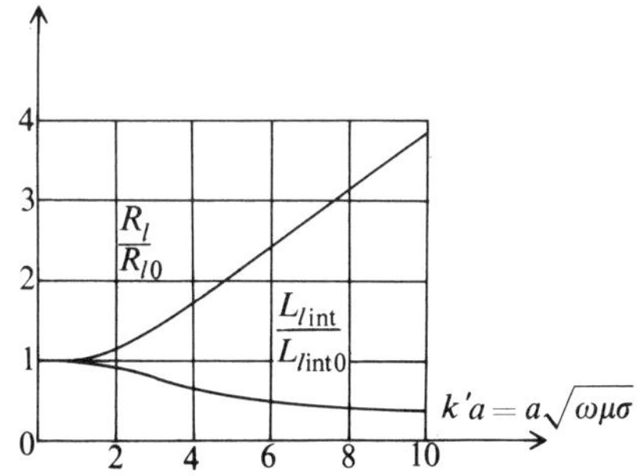


Figure 14.G5

***14.G5.** Determine the resistance and the internal inductance per unit length of the round wire analyzed in Section 14.4. Make use of Eq. (c) and (d) of Problem 14.G3.

*Solution.* From Fig. 14.7 we see that, on the surface of the conductor, the vector $\mathbf{E} \times \mathbf{H}^*$ is directed *into* the conductor, that is, in the direction opposite to an outward, positive normal.

According to Eqs. (14.4.18) and (14.4.19) and Eqs. (c) and (d) of Problem 14.G3 we have

$$R_l = +\frac{1}{|I|^2}\,\mathrm{Re}\left\{\frac{J(ka)}{\sigma}\,H^*(ka)\,2\pi a\right\} = \frac{k'}{2\pi a\sigma}\,\mathrm{Re}\{f(k'a)\}, \tag{a}$$

$$L_{l\,\mathrm{int}} = \frac{k'}{2\pi a\sigma\omega}\,\mathrm{Im}\{f(k'a)\}, \tag{b}$$

where $k' = \sqrt{\omega\mu\sigma}$, and

$$f(k'a) = j\,\frac{ber\,k'a + j\,bei\,k'a}{ber'\,k'a + j\,bei'\,k'a}$$

$$= \frac{ber\,k'a\,bei'\,k'a - bei\,k'a\,ber'\,k'a}{(ber'\,k'a)^2 + (bei'\,k'a)^2} + j\,\frac{ber\,k'a\,ber'\,k'a + bei\,k'a\,bei'\,k'a}{(ber'\,k'a)^2 + (bei'\,k'a)^2}. \tag{c}$$

For very small values of the parameter $k'a = a\sqrt{\omega\mu\sigma}$, Eqs. (a) and (b) must become the same as those valid for the steady current case. Substituting into Eq. (c) the first terms of series in Eqs. (14.4.11), (14.4.12), (14.4.16), and (14.4.17) we easily find that then

$$R_l \simeq \frac{1}{a^2\pi\sigma} = R_{l0}, \qquad L_{l\,\mathrm{int}} \simeq \frac{\mu}{8\pi} = L_{l\,\mathrm{int}\,0}. \tag{d}$$

Figure 14.G5 gives the ratios $R_l/R_{l0}$ and $L_{l\,\mathrm{int}}/L_{l\,\mathrm{int}\,0}$ as functions of $k'a$. Note that $L_{l\,\mathrm{int}}$ is always smaller than $L_{l\,\mathrm{int}\,0}$, and that $R_l$ is always larger than $R_{l0}$.

For very large values of $k'a$, that is, for frequencies high enough, the skin effect is very pronounced. The current is concentrated in a very thin layer near the surface of the conductor. The distribution of the current can be determined by considering the behavior of the *bei*, *ber*, *ber'* and *bei'* functions for large values of the argument. The same result can be obtained also in a simpler way. Considering the current layer to be approximately plane, we make use of Eq. (14.2.19):

$$J(r) \simeq J(ka)e^{k(r-a)}e^{jk(r-a)}, \qquad k = \sqrt{\omega\mu\sigma/2}, \qquad ka \gg 1. \tag{e}$$

According to Eq. (14.4.1) we know that

$$H(r) = \frac{1}{j\omega\mu\sigma}\,\frac{dJ(kr)}{dr} = \frac{(1+j)k}{j\omega\mu\sigma}\,J(ka)e^{k(r-a)}e^{jk(r-a)}. \tag{f}$$

The constant $J(ka)$ can be expressed in terms of the total current $I$ through the wire by using Ampère's circuital law:

$$2\pi a H(a) = I, \qquad \text{from which} \qquad J(ka) = \frac{j\omega\mu\sigma I}{2\pi a(1+j)k} = \frac{(1+j)kI}{2\pi a}.$$

So we have that

$$E(a) = J(a)/\sigma = (1+j)kI/2\pi a\sigma, \qquad H(a) = I/2\pi a,$$

from which

$$R_l + j\omega L_{l\,\mathrm{int}} = -\frac{1}{|I|^2}\left[-E(a)H^*(a)2\pi a\right] = \frac{k}{2\pi a\sigma} + j\,\frac{k}{2\pi a\sigma}. \tag{g}$$

The resistance is seen to be the same as if the total current is distributed uniformly in a thin layer, the thickness of which is equal to the skin depth $\delta = 1/k$. The total current is *not* distributed like that, but, remembering that $\delta = (2/\omega\mu\sigma)^{1/2} = 1/k$, we can easily recall formula (g).

***14.G6.** A long straight round wire, of radius $a$, permeability $\mu$, and conductivity $\sigma$, is situated in a uniform external magnetic field parallel to the wire. The magnetic field varies in time sinusoidally, with an angular frequency $\omega$. The amplitude of the magnetic field intensity on the surface of the wire is $H_0$. Determine the distribution of the magnetic field and of the current density in the wire.

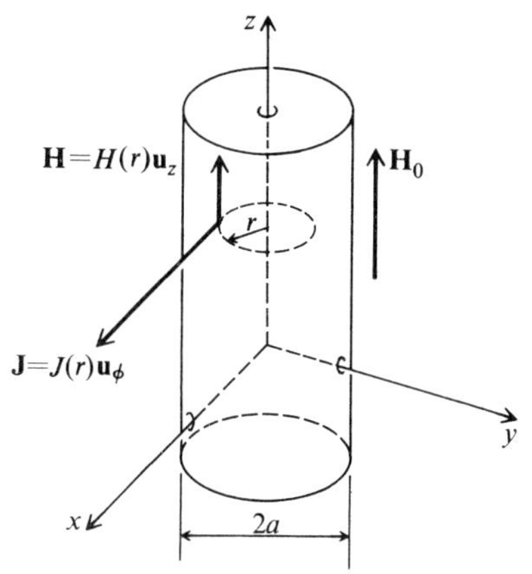


**Figure 14.G6**

*Solution.* Figure 14.G6 shows the wire with the $z$ axis of the cylindrical coordinate system coinciding with the wire axis. By symmetry, only the $z$ component of the magnetic field intensity and the $\phi$ component of the current density exist. Let us write the two Maxwell's curl equations (13.3.24) and (13.3.25), ignoring the displacement current density. The expression for the curl in the cylindrical coordinate system is derived in Problem 8.G20. The result is

$$\frac{1}{r}\frac{d}{dr}(rE) = -j\omega\mu H, \qquad -\frac{dH}{dr} = \sigma E = J. \tag{a}$$

Eliminating $E$ from these equations, we obtain

$$\frac{d^2H}{dr^2} + \frac{1}{r}\frac{dH}{dr} - j\omega\mu\sigma H = 0. \tag{b}$$

This equation is exactly the same as Eq. (14.4.3), with $H$ instead of $J$. Hence the resulting Eq. (14.4.13) also applies:

$$H(k'r) = H(0)\,(ber\; k'r + j\, bei\; k'r), \qquad k' = \sqrt{\omega\mu\sigma}. \tag{c}$$

According to Eq. (a) the current density is given by

$$J(k'r) = -k'H(0)\,(ber'\; k'r + j\, bei'\; k'r). \tag{d}$$

The constant $H(0)$ can be expressed in terms of the amplitude $H_0$ of the external field.

Since $H(k'a) = H_0$, from Eq. (c)

$$H(0) = \frac{H_0}{ber\, k'a + j\, bei\, k'a},$$

so that

$$H(k'r) = H_0 \frac{ber\, k'r + j\, bei\, k'r}{ber\, k'a + j\, bei\, k'a}, \tag{e}$$

$$J(k'r) = -k'H_0 \frac{ber'\, k'r + j\, bei'\, k'r}{ber\, k'a + j\, bei\, k'a}. \tag{f}$$

***14.G7.** Determine the distribution of the current and of the magnetic field intensity in the wire from the preceding problem for the special case when $k'a \ll 1$. Hence compute the average power transformed into heat per unit length of the wire.

*Solution.* Using only the first terms of the series in Eqs. (14.4.11), (14.412), (14.4.16), and (14.4.17), Eqs. (e) and (f) of the preceding problem become

$$H(k'r) \simeq H_0 \frac{1 + j(k'r)^2/2^2}{1 + j(k'a)^2/2^2} \simeq H_0,$$

$$J(k'r) \simeq -k'H_0 \frac{-4(k')^3/2^2 \cdot 4^2 + j2(k'r)/2^2}{1 + j(k'a)^2/2^2} \simeq -jk'H_0 \frac{k'r}{2}.$$

The power transformed into heat per unit length of the wire can be determined by integrating the expression $|J(k'r)|^2/2\sigma$ over the volume of unit length of the wire:

$$P_{Jl} = \int_0^a \frac{|J(k'r)|^2}{2\sigma} 2\pi r\, dr \simeq \frac{\pi k'^4 H_0^2 a^4}{16\sigma} = \frac{\pi\sigma\omega^2\mu^2 H_0^2 a^4}{16}.$$

This process is used for heating and melting pieces of metal by strong alternating magnetic fields. Furnaces based on this principle are known as *induction furnaces*.

***14.G8.** A ferromagnetic core of a solenoidal coil is made of a bunch of thin, mutually insulated round wires. The angular frequency of the sinusoidal current in the coil is $\omega$; the permeability of the wires is $\mu$, and their conductivity $\sigma$. Determine approximately the radius $a$ of the wires so that the magnetic field is distributed practically uniformly over their cross-section. Why is this condition important in practice?

***14.G9.** Determine the resistance and internal inductance per unit length of the strip line shown in Fig. 14.11, if the width of the strip (in the $x$ direction) is $a$. Use Eq. (14.5.2) for the current distribution and Eqs. (c) and (d) of Problem 14.G3, and take *both* strips into account. What is the external inductance of such a line per unit length?

*Answer.* For $kd \ll 1$ the resistance of the line per unit length is $2/\sigma ad$.

## Numerical problems

**14.N1.** A thin copper strip, of thickness $d = 1$ mm, carries a sinusoidal current of frequency $f$. By reference to Fig. 14.1 and Eq. (14.2.16), find the ratio $|J_z(kd/2)|/J_z(0)$ for the following frequencies: 50 Hz, $10^4$ Hz, and $10^6$ Hz.

**14.N2.** Suppose that the skin effect phenomena are to be analyzed in sea water and dry earth. Determine approximately the highest frequency for which the analysis of Section 14.2

would apply in these cases. Refer to Table 6.1, and assume that $\varepsilon_{\text{earth}} \simeq 10\varepsilon_0$ and $\varepsilon_{\text{sea water}} \simeq 81\varepsilon_0$.

**14.N3.** Check the values of the skin depth given in Table 14.1.

***14.N4.** By reference to Problem 14.G3, sketch the ratio $R_l/R_{l0}$ and $L_{l\,\text{int}}/L_{l\,\text{int}\,0}$, if $R_{l0}$ and $L_{l\,\text{int}\,0}$ correspond to $kd/2 \ll 1$.

**14.N5.** Determine approximately the thickness of an iron sheet with $\mu_r \simeq 1000$ so that, at the frequency $f = 60$ Hz, the distribution of the magnetic field in the sheet is practically uniform. What are the losses per unit volume of such a sheet if $B_0 = 0.1$ T?

***14.N6.** By reference to Problem 14.G5, find the resistance and internal inductance per unit length of a copper and an aluminum wire, of radius $a = 0.5$ mm, for the following frequencies: 60 Hz, 500 Hz, $10^5$ Hz, and $10^8$ Hz.

**14.N7.** A bunch of $N = 1000$ iron wires, of radius $a = 0.1$ mm and length 10 cm, is situated in a uniform external magnetic field of amplitude $B_0 = 0.1$ T and frequency $f = 60$ Hz. Assuming $\mu_r = 1000$ and $\sigma = 10^7$ S/m, find the average power losses in the bunch. Vector $B_0$ is parallel to the wires (see Problem 14.G7).

**Suggested reading.** The list progresses towards more advanced texts. For full bibliographical description see Bibliography.

1. WALSH, J. B., *Electromagnetic Theory and Engineering Applications*, Sections 6–8 to 6–11 and Chapter 8.
2. CARTER, G. W., *The Electromagnetic Field in Its Engineering Aspects*, Chapter 11.
3. RAMO, S., J. R. WHINNERY, and T. VAN DUZER, *Fields and Waves in Communication Electronics*, Chapter 5.
4. SMYTHE, W. R., *Static and Dynamic Electricity*, Chapter X.

CHAPTER 15

# UNIFORM PLANE ELECTROMAGNETIC WAVES

## 15.1 THE CONCEPT OF ELECTROMAGNETIC WAVE

One of the most brilliant achievements of Maxwell's theory of electromagnetism was the prediction of the existence of electromagnetic waves. In the present chapter the simplest types (and in a sense idealized types) of these waves, known as the plane waves, will be analyzed. Although very simple, they are of considerable practical importance as approximations to certain wave types frequently encountered.

In the formal analysis of plane electromagnetic waves their real sources, that is, time-varying charges and currents, do not explicitly appear. Only in a later chapter shall we consider some systems for radiating electromagnetic waves. But an appreciation of the physical meaning of the term "electromagnetic waves" is scarcely possible without some insight into the process of radiation. For this reason, a note on how electromagnetic waves are created is presented first.

Let us begin by considering a very long two-wire line situated in a vacuum (see Fig. 15.1), with current $i(t) = I \cos \omega t$ at the short-circuited end, and suppose that along the $z$ axis fixed point charges $Q$ are placed, as indicated by the black dots. Let us determine the forces on these charges as functions of time and distances $z$ from the line end.

Since the charges are at rest, they can be acted on only by the electric field. There are practically no charges at the line end, but the current is time-varying, so that the component of the electric field exists, as the result of the accelerated charges. We shall consider only the field produced by the current in the short-circuited end.

According to Eq. (13.4.31) the vector potential has an $x$ component only:

$$A_x = \frac{\mu_0}{4\pi} \frac{I \Delta l}{z} \cos \omega(t - z/c_0), \qquad c_0 = (\varepsilon_0 \mu_0)^{-1/2}. \tag{15.1.1}$$

From Eq. (13.4.17) we have that $E_y = E_z = 0$, and

$$E_x = -\frac{\partial A_x}{\partial t} = \frac{\mu_0}{4\pi} \frac{\omega I \Delta l}{z} \sin \omega(t - z/c_0). \tag{15.1.2}$$

The forces on the charges are simply $F_x = QE_x$. Note that they have different

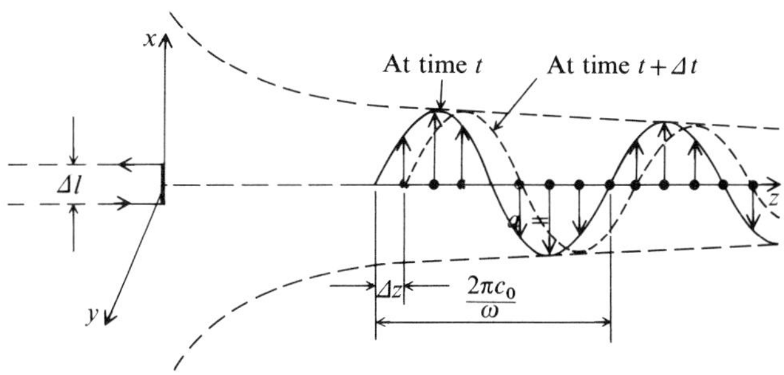


**Fig. 15.1** Electric wave produced by a small current element.

magnitudes because of the different distances $z$ of the charges from the end of the line. In addition, at the instant considered, the *directions* of the forces are also generally different, as the result of retardation, as indicated in Fig. 15.1. This situation does not occur in the case of static and quasi-static fields.

If we now consider a somewhat later instant, say $(t + \Delta t)$, we shall find that the electric field intensity at all points will be somewhat different. However, the value of the sine function at *all* points will be the same as the value it had at the instant $t$ at points for $\Delta z = c_0 \Delta t$ to the left, since

$$\sin \omega[t + \Delta t - (z + \Delta z)/c_0] = \sin \omega(t - z/c_0), \qquad \text{if} \qquad \Delta z = c_0 \Delta t.$$

The picture of the electric field at the instant $(t + \Delta t)$ is obtained simply by moving this picture at the instant $t$ to the right for $\Delta z = c_0 \Delta t$. This is shown in Fig. 15.1 in a dashed line. Now, the instant $t$ was completely arbitrary, so the electric field intensity varies at all points just as if the sinusoidal wave indicated moves to the right with a velocity $c_0 = \Delta z/\Delta t$. *Thus, changes of the electric field in space and time can be visualized as a sinusoidal "wave" moving with a constant velocity away from the source.* This is referred to as the sinusoidal electric wave. It can be shown that the magnetic field at large distances from the line end has exactly the same behavior. In fact, the electric and magnetic fields are only two parts of a unique field of force of charges moving in the end of the line. We can visualize the changes of the two fields in space and time as two simultaneously existing travelling waves. This two-wave system is referred to as the *electromagnetic wave*.

Two extremely important properties of all electromagnetic waves can now be easily deduced. The first is that an electromagnetic wave, once created, has no connection whatsoever with the system of charges and currents that produced it. We can best understand this by the following reasoning. Suppose the current in the line under consideration is suddenly turned off. The information that this has happened can never reach the waves produced earlier, because the end of a wave moves with the same velocity as the rest of it. The waves, therefore, go on propagating unaware, so to speak, that no successors are following them. In other words, once created, they have no further connection with their source.

The other property of electromagnetic waves is that they carry a certain amount of energy. This can be ascertained as follows. If a charge is situated in an electromagnetic wave, it will be acted upon by the electric force, and the work done on the charge by the electric force may be converted into mechanical energy or heat, or into some other form of energy. Since the electromagnetic waves have no connection with the sources which produced them, this energy must be supplied by the wave itself. In other words, the electromagnetic waves must carry a certain amount of energy. This fact is their most important property; without it their existence could not be detected, nor would the transmission of any type of information by means of electromagnetic waves be possible.

Electromagnetic waves are produced whenever charges and currents are time-varying. We can therefore imagine an unlimited number of electromagnetic wave types, but in this chapter we shall consider only the simplest case, the uniform plane electromagnetic waves. The importance of understanding the properties of these special waves cannot be overemphasized. Besides being the most simple type of waves, they represent approximations of waves at large distances from any radiating system.

## 15.2 UNIFORM PLANE ELECTROMAGNETIC WAVES IN HOMOGENEOUS UNBOUNDED DIELECTRICS

Let us consider an unbounded homogeneous dielectric, of permittivity $\varepsilon$ and permeability $\mu$, with no charges or currents at any point. All possible electromagnetic fields in that case must satisfy the following specialized set of general Maxwell's equations (13.3.6) to (13.3.9):

$$\text{curl}\,\mathbf{E} = -\mu\frac{\partial\mathbf{H}}{\partial t}, \qquad \text{div}\,\mathbf{E} = 0, \tag{15.2.1}$$

$$\text{curl}\,\mathbf{H} = \varepsilon\frac{\partial\mathbf{E}}{\partial t}, \qquad \text{div}\,\mathbf{H} = 0. \tag{15.2.2}$$

Eliminating **H** from the first set and **E** from the second set we have

$$\text{curl}\,(\text{curl}\,\mathbf{E}) = -\varepsilon\mu\frac{\partial^2\mathbf{E}}{\partial t^2}, \qquad \text{div}\,\mathbf{E} = 0, \tag{15.2.3}$$

$$\text{curl}\,(\text{curl}\,\mathbf{H}) = -\varepsilon\mu\frac{\partial^2\mathbf{H}}{\partial t^2}, \qquad \text{div}\,\mathbf{H} = 0. \tag{15.2.4}$$

These equations are valid generally. Both pairs can each be compressed into one equation, using the identity in Eq. (8.9.10):

$$\text{div}\,(\text{grad}\,\mathbf{E}) - \varepsilon\mu\frac{\partial^2\mathbf{E}}{\partial t^2} = 0, \tag{15.2.5}$$

$$\text{div}\,(\text{grad}\,\mathbf{H}) - \varepsilon\mu\frac{\partial^2\mathbf{H}}{\partial t^2} = 0, \tag{15.2.6}$$

with "div grad" as defined by Eq. (8.9.11). If we use the Laplacian operator acting formally on vector functions **E** and **H**, we can write this in a still simpler form:

$$\nabla^2\mathbf{E} - \varepsilon\mu\frac{\partial^2\mathbf{E}}{\partial t^2} = 0, \qquad \nabla^2\mathbf{H} - \varepsilon\mu\frac{\partial^2\mathbf{H}}{\partial t^2} = 0. \tag{15.2.7}$$

Equations of this type are referred to as *wave equations.*

The *uniform plane wave* is defined as follows. The electric and magnetic field intensities are assumed to be constant at all points of any plane perpendicular to a given direction. If we let this direction coincide with the $z$-axis of a rectangular coordinate system, then vectors **E** and **H** can only be functions of time and the $z$-coordinate. There is no lack of generality if two components of **E**, say $E_x$ and $E_z$, only are assumed to exist, since orientation of the coordinate system is arbitrary.

Let us then analyze possible types of electromagnetic waves having only the components $E_x(z, t)$ and $E_z(z, t)$. From the equation div $\mathbf{E} = 0$ it follows that $\partial E_z/\partial z = 0$. Hence $E_z$ does not vary with $z$, and therefore cannot represent a traveling wave. Thus the $x$ component $E_x(z, t)$ is the only possible component. According to Eq. (8.9.9) the first of the wave equations (Eq. 15.2.7) becomes

$$\frac{\partial^2 E_x}{\partial z^2} - \varepsilon\mu\frac{\partial^2 E_x}{\partial t^2} = 0. \tag{15.2.8}$$

By substitution it is not difficult to prove that the solution of this equation is

$$E_x(z, t) = E_1 f_1(t - z/c) + E_2 f_2(t + z/c), \qquad c = (\varepsilon\mu)^{-1/2}, \tag{15.2.9}$$

$E_1$ and $E_2$ being constants, and $f_1$ and $f_2$ *any* well-behaved (that is, differentiable) functions of the variables $(t \pm z/c) = (t \pm z\sqrt{\varepsilon\mu})$. These solutions represent waves traveling in opposite directions, a subject which is demonstrated below.

Let us first consider the term $f_1(t - z/c)$, and suppose that at an instant $t$ the function $f_1(t - z/c)$ has the form shown in solid line in Fig. 15.2. At a somewhat later instant, say $(t + \Delta t)$, this function will have exactly the same values as earlier,

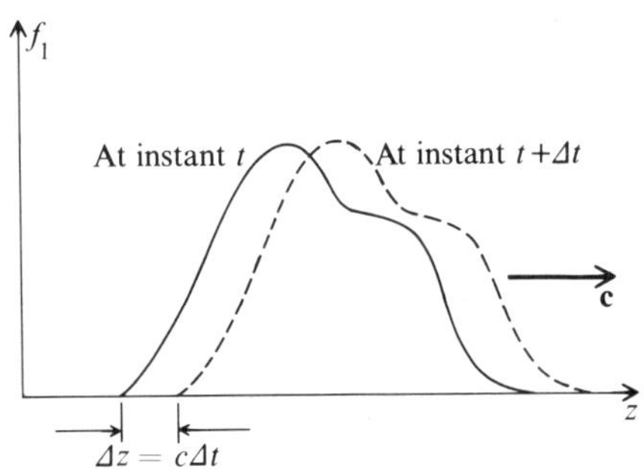


**Fig. 15.2** A function $f_1(t-z/c)$ at two close instants of time.

but now at points $(z + c\,\Delta t)$, since

$$f_1[(t + \Delta t) - (z + \Delta z)/c] = f_1(t - z/c), \qquad \text{if} \qquad \Delta z = c\,\Delta t.$$

This means that the waveform shown in Fig. 15.2 moves to the right with a velocity $c = \Delta z/\Delta t = (\varepsilon\mu)^{-1/2}$. We can easily show that $f_2(t + z/c)$ represents a waveform traveling in the negative $z$ direction with velocity $c$. Which one exists depends upon where the sources of the wave are. We shall assume that all sources are far to the left and that, unless explicitly mentioned, this other component does not exist. So we assume that

$$E_x(z, t) = E_1 f_1(t - z/c), \qquad c = (\varepsilon\mu)^{-1/2}. \tag{15.2.10}$$

The corresponding components of **H** can be now determined from the first part of Eq. (15.2.1). According to Eq. (8.8.8) for the curl in a rectangular coordinate system, we obtain

$$-\mu\frac{\partial H_x}{\partial t} = 0, \qquad -\mu\frac{\partial H_y}{\partial t} = \frac{\partial E_x}{\partial z} = E_1\frac{\partial f_1}{\partial z}, \qquad -\mu\frac{\partial H_z}{\partial t} = 0,$$

so that only the $H_y$ component exists. Now,

$$\frac{\partial f_1(t - z/c)}{\partial t} = \frac{\partial f_1(t - z/c)}{\partial(t - z/c)}\frac{\partial(t - z/c)}{\partial t} = \frac{\partial f_1(t - z/c)}{\partial(t - z/c)},$$

$$\frac{\partial f_1(t - z/c)}{\partial z} = \frac{\partial f_1(t - z/c)}{\partial(t - z/c)}\frac{\partial(t - z/c)}{\partial z} = -\frac{1}{c}\frac{\partial f_1(t - z/c)}{\partial(t - z/c)},$$

from which

$$\frac{\partial f_1}{\partial z} = -\frac{1}{c}\frac{\partial f_1}{\partial t}.$$

Omitting a possible term constant in time we therefore have

$$H_y(z, t) = \frac{1}{\mu c}E_1 f_1(t - z/c) = \frac{1}{\mu c}E_x(z, t).$$

As $c = (\varepsilon\mu)^{-1/2}$ this becomes

$$\boxed{H_y(z, t) = \sqrt{\frac{\varepsilon}{\mu}}\,E_x(z, t)}\,. \tag{15.2.11}$$

Since the unit of $H$ is ampère per meter and the unit of $E$ volt per meter, $(\mu/\varepsilon)^{1/2}$ has the dimension of resistance. It is referred to as the *intrinsic impedance* of the medium. The intrinsic impedance is usually designated by Greek letter $\eta$ (eta):

$$\boxed{\eta = \sqrt{\mu/\varepsilon}} \qquad \text{(the definition of intrinsic impedance).} \tag{15.2.12}$$

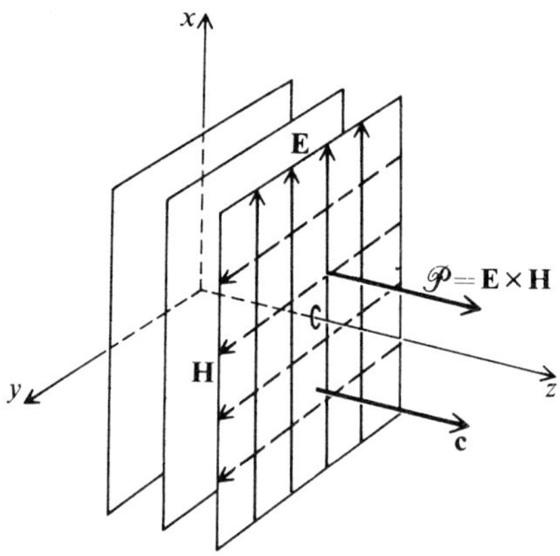


**Fig. 15.3** Vectors **E**, **H**, and $\mathscr{P} = \mathbf{E} \times \mathbf{H}$ of a plane electromagnetic wave.

Perhaps the most important case is that of waves propagating in a vacuum. Then

$$\boxed{\eta_0 = \sqrt{\mu_0/\varepsilon_0} \simeq 120\,\pi \simeq 377\,\Omega} \quad \text{(intrinsic impedance of a vacuum)}, \tag{15.2.13}$$

$$\boxed{c_0 = 1/\sqrt{\varepsilon_0\mu_0} \simeq 3 \times 10^8 \text{ m/s}} \quad \text{(velocity of propagation of plane waves in a vacuum)}. \tag{15.2.14}$$

*The velocity of propagation of plane waves in a vacuum is equal to the velocity of light*. This result was the basis of Maxwell's electromagnetic theory of light.

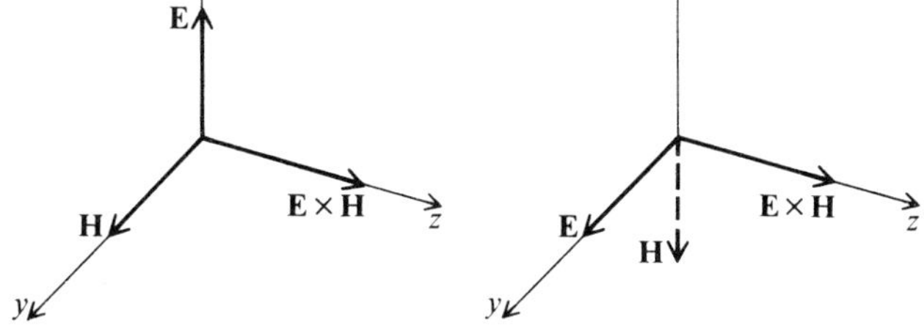


**Fig. 15.4** Relation between components of vectors **E** and **H** of a plane wave.

Finally, note that the cross product $\mathbf{E}_x \times \mathbf{H}_y$ of the plane wave components is in the direction of the wave propagation (see Fig. 15.3). This is exactly what we would expect, since the Poynting vector $\mathscr{P} = \mathbf{E} \times \mathbf{H}$ should be in the direction of the flow of electromagnetic energy. Thus the component $-H_x$ corresponds to the component $E_y$ (see Fig. 15.4).

**15.2.1 Harmonic (sinusoidal) plane waves.** Both practically and theoretically the most important type of plane waves are those varying with time according to the simple harmonic (sinusoidal) law. Waves used for broadcasting, television transmission, radar, etc. are very nearly sinusoidal, and any waveform can be obtained theoretically by a suitable combination of sinusoidal waves.

So instead of $f_1(t - z/c)$ let us substitute into Eqs. (15.2.10) and (15.2.11) $\cos \omega(t - z/c)$:

$$E_x(z, t) = E \cos \omega(t - z/c), \qquad H_y(z, t) = \frac{E}{\eta} \cos \omega(t - z/c). \qquad (15.2.15)$$

Much more often these expressions are written in the form

$$E_x(z, t) = E \cos (\omega t - \beta z), \qquad H_y(z, t) = \frac{E}{\eta} \cos (\omega t - \beta z), \qquad (15.2.16)$$

where

$$\boxed{\beta = \omega/c = \omega\sqrt{\varepsilon\mu}} \qquad \text{(definition of the phase constant).} \qquad (15.2.17)$$

The constant $\beta$ is referred to as the *phase constant*.

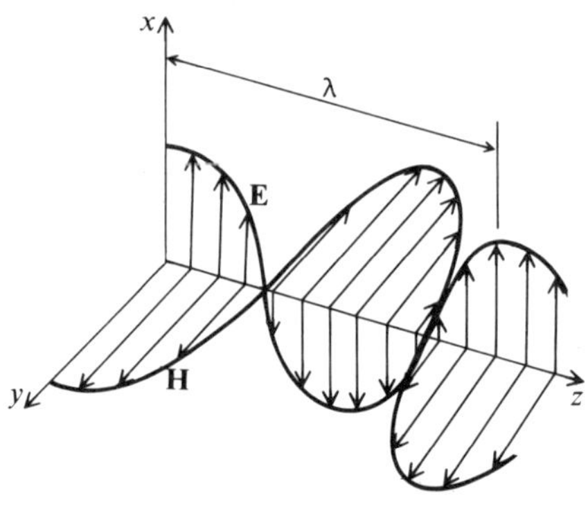


**Fig. 15.5** A graphical representation of the plane wave.

Figure 15.5 shows a harmonic plane wave, represented by Eq. (15.2.16). This figure is relatively simple, but it can also be misleading, because the fact that $E$ and $H$ are constant in the planes perpendicular to the $z$-axis is not clear from the figure. A more realistic sketch (though also a more complicated one) of a plane wave is shown in Fig. 15.6.

The distance between two successive maxima or minima of a periodic plane wave is known as the *wavelength* $\lambda$. In the case of sinusoidal waves $\beta\lambda = 2\pi$, from

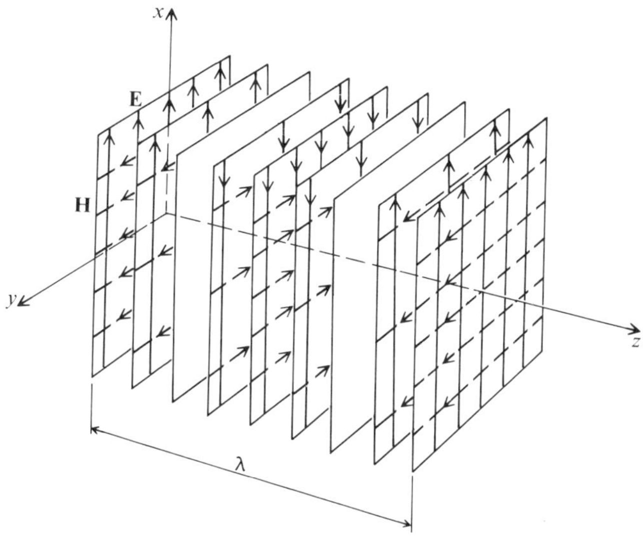


**Fig. 15.6** An alternative representation of the plane wave.

which

$$\boxed{\lambda = 2\pi/\beta = 2\pi/(\omega/c) = c/f} \qquad \text{(the definition of wavelength).} \quad (15.2.18)$$

Note that it is possible to define the wavelength only for periodic waves.

**15.2.2 The polarization of plane waves.** In engineering practice we most often deal with plane, or nearly plane, waves of the type described. That is, only one component of **E** and one component of **H** exists, their cross product **E** × **H** being in the direction of the wave propagation. But, again, any number of plane waves propagating in the same direction form a plane wave, so a plane wave can be a very complicated structure. Fortunately, we have not usually to deal with such complicated waves. We shall analyze only one type which is slightly more complicated than the simple plane wave already considered. This is the plane wave obtained when two plane waves of the same frequency, but with different orientations, amplitudes, and phases of vectors **E** and **H**, propagate in the same direction.

We will consider two plane waves, one having only the $x$ component and the other only the $y$ component of **E**, and suppose that

$$E_x(z, t) = E_x \cos(\omega t - \beta z),$$
$$E_y(z, t) = E_y \sin(\omega t - \beta z).$$

In this simple case

$$\left[\frac{E_x(z, t)}{E_x}\right]^2 + \left[\frac{E_y(z, t)}{E_y}\right]^2 = 1.$$

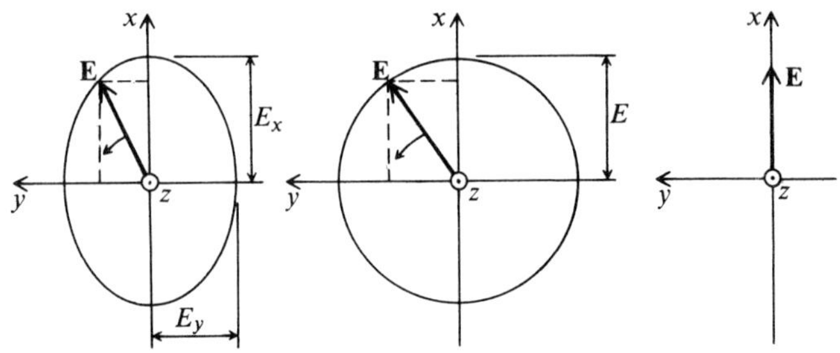


**Fig. 15.7** Illustration of elliptic, circular, and linear polarizations.

For a fixed $z$ this is the equation of an *ellipse* with semiaxes $E_x$ and $E_y$. This means that the tip of the total electric field intensity vector $\mathbf{E}_t(z, t) = E_x(z, t)\mathbf{u}_x + E_y(z, t)\mathbf{u}_y$ for any fixed $z$ describes this ellipse. Such a wave is said to be *elliptically polarized.*

In the special case when $E_x = E_y = E$, the tip of the total electric field intensity vector $\mathbf{E}_t(z, t)$ for a fixed $z$ describes a circle of radius $E$. This is said to be a plane wave with a *circular polarization.*

If either $E_x$ or $E_y$ is zero, we have a simple wave with only one component of the electric field intensity. This is referred to as *linear polarization.* The three types of polarization are shown in Fig. 15.7, while Fig. 15.8 shows an elliptically polarized wave in a way analogous to Fig. 15.6.

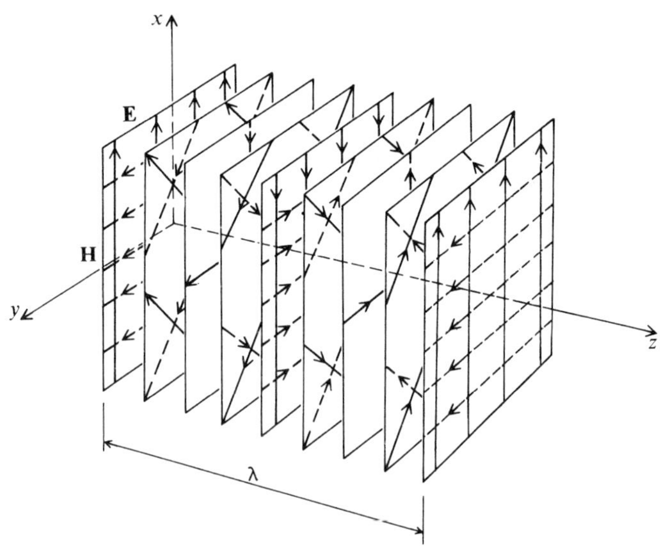


**Fig. 15.8** A representation of an elliptically polarized plane wave.

**15.2.3 The complex representation of plane waves.** If we deal with harmonic plane waves, we can devise expressions describing the components of the wave in the complex form. We recall that by definition the complex amplitude $E_{xc}$ of, say, the electric field intensity vector is defined by

$$E_x(z, t) = \text{Re}\,\{E_{xc}(z)e^{j\omega t}\}. \tag{15.2.19}$$

Comparing this with Eq. (15.2.16) we see that

$$\boxed{E_{xc}(z) = Ee^{-j\beta z}}\,. \tag{15.2.20}$$

In general, $E$ can also be complex. For example, if in Eq. (15.2.16) instead of $\cos(\omega t - \beta z)$ we have $\cos(\omega t - \beta z + \theta)$, the complex form of the expression would be $Ee^{j\theta}e^{-j\beta z} = E_c e^{-j\beta z}$. The complex expression for the magnetic field of a plane wave is of an analogous form. Thus, in complex notation, the presence of the factor $e^{-j\beta z}$ implies a wave propagating in the positive $z$ direction.

*Problems referring to Section 15.2:* 15.T1 to 15.T5, 15.G1 to 15.G6, 15.N1 to 15.N2.

## 15.3 PLANE WAVES IN IMPERFECT DIELECTRICS AND IN GOOD CONDUCTORS

**15.3.1 Harmonic plane waves in imperfect dielectrics.** Let us consider a homogeneous imperfect dielectric, of permittivity $\varepsilon$, permeability $\mu$, and conductivity $\sigma$, and assume that no charges exist inside the medium. Equations (15.2.1) are therefore also valid in this case. In Eq. (15.2.2) we have to add the term $\sigma\mathbf{E}$ to $\varepsilon\partial\mathbf{E}/\partial t$, so that, instead of Eqs. (15.2.5) and (15.2.6), we obtain

$$\text{div}\,(\text{grad}\,\mathbf{E}) - \mu\sigma\frac{\partial\mathbf{E}}{\partial t} - \varepsilon\mu\frac{\partial^2\mathbf{E}}{\partial t^2} = 0, \tag{15.3.1}$$

$$\text{div}\,(\text{grad}\,\mathbf{H}) - \mu\sigma\frac{\partial\mathbf{H}}{\partial t} - \varepsilon\mu\frac{\partial^2\mathbf{H}}{\partial t^2} = 0. \tag{15.3.2}$$

These equations are still more formidable than Eqs. (15.2.5) and (15.2.6), but, fortunately, if we write them in the complex form they are of the same *form* as the complex form of the earlier equations. When we take the equation for the vector $\mathbf{E}$ only, assuming $\mathbf{E}$ in the form $\mathbf{E}_c e^{j\omega t}$, Eqs. (15.2.5) and (15.3.1) become

$$\begin{aligned} \text{div}\,(\text{grad}\,\mathbf{E}_c) + \omega^2\varepsilon\mu\mathbf{E}_c &= 0,\\ \text{div}\,(\text{grad}\,\mathbf{E}_c) + (\omega^2\varepsilon\mu - j\omega\mu\sigma)\mathbf{E}_c &= 0. \end{aligned} \tag{15.3.3}$$

If we rearrange the last of these equations as follows

$$\text{div}\,(\text{grad}\,\mathbf{E}_c) + \omega^2\varepsilon_c\mu\mathbf{E}_c = 0, \tag{15.3.4}$$

in which

$$\boxed{\varepsilon_c = \varepsilon - j\sigma/\omega} \quad \text{(the definition of complex permittivity)}, \tag{15.3.5}$$

we see that the two equations are formally identical. The complex quantity $\varepsilon_c$ is referred to as the *complex permittivity*. For $\sigma = 0$ it becomes simply $\varepsilon$, and Eq. (15.3.4) reduces to Eq. (15.3.3). Thus Eq. (15.3.4) can be regarded as the general wave equation (in complex form).

The plane wave solution of Eq. (15.3.3) is given by Eq. (15.2.20), and the plane wave solution of Eq. (15.3.4) is obviously of the same form, except that $\varepsilon$ should be replaced by $\varepsilon_c$. According to Eq. (15.2.17), this means that the solution is obtained by simply changing $j\beta$ in Eq. (15.2.20) to

$$\boxed{\gamma_c = j\omega\sqrt{\varepsilon_c\mu} = j\omega\sqrt{(\varepsilon - j\sigma/\omega)\mu} = \alpha + j\beta} \quad \text{(the definition of complex propagation constant)}, \tag{15.3.6}$$

where

$$\alpha = \text{Re}\,\{j\omega\sqrt{(\varepsilon - j\sigma/\omega)\mu}\}, \qquad \beta = \text{Im}\,\{j\omega\sqrt{(\varepsilon - j\sigma/\omega)\mu}\}. \tag{15.3.7}$$

The complex constant $\gamma_c$ is referred to as the *complex propagation constant*. As before, $\beta$ is known as the *phase constant*, and $\alpha$ is termed the *attenuation constant*. With Eqs. (15.3.6) and (15.2.20) the simple plane wave solution of Eq. (15.3.4) is

$$E_{xc}(z) = E_c e^{-(\alpha + j\beta)z} = E_c e^{-\alpha z} e^{-j\beta z}. \tag{15.3.8}$$

The corresponding magnetic field is given by

$$H_{yc}(z) = \frac{E_c}{\eta_c} e^{-\alpha z} e^{-j\beta z}, \tag{15.3.9}$$

with

$$\eta_c = \sqrt{\frac{\mu}{\varepsilon_c}} = \sqrt{\frac{\mu}{\varepsilon - j\sigma/\omega}}. \tag{15.3.10}$$

Since $\eta_c$ is complex, the electric and magnetic fields do not have the same phase. Let $\eta_c = |\eta_c| \exp \theta_\eta$, and $E_c$ be real; the instantaneous values of $E_x(z, t)$ and $H_y(z, t)$ are then

$$E_x(z, t) = \text{Re}\,\{E_{xc}(z)e^{j\omega t}\} = Ee^{-\alpha z}\cos(\omega t - \beta z), \tag{15.3.11}$$

$$H_y(z, t) = \text{Re}\,\{H_{yc}(z)e^{j\omega t}\} = \frac{E}{|\eta_c|} e^{-\alpha z}\cos(\omega t - \beta z - \theta_\eta). \tag{15.3.12}$$

The amplitudes of both $E_x(z, t)$ and $H_y(z, t)$ decrease exponentially as the wave propagates in the $z$ direction. The measure of the rate at which they decrease is the attenuation constant $\alpha$. The amplitudes of $E_x(z, t)$ and $H_y(z, t)$ at the plane $(z + 1/\alpha)$ are exactly $e = 2.718\ldots$ times smaller than at the plane $z$. Thus, the larger $\alpha$ is, the larger the attenuation of the wave.

The attenuation between two points $\Delta z$ apart is said to be 1 *neper*, if the natural logarithm of the ratio of the wave amplitudes at the starting point and at the end point is unity. Thus for $\Delta z = d$ we have

$$\text{attenuation in nepers} = \ln \frac{Ee^{-\alpha z}}{Ee^{-\alpha(z+d)}} = \ln e^{\alpha d} = \alpha d. \qquad (15.3.13)$$

Consequently, the attenuation constant is measured in *nepers per meter* (abbreviation, Np/m). We must note that, according to Eq. (15.3.7), the attenuation constant is not only a function of $\varepsilon$, $\mu$ and $\sigma$; it is also a function of frequency. This means that waves of different frequencies have different attenuation.

The velocity of propagation of plane waves in lossy dielectrics is again $\omega/\beta$, as we can see from Eq. (15.3.11) and (15.3.12). But $\beta$ is now *not* equal to $\omega\sqrt{\varepsilon\mu}$; its value is given by Eq. (15.3.7). It is also a function of frequency, a phenomenon which is not present in the case of perfect dielectrics. If a signal consists of waves having different frequencies (which in practice is always so) it "disperses" when propagating through such a medium. Waves of lower frequency have different velocities from those of higher frequencies, and the original signal will be "smeared". The media in which the velocity of propagation is a function of the wave frequency are therefore termed *dispersive*, and it is clear that a lossy dielectric is a dispersive medium.

The conductivity of lossy dielectrics is usually very small, that is, $\sigma/\omega$ is much smaller than $\varepsilon$. In such cases the complex propagation constant of Eq. (15.3.6) can be written in the following approximate form:

$$\gamma_c = j\omega\sqrt{\varepsilon\mu}\sqrt{1 - j\frac{\sigma}{\omega\varepsilon}} \simeq j\omega\sqrt{\varepsilon\mu}\left[1 - j\frac{\sigma}{2\omega\varepsilon} + \frac{1}{8}\left(\frac{\sigma}{\omega\varepsilon}\right)^2\right].$$

Hence, according to Eq. (15.3.7), we have

$$\alpha \simeq \omega\sqrt{\varepsilon\mu}\,\frac{\sigma}{2\omega\varepsilon} = \frac{\sigma}{2}\sqrt{\frac{\mu}{\varepsilon}} \qquad (\sigma \ll \omega\varepsilon), \qquad (15.3.14)$$

$$\beta \simeq \omega\sqrt{\varepsilon\mu}\left[1 + \frac{1}{8}\left(\frac{\sigma}{\omega\varepsilon}\right)^2\right] \qquad (\sigma \ll \omega\varepsilon). \qquad (15.3.15)$$

The velocity of propagation is

$$v = \frac{\omega}{\beta} \simeq \frac{1}{\sqrt{\varepsilon\mu}}\left[1 - \frac{1}{8}\left(\frac{\sigma}{\omega\varepsilon}\right)^2\right] \qquad (\sigma \ll \omega\varepsilon). \qquad (15.3.16)$$

The complex intrinsic impedance of the medium in this case is

$$\eta_c = \sqrt{\frac{\mu}{\varepsilon_c}} = \sqrt{\frac{\mu}{\varepsilon(1 - j\sigma/\omega\varepsilon)}}$$

$$= \sqrt{\frac{\mu}{\varepsilon}}\left(1 - j\frac{\sigma}{\omega\varepsilon}\right)^{-1/2} \simeq \sqrt{\frac{\mu}{\varepsilon}}\left(1 + j\frac{\sigma}{2\omega\varepsilon}\right). \qquad (15.3.17)$$

**15.3.2 Harmonic plane waves in good conductors.** Let us consider a good conducting medium, that is, let us suppose that $\sigma \gg \omega\varepsilon$. The general conclusions from the preceding section are also formally valid in this case, but can be considerably simplified. The propagation constant of Eq. (15.3.6) becomes simply

$$\gamma_c \simeq j\omega\sqrt{-j\sigma\mu/\omega} = j\sqrt{-j}\sqrt{\omega\mu\sigma} = (1 + j)\sqrt{\omega\mu\sigma/2}$$

$$= (1 + j)\sqrt{\pi f\mu\sigma}, \qquad (15.3.18)$$

since $j\sqrt{-j} = e^{j\pi/2}e^{-j\pi/4} = e^{j\pi/4} = (1 + j)/\sqrt{2}$, and $\omega = 2\pi f$. According to Eq. (15.3.7), the attenuation and phase constants in this case are equal:

$$\alpha = \beta = \sqrt{\pi f\mu\sigma} \qquad \text{(a good conductor).} \qquad (15.3.19)$$

Waves propagating through good conductors are attenuated very rapidly; for example, in copper ($\sigma \simeq 5.8 \times 10^7$ S/m, $\mu \simeq \mu_0$), at frequency $f = 10^6$ Hz, we find that $\alpha \simeq 15{,}150$ Np/m. Thus, after travelling a distance of 0.5 mm $= 0.5 \times 10^{-3}$ m, the wave has an amplitude $e^{7.575} \simeq 1950$ times smaller. Consequently, the wave, together with the currents it induces in a conductor, is restricted to a thin layer near the surface, and we recognize the skin effect phenomenon discussed in the preceding chapter.

For practical applications at high frequencies it is important to know the average power losses per unit area of a conductor for a given amplitude of the wave at the surface. We shall use this information for determining the attenuation along transmission lines and waveguides. Suppose that at the plane surface of a conductor the amplitude of the magnetic field is $H_0$, and that the wave propagates into the conductor (see Fig. 15.9). The average power absorbed from the wave per unit area of the conductor is given as the real part of the complex Poynting

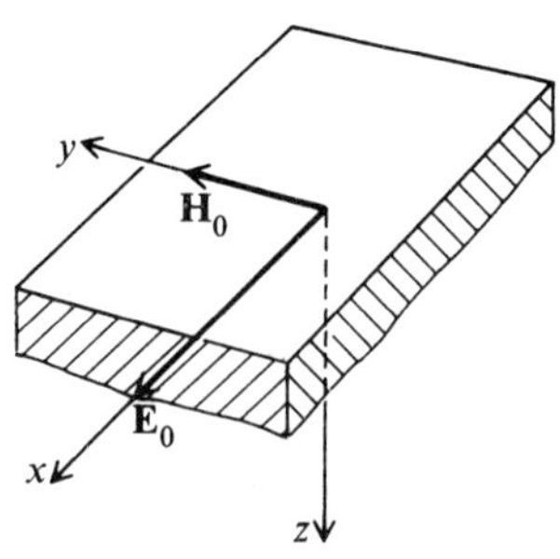


**Fig. 15.9** A plane wave propagating into a conductor.

vector (see Problem 13.G7):

$$\frac{P}{S} = \text{Re}\,\{\tfrac{1}{2}E_0 H_0^*\}.$$

Now, for a good conductor the intrinsic impedance of Eq. (15.3.10) is simply $\eta_c \simeq \sqrt{j}\,\sqrt{\omega\mu/\sigma}$. So we have

$$\frac{P}{S} = \text{Re}\,\{\tfrac{1}{2}\eta_c H_0 H_0^*\} = \text{Re}\left\{\tfrac{1}{2}(1 + j)\,|H_0|^2 \sqrt{\frac{\omega\mu}{2\sigma}}\right\} = \tfrac{1}{2}\,|H_0|^2 \sqrt{\frac{\omega\mu}{2\sigma}}\,.$$

The square root on the right has the dimension of resistance. It is usually referred to as the *surface resistance*, and designated by $R_s$:

$$\boxed{R_s = \sqrt{\frac{\omega\mu}{2\sigma}} = \sqrt{\frac{\pi f \mu}{\sigma}}} \qquad \text{(the definition of surface resistance).} \tag{15.3.20}$$

With this definition, the average power absorbed from the wave per unit area of a good conductor is simply

$$\boxed{\frac{dP}{dS} = \tfrac{1}{2}R_s\,|H_0|^2}\,. \tag{15.3.21}$$

This is an important formula, which is strictly valid only for plane conducting surfaces. But it obviously represents an excellent approximation whenever the radius of curvature of the surface is much larger than the skin depth. In practice this is almost always the case at very high frequencies. The average losses are then obtained by integrating Eq. (15.3.21) over the surface of the conductor, since $|H_0|$ usually varies at different points.

*Problems referring to Section 15.3:* 15.T6, 15.N3 to 15.N4.

## 15.4 REFLECTION AND REFRACTION OF PLANE WAVES

**15.4.1 Introduction.** In practice, plane, or almost plane, waves never propagate through unbounded homogeneous media. Whenever an object with properties different from the surrounding homogeneous medium is present in the electromagnetic field of a plane wave, the foregoing considerations are not valid. The electric field of the wave will induce either conduction or polarization currents in the object, depending upon whether it is a conductor or a dielectric. The total wave is the sum of the initial, original wave, and a secondary wave radiated by the currents and charges induced on the object. The determination of this resultant wave is very difficult even for objects of the simplest shapes.

One class of problems of great practical importance is, however, relatively easy to analyze. These are the cases when a plane wave is incident on a *plane* interface of two different homogeneous media, when again the secondary waves radiated by the currents induced on the boundary surface are plane. It follows that the resultant waves in both media are the resultants of two plane waves.

Strictly speaking, harmonic *plane* waves incident on *plane* boundary surfaces are never encountered in practice. But *almost* harmonic and *almost* plane waves incident on an *almost* plane surface are of very frequent occurrence. Hence the great practical importance of the idealized cases we shall consider.

**15.4.2 Perpendicular incidence on a perfectly conducting plane.** A perfect conductor does not exist, but all good conductors, like silver, copper, and aluminum, can be regarded as approximately perfect.

Let us suppose that a very large, perfectly conducting slab is in the path of a uniform plane wave propagating through a perfect dielectric. We assume first that the conducting surface is perpendicular to the direction of the propagation of the wave, that is, parallel to both vectors $\mathbf{E}_i$ and $\mathbf{H}_i$ of the incident wave (see Fig. 15.10). The electric field of the wave will induce currents on the surface of the conductor, which in turn will give rise to a secondary wave. The resultant wave is the sum of the incident and secondary waves.

To determine the secondary wave we make use of boundary conditions. According to Eq. (13.4.7), the tangential component of the *total* electric field intensity at the surface of a perfect conductor must be zero. Hence it follows that at the conductor surface the electric field intensity of the secondary wave must be of equal amplitude to that of the incident wave, and opposite to it (see Fig. 15.10). The secondary magnetic field produced by the induced surface currents to the right of the surface must be of the same magnitude as $H_i$ but in the opposite direction. Otherwise the magnetic field would exist inside the perfect conductor, which is impossible.

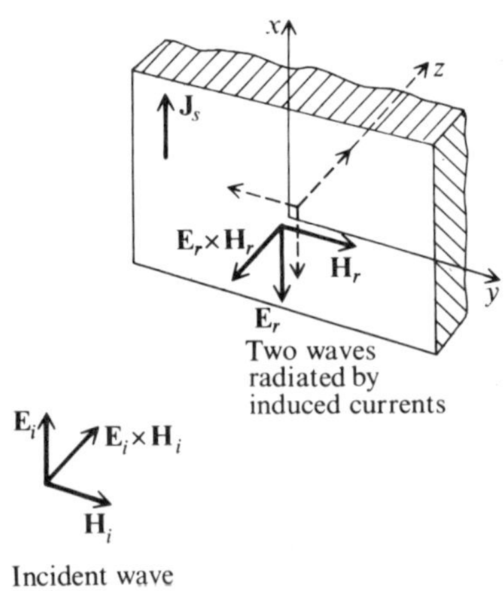


**Fig. 15.10** Perpendicular incidence on a perfectly conducting plane.

The total field is equal to the sum of the incident and the secondary waves, and, by assumption, this sum inside the conductor is zero. To the left of the conductor surface we have two plane waves, propagating in opposite directions. The secondary wave is really due to the induced currents in the slab, but once we understand this, it is customary for us to refer to the secondary wave as the *reflected wave*. To have a clear physical picture of what is going on, we must always remember that the reflected wave is created by the currents induced on the conductor surface.

Now we know that two traveling waves exist to the left of the conductor surface:

$$\mathbf{E}_i(z) = E_i e^{-j\beta z}\mathbf{u}_x, \qquad \mathbf{H}_i(z) = H_i e^{-j\beta z}\mathbf{u}_y \qquad \text{(the incident wave)},$$
$$\mathbf{E}_r(z) = -E_i e^{+j\beta z}\mathbf{u}_x, \qquad \mathbf{H}_r(z) = H_i e^{+j\beta z}\mathbf{u}_y \qquad \text{(the reflected wave)}.$$

Their sum is

$$\mathbf{E}_{\text{total}}(z) = \mathbf{E}_i(z) + \mathbf{E}_r(z) = E_i(e^{-j\beta z} - e^{+j\beta z})\,\mathbf{u}_x = -2jE_i \sin\beta z\,\mathbf{u}_x, \tag{15.4.1}$$
$$\mathbf{H}_{\text{total}}(z) = \mathbf{H}_i(z) + \mathbf{H}_r(z) = H_i(e^{-j\beta z} + e^{+j\beta z})\,\mathbf{u}_y = 2H_i \cos\beta z\,\mathbf{u}_y. \tag{15.4.2}$$

The instantaneous value of the total wave is therefore

$$\mathbf{E}_{\text{total}}(z, t) = \text{Re}\,\{\mathbf{E}_{\text{total}}(z)e^{j\omega t}\} = 2E_i \sin\beta z \sin\omega t\,\mathbf{u}_x, \tag{15.4.3}$$
$$\mathbf{H}_{\text{total}}(z, t) = \text{Re}\,\{\mathbf{H}_{\text{total}}(z)e^{j\omega t}\} = 2H_i \cos\beta z \cos\omega t\,\mathbf{u}_y. \tag{15.4.4}$$

The resultant wave is of a peculiar form. It is not a traveling wave, since the planes in which $\mathbf{E}_{\text{total}}$ and $\mathbf{H}_{\text{total}}$ are zero are fixed: $\mathbf{E}_{\text{total}}$ is zero in the planes defined by $\beta z = -n\pi$, $n = 0, 1, 2, \ldots$; $\mathbf{H}_{\text{total}}$ is zero in the planes defined by $\beta z = -(n + 1)\pi/2$, $n = 0, 1, 2, \ldots$. In a traveling wave the whole waveform moves with a certain velocity. In the present case a wave exists, but, in a sense, it is fixed.

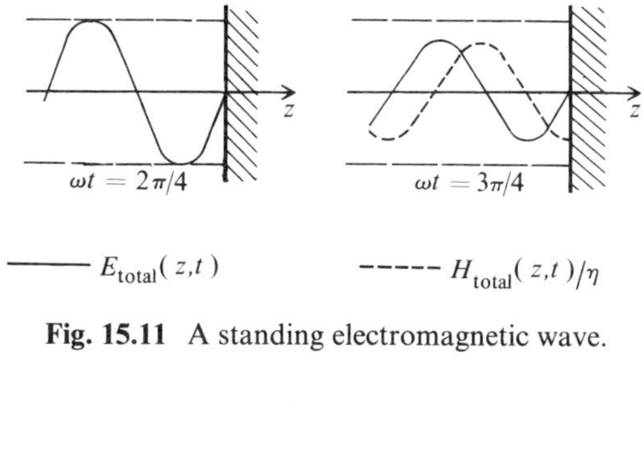


**Fig. 15.11** A standing electromagnetic wave.

Such waves are known as *standing waves*. Figure 15.11 shows a sequence of values of $\mathbf{E}_{\text{total}}(z, t)$ and $\mathbf{H}_{\text{total}}(z, t)/\eta$, as given by Eqs. (15.4.3) and (15.4.4).

**15.4.3 Oblique incidence on a perfectly conducting plane.** The general philosophy for determining the reflected wave described in the preceding section can also be applied in this case. The charges and currents induced on the plane by the incident wave give rise to a secondary wave, which, on the conductor side, must exactly cancel the incident wave, since a varying electromagnetic field cannot exist in a perfect conductor. This determines uniquely the secondary wave radiated to both sides of the interface. As before, the secondary wave radiated into the dielectric is called the reflected wave.

Now let us consider a plane wave which is incident at an angle $\theta$ with respect to the normal to the conductor surface (see Fig. 15.12). The plane perpendicular to the interface and parallel to the direction of the wave is known as the *plane of incidence*, a concept which we shall use frequently. In Fig. 15.12 the plane of incidence coincides with the plane $x = 0$. Denote by $\mathbf{E}_s$ and $\mathbf{H}_s$ the vector components of the secondary wave created by the currents and charges induced on the conductor surface. Below the surface (inside the conductor) the secondary wave must exactly cancel the incident wave, as indicated in the figure by dashed arrows. But, by symmetry, these charges and currents must also radiate a wave symmetric to $\mathbf{E}_s$, $\mathbf{H}_s$ with respect to the surface, which is the reflected wave. In the figure it is designated by $\mathbf{E}_r$, $\mathbf{H}_r$. Obviously, the *direction of propagation of the reflected wave is in the plane of incidence and makes exactly an angle $\theta$ with the normal to the plane*, as shown.

It is not easy to determine the reflected and the resultant waves for arbitrary directions of $\mathbf{E}_i$ and $\mathbf{H}_i$ (which of course are mutually perpendicular and also perpendicular to the direction of propagation). Therefore we will consider two special

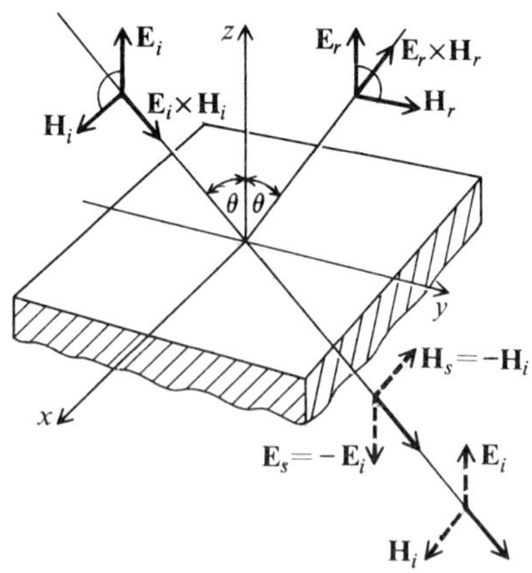


**Fig. 15.12** Oblique incidence on a perfectly conducting plane.

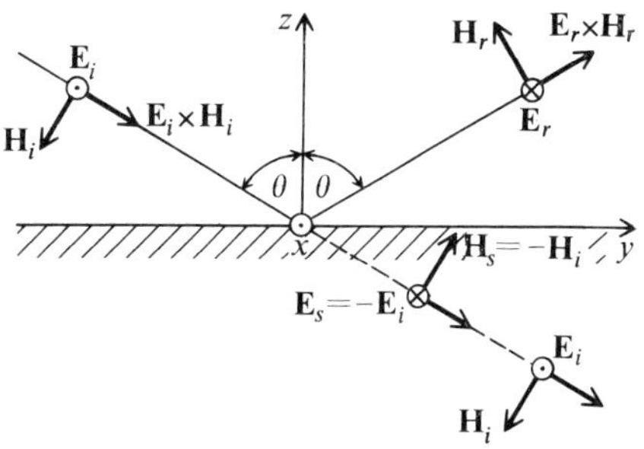


**Fig. 15.13** Horizontally polarized wave obliquely incident on a perfectly conducting plane.

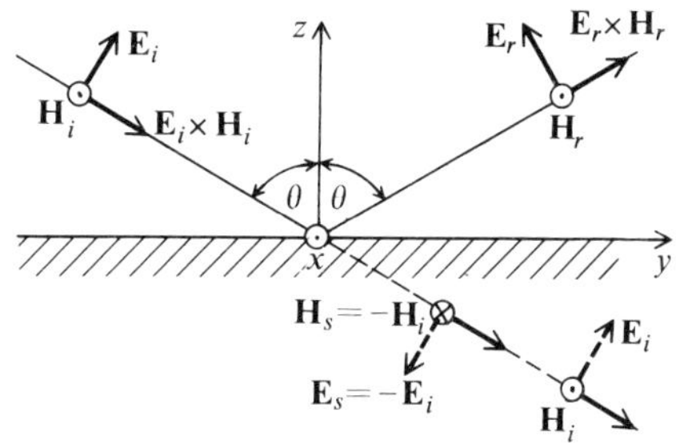


**Fig. 15.14** Parallel polarized wave obliquely incident on a perfectly conducting plane.

cases, from which any wave can be obtained by superposition. The first is when $\mathbf{E}_i$ is perpendicular to the plane of incidence. It is then said that the wave is *perpendicularly* or *horizontally* polarized. The second case is when $\mathbf{E}_i$ is parallel to the plane, when the polarization is said to be *parallel*.

Figure 15.13 shows a horizontally polarized incident wave with its corresponding reflected wave. According to the above general reasoning, the amplitudes of the two field vectors must be equal; their direction is shown in the figure. We have to determine the relative phase of the reflected wave with respect to the incident wave, so that at any instant, for $z = 0$, $\mathbf{E}_i + \mathbf{E}_r = 0$. The resultant wave can now be determined relatively easily.

A parallel polarized incident wave with its corresponding reflected wave is shown in Fig. 15.14. Again the amplitudes of the components of the two waves are equal; their directions are shown in the figure. We have to determine the phase of the reflected wave with respect to the incident wave, so that at $z = 0$ no parallel component of the electric field intensity vector exists. Their phases are therefore equal.

**15.4.4 Perpendicular incidence on the plane surface of a dielectric.** Let us consider two homogeneous dielectric media which have a common plane interface, and suppose that an incident plane wave propagating through dielectric 1 (of parameters $\varepsilon_1$ and $\mu_1$) impinges on the interface (see Fig. 15.15). As the interface is discontinuous in $\varepsilon$ and $\mu$, it will give rise to secondary plane waves propagating away from it in both directions. The secondary wave propagating to the left through medium 1 is again termed the reflected wave, and its electric and magnetic fields are designated by the subscript "$r$". So in medium 1 there are two waves propagating in opposite directions, the incident wave and the reflected wave.

In medium 2 both the incident and the secondary waves exist, propagating to the right; they constitute a single plane wave propagating in the positive $z$ direction.

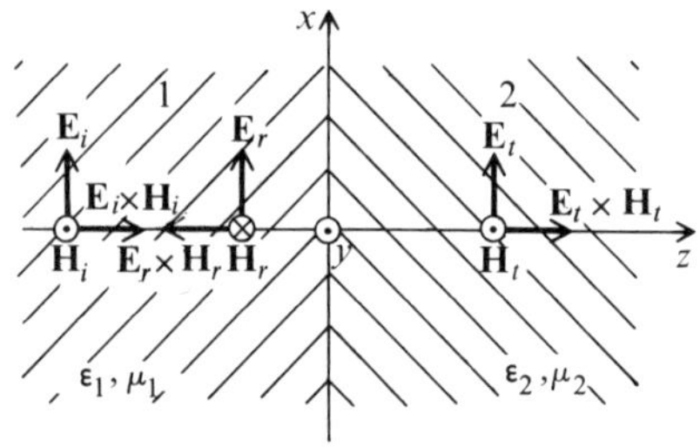

**Fig. 15.15** A plane wave incident on a plane boundary surface between two dielectrics.

The wave in medium 2 is referred to as the *transmitted* wave. The electric and magnetic fields of the transmitted wave are designated by the subscript "$t$".

The relations between the amplitudes of the components of the three waves are obtained from the boundary conditions. Using the complex notation, let

$$\mathbf{E}_i(z) = E_i e^{-j\beta_1 z}\mathbf{u}_x, \qquad \mathbf{H}_i(z) = \frac{E_i}{\eta_1} e^{-j\beta_1 z}\mathbf{u}_y, \tag{15.4.5}$$

$$\mathbf{E}_r(z) = E_r e^{+j\beta_1 z}\mathbf{u}_x, \qquad \mathbf{H}_r(z) = -\frac{E_r}{\eta_1} e^{+j\beta_1 z}\mathbf{u}_y, \tag{15.4.6}$$

$$\mathbf{E}_t(z) = E_t e^{-j\beta_2 z}\mathbf{u}_x, \qquad \mathbf{H}_t(z) = \frac{E_t}{\eta_2} e^{-j\beta_2 z}\mathbf{u}_y. \tag{15.4.7}$$

According to the boundary conditions in Eqs. (13.4.1) and (13.4.3), we find that, for $z = 0$, the following relations must hold:

$$\mathbf{E}_i(0) + \mathbf{E}_r(0) = \mathbf{E}_t(0), \qquad \text{or} \qquad E_i + E_r = E_t, \tag{15.4.8}$$

$$\mathbf{H}_i(0) + \mathbf{H}_r(0) = \mathbf{H}_t(0), \qquad \text{or} \qquad (E_i - E_r)/\eta_1 = E_t/\eta_2. \tag{15.4.9}$$

From these equations

$$E_r = \frac{\eta_2 - \eta_1}{\eta_1 + \eta_2} E_i, \qquad E_t = \frac{2\eta_2}{\eta_1 + \eta_2} E_i. \tag{15.4.10}$$

The ratio $E_r/E_i$ is referred to as the *reflection coefficient*, and we shall designate it by $\mathscr{R}$. Similarly, $E_t/E_i$ is known as the *transmission coefficient*, $\mathscr{T}$. So

$$\mathscr{R} = \frac{E_r}{E_i} = \frac{\eta_2 - \eta_1}{\eta_1 + \eta_2} \qquad \text{(the reflection coefficient)}, \tag{15.4.11}$$

$$\mathscr{T} = \frac{E_t}{E_i} = \frac{2\eta_2}{\eta_1 + \eta_2} \qquad \text{(the transmission coefficient)}. \tag{15.4.12}$$

The wave in medium 2 is a wave traveling in the positive $z$ direction; in medium 1 there are two waves, traveling in opposite directions. Since one part of the wave is transmitted to region 2, one part of the total wave in medium 1 must be a traveling wave; the other part is of the standing-wave type. According to Eqs. (15.4.5), (15.4.6), and (15.4.11), we have

$$\mathbf{E}_1(z) = \mathbf{E}_i(z) + \mathbf{E}_r(z) = E_i e^{-j\beta_1 z}(1 + \mathscr{R}e^{+2j\beta_1 z})\mathbf{u}_x. \qquad (15.4.13)$$

If we recall that $e^{\pm jx} = \cos x \pm j \sin x$, so that $|e^{\pm jx}| = 1$ and $e^{\pm j2n\pi} = 1$, then, provided that $\mathscr{R} > 0$, i.e. $\eta_2 > \eta_1$, the maximum values of $E_1(z)$ will occur for the values of $z$ determined by

$$2\beta_1 z_{\max} = -2n\pi, \quad \text{or} \quad z_{\max} = -\frac{n\pi}{\beta_1} = -\frac{n\lambda_1}{2}, \quad n = 0, 1, 2, \ldots \qquad (15.4.14)$$

since $\beta_1 = 2\pi/\lambda_1$. Similarly, since $e^{\pm j(2n+1)\pi} = -1$, provided that $\mathscr{R} > 0$ minimum values of $E_1(z)$ are obtained for the following values of $z$:

$$2\beta_1 z_{\min} = -(2n + 1)\pi, \quad \text{or}$$

$$z_{\min} = -\frac{(2n + 1)\pi}{2\beta_1} = -\frac{(2n + 1)\lambda_1}{4}, \quad n = 0, 1, 2, \ldots \qquad (15.4.15)$$

If $\mathscr{R} < 0$ (i.e. if $\eta_2 < \eta_1$), the positions of the maxima and minima are interchanged. The ratio

$$s = \frac{|E_1(z)|_{\max}}{|E_1(z)|_{\min}} = \frac{1 + |\mathscr{R}|}{1 - |\mathscr{R}|} \quad \text{(the definition of the standing-wave ratio)} \qquad (15.4.16)$$

is known as the *standing-wave ratio*. Let us remember: the larger the reflected wave, the larger the standing-wave ratio. Figure 15.16 shows the standing-wave pattern for $s = 2$.

Standing waves are usually undesirable, since we wish to transmit all the energy of the wave to the other region, but in some instances they are very useful. For example, suppose that we wish to determine the intrinsic impedance $\eta_2$ of medium 2, $\eta_1$ being known. In these circumstances we can make use of the standing wave pattern. Firstly, if the maximum occurs at the interface, we conclude that $\eta_2 > \eta_1$. Secondly, we can measure the standing wave ratio $s$. Both measurements can be made, for example, by using a thin opened coaxial cable, with the

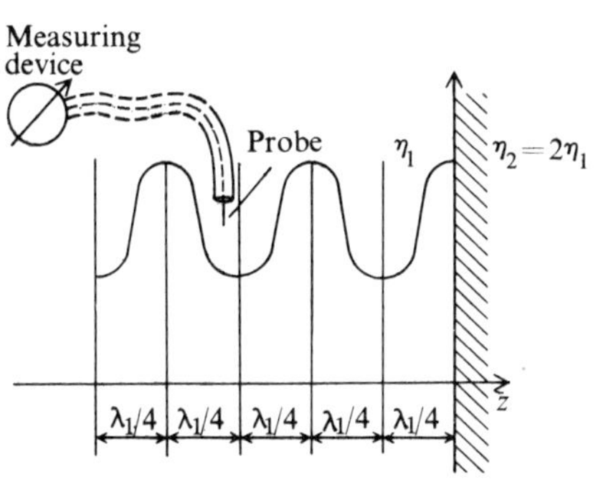


**Fig. 15.16** Standing wave pattern for perpendicular incidence and for $s = 2$.

inner conductor slightly protruding, the other end of the cable being connected to a convenient measuring device (see Fig. 15.16). The electromotive force induced in the protruding end is proportional to the electric field in which it is immersed. We can therefore move the probe, detect the largest and smallest indications of the measuring device, and determine their ratio $s$. $\mathscr{R}$ is then obtained from Eq. (15.4.16), and $\eta_2$ from Eq. (15.4.11). A similar technique is used for measuring various quantities on transmission lines and waveguides.

It is important to note that the foregoing analysis is also formally valid if one or both of the two media have nonzero conductivities. The only difference is that $\eta_1$ and $\eta_2$ are then complex, and in consequence the reflection and transmission coefficients are also complex. This only modifies the above analysis quantitatively. For example, if $\sigma_2 \gg \omega\varepsilon_2$ (medium 2 being a very good conductor), and in addition $|\eta_2| \ll |\eta_1|$, we find that $\mathscr{R} = -1$, which we said was true for a perfect conductor (Section 15.4.2).

**15.4.5 Oblique incidence on the plane surface of a dielectric.** Let us consider a plane wave propagating through a dielectric 1 (of parameters $\varepsilon_1$ and $\mu_1$), obliquely incident at an angle $\theta_1$ on a plane interface between dielectric 1 and dielectric 2 (of parameters $\varepsilon_2$ and $\mu_2$). In medium 1 both the incident and the reflected waves are then present. In medium 2 a single traveling (transmitted) wave exists, which we often also refer to as the *refracted* wave.

Figure 15.17 shows diagrammatically the three waves by means of a set of equiphase planes representing the crests of the waves. Certain boundary conditions must be satisfied at the interface. Suppose that at an instant of time they are satisfied at points A, B, C, ... These points are moving along the $y$-axis, but the relations between the field vectors of the three waves must not change, since such a change would violate the boundary conditions. This means that the whole picture shown in the figure must move along the $y$-axis unchanged in time. Note that the velocity of propagation of both incident and reflected waves is $v_1$, and that the velocity of the transmitted (refracted) wave is $v_2$. From the figure it follows that the mutual relationship between the equiphase planes will remain unchanged

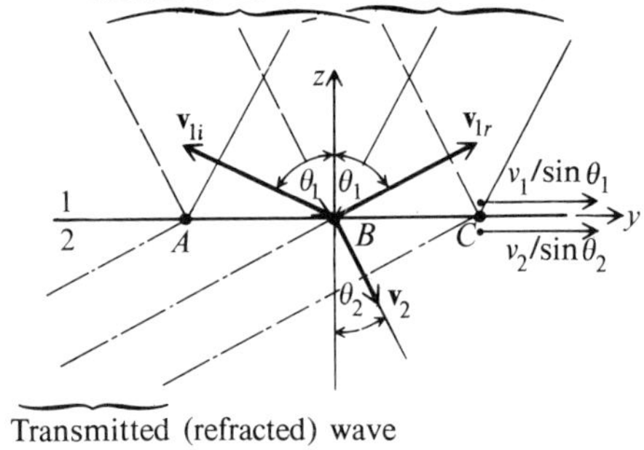


**Fig. 15.17** Oblique incidence of a plane wave on the interface between two dielectrics.

only if $\theta_r = \theta_1$, and $v_1/\sin\theta_1 = v_2/\sin\theta_2$, or

$$\boxed{\frac{\sin\theta_1}{\sin\theta_2} = \frac{v_1}{v_2}} \qquad \text{(Snell's law).} \qquad (15.4.17)$$

This relation is known as *Snell's law*. The ratio $v_1/v_2$ is referred to as the *index of refraction* $n_{12}$ for the two media. Usually $\mu_1 = \mu_2 = \mu_0$, and the index of refraction is simply

$$n_{12} = v_1/v_2 = \sqrt{\varepsilon_2/\varepsilon_1} \qquad \text{(index of refraction when } \mu_1 = \mu_2 = \mu_0). \qquad (15.4.18)$$

According to Snell's law, $\theta_2 < \theta_1$ if $\varepsilon_2 > \varepsilon_1$. The wave is refracted towards the normal to the surface; this is true when medium 1 is air and medium 2 any dielectric. A refracted wave exists for all angles of incidence of the incident wave.

If $\varepsilon_1 > \varepsilon_2$, however, $\theta_2 > \theta_1$. So, for a certain angle $\theta_1 = \theta_T$ the angle $\theta_2$ will be exactly $\pi/2$. Since $\theta_2$ cannot be larger than $\pi/2$, for $\theta_1 > \theta_T$ there is no refracted (transmitted) wave, and the wave is said to be *totally reflected.* The angle $\theta_1 = \theta_T$ of the total reflection is easily obtained from Snell's law. We let $\theta_2 = \pi/2$, and obtain

$$\sin\theta_T = v_1/v_2 = \sqrt{\varepsilon_2/\varepsilon_1} \qquad \text{(the angle of total reflection).} \qquad (15.4.19)$$

The phenomenon of total reflection is used extensively in practice, particularly in optics (e.g. optical prisms). Another application of total reflection is in what are called dielectric waveguides, where an electromagnetic wave, which is totally reflected back from the surface of a dielectric slab situated in air to the slab, "bounces" between the two faces of the slab and propagates along it.

The foregoing conclusions are valid for any polarization of the incident wave. Any wave can be represented as a sum of perpendicularly (or horizontally)

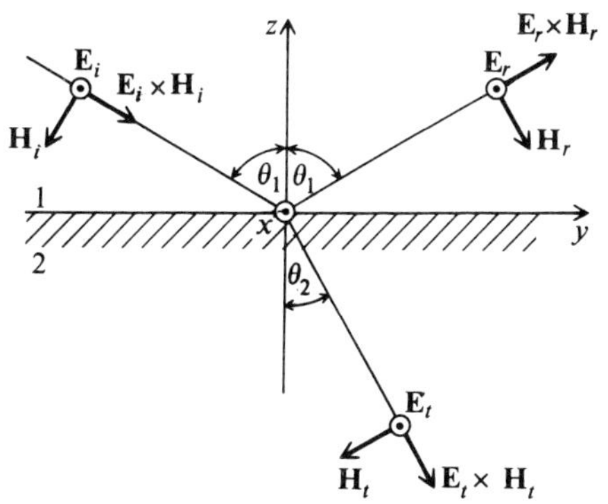


**Fig. 15.18** Horizontally polarized wave obliquely incident on the interface between two dielectrics.

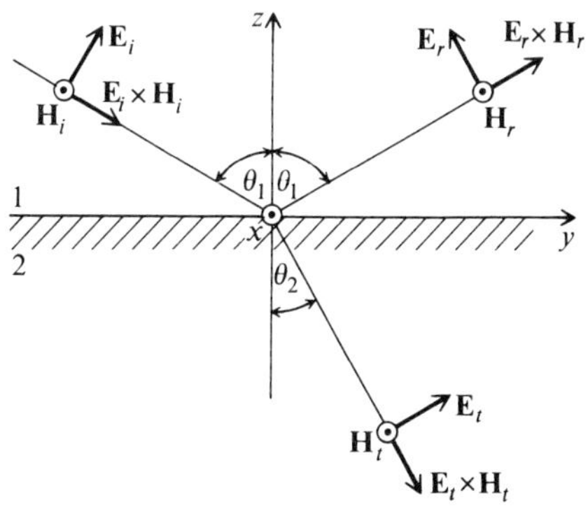


**Fig. 15.19** Parallel polarized wave obliquely incident on the interface between two dielectrics.

polarized and parallel polarized waves. Figure 15.18 illustrates the first case, Fig. 15.19 the second. If we denote by $E_i$, $E_r$, and $E_t$ the complex amplitudes of the electric field intensity of the three waves at the boundary surface, then, according to the general boundary conditions in Eqs. (13.4.1) and (13.4.3) and Figs. 15.18 and 15.19, we have

$$E_i + E_r = E_t, \qquad (H_i - H_r)\cos\theta_1 = H_t\cos\theta_2 \qquad \text{(horizontal polarization)},$$
$$(E_i - E_r)\cos\theta_1 = E_t\cos\theta_2, \qquad H_i + H_r = H_t \qquad \text{(parallel polarization)}.$$

Noting that $H_i = E_i/\eta_1$, $H_r = E_r/\eta_1$ and $H_t = E_t/\eta_2$, we find the coefficients of reflection and refraction:

$$\mathscr{R}_h = \left(\frac{E_r}{E_i}\right)_h = \frac{\eta_2\cos\theta_1 - \eta_1\cos\theta_2}{\eta_2\cos\theta_1 + \eta_1\cos\theta_2} \qquad \text{(horizontal polarization)}, \qquad (15.4.20)$$

$$\mathscr{T}_h = \left(\frac{E_t}{E_i}\right)_h = \frac{2\eta_2\cos\theta_1}{\eta_2\cos\theta_1 + \eta_1\cos\theta_2} \qquad \text{(horizontal polarization)}, \qquad (15.4.21)$$

$$\mathscr{R}_p = \left(\frac{E_r}{E_i}\right)_p = \frac{\eta_1\cos\theta_1 - \eta_2\cos\theta_2}{\eta_1\cos\theta_1 + \eta_2\cos\theta_2} \qquad \text{(parallel polarization)}, \qquad (15.4.22)$$

$$\mathscr{T}_p = \left(\frac{E_t}{E_i}\right)_p = \frac{2\eta_2\cos\theta_1}{\eta_1\cos\theta_1 + \eta_2\cos\theta_2} \qquad \text{(parallel polarization)}. \qquad (15.4.23)$$

These expressions are completely general. If the two media are perfect dielectrics, $\eta_1$ and $\eta_2$, and hence also coefficients of reflection and transmission, are real. This means that the field components are either in phase or shifted for 180°. If the two media have nonzero conductivities, $\eta_1$ and $\eta_2$, and consequently $\mathscr{R}_h, \ldots, \mathscr{T}_p$, are complex, and the field components of the three waves are not in phase.

We shall consider only the case where the two media are perfect dielectrics. Usually in these cases $\mu_1 = \mu_2 = \mu_0$, so that, according to Eq. (15.4.18),

$$\eta_1/\eta_2 = \sqrt{\mu_1/\varepsilon_1}\sqrt{\varepsilon_2/\mu_2} = \sqrt{\varepsilon_2/\varepsilon_1} = n_{12}. \qquad (15.4.24)$$

So (15.4.20) to (15.4.23) become

$$\mathscr{R}_h = \frac{\cos\theta_1 - n_{12}\cos\theta_2}{\cos\theta_1 + n_{12}\cos\theta_2}, \qquad \mathscr{T}_h = \frac{2\cos\theta_1}{\cos\theta_1 + n_{12}\cos\theta_2}, \qquad (15.4.25)$$

$$\mathscr{R}_p = \frac{n_{12}\cos\theta_1 - \cos\theta_2}{n_{12}\cos\theta_1 + \cos\theta_2}, \qquad \mathscr{T}_p = \frac{2\cos\theta_1}{n_{12}\cos\theta_1 + \cos\theta_2}. \qquad (15.4.26)$$

According to Snell's law in Eq. (15.4.17), $n_{12} = \sin\theta_1/\sin\theta_2$. Therefore $\mathscr{R}_h$ can never be zero, since this would require $n_{12} = \cos\theta_1/\cos\theta_2$, which is not possible. But $\mathscr{R}_p$ *can* be zero, if $\cos\theta_2 = n_{12}\cos\theta_1$. It is convenient to determine $\theta_1$ in terms of $n_{12}$ only. Since $n_{12}\cos\theta_1 = \cos\theta_2$ and $\sin\theta_1 = n_{12}\sin\theta_2$ we have

$$\sin\theta_1\cos\theta_1 = \sin\theta_2\cos\theta_2, \qquad \text{or} \qquad \sin 2\theta_1 = \sin 2\theta_2.$$

Now, $\theta_1 \neq \theta_2$, except in the trivial case when $n_{12} = 1$. Recalling that $\sin(180° - 2\theta_2) = \sin 2\theta_2$, we then have $2\theta_2 = 180° - 2\theta_1$. So $\theta_2 = 90° - \theta_1$, and $\cos\theta_2 = \sin\theta_1$. The condition $\cos\theta_2 = n_{12}\cos\theta_1$ therefore becomes

$$\tan\theta_1 = n_{12} \quad \text{(the Brewster angle; no reflected wave if parallel polarization).} \tag{15.4.27}$$

Angle $\theta_1$ thus determined is referred to as the *Brewster angle*. If a composite wave having both parallel and horizontally polarized components is incident on the interface at the Brewster angle, a horizontally polarized reflected component only exists. If another slab of the dielectric is introduced in the path of this reflected wave so that the wave impinges on it with *parallel* polarization and at the Brewster angle, the reflected wave disappears completely. This fact has been known in optics for a long time.

*Problems referring to Section 15.4:* 15.T7 to 15.T12, 15.G7 to 15.G11, 15.N5 to 15.N8.

## 15.5 THE PROPAGATION OF PLANE WAVES IN IONIZED MEDIA*

The upper layer of the atmosphere, roughly between 50 km and 500 km above the surface of the earth, is known to consist of rarified ionized gases. This ionized layer around the earth is known at the *ionosphere*, and it has a considerable influence on the propagation of electromagnetic waves. For analyzing radio communications with satellites, or between two points at the surface of the earth if the ionosphere is in the path of the electromagnetic wave, it is therefore necessary to have some knowledge of the mechanism of propagation through ionized gases.

Let us consider a plane electromagnetic wave in a homogeneous ionized gas, and let $N$ be the number of charged particles per unit volume, $Q$ their charge, and $m$ their mass. If we assume that the electric field intensity of the wave at a given point varies according to the law $\mathbf{E}(t) = \mathbf{E}\sin\omega t$, then the equation of motion of each charge is

$$m\frac{d\mathbf{v}}{dt} = Q\mathbf{E}\sin\omega t,$$

and the velocity of the charge due to the electric field of the wave

$$\mathbf{v} = -\frac{Q}{\omega m}\mathbf{E}\cos\omega t.$$

The density of the convection current produced by the moving charges is

$$\mathbf{J} = NQ\mathbf{v} = -\frac{NQ^2}{\omega m}\mathbf{E}\cos\omega t.$$

* This section may be omitted without loss of continuity.

So Maxwell's equation (13.3.7) becomes

$$\text{curl } \mathbf{H} = \mathbf{J} + \varepsilon_0 \frac{\partial \mathbf{E}}{\partial t} = \omega \left( \varepsilon_0 - \frac{NQ^2}{\omega^2 m} \right) \mathbf{E} \cos \omega t.$$

The electromagnetic waves in an ionized medium can thus be analyzed as in the case of a perfect dielectric if we replace the permittivity $\varepsilon$ of the dielectric by

$$\varepsilon' = \varepsilon_0 \left( 1 - \frac{NQ^2}{\varepsilon_0 \omega^2 m} \right) = \varepsilon_0 \left( 1 - \frac{f_c^2}{f^2} \right), \qquad (15.5.1)$$

where

$$f_c^2 = \frac{NQ^2}{4\pi^2 \varepsilon_0 m}. \qquad (15.5.2)$$

Accordingly, the phase velocity of the plane wave in the ionized medium is

$$v_p = \frac{1}{\sqrt{\mu_0 \varepsilon'}} = \frac{c_0}{\sqrt{1 - f_c^2/f^2}}. \qquad (15.5.3)$$

If $f > f_c$, the radical is positive, but smaller than unity, so that $v_p > c_0$. The velocity $v_p$ is however not the velocity of the propagation of the *energy* carried by the plane wave. It is a purely geometrical velocity, like the velocity of points A, B, and C in Fig. 15.17, which is also greater than the velocity of light in a vacuum, if medium 1 is a vacuum. A more detailed explanation of the concept of phase velocity is presented in Section 16.3.4.

If $f < f_c$, the radical is negative, and the phase velocity is imaginary, which means that waves of frequencies less than $f_c$ cannot propagate through the ionized medium. The frequency $f_c$ is therefore sometimes referred to as the cutoff fre-

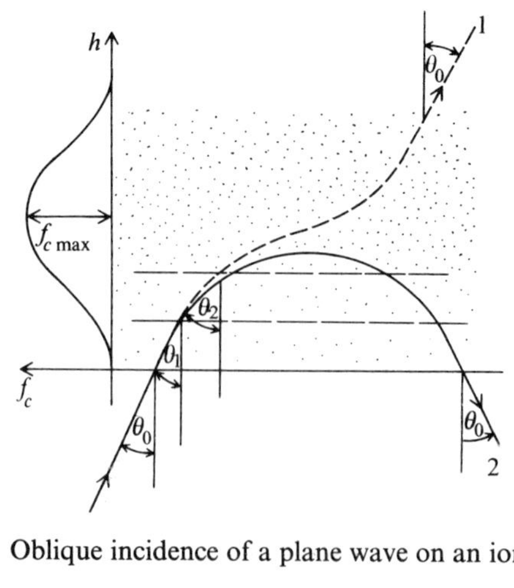


**Fig. 15.20** Oblique incidence of a plane wave on an ionized layer.

quency. If we ignore collisions of charged particles with neutral atoms, the wave cannot transmit any energy to the ionized medium. Hence, if a wave of frequency $f < f_c$ impinges perpendicularly on an ionized layer having a cutoff frequency of $f_c$, it is totally reflected.

Let us now consider an ionized layer in which the ionization varies from zero to a certain maximum value inside the layer, and suppose that a cutoff frequency of $f_{c\max}$ corresponds to the maximum ionization. Let a wave of frequency $f$ impinge on the layer at an angle $\theta_0$ with respect to the normal to the layer (Fig. 15.20). We can then formally apply Snell's law from Eq. (15.4.17):

$$\sin\theta_0/\sin\theta_1 = \sqrt{\varepsilon_1'/\varepsilon_0}, \qquad \sin\theta_1/\sin\theta_2 = \sqrt{\varepsilon_2'/\varepsilon_1'}, \ldots,$$

where $\varepsilon_1', \varepsilon_2', \ldots,$ are the equivalent permittivities at successive points of the trajectory of the wave propagating through the layer. Therefore angle $\theta'$, corresponding to the point at which the equivalent permittivity is $\varepsilon'$, is determined by $\sin\theta'\sqrt{\varepsilon'} = \sin\theta_0\sqrt{\varepsilon_0}$. According to Eq. (15.5.1) we thus obtain

$$\sin\theta'\sqrt{1 - f_c^2/f^2} = \sin\theta_0. \tag{15.5.4}$$

If frequency $f$ and angle $\theta_0$ are such that $\theta' < \pi/2$ at the point where the cutoff frequency is maximal, the wave will be bent, but will pass through the layer, as shown by path 1 in Fig. 15.20. If frequency $f$ and the incident angle $\theta_0$ are such that $\theta' = \pi/2$ at a certain point, the wave is totally reflected (path 2). In that case the wave turns back at the height where the ionization is such, that

$$\sqrt{1 - f_c^2/f^2} = \sin\theta_0,$$

from which

$$f_c = f\cos\theta_0. \tag{15.5.5}$$

If $\theta_0 = 0, f_c = f$ as before.

The ionization of the ionosphere varies greatly during the twenty-four hours of a day and during a year. Roughly, the cutoff frequency for perpendicular (vertical) incidence ranges between 3 MHz and 8 MHz. From Eq. (15.5.5) we

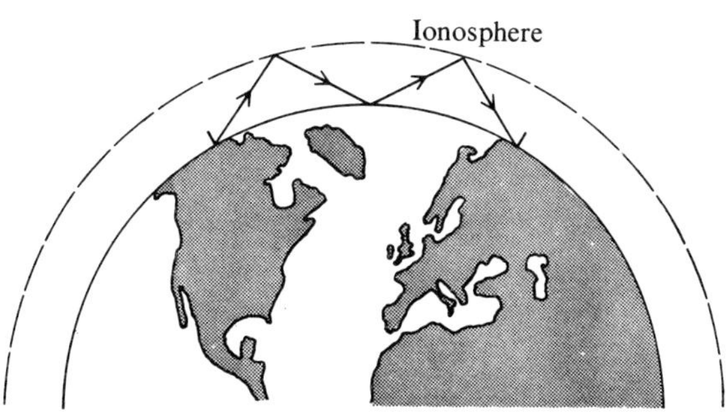


**Fig. 15.21** Ionosphere and surface of the earth acting as mirrors for electromagnetic waves.

know that frequencies can be still higher if the incidence is not perpendicular; but no wave, having a frequency less than, say, 3 MHz, can leave the earth, nor can it reach the surface of the earth from outer space. [The exceptions are waves of very low frequencies, for which the earth's ionosphere is not sufficiently thick (in terms of the wavelength) to produce complete reflection.]

So, if we wish to communicate with a satellite above the ionosphere, we must use frequencies high enough to ensure communication for almost all the angles of incidence. Conversely, if we wish to use the ionosphere and the surface of the earth as reflectors of radio waves for communicating between distant points at the surface of the earth, we use frequencies low enough that the wave does not "escape" to the outer world (Fig. 15.21).

*Problems referring to Section 15.5:* 15.T13 to 15.T14, 15.G12 to 15.G13, 15.N9.

## PROBLEMS

### Test problems

**15.T1.** Write the wave equations (15.2.7) in complex form.

**15.T2.** Prove that both $f_1(t - z/c)$ and $f_2(t + z/c)$ are solutions of the wave equation (15.2.8).

**15.T3.** Prove that $f_2(t + z/c)$ represents a wave travelling with a velocity $c$ in the direction opposite to the $z$ axis.

**15.T4.** What is the ratio of the wavelengths of a sinusoidal plane wave of frequency $f$, if it propagates through perfect dielectrics, which have permittivities $\varepsilon_1$ and $\varepsilon_2$ and permeabilities $\mu_0$?

**15.T5.** Deduce Eq. (15.2.20) from the complex form of Eq. (15.2.8).

**15.T6.** From Eq. (15.3.9) and (15.3.10) deduce the range of phase difference between the electric and magnetic field intensities of a plane wave propagating through conducting media.

**15.T7.** A small conducting closed loop is situated in the standing wave shown in Fig. 15.11. How must the loop be oriented and where must it be positioned in order to obtain the maximum induced current in it?

**15.T8.** Prove that the average value of the Poynting vector of the standing wave shown in Fig. 15.11 is zero.

**15.T9.** Discuss the possibility of eliminating the reflected wave in the case of perpendicular incidence by a suitable choice of the two media.

**15.T10.** What is the magnitude of the reflection coefficient if the standing-wave ratio is 1 and 1.1?

**15.T11.** Imagine that you are watching fish in a large aquarium through a thin glass wall. Discuss qualitatively the relation between the real positions of fishes and their positions as you see them.

**15.T12.** With reference to Eqs. (15.4.25) and (15.4.26), explain why $\mathscr{R}_p$ can be zero, but $\mathscr{R}_h$ can never be zero if $\mu_1 = \mu_2$.

**15.T13.** If an ionized layer consists of free electrons and positive ions of some gas in equal number per unit volume, which one of the two components primarily determines the cutoff frequency?

**15.T14.** Is an ionized medium a dispersive medium?

## General problems

**15.G1.** A circular conducting loop of radius $a$ is situated in a vacuum in the electromagnetic field of a plane wave, of wavelength $\lambda$ ($\lambda \gg a$) and the amplitude of the electric field intensity $E$. How should the loop be positioned in order that the electromotive force induced in it shall be maximal? Determine the amplitude of the induced electromotive force in that case.

*Solution.* According to Faraday's law of electromagnetic induction, the maximal electromotive force is induced in the loop if it is perpendicular to the lines of vector **B** (that is, of **H**). Suppose that in the region where the loop is situated the magnetic field of the wave varies according to the law $H(t) = H \cos \omega t$. The electromotive force induced in the loop is then

$$\mathscr{E}(t) = -a^2\pi\mu_0 \frac{dH(t)}{dt} = a^2\pi\omega\mu_0 H \sin \omega t. \tag{a}$$

Now, instead of $\omega$ and $H$, we know $\lambda$ and $E$. But $\omega = 2\pi c_0/\lambda$ and $H = E/\eta_0$. So the amplitude $\mathscr{E}$ of the electromotive force $\mathscr{E}(t)$ is

$$\mathscr{E} = \frac{2a^2\pi^2 c_0\mu_0}{\lambda\eta_0} E = \frac{2a^2\pi^2}{\lambda} E,$$

since $\eta_0 = c_0\mu_0$.

**15.G2.** A rectangular conducting loop with sides $a$ and $b$ is situated in a vacuum in the electromagnetic field of a sinusoidal plane wave. The amplitude of the electric field intensity of the wave is $E$, and its wavelength is $\lambda$ ($\lambda \gg a, b$). The loop is oriented as shown in Fig. 15.G2. Determine the amplitude of the induced electromotive force in the loop both using Faraday's law and calculating it as the line integral of **E** around the loop.

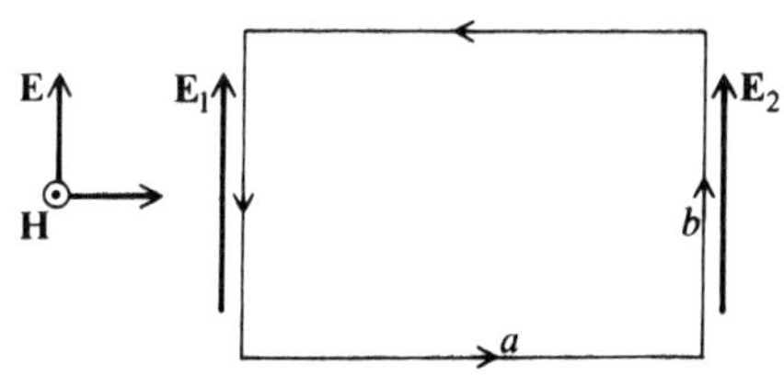


**Figure 15.G2**

*Solution.* Suppose that in the region where the loop is situated $H(t) = H \cos \omega t$. Using Faraday's law we then obtain

$$\mathscr{E}(t) = -ab\mu_0 \frac{dH(t)}{dt} = ab\mu_0\omega H \sin \omega t. \tag{a}$$

Now, $-d\Phi/dt$ in Faraday's law stands in fact for the line integral of **E** around the loop. So, according to Fig. 15.G2, it must also be

$$\mathscr{E}(t) = [-E_1(t) + E_2(t)]b. \tag{b}$$

In applying Faraday's law we assumed that $H(t)$ approximately equalled $H\cos\omega t$ at all points on the loop. A similar approximation for $E(t)$ would result in $\mathscr{E}(t) = 0$, so we must take into account the difference between $E_1(t)$ and $E_2(t)$ which really exists. Let $E_1(t) = E\cos\omega t$. Then

$$E_2(t) = E\cos(\omega t - \beta a) = E\cos(\omega t - 2\pi a/\lambda).$$

Now, since $a \ll \lambda$, we have that $\cos(2\pi a/\lambda) \simeq 1$ and $\sin(2\pi a/\lambda) \simeq 2\pi a/\lambda$. So

$$\begin{aligned}\mathscr{E}(t) &= bE[-\cos\omega t + \cos(\omega t - 2\pi a/\lambda)]\\ &= bE\left(-\cos\omega t + \cos\omega t\cos\frac{2\pi a}{\lambda} + \sin\omega t\sin\frac{2\pi a}{\lambda}\right) \simeq bE\frac{2\pi a}{\lambda}\sin\omega t.\end{aligned}$$

This is the same as Eq. (a), because $H = E/\eta_0$ and $\omega = 2\pi c_0/\lambda$.

**15.G3.** A sinusoidal plane wave, of frequency $f$ and average value of the Poynting vector $\mathscr{P}$, propagates through distilled water ($\mu = \mu_0$, $\varepsilon = \varepsilon_r\varepsilon_0$). Find the amplitude of the electromotive force induced in a circular conducting loop of radius $a$, oriented so that the electromotive force induced is maximal. What condition must be met in order that the current in the loop can be considered as quasi-stationary?

*Answer.* For $f = 10^6/3$ Hz, $\mathscr{P} = 40\pi/3\ \mathrm{W/m^2}$, $\varepsilon_r = 81$, and $a = 10$ cm, the amplitude of the electromotive force induced in the loop is $\mathscr{E} = 8\pi^3\sqrt{2}/3$ mV.

**15.G4.** A sinusoidal plane wave propagating in a vacuum impinges perpendicularly on a plane absorbing screen. The screen absorbs the energy of the wave almost completely. The amplitude of the electric field intensity of the wave is $E$. Determine the energy transformed per unit area of the screen into heat in time $t$.

*Solution.* We can obtain the result by noting that the average Poynting vector represents the power flow per unit area perpendicular to the direction of propagation of the wave. So

$$W_s = \mathscr{P}_{\mathrm{av}}t = \tfrac{1}{2}EHt = \tfrac{1}{2}E^2t\sqrt{\varepsilon_0/\mu_0}. \tag{a}$$

The same result can be obtained by following another line of reasoning. In time $t$ the wave travels for a distance $c_0t$. If the screen absorbs the wave completely, as assumed, the total energy delivered to the screen in time $t$ equals the energy stored in the wave in a long parallelepiped of base of unit area and "height" $c_0t$. Suppose that the expressions for the wave components are

$$E(z,t) = E\cos(\omega t - \beta z), \qquad H(z,t) = \frac{E}{\eta_0}\cos(\omega t - \beta z).$$

Then at, say, the instant $t = 0$, the density of the energy (electric plus magnetic) along the wave is given by

$$\frac{dW}{dv} = \tfrac{1}{2}\varepsilon_0E^2\cos^2\beta z + \tfrac{1}{2}\mu_0\frac{E^2}{\eta_0^2}\cos^2\beta z = \varepsilon_0E^2\cos^2\beta z.$$

The average density of the energy is

$$\left(\frac{dW}{dv}\right)_{\mathrm{av}} = \frac{1}{\lambda}\int_{z=-\lambda}^{0}\frac{dW}{dv}\,dz = \frac{\varepsilon_0E^2}{\lambda}\int_{z=-\lambda}^{0}\cos^2\beta z\,dz = \tfrac{1}{2}\varepsilon_0E^2.$$

So we obtain

$$W_s = \left(\frac{dW}{dv}\right)_{\text{av}} c_0 t = \tfrac{1}{2}\varepsilon_0 E^2 c_0 t = \tfrac{1}{2}E^2 t\sqrt{\frac{\varepsilon_0}{\mu_0}},$$

as before.

**15.G5.** It was found by measurement that the radiation from the sun on the earth transfers in time $t$ an energy $W_s$ per unit area to a "black" screen perpendicular to the rays. This radiation is composed of many components of different frequencies and different polarizations. Approximating the total wave by a linearly polarized plane wave, find the values of $E$ and $H$ of this equivalent wave.

*Answer.* For $W_s/t = 1350\ \text{W/m}^2$, $E \simeq 1000$ V/m and $H \simeq 2.68$ A/m.

**15.G6.** Two plane waves of equal frequencies and phases propagate in the same direction. Both are circularly polarized, but in opposite directions. The amplitudes of the electric field intensities of the two waves are $E_1$ and $E_2$. Find the polarization of the resultant wave in terms of $E_1$ and $E_2$.

*Solution.* Let the two waves propagate in the direction of the $z$ axis, as shown in Fig. 15.G6. The two circularly polarized waves can be represented as pairs of linearly polarized waves:

$$E_{1x}(z, t) = E_1 \cos(\omega t - \beta z), \qquad E_{1y}(z, t) = E_1 \sin(\omega t - \beta z),$$
$$E_{2x}(z, t) = E_2 \cos(\omega t - \beta z), \qquad E_{2y}(z, t) = -E_2 \sin(\omega t - \beta z).$$

So for components of the total wave we have

$$E_{x\,\text{total}} = (E_1 + E_2)\cos(\omega t - \beta z),$$
$$E_{y\,\text{total}} = (E_1 - E_2)\sin(\omega t - \beta z).$$

If $E_1 \neq E_2$ the total wave is polarized elliptically. If $E_1 = E_2$ it becomes linearly polarized. Circular polarization is obtained only if $E_1 = 0$ or $E_2 = 0$.

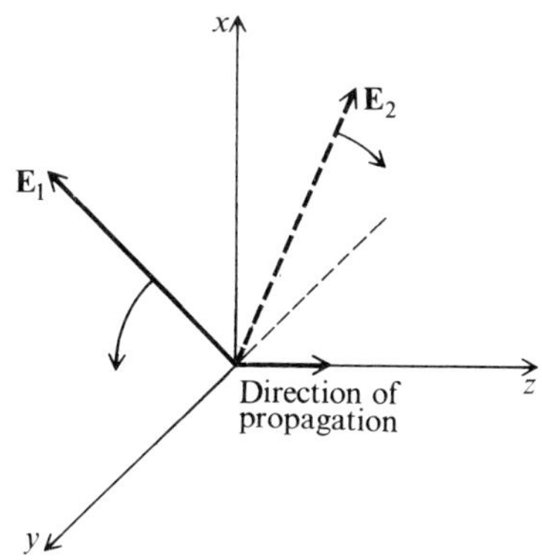


**Figure 15.G6**

**15.G7.** A plane wave, of frequency $f$ and amplitude of the electric field intensity $E$, impinges perpendicularly on a plane, very good conducting block, of conductivity $\sigma$ and permeability $\mu$. Find the average approximate value of the power absorbed by the block per unit area.

*Solution*. According to Eq. (15.3.21), the only quantity necessary for determining the power is the amplitude $H_0$ of the total magnetic field at the conductor surface. Since the conductor is assumed to be very good, $H_0$ is approximately the same as if the conductor were perfect. According to Eq. (15.4.4), $H_0 = 2H = 2E/\eta_0$, so that the power absorbed per unit area is

$$P_s = \tfrac{1}{2}R_s\,|H_0|^2 = \frac{2E^2R_s}{\eta_0^2} = 2E^2\,\frac{\varepsilon_0}{\mu_0}\sqrt{\frac{\pi\mu f}{\sigma}}\,.$$

**15.G8.** A linearly polarized plane wave, of wavelength $\lambda_0$ in a vacuum, impinges perpendicularly on the plane surface of water ($\mu = \mu_0$, $\varepsilon = \varepsilon_r\varepsilon_0$). The amplitude of the electric field intensity of the wave in air is $E_0$. Determine the amplitude $\mathscr{E}$ of the electromotive force induced in a small conducting loop of area $S$ situated in water, if it is oriented so that the induced electromotive force is maximal.

*Answer*. For $E_0 = 100$ mV/m (i.e. 0.100 V/m), $\lambda_0 = 300$ m, $\varepsilon_r = 81$, and $S = 0.1\ \text{m}^2$, $\mathscr{E} = 0.377$ mV.

**15.G9.** A plane wave propagating in dielectric 1 impinges perpendicularly on a dielectric slab 2, of thickness $d$, permittivity $\varepsilon_2$ and permeability $\mu_2$. To the right of the slab is a dielectric 3, of permittivity $\varepsilon_3$ and permeability $\mu_3$ (see Fig. 15.G9). Determine the ratio of the amplitudes of the reflected and incident waves in dielectric 1.

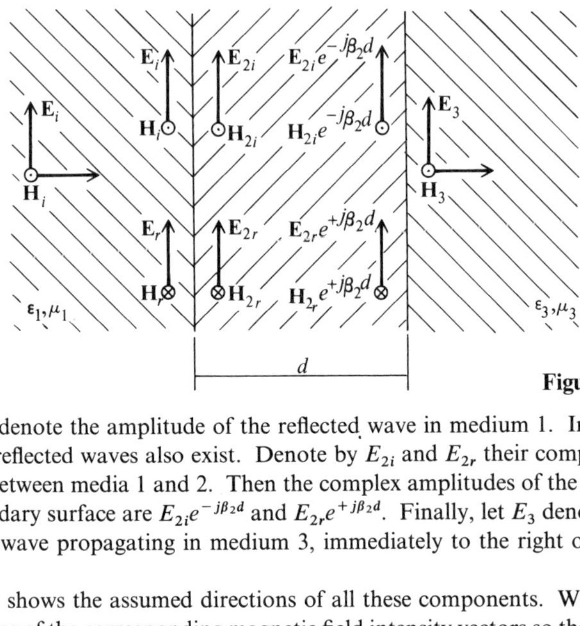


**Figure 15.G9**

*Solution*. Let $E_r$ denote the amplitude of the reflected wave in medium 1. In medium 2 the incident and the reflected waves also exist. Denote by $E_{2i}$ and $E_{2r}$ their complex amplitudes on the interface between media 1 and 2. Then the complex amplitudes of the two waves next to the other boundary surface are $E_{2i}e^{-j\beta_2 d}$ and $E_{2r}e^{+j\beta_2 d}$. Finally, let $E_3$ denote the complex amplitude of the wave propagating in medium 3, immediately to the right of the boundary surface.

Figure 15.G9 shows the assumed directions of all these components. We determine the indicated directions of the corresponding magnetic field intensity vectors so that we obtain the correct directions of the Poynting vector. The boundary conditions require equality of the total tangential components of vectors **E** and **H**. Recalling that $H = E/\eta$ (with appropriate subscripts) we obtain

$$E_i + E_r = E_{2i} + E_{2r}, \qquad (E_i - E_r)/\eta_1 = (E_{2i} - E_{2r})/\eta_2, \tag{a}$$

$$E_{2i}e^{-j\beta_2 d} + E_{2r}e^{j\beta_2 d} = E_3, \qquad (E_{2i}e^{-j\beta_2 d} - E_{2r}e^{j\beta_2 d})/\eta_2 = E_3/\eta_3. \tag{b}$$

After a somewhat lengthy, but elementary, calculation we find that

$$E_r/E_i = (\eta_e - \eta_1)/(\eta_e + \eta_1), \tag{c}$$

where

$$\eta_e = \eta_2 \frac{\eta_3 \cos \beta_2 d + j\eta_2 \sin \beta_2 d}{\eta_2 \cos \beta_2 d + j\eta_3 \sin \beta_2 d}. \tag{d}$$

Comparing Eq. (c) with Eq. (15.4.11) we see that $\eta_e$ can be regarded as an equivalent intrinsic impedance of media 2 and 3 with respect to medium 1.

**15.G10.** Starting from Eqs. (c) and (d) of the preceding problem, determine $d$ so that there is no reflected wave in medium 1.

*Solution.* According to Eq. (c) of the preceding problem the necessary and sufficient condition for the disappearance of the reflected wave is that $\eta_e = \eta_1$, so $\eta_e$ must be real. According to Eq. (d), this could be so if $\cos \beta_2 d = 0$, that is, if $\beta_2 d = \pi/2$, from which

$$d = \lambda_2/4, \tag{a}$$

or if $\sin \beta_2 d = 0$, that is, if $\beta_2 d = \pi$, from which

$$d = \lambda_2/2. \tag{b}$$

In these equations only the smallest value of $d$ is determined.

In the case of Eq. (a), Eq. (d) of the preceding problem becomes $\eta_e = \eta_2^2/\eta_3$, so that the reflected wave disappears if

$$\eta_2 = \sqrt{\eta_1\eta_3}, \qquad \text{with} \qquad d = \lambda_2/4. \tag{c}$$

This equation is used in optics for determining the thickness and relative permittivity of a thin transparent layer over lenses, which are designed to minimize the reflection of electromagnetic waves from the optical region.

In the case of Eq. (b), Eq. (d) of the preceding problem becomes $\eta_e = \eta_3$. So the reflected wave disappears if

$$\eta_1 = \eta_3, \qquad \text{with} \qquad d = \lambda_2/2. \tag{d}$$

This is also an important result. For example, if a radar antenna is to be protected from atmospheric influences, it can be covered with a dome of some dielectric material. If its thickness is approximately $\lambda/2$ (or, of course, $n\lambda/2$, $n = 1, 2, \ldots$), $\lambda$ denoting the wavelength in the dielectric, it represents an almost transparent cover for the electromagnetic waves of the wavelength $\lambda$ used by the radar. Such domes are referred to as *radomes*. A familiar, although perhaps not recognized, radome is the (usually black) dome at the front of some airplanes.

**15.G11.** Electromagnetic waves propagate through a medium unattenuated if its conductivity is practically zero. Sea water is a relatively good conductor ($\sigma \simeq 4$ S/m) at low frequencies. It is not "transparent" for radio waves, but it is quite reasonably transparent for electromagnetic waves from the optical region. Explain this phenomenon qualitatively. [*Hint:* note that conductivity at high frequencies is not a constant. Free charge carriers become too "lazy" to follow rapid changes of the field because of their nonzero mass. This at least is the standpoint of the classical theory.]

**15.G12.** A plane electromagnetic wave impinges perpendicularly on an ionized layer consisting of free electrons. The number of electrons per unit volume is given by

$$N = N_{max} - \frac{N_{max}}{h^2}(h - z)^2 \qquad \text{for} \qquad 0 \le z \le 2h,$$

$$N = 0 \qquad \text{for} \qquad z \ge 2h, \qquad \text{and} \qquad z \le 0,$$

where $N_{max}$ is a constant (see Fig. 15.G12). If frequency $f$ of the wave is lower than the cutoff frequency corresponding to $z = h$, determine the depth $z$ to which the wave penetrates.

*Solution.* Substituting into Eq. (15.5.2) $Q = e = 1.60 \times 10^{-19}$ C, $m = 9.1 \times 10^{-31}$ kg, and the value of $\varepsilon_0$, we find that the cutoff frequency for the ionized layer consisting of free electrons is

$$f_c = \sqrt{80.5\,N}\ \text{Hz}. \tag{a}$$

So the plane from which the wave is reflected is determined by

$$f^2 = f_c^2 = 80.5\left[N_{max} - \frac{N_{max}}{h^2}(h - z)^2\right],$$

from which

$$z = h\left(1 - \sqrt{1 - \frac{f^2}{80.5\,N_{max}}}\right).$$

The other sign of the square root is impossible, since the wave impinges on the layer from the left.

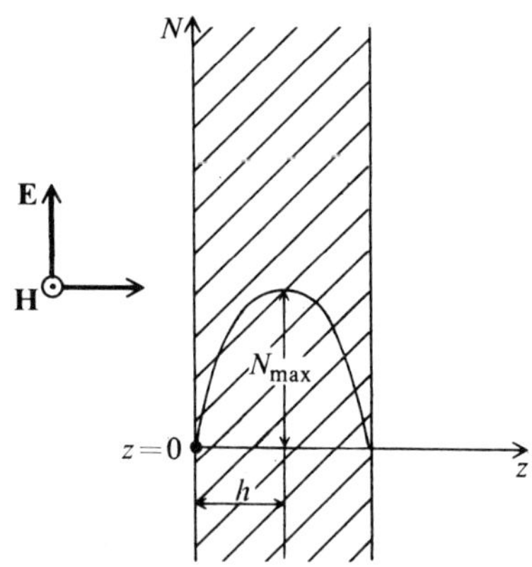


**Figure 15.G12**

**15.G13.** A plane wave of frequency $f$ impinges perpendicularly on an ionized layer consisting of free electrons. The number of electrons per unit volume of the layer is given approximately by

$$N = n^2(h - H) \qquad \text{for} \qquad h \ge H, \qquad \text{and} \qquad N = 0 \qquad \text{for} \qquad h \le H,$$

where $n$ and $H$ are constants. Determine the height $h$ reached by the wave.

*Answer.* $h = H + f^2/80.5n^2$.

## Numerical problems

**15.N1.** Find the free-space wavelengths of plane electromagnetic waves of the following frequencies: 60 Hz, 1000 Hz, $10^5$ Hz, $10^8$ Hz, and $10^{10}$ Hz.

**15.N2.** Repeat the preceding problem if the plane wave propagates in water ($\mu = \mu_0$, $\varepsilon_r = 81$, $\sigma \simeq 0$).

**15.N3.** Determine the complex propagation constant, and its real and imaginary parts, for salt water ($\varepsilon_r = 81$, $\sigma = 4$ S/m, $\mu = \mu_0$) and moist earth ($\varepsilon_r = 10$, $\sigma = 10^{-2}$ S/m, $\mu = \mu_0$), at a frequency $f = 10^6$ Hz.

**15.N4.** Find the real and imaginary parts of the complex intrinsic impedance for the two cases of the preceding problem.

**15.N5.** A plane wave, of frequency $f = 10^8$ Hz and of amplitude of the electric field intensity $E = 1$ V/m, propagates through air, and impinges perpendicularly on a plane copper sheet. Determine the average power losses per unit area of the sheet.

**15.N6.** Find the reflection and transmission coefficients for the interface between air and fresh water ($\varepsilon_r = 81$, $\sigma = 0$), in the case of perpendicular incidence.

**15.N7.** A plane wave is perpendicularly incident on the interface between air and a dielectric having $\mu = \mu_0$ and an unknown permittivity $\varepsilon$. The measured standing-wave ratio in air is 1.8. Determine $\varepsilon$.

**15.N8.** Determine the angle of total reflection for fresh water and air at radio frequencies (for water at radio frequencies $\varepsilon_r \simeq 81$).

**15.N9.** Prove that, for an ionized layer consisting of electrons, $f_c = \sqrt{80.5\,N}$, where $N$ is the number of electrons per unit volume.

**Suggested reading.** The list progresses towards more advanced texts. For full bibliographical description see Bibliography.

1. Kraus, J. D., *Electromagnetics*, Chapters 9 and 10.
2. Bohn, E. V., *Introduction to Electromagnetic Fields and Waves*, Chapter 9.
3. Jordan, E. C., and K. G. Balmain, *Electromagnetic Waves and Radiating Systems*, Chapters 5 and 17.

CHAPTER 16

# GUIDED ELECTROMAGNETIC WAVES

## 16.1 INTRODUCTION

In the preceding chapter we considered electromagnetic waves propagating through space without referring to their sources or their destination, and we concluded that they are able to transport energy. In practical applications, electromagnetic waves are in fact used precisely for the purpose of transporting energy. For example, the antenna of a broadcasting or television station radiates energy carrying pieces of information towards all possible receivers in the vicinity. Thus, an important application of electromagnetic waves is for the convenience of anyone desiring to receive them in a certain region. Obviously, a large part of the energy of such waves is inevitably lost.

Another, and perhaps more important, application of electromagnetic waves is in the transmission of electromagnetic energy from a source to a *single* receiver. In these cases it is obviously desirable to transmit energy with the smallest possible losses to that particular receiver, and only to that one. Electromagnetic waves propagating through free space are often unsuitable. Some systems for guiding the energy of the wave along a desired route have to be devised. All such systems can logically be called guides of electromagnetic waves. Electromagnetic waves propagating along these guides are known as *guided electromagnetic waves*.

For example, in power stations, other types of energy (chemical, hydro, nuclear) are transformed by a suitable process into electric energy, and this electric energy is then guided to various receivers by means of power *transmission lines*. Theoretically, the power station could just as well radiate all that energy and let the consumers "hunt" for as much of it as they needed from the air. But this process would be extremely uneconomic and, in fact, practically impossible.

Or let us suppose that we wish to transmit a very private message from New York to a friend in San Francisco by means of electromagnetic waves. Of course we can't use a powerful transmitter creating electromagnetic waves over the whole area of the USA; secrecy would hardly be the word if we did. Instead, we use electromagnetic waves propagating along telephone lines and radio links, with appropriate devices for regenerating the strength of our message, weakened by the long trip.

Again, let us suppose that we desire to transmit the extremely high electromagnetic energy created in pulses by a radar transmitter to its antenna. We very often use highly conductive metal tubes, and transmit the pulses of energy *through* them. This turns out to be the most suitable method with respect to losses along the guiding system. Such tubes are referred to as the *waveguides*. There are still other types of guides for electromagnetic waves, but we shall restrict our study to the two types mentioned.

Types of electromagnetic waves propagating along transmission lines and waveguides are quite different. In the case of transmission lines, electromagnetic waves having both field vectors **E** and **H** in practically transversal planes are usually used, and are referred to as *t*ransverse *e*lectro*m*agnetic waves, or simply *TEM* waves. The plane waves which we considered earlier are of the *TEM* type; the only important difference is that now **E** and **H** are *not* constant in a transverse plane. So *TEM* waves along transmission lines are not uniform.

There are many types of transmission lines, but we shall consider only those having two parallel cylindrical conductors. The types most often encountered are the two-wire line, the coaxial line (coaxial cable), and the strip line. Their characteristics, such as capacitance, inductance, conductance, and resistance per unit length, were determined in the preceding chapters.

Waveguides can be in the form of metal tubes or dielectric rods of varying cross-sections. All electromagnetic waves propagating through them have one thing in common; they are *not* of the *TEM* type. In addition, only waves which have wavelengths *smaller* than a certain critical wavelength can propagate through waveguides, that critical wavelength being of the same order as their transverse dimensions. Waveguides can therefore be used only at very high frequencies, for only then are the cross-sectional dimensions necessary for wave propagation of a practically reasonable order of magnitude.

## 16.2 TRANSMISSION LINES

**16.2.1 Introduction.** Transmission lines most often encountered in practice are in the form of two, very nearly parallel, cylindrical conductors. The conductors can never be perfect, nor can the dielectric surrounding them. Real transmission lines are therefore extremely difficult to analyze exactly.

Many times in this text we have emphasized that all the solutions which we find for any real system are of necessity approximate, although some of them can be regarded as exact for all practical purposes. Some, however, are poorer approximations. But, as engineers, we do not necessarily desire to know the exact solution. A simple approximate solution is usually to be preferred to a more accurate, but much more complicated one.

The approximation we are about to describe has been proved to be an excellent one in the case of transmission lines. For the purposes of this approximation, we consider a transmission line as an apparently complicated collection of very small resistors, capacitors, and coils, instead of as a boundary value problem (see Fig. 16.1). In fact, in the case of real transmission lines these "elements" should be considered as to some extent overlapping.

Systems of the type shown in Fig. 16.1 are referred to as having *distributed parameters*. They can be analyzed by using the basic laws of circuit theory, which was shown to represent, in certain circumstances, an excellent approximation to the field theory. So, instead of the electric and magnetic field intensities at all points where the electromagnetic field of the line exists, we consider as unknowns the voltage between the two conductors and the current intensity through them.

**16.2.2 The approximate analysis of wave propagation along real (lossy) lines.** Let us suppose that a transmission line is characterized by the following parameters per unit length: resistance $R$ (taking both conductors into account), inductance $L$, capacitance $C$, and conductance $G$.* Then, following the approximate representation of the line shown in Fig. 16.1, we can approximate a small section $\Delta z$ of the line by means of the simple network shown in Fig. 16.2. The parameters $R$, $L$, $C$, and $G$ are functions of the line geometry, and of the particular materials used for the conductors and the dielectric of the line.

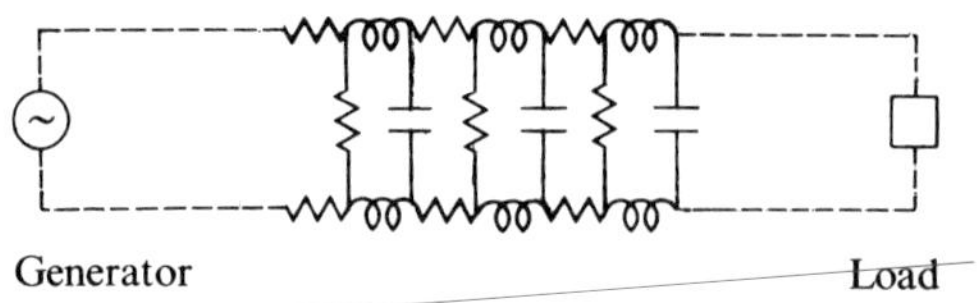


**Fig. 16.1** Circuit theory approximation of a transmission line.

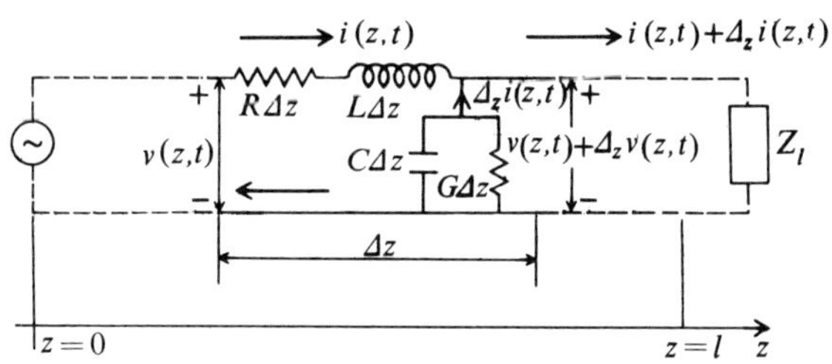


**Fig. 16.2** Circuit theory approximation of a small section of a transmission line.

The voltage $v(z, t)$ between the line conductors, as well as the current $i(z, t)$ through the line, vary not only with time, but also along the line. According to the figure

$$v(z, t) - R\Delta z\, i(z, t) - L\Delta z \frac{\partial i(z, t)}{\partial t} = v(z, t) + \Delta_z v(z, t),$$

$$i(z, t) - G\Delta z[v(z, t) + \Delta_z v(z, t)] - C\Delta z \frac{\partial}{\partial t}[v(z, t) + \Delta_z v(z, t)]$$
$$= i(z, t) + \Delta_z i(z, t).$$

* In this chapter we shall deal only with parameters per unit length. We omit the subscript "$l$" (= per unit length) for simplicity.

After some elementary manipulations, and ignoring terms with $(\Delta z)^2$, we obtain

$$-\frac{\Delta_z v(z, t)}{\Delta z} = Ri(z, t) + L\frac{\partial i(z, t)}{\partial t},$$

$$-\frac{\Delta_z i(z, t)}{\Delta z} = Gv(z, t) + C\frac{\partial v(z, t)}{\partial t}.$$

If $\Delta z$ is sufficiently small, these equations become

$$-\frac{\partial v(z, t)}{\partial z} = Ri(z, t) + L\frac{\partial i(z, t)}{\partial t}, \tag{16.2.1}$$

$$-\frac{\partial i(z, t)}{\partial z} = Gv(z, t) + G\frac{\partial v(z, t)}{\partial t}. \tag{16.2.2}$$

The most important case in practice is when $i(z, t)$ and $v(z, t)$ vary *with time* according to the simple sinusoidal law. We then make use of complex notation and write

$$i(z, t) = \mathrm{Re}\{I(z)e^{j\omega t}\}, \qquad v(z, t) = \mathrm{Re}\{V(z)e^{j\omega t}\}, \tag{16.2.3}$$

where $I(z)$ and $V(z)$ are *complex* amplitudes of current and voltage along the line. Equations (16.2.1) and (16.2.2) become

$$\frac{dV(z)}{dz} = -(R + j\omega L)I(z), \tag{16.2.4}$$

$$\frac{dI(z)}{dz} = -(G + j\omega C)V(z). \tag{16.2.5}$$

Differentiating the first equation with respect to $z$ and using the value of $dI(z)/dz$ from the second, we obtain

$$\frac{d^2V(z)}{dz^2} = (R + j\omega L)(G + j\omega C)V(z). \tag{16.2.6}$$

By an analogous procedure

$$\frac{d^2I(z)}{dz^2} = (R + j\omega L)(G + j\omega C)I(z). \tag{16.2.7}$$

Let

$$\boxed{(R + j\omega L)(G + j\omega C) = \gamma^2 = (\alpha + j\beta)^2}\,. \tag{16.2.8}$$

We refer to $\gamma$ as the *complex propagation constant*, $\alpha$ as the *attenuation constant*, and

$\beta$ as the *phase constant*. With Eq. (16.2.8), Eqs. (16.2.6) and (16.2.7) become

$$\frac{d^2V(z)}{dz^2} = \gamma^2 V(z), \tag{16.2.9}$$

$$\frac{d^2I(z)}{dz^2} = \gamma^2 I(z). \tag{16.2.10}$$

The solution of both equations is the same:

$$V(z) = V_i e^{-\gamma z} + V_r e^{\gamma z}, \tag{16.2.11}$$

$$I(z) = I_i e^{-\gamma z} + I_r e^{\gamma z}, \tag{16.2.12}$$

where $V_i$, $V_r$, $I_i$, and $I_r$ are constants. Comparing these equations with Eqs. (15.3.8) to (15.3.12), we see that terms with $e^{-\gamma z}$ represent attenuated voltage and current waves traveling in the positive $z$ direction, and we can readily conclude that terms with $e^{\gamma z}$ represent attenuated voltage and current waves propagating in the opposite direction. So voltage and current along a line can in general be considered as the sums of two oppositely traveling attenuated waves.

The four constants $V_i, \ldots, I_r$ are not independent. According to Eqs. (16.2.4), (16.2.11), and (16.2.12), we have

$$-\gamma V_i e^{-\gamma z} + \gamma V_r e^{\gamma z} = -(R + j\omega L)(I_i e^{-\gamma z} + I_r e^{\gamma z}).$$

This equation can be satisfied for any $z$ only if the coefficients of $e^{-\gamma z}$ and $e^{\gamma z}$ on the two sides are equal. So we obtain

$$\begin{aligned} \frac{V_i}{I_i} &= \frac{R + j\omega L}{\gamma} = \sqrt{\frac{R + j\omega L}{G + j\omega C}}, \\ \frac{V_r}{I_r} &= -\frac{R + j\omega L}{\gamma} = -\sqrt{\frac{R + j\omega L}{G + j\omega C}}. \end{aligned} \tag{16.2.13}$$

The square root on the right of these expressions represents a certain impedance, which is referred to as the *characteristic impedance* $Z_0$ of the line:

$$\boxed{Z_0 = \sqrt{\frac{R + j\omega L}{G + j\omega C}}} \quad \text{(the definition of characteristic impedance).} \tag{16.2.14}$$

With Eqs. (16.2.13) and (16.2.14) the voltage and current along the line can be represented as

$$\boxed{V(z) = V_i e^{-\gamma z} + V_r e^{\gamma z}} \tag{16.2.15}$$

(fundamental transmission line equations).

$$\boxed{I(z) = \frac{1}{Z_0}(V_i e^{-\gamma z} - V_r e^{\gamma z})} \tag{16.2.16}$$

These equations can be regarded as fundamental transmission line equations. We can find the constants $V_i$ and $V_r$ from any set of two boundary conditions. For example, voltage and current at $z = 0$, or at $z = l$ ($l$ being the length of the line), may be specified, or the terminating impedance $Z_l$ may be given along with some other information. We shall not, however, consider this in detail. Some exercise in determining $V_i$ and $V_r$ is given in Problem 16.T2.

There are, however, two important cases. The first is the case of an infinitely long line, where reflected wave does not exist, so that

$$V(z) = V_i e^{-\gamma z}, \qquad I(z) = \frac{V_i}{Z_0} e^{-\gamma z}, \tag{16.2.17}$$

from which

$$Z_0 = \frac{V(z)}{I(z)} \quad \text{(infinitely long line).} \tag{16.2.18}$$

*The characteristic impedance of any line is equal to the impedance of a very long section of that line.*

**Table 16.1.** Parameters of some transmission lines at high frequencies

| Parameter | Coaxial line 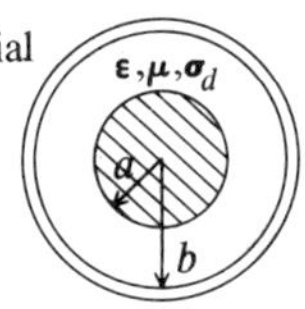 | Two–wire line ($d \gg 2a$) 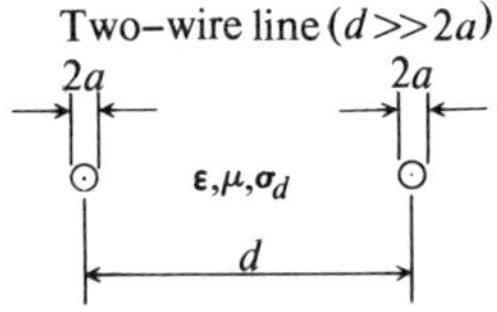 | Strip line ($b \gg a$) 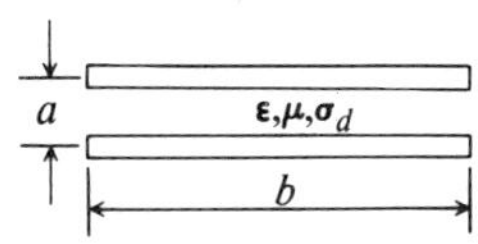 |
|---|---|---|---|
| $C$ | $\dfrac{2\pi\varepsilon}{\ln b/a}$ | $\dfrac{\pi\varepsilon}{\ln d/a}$ | $\varepsilon \dfrac{b}{a}$ |
| $G$ | — | $\dfrac{\sigma_d}{\varepsilon} C$ | — |
| $L_{ext}$* | $\dfrac{\mu}{2\pi} \ln \dfrac{b}{a}$ | $\dfrac{\mu}{\pi} \ln \dfrac{d}{a}$ | $\mu \dfrac{a}{b}$ |
| $R$ | $\dfrac{R_s}{2\pi}\left(\dfrac{1}{a} + \dfrac{1}{b}\right)$ | $\dfrac{R_s}{\pi a}$ | $\dfrac{2R_s}{b}$ |

*$L_{int} = R/\omega$ in all three cases.

The other case, which is even more important in practice, is when $Z_l = Z_0$. Then $V(l) = Z_0 I(l)$, so that, from Eqs. (16.2.15) and (16.2.16), we obtain

$$V_i e^{-\gamma l} + V_r e^{\gamma l} = V_i e^{-\gamma l} - V_r e^{\gamma l},$$

from which $V_r = 0$. *There is no reflected wave on a line terminated in its characteristic impedance:* it is said that the line is *matched.* A matched line is most efficient for transmitting a signal from the sending to the receiving end. Since no reflected wave exists, the total energy of the signal that reaches the terminating impedance is absorbed by it.

Starting from the basic equations (16.2.8), (16.2.14), (16.2.15), and (16.2.16), we can determine the distribution of current and voltage along any transmission line, provided that the line parameters $R$, $L$, $C$ and $G$ are known. These parameters for coaxial, two-wire and strip lines are summarized in Table 16.1, which is only valid for high frequencies. There are also many other associated problems that could be analyzed, and we shall consider some of these when we deal with the case of lines with very small, theoretically zero, losses. For a more detailed treatment of transmission lines the reader is referred to more specialized texts.

**16.2.3 Lines with negligible losses ("lossless" lines).** In some important cases losses due to the resistivity of line conductors and the conductivity of the surrounding dielectric can be ignored in the first approximation. This is true particularly if the lines are used at very high frequencies and are reasonably short. Such lines are referred to as "lossless" lines. The analysis of such lines is considerably simpler than that of lossy lines.

Assuming that, in the basic transmission line equations (16.2.8), (16.2.14), (16.2.15), and (16.2.16), $R = 0$ and $G = 0$, we have

$$\boxed{\alpha = 0, \qquad \beta = \omega\sqrt{LC}} \qquad \text{(lossless lines)}, \tag{16.2.19}$$

$$\boxed{Z_0 = \sqrt{L/C}} \qquad \text{(lossless lines)}, \tag{16.2.20}$$

$$\boxed{V(z) = V_i e^{-j\beta z} + V_r e^{j\beta z}} \qquad \text{(lossless lines)}, \tag{16.2.21}$$

$$\boxed{I(z) = \frac{1}{Z_0}(V_i e^{-j\beta z} - V_r e^{j\beta z})} \qquad \text{(lossless lines)}. \tag{16.2.22}$$

From Table 16.1 we see that, if all three types of line are considered to be lossless ($R = 0$, $G = 0$, *and* $L_{\text{int}} = 0$),

$$\boxed{\beta = \omega\sqrt{\varepsilon\mu}} \qquad \text{(lossless lines)}. \tag{16.2.23}$$

This can be shown to be true for all lossless lines. So we learn that the velocity of propagation along such lines equals the velocity of propagation of the plane waves, $c = \omega/\beta = (\varepsilon\mu)^{-1/2}$. Thus *wavelengths measured along a lossless line correspond to the wavelengths of the plane waves propagating in the dielectric of the line.*

Let us consider a line terminated by an arbitrary terminating impedance $Z_l$, and let $I(l)$ be the complex amplitude of the current through $Z_l$. According to Eqs. (16.2.21) and (16.2.22)

$$Z_l I(l) = V_i e^{-j\beta l} + V_r e^{j\beta l},$$
$$Z_0 I(l) = V_i e^{-j\beta l} - V_r e^{j\beta l},$$

from which

$$V_i = \tfrac{1}{2}(Z_l + Z_0)I(l)e^{j\beta l}, \qquad V_r = \tfrac{1}{2}(Z_l - Z_0)I(l)e^{-j\beta l}.$$

Substituting these values of $V_i$ and $V_r$ into Eqs. (16.2.21) and (16.2.22), we obtain

$$V(z) = \frac{I(l)}{2}\left[(Z_l + Z_0)e^{j\beta(l-z)} + (Z_l - Z_0)e^{-j\beta(l-z)}\right],$$

$$I(z) = \frac{I(l)}{2Z_0}\left[(Z_l + Z_0)e^{j\beta(l-z)} - (Z_l - Z_0)e^{-j\beta(l-z)}\right].$$

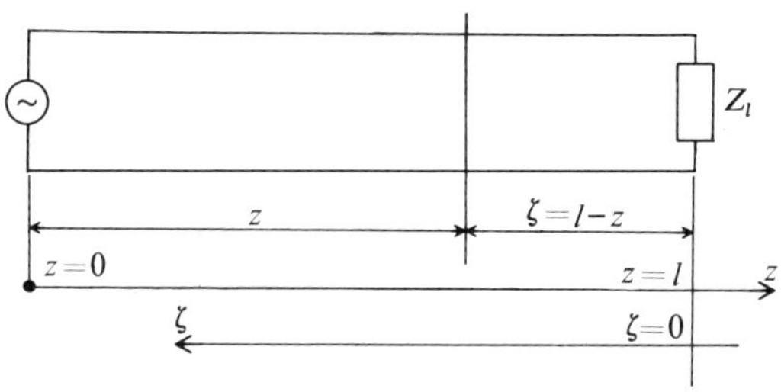


**Fig. 16.3** Coordinates used for analysing transmission lines.

It is very convenient to introduce a new variable, the distance $\zeta = l - z$ from the load (see Fig. 16.3). With this, and for simplicity designating the current $I(l)$ through the load by $I_l$, we transform the above equations into

$$V(\zeta) = \frac{I_l}{2}\left[(Z_l + Z_0)e^{j\beta\zeta} + (Z_l - Z_0)e^{-j\beta\zeta}\right], \tag{16.2.24}$$

$$I(\zeta) = \frac{I_l}{2Z_0}\left[(Z_l + Z_0)e^{j\beta\zeta} - (Z_l - Z_0)e^{-j\beta\zeta}\right]. \tag{16.2.25}$$

If we recall that

$$e^{j\beta\zeta} + e^{-j\beta\zeta} = 2\cos\beta\zeta, \qquad e^{j\beta\zeta} - e^{-j\beta\zeta} = 2j\sin\beta\zeta,$$

we can simplify these equations to

$$V(\zeta) = I_l(Z_l \cos\beta\zeta + jZ_0 \sin\beta\zeta), \tag{16.2.26}$$

$$I(\zeta) = \frac{I_l}{Z_0}(Z_0 \cos\beta\zeta + j\,Z_l \sin\beta\zeta). \tag{16.2.27}$$

These equations describe the distributions of voltage and current along the line in terms of the conditions at the receiving (load) end. In particular, the impedance of the line terminated by $Z_l$ at any distance $\zeta$ from the load and looking towards the load is

$$Z(\zeta) = \frac{V(\zeta)}{I(\zeta)} = Z_0 \frac{Z_l + jZ_0 \tan\beta\zeta}{Z_0 + jZ_l \tan\beta\zeta}. \tag{16.2.28}$$

Two important special cases are the open-circuited and the short-circuited lines. With $Z_l \to \infty$ and $Z_l = 0$, respectively, Eq. (16.2.28) becomes

$$Z(\zeta) = -jZ_0/\tan\beta\zeta \qquad \text{(open-circuited line)}, \tag{16.2.29}$$

$$Z(\zeta) = jZ_0 \tan\beta\zeta \qquad \text{(short-circuited line)}. \tag{16.2.30}$$

Figure 16.4 shows the ratio $Z(\zeta)/j$ for both cases. We should note that the input impedance of both open-circuited and short-circuited lossless lines is purely imaginary, so they may replace any capacitor or any coil. We can also see that each can behave as a short or as an open circuit, if only the length of the line is appropriately chosen.

Finally, recalling that in Eq. (16.2.24) the term with $e^{j\beta\zeta}$ represents the incident voltage wave and that with $e^{-j\beta\zeta}$ the reflected voltage wave, we find that the voltage reflection coefficient $\mathscr{R} = V_r/V_i$ is given by

$$\boxed{\mathscr{R} = \frac{Z_l - Z_0}{Z_l + Z_0}} \qquad \text{(the voltage reflection coefficient).} \tag{16.2.31}$$

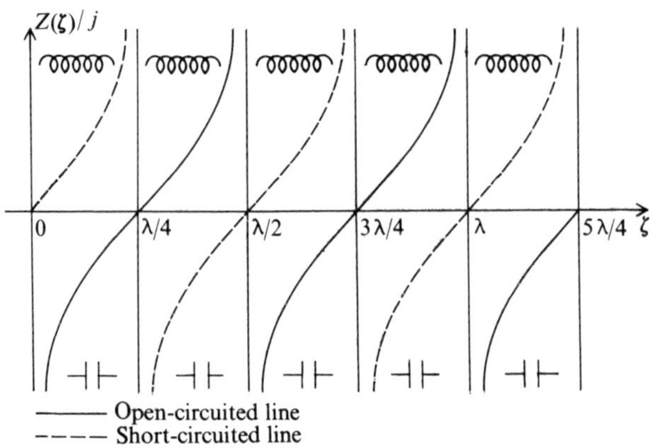


**Fig. 16.4** Input impedance of open-circuited and short-circuited lossless lines as a function of their length.

Note that it is of the same form as the reflection coefficient in Eq. (15.4.11) in the case of a plane wave perpendicularly incident on a plane interface between two dielectric media. Using $\mathscr{R}$, we can rewrite Eq. (16.2.24) as

$$V(\zeta) = \tfrac{1}{2}I_l(Z_l + Z_0)e^{j\beta\zeta}(1 + \mathscr{R}e^{-j2\beta\zeta}). \tag{16.2.32}$$

Since $|e^{-j2\beta\zeta}| = 1$, the ratio of the maximum and minimum amplitudes of the voltage along the line is

$$\frac{V_{\max}}{V_{\min}} = s = \frac{1 + |\mathscr{R}|}{1 - |\mathscr{R}|} \qquad \text{(definition of the voltage standing-wave ratio).} \tag{16.2.33}$$

This ratio is referred to as the voltage *standing-wave ratio*, which is sometimes shortened to VSWR. As we can see from Eq. (16.2.33), $1 \leq s \leq \infty$.

So, if $|\mathscr{R}| \neq 0$, a standing wave along the line is superimposed on the progressive voltage wave towards the load, just as in the case of plane waves perpendicularly incident on a plane interface. And, as in connection with Eq. (15.4.13), we can conclude that, if $Z_l$ is real (remember that for lossless lines $Z_0$ is always real), the voltage maximum or minimum occurs at the load. The maximum is obtained if $Z_l > Z_0$, and the minimum if $Z_l < Z_0$. If $Z_l$ is complex, the voltage across $Z_l$ is neither maximum nor minimum. The knowledge of the position of voltage maxima and minima and of $|\mathscr{R}|$ is sufficient for determining the load impedance $Z_l$. This represents the basis of an important experimental method of determining impedances at very high frequencies (see Problem 16.G4).

The input impedance given in Eq. (16.2.28) looking toward the load, at any distance $\zeta$ from the load, can also be expressed in terms of $Z_0$ and the complex reflection coefficient $\mathscr{R}$ only. For that purpose we first modify Eq. (16.2.25) to a form analogous to Eq. (16.3.32). Dividing the two equations we obtain

$$Z(\zeta) = \frac{V(\zeta)}{I(\zeta)} = Z_0\frac{1 + \mathscr{R}e^{-j2\beta\zeta}}{1 - \mathscr{R}e^{-j2\beta\zeta}}. \tag{16.2.34}$$

This might seem to be just juggling around the equations that we have; the last expression, however, is important for the graphical solution of some problems.

Using the foregoing equations, we can solve any problem involving transmission lines analytically. For example, from Eq. (16.2.28) we can determine the input impedance of any transmission line terminated with any impedance, or we can determine the necessary length of a line to have a desired input impedance if terminated by a certain load. Many other more or less similar problems crop up in practice. Although the solution is relatively simple conceptually, it is usually very lengthy, because complex rather than real numbers are involved. In most cases an approximate, but usually sufficiently accurate solution can be obtained by using certain graphical methods. Perhaps the best known and most widely used is the one which makes use of the *Smith chart*.*

---

* P. H. Smith, "Transmission-line Calculator", *Electronics*, **12**, January 1939, p. 29; "An Improved Transmission-line Calculator", *Electronics*, **17**, January 1944, p. 130.

**16.2.4 The graphical solution of line problems using the Smith chart.*** Essentially, the Smith chart enables us to make a direct determination of the complex reflection coefficient $\mathscr{R}$ given by Eq. (16.2.31), corresponding to a given load impedance $Z_l$ and the characteristic impedance $Z_0$ of the line. Conversely, if $\mathscr{R}$ is determined experimentally, we can read directly from the chart the load impedance $Z_l$, if $Z_0$ is known. However, the usefulness of the Smith chart far surpasses these two relatively simple tasks. To understand how to use it properly, we must understand how it is constructed.

Let us first divide the numerator and denominator on the right of Eq. (16.2.31) by $Z_0$. We thus express the reflection coefficient by the ratio $Z_l/Z_0$ instead of by $Z_l$ and $Z_0$ separately. The ratio $Z_l/Z_0$ is known as the load impedance *normalized* with respect to the characteristic impedance of the line. The normalized complex impedance is usually designated by $z$, and its real and imaginary parts by $r$ and $x$ respectively. So we have

$$\boxed{\frac{Z_l}{Z_0} = \frac{R_l + jX_l}{Z_0} = z = r + jx} \quad \text{(definition of the normalized load impedance)}, \tag{16.2.35}$$

and

$$\mathscr{R} = \frac{z - 1}{z + 1}.$$

Solving for $z$ this equation becomes

$$z = \frac{1 + \mathscr{R}}{1 - \mathscr{R}}, \tag{16.2.36}$$

or

$$r + jx = \frac{(1 + \mathscr{R}_r) + j\mathscr{R}_i}{(1 - \mathscr{R}_r) - j\mathscr{R}_i}.$$

$\mathscr{R}_r$ and $\mathscr{R}_i$ respectively designate the real and imaginary parts of the complex reflection coefficient $\mathscr{R}$. The last equation represents two real equations, since the real and imaginary parts of any complex equation must be equal. Now,

$$\frac{(1 + \mathscr{R}_r) + j\mathscr{R}_i}{(1 - \mathscr{R}_r) - j\mathscr{R}_i} = \frac{[(1 + \mathscr{R}_r) + j\mathscr{R}_i][(1 - \mathscr{R}_r) + j\mathscr{R}_i]}{(1 - \mathscr{R}_r)^2 + \mathscr{R}_i^2} = \frac{(1 - \mathscr{R}_r^2 - \mathscr{R}_i^2) + j2\mathscr{R}_i}{(1 - \mathscr{R}_r)^2 + \mathscr{R}_i^2}.$$

So we have

$$r = \frac{1 - \mathscr{R}_r^2 - \mathscr{R}_i^2}{1 - 2\mathscr{R}_r + \mathscr{R}_r^2 + \mathscr{R}_i^2}, \qquad x = \frac{2\mathscr{R}_i}{1 - 2\mathscr{R}_r + \mathscr{R}_r^2 + \mathscr{R}_i^2},$$

* This section can be omitted without loss of continuity.

from which, after some manipulation, we find that

$$\left(\mathscr{R}_r - \frac{r}{r+1}\right)^2 + \mathscr{R}_i^2 = \left(\frac{1}{r+1}\right)^2, \tag{16.2.37}$$

$$(\mathscr{R}_r - 1)^2 + \left(\mathscr{R}_i - \frac{1}{x}\right)^2 = \left(\frac{1}{x}\right)^2. \tag{16.2.38}$$

Each of these two equations represents a family of circles with $r$ and $x$, respectively, as the parameters. Note that, if $Z_0$ is real, by the definition in Eq. (16.2.31) $|\mathscr{R}| \leq 1$. So only those parts of these circles are of interest which are within the unit circle

$$|\mathscr{R}|^2 = \mathscr{R}_r^2 + \mathscr{R}_i^2 = 1 \tag{16.2.39}$$

in the complex $\mathscr{R}$ plane.

Construction of the circles represented by Eqs. (16.2.37) and (16.2.38) for different values of $r$ and $x$ is simple. Note that the radii of the $r$-circles are $1/(r+1)$, and that the coordinates of their centers are $\mathscr{R}_r = r/(r+1)$ and $\mathscr{R}_i = 0$. Similarly, the radii of the $x$-circles are $1/x$, and the coordinates of their centers $\mathscr{R}_i = 1/x$ and $\mathscr{R}_r = 1$.

Figure 16.5 shows parts of several of these circles lying within the unit circle $|\mathscr{R}| = 1$. A much more elaborate chart, with some additional details, the use of which will be explained later, is given in Fig. 16.6.

The information that can be obtained directly from such a chart is the complex reflection coefficient $\mathscr{R}$ corresponding to a certain normalized load impedance $z$. For example, for $Z_0 = 50\Omega$ and $Z_l = (40 + j60)\Omega$, we have $z = Z_l/Z_0 = (0.8 + j1.2)$. So $r = 0.8$, and $x = +1.2$. In the crude diagram of Fig. 16.5

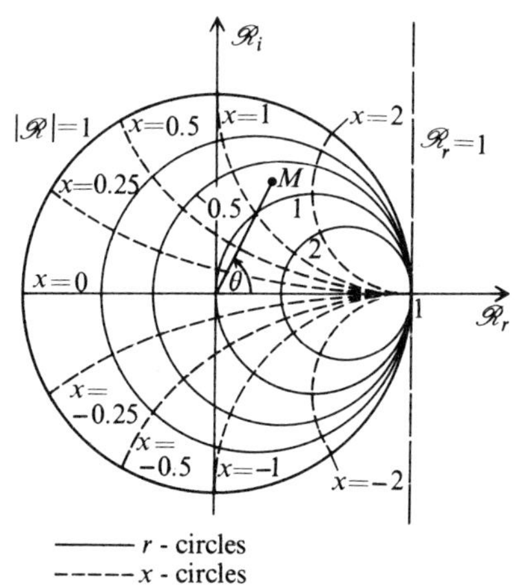


**Fig. 16.5** A crude Smith's chart.

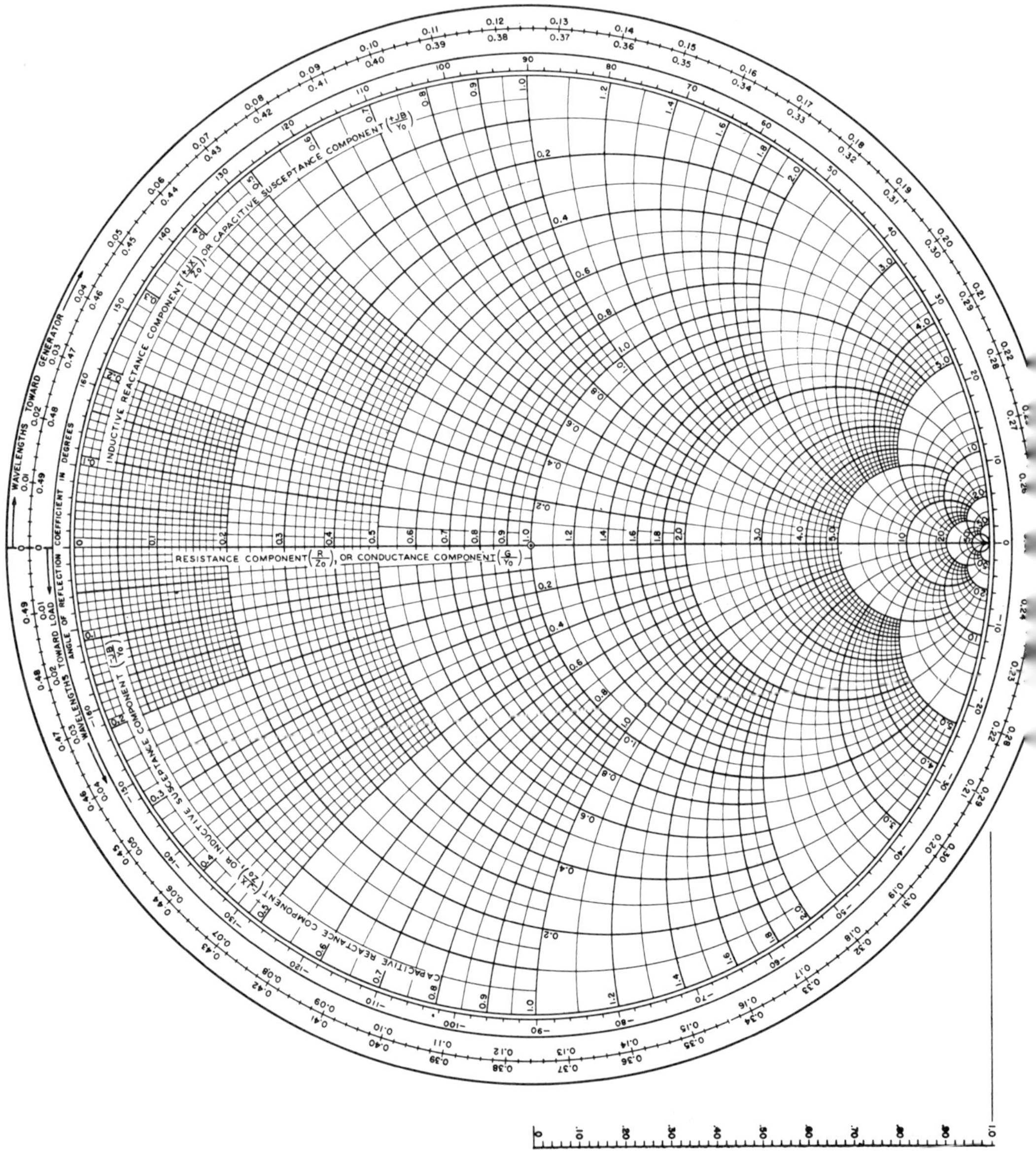


**Fig. 16.6** The Smith chart. (Reproduced by permission of Kay Electric Company, Pine Brook, N. J.).

this corresponds to $|\mathscr{R}| \simeq 0.6$, and the angle of $\mathscr{R}$, $\theta_{\mathscr{R}} \simeq 60°$. According to the more elaborate chart shown in Fig. 16.6 we find that $|\mathscr{R}| \simeq 0.57$ and $\theta_{\mathscr{R}} \simeq 66°$. The magnitude of $\mathscr{R}$ is obtained by first measuring the distance of the point $M$ from the chart center (point $\mathscr{R}_r = \mathscr{R}_i = 0$), using a compass. Below the chart a linear scale is provided; by using this we can obtain the distance measured by the compass in terms of the chart radius (unit circle in the complex $\mathscr{R}$ plane). For easy reading of the angle $\theta_{\mathscr{R}}$, an angle scale marked "angle of reflection coefficient in degrees" is provided around the main chart. So we need only draw a straight line from the origin through $M$ to determine its intersection with the angle scale.

The converse problem, of determining the normalized load impedance for a given (say, experimentally determined) reflection coefficient, is equally simple. For example, according to Fig. 16.6 for $\mathscr{R} = 0.8e^{-j45°}$, that is, $|\mathscr{R}| = 0.8$ and $\theta_{\mathscr{R}} = -45°$, we obtain $z \simeq 0.75 - j2.20$. So if, say, $Z_0 = 60\Omega$, the load impedance is $Z_l = zZ_0 \simeq (45.0 - j132)\Omega$.

In addition to the two foregoing basic and simple applications of the Smith chart, there are several more sophisticated ones. Perhaps the most important is that of determining the input impedance of a line of a given length and characteristic impedance, terminated by a given load impedance. This problem can be solved by using Eq. (16.2.28), but an approximate solution using the Smith chart is much faster. For that purpose we first normalize $Z(\zeta)$ given by Eq. (16.2.34) with respect to $Z_0$:

$$z(\zeta) = \frac{Z(\zeta)}{Z_0} = \frac{1 + \mathscr{R}e^{-j2\beta\zeta}}{1 - \mathscr{R}e^{-j2\beta\zeta}}. \tag{16.2.40}$$

For $\zeta = 0$, $z(\zeta)$ is identical with our earlier $z$, given by (16.2.36). But $z(\zeta)$ is of *exactly* the same form as $z$ if we denote $\mathscr{R}e^{-j2\beta\zeta}$ by, say, $\mathscr{R}_\zeta$:

$$z(\zeta) = \frac{1 + \mathscr{R}_\zeta}{1 - \mathscr{R}_\zeta}, \qquad \mathscr{R}_\zeta = \mathscr{R}e^{-j2\beta\zeta}. \tag{16.2.41}$$

Since $|\mathscr{R}| \leq 1$ and $|e^{-j2\beta\zeta}| = 1$, $|\mathscr{R}_\zeta| \leq 1$. So exactly the same chart applies to $z(\zeta)$ and to $\mathscr{R}_\zeta$ as to $z$ and $\mathscr{R}$. Now, if we know $z$ (i.e. $Z_l$ and $Z_0$), we can locate the point in the Smith chart which determines $\mathscr{R}$ directly. To obtain $z(\zeta)$, however, $\mathscr{R}_\zeta = \mathscr{R}e^{-j2\beta\zeta}$ is needed rather than $\mathscr{R}$. But multiplying a complex number by $e^{-j2\beta\zeta}$ implies changing its angle for $-2\beta\zeta$, leaving its magnitude constant, which means that we simply have to rotate $\mathscr{R}$ (corresponding to $z$) for $2\beta\zeta$, in the *clockwise* (negative) direction. Thus we obtain $\mathscr{R}_\zeta$, and can read $z(\zeta)$ directly from the chart. Since

$$2\beta\zeta = 2\frac{2\pi}{\lambda}\zeta = \frac{2\zeta}{\lambda}2\pi, \tag{16.2.42}$$

it follows that an angle of rotation $2\pi$ corresponds to $\zeta = \lambda/2$. This must be so, since from Eq. (16.2.28) we know that $Z(\zeta) = Z(\zeta + \lambda/2)$. To facilitate the rotation of $\mathscr{R}$ for the proper angle, an additional scale around the Smith chart is provided, with 0.5 (wavelengths) corresponding to one complete revolution around the unit circle $|\mathscr{R}| = 1$ In the chart shown in Fig. 16.6 this wavelength scale is designated by "wavelengths toward generator." For some applications the same scale in the opposite (counterclockwise) direction is also useful. This is also provided, and is designated by "wavelengths toward load".

As an example, let us consider a line of characteristic impedance $Z_0 = 60\Omega$ and of length $\zeta = 0.40\lambda$ at the frequency used. Let us suppose that $Z_l = (90 - j60)\Omega$, and that we wish to determine the input impedance of the line thus terminated, using the Smith chart.

First we have that $z = Z_l/Z_0 = 1.5 - j1$, and we start with this value in the chart. This point has to be rotated in a clockwise direction for 0.4 units of the wavelength scale. Therefore we draw a straight line from the center of the chart through the point $z = 1.5 - j1$. The intersection of this line and the "wavelengths toward generator" scale is at 0.308 on the scale. We add 0.40 to this, and get 0.708. This is for 0.208 further than point 0.000 on the scale, that is, at the point 0.208. We draw a straight line from the 0.208 point of the wavelength scale towards the chart center, and measure along this line from the center the distance of the point $z = 1.5 - j1$ from the center. The point thus found determines $z(\zeta) = z(0.4\lambda)$. From the chart we find that $z(0.4\lambda) \simeq (1.83 + j0.95)$. So the input impedance to a $0.4\lambda$ line where $Z_0 = 60\Omega$ and terminated with $Z_l = (90 - j60)\Omega$ is $Z(0.4\lambda) = Z_0 z(0.4\lambda) \simeq (110 + j57.0)\Omega$.

These few simple examples are only illustrations of some of the simplest applications of the Smith chart. Several other examples will be found among the problems at the end of the chapter. The applications of the Smith chart are much more diverse than these examples suggest. It can be used also for analyzing plane waves perpendicularly incident on a plane boundary surface, for analyzing lossy lines, and for many other problems. However, once the logical basis of the chart is understood, there should be no difficulty whatsoever in applying it to other situations.

**16.2.5 Conclusion.** Transmission lines are used for many different purposes, and in many different forms. Power transmission lines, telephone lines, and lines delivering electromagnetic energy to antennas (feeders), are perhaps the most important, but by no means the only types. They all have different, specific requirements imposed upon them. For power lines distorsion of the "signal" is irrelevant, but lines with the smallest possible losses are desirable. Telephone lines should be as free from distortion as possible, attenuation being of secondary importance, since low-power signals can easily be amplified. All these and many other practical questions necessarily remain unmentioned in this short introduction.

*Problems referring to Section 16.2:* 16.T1 to 16.T8, 16.G1 to 16.G7, 16.N1 to 16.N9.

## 16.3 WAVEGUIDES

**16.3.1 Introduction.** Transmission lines are defined as guiding systems which usually have two conductors (but there may also be more, as in three-phase power lines). The currents in the two conductors at any cross-section are of equal magnitudes but in opposite directions at any instant, and the electric and magnetic field vectors are essentially in planes transverse to the line direction. The waves propagating along the transmission lines are therefore said to be transverse electromagnetic (*TEM*) waves.

Waveguides are guiding systems in the form of highly conductive tubes or dielectric rods of arbitrary cross-sections, and electromagnetic energy is transported *through* such systems. There are many types of waveguides, and the problems associated with them are still more numerous. In this short introduction one type only will be analyzed. This is the type known as the rectangular waveguide. It takes the form of a tube of rectangular cross-section and is very often used in practice.

As usual, it is easier to analyze theoretically an idealized system rather than the real one. The walls of the rectangular waveguide can never be made of a perfect conductor, but we shall nevertheless consider them to be perfect, as otherwise the solution is extremely involved. For most practical applications only a small simple correction of the solution of this idealized case is necessary. In addition, waveguides are never perfectly straight or infinitely long. We consider this case notwithstanding, simply because we do not know how to solve a more complicated one. In spite of all drawbacks, the picture we obtain of what is going on in a waveguide is very close to what actually happens in practice.

Waveguides, and the many different structures derived from them, are the basic building blocks of any microwave system. The engineering problems relating to such structures are very diverse, but, in an introductory text such as this, we can deal only with some of the basic ideas. In consequence only a few of the most important concepts pertaining to waveguides will be presented.

**16.3.2 The propagation of electromagnetic waves in rectangular waveguides.** Let us take the tube of rectangular cross-section shown in Fig. 16.7. We assume that the waveguide is infinitely long, that its walls are made of a perfect conductor, and that the dielectric inside it is also perfect, of permittivity $\varepsilon$ and permeability $\mu$. Solutions of Maxwell's equations are needed for representing the sinusoidally time-varying electromagnetic waves traveling through the waveguide in the positive $z$ direction.

We have seen that, in complex notation, a wave traveling in the positive $z$ direction depends on the $z$ coordinate as $e^{-j\beta z}$. When attenuation is also present, we insert $\gamma = \alpha + j\beta$ instead of $j\beta$. In general, therefore, a factor $e^{-\gamma z}$ designates the propagation in the positive $z$ direction. So we assume that the (complex) amplitudes of the field vectors are of the form

$$\mathbf{E} = \mathbf{E}(x, y)\, e^{-\gamma z}, \qquad \mathbf{H} = \mathbf{H}(x, y)\, e^{-\gamma z}. \tag{16.3.1}$$

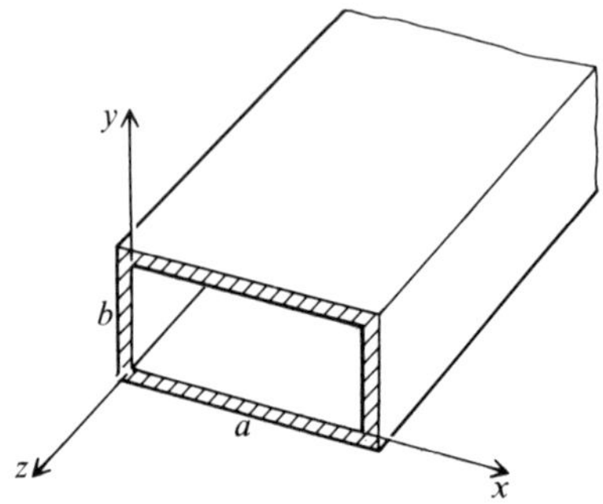


**Fig. 16.7** A rectangular waveguide.

$\mathbf{E}(x, y)$ and $\mathbf{H}(x, y)$ can have all three components, but they are functions of $x$ and $y$ only. Since

$$\frac{\partial}{\partial z}\left[E_x(x, y)\, e^{-\gamma z}\right] = -\gamma E_x(x, y)\, e^{-\gamma z}, \tag{16.3.2}$$

and similarly for all other components, whenever a derivative with respect to $z$ appears, we can replace it by the factor $-\gamma$.

The types of waves in waveguides are divided into two groups. If $E_z = 0$, the electric field intensity vector is in the transverse plane, so that the waves with $E_z = 0$ are referred to as transverse electric, or *TE*, waves. Similarly, waves with $H_z = 0$ are known as transverse magnetic, or *TM*, waves. Both wave types can exist in a waveguide independently or simultaneously, but we shall restrict our attention to *TE* waves which are practically more important, that is, we shall assume $E_z = 0$.

With these remarks and omitting the common factor $e^{-\gamma z}$, the two Maxwell's curl equations in complex form [Eqs. (13.3.24) and (13.3.25)] for the electromagnetic field inside the waveguide become

$$\gamma E_y = -j\omega\mu\, H_x, \tag{16.3.3}$$

$$-\gamma E_x = -j\omega\mu\, H_y, \tag{16.3.4}$$

$$\frac{\partial E_y}{\partial x} - \frac{\partial E_x}{\partial y} = -j\omega\mu H_z, \tag{16.3.5}$$

$$\frac{\partial H_z}{\partial y} + \gamma H_y = j\omega\varepsilon E_x, \tag{16.3.6}$$

$$-\gamma H_x - \frac{\partial H_z}{\partial x} = j\omega\varepsilon E_y, \tag{16.3.7}$$

$$\frac{\partial H_y}{\partial x} - \frac{\partial H_x}{\partial y} = 0. \tag{16.3.8}$$

Combining Eqs. (16.3.3) and (16.3.7) we obtain

$$-\gamma^2 H_x - \gamma \frac{\partial H_z}{\partial x} = \omega^2 \varepsilon\mu H_x$$

from which

$$H_x = -\frac{\gamma}{k^2}\frac{\partial H_z}{\partial x}, \tag{16.3.9}$$

where

$$k^2 = \gamma^2 + \omega^2\varepsilon\mu. \tag{16.3.10}$$

From Eqs. (16.3.4) and (16.3.6), we similarly obtain

$$H_y = -\frac{\gamma}{k^2}\frac{\partial H_z}{\partial y}. \tag{16.3.11}$$

Hence, according to Eqs. (16.3.3) and (16.3.4) it follows that

$$E_x = -\frac{j\omega\mu}{k^2}\frac{\partial H_z}{\partial y}, \tag{16.3.12}$$

$$E_y = \frac{j\omega\mu}{k^2}\frac{\partial H_z}{\partial x}. \tag{16.3.13}$$

So, if we determine $H_z$, all other components can easily be determined. We obtain the differential equation for $H_z$ by substituting $E_x$ and $E_y$, given by the last two equations, into Eq. (16.3.5):

$$\frac{\partial^2 H_z}{\partial x^2} + \frac{\partial^2 H_z}{\partial y^2} + k^2 H_z = 0. \tag{16.3.14}$$

To solve for $H_z$, we assume that $H_z$ (a function of $x$ and $y$ only) can be represented as a product of two functions of $x$ and $y$ only,

$$H_z = H_z(x, y) = X(x)Y(y). \tag{16.3.15}$$

With $H_z$ thus assumed, Eq. (16.3.14) becomes

$$\frac{1}{X}\frac{d^2X}{dx^2} + \frac{1}{Y}\frac{d^2Y}{dy^2} + k^2 = 0. \tag{16.3.16}$$

The first term is a function of $x$, and the second of $y$ alone. So the equation can be satisfied for any $x$ and $y$ only if both are equal to a constant:

$$d^2X/dx^2 = -k_x^2 X, \qquad d^2Y/dy^2 = -k_y^2 Y, \tag{16.3.17}$$

where

$$k_x^2 + k_y^2 = k^2 = \gamma^2 + \omega^2\varepsilon\mu. \tag{16.3.18}$$

Solutions of Eq. (16.3.17) are of the same form

$$X(x) = A_x \sin k_x x + B_x \cos k_x x,$$
$$Y(y) = A_y \sin k_y y + B_y \cos k_y y,$$

and $H_z(x, y)$ becomes

$$H_z = H_z(x, y) = (A_x \sin k_x x + B_x \cos k_x x)(A_y \sin k_y y + B_y \cos k_y y). \quad (16.3.19)$$

$A_x$, $B_x$, $A_y$, and $B_y$ are constants which have to be determined together with constants $k_x$ and $k_y$ so that the boundary conditions on the waveguide walls are satisfied. By reference to Fig. 16.7 and to Eq. (13.4.7), we find that these conditions are

$$E_x(x, 0) = 0, \qquad E_x(x, b) = 0, \quad (16.3.20)$$
$$E_y(0, y) = 0, \qquad E_y(a, y) = 0. \quad (16.3.21)$$

From Eqs. (16.3.9) to (16.3.13) we can see that, if these conditions are satisfied, then the normal components of **H** are also zero on the walls, as required by Eq. (13.4.9).

From Eqs. (16.3.12), (16.3.19), and (16.3.20) we find that it must be

$$(A_x \sin k_x x + B_x \cos k_x x)(A_y k_y \cos k_y y - B_y k_y \sin k_y y) = 0 \text{ for } y = 0 \text{ and } y = b.$$

This is possible only if $A_y = 0$ and $k_y = n\pi/b$, $n = 0, 1, 2, \ldots$. The boundary conditions of Eq. (16.3.21) are similarly found to be satisfied only if $A_x = 0$ and $k_x = m\pi/a$, $m = 0, 1, 2, \ldots$ So (16.3.19) becomes

$$H_z = H_z(x, y) = H_0 \cos \frac{m\pi}{a} x \cos \frac{n\pi}{b} y, \quad m, n = 0, 1, 2, \ldots. \quad (16.3.22)$$

$H_0$ stands for $B_x B_y$. It is determined by the intensity of the wave propagating through the waveguide. We must note that the boundary conditions could not be satisfied if $H_z = 0$, which we can easily see from Eqs. (16.3.3) to (16.3.8). If $H_z = 0$ these equations simply determine a plane wave propagating in the positive $z$ direction, for which **E** and **H** are *constant* in the transverse planes. *So TEM waves cannot propagate along waveguides.*

Using Eq. (16.3.9) to (16.3.13), we can easily calculate all the other components of an electromagnetic wave of the *TE* type. All the components have numbers $m$ and $n$ as parameters. An infinite number of *TE* wave types, corresponding to different pairs of numbers $m$ and $n$, can propagate through the rectangular waveguide. The general $mn$ wave is designated by the expression $TE_{mn}$, and we can have particular types numbered $TE_{10}$, $TE_{01}$, $TE_{12}$, $TE_{45}$, etc. Note that $m$ and $n$ cannot simultaneously be zero. Then $H_z = H_0 =$ constant, so that, according to Eqs. (16.3.9) to (16.3.16), all other components are zero. We suggest that the reader should derive complete expressions for the general $TE_{mn}$ wave as an exercise.

Since $k_x = m\pi/a$ and $k_y = n\pi/b$, the propagation constant $\gamma$ is, according to

Eq. (16.3.18), given by

$$\gamma = \sqrt{\left(\frac{m\pi}{a}\right)^2 + \left(\frac{n\pi}{b}\right)^2 - \omega^2\varepsilon\mu} = j\omega\sqrt{\varepsilon\mu}\sqrt{1 - \frac{f_c^2}{f^2}}, \qquad (16.3.23)$$

where

$$f_c = \frac{\omega_c}{2\pi} = \frac{1}{2\sqrt{\varepsilon\mu}}\sqrt{\left(\frac{m}{a}\right)^2 + \left(\frac{n}{b}\right)^2}. \qquad (16.3.24)$$

If frequency $f$ of the wave is larger than $f_c$, the propagation constant is purely imaginary, which is precisely the condition for an unattenuated wave propagation in the positive $z$ direction. But if $f < f_c$, the propagation constant is real. So, instead of the common factor of the type $e^{-j\beta z}$, we have $e^{-\alpha z}$, which means that the wave is attenuated in spite of the fact that there are no losses in the guiding system. Consequently, waves with frequency $f$ *lower* than $f_c$ given by Eq. (16.3.24) cannot propagate through the waveguide. The particular frequency $f_c$ is referred to as the *cutoff frequency*.

The cutoff frequencies of different $TE_{mn}$ waves are also different. Assuming that $a > b$, $TE_{10}$ wave ($m = 1, n = 0$) has the lowest cutoff frequency:

$$(f_c)_{TE_{10}} = \frac{1}{2a\sqrt{\varepsilon\mu}} = \frac{c}{2a}. \qquad (16.3.25)$$

The free-space wavelength corresponding to this frequency is $\lambda_c = 2a$. The next lowest cutoff frequency is that of the $TE_{01}$ type, being equal to $c/2b$. *So, in the range $c/2a < f < c/2b$, no other wave can propagate through the waveguide except $TE_{10}$.* This is an extremely important property. If the waveguide is bent or twisted (which in practice it is always necessary to do) the currents and charges induced in these sections may give rise to other wave types, but, nevertheless, only the $TE_{10}$ wave type will propagate further. So, if we inject a $TE_{10}$ wave of a proper frequency at one end, no matter what we do with the waveguide, only this type will appear at the other end. The $TE_{10}$ wave is known as the *dominant* wave.

Of all $TE$ waves, the $TE_{10}$ type is almost the only wave type used in rectangular waveguides. It is worth mentioning that, in addition to the foregoing important property, the $TE_{10}$ type requires the minimum cross-sectional dimensions of a waveguide, and has the smallest attenuation of all $TE$ (and also $TM$) types (for a given frequency). No wonder that it is the favorite of engineers. For our part, we shall restrict our attention to this important wave.

**16.3.3 The properties of $TE_{10}$ wave.** According to Eqs. (16.3.1), (16.3.9) to (16.3.13), and (16.3.22), the complex components of a $TE_{10}$ wave (with $j\beta$ instead of $\gamma$) are

$$E_y(x, y, z) = -j\omega\mu\frac{a}{\pi}H_0 \sin\frac{\pi}{a}x\, e^{-j\beta z}, \qquad (16.3.26)$$

$$H_x(x, y, z) = j\beta \frac{a}{\pi} H_0 \sin \frac{\pi}{a} x \, e^{-j\beta z}, \tag{16.3.27}$$

$$H_z(x, y, z) = H_0 \cos \frac{\pi}{a} x \, e^{-j\beta z}. \tag{16.3.28}$$

The phase constant $\beta$ for the $TE_{10}$ type is given by Eqs. (16.3.23) and (16.3.25):

$$\beta = \omega\sqrt{\varepsilon\mu}\sqrt{1 - f_c^2/f^2}, \qquad f_c = \frac{1}{2a\sqrt{\varepsilon\mu}} < f. \tag{16.3.29}$$

The wavelength $\lambda_z$ of the wave propagating through the guide *measured along the z axis* is obtained from the relation $\beta = 2\pi/\lambda_z$:

$$\lambda_z = \frac{2\pi}{\beta} = \frac{c}{f\sqrt{1 - f_c^2/f^2}} = \frac{\lambda}{\sqrt{1 - f_c^2/f^2}}. \tag{16.3.30}$$

($\lambda$ designates the wavelength of plane wave of the same frequency.)

The velocity of propagation of the wave along the waveguide is obtained as

$$v_p = f\lambda_z = \frac{c}{\sqrt{1 - f_c^2/f^2}}, \qquad c = \frac{1}{\sqrt{\varepsilon\mu}}. \tag{16.3.31}$$

If the dielectric in the waveguide is a vacuum (air), the velocity $v_p$ is larger than the velocity $c_0$ of light in a vacuum! However, $v_p$ is only a geometrical velocity, showing how fast the *phase* of the wave propagates along the guide, and is referred to as the *phase* velocity. We shall see in the next section that energy propagates through the waveguide with a different velocity, $v_g$, known as the *group* velocity. Note that $v_p$ is a function of frequency, so a waveguide is a dispersive medium.

Figure 16.8 shows the distribution of the electric and magnetic fields of a $TE_{10}$ wave. We have to imagine this distribution moving with the velocity $v_p$ along the guide.

For some applications of waveguides it is necessary to know the distribution of the currents induced on the walls of the guide, and this can be determined from the

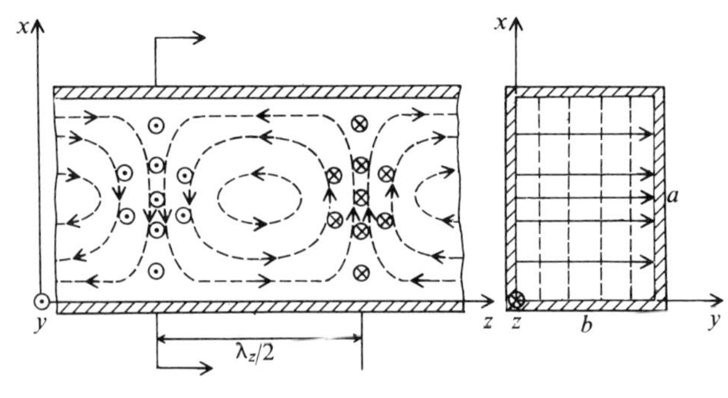


**Fig. 16.8** $TE_{10}$ wave in rectangular waveguide.

boundary condition of Eq. (13.4.11) by using Eqs. (16.3.27) and (16.3.28). Figure 16.9 shows the current distribution on the waveguide walls. Again we have to imagine this distribution moving with the velocity $v_p$ along the guide.

We can find the average power transmitted through the guide by integrating the complex Poynting vector over a cross-section of the guide. Since the complex Poynting vector in the $z$ direction is

$$\mathscr{P} = \tfrac{1}{2}\mathbf{E}_{\text{transverse}} \times \mathbf{H}^*_{\text{transverse}} = -\tfrac{1}{2} E_y H_x^* \mathbf{u}_z,$$

according to Eqs. (16.3.26) and (16.3.27) we find that the average power transmitted through the guide is

$$(P_{\text{av}})_{TE_{10}} = \int_0^a \int_0^b \text{Re}\{-\tfrac{1}{2}E_y H_x^*\}\, dx\, dy = \frac{\omega\mu\beta a^2 H_0^2}{2\pi^2} \int_0^a \int_0^b \sin^2 \frac{\pi}{a} x\, dx\, dy,$$

from which

$$(P_{\text{av}})_{TE_{10}} = \frac{\omega\mu\beta a^3 b H_0^2}{4\pi^2}.$$

According to Eq. (16.3.29), this can be written in the form

$$(P_{\text{av}})_{TE_{10}} = \frac{ab}{4}\sqrt{\frac{\mu}{\varepsilon}}\frac{f^2}{f_c^2}\sqrt{1 - f_c^2/f^2}\, H_0^2. \tag{16.3.32}$$

This short introduction to waveguides would be incomplete without mentioning how the waves propagating through them are excited. There are various methods of exciting the desired wave type in a waveguide. Figure 16.10 shows one

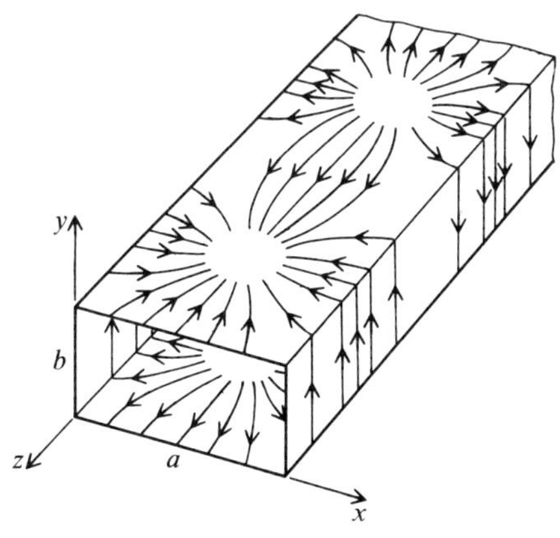


**Fig. 16.9** Distribution of currents on the walls of a rectangular waveguide with $TE_{10}$ wave.

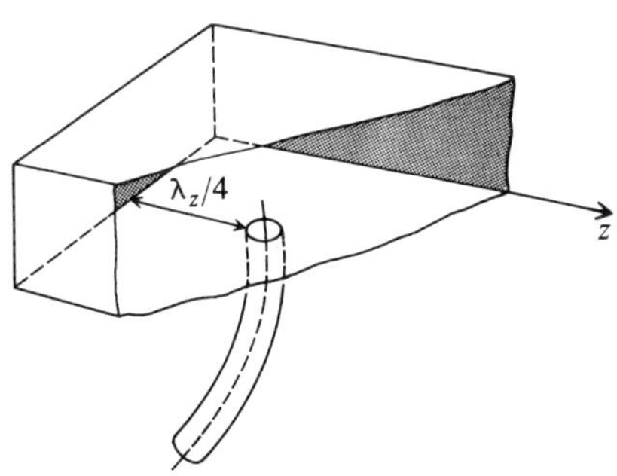


**Fig. 16.10** A method of exciting the $TE_{10}$ wave.

method of initiating the $TE_{10}$ wave. The end of the waveguide is closed with a conducting plate, and at a distance of approximately $\lambda_z/4$ from the end plate a small hole is drilled at the middle of the waveguide. The outer conductor of a coaxial cable is fastened to the guide, and the inner conductor protrudes slightly into the guide through the hole. Obviously, the wave produced in the excitation region is of a complicated structure, but, if the frequency of the wave is chosen appropriately, only the $TE_{10}$ wave will eventually propagate out of this mess.

**16.3.4 Phase velocity and group velocity.** For some practical applications it is necessary to know the velocity of a *signal* through a waveguide. As mentioned above, this is different from the phase velocity $v_p = \omega/\beta$. We can obtain the general expression for the group, that is, the signal velocity valid for any dispersive medium, as follows.

Note first that a purely sinusoidal wave cannot transmit any information other than "I am present." If sinusoidal waves are to transmit a signal they must be modified in one of several ways. But, whatever the method, the result is a package of waves of frequencies which are usually very close to the fundamental frequency (frequency of the wave without a signal). Our interest is in the velocity of this package, since it carries the information.

As an example of the simplest "frequency package", let us take two waves of angular frequencies $\omega_1$ and $\omega_2$, which are very close to the fundamental frequency $\omega$. According to Eq. (16.3.29), the phase constants $\beta_1$ and $\beta_2$ of the two waves are also somewhat different. So we have to add two waves, in the forms $f_1(t, z) = \cos(\omega_1 t - \beta_1 z)$ and $f_2(t, z) = \cos(\omega_2 t - \beta_2 z)$. According to the basic trigonometric relation

$$\cos\theta_2 + \cos\theta_1 = 2\cos\frac{\theta_2+\theta_1}{2}\cos\frac{\theta_2-\theta_1}{2},$$

the sum of the two waves is

$$f_2(t,z) + f_1(t,z) = 2\cos\left(\frac{\omega_2+\omega_1}{2}t - \frac{\beta_2+\beta_1}{2}z\right)\cos\left(\frac{\omega_2-\omega_1}{2}t - \frac{\beta_2-\beta_1}{2}z\right).$$

Provided that $\omega_2$ and $\omega_1$ are almost equal to the fundamental frequency $\omega$, we can write

$$(\omega_2 + \omega_1)/2 \simeq \omega, \qquad (\beta_2 + \beta_1)/2 \simeq \beta, \qquad \omega_2 - \omega_1 = \Delta\omega, \qquad \beta_2 - \beta_1 = \Delta\beta,$$

so that the sum of the two waves becomes

$$f_2(t, z) + f_1(t, z) = 2\cos[(\Delta\omega t - \Delta\beta z)/2]\cos(\omega t - \beta z). \tag{16.3.33}$$

Since $\Delta\omega \ll \omega$, this represents a signal in the form of a wave $\cos(\omega t - \beta z)$, with a varying amplitude $2\cos[(\Delta\omega t - \Delta\beta z)/2]$ (see Fig. 16.11). The sinusoidal wave inside the envelope propagates with the velocity $v_p = \omega/\beta$. But the wave packages enclosed by the envelope are carriers of energy, since no energy can be transmitted

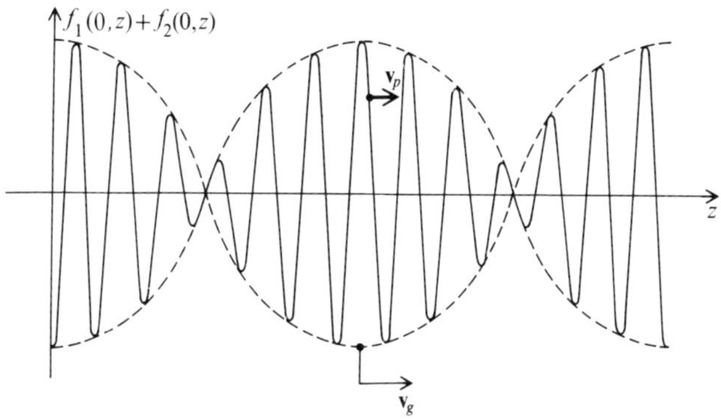


**Fig. 16.11** Sum of two sinusoidal functions of somewhat different frequencies.

through the zero values of the wave. The velocity of the envelope is $\Delta\omega/\Delta\beta$, which in the limiting case becomes

$$\boxed{v_g = \frac{d\omega}{d\beta} = \frac{1}{d\beta/d\omega}} \qquad \text{(group velocity in a dispersive medium).} \tag{16.3.34}$$

This expression for the group (signal) velocity is valid for all dispersive media, provided that the frequency spectrum of the signal is very narrow.

For example, in the case of the rectangular waveguide, $\beta$ is given by Eq. (16.3.29). By using Eq. (16.3.34) it is not difficult to discover that the signal velocity in the waveguide is

$$v_g = \frac{1}{\sqrt{\varepsilon\mu}}\sqrt{1 - f_c^2/f^2} = c\sqrt{1 - f_c^2/f^2}\,. \tag{16.3.35}$$

So, when the phase velocity is of the form of Eq. (16.3.31), *and in that case only*, the product $v_p v_g = c^2$. This is a useful relation, for if we memorize either $v_p$ or $v_g$ we can easily determine the other. Note that the phase velocity in ionized media is also of the form of Eq. (16.3.31), so that Eq. (16.3.35) is valid for that case also.

*Problems referring to Section 16.3:* 16.T9 to 16.T15, 16.G8 to 16.G15, 16.N10 to 16.N13.

## 16.4 WAVE ATTENUATION ALONG GUIDING SYSTEMS WITH SMALL LOSSES

In the preceding sections we considered "lossless" guiding systems, such as "lossless" transmission lines and waveguides; but lossless systems do not exist in reality. At least there are some very small, but nonzero, losses. In this section we shall present an approximate method for determining the attenuation of waves propagating along real, low-loss systems.

Provided that the losses are very small, it is logical to assume that for any sufficiently short section of a real, lossy guiding system the field configuration is practically the same as in idealized, lossless guiding system. Experiments confirm the correctness of this assumption without exceptions, so we can use the results derived for lossless systems as our starting point.

Next we assume that the only effect of the losses is a small exponential attenuation of the wave amplitude as it travels along the system. Hence we multiply the two field vectors, or current and voltage, found for the idealized system by $e^{-\alpha z}$. The quantities thus obtained are certainly much better approximations to true quantities, but the attenuation constant $\alpha$ is unknown. As a general rule it can be determined as follows.

If the amplitudes of the two field vectors, or of current and voltage, diminish along the system as $e^{-\alpha z}$, the average power transmitted diminishes as $e^{-2\alpha z}$. Thus we know that the average value of the power transmitted can be written in the form

$$P_{av}(z) = P_{av}(0)e^{-2\alpha z}. \tag{16.4.1}$$

That part of the power lost in the system represents Joule's losses. If we denote these losses per unit length by $dP_J(z)/dz$, then, according to the law of conservation of energy,

$$-\frac{dP_{av}(z)}{dz} = \frac{dP_J(z)}{dz}. \tag{16.4.2}$$

The minus sign is necessary since we know that $P_{av}(z)$ decreases with $z$. From these two equations we have

$$-[-2\alpha P_{av}(0)e^{-2\alpha z}] = dP_J(z)/dz,$$

from which, noting that $P_{av}(0)e^{-2\alpha z} = P_{av}(z)$,

$$\boxed{\alpha = \frac{dP_J(z)/dz}{2P_{av}(z)}} \quad \text{(the attenuation constant for low-loss guiding systems).} \tag{16.4.3}$$

This simple formula for the attenuation constant is applicable to all low-loss guiding systems. $P_{av}(z)$ is usually determined by using Poynting's theorem. We can find $P_J(z)$ from the known distribution of the magnetic field in the idealized case and Eq. (15.3.21).

As an example of the determination of the attenuation constant by using Eq. (16.4.3), let us take the strip-line shown in Fig. 16.12. We ignore the losses in the dielectric, and assume that the line terminates in its characteristic impedance. If we ignore the fringing effect, we find that the magnetic field exists only between the strips, and that its amplitude at $z = 0$ is $H_0 = I/b$. So, according to Eq. (15.3.21), the losses in the two crosshatched sections of the strips are

$$dP_J(0) = 2 \times \tfrac{1}{2}R_s\frac{I^2}{b^2}\,b\,dz = R_s\frac{I^2}{b}\,dz.$$

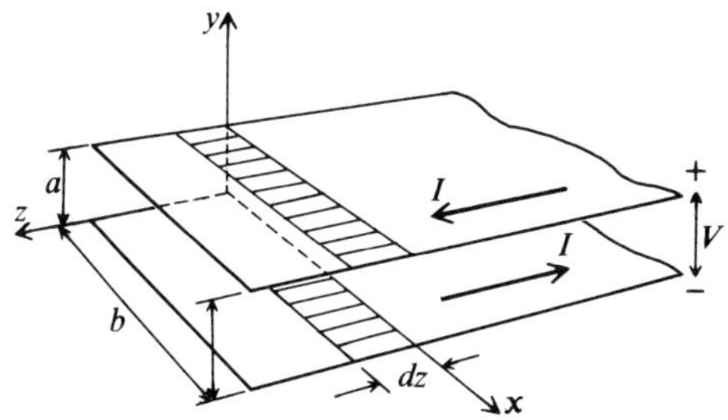


**Fig. 16.12** Determination of attenuation constant of a strip line.

The average power transmitted across the cross-section $z = 0$ is

$$P_{\text{av}}(0) = \tfrac{1}{2}VI = \tfrac{1}{2}Z_0I^2.$$

So we obtain

$$\alpha = \frac{dP_J(z)/dz}{2P_{\text{av}}(z)}\bigg|_{z=0} = \frac{R_s}{bZ_0}.$$

The characteristic impedance $Z_0$ of the line can be obtained from Eq. (16.2.20) and Table 16.1.

*Problems referring to Section 16.4:* 16.G16 to 16.G21, 16.N14.

## PROBLEMS

### Test problems

**16.T1.** The characteristic impedance from Eq. (16.2.14) of a line is real if $R = 0$ and $G = 0$. Can it be real for some other relation between $R$, $L$, $G$, and $C$ for all frequencies?

**16.T2.** Determine $V_i$ and $V_r$ in Eqs. (16.2.15) and (16.2.16) in terms of the sending-end voltage $V(0)$ and current $I(0)$.

**16.T3.** Sketch the distribution of the voltage and current near the end of a lossless line if it is (1) open-circuited, and (2) short-circuited.

**16.T4.** What are the input impedances to lossless lines of lengths $\lambda/4$ and $\lambda/2$, if they are (1) open-circuited, and (2) short-circuited?

**16.T5.** Prove that it is possible to obtain the characteristic impedance of any lossless line by measuring the input impedance of a section of the line when it is open-circuited and when it is short-circuited.

**16.T6.** Prove that a resistor of resistance $R$ can always be matched to a lossless line which has the characteristic impedance $Z_0 \neq R$ by connecting the resistor to the line by a $\lambda/4$ section of another lossless line with an appropriate characteristic impedance. Determine the necessary characteristic impedance of this *quarter-wavelength transformer* in terms of $Z_0$ and $R$.

**16.T7.** Using the Smith chart, determine the complex reflection coefficients on a 60Ω line, if it is terminated by (1) 80Ω, (2) $(30 - j40)\Omega$ and (3) $(40 + j90)\Omega$.

**16.T8.** Using the Smith chart, determine the terminating impedance of a 70Ω line, if it was found experimentally that the complex voltage reflection coefficient is (1) 0.8, (2) $0.2e^{-j45°}$ and (3) $0.5e^{j60°}$.

***16.T9.** Prove that for $H_z = 0$ the system of Eqs. (16.3.3) to (16.3.8) represents the equations of plane waves traveling along the $z$ axis in opposite directions.

***16.T10.** Determine the expressions for $E_x$, $E_y$, $H_x$, and $H_y$ components of the $TE_{mn}$ type. From these general expressions deduce these components for $TE_{10}$, $TE_{01}$, and $TE_{11}$ waves.

**16.T11.** Find the cutoff frequencies and corresponding free-space wavelengths for $TE_{01}$, $TE_{11}$, and $TE_{22}$ waves.

**16.T12.** A signal in the form of a package of $TE_{10}$ waves in the frequency region $f_1 \leq f \leq f_2$ is emitted at one end of a waveguide. Discuss what will happen with the signal if (1) $f_2$–$f_1$ is not small, and (2) if the length of the waveguide is not reasonable.

**16.T13.** With reference to Fig. 16.9, discuss where we can make narrow slots in a waveguide so as not to influence the propagation of the dominant $TE_{10}$ mode through it.

**16.T14.** From Eqs. (16.3.26) to (16.3.28), write the expressions for the *instantaneous* values of the three field components of a $TE_{10}$ wave.

**16.T15.** Two signals, one having a frequency spectrum in the vicinity of frequency $f_1$, the other in the vicinity of frequency $f_2$, propagate through a rectangular waveguide in the form of dominant $TE_{10}$ waves. If $f_1 > f_2$, which of the two is the faster?

## General problems

**16.G1.** The expressions for $C$, $G$, and $L_{\text{ext}}$ given in Table 16.1 (p. 549) were derived earlier, but the expressions for $R$ and $L_{\text{int}}$ were not derived. Show that they are correct by using Eq. (15.3.21).

*Solution.* Consider the coaxial cable of Table 16.1. According to Eq. (15.3.21), for determining losses per unit length of the cable at high frequencies, it is necessary to determine the tangential component of the magnetic field intensity in addition to $R_s$. Suppose that the amplitude of the current through the cable is $I$. Then, according to Ampère's circuital law,

$$H(a) = I/2\pi a,$$

$$H(b) = I/2\pi b.$$

As the average losses per unit length of the cable are

$$P_l = \tfrac{1}{2}R_s[H(a)]^2\, 2\pi a + \tfrac{1}{2}R_s[H(b)]^2\, 2\pi b,$$

we have

$$P_l = \frac{R_s I^2}{2}\left(\frac{1}{2\pi a} + \frac{1}{2\pi b}\right).$$

The energy definition of the resistance $R$ per unit length is

$$P_l = \tfrac{1}{2}RI^2.$$

If we compare the last two equations, the expression for $R$ of the coaxial cable appearing in Table 16.1 immediately follows.

It is left to the reader to check in a similar way the resistances per unit length of the other two line types.

In deriving Eq. (15.3.21) we saw that the real and imaginary parts of the complex Poynting vector were equal, so the average values of the active and reactive power per unit area of a good conductor are also equal. Since the positive reactive power in terms of the internal inductance $L_{\text{int}}$ per unit length is $L_{\text{int}}\omega I^2/2$, we discover that at high frequencies $L_{\text{int}} = R/\omega$.

**16.G2.** A lossless line, of a characteristic impedance $Z_{01}$ and length $l_1$, is terminated by an impedance $Z_l$. This line serves as a load for another lossless line, of characteristic impedance $Z_{02}$ and length $l_2$. The dielectric in both lines is air, and the angular frequency of the current is $\omega$. Determine general expressions for the input impedance of the second line, the reflection coefficient on both lines, and the voltage standing-wave ratio.

**16.G3.** Prove that for all *lossless* lines $Z_0 = \sqrt{\varepsilon\mu}/C$ and $L = \varepsilon\mu/C$. These relations show that the capacitance $C$ per unit length, which is usually very easily determined, is a parameter sufficient for describing lossless lines. Hence determine the characteristic impedances of the lines from Table 16.1.

*Proof.* First, from Eqs. (16.2.19) and (16.2.23) it follows that $LC = \varepsilon\mu$. So we obtain directly that $L = \varepsilon\mu/C$. Substituting this into Eq. (16.2.20) we find indeed that

$$Z_0 = \sqrt{\varepsilon\mu}/C. \tag{a}$$

Hence, according to Table 16.1, we obtain the characteristic impedances of the three line types as

coaxial line:
$$Z_0 = \frac{1}{2\pi}\sqrt{\frac{\mu}{\varepsilon}}\ln\frac{b}{a}, \tag{b}$$

two-wire line:
$$Z_0 = \frac{1}{\pi}\sqrt{\frac{\mu}{\varepsilon}}\ln\frac{d}{a}, \tag{c}$$

strip line:
$$Z_0 = \frac{a}{b}\sqrt{\frac{\mu}{\varepsilon}}. \tag{d}$$

**16.G4.** Figure 16.G4a shows what is called the slotted coaxial line. Slotted lines are used for measuring impedances at very high frequencies. Explain how an unknown impedance can be measured with such a line. Assume that the generator has a fixed, but unknown frequency, and that the dielectric in the slotted line is air, and its characteristic impedance $Z_0$. For simplicity it is assumed that the graduation of the slot scale starts from the line end. Usually only the slot is calibrated, and the position of the end is obtained by extrapolating the standing-wave pattern.

*Solution.* Note first that the slot is made *along* the current-flow lines in the outer cable conduc-

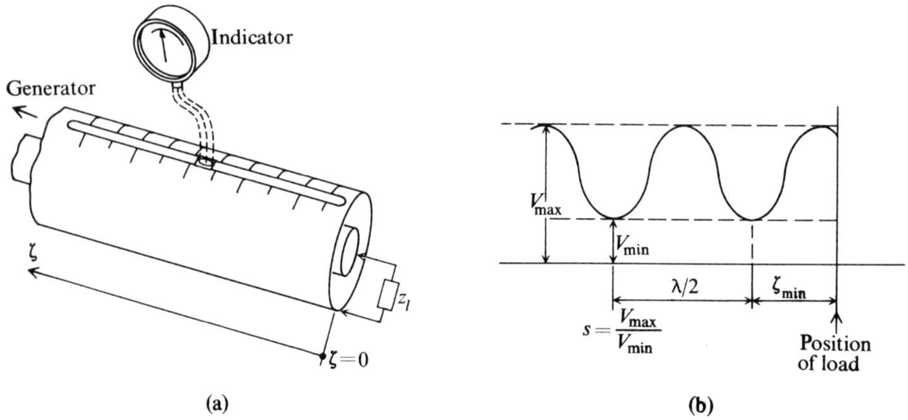


**Figure 16.G4**

tor, so it only slightly affects the distribution of current and voltage in the cable. The probe can measure only the relative magnitude of the electric field, and hence also that of the voltage, at a certain cross-section of the line. This is, however, sufficient for measuring the terminating impedance of the line.

The wavelength measured along the line is practically equal to the free-space wavelength $\lambda_0 = c_0/f$. We can measure it simply by moving the probe back and forth and determining the two successive minima, while the line is either open-circuited or short-circuited. In both cases a standing-wave pattern exists, with two successive minima at a distance $\lambda_0/2$ apart. To measure $\lambda_0$ the line is usually short-circuited, to eliminate radiation from the open end, which would represent a certain, unknown load. We seek minima rather than maxima, because they are sharp and can be detected more precisely. Thus we can determine $\lambda_0$, and from it the frequency $f$ of the generator.

Suppose we now replace the short circuit by an element the impedance of which we wish to measure. By moving the probe back and forth we determine the maximum and minimum readings of the indicator, which gives us the voltage standing-wave ratio $s$ (see Fig. 16.G4b). Then we locate as precisely as possible the distance $\zeta_{min}$ of the first minimum from the load in terms of wavelength, which was determined earlier by measurements on the short-circuited line. From Eq. (16.2.32) we find that the voltage minimum occurs when $\mathscr{R}e^{-j2\beta\zeta}$ is real and negative, that is, equal to $-|\mathscr{R}|$. So the impedance $Z(\zeta)$ given by Eq. (16.2.34) for $\zeta = \zeta_{min}$ is real, and equal to

$$Z(\zeta_{min}) = Z_0 \frac{1 - |\mathscr{R}|}{1 + |\mathscr{R}|} = \frac{Z_0}{s}. \tag{a}$$

As we now know $Z(\zeta_{min})$ and $\zeta_{min}$ (in terms of wavelength), we use the Smith chart for determining $Z_l$.

As a numerical example, let us consider a $Z_0 = 60\Omega$ slotted line, and suppose we found experimentally that $s = V_{max}/V_{min} = 2$ and that the first minimum is at a distance $\zeta_{min} = 0.186\lambda$ from the load. The normalized impedance at $\zeta_{min}$ is

$$z(\zeta_{min}) = Z(\zeta_{min})/Z_0 = 1/s = 0.5.$$

We start with this value in the Smith chart ($R/Z_0 = 0.5$, $X/Z_0 = 0$), and to obtain the load impedance we simply move this point along the circle centered at the center of the chart for 0.186, measured along the scale "wavelengths toward load". Thus we find that the normalized load impedance is $z_l \simeq 1.37 - j0.72$, so that $Z_l = Z_0 z_l \simeq (82.2 - j43.2)\Omega$.

The terminating impedance can also be determined by measuring the position $\zeta_{max}$ of the first maximum from the load. We leave this as an exercise for the reader. In practice, the latter method is rarely used, because minima are always sharper than maxima.

**16.G5.** Trace the process of solving Problem 16.G2 by means of the Smith chart.

**16.G6.** The reciprocals of complex numbers can be determined easily from the Smith chart. Starting from Eq. (16.2.36), deduce how this can be done.

*Solution.* According to Eq. (16.2.36), the reciprocal $1/z$ of a complex number $z$ is given by

$$\frac{1}{z} = \frac{1 - \mathscr{R}}{1 + \mathscr{R}} = \frac{1 + \mathscr{R}e^{j\pi}}{1 - \mathscr{R}e^{j\pi}}.$$

So, for obtaining $1/z$ we have first to enter in the chart with $z$. This locates the complex number $\mathscr{R}$ in the complex $\mathscr{R}$ plane. To $1/z$ corresponds $\mathscr{R}$ of the same *magnitude*, but the phase of the new $\mathscr{R}$ is for $\pi$ larger. That is, we simply draw a straight line through $z$ and the origin, and find in the diagram the symmetric point to $z$, which determines $1/z$.

For example, let $z = 0.3 + j0.5$. The symmetric point to this point with respect to the chart center is $1/z = 0.88 - j1.47$.

This simple conclusion has important implications. It shows that we can also use the Smith chart as an admittance (reciprocal of impedance) chart, which is particularly valuable when we have two or more lines connected in parallel, since their equivalent input admittance simply equals the sum of their admittances. The next problem illustrates an application of the Smith chart as an admittance chart.

**16.G7.** A fixed, known complex impedance $Z_l$ has to be connected to a lossless line having the characteristic impedance $Z_0$. Show that it is possible to eliminate the reflected wave along the line if an appropriate length of the same line, assumed to be short-circuited, is connected at an appropriate place on the line near $Z_l$ (see Fig. 16.G7).

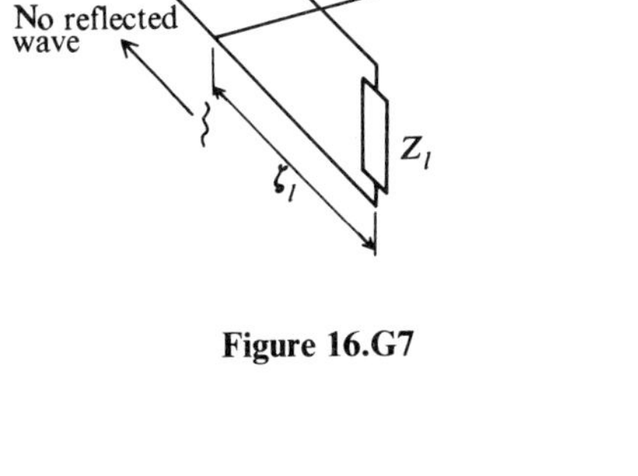


**Figure 16.G7**

*Solution.* We note first that the impedance, and hence also the admittance, of the short-circuited stub can only be imaginary. Next we note that, as we move along the line away from the load, the input impedance to the portion of the line terminated by the load varies. At a certain point its real part is equal to $Z_0$, so it will be matched to the left of that point if we compensate the imaginary part of that impedance by the stub.

Although the idea of how the matching can be effected is simple, it is a lengthy process to find $\zeta_l$ and $\zeta_s$ without the aid of the Smith chart. By using the Smith chart, however, we can do it easily. We simply locate the normalized load impedance on the chart, and then the normalized load admittance. We next move this point along the circle centered at the chart center until it coincides with the $R/Z_0 = 1$ circle. There will always be two solutions, and either can be adopted. A certain imaginary part of the admittance corresponds to that point. The imaginary admittance of the stub connected at that point must have equal magnitude, but be of the opposite sign. We can find both the length of the stub $\zeta_s$ and the distance $\zeta_l$ from the load from the chart.

As an example, consider a $Z_0 = 50\Omega$ line terminated by an impedance $Z_l = (65 + j50)\Omega$. Thus $z_l = (1.3 + j1)$. We locate this point on the chart, and determine the symmetrical point with respect to the origin. Thus we obtain $1/z_l \simeq (0.48 - j0.37)$. We move this point around the circle $|\mathscr{R}| =$ constant in a clockwise direction (towards the generator) to the point where it intersects the circle $G/Y_0 = 1$. At this point we find that $1/z(\zeta_l) = (1 + j0.92)$. By adding a stub having $1/z(\zeta_s) = -j0.92$, we discover that the total normalized admittance at this point is 1. So the line is matched to the left of that point.

The distances $\zeta_l$ and $\zeta_s$ are obtained as follows. First, we note that the starting point, $1/z_l$, corresponds approximately to 0.431 on the "wavelengths toward generator" scale. The end point, $1/z(\zeta_l) = (1 + j0.92)$, corresponds approximately to 0.159 on the same scale, so the total distance $\zeta_l$ is $[0.500 - 0.431) + 0.159] = 0.228$ wavelengths.

The point $1/z_s = -j0.92$ is on the $G/Y_0 = 0$ circle, and we have to go round this circle to the point where $B/Y_0 = \infty$ (the line is short-circuited) in a counterclockwise direction (toward the load). On the scale "wavelengths toward load" we find that there are $(0.250 - 0.118) = 0.132$ wavelengths to go round, so $\zeta_s = 0.132\lambda$.

**16.G8.** The maximum average power that can be transmitted through a waveguide is limited by the properties of the dielectric inside it. The largest amplitude of the electric field intensity must be less than the dielectric strength of the dielectric. Find the maximum value of the average power in the case of an air-filled waveguide for a $TE_{10}$ wave. In particular, determine this power if $a = 2$ cm, $b = 1$ cm, and $f = 10$ GHz (1 GHz $= 10^9$ Hz).

*Solution.* According to Eq. (16.3.26), the largest amplitude of the electric field intensity is for $x = a/2$. So it must be

$$E_{y\,\max} = \omega\mu \frac{a}{\pi} H_0 \leq E_{\text{diel.str.}},$$

from which

$$H_{0\,\max} = \pi E_{\text{diel.str.}}/\omega\mu a.$$

Substituting this value of $H_0$ into Eq. (16.3.32) we find that the maximum average power that can be transmitted through the waveguide using a $TE_{10}$ wave is

$$P_{\text{av max}} = \frac{ab}{4}\sqrt{\frac{\mu}{\varepsilon}}\frac{f^2}{f_c^2}\sqrt{1 - f_c^2/f^2}\left(\frac{\pi E_{\text{diel.str.}}}{\omega\mu a}\right)^2. \tag{a}$$

In the particular case when $\varepsilon = \varepsilon_0$, $\mu = \mu_0$, $a = 2$ cm, $b = 1$ cm, $E_{\text{diel.str.}} = 30$ kV/cm, and $f = 10^{10}$ Hz, by using Eq. (16.3.25) we obtain $f_c = 7.5$ GHz. From Eq. (a) we then find that $P_{\text{av max}} \simeq 800$ kW.

**16.G9.** The dominant $TE_{10}$ mode, transmitting an average power $P_{\text{av}}$, propagates through a rectangular waveguide of sides $a$ and $b$. Find the largest amplitude of the electric field intensity in the waveguide.

*Answer.*

$$E_{\max} = 4\mu f_c \sqrt{\sqrt{\frac{\varepsilon}{\mu}} \frac{aP_{\text{av}}}{b\sqrt{1 - f_c^2/f^2}}}.$$

**16.G10.** In a rectangular waveguide, with sides $a$ and $b$, two signals are simultaneously emitted. The frequency spectrum of one is in the immediate vicinity of frequency $f_1$, that of the other in the vicinity of $f_2$. Both propagate as $TE_{10}$ waves. Find the time interval $\Delta t$ between the arrival of the two signals at the receiving point, which is at a distance $L$ from the sending point, if the dielectric is air.

*Answer.* For $f_1 = 10$ GHz, $f_2 = 12$ GHz, $a = 2$ cm, $b = 1$ cm, and $L = 10$ m, $\Delta t = 7.69 \times 10^{-9}$ s.

**16.G11.** A rectangular waveguide of sides $a$ and $b$ is closed at one end with a "perfectly" conducting plate (see Fig. 16.G11). The other end is very far to the left. If a $TE_{10}$ wave arrives from the source situated at the far left, determine the resultant electromagnetic field in the waveguide. (Such a waveguide is sometimes referred to as a *short-circuited* waveguide.)

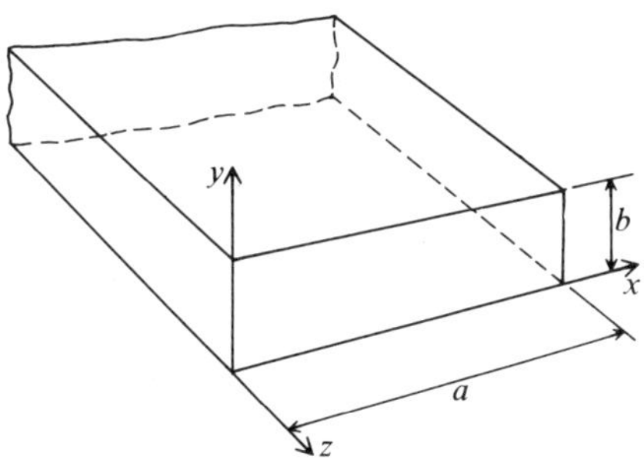


**Figure 16.G11**

*Solution.* The resultant electromagnetic field in the short-circuited waveguide is the sum of the incident wave and the reflected wave. The components of the incident wave are given by Eqs. (16.3.26) to (16.3.28):

$$E_{yi} = -j\omega\mu \frac{a}{\pi} H_0 \sin \frac{\pi}{a} x \, e^{-j\beta z}, \tag{a}$$

$$H_{xi} = j\beta \frac{a}{\pi} H_0 \sin \frac{\pi}{a} x \, e^{-j\beta z}, \tag{b}$$

$$H_{zi} = H_0 \cos \frac{\pi}{a} x \, e^{-j\beta z}. \tag{c}$$

The components of the reflected wave are determined from the boundary conditions at $z = 0$ (at the "short-circuit"). At $z = 0$ the boundary conditions are

$$(E_{yi} + E_{yr})_{z=0} = 0, \qquad (H_{zi} + H_{zr})_{z=0} = 0.$$

In addition, $-j\beta z$ in Eqs. (a) to (c) has to be replaced by $+j\beta z$, since the reflected wave propagates in the opposite direction. Thus we obtain

$$E_{yr} = j\omega\mu \frac{a}{\pi} H_0 \sin \frac{\pi}{a} x\, e^{j\beta z}, \tag{d}$$

$$H_{xr} = j\beta \frac{a}{\pi} H_0 \sin \frac{\pi}{a} x\, e^{j\beta z}, \tag{e}$$

$$H_{zr} = -H_0 \cos \frac{\pi}{a} x\, e^{j\beta z}. \tag{f}$$

It follows that the components of the resultant electromagnetic field are

$$E_y = E_{yi} + E_{yr} = j\omega\mu \frac{a}{\pi} H_0 \sin \frac{\pi}{a} x\, (e^{j\beta z} - e^{-j\beta z})$$

$$= -2\omega\mu \frac{a}{\pi} H_0 \sin \frac{\pi}{a} x \sin \beta z, \tag{g}$$

$$H_x = H_{xi} + H_{xr} = 2j\beta \frac{a}{\pi} H_0 \sin \frac{\pi}{a} x \cos \beta z, \tag{h}$$

$$H_z = H_{zi} + H_{zr} = -2jH_0 \cos \frac{\pi}{a} x \sin \beta z. \tag{i}$$

The factor $e^{\pm j\beta z}$, characteristic for traveling waves, is not present. The resultant wave is a kind of a standing wave, which was to be expected, since no energy can be transmitted to the perfectly conducting end plate.

**16.G12.** From the result of the preceding problem deduce that an electromagnetic field can exist inside a cavity in the form of a rectangular parallelepiped. Analyze qualitatively the distribution of the charges and currents on the walls of such a *resonant cavity*.

*Solution.* From Eqs. (g) and (i) of the preceding problem we see that $E_y$ and $H_z$ are zero for $z$ determined by $\beta z = -p\pi, p = 1, 2, \ldots$. Since $\beta = 2\pi/\lambda_z$, we find that $E_y$ and $H_z$ are zero at

$$z = -\frac{p\pi}{2\pi} \lambda_z = -p\frac{\lambda_z}{2}.$$

If we insert in such a plane a very thin, perfectly conducting foil, nothing will be changed in the waveguide. The necessary boundary conditions ($E_t = 0$, $H_n = 0$) are automatically satisfied. Thus we obtain a box with an oscillating electromagnetic field inside it, that is, a resonant cavity. If the foil is placed at $z = -p\lambda_z/2$, we can say that the standing wave in the box is the $TE_{10p}$ wave.

Let us consider the simplest, $TE_{101}$ wave. Let $\lambda_z/2 = d$ (see Fig. 16.G12a). Then we have

$$\beta = \frac{2\pi}{\lambda_z} = \frac{\pi}{d}, \tag{a}$$

so that Eqs. (g) to (i) of the preceding problem become

$$E_y = -2\omega\mu \frac{a}{\pi} H_0 \sin\frac{\pi}{a} x \sin\frac{\pi}{d} z, \tag{b}$$

$$H_x = 2j\frac{a}{d} H_0 \sin\frac{\pi}{a} x \cos\frac{\pi}{d} z, \tag{c}$$

$$H_z = -2jH_0 \cos\frac{\pi}{a} x \sin\frac{\pi}{d} z. \tag{d}$$

The other components are zero.

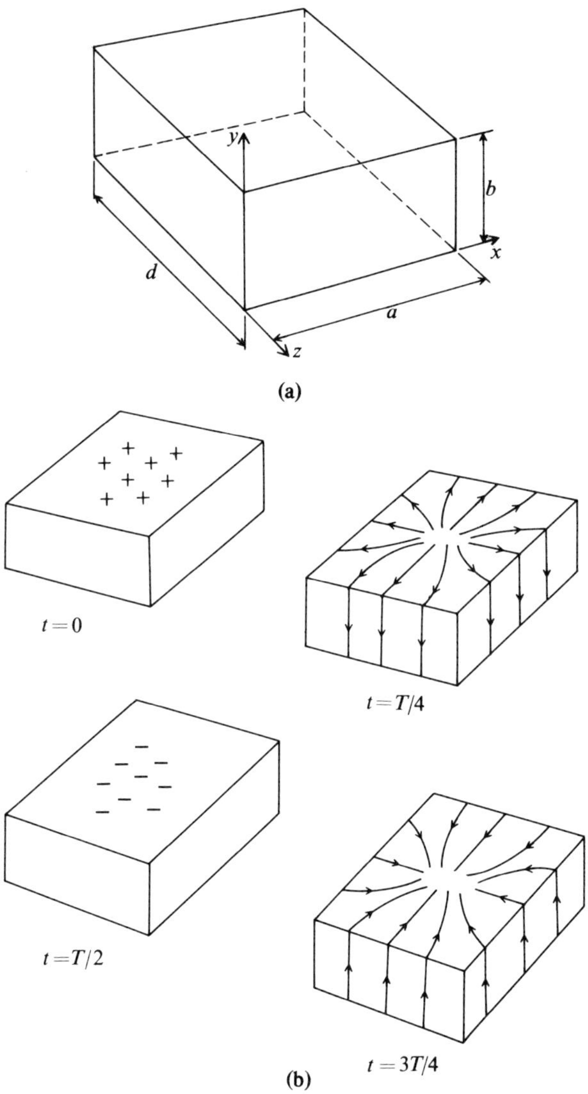


**Figure 16.G12**

From these equations we can easily deduce how the electromagnetic oscillations in the cavity are maintained. There are induced electric charges on the upper and lower cavity faces, since the normal component $E_y$ of the electric field intensity exists. Surface currents in the $y$ direction appear only on the side faces, since at these surfaces $H_x$ and $H_z$ components exist, $E_y$ being zero. So the oscillations of the electromagnetic field inside the cavity are accompanied by charges and currents on its walls.

According to Eqs. (b), (c), and (d), the electric and magnetic fields are shifted in phase for $\pi/2$, which is comprised in the factor $j = e^{j\pi/2}$, so that instants exist with no currents on the walls, as well as instants with no charges. These instants occur after intervals of $T/4$, where $T = 1/f$ is the period of the oscillations. Figure 16.G12b shows the distribution of charges and currents during one period, starting at the instant when the upper face carries the maximum positive charge.

***16.G13.** Determine the total energy stored inside the cavity of the preceding problem.

*Solution.* Since no energy can be lost, the conductor and the dielectric being perfect, the *total* energy must be constant in time. It oscillates from the energy stored completely in the electric field at instants when $H = 0$ at all points, to the energy stored completely in the magnetic field at instants when $E = 0$ at all points. So it is sufficient to determine the energy in any one of the two extreme cases.

The energy stored in the electric field when $H = 0$ is

$$W = W_{e\,\max} = \int_v \tfrac{1}{2}\varepsilon E_{\max}^2 \, dv = \int_0^a \int_0^b \int_{-d}^0 \tfrac{1}{2}\varepsilon \,|E_y|^2 \, dx \, dy \, dz.$$

According to Eq. (b) of the preceding problem, we obtain, after performing integrations,

$$W = W_{e\,\max} = \frac{1}{2\pi^2} \varepsilon\mu^2 a^3 b d \omega^2 H_0^2. \tag{a}$$

(Note that $\omega$ is determined by the cavity dimensions.) Since for a $TE_{101}$ wave

$$d = \frac{\lambda_z}{2} = \frac{\lambda}{2\sqrt{1 - f_c^2/f^2}} \quad \text{and} \quad f_c = \frac{1}{2a\sqrt{\varepsilon\mu}},$$

the resonant frequency of the cavity is

$$f^2 = f_{\text{res}}^2 = \frac{1}{4\varepsilon\mu} \frac{a^2 + d^2}{a^2 d^2}. \tag{b}$$

***16.G14.** If the surface resistance of the cavity from Problem 16.G12 is $R_s$, determine the average power losses $P_J$ in the walls. [*Hint:* use Eq. (15.3.21).]

*Answer.*

$$P_J = R_s H_0^2 \left( 2\frac{a^3 b}{d^2} + \frac{a^3}{d} + ad + 2bd \right).$$

***16.G15.** The *quality factor Q* of a resonant cavity is defined as

$$Q = 2\pi \frac{f_{\text{res}} W}{P_J}.$$

Find the $Q$ factor for the rectangular cavity from the preceding problem for a $TE_{101}$ wave.

*Answer.* For $a = b = d$, $\varepsilon = \varepsilon_0$, $\mu = \mu_0$, $\sigma = 5.6 \times 10^7$ S/m, and $f = 10^9$ Hz, $Q \simeq 33{,}300$. Note that the $Q$ factor of ordinary resonant circuits is of the order of a few hundreds, and of resonant lines a few thousands.

**16.G16.** Find the attenuation constant of the strip line shown in Fig. 16.12, if the conductivity of the strips is assumed to be infinite, but the dielectric is not perfect. The parameters of the dielectric are $\varepsilon$, $\mu_0$, and $\sigma$.

*Answer.* $\alpha = 0.5\sigma\sqrt{\mu_0/\varepsilon}$.

**16.G17.** Find the attenuation constant of a coaxial line with conductor radii $a$ and $b$. Assume a perfect dielectric and imperfect conductors, having a resistance $R$ per unit length. The line is terminated in its characteristic impedance $Z_0$.

*Answer.* $\alpha = R/2Z_0$.

**16.G18.** If the outer radius $b$ of the coaxial line in the preceding problem is fixed, find the radius of the inner conductor so that the attenuation is minimal.

*Solution.* According to the answer to the preceding problem, Table 16.1 and Eq. (b) of Problem 16.G3, the attenuation constant can be written in the form

$$\alpha = \sqrt{\frac{\varepsilon}{\mu}}\,\frac{R_s}{2}\,\frac{1/a + 1/b}{\ln b/a}.$$

The required value of $a$ is obtained from the condition $\partial\alpha/\partial a = 0$. Thus we obtain the following equation:

$$1 + \frac{b}{a} = \frac{b}{a}\ln\frac{b}{a}, \tag{a}$$

which is a transcendental equation. Using some approximate method, say by determining the intersection of curves $y = (1 + x)$ and $y = x \ln x$, we find that $x = b/a \simeq 3.6$. According to Eq. (b) of Problem 16.G3, the characteristic impedance of such a cable is

$$Z_0 \simeq 0.204\sqrt{\mu/\varepsilon}.$$

If the dielectric is air, this yields $Z_0 \simeq 77\Omega$. Coaxial lines for high frequencies and large power are almost always designed so that $\alpha$ is minimal.

**16.G19.** Suppose that a homogeneous dielectric in a coaxial line of radii $a$ and $b$ has a very small, but nonzero conductivity $\sigma$. The losses in the conductors can be ignored. The permittivity and permeability of the dielectric are $\varepsilon$ and $\mu$. Find the attenuation constant, assuming that the line is terminated in its characteristic impedance.

*Answer.* $\alpha = 0.5\sigma\sqrt{\mu/\varepsilon}$.

**16.G20.** Find the attenuation constant of a two-wire line with conductors having a surface resistance $R_s$. The radius of the conductors is $a$, the distance between them $d$ ($d \gg a$), and the dielectric can be regarded as perfect. The line is terminated in its characteristic impedance $Z_0$.

*Answer.* $\alpha = R/2Z_0$, where $R$ is the resistance of the line per unit length.

***16.G21.** Prove that the attenuation constant of a $TE_{10}$ wave in a waveguide, with a perfect dielectric but having walls of surface resistance $R_s$, is

$$\alpha_{TE_{10}} = \sqrt{\frac{\varepsilon}{\mu}} \frac{R_s}{b\sqrt{1 - f_c^2/f^2}} \left(1 + \frac{2b}{a} \frac{f_c^2}{f^2}\right).$$

## Numerical problems

**16.N1.** Determine all the parameters in Table 16.1 if the conductors are made of copper, if the dielectric is air and if frequency $f = 10^8$ Hz. For a coaxial cable, let $a = 1$ mm and $b = 3.6$ mm; for a two-wire line, let $a = 2$ mm and $d = 50$ cm; and for a strip-line, let $a = 1$ cm and $b = 10$ cm.

**16.N2.** Find the characteristic impedances of the three lines from the preceding problem if they are considered to be lossless.

**16.N3.** Find the input impedance of an open-circuited lossless line having $Z_0 = 70\Omega$, if its length is 1 m, at frequency $f = 100$ MHz. The dielectric is air.

**16.N4.** If $Z_0 = 70\Omega$ and $Z_l = (80 - j10)\Omega$, use the Smith chart to determine the reflection coefficient.

**16.N5.** If $Z_0 = 50\Omega$ and $\mathcal{R} = 0.7e^{j30^\circ}$, use the Smith chart to find the terminating impedance $Z_l$.

**16.N6.** A line with $Z_0 = 60\Omega$ and of length $0.210\lambda$ is terminated with impedance (1) $Z_l = (60 + j30)\Omega$, (2) $Z_l = (100 - j40)\Omega$, and (3) $Z_l = (30 - j40)\Omega$. Determine the input impedances of the line in all three cases.

**16.N7.** If in Problem 16.G4, the measurements on a $Z_0 = 50\Omega$ slotted line yielded $s = 1.8$ and $\zeta_{\min} = 0.1\lambda$, find the value of $Z_l$.

**16.N8.** Using the Smith chart, determine the reciprocals of numbers $(3 - j2)$, $(0.5 + j2)$, and $(4 - j0.5)$. (See Problem 16.G6.)

**16.N9.** Referring to Problem 16.G7, find the position and length of a short-circuited stub matching an impedance $Z_l = (40 - j30)\Omega$ on a line with $Z_0 = 60\Omega$.

**16.N10.** Determine the cutoff frequencies of an air-filled rectangular waveguide with $a = 2.5$ cm and $b = 1.25$ cm, for the following wave types: $TE_{01}$, $TE_{10}$, $TE_{11}$, $TE_{12}$, and $TE_{22}$.

**16.N11.** Find the wavelength $\lambda_z$ measured along a waveguide with $a = 2$ cm, $b = 1$ cm, and the dielectric air, if the wave is of the $TE_{10}$ type, for frequencies (1) 8 GHz and (2) 10 GHz.

**16.N12.** Determine the phase and group velocities for both cases in the preceding problem.

***16.N13.** Referring to Problem 16.G13, determine the maximum possible energy stored in a cavity with $a = b = d = 10$ cm at a resonant frequency corresponding to the $TE_{101}$ wave. The dielectric is air, for which $E_{\text{diel.str.}} = 3 \times 10^6$ V/m. (See Problem 16.G13.)

**16.N14.** Find the attenuation constant of the strip-line discussed in Section 16.14, and of the lines of Problems 16.G17 and 16.G20, for the numerical values in Problem 16.N1.

**Suggested reading.** The list progresses towards more advanced texts. For full bibliographical description see Bibliography.

1. WALSH, J. B., *Electromagnetic Theory and Engineering Applications,* Chapter 9.
2. HOLT, C. A., *Introduction to Electromagnetic Fields and Waves*, Chapters 20, 21 and 22.
3. KRAUS, J. D., *Electromagnetics*, Chapters 11 and 12.
4. RAMO, S., J. R. WHINNERY, and T. VAN DUZER, *Fields and Waves in Communication Electronics*, Chapters 7 and 8.

CHAPTER 17

# RADIATION AND RECEPTION OF ELECTROMAGNETIC WAVES

## 17.1 INTRODUCTION

The sources of any electromagnetic field are electric currents and charges. If the charges and currents are time-varying, they may produce electromagnetic waves propagating away from their sources and never returning to them. This process is known as the radiation of electromagnetic waves.

Theoretically, any system of varying charges and currents radiates a certain amount of energy. In many cases in actual practice the radiation either can be ignored, or is intentionally suppressed. For example, radiation from 60 Hz power transmission lines exists in theory, but the power radiated is so small that it practically cannot be detected. In the case of high-frequency transmission lines, coaxial cables are generally used because they do not radiate at any frequency.

Conversely, in many practical cases, a configuration of charges and currents which radiates considerable energy in the surrounding space is desirable. A whole complicated science of designing and analyzing radiating systems has been developed, which, essentially, is part of the theory of electromagnetic fields. However, it is highly specialized and the problems encountered are very delicate. In an introductory text on electromagnetic fields such as this it is possible to present only a few of the most basic and simple types of radiating systems.

Systems designed for the specific purpose of effectively radiating electromagnetic waves are called *antennas*, and there are many varied types of antenna. Their exact analysis is invariably extremely complicated, and in most cases only rough approximate theories have been used. A certain theoretical insight followed by delicate experimental research is a typical procedure for analyzing new types.

Once the electromagnetic wave has been created by an antenna, it travels through space towards the place or places where the message impressed in the wave has to be extracted from it. The distance between the transmitting antenna and the point of reception of the signal can be tremendous; radio communication, for instance, between two points hundreds of thousands of miles apart has been achieved. In these cases, special antennas which radiate very narrow beams are used. The signal is weak, and a system capable of catching as much as possible of the energy of the transmitted wave is needed. Even in the case of receivers closer to the transmitting antenna it is desirable to catch as much as possible of the radiated energy, for example, to ensure a good quality television picture, to give good reproduction of messages in sound, etc.

How is the energy extracted from the wave? For receiving purposes, structures similar to transmitting antennas are used; these are referred to as the *receiving* antennas. Their diversity is very great, but they all have one characteristic in common. The wave induces electromotive forces in them, that is, they behave as generators. The analysis of receiving antennas is even more difficult than that of transmitting antennas.

Because of the extreme complexity of antenna problems, we shall restrict our attention to analyzing only a few simple antenna types, but, although these types are simple, they are of considerable practical importance.

## 17.2 THE ELECTRIC DIPOLE ANTENNA (THE HERTZIAN DIPOLE)

**17.2.1 The electromagnetic field of an oscillating dipole.** The electric dipole antenna, or the Hertzian dipole, is the most simple of all radiating systems. It consists of a straight, very thin conductor terminated by two small conducting spheres or disks. A generator of time-varying electromotive force is connected somewhere along the wire. We shall assume that the whole system is isolated and situated in a vacuum (see Fig. 17.1).

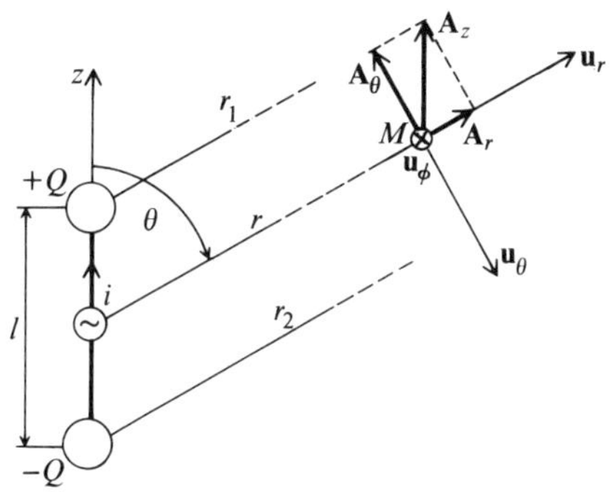


**Fig. 17.1** The Hertzian dipole.

If the two end spheres have a capacitance much larger than that of the wire, practically no charges will be distributed along the wire. If the length $l$ of the wire is much smaller than the wavelength of the wave radiated, we can assume that the current along the wire is constant at any instant.

The time-varying electromotive force of the generator to which the dipole is connected may be any function of time. The most important practical case is when it is sinusoidally time-varying, and this is the only case we shall consider. We can then simplify the analysis by considering the complex amplitudes of the field vectors rather than their instantaneous values.

Let us suppose that the angular frequency of the current in the dipole is $\omega$, and its amplitude $I$. From Fig. 17.1 we have $i(t) = dQ(t)/dt$, which in complex notation becomes simply $I = j\omega Q$, or $Q = -jI/\omega$. To determine the electric and magnetic field intensities at the point $M$, we first determine the retarded potentials at that point. The general complex forms of the electric scalar and magnetic vector potentials are given by Eqs. (13.4.32) and (13.4.33). As in the case of the waves considered earlier, we denote $\omega/c$ by $\beta$, and these expressions in the present case become

$$V = \frac{-jI}{4\pi\varepsilon_0\omega}\left(\frac{e^{-j\beta r_1}}{r_1} - \frac{e^{-j\beta r_2}}{r_2}\right), \tag{17.2.1}$$

$$\mathbf{A} = A_z\mathbf{u}_z = \frac{\mu_0 Il}{4\pi}\frac{e^{-j\beta r}}{r}\mathbf{u}_z. \tag{17.2.2}$$

Now, if $\Delta r$ is small, by the very definition of the derivative we have for any function $f(r)$

$$f(r - \Delta r) - f(r + \Delta r) = -[f(r + \Delta r) - f(r - \Delta r)] \simeq -\frac{df}{dr}2\,\Delta r. \tag{17.2.3}$$

In Eq. (17.2.1) $r \gg l$, so that

$$r_1 \simeq r - (l\cos\theta)/2, \qquad r_2 \simeq r + (l\cos\theta)/2.$$

Let $(l\cos\theta)/2 = \Delta r$; then, according to Eq. (17.2.3), the expression for the scalar potential becomes

$$V \simeq \frac{+jI}{4\pi\varepsilon_0\omega}\frac{d}{dr}\left(\frac{e^{-j\beta r}}{r}\right)l\cos\theta = \frac{-jIl\cos\theta}{4\pi\varepsilon_0\omega}\left(\frac{1}{r^2} + \frac{j\beta}{r}\right)e^{-j\beta r}. \tag{17.2.4}$$

Having found the expressions for the electric scalar and magnetic vector potentials, we can easily find the magnetic and electric field intensities. Due to symmetry, the most natural coordinate system for considering the field is the spherical. The potential $V$ in Eq. (17.2.4) is expressed in terms of the spherical coordinates $r$ and $\theta$. The spherical components of $\mathbf{A}$ are (see Fig. 17.1):

$$\left.\begin{aligned} A_r &= A_z\cos\theta = \frac{\mu_0 Il\cos\theta}{4\pi}\frac{e^{-j\beta r}}{r}, \\ A_\theta &= -A_z\sin\theta = -\frac{\mu_0 Il\sin\theta}{4\pi}\frac{e^{-j\beta r}}{r}, \\ A_\phi &= 0. \end{aligned}\right\} \tag{17.2.5}$$

Note that neither $\mathbf{A}$ nor $V$ is a function of coordinate $\phi$. So, according to the expression for curl in the spherical coordinate system (see Problem 8.G21), the relation $\mathbf{H} = \text{curl}\,\mathbf{A}/\mu_0$ yields

$$H_r = 0, \qquad H_\theta = 0, \qquad H_\phi = \frac{Il\sin\theta}{4\pi}\left(\frac{j\beta}{r} + \frac{1}{r^2}\right)e^{-j\beta r}. \tag{17.2.6}$$

The electric field intensity vector is obtained from $V$ and $\mathbf{A}$ as $\mathbf{E} = -\operatorname{grad} V - \partial\mathbf{A}/\partial t$. In the complex notation this simplifies to

$$\mathbf{E} = -\operatorname{grad} V - j\omega\mathbf{A}. \tag{17.2.7}$$

So, by reference to the expression for the gradient in the spherical coordinate system (see Problem 1.G12), we have, according to Eqs. (17.2.4), (17.2.5), and (17.2.7),

$$E_r = \frac{Il\cos\theta}{2\pi}\sqrt{\frac{\mu_0}{\varepsilon_0}}\left(\frac{1}{r^2} - \frac{j}{\beta r^3}\right)e^{-j\beta r}, \tag{17.2.8}$$

$$E_\theta = \frac{Il\sin\theta}{4\pi}\sqrt{\frac{\mu_0}{\varepsilon_0}}\left(\frac{j\beta}{r} + \frac{1}{r^2} - \frac{j}{\beta r^2}\right)e^{-j\beta r}, \tag{17.2.9}$$

$$E_\phi = 0. \tag{17.2.10}$$

Whenever we are dealing with radiation we consider only those points which are at great distances from the dipole. The electromagnetic field at great distances from an antenna is referred to as the *radiation field*, the *far-field*, or the *far-zone field*. From Eqs. (17.2.6) and (17.2.8) to (17.2.10), we see that the far-field components of the electromagnetic field created by a Hertzian dipole are

$$\boxed{E_\theta = \frac{j\beta Il\sin\theta}{4\pi r}\sqrt{\frac{\mu_0}{\varepsilon_0}}\,e^{-j\beta r}} \qquad \text{(Hertzian dipole, far-field)}, \tag{17.2.11}$$

and $$H_\phi = \frac{j\beta Il\sin\theta}{4\pi r}e^{-j\beta r} \qquad \text{(Hertzian dipole, far-field)}. \tag{17.2.12}$$

All other components are either zero or negligibly small. Note the simple relation between $E_\theta$ and $H_\phi$:

$$\boxed{\frac{E_\theta}{H_\phi} = \sqrt{\frac{\mu_0}{\varepsilon_0}}} \qquad \text{(far-field of all antennas)}, \tag{17.2.13}$$

as in the case of plane waves. In fact, any small portion of the large spherical waves in the far-field of the dipole can be approximated by a plane wave, so that this result in a sense is logical. We also expect the wavelength in the far-field to be the same as for plane waves. We can easily check whether this is true. The factor $e^{-j\beta r}$ is unaltered if we substitute $(r + 2\pi/\beta)$ for $r$, since

$$e^{-j\beta(r+2\pi/\beta)} = e^{-j\beta r}e^{-j2\pi} = e^{-j\beta r}.$$

So the wavelength of the spherical wave at great distances from the dipole is

$$\boxed{\lambda = \frac{2\pi}{\beta} = \frac{2\pi}{\omega\sqrt{\varepsilon_0\mu_0}} = \frac{c_0}{f}}. \tag{17.2.14}$$

The Hertzian dipole is an extremely important radiating structure. Any radiating system can always be subdivided into a large number of Hertzian dipoles, provided that the current distribution of the system is known. Hence relations of the form of Eqs. (17.2.13) and (17.2.14) are true for all radiating systems.

**17.2.2 Power radiated and the radiation resistance of the Hertzian dipole.** Spherical waves radiated by an antenna carry away a certain amount of energy, which is of course supplied by the generator to which the antenna is connected. The average power radiated by an antenna is best calculated by determining the flux of the time-average Poynting vector through a very large sphere enclosing the antenna. To determine the power radiated, therefore, it is necessary to know only the components of the far-field. The integration involved is very complicated even for the simplest antennas, but the result can always be determined by numerical integration. The Hertzian dipole is one of the rare examples where this integration can be performed relatively easily.

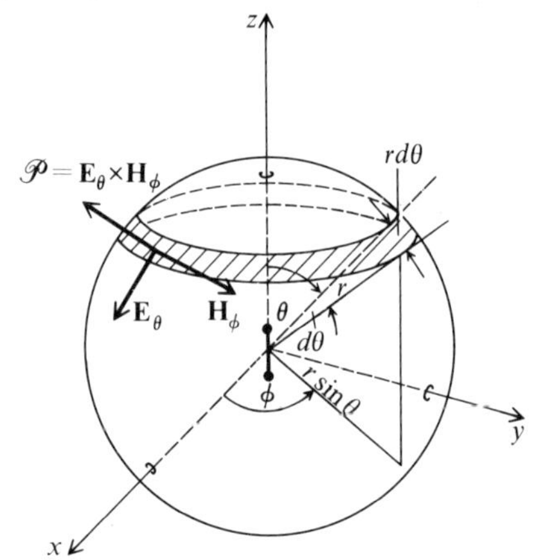


**Fig. 17.2** Determination of power radiated by the Hertzian dipole.

Figure 17.2 shows a large sphere, of radius $r$, centered at the dipole. According to Eqs. (17.2.11) and (17.2.12), the (complex) Poynting vector is perpendicular to the sphere at all points and is real. It is given by

$$\mathscr{P} = \tfrac{1}{2}\mathbf{E}_\theta \times \mathbf{H}_\phi^* = \frac{\beta^2 I^2 l^2 \sin^2\theta}{32\pi^2 r^2}\sqrt{\frac{\mu_0}{\varepsilon_0}}\,\mathbf{u}_r.$$

Since $\mathscr{P}$ is a function of $\theta$, the most suitable surface element is the ring shown crosshatched in Fig. 17.2. The area of the ring is $dS = 2\pi r \sin\theta\, r\, d\theta$, so that the average power radiated by the dipole is

$$P_{\text{av}} = \oint_{\text{sphere}} \mathscr{P}\cdot d\mathbf{S} = \int_0^\pi \frac{\beta^2 I^2 l^2 \sin^2\theta}{32\pi^2 r^2}\sqrt{\frac{\mu_0}{\varepsilon_0}}\, 2\pi r^2 \sin\theta\, d\theta,$$

or

$$P_{av} = \frac{\beta^2 I^2 l^2}{16\pi} \sqrt{\frac{\mu_0}{\varepsilon_0}} \int_0^\pi \sin^3 \theta \, d\theta.$$

Now

$$\int_0^\pi \sin^3 \theta \, d\theta = -\int_{\theta=0}^{\pi} (1 - \cos^2 \theta) \, d(\cos \theta) = -\int_1^{-1} (1 - x^2) \, dx = \tfrac{4}{3}.$$

So we have

$$P_{av} = \frac{\beta^2 I^2 l^2}{12\pi} \sqrt{\frac{\mu_0}{\varepsilon_0}}. \tag{17.2.15}$$

As $\beta = 2\pi/\lambda$ and $\sqrt{\mu_0/\varepsilon_0} = 120\pi$, this expression can be written in the form

$$P_{av} = \tfrac{1}{2}[80(\pi l/\lambda)^2] I^2. \tag{17.2.16}$$

Since $I$ is the *amplitude* of the dipole current, we can see that the power radiated can be *formally* regarded as dissipated in a resistor of resistance

$$\boxed{R_{rad} = 80(\pi l/\lambda)^2} \quad \text{(the radiation resistance of the Hertzian dipole).} \tag{17.2.17}$$

This is known as the *radiation resistance* of the Hertzian dipole. The radiation resistance of any antenna is defined by the relation

$$\boxed{P_{av\ radiated} = \tfrac{1}{2} R_{rad} I^2} \quad \text{(definition of the radiation resistance).} \tag{17.2.18}$$

It should be noted that the radiation resistance is *not* any real resistance, but a measure of how much power is radiated by an antenna, if the current at a given point along the antenna is $I$.

A very important conclusion can be drawn from Eqs. (17.2.17) and (17.2.18). It is that *a measure of radiation efficiency of an antenna is its length measured in wavelengths.* Only little power can be radiated by "electrically" very short antennas, and, since the physical dimensions of antennas are restricted in practice, high frequencies must be used for efficient radiation. The lowest frequencies used are of the order of tens of thousands Hz. Note, however, that Eq. (17.2.17) is not valid generally, but only for the Hertzian dipole. The radiation resistance is quite different for longer antennas.

**17.2.3 The directional properties of an antenna and of the Hertzian dipole.** From Eq. (17.2.11) we can see that the Hertzian dipole does not radiate uniformly in all directions. The strongest radiation is in the plane $\theta = \pi/2$, normal to the dipole, and radiation is zero in the direction of the dipole, $\theta = 0$. Unequal radiation in different directions is a common and important characteristic of *all* radiating systems. In practice, antennas with diverse radiation distributions in space are used. For example, a broadcasting antenna usually radiates equally in all directions in the horizontal plane, but antennas used for directive radio communications between two points have highly directive radiation properties.

Several quantities are used to describe the directional properties of antennas, but we shall mention only four of them. Perhaps the simplest way of visualizing directional properties is to consider the *radiation pattern of the electric field intensity*. This is a plot of the magnitude of vector **E** in the far zone as a function of direction. For convenience, the magnitude of **E** is often normalized with respect to its magnitude in the direction of maximal radiation. The radiation pattern is then said to be a *normalized* or *relative* radiation pattern of the electric field intensity.

A complete radiation pattern is a three-dimensional surface and is therefore difficult to represent or to visualize. Often only one or two cross-sections of the pattern are used, giving a sufficient indication of its general shape. As an example, Figure 17.3 shows the relative radiation pattern of the Hertzian dipole in a plane containing the dipole. We can obtain the complete, three-dimensional pattern by rotating the pattern shown about the $z$ axis. So it is in the form of a doughnut of zero inner radius.

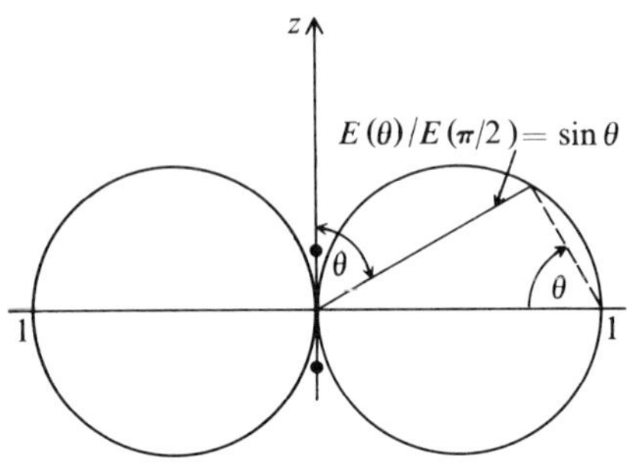


**Fig. 17.3** Relative radiation pattern of the Hertzian dipole.

Instead of the radiation pattern of the electric field intensity, the pattern of the average Poynting vector as a function of direction is also sometimes used. Since in the far-zone $\mathscr{P}_{\text{av}} = E_\theta H_\phi/2 = E_\theta^2/2\eta_0$, the Poynting vector, or power pattern is proportional to the square of the pattern of the field intensity.

Another concept used for characterizing the directive properties of antennas is the *directive gain* of an antenna, which again is a function of direction. It is usually designated by $g_d(\theta)$, or by $g_d(\theta, \phi)$ if the radiation is not symmetrical with respect to the $z$ axis. To define the directive gain of an antenna we must first define the concept of an imaginary, but useful *isotropic* antenna. The isotropic antenna is an imaginary antenna which radiates equally in all directions, so that its radiation field is only a function of the distance $r$ from the antenna. If we denote the electric field intensity of the isotropic antenna by $E_{is}(r)$, the directive gain of an antenna is then defined as

$$g_d(\theta, \phi) = \left[\frac{\text{power radiated by isotropic antenna}}{\text{power radiated by real antenna}}\right]_{\text{for } E(r,\theta,\phi) = E_{is}(r)}$$

$$\text{(definition of directive gain).} \qquad (17.2.19)$$

The directive gain of the isotropic antenna is equal to 1.

Using this definition, we can find the directive gain of any antenna in any direction. As an example, we can find the directive gain of the Hertzian dipole with respect to the isotropic antenna as reference, as explained in the next paragraph.

Let us suppose that the electric field intensity of the isotropic antenna at a distance $r$ is $E_{is}(r)$. Then it radiates the average power

$$P_{\text{av is}} = \tfrac{1}{2}E_{is}(r)H_{is}(r)4\pi r^2 = \frac{1}{2}\sqrt{\frac{\varepsilon_0}{\mu_0}}\,E_{is}^2(r)4\pi r^2. \qquad (17.2.20)$$

Since $E_\theta(r, \theta, \phi)$ should be equal to $E_{is}(r)$, by the definition in Eq. (17.2.19), according to Eqs. (17.2.11) and (17.2.18)

$$P_{\text{av radiated}} = \tfrac{1}{2}R_{\text{rad}}\frac{[4\pi r E_{is}(r)]^2}{\beta^2 l^2 \sin^2\theta}\frac{\varepsilon_0}{\mu_0}.$$

Thus the directive gain of the Hertzian dipole is

$$g_d(\theta) = \sqrt{\frac{\mu_0}{\varepsilon_0}}\frac{\beta^2 l^2 \sin^2\theta}{4\pi R_{\text{rad}}} \qquad \text{(Hertzian dipole).}$$

Taking into account that $\beta = 2\pi/\lambda$, $\sqrt{\mu_0/\varepsilon_0} = 120\pi$, and $R_{\text{rad}} = 80(\pi l/\lambda)^2$, this becomes

$$g_d(\theta) = 120\pi\frac{4\pi^2 l^2 \sin^2\theta}{4\pi\lambda^2 80\pi^2 l^2/\lambda^2} = 1.5\sin^2\theta \qquad \text{(Hertzian dipole).} \quad (17.2.21)$$

The maximum value of the directive gain is known as *directivity* $D$. According to Eq. (17.2.21), the maximum value of the directive gain of the Hertzian dipole, that is, its directivity $D$, is

$$D = g_d(\pi/2) = g_{d\,\max} = 1.5 \qquad \text{(Hertzian dipole).} \qquad (17.2.22)$$

*Problems referring to Section 17.2:* 17.T1 to 17.T2, 17.G1, 17.N1 to 17.N4.

## 17.3 THE THIN LINEAR ANTENNA

**17.3.1 Current distribution along thin linear antennas.** A frequently used type of antenna is the thin straight center-fed conductor (see Fig. 17.4). If the current distribution along such an antenna is known, finding the radiation field of the

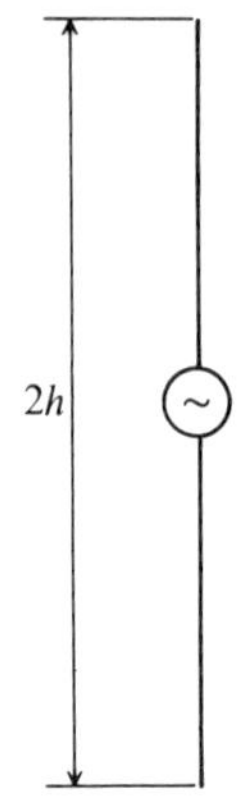


Fig. 17.4 A thin linear antenna.

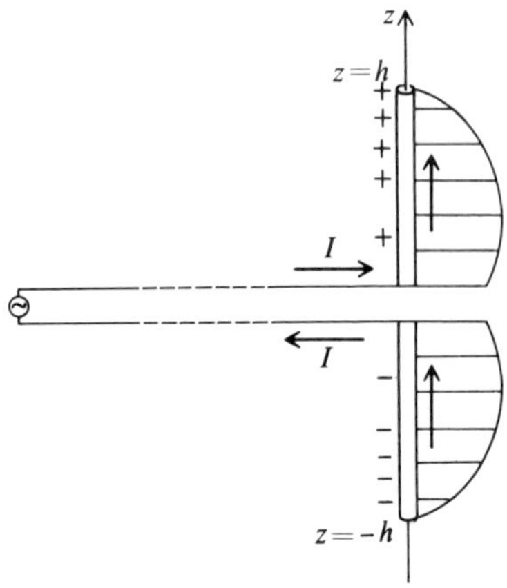


**Fig. 17.5** A thin linear antenna fed by a two-wire line.

antenna reduces to determining the field created by a chain of electric dipoles. Unfortunately, determining the exact current distribution is the most difficult (and generally unsolved) problem of the antenna theory. To understand, at least qualitatively, the difficulties implicit in this problems, let us consider the antenna fed by a two-wire line, as shown in Fig. 17.5. Both charges and current are present along the antenna, and, if we assume for simplicity that the antenna is made of a perfect conductor, the boundary conditions require that the electric field intensity shall have a zero tangential component at all points. The total electric field intensity is due to all the charges and currents of the system, and it is obtained as a complicated integral of these quantities, which are assumed to be unknown functions. Equations in which the unknown appears under the integral sign are known as the *integral equations*, for the solving of which there is no simple method. Even in the simplest cases, the procedure is exceptionally involved.

Instead of searching, therefore, for an exact solution, we are again compelled to look for a good approximate one. For thin linear antennas, a very simple approximate current distribution is sufficiently accurate for many purposes. Referring to Fig. 17.5, that distribution is of the form

$$I_z(z) = I_m \sin \beta(h - |z|), \qquad \beta = 2\pi/\lambda, \tag{17.3.1}$$

where $I_m$ is the largest amplitude of the current along the antenna. Such a distribution for $h \simeq \lambda/3$ is shown in Fig. 17.5. Again we should note that this current distribution is *not* exact. It is to be expected, therefore, that only those properties of linear antennas which depend mainly on the general shape of the current distribution can be calculated with reasonable accuracy by using this distribution. The properties concerned are the radiation pattern, the radiated power, the radiation resistance, and the antenna directive gain. The experimental evidence has shown that all these quantities, as computed by means of the sinusoidal current distribution, are very good indicators of what we can expect in practice.

If sinusoidal distribution is used for determining the properties of antennas which critically depend on the exact current distribution, a poor degree of accuracy in the results is to be expected, and this is indeed the case with an important parameter of the linear antenna—its input impedance. To determine the input impedance exactly, we must know the exact current at the antenna terminals, which, as explained above, is the most difficult problem of the antenna theory. In this short introduction we shall not deal with this problem of impedance, but only with the radiation properties of antennas. For this purpose, the sinusoidal current distribution is a sufficiently accurate approximation.

**17.3.2 The radiation field of thin linear antennas with a sinusoidal current distribution.** Let us consider a thin, center-fed straight antenna of length $2h$, shown in Fig. 17.6, and assume that the amplitude of the sinusoidally time-varying current along the antenna is given by Eq. (17.3.1). Thus, we know the current intensity in any element $dz$ of the antenna. Each element can be considered as an elemental Hertzian dipole, although, at first glance, we could argue that this is not so, since there are no charges at the ends. But in the components of the radiation field of the dipole, given by Eqs. (17.2.11) and (17.2.12), only the current amplitude $I$ appears. The charges at the ends are implicitly incorporated by virtue of the continuity equation. Thus, even if we take the case of the field of a current element, these charges are automatically incorporated in the solution if it is determined correctly from Maxwell's equations.

Each current element creates a far-field having only the $dE_\theta$ and $dH_\phi$ components. Since $r \gg h$, the resultant far-field will also have only the $E_\theta$ and $H_\phi$ components, related by Eq. (17.2.13). We shall therefore determine only the resultant $E_\theta$ component.

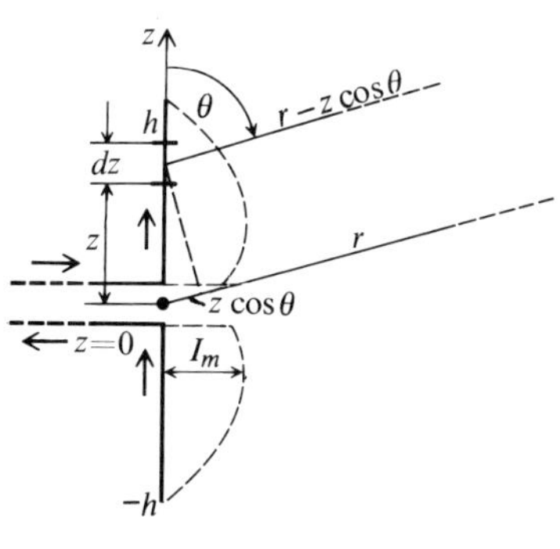


**Fig. 17.6** Determination of the far-zone field of a linear antenna with sinusoidal current distribution.

According to Eq. (17.2.11) and by reference to Fig. 17.6

$$E_\theta = \frac{j\beta}{4\pi}\sqrt{\frac{\mu_0}{\varepsilon_0}}\int_{-h}^{h}\frac{I(z)\sin\theta\,dz}{r - z\cos\theta}e^{-j\beta(r - z\cos\theta)}. \tag{17.3.2}$$

Now, $(r - z\cos\theta)$ in the denominator of the integrand determines the amplitude of the field created by an element positioned at $z$, and since $r \gg z$, we can in this case change $(r - z\cos\theta)$ simply to $r$. For example, if $h = 1$ m and $r = 10$ km, in practice it is irrelevant whether we have 1/10,001 or 1/10,000. This can*not* be done, however, with the expression $(r - z\cos\theta)$ in the exponent. The exponent determines the relative phase of the individual elemental waves, and its exact value is absolutely necessary if the correct resultant field is to be obtained. For example, if $\beta z\cos\theta = \pi/2$, then the field created by the element at $z$ is zero at an instant when the field created by an element near $z = 0$ is maximal. If we neglected $\beta z\cos\theta$ in the exponent, both would be maximal at the same time, which is obviously far from the truth.

In the light of these remarks and by the use of $I(z)$ given by Eq. (17.3.1), Eq. (17.3.2) becomes

$$E_\theta = \frac{j\beta I_m\sin\theta}{4\pi r}\sqrt{\frac{\mu_0}{\varepsilon_0}}e^{-j\beta r}\int_{-h}^{h}\sin\beta(h - |z|)e^{j\beta z\cos\theta}\,dz. \tag{17.3.3}$$

The integral can be evaluated as follows. We begin with

$$\begin{aligned} C &= \int_{-h}^{h}\sin\beta(h - |z|)e^{j\beta z\cos\theta}\,dz \\ &= \int_{-h}^{h}\sin\beta(h - |z|)\left[\cos(\beta z\cos\theta) + j\sin(\beta z\cos\theta)\right]dz. \end{aligned}$$

Note that $\sin\beta(h - |z|)$ is an even function of $z$, since $|-z| = |z|$. The first function in the brackets is also even, but the second is odd [that is, $\sin(-\beta z\cos\theta) = -\sin(\beta z\cos\theta)$]. By reference to the sketches in Fig. 17.7 the integral becomes

$$C = 2\int_0^h \sin\beta(h - z)\cos(\beta z\cos\theta)\,dz.$$

Using the trigonometric identity,

$$\sin\alpha_1\cos\alpha_2 = \tfrac{1}{2}[\sin(\alpha_1 + \alpha_2) + \sin(\alpha_1 - \alpha_2)],$$

we can easily solve the last integral. The result is

$$C = \frac{2}{\beta\sin^2\theta}[\cos(\beta h\cos\theta) - \cos\beta h],$$

so that the expression for the far-zone electric field of Eq. (17.3.3) becomes

$$E_\theta = \frac{jI_m}{2\pi r}\sqrt{\frac{\mu_0}{\varepsilon_0}}e^{-j\beta r}\frac{\cos(\beta h\cos\theta) - \cos\beta h}{\sin\theta}. \tag{17.3.4}$$

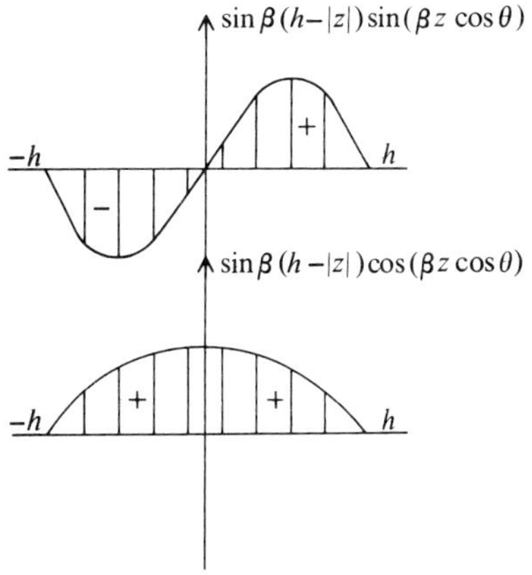


**Fig. 17.7** Graphs of functions $\sin \beta(h - |z|) \sin (\beta z \cos \theta)$ and $\sin \beta(h - |z|) \cos (\beta z \cos \theta)$.

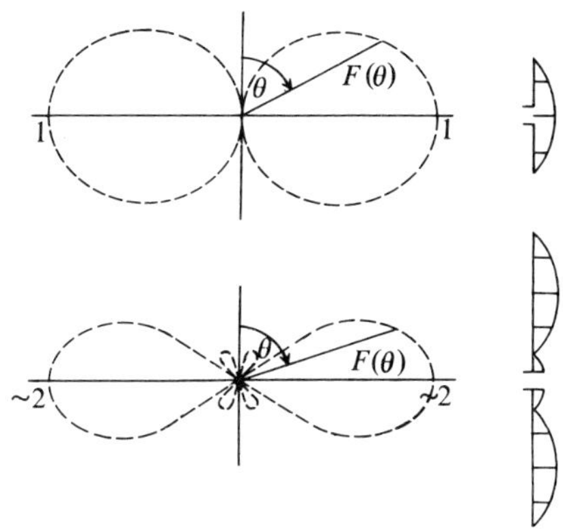


**Fig. 17.8** Radiation characteristic of the half-wave dipole (above) and of a dipole somewhat longer than the wavelength (below).

The expression

$$F(\theta) = \frac{\cos (\beta h \cos \theta) - \cos \beta h}{\sin \theta} \tag{17.3.5}$$

is sometimes referred to as the *radiation characteristic* of the antenna considered. Obviously, it is proportional to the relative radiation pattern of the electric field intensity.

If only the magnitude of $E_\theta$ is of interest, which is often the case, Eq. (17.3.4) can be written in a simpler form. With $\sqrt{\mu_0/\varepsilon_0} = 120\pi$, it becomes

$$|E_\theta| = \frac{I_m}{2\pi r}\sqrt{\frac{\mu_0}{\varepsilon_0}}\,F(\theta) = \frac{60 I_m}{r}\,F(\theta) \qquad \text{(antenna in air, sinusoidal current).} \tag{17.3.6}$$

The radiation characteristic $F(\theta)$ is often decisive in adopting an antenna for a specific purpose. Almost any book on antennas gives graphs of $F(\theta)$ for many lengths $h$ of linear antennas. Figure 17.8 shows the radiation characteristics for the half-wave dipole ($2h = \lambda/2$) and the antenna where $h$ is somewhat larger than $\lambda/2$. The latter is very much used as a medium-wave broadcasting antenna, because of its flat characteristic and for some other reasons.

For the important half-wave dipole the radiation characteristic of Eq. (17.3.5) becomes

$$F(\theta) = \frac{\cos\left(\dfrac{\pi}{2}\cos\theta\right)}{\sin\theta} \qquad \text{(the half-wave dipole).} \tag{17.3.7}$$

The receiving point is most often in the plane $\theta = \pi/2$, where the field is strongest. In that case

$$F(\pi/2) = 1 \qquad \text{(the half-wave dipole).} \tag{17.3.8}$$

**17.3.3 Power radiated, the radiation resistance and the directive gain of antennas with sinusoidal current distribution.** As we know the electric field intensity in the far-zone of antennas with sinusoidal current distribution, we can find the average power radiated in the same way as for the Hertzian dipole. By reference to Fig. 17.2, and Eqs. (17.2.13) and (17.3.6), we obtain

$$P_{av} = \int_0^\pi \frac{1}{2}\sqrt{\frac{\mu_0}{\varepsilon_0}}\frac{I_m^2}{4\pi^2 r^2} F^2(\theta) 2\pi r^2 \sin\theta \, d\theta,$$

from which

$$P_{av} = \tfrac{1}{2} R_{rad} I_m^2, \tag{17.3.9}$$

where the radiation resistance

$$R_{rad} = \frac{1}{2\pi}\sqrt{\frac{\mu_0}{\varepsilon_0}}\int_0^\pi F^2(\theta)\sin\theta\, d\theta. \tag{17.3.10}$$

Note that $R_{rad}$ thus defined refers to the amplitude $I_m$ of the *largest* current intensity along the antenna, and *not* to that at the antenna terminals. $F(\theta)$ is given by Eq. (17.3.5). The integral in $R_{rad}$ cannot be expressed in terms of elementary functions, and will not be solved here. Figure 17.9 shows $R_{rad}$ as a function of the antenna half-length $h$. For example, the often used half-wave dipole has a radiation resistance

$$R_{rad} \simeq 73\Omega \qquad \text{(the half-wave dipole).} \tag{17.3.11}$$

Finally, the directive gain of any linear antenna with sinusoidal current distribution can easily be obtained from the defining formula in Eq. (17.2.19), using Eq. (17.2.20), (17.3.6), and (17.3.9). From Eqs. (17.3.6) and (17.3.9) we first have

$$P_{av} = \frac{1}{2}\frac{\varepsilon_0}{\mu_0}\frac{4\pi^2 r^2 |E_\theta|^2 R_{rad}}{F^2(\theta)}$$

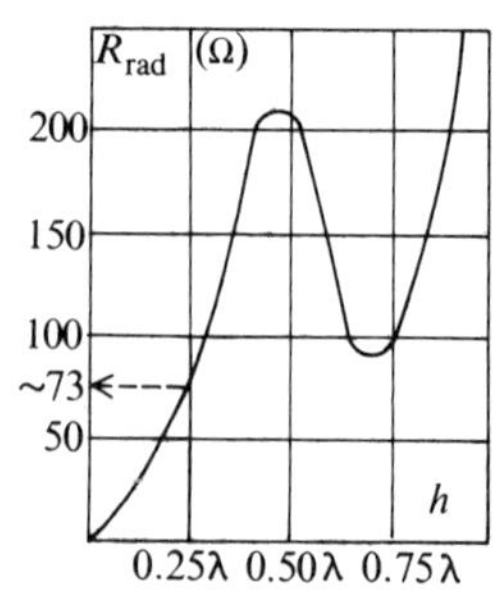


**Fig. 17.9** Radiation resistance of linear antennas with sinusoidal current distribution.

Since $|E_\theta|$ must be equal to $E_{is}$, from Eqs. (17.2.19) and (17.2.20) we obtain

$$g_d(\theta) = \sqrt{\frac{\mu_0}{\varepsilon_0}}\frac{F^2(\theta)}{\pi R_{rad}} \quad \text{(the directive gain of linear antennas with sinusoidal current distribution).} \tag{17.3.12}$$

For example, according to Eqs. (17.3.8) and (17.3.11) the half-wave dipole has its directivity $D$ equal to

$$D = g_d(\pi/2) = 120\pi\frac{F^2(\pi/2)}{\pi R_{rad}} = \frac{120}{73} \simeq 1.64 \quad \text{(the half-wave dipole).} \tag{17.3.13}$$

Instead of the isotropic antenna either the Hertzian dipole ($D = 1.5$) or the half-wave dipole is often used as the reference antenna, and the directive gain is defined with respect to the *directivity* of one of them. Recalling that the directivity of the isotropic antenna is unity, we can obtain the directive gain with respect to the Hertzian dipole or the half-wave dipole simply by dividing the directive gain with respect to the isotropic radiator by 1.5 or 1.64, respectively. Because of these three possibilities as to the reference antenna adopted for analysis, it is always necessary to specify which one is being used.

*Problems referring to Section 17.3:* 17.T3 to 17.T5, 17.G2 to 17.G3, 17.N5 to 17.N7.

## 17.4 ELEMENTS OF LINEAR ARRAYS

In many cases in practice a very directive radiation pattern is required. We have seen that the radiation pattern of a single linear antenna is toroid-like, that is, that the radiation field is independent of the $\phi$ coordinate. We shall now show that a radiation pattern which also has directive properties in coordinate $\phi$ can be obtained by combining two or more linear antennas.

Let us consider the simplest case of two parallel and identical thin linear antennas at a distance $d$ apart (see Fig. 17.10), and assume that the complex amplitudes of the currents at the current maxima of the two antennas are

$$I_1 = I_0, \qquad \text{and} \qquad I_2 = I_0 e^{j\delta}, \tag{17.4.1}$$

that is, that their magnitudes are equal, but that they differ in phase. The more general case, when the magnitudes of the currents are also different, will not be

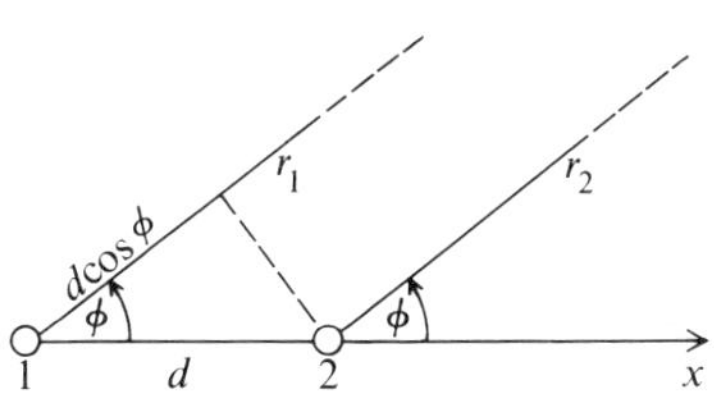


**Fig. 17.10** An array of two parallel identical dipoles.

considered here. In addition, we shall consider only points in the plane $\theta = \pi/2$.

Since the amplitudes of the two currents at the current maxima are equal (remember that $|e^{j\delta}| = 1$), according to Eq. (17.3.6) the *amplitudes* of the electric far-field intensity created by the two antennas are also equal. But their *phases* are *not* equal. There are two sources of the phase difference: the first is that there is a phase difference $\delta$ between the currents themselves, the second is that the distances between the two antennas and a receiving point are generally also different. For example, in the direction shown in Fig. 17.10 the two distances differ for $r_1 - r_2 = d \cos \phi$. So the wave from antenna 2 arrives at the receiving point for $\Delta t = d \cos \phi / c_0$ earlier than that from antenna 1.

Now, according to the assumed phase difference between the currents represented by Eq. (17.4.1), the wave from antenna 2 also advances the wave from antenna 1 because the current $I_2$ advances the current $I_1$ for angle $\delta$. So the total phase *advance* of the electric far-field, due to antenna 2 with respect to that due to antenna 1, is $(\delta + \beta d \cos \phi)$.

If we denote by $E_0$ the complex amplitude of the electric field due to antenna 1, then the complex amplitude $E_\theta$ of the resultant field is, according to the foregoing,

$$E_\theta = E_{\theta 1} + E_{\theta 2} = E_0 + E_0 e^{j(\delta + \beta d \cos \phi)}.$$

This can be written in the form

$$E_\theta = E_0\{[1 + \cos(\delta + \beta d \cos \phi)] + j \sin(\delta + \beta d \cos \phi)\}.$$

Except in rare instances we are interested only in the real amplitude of the total electric field intensity $E_\theta$, and, according to the foregoing equation it is given by

$$|E_\theta| = E_0 \sqrt{[1 + \cos(\delta + \beta d \cos \phi)]^2 + \sin^2(\delta + \beta d \cos \phi)},$$

or

$$|E_\theta| = E_0 \sqrt{2[1 + \cos(\delta + \beta d \cos \phi)]} \qquad \text{(valid in plane } \theta = \pi/2\text{)}. \qquad (17.4.2)$$

So the radiation pattern is a function of both the distance between the antennas (measured in wavelengths) and the phase difference $\delta$ between the two currents.

Numerous graphs of the function $|E_\theta|/E_0$ can be found in any book on antennas. Figure 17.11 gives three of them, for the values of $\delta$ and $d$ indicated.

Various combinations of more than two thin parallel antennas are also used in practice. Assuming a sinusoidal current distribution, the determination of the radiation pattern is not usually complicated, provided that the relative phases of the currents in the elements are known.

Another type of linear array has what are called *parasitic* elements, which are simply short-circuited antennas, that is, antennas which are not connected to a generator. The currents in the parasitic elements are induced by currents in driven elements. The resultant radiation pattern is the result of all these currents.

An important and familiar type of a linear array with parasitic elements is the

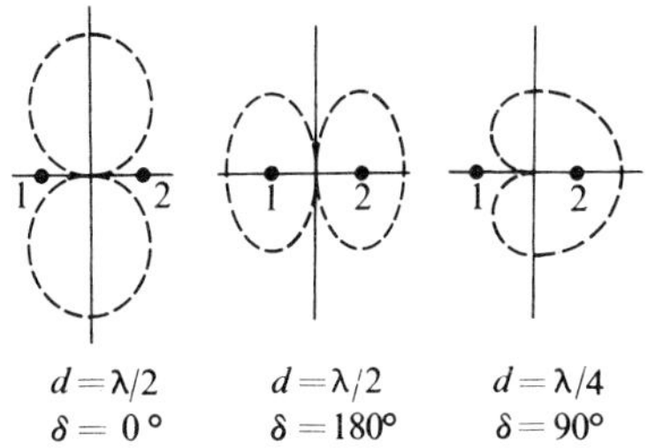


**Fig. 17.11** Radiation patterns of two parallel identical dipoles with sinusoidal current distribution, in the plane $\theta = \pi/2$.

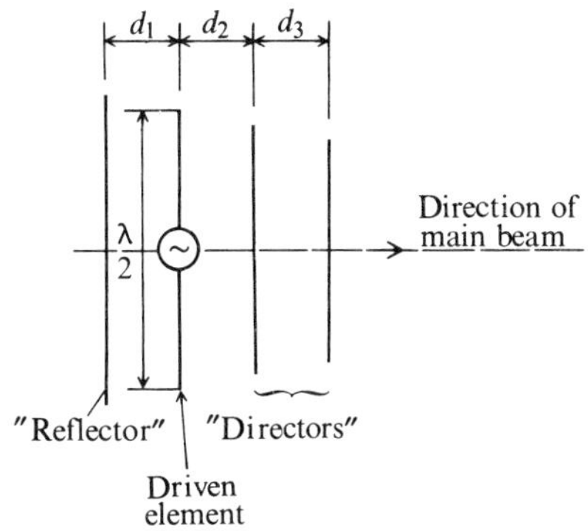


**Fig. 17.12** An Uda-Yagi array.

Uda-Yagi array shown in Fig. 17.12. Even an approximate analysis of this type of array is extremely involved. The distances $d_1$, $d_2$, $d_3$, . . . between the elements of the array are mainly determined experimentally. The driven element is approximately $\lambda/2$ long, the "reflector" is slightly longer, and the "directors" are slightly shorter. The number of directors is not necessarily two as shown; arrays with only one director, or several of them, are also used. The Uda-Yagi antenna is most often used as a receiving antenna.

*Problems referring to Section 17.4:* 17.T6 to 17.T8, 17.G4.

## 17.5 ANTENNAS IN THE PRESENCE OF A PLANE CONDUCTING SURFACE

In practice, antennas are very often in the presence of a large, approximately plane conducting surface, which is usually the surface of the earth. The electromagnetic field of antennas induce currents and charges on the conducting surface. The total electromagnetic field is due to all the currents and charges, that is, to those in the antennas themselves and to the secondary, induced currents and charges on the conducting surface.

Exact determination of the resultant field in the general case of an antenna above an imperfectly conducting earth is extremely involved. But, if the earth is regarded to be an approximately perfect conductor, the secondary field, due to the induced currents and charges, can be easily determined.

Figure 17.13 shows an antenna above a perfectly conducting plane. The boundary conditions require that the tangential component of the electric field intensity and the normal component of the magnetic flux density at points of the

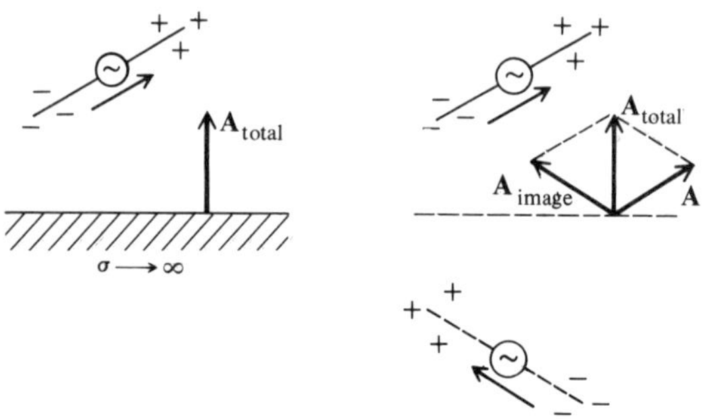


**Fig. 17.13** An antenna above a perfectly conducting plane, and its image.

plane are zero. Since

$$\mathbf{E} = -\text{grad } V - \frac{\partial \mathbf{A}}{\partial t} \qquad \text{and} \qquad \mathbf{B} = \text{curl } \mathbf{A},$$

both boundary conditions are satisfied if on the plane $V = \text{constant} = 0$, and if **A** is perpendicular to it.

Having the expressions in Eq. (13.4.32) and (13.4.33) for the retarded potentials in mind, we can see from Fig. 17.13 that the required boundary conditions will be satisfied if the plane is removed and the image antenna introduced instead. (Note the direction of the current in the image antenna.) By the uniqueness theorem, the secondary field radiated by the induced charges and currents on the perfectly conducting plane is the same as that due to the image antenna.

As an example, let us consider the vertical, bottom-fed antenna above the surface of the earth, shown in Fig. 17.14. The radiation field exists in the upper half space only, where it is equal to the field created by the *symmetrical* antenna in the figure. The radiation pattern can therefore be readily determined.

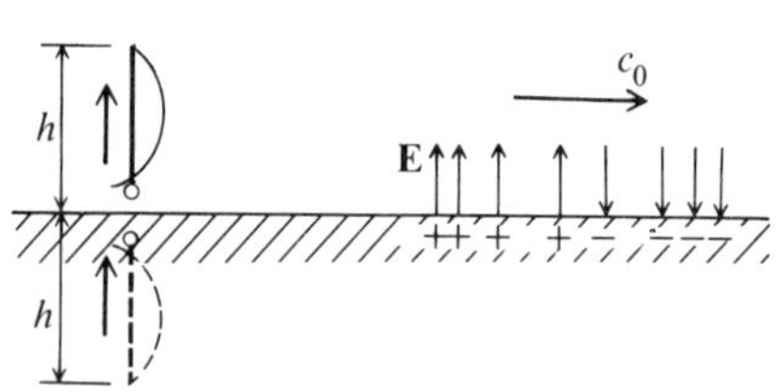


**Fig. 17.14** A vertical antenna above the surface of the earth, assumed to be a perfect conductor.

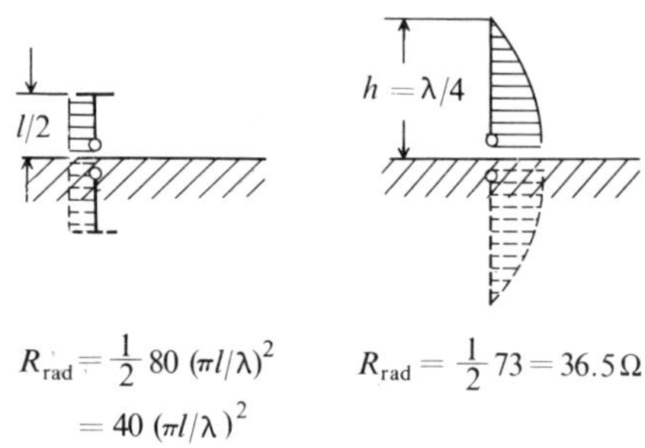


**Fig. 17.15** Short top-loaded antenna and quarter-wave antenna above the surface of the earth.

The wave propagating along the surface of the earth is accompanied by induced charges and currents. In the real case the earth is not a very good conductor; as a result, considerable attenuation of this surface wave takes place.

The radiation pattern in the upper half space, where the field exists, is identical with that created by a symmetrical antenna, but the power radiated is obviously only half of that radiated by the symmetrical dipole. According to Eq. (17.2.18), this means that the radiation resistance of a vertical, bottom-fed antenna above a perfectly conducting plane is equal to exactly *one half* of the radiation resistance of the symmetric antenna in free space. This is illustrated in Fig. 17.15 for the case of a short, top-loaded antenna and a quarter-wave antenna. The radiation resistances in the two cases for the symmetrical antenna are given by Eqs. (17.2.17) and (17.3.11), respectively.

*Problems referring to Section 17.5:* 17.T9 and 17.T10, 17.G5 to 17.G7.

### 17.6 THE ELEMENTS OF APERTURE ANTENNAS

The thin linear antenna considered in the preceding sections belongs to a frequently-used antenna type, but many other types are also used. We shall restrict our attention to one more class of radiating systems, to what are called aperture antennas, which are used as radiating structures in the microwave region. Because of the relative complexity of their analysis, we shall give only a qualitative description of their functioning.

In general, the term "aperture antennas" usually implies highly conductive structures which have physical apertures through which electromagnetic waves are radiated. One example of an aperture antenna is a rectangular waveguide opened at one end. Part of the energy propagating through the guide will leave it, so that the open end of the waveguide behaves as a radiating structure. To promote radiation from the waveguide end and to obtain a more directive pattern, a transition between the waveguide and free space is made in the form of horns of many different shapes. Two of these are shown in Fig. 17.16. Electromagnetic horns

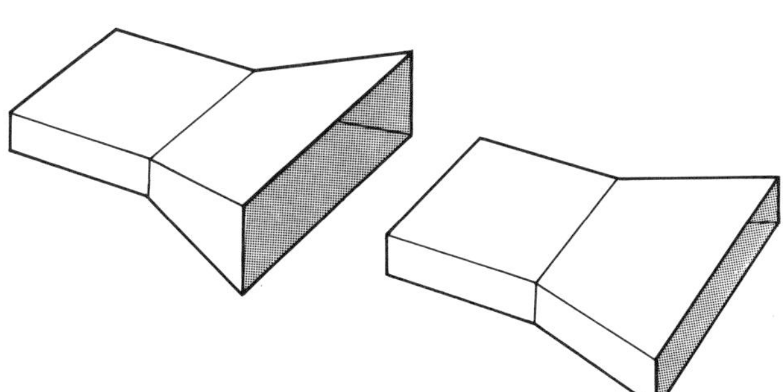

**Fig. 17.16** Two types of electromagnetic horns.

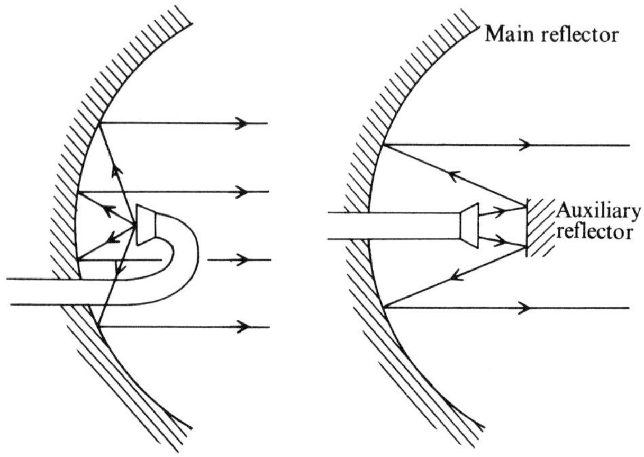


**Fig. 17.17** Electromagnetic horn combined with a reflector.

are most often used in connection with approximately parabolic reflectors, as in Fig. 17.17.

Another type of aperture antennas are known as the *slot antennas*, which are obtained whenever a slot is made in a structure inside which a high-frequency time-varying electromagnetic field exists. A certain amount of electromagnetic energy will leak through the slot, so that it behaves as a source of electromagnetic waves. One example of the slot antenna is a slot in a waveguide. Slot antennas are also used in aircraft. The slot is made in the shell of the aircraft and then filled with a suitable dielectric, and an antenna is obtained which does not require changing the aerodynamic shape of the plane.

The radiation from all types of aperture antennas is due to an *unknown* distribution of time-varying charges and currents on the conducting surface in the vicinity of the aperture. The exact analysis of any aperture antenna is therefore practically impossible. To analyze the radiation from the apertures we have to devise some approximate method simpler than that of determining the exact distribution of the currents and charges in the system. We usually start by *postulating* some reasonable distribution of the electromagnetic field over the area of the aperture. It can be shown that the electromagnetic field specified over the aperture surface uniquely determines the radiation from the aperture. Obviously, because of the approximate field distribution postulated, only an approximate radiation field from the aperture can be obtained, but it is usually sufficiently accurate.

## 17.7 THE RECEIVING ANTENNA

**17.7.1 Introduction.** The task of a receiving antenna is to extract energy from an electromagnetic wave, together with the signal carried by the wave, and to deliver it to a receiver. A qualitative explanation of the mechanism which enables a receiving antenna to extract energy from the wave is simple. The time-varying electric field of the wave, in which the antenna is immersed, induces currents and charges in the body of the conducting antenna. A certain voltage is thus developed across the antenna terminals, which varies with time according to the same law as the wave. This voltage represents the input to a receiver, in which it is usually amplified to a desired level, and then converted into the desired form (sound, picture, etc.).

The geometrical shape of the receiving antennas used in practice is less diverse than that of transmitting antennas. But any transmitting antenna can be used as a receiving antenna, and vice versa. For example, the aperture antennas shown in Figs. 17.16 and 17.17 are used as receiving antennas as well as the Hertzian dipole or the thin linear antenna. It is important to understand, however, that the working conditions in the receiving and transmitting cases are not equal. In the transmitting case, the impressed external electric field acts only between the antenna terminals. In the receiving case, the external electric field, the field of the wave, exists over the entire surface of the antenna. The resulting current and charge distribution in the two cases will obviously be very different. This might give an impression that no knowledge of the transmitting properties of an antenna can be used for analyzing its receiving properties, but fortunately this is not the case. It can be shown that the two most important characteristics of any receiving antenna in practice, its directive properties and its impedance, are identical to those in the transmitting case. In this short introduction to the receiving antenna, however, we shall consider only the electromotive force of the two simplest receiving antennas—the small loop antenna and the Hertzian dipole.

**17.7.2 The small loop and the Hertzian dipole as receiving antennas.** Let us consider a small conducting loop in the electromagnetic field radiated by a distant antenna. By the term "small" we imply that the wavelength of the wave is much larger than the dimensions of the loop. The electric field intensities at points along the loop differ slightly, and, as a result, the line integral of **E** is not zero. A time-varying electromotive force is thus induced in the loop. The amplitude of the electromotive force is a function of the orientation of the loop with respect to **E**.

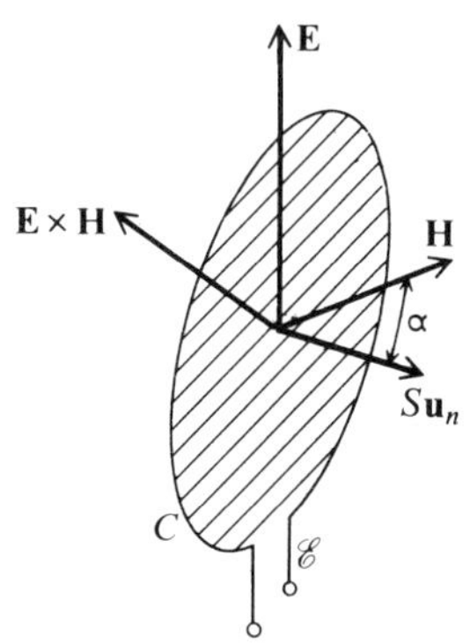


**Fig. 17.18** A loop receiving antenna.

For example, it is evident that if the plane of the loop is perpendicular to **E**, or perpendicular to the direction of wave propagation, the induced electromotive force is zero.

Figure 17.18 shows such a loop and the directions of the field vectors. The electromotive force induced in the loop is

$$\mathscr{E}(t) = \oint_C \mathbf{E} \cdot d\mathbf{l}.$$

Instead of calculating the line integral, we can, by using the Stokes theorem and the curl equation, curl $\mathbf{E} = -\partial \mathbf{B}/\partial t$, transform this into

$$\mathscr{E}(t) = \oint_C \mathbf{E} \cdot d\mathbf{l} = \int_S \text{curl}\, \mathbf{E} \cdot d\mathbf{S} = -\frac{\partial}{\partial t} \int_S \mathbf{B} \cdot d\mathbf{S}.$$

Since the loop is assumed to be small, **B** is practically constant over its area $S$, so that we can write

$$\mathscr{E}(t) = -\mu S \mathbf{u}_n \cdot \frac{\partial \mathbf{H}}{\partial t}. \tag{17.7.1}$$

At the point considered let $\mathbf{H}(t) = \mathbf{H} \cos \omega t$. Then

$$\mathscr{E}(t) = \mu S \omega H \cos \alpha \sin \omega t, \tag{17.7.2}$$

where $\alpha$ is the angle between **H** and $\mathbf{u}_n$ (see Fig. 17.18). The electromotive force is thus maximal when the loop is perpendicular to **H**, in which case its plane coincides with the direction of propagation of the wave. $\mathscr{E}$ is zero if $\alpha = \pi/2$, that is, when the loop is perpendicular to the direction of propagation. Either can be used for determining the direction of the wave, and to locate, say, a clandestine broadcasting station. Usually the condition $\alpha = \pi/2$ is used, since the zero of $\mathscr{E}$ can be detected more accurately than its maximum. With two or several search loops, the location of the transmitting antenna can be determined with a high accuracy (see

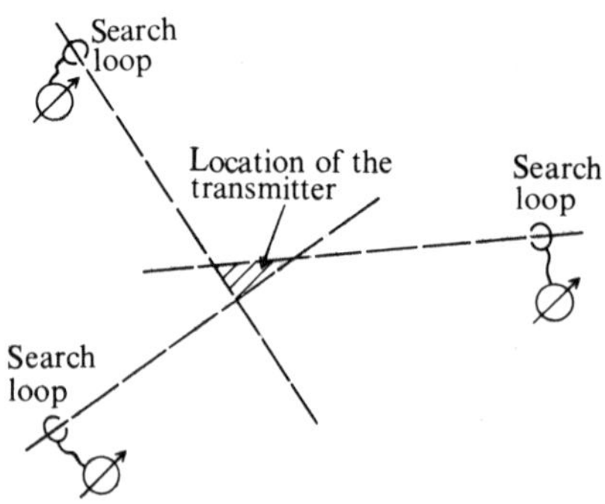


**Fig. 17.19** Locating a transmitter using search loop antennas.

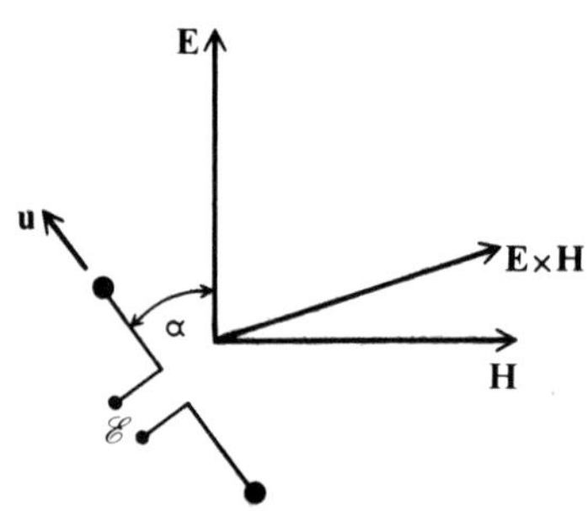


**Fig. 17.20** Hertzian dipole as a receiving antenna.

Fig. 17.19). The loop antenna is also very much used as a room television antenna for ultra-high frequencies, and in radio receivers. The ferrite antenna in radio receivers is a kind of loop antenna.

Let us now consider a Hertzian dipole situated in the electromagnetic field radiated by an antenna. Since its length $l$ is very small when compared with the wavelength of the wave, the electromotive force induced in it is, as shown in Fig. 17.20, simply

$$\mathscr{E} = \mathbf{E}\cdot\mathbf{u}\,l = El\cos\alpha \quad (l \ll \lambda). \tag{17.7.3}$$

$\mathscr{E}$ is, of course, maximal when the dipole is parallel to **E**, that is, $\alpha = 0$. Hertzian, or almost Hertzian dipoles (often only one half of a dipole) are used as the receiving antennas of television receivers, small portable radios, car radios, etc.

*Problems referring to Section 17.7:* 17.T11 and 17.T12, 17.G8 to 17.G11, 17.N8.

## PROBLEMS

### Test problems

**17.T1.** Assuming that the dipole shown in Fig. 17.1 has no end spheres, discuss the current distribution if $l \ll \lambda$.

**17.T2.** Write Eqs. (17.2.6), (17.2.8), and (17.2.9) for points very close to the dipole ($r \ll \lambda$). Is the result obtained logical? [Note that $I = j\omega Q$ and $\beta = \omega\sqrt{\varepsilon_0\mu_0}$.]

**17.T3.** If the current intensity through antenna terminals is zero, no power can be delivered to the antenna. With this in mind, discuss the usefulness of the approximate sinusoidal distribution for analyzing the *input impedance* of a full-wave dipole ($2h = \lambda$).

**17.T4.** Sketch the radiation characteristic $F(\theta)$ for $h = \lambda/2$ and $h = 3\lambda/4$.

**17.T5.** If the radiation resistance $R_{\text{rad}}$ (with reference to the current maximum) and the average power radiated are known, rather than $I_m$, what is the form of Eq. (17.3.6)?

**17.T6.** Sketch $|E_\theta|/E_0$ given by Eq. (17.4.2) for $\delta = 0$ and $d = \lambda$.

**17.T7.** If in Fig. 17.T7 $I_1 = I_2 = I_3$, determine the minimum value of $d$ so that the radiation of the array is maximal in the direction $\phi = 0$ and $\phi = \pi$. In which other directions is it also maximal?

**17.T8.** Given that in Fig. 17.T7 $I_1 = I$, $I_2 = Ie^{-j\delta}$, and $I_3 = Ie^{-j2\delta}$, determine $\delta$ so that the radiation of the array is maximal in the direction (1) $\phi = 0$, and (2) $\phi = \pi$. The distance $d$ is fixed, and the wavelength of the radiated wave is $\lambda$.

**17.T9.** A linear antenna resides in air, parallel to the conducting surface of the earth. Assuming that the earth is a perfect conductor, determine the minimum necessary height of

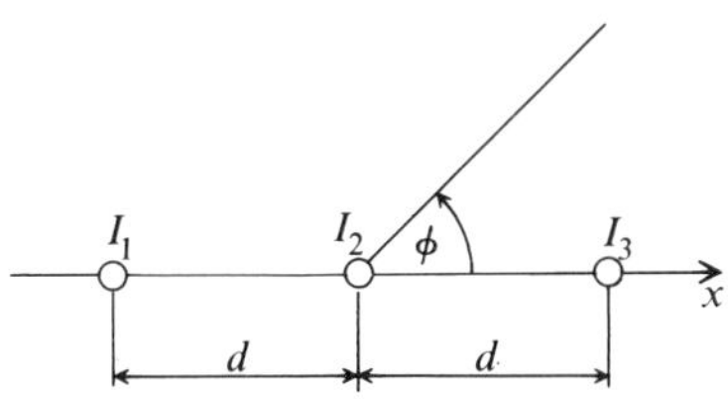


**Figure 17.T7**

the antenna so that the radiation in the direction perpendicular to the surface of the earth is (1) zero, and (2) the maximum possible.

**17.T10.** With reference to Fig. 17.15, note that for the antennas shown only half of the voltage of the symmetric antenna is required to obtain the same intensity of current. Starting from this fact prove that the radiation resistances of the antennas shown are equal to exactly one half of those for symmetric antennas.

**17.T11.** Is it possible to locate a transmitting antenna using short dipoles instead of loops as receiving antennas?

**17.T12.** On many short antennas there is a small conducting ball at each end. What is the ball for?

### General problems

**17.G1.** A Hertzian dipole, of length $l$ and wire radius $a$, is made of copper. If the frequency of the radiated wave is $f$, find the ratio of average power radiated to the total average power delivered to the dipole. (This ratio is a measure of the efficiency of the antenna.)

*Solution.* Since there are some losses in the copper wire of the dipole, a part of the power delivered to the dipole is converted into heat in the dipole itself. Designating the amplitude of the dipole current by $I$ and the resistance of the wire by $R_w$, the average power transformed into heat is simply

$$P_J = \tfrac{1}{2}R_w I^2.$$

According to Eq. (17.2.18) the power radiated is

$$P_{\text{rad}} = \tfrac{1}{2}R_{\text{rad}} I^2.$$

The total average power delivered to the dipole is thus

$$P_{\text{total}} = P_J + P_{\text{rad}} = \tfrac{1}{2}(R_w + R_{\text{rad}})I^2,$$

so that

$$\frac{P_{\text{rad}}}{P_{\text{total}}} = \frac{R_{\text{rad}}}{R_w + R_{\text{rad}}} = \frac{1}{1 + R_w/R_{\text{rad}}}. \tag{a}$$

Now, $R_{\text{rad}}$ is given by Eq. (17.2.17). $R_w$ is *not* the resistance $l/\sigma a^2\pi$ of the wire for the direct current, but rather its high-frequency resistance. We can readily determine the resistance according to Eq. (15.3.21). Since the amplitude of the magnetic field intensity on the wire surface is approximately $H_0 = I/2\pi a$, we have

$$P_J = \tfrac{1}{2}R_s\left(\frac{I}{2\pi a}\right)^2 2\pi a l = \frac{1}{2}\frac{R_s l}{2\pi a} I^2,$$

from which

$$R_w = \frac{R_s}{2\pi a} l. \tag{b}$$

(See also Problem 17.N4.)

***17.G2.** Repeat the preceding problem for a linear antenna of length $2h$ with a sinusoidal current distribution.

*Answer.* $R_w = \dfrac{R_s}{2\pi a}\left(h - \dfrac{\lambda}{4\pi}\sin\dfrac{4\pi h}{\lambda}\right).$

**17.G3.** Two half-wave dipoles are at right angles to each other and the complex amplitudes of their currents are $I_m$ and $jI_m$ (see Fig. 17.G3). Determine the resultant electric field intensity in the far-zone, in the direction perpendicular to the plane of the dipoles.

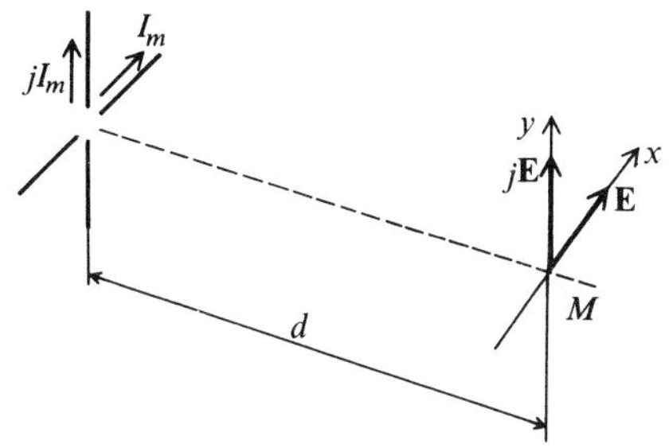


**Figure 17.G3**

*Solution.* At a distant point $M$ in Fig. 17.G3, the electric field intensity vectors have the directions shown. They are of equal amplitudes, but their phases differ for $\pi/2$ due to the phase difference of the currents in the two antennas. So if, say, $E_x = E\cos\omega t$, then $E_y = E\cos(\omega t + \pi/2) = E\sin\omega t$. The total electric field intensity vector is of a constant amplitude,

$$E_{\text{total}} = \sqrt{E_x^2 + E_y^2} = E.$$

The angle $\alpha$ which $\mathbf{E}_{\text{total}}$ makes with the $x$ axis is determined from

$$\tan\alpha = \frac{E_y}{E_x} = \frac{\sin\omega t}{\cos\omega t} = \tan\omega t,$$

so that $\alpha = \omega t$. The resultant electric field intensity vector $\mathbf{E}_{\text{total}}$ rotates at $M$ with an angular velocity $\omega$. The wave is circularly polarized. (Such a combination of antennas is referred to as the *turnstile* type of antenna.)

**17.G4.** Starting from Eq. (17.4.2), plot the radiation pattern, that is, the ratio $|E_\theta|/E_0$ of the two antennas shown in Fig. 17.10, in the plane $\theta = \pi/2$, for the following pairs of values of $d$ and $\delta$: (1) $d = \lambda/4$, $\delta = 90°$; (2) $d = \lambda/4$, $\delta = -90°$; (3) $d = \lambda/2$, $\delta = 0°$; (4) $d = \lambda/2$, $\delta = 180°$; (5) $d = \lambda$, $\delta = 0°$. In all cases determine $|E_\theta|/E_0$ for the following values of $\phi$ only: $\phi = 0°, 45°, 90°, 135°$ and $180°$. Using these values, make a free-hand plot of the diagrams.

**17.G5.** A half-wave dipole with current $I_m$ is placed near a conducting 90° corner, as shown in Fig. 17.G5a. Assuming in the first approximation that the corner is infinitely large and perfectly conducting, determine the resultant electric field at the distant point $M$ shown in the figure.

*Solution.* The electromagnetic field of the system is due to the current in the dipole, and the currents induced on the surface of the corner. The distribution of the latter is unknown. However, by the method of images we find that the field created by the induced currents is the same as the field of the *three* image antennas positioned as shown in Fig. 17.G5b, the corner assumed have been removed. We can easily check that the boundary conditions on the corner surfaces are satisfied.

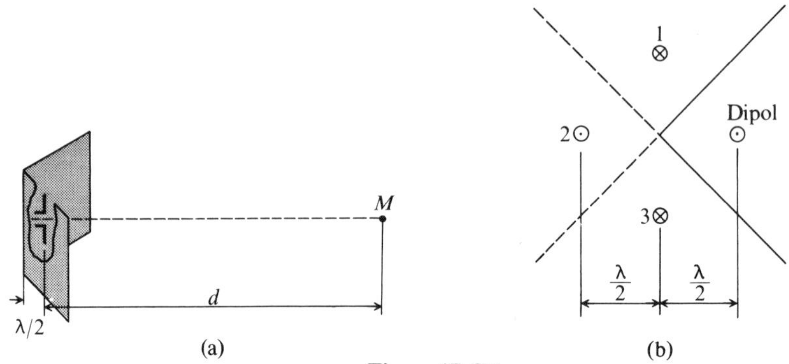


Figure 17.G5

Thus the total electric field intensity at $M$ is the sum of four electric field intensities. All four are parallel, but to add them properly we must take their relative phases into account. We can readily see from Fig. 17.G5b that the phases of all four are equal at $M$. So, according to Eqs. (17.3.6) and (17.3.8), we have that

$$E_{\text{total}} = 4E_\theta = 240\ I_m/r.$$

**17.G6.** A symmetric linear antenna resides at a height $h$ parallel to the plane surface of the earth. Assuming in the first approximation that the earth is a perfect conductor, determine the smallest value of $h$ so that the radiation is maximal at an angle $\theta$ as shown in Fig. 17.G6.

*Answer.* $h = \lambda/(4 \cos \theta)$, where $\lambda$ is the wavelength of the radiated wave.

**17.G7,** A short vertical transmitting antenna of height $h$ has a conducting plate at the top, so that the current along the antenna is practically uniform, of amplitude $I$. At the receiving point, at a distance $d$ from the antenna, only the wave reflected by the ionosphere arrives, as shown in Fig. 17.G7. The ionosphere can be approximated by a perfectly conducting plane at a height $h_i$ above the surface of the earth. Assuming the earth at both the transmitting and receiving regions to be a perfect conductor, determine the amplitude of the electric field intensity at the receiving point. The wavelength of the radiated wave is $\lambda$.

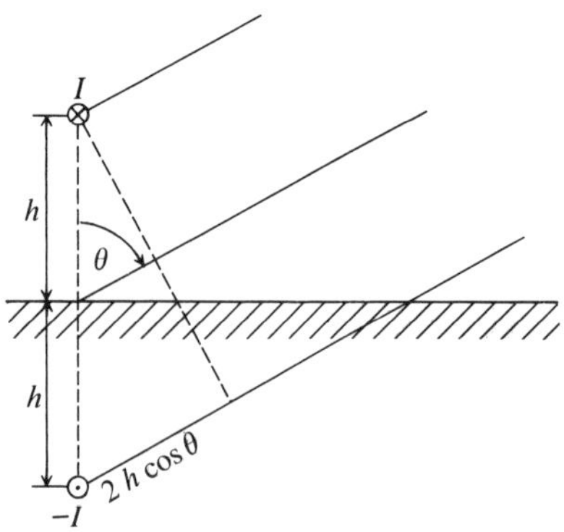


Figure 17.G6

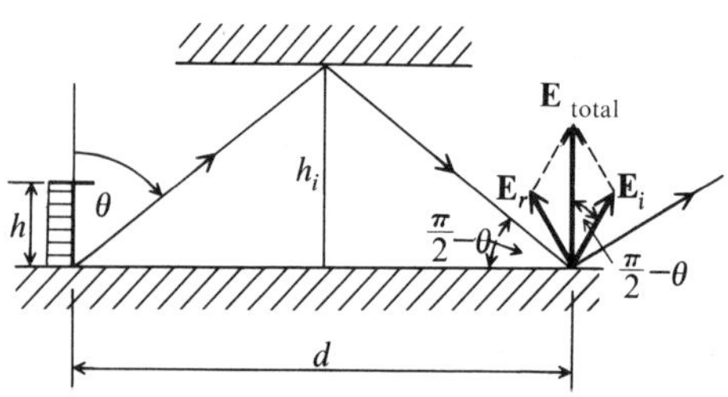


Figure 17.G7

*Solution.* The earth at the transmitting place is assumed to be a perfect conductor, so the wave radiated by the antenna and the currents induced on the surface of the earth is the same as that radiated by the antenna and its image, that is, by a Hertzian dipole of length $2h$. The wave toward the receiving point makes an angle $\theta$ with the antenna, and the distance it covers is $2\sqrt{h_i^2 + (d/2)^2}$. According to Eq. (17.2.11), the amplitude of the electric field intensity of the *incident* wave at the receiving point is

$$E_i = \frac{\beta I 2h \sin\theta}{4\pi 2\sqrt{h_i^2 + (d/2)^2}}\sqrt{\frac{\mu_0}{\varepsilon_0}} = \frac{hI\sin\theta}{2\lambda\sqrt{h_i^2 + (d/2)^2}}\sqrt{\frac{\mu_0}{\varepsilon_0}}.$$

The total electric field intensity at the receiving point is the sum of the incident wave and the reflected wave. The direction and amplitude of the reflected wave are determined from the condition that the tangential component of the resultant electric field intensity vector on the surface of the earth is zero. This must be so if the earth is assumed to be a perfect conductor. Hence, according to Fig. 17.G7, we find that the amplitude of the total electric field intensity vector, which is perpendicular to the surface of the earth, is

$$E_{\text{total}} = 2E_i \cos\left(\frac{\pi}{2} - \theta\right) = \frac{hI\sin^2\theta}{\lambda\sqrt{h_i^2 + (d/2)^2}}\sqrt{\frac{\mu_0}{\varepsilon_0}}.$$

Finally, since

$$\sin^2\theta = \frac{(d/2)^2}{h_i^2 + (d/2)^2},$$

and with $\sqrt{\mu_0/\varepsilon_0} = 120\pi$, $E_{\text{total}}$ becomes

$$E_{\text{total}} = \frac{30hId^2\pi}{\lambda[h_i^2 + (d/2)^2]^{3/2}}.$$

**17.G8.** A quarter-wave vertical antenna serves as a transmitting antenna above the surface of the earth, fed with an average power $P_{\text{av}}$. The receiving antenna is a thin loop, of area $S$ and with $N$ turns of thin wire, at a distance $d$ ($d \gg \lambda$) from the transmitting antenna. The earth can be regarded to be approximately a perfect conductor. If the loop is positioned so that the electromotive force induced in it is maximal and if the frequency of the radiated wave is $f$, determine the amplitude of the electromotive force induced in the loop.

*Solution.* According to Eq. (17.7.2), for determining the amplitude $\mathscr{E}_m$ of the electromotive force induced in the loop it is necessary to determine the magnetic field intensity in the vicinity of the loop. The earth being assumed to be a perfect conductor, the quarter-wavelength transmitting antenna creates near the loop the same field as a half-wave dipole with the same current. Since the average power $P_{\text{av}}$ radiated rather than the current amplitude $I_m$ is known, we first determine $I_m$ according to Eq. (17.3.9) and Fig. 17.15:

$$I_m = \sqrt{2P_{\text{av}}/R_{\text{rad}}} = \sqrt{2P_{\text{av}}/36.5}.$$

Next, according to Eqs. (17.2.13), (17.3.6), and (17.3.8), the magnitude of the magnetic field intensity $H$ at the receiving point is

$$H = \frac{I_m}{2\pi d} = \frac{1}{2\pi d}\sqrt{\frac{2P_{\text{av}}}{36.5}}.$$

So the amplitude of the induced electromotive force is

$$\mathscr{E}_m = \omega\mu_0 NSH = \frac{\mu_0 NSf}{d}\sqrt{\frac{2P_{av}}{36.5}}.$$

**17.G9.** Suppose that the transmitting antenna in Fig. 17.G7 is a vertical quarter-wave antenna rather than the short, top-loaded antenna shown. Let the quarter-wave antenna be fed with an average power $P$, and let the receiving antenna be a loop of area $S$ with $N$ turns of thin wire. Determine the amplitude of the electromotive force induced in the loop if it is positioned so that the electromotive force is maximal. The wavelength of the radiated wave is $\lambda$.

*Answer.* For $P = 1$ kW, $d = 200$ km, $h_i = 100$ km, $S = 2500$ cm$^2$, $N = 10$, and $\lambda = 100$ m, $\mathscr{E}_m = 0.311$ mV.

**17.G10.** The transmitting antenna is a half-wave dipole, with current $I_m$, residing parallel to a large planar conducting sheet. The receiving antenna is a Hertzian dipole of length $l$. The system is shown in Fig. 17.G10, with $D \gg d$. Find first the smallest value of $d$ for which the field at the receiving point is maximal, then find the amplitude $\mathscr{E}_m$ of the electromotive force induced in the receiving dipole.

*Answer.* $\mathscr{E}_m = 120\, I_m l/D$.

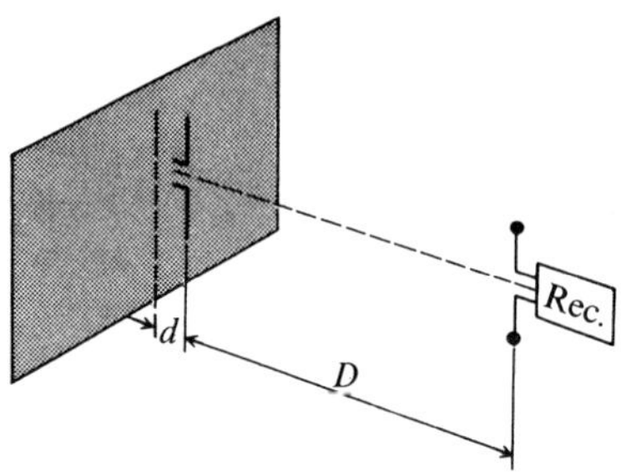


**Figure 17.G10**

**17.G11.** A Hertzian dipole of length $l$ is used as the receiving antenna in the system shown in Fig. 17.G3. The dipole is in the $x$-$y$ plane. Find the amplitude $\mathscr{E}_m$ of the electromotive force induced in the dipole as a function of the angle $\alpha$ between the dipole and the $x$ axis.

*Answer.* $\mathscr{E}_m = El$, independent of $\alpha$.

### Numerical problems

**17.N1.** A Hertzian dipole of length $l = 1$ m is fed with a current of amplitude $I = 1$ A and of frequency $f = 10^6$ Hz. Find $E_\theta$ and $H_\phi$ in the equatorial plane (plane $\theta = \pi/2$) of the dipole, at a distance $r = 10$ km.

**17.N2.** Find the radiation resistance of a Hertzian dipole of length $l = 2$ m at a frequency $f = 0.5$ MHz.

**17.N3.** If the average power radiated by the dipole from the preceding problem is $P = 1$ kW, find the amplitude of the current in the dipole.

**17.N4.** The radius of the copper wire of the dipole of Problem 17.N2 is $a = 1.5$ mm. By reference to Problem 17.G1, find $P_{rad}/P_{total}$ at the frequency given.

**17.N5.** A half-wave dipole is fed with the average power $P = 10$ kW. Find the current amplitude at the antenna terminals.

**17.N6.** It is desired to have at a distance $d = 10$ km from a half-wave dipole, in its equatorial plane, an electric field intensity of amplitude $E = 1$ V/m. Find the power which must be radiated by the dipole, and the current amplitude at its terminals.

**17.N7.** The copper wire of a half-wave dipole has a radius $a = 5$ mm. By reference to Problem 17.G2, find the ratio $P_{\text{rad}}/P_{\text{total}}$ at a frequency $f = 10^8$ Hz.

**17.N8.** Find the amplitude of the electromotive force induced in the loop of Problem 17.G8, if $N = 10$, $S = 0.1\ \text{m}^2$, $f = 10^6$ Hz, $d = 100$ km, and $P_{\text{av}} = 100$ kW.

**Suggested reading.** The list progresses towards more advanced texts. For full bibliographical description see Bibliography.

1. KRAUS, J. D., *Electromagnetics*, Chapter 13.

2. BOHN, E. V., *Introduction to Electromagnetic Fields and Waves*, Chapter 11.

3. JORDAN, E. C., and K. G. BALMAIN, *Electromagnetic Waves and Radiating Systems*, Chapters 10 to 14.

APPENDIX 1

# THE MKSA SYSTEM OF UNITS

By inspecting the number of relations between physical quantities and physical constants encountered in the basic theory of electromagnetic fields, and the number of these quantities and constants, we reach the following conclusion: there are *four* relations less than the number of physical quantities and constants put together. This means that, if *any* set of four independent quantities and constants is adopted as fundamental, a sufficient number of relations is available to express all others in terms of these four. This is obviously not the only possibility. For example, one of the physical constants can be arbitrarily considered as a pure number, and some *three* independent quantities considered as fundamental. Several systems of fundamental quantities are used in engineering and physics. In electrical engineering, the fundamental system of four quantities, length (abbreviation $L$), mass ($M$), time ($T$), and one electrical quantity or physical constant is today generally adopted. As the electrical quantity, the current intensity ($I$) is most often used, but the charge or permeability of free space is also sometimes adopted. All other quantities can be expressed in terms of these fundamental quantities. The expression describing a quantity in terms of fundamental quantities is referred to as its *dimension*.

The *unit* of a physical quantity may be any convenient magnitude of that quantity. According to the particular branch of engineering or physics, various units are adopted for the same physical quantity. The criterion is most often that the unit is close to the usual order of magnitude of the physical quantity considered. It is convenient to adopt units for the four (or possibly less) fundamental quantities, and then to derive units of all other quantities by using the relations among them. A *consistent* system of units is said to result.

There are several consistent systems of units. In electrical engineering what is known as the MKSA system is today practically standard.* The fundamental units in this system are meter, kilogram, second and ampere. The exact definitions of these units officially adopted by many countries (the United States among them)

* *IEEE Spectrum*, March 1966, pp. 169–173.

may be found elsewhere, but they are very close or equal to the following units. The meter is approximately equal to the distance between two marks on a bar of platinum-iridium, which is kept in Paris as the international standard of length. The kilogram is the mass of the international standard of mass, a lump of platinum-iridium, also kept in Paris. The second is approximately 1/86,400 (that is $1/60 \cdot 60 \cdot 24$) of a mean solar day. Finally, the ampere is equal to the current intensity which, if maintained in two straight parallel conductors of infinite length and negligible cross-section, situated in a vacuum at a distance of one meter, will produce between the two conductors a force of $2 \times 10^{-7}$ newtons per meter of the conductors. According to Eq. (12.3.6), this implicity assigns to $\mu_0$ the value of *exactly* $4\pi \times 10^{-7}$ henries/meter.

The MKSA units of all other quantities in terms of the four fundamental units are obtained by using the mathematical relations existing among them. These derived units are often given separate names. For convenience, we give in Table A1 the MKSA units of some of the quantities most often encountered in connection with electromagnetic theory, together with their internationally adopted names and abbreviations.

**Table A1.** List of fundamental and derived units in the MKSA system

| Physical quantity | Symbol | Unit (and its abbreviation) | Dimension |
|---|---|---|---|
| | | *Fundamental quantities* | |
| Length, distance | $r, l, \ldots$ | meter (m) | $L$ |
| Mass | $m$ | kilogram (kg) | $M$ |
| Time | $t$ | second (s) | $T$ |
| Current intensity | $I, i$ | ampere (A) | $I$ |
| | | *Some derived mechanical quantities* | |
| Area | $S$ | meter$^2$ (m$^2$) | $L^2$ |
| Volume | $v$ | meter$^3$ (m$^3$) | $L^3$ |
| Velocity | $v$ | meter/second (m/s) | $LT^{-1}$ |
| Force | $F$ | newton (N) | $MLT^{-2}$ |
| Moment (torque) | $T$ | newton-meter (N-m) | $ML^2T^{-2}$ |
| Energy and work | $W$ | joule (J) | $ML^2T^{-2}$ |
| Power | $P$ | watt (W) | $ML^2T^{-3}$ |
| Frequency | $f$ | hertz (Hz) | $T^{-1}$ |
| Angular frequency | $\omega$ | radian/second (rad/s) | $T^{-1}$ |
| | | *Derived quantities of the static electric field* | |
| Charge | $Q, q$ | coulomb (C) | $TI$ |
| Permittivity | $\varepsilon$ | farad/meter (F/m) | $M^{-1}L^{-3}T^4I^2$ |
| Electric field intensity | $E$ | volt/meter (V/m) | $MLT^{-3}I^{-1}$ |
| Volume charge density | $\rho$ | coulomb/meter$^3$ (C/m$^3$) | $L^{-3}TI$ |

| Physical quantity | Symbol | Unit (and its abbreviation) | Dimension |
|---|---|---|---|
| Surface charge density | $\rho_s$ | coulomb/meter$^2$ (C/m$^2$) | $L^{-2}TI$ |
| Potential | $V$ | volt (V) | $ML^2T^{-3}I^{-1}$ |
| Polarization | $P$ | coulomb/meter$^2$ (C/m$^2$) | $L^{-2}TI$ |
| Electric displacement | $D$ | coulomb/meter$^2$ (C/m$^2$) | $L^{-2}TI$ |
| Capacitance | $C$ | farad (F) | $M^{-1}L^{-2}T^4I^2$ |
| | *Derived quantities of the steady current field* | | |
| Current density | $J$ | ampere/meter$^2$ (A/m$^2$) | $L^{-2}I$ |
| Electromotive force | $\mathscr{E}$ | volt (V) | $ML^2T^{-3}I^{-1}$ |
| Resistance | $R$ | ohm ( Ω) | $ML^2T^{-3}I^{-2}$ |
| Conductance | $G$ | siemens (S) or mho | $M^{-1}L^{-2}T^3I^2$ |
| Resistivity | $\rho$ | ohm-meter ( Ω-m) | $ML^3T^{-3}I^{-2}$ |
| Conductivity | $\sigma$ | siemens/meter (S/m) | $M^{-1}L^{-3}T^3I^2$ |
| | *Derived quantities of the static magnetic field* | | |
| Permeability | $\mu$ | henry/meter (H/m) | $MLT^{-2}I^{-2}$ |
| Magnetic flux density | $B$ | tesla (T) | $MT^{-2}I^{-1}$ |
| Magnetic flux | $\Phi$ | weber (Wb) | $ML^2T^{-2}I^{-1}$ |
| Magnetic vector potential | $A$ | weber/meter (Wb/m) | $MLT^{-2}I^{-1}$ |
| Magnetic scalar potential | $V_m$ | ampere (A) | $I$ |
| Magnetization | $M$ | ampere/meter (A/m) | $L^{-1}I$ |
| Magnetic field intensity | $H$ | ampere/meter (A/m) | $L^{-1}I$ |
| Reluctance | $R_m$ | ampere/weber (A/Wb) | $M^{-1}L^{-2}T^2I^2$ |
| Inductance | $L$ | henry (H) | $ML^2T^{-2}I^{-2}$ |

APPENDIX 2

# THE HELMHOLTZ THEOREM

Essentially, the Helmholtz theorem states that a vector function is determined uniquely if the values of its curl and its divergence are known at all points.

Let us consider a vector function $\mathbf{F}$ and suppose that not $\mathbf{F}$ but curl $\mathbf{F}$ and div $\mathbf{F}$ are known at all points:

$$\text{curl}\,\mathbf{F} = \mathbf{G}, \qquad \text{div}\,\mathbf{F} = \mathbf{g}. \tag{a}$$

$\mathbf{G}$ and $\mathbf{g}$ are arbitrary *known* functions of position. Let us assume that $\mathbf{F}$ is *not* determined uniquely by Eqs. (a), contrary to the Helmholtz theorem, and that also

$$\text{curl}\,(\mathbf{F} + \mathbf{F}_0) = \mathbf{G} \qquad \text{and} \qquad \text{div}\,(\mathbf{F} + \mathbf{F}_0) = \mathbf{g}, \tag{b}$$

where $\mathbf{F}_0$ is some function of position. We shall prove that if Eqs. (a) are satisfied, then $\mathbf{F}_0 = 0$.

According to Eqs. (a) and (b) it follows that

$$\text{curl}\,\mathbf{F}_0 = 0 \qquad \text{and} \qquad \text{div}\,\mathbf{F}_0 = 0. \tag{c}$$

Let us assume, however, that div $\mathbf{F}_0 = g_0$, where $g_0$ is an arbitrary function of position, although we know that $g_0 = 0$. So instead of Eqs. (c) we consider the case

$$\text{curl}\,\mathbf{F}_0 = 0, \qquad \text{div}\,\mathbf{F}_0 = g_0, \tag{d}$$

with $g_0 = 0$ in the final result. From the first equation

$$\mathbf{F}_0 = -\text{grad}\,V, \tag{e}$$

where $V$ is an arbitrary scalar function of position. Substituting the value of $\mathbf{F}_0$ given by Eq. (e) into the second of Eqs. (d), an equation with $V$ as unknown is obtained:

$$\text{div grad}\,V = -g_0. \tag{f}$$

This is Poisson's equation (1.12.1), except that $\rho/\varepsilon_0$ is replaced by $g_0$. Hence the solution in the form of the volume integral in Eq. (1.6.12) applies, so that

$$V = \int_v \frac{g_0\,dv}{4\pi r} = \int_v \frac{\text{div}\,\mathbf{F}_0\,dv}{4\pi r}. \tag{g}$$

The integration is to be performed over all space. Now, if div $\mathbf{F}_0 = 0$ everywhere, then $V = 0$ at all points as well. From Eq. (e) it then follows that $\mathbf{F}_0 = 0$. A vector function is therefore defined uniquely if its curl and its divergence are specified at all points. In particular, if both the curl and the divergence of a vector function are zero at all points, it is equal to zero (or possibly to a constant) everywhere.

APPENDIX 3

# SUMMARIZED VECTOR IDENTITIES

## ALGEBRAIC IDENTITIES

1. $\mathbf{A} + \mathbf{B} = \mathbf{B} + \mathbf{A}$ (0.3.2)
2. $(\mathbf{A} + \mathbf{B}) + \mathbf{C} = \mathbf{A} + (\mathbf{B} + \mathbf{C})$ (0.3.3)
3. $\mathbf{A} \cdot \mathbf{B} = AB \cos \alpha$ ($\alpha$—angle between **A** and **B**) (0.3.7)
4. $\mathbf{A} \cdot \mathbf{A} = \mathbf{A}^2 = A^2$ (0.3.7a)
5. $\mathbf{A} \cdot \mathbf{B} = \mathbf{B} \cdot \mathbf{A}$ (0.3.8)
6. $\mathbf{A} \cdot (\mathbf{B} + \mathbf{C}) = \mathbf{A} \cdot \mathbf{B} + \mathbf{A} \cdot \mathbf{C}$ (0.3.9)
7. $\mathbf{A} \times \mathbf{B} = AB \sin \alpha \, \mathbf{u}_n$ ($\mathbf{u}_n$—unit vector normal to both **A** and **B**; its direction is determined by the right-hand screw rule) (0.3.12)
8. $\mathbf{B} \times \mathbf{A} = -\mathbf{A} \times \mathbf{B}$ (0.3.13)
9. $\mathbf{A} \times (\mathbf{B} + \mathbf{C}) = \mathbf{A} \times \mathbf{B} + \mathbf{A} \times \mathbf{C}$ (0.3.14)
10. $\mathbf{A} \cdot (\mathbf{B} \times \mathbf{C}) = \mathbf{C} \cdot (\mathbf{A} \times \mathbf{B}) = \mathbf{B} \cdot (\mathbf{C} \times \mathbf{A})$ (0.3.15)
11. $\mathbf{A} \times (\mathbf{B} \times \mathbf{C}) = (\mathbf{A} \cdot \mathbf{C})\mathbf{B} - (\mathbf{A} \cdot \mathbf{B})\mathbf{C}$ (0.3.17)
12. $\mathbf{u}_A = \dfrac{\mathbf{A}}{A}$ (unit vector in direction of vector **A**) (0.4.1)

## IDENTITIES INVOLVING DIFFERENTIAL OPERATORS

Most identities are given in both their explicit form and the short form using the del operator. **A** and **B** designate any two vector functions, $V$ and $W$ any two scalar functions.

### Identities involving gradient

13. $\mathbf{u}_l \cdot \text{grad } V = dV/dl$, or $\mathbf{u}_l \cdot \nabla V = dV/dl$ (1.7.3)
14. $\text{grad } (V + W) = \text{grad } V + \text{grad } W$, or $\nabla(V + W) = \nabla V + \nabla W$
15. $\text{grad } (VW) = V \text{ grad } W + W \text{ grad } V$, or $\nabla(VW) = V\nabla W + W\nabla V$
16. $\nabla(\mathbf{A} \cdot \mathbf{B}) = (\mathbf{A} \cdot \nabla)\mathbf{B} + (\mathbf{B} \cdot \nabla)\mathbf{A} + \mathbf{A} \times (\nabla \times \mathbf{B}) + \mathbf{B} \times (\nabla \times \mathbf{A})$

17. $\displaystyle\int_v \text{grad } V\, dv = \oint_S V\, d\mathbf{S}, \qquad \text{or} \qquad \int_v \nabla V\, dv = \oint_S V\, d\mathbf{S}$

18. $\displaystyle\int_S \mathbf{u}_n \times \text{grad } V\, dS = \oint_C V\, d\mathbf{l}, \qquad \text{or} \qquad \int_S \mathbf{u}_n \times \nabla V\, dS = \oint_C V\, d\mathbf{l}$

**Identities involving divergence**

19. $\displaystyle\text{div } \mathbf{A} = \nabla \cdot \mathbf{A} = \lim_{\Delta v \to 0} \frac{\oint_{\Delta S} \mathbf{A} \cdot d\mathbf{S}}{\Delta v}$ (1.11.4)

20. $\displaystyle\int_v \text{div } \mathbf{A}\, dv = \oint_S \mathbf{A} \cdot d\mathbf{S}$ (divergence theorem) (1.11.10)

21. $\text{div } (\mathbf{A} + \mathbf{B}) = \text{div } \mathbf{A} + \text{div } \mathbf{B}, \qquad \text{or} \qquad \nabla \cdot (\mathbf{A} + \mathbf{B}) = \nabla \cdot \mathbf{A} + \nabla \cdot \mathbf{B}$

22. $\text{div } (V\mathbf{A}) = V \text{div } \mathbf{A} + \mathbf{A} \cdot \text{grad } V, \qquad \text{or} \qquad \nabla \cdot (V\mathbf{A}) = V\nabla \cdot \mathbf{A} + \mathbf{A} \cdot \nabla V$

23. $\text{div } (\mathbf{A} \times \mathbf{B}) = \mathbf{B} \cdot \text{curl } \mathbf{A} - \mathbf{A} \cdot \text{curl } \mathbf{B},$
or $\nabla \cdot (\mathbf{A} \times \mathbf{B}) = \mathbf{B} \cdot (\nabla \times \mathbf{A}) - \mathbf{A} \cdot (\nabla \times \mathbf{B})$ (12.1.7)

24. $\text{div } (\text{curl } \mathbf{A}) = \nabla \cdot (\nabla \times \mathbf{A}) = 0$ (8.9.1)

25. $\text{div } (\text{grad } V) = \nabla \cdot (\nabla V) = \nabla^2 V =$ Laplacian of $V$ (1.12.5)

26. $\text{div } [\text{grad } (V + W)] = \text{div } (\text{grad } V) + \text{div } (\text{grad } W)$
or $\nabla^2(V + W) = \nabla^2 V + \nabla^2 W$ (1.13.6)

**Identities involving curl**

27. $\displaystyle\mathbf{u}_n \cdot \text{curl } \mathbf{A} = \mathbf{u}_n \cdot (\nabla \times \mathbf{A}) = \lim_{\Delta S \to 0} \frac{\oint_{\Delta C} \mathbf{A} \cdot d\mathbf{l}}{\Delta S}, \qquad \Delta\mathbf{S} = \mathbf{u}_n\, \Delta S$ (8.8.4)

28. $\displaystyle\int_S \text{curl } \mathbf{A} \cdot d\mathbf{S} = \oint_C \mathbf{A} \cdot d\mathbf{l}$ (Stokes' theorem) (8.8.14)

29. $\text{curl } (\mathbf{A} + \mathbf{B}) = \text{curl } \mathbf{A} + \text{curl } \mathbf{B},$
or $\nabla \times (\mathbf{A} + \mathbf{B}) = \nabla \times \mathbf{A} + \nabla \times \mathbf{B}$

30. $\text{curl } (V\mathbf{A}) = \text{grad } V \times \mathbf{A} + V \text{curl } \mathbf{A},$
or $\nabla \times (V\mathbf{A}) = \nabla V \times \mathbf{A} + V\nabla \times \mathbf{A}$

31. $\nabla \times (\mathbf{A} \times \mathbf{B}) = \mathbf{A}\nabla \cdot \mathbf{B} - \mathbf{B}\nabla \cdot \mathbf{A} + (\mathbf{B} \cdot \nabla)\mathbf{A} - (\mathbf{A} \cdot \nabla)\mathbf{B}$

32. $\text{curl } (\text{grad } V) = \nabla \times (\nabla V) = 0$ (8.9.3)

33. $\text{curl } (\text{curl } \mathbf{A}) = \text{grad } (\text{div } \mathbf{A}) - \text{div } (\text{grad } \mathbf{A}),$
or $\nabla \times (\nabla \times \mathbf{A}) = \nabla(\nabla \cdot \mathbf{A}) - \nabla^2 \mathbf{A}$ (definition of operation div grad or $\nabla^2$ acting on a vector function) (8.9.10)

34. $\displaystyle\int_v \text{curl } \mathbf{A}\, dv = \oint_S \mathbf{u}_n \times \mathbf{A}\, dS$

APPENDIX 4

# GRADIENT, DIVERGENCE, LAPLACIAN, AND CURL IN RECTANGULAR, CYLINDRICAL, AND SPHERICAL COORDINATE SYSTEMS

**Rectangular coordinate system**

$$\text{grad } V = \nabla V = \frac{\partial V}{\partial x}\mathbf{u}_x + \frac{\partial V}{\partial y}\mathbf{u}_y + \frac{\partial V}{\partial z}\mathbf{u}_z$$

$$\text{div } \mathbf{A} = \nabla \cdot \mathbf{A} = \frac{\partial A_x}{\partial x} + \frac{\partial A_y}{\partial y} + \frac{\partial A_z}{\partial z}$$

$$\text{div (grad } V) = \nabla^2 V = \frac{\partial^2 V}{\partial x^2} + \frac{\partial^2 V}{\partial y^2} + \frac{\partial^2 V}{\partial z^2}$$

$$\text{curl } \mathbf{A} = \nabla \times \mathbf{A} = \left(\frac{\partial A_z}{\partial y} - \frac{\partial A_y}{\partial z}\right)\mathbf{u}_x + \left(\frac{\partial A_x}{\partial z} - \frac{\partial A_z}{\partial x}\right)\mathbf{u}_y + \left(\frac{\partial A_y}{\partial x} - \frac{\partial A_x}{\partial y}\right)\mathbf{u}_z$$

**Cylindrical coordinate system**

$$\text{grad } V = \nabla V = \frac{\partial V}{\partial r}\mathbf{u}_r + \frac{1}{r}\frac{\partial V}{\partial \phi}\mathbf{u}_\phi + \frac{\partial V}{\partial z}\mathbf{u}_z$$

$$\text{div } \mathbf{A} = \nabla \cdot \mathbf{A} = \frac{1}{r}\frac{\partial}{\partial r}(rA_r) + \frac{1}{r}\frac{\partial A_\phi}{\partial \phi} + \frac{\partial A_z}{\partial z}$$

$$\text{div (grad } V) = \nabla^2 V = \frac{1}{r}\frac{\partial}{\partial r}\left(r\frac{\partial V}{\partial r}\right) + \frac{1}{r^2}\frac{\partial^2 V}{\partial \phi^2} + \frac{\partial^2 V}{\partial z^2}$$

$$\text{curl } \mathbf{A} = \nabla \times \mathbf{A} = \left(\frac{1}{r}\frac{\partial A_z}{\partial \phi} - \frac{\partial A_\phi}{\partial z}\right)\mathbf{u}_r + \left(\frac{\partial A_r}{\partial z} - \frac{\partial A_z}{\partial r}\right)\mathbf{u}_\phi + \frac{1}{r}\left[\frac{\partial}{\partial r}(rA_\phi) - \frac{\partial A_r}{\partial \phi}\right]\mathbf{u}_z$$

**Spherical coordinate system**

$$\text{grad } V = \nabla V = \frac{\partial V}{\partial r}\mathbf{u}_r + \frac{1}{r}\frac{\partial V}{\partial \theta}\mathbf{u}_\theta + \frac{1}{r\sin\theta}\frac{\partial V}{\partial \phi}\mathbf{u}_\phi$$

$$\text{div } \mathbf{A} = \nabla\cdot\mathbf{A} = \frac{1}{r^2}\frac{\partial}{\partial r}(r^2 A_r) + \frac{1}{r\sin\theta}\frac{\partial}{\partial\theta}(A_\theta \sin\theta) + \frac{1}{r\sin\theta}\frac{\partial A_\phi}{\partial\phi}$$

$$\text{div (grad } V) = \nabla^2 V = \frac{1}{r^2}\frac{\partial}{\partial r}\left(r^2\frac{\partial V}{\partial r}\right) + \frac{1}{r^2\sin\theta}\frac{\partial}{\partial\theta}\left(\sin\theta\frac{\partial V}{\partial\theta}\right) + \frac{1}{r^2\sin^2\theta}\frac{\partial^2 V}{\partial\phi^2}$$

$$\text{curl } \mathbf{A} = \nabla\times\mathbf{A} = \frac{1}{r\sin\theta}\left[\frac{\partial}{\partial\theta}(A_\phi\sin\theta) - \frac{\partial A_\theta}{\partial\phi}\right]\mathbf{u}_r$$

$$+ \frac{1}{r}\left[\frac{1}{\sin\theta}\frac{\partial A_r}{\partial\phi} - \frac{\partial}{\partial r}(rA_\phi)\right]\mathbf{u}_\theta + \frac{1}{r}\left[\frac{\partial}{\partial r}(rA_\theta) - \frac{\partial A_r}{\partial\theta}\right]\mathbf{u}_\phi$$

APPENDIX 5

# TABLE OF SOME IMPORTANT CONSTANTS

| | |
|---|---|
| Permittivity of free space | $\varepsilon_0 = 8.854 \times 10^{-12} \simeq \dfrac{10^{-9}}{36\pi}$ farad/meter |
| Permeability of free space | $\mu_0 = 4\pi \times 10^{-7}$ henry/meter |
| Velocity of light in vacuum | $c_0 = (\varepsilon_0\mu_0)^{-1/2} = 2.998 \times 10^8 \simeq 3 \times 10^8$ m/s |
| Intrinsic impedance of vacuum | $\eta_0 = (\mu_0/\varepsilon_0)^{1/2} = 376.7 \simeq 120\pi$ ohm |
| Charge of electron | $e = 1.602 \times 10^{-19}$ coulomb |
| Mass of electron | $m_e = 9.107 \times 10^{-31}$ kilogram |
| Charge/mass of electron | $e/m_e = 1.759 \times 10^{11}$ coulomb/kilogram |
| Mass of proton | $m_p = 1.673 \times 10^{-27}$ kilogram |

**Location of tables of physical properties of materials**

# BIBLIOGRAPHY

During the preparation of the present text the author consulted the following works on electromagnetic fields. Those which the author found particularly useful are designated by an asterisk.

## INTRODUCTORY LEVEL


ATTWOOD, S. S., *Electric and Magnetic Fields*, John Wiley & Sons, New York, 1958.

BENUMOF, R., *Concepts in Electricity and Magnetism*, Holt, Rinehart and Winston, New York, 1961.

BOAST, W. B., *Vector Fields*, Harper & Row, New York, 1964.

BRADSHAW, M. D. and W. J. BYATT, *Introductory Engineering Field Theory*, Prentice-Hall, Englewood Cliffs, N. J., 1967.

*CARTER, G. W., *The Electromagnetic Field in Its Engineering Aspects*, American Elsevier, New York, 1967.

CHESTON, W. B., *Elementary Theory of Electric and Magnetic Fields*, John Wiley & Sons, New York, 1964.

CORSON, D. R. and P. LORRAIN, *Introduction to Electromagnetic Fields and Waves*, W. H. Freeman, San Francisco, 1962.

COULSON, C. A., *Electricity*, Oliver and Boyd, Edinburgh, 1953.

HARRINGTON, R. F., *Introduction to Electromagnetic Engineering*, McGraw-Hill Book Company, New York, 1958.

*HAYT, W. H., Jr., *Engineering Electromagnetics*, McGraw-Hill Book Company, New York, 1967.

HOLT, C. A., *Introduction to Electromagnetic Fields and Waves*, John Wiley & Sons, New York, 1966.

*KRAUS, J. D., *Electromagnetics*, McGraw-Hill Book Company, New York, 1953.

NEUMANN, L. R. and P. L. KALANTAROF, *Theoretical Foundations of Electrical Engineering*, parts I and III, Naučna Knjiga, Belgrade, 1951–1952 (in Serbo-Croatian, translated from Russian).

OWEN, G. E., *Introduction to Electromagnetic Theory*, Allyn and Bacon, Boston, 1963.

Popović, B. D., *A Collection of Problems in Electromagnetism* (with solutions), Gradjevinska Knjiga, Belgrade, 1966 (in Serbo-Croatian).

Reitz, J. R. and F. J. Milford, *Foundations of Electromagnetic Theory*, Addison-Wesley Publishing Company, Reading, Mass., 1960.

Rogers, W. E., *Introduction to Electric Fields*, McGraw-Hill Book Company, New York, 1954.

Schelkunoff, S. A., *Electromagnetic Fields*, Blaisdell Publishing Company, New York, 1963.

Seely, S., *Introduction to Electromagnetic Fields*, McGraw-Hill Book Company, New York, 1958.

Skilling, H. H., *Fundamentals of Electric Waves*, John Wiley & Sons, New York, 1964.

*Walsh, J. B., *Electromagnetic Theory and Engineering Applications*, The Ronald Press Company, New York, 1960.

## INTERMEDIATE LEVEL

Bewley, L. V., *Flux Linkages and Electromagnetic Induction*, Dover Publications, New York, 1964.

Bewley, L. V., *Two-Dimensional Fields in Electrical Engineering*, The Macmillan Company, New York, 1948.

*Bohn, E. V., *Introduction to Electromagnetic Fields and Waves*, Addison-Wesley Publishing Company, Reading, Mass., 1968.

Della Torre, E. and C. V. Longo, *The Electromagnetic Field*, Allyn and Bacon, Boston, 1969.

Fano, R. M., Chu, L. J., and R. B. Adler, *Electromagnetic Fields, Energy and Forces*, John Wiley & Sons, New York, 1960.

Hallén, E., *Electromagnetic Theory*, John Wiley & Sons, New York, 1962.

Harnwell, G. P., *Principles of Electricity and Electromagnetism*, McGraw-Hill Book Company, New York, 1949.

Johnson, C. C., *Field and Wave Electrodynamics*, McGraw-Hill Book Company, New York, 1965.

King, R. W. P., *Fundamental Electromagnetic Theory*, Dover Publications, New York, 1963.

Küpfmüller, K., *Einfürung in die theoretische Elektrotechnik*, Springer, Berlin, 1962.

Langmuir, R. V., *Electromagnetic Fields and Waves*, McGraw-Hill Book Company, New York, 1961.

Mason, M. and W. Weaver, *The Electromagnetic Field*, Dover Publications, New York, 1929.

Moon, P. and D. E. Spencer, *Field Theory for Engineers*, D. Van Nostrand Company, Princeton, N. J., 1961.

Maxwell, J. C., *A Treatise on Electricity and Magnetism*, Vol. 1 and 2, Dover Publications, New York, 1954.

Plonsey, R. and R. E. Collin, *Principles and Applications of Electromagnetic Fields*, McGraw-Hill Book Company, New York, 1961.

RAMO, S., WHINNERY, J. R. and T. VAN DUZER, *Fields and Waves in Communication Electronics*, John Wiley & Sons, New York, 1965.

*SCOTT, W. T., *The Physics of Electricity and Magnetism*, John Wiley & Sons, New York, 1966.

SILVESTER, P., *Modern Electromagnetic Fields*, Prentice-Hall, Englewood Cliffs, N. J., 1968.

SLATER, J. C. and N. H. FRANK, *Electromagnetism*, McGraw-Hill Book Company, New York, 1947.

*SURUTKA, J. V., *Electromagnetics*, Gradjevinska Knjiga, Belgrade, 1966 (in Serbo-Croatian).

*TAMM, I. E., *Fundamentals of Electrical Theory*, Science Publishing House, Moscow, 1966 (in Russian).

*WEBER, E., *Electromagnetic Fields, Theory and Applications*, Volume I—*Mapping of Fields*, John Wiley & Sons, New York, 1957.

*WEISS, A. VON and H. KLEINWÄCHTER, *Übersicht über die theoretische Electrotechnik*, part II, *Ausgewählte Kapitel und Aufgaben*, Geest & Portig, Leipzig, 1956.

## ADVANCED LEVEL

ABRAHAM, M. and R. BECKER, *The Classical Theory of Electricity and Magnetism*, Hafner Publishing Company, New York, 1951.

BATYGIN, V. V. and I. N. TOPTYGIN, *Problems in Electrodynamics*, Academic Press, New York, 1964.

BLEANEY, B. I. and B. BLEANEY, *Electricity and Magnetism*, Clarendon Press, Oxford, 1965.

DURAND, E., *Electrostatique et magnétostatique*, Masson, Paris, 1953.

ELLIOT, R. S., *Electromagnetics*, McGraw-Hill Book Company, New York, 1966.

JACKSON, J. D., *Classical Electrodynamics*, John Wiley & Sons, New York, 1962.

*JORDAN, E. C. and K. G. BALMAIN, *Electromagnetic Waves and Radiating Systems*, Prentice-Hall, Englewood Cliffs, N. J., 1968.

LANDAU, L. and E. LIFCHITZ, *The Classical Theory of Fields*, Addison-Wesley Publishing Company, Reading, Mass., 1951.

PANOFSKY, W. K. H. and M. PHILLIPS, *Classical Electricity and Magnetism*, Addison-Wesley Publishing Company, Reading, Mass., 1956.

SMYTHE, W. R., *Static and Dynamic Electricity*, McGraw-Hill Book Company, New York, 1968.

SOMMERFELD, A., *Electrodynamics*, Academic Press, New York, 1952.

STRATTON, J. A., *Electromagnetic Theory*, McGraw-Hill Book Company, New York, 1941.

VAN BLADEL, J., *Electromagnetic Fields*, McGraw-Hill Book Company, New York, 1964.

WEEKS, W. L., *Electromagnetic Theory for Engineering Applications*, John Wiley & Sons, New York, 1964.

# INDEX

# INDEX

ABCDE7987654321